Thrustbelts
Structural Architecture, Thermal Regimes
and Petroleum Systems

Thrustbelts form along tectonically active plate margins and internally within plates through reactivation of pre-existing crustal weaknesses. These regions are likely to be productive sources of hydrocarbons well into the future. Finding new resources will include searching for deeper targets, smaller and unconventional traps, and by-passed resources. Many new technical tools are enabling new discoveries, or the more efficient recovery of known reserves.

Thrustbelts provides a comprehensive account of thrust systems, including orogenic thrustbelts, transpressional ranges and accretionary prisms, and discusses both thin-skin thrust systems and thick-skin inversion structures. It includes major sections on the basic concepts, definitions and mechanics of thrust systems, the roles of stratigraphy, syn-tectonic deposition, erosion and fluid flow in determining structural style, the origins and nature of evolving thermal regimes in thrustbelts, and a thorough analysis of petroleum systems and hydrocarbon plays in thrustbelts. Wherever possible, in-depth treatment of specific analytical techniques is included, providing the reader with sufficient practical, mathematical and physical background to understand the principles. Throughout the book, case studies are worked through with a discussion of the potential applications of the technique, possible limitations and future developments. A comprehensive database of thrustbelts is available to download.

This book will be an invaluable resource for research scientists, oil company managers and students.

MICHAL NEMČOK is a research professor at the Energy and Geoscience Institute of the University of Utah. He has performed structural and hydrocarbon analyses of many thrustbelts around the world including the Andes of Bolivia, the Balkans, the Carpathians, the Eastern Alps, the Greater Caucasus and the Variscides of Wales. Together with Steven Schamel and other colleages from EGI, he has undertaken a study entitled 'Systematics of hydrocarbon exploration and production in thrustbelts', which became a nucleus for this book.

STEVEN SCHAMEL was formerly a research professor and associate director of the Petroleum Research Center at the University of Utah. His field studies have also taken him to many of the world's thrustbelts including the Appalachians, the Rocky Mountains, the Andes of Colombia and Venezuela and the northern Urals. Together with Michal Nemčok he has studied the transpressional tectonics in the southern San Joaquin basin and taught short courses focused on petroleum systems of thrustbelts.

RODNEY A. GAYER is a structural geologist and teacher with specialist knowledge of orogenic tectonics and coal geology. His initial research was in thrust tectonics in the Spitsbergen and Scandinavian Caledonides, where with Brian Harland he defined the Iapetus Ocean. He later diversified into the Variscan orogen of Europe. Together with Michal Nemčok he has investigated the role of inversion tectonics in the development of the Bristol Channel Mesozoic basin.

Thrustbelts

Structural Architecture, Thermal Regimes and Petroleum Systems

MICHAL NEMČOK
University of Utah

STEVEN SCHAMEL
GeoX Consulting Inc

ROD GAYER
Cardiff University

CAMBRIDGE
UNIVERSITY PRESS

CAMBRIDGE UNIVERSITY PRESS
Cambridge, New York, Melbourne, Madrid, Cape Town, Singapore,
São Paulo

CAMBRIDGE UNIVERSITY PRESS
The Edinburgh Building, Cambridge CB2 2RU, UK

Published in the United States of America by Cambridge University
Press, New York

www.cambridge.org
Information on this title: www.cambridge.org/9780521822947

© M. Nemčok, S. Schamel and R. Gayer 2005

First published 2005

Printed in the United Kingdom at the University Press, Cambridge

A record for this publication is available from the British Library

ISBN-13 978-0-521-82294-7 hardback
ISBN-10 0-521-82294-7 hardback

To Renata, Ivana and Ján

Contents

Preface

This book aims to provide a comprehensive understanding of thrustbelts as a whole. The aim is to synthesize existing information devoted to specific aspects of these important hydrocarbon habitats. The book assembles this information in one volume, in a manner that permits the knowledge to be used to assess the risks of exploring and operating in these settings.

The plan for this book originated with a project called *Systematics of Hydrocarbon Exploration and Production in Thrustbelts*, which summarized various aspects of exploration and production in thrustbelt settings provided by a large and diverse literature, and addressed gaps in knowledge.

This synthesis is completed from results of personal, long-term research on thrustbelts, numerical validations of various concepts and extensive tables documenting various factors influencing structural styles, thermal regimes and petroleum systems, as well as rates of various modern geological processes. The book contains an enclosed database on characteristic features of existing hydrocarbon fields in thrustbelts, which serves as further documentation. This book should have value to a broad range of readers, from geology students to exploration managers searching for the character of producing fields in analogous settings. The book is divided into four parts.

Part I defines the scope of the book, with the thrustbelt being defined broadly enough to include conventional thrustbelts, transpressional ranges, toe thrusts and accretionary prisms. It describes fundamental structural styles and variation of styles in thrustbelts, illustrated by worldwide thrustbelt data, including their location, age, tectonic character and vergency. The text follows with descriptions of the thrust wedge development, covering the two-dimensional frictional Coulomb wedge model, its limitations, its extension to three dimensions and additions to handle the brittle–ductile transition. The section on wedge mechanics is followed by thrust sheet mechanics, focusing on the rheological-stress control of the resultant thrust structures and the energy balance behind the chance that the active thrust sheet witnesses the propagation of a new thrust sheet, whether it experiences reactivation of existing faults, whether it undergoes further movements or whether it experiences internal deformation. Sandbox modelling and examples from the Taiwan thrustbelt and the West Carpathians, for example, document the role of the energy balance on complex thrust fault activity, including numerous out-of-sequence thrusting events. Subsequent chapters of the first part define thin- and thick-skin structures in thrustbelts, stress control on development, the importance related to hydrocarbon accumulation and the potential translation under subsequent shortening. These chapters are illustrated with geological cross sections, outcrop photographs and seismic data. The part concludes with descriptions of available methods for determination of thrust timing and related deformation rates.

Part II focuses on the importance of various factors controlling the structural architecture of thrustbelts. It was written with specific questions in mind. How do pre-existing structures affect the evolving structural style? What is the role of sedimentary rheology in the evolving structural style and which geological factors are most important in this role? How can knowledge of these factors be used to constrain geometric interpretations of structures? How do fluids influence the structural architecture, what are the fluid mechanisms in thrustbelts and how does one determine fluid sources, sinks and overpressure-buildup mechanisms?

Part II draws from physical phenomena described in Part I and analyses the importance of large-scale influencing factors. Sedimentary rheology is broken down into factors such as the relative strength of the rocks involved, rock lithification stages and how rocks are organized in the stratigraphic package by means of individual layer strengths, layer thicknesses, layer patterns and friction along their contacts, illustrated by extensive rock mechanics tables and factor interaction calculations. The text also focuses on rock layers undergoing deformation, how this may control reservoir horizons and trap geometries that could result from various combinations of rheologies and stress regimes. Fault-prone sequences become shortened by fault-bend folds, basement uplifts or inverted grabens. Sequences with upper weak layers and underlying stronger layers enhance the potential for passive roof duplexes, whereas little strength difference combined with high basal friction enhances foreland-vergent structures. Fold-prone sequences undergo detachment folding. Examples such as those from the Bolivian sub-Andean thrustbelt document how orogen strike-parallel changes in lithostratigraphy of accreted sediments can account for varying structural architecture under the same stress regime. A simple stress control argument indicates why the large number of interpreted fault-propagation folds may be faulted detachment folds. Numerical models and global examples document controls on the relative locations of detachment faults.

The text describes how the combination of rheology and stress in control of structural architecture becomes complicated in the upper 10–15 km of the crust, which contains pre-existing structures. It also explains how syntectonic deposition and erosion introduce further complexity. Modern-day rates of deposition and erosion from various thrustbelt settings are illustrated in extensive tabulated databases. The contribution of erosion and deposition resides not only in their effect on the thrustbelt itself, but also on the sediments to be accreted later, which in turn affect the thrustbelt development during accretion. Thick foreland sediments enhance the width and advance distance of the thrustbelt whereas thin sediments promote its internal deformation. Variable distribution of syn-tectonic sediments, together with variable sedimentary taper of the foreland basin, may produce a spectrum of structural styles, due to varying thrust spacing, various thrust trajectories and strength contrasts.

The text on fluid flow identifies the topography-driven/compaction-driven fluid flows as the main fluid flow mechanisms in thrustbelts. Although thrustbelts contain a variety of fluid sources, only compaction-released fluids, release of the structurally bound water from smectite, fluids produced by the gypsum-to-anhydrite transformation and hydrocarbon expulsion account for overpressure generation controlling the structural architecture. Fault cores sandwiched between high-fracture density damage zones are the most likely fluid migration paths in active thrustbelts with a tendency to collapse after their movement stops. The fluid flow along main detachment faults and ramps typically has a transient character.

Part III focuses on the importance of various factors controlling thermal regimes in thrustbelts. It addresses the following questions. What are the effects of pre-orogenic heat flow on subsequent thermal regimes? What are the effects of deformation on thermal regimes? What are the roles of stratigraphic development in thermal regimes, via thermal conductivity, specific heat and radioactive heat production? How does stratigraphic distribution affect the thermal regime? What are the roles of various fluid flow mechanisms in perturbing thermal regimes? How can the advective component of the heat transport be recognized? What are the roles of deposition and erosion in thermal regimes?

Part III builds on extensive tables of modern movement rates of plates and thrust sheets, and data sets on thermal conductivity, thermal diffusivity, specific heat capacity and heat production of rocks involved in thrustbelts. The review of natural slip rates for thrust sheets shows a characteristic range of 0.3–$4.3 \, mm \, yr^{-1}$, documented by minimum values from the Wyoming thrustbelt, the Perdido foldbelt in the Gulf of Mexico and the Argentinean Precordillera, by intermediate rates from the Pyrenees and the North Apennines, and by maximum rates from the San Joaquin basin and the Swiss Molasse basin.

Part III identifies, also drawing from new numerical modelling, the order of importance of various factors on thermal regimes. The list starts with factors as important as the presence or absence of syn-tectonic deposition/erosion, the pre-tectonic heat flow and the presence of critical fluid flow regimes. The part follows with factors such as the thermal blanketing potential of the uppermost layers of the accreted sequence, the slip rate, the accreted sequence lithology, the basal frictional heat and the radiogenic heat. The list ends with the lithology of detachment horizons and internal strain heating. Discussed perturbations in thrustbelts indicate that the thrust displacement of rock layers characterized by different heat production produces vertical and lateral thermal gradients.

Among fluid flow mechanisms, only the topography-driven and compaction-/compression-driven fluid flows are capable of affecting the thermal regime of a thrustbelt more than locally. The impact of both mechanisms is important because the flow rates can be greater than $10 \, m \, yr^{-1}$. While the topography-driven flow is generally robust enough to achieve such flow rates, the compaction-/compression-driven flow usually requires a flow enhancement, such as flow focusing by faults. The advective component of the heat transport in thrustbelts is recognized either by analytical calculations or by analysis of maturation data. The surface heat flow in thrustbelts also responds quickly to deposition and erosion, but recovers slowly after the end of activity. Whereas the erosion increases the surface heat flow, the heat flow is significantly depressed by deposition rates equal to or greater than $0.1 \, mm \, yr^{-1}$. The deposition can also depress the heat flow due to overpressure development. Examples (e.g., Kura basin in Azerbaijan) document that this phenomenon is caused by retarded heat transfer through undercompacted sediments.

Part IV focuses on the importance of various factors controlling petroleum systems in thrustbelts. It emphasizes the following questions.

What factors control the deposition and quality of source rocks in thrustbelts? What factors control the initiation and termination of hydrocarbon expulsion? What factors impact hydrocarbon migration, and how? What types of traps dominate in thrustbelts? What kinds of lithological seals are typical in thrustbelts and what factors control their sealing quality? What types of fault seals are typical in thrustbelts and what factors control their sealing quality? What factors enhance and destroy reservoir rocks, and how? What is the optimal timing for operation of the petroleum system?

Part IV discusses the critical presence of quality source rocks in correct stratigraphic and structural positions to have reached maturity near the close of, or following, thrusting. For example, burial beneath about 2 km of sediments shed into the Green River basin from the flanking Paleocene–Eocene Rocky Mountain uplifts was responsible for the post-thrust hydrocarbon generation from the Lower Cretaceous foreland basin source rocks in the Late Cretaceous Wyoming–Utah thrustbelt. However, source rock distributions in thrustbelts show a large variety of depositional settings and source rock quality and the magnitude of hydrocarbon reserves also vary. Generally, the deposition of source rocks is independent of the tectonic events that led directly to thrusting, but specific depositional settings appear to favour both quality source rocks and eventual contractional tectonics. These include eustatically flooded passive margin basins, silled pull-apart basins and syn-orogenic foreland basins. Typically, oil-prone black shales are deposited in the distal portions of the basins and gas-prone, coaly rocks characterize the proximal parts of the basins.

Rapid burial associated with thrust imbrication, followed by rapid post-thrusting rebound and erosion, perturb thermal gradients and complicate source rock maturation. Thrusts and associated fracture systems serve as conduits for migration of hydrocarbons and connate waters from the hydrocarbon kitchens to traps, and for the inward flux of meteoric waters that in time flush or degrade pooled hydrocarbons. Rock deformation and fluid migration during thrusting both enhance and degrade the quality of reservoirs. The majority of hydrocarbon fields reside in broad, simple anticlines in parts of thrustbelts where overall shortening and internal strains are relatively small. These folds are dominantly detachment and fault-propagation folds with a large radius of curvature in cross section (e.g., the Zagros foldbelt in Iran, the sub-Andean thrustbelt in Bolivia and Argentina, and the Wyoming thrustbelt). Additional traps are located in fault-seal-dependent footwall traps and sub-thrust autochthonous and para-autochthonous strata beneath the leading edges of thrustbelts.

Part IV also examines the worldwide distribution of oil and gas resources and the interplay of factors in thrustbelt settings to generate, entrap and preserve hydrocarbons. Many thrustbelts host hydrocarbons. A few are the site of world-class oil and gas accumulations. About a dozen thrustbelts hold the lion's share of hydrocarbon resources within this habitat. The strong asymmetry in the global distribution of hydrocarbon reserves in thrustbelts is examined here.

There is good reason to believe that thrustbelts will be productive sources of hydrocarbons well into the future. The opportunities include pure frontier plays, but extending exploration into less mature portions of established petroleum provinces is a safer path. This includes searching for deeper targets, smaller and unconventional traps, and by-passed resources. Many new technical tools are now available to assist in the discovery of new oil and gas, or in the more efficient recovery of known reserves. This book closes by pointing to likely directions of continued hydrocarbon exploration and development in thrustbelts. It seems clear that as world consumption of, and thus demand for, hydrocarbons continues to rise (BP, 2004), the 'real' price of the product will inevitable increase, thus funding the advances in technology that will be required for exploration and production of the future. We hope that this book will serve as a useful companion for those involved in this endeavour.

Acknowledgments

We wish to thank all those who contributed to the progress of this book, including the work on the earlier report that preceded it. Elf, Enterprise, OXY, Repsol YPF and Texaco funded the original research project. Ivan Vrúbel, Peter Ostrolúcky, Chelsea Christensen, Lubomil Pospíšil and Ray Levey helped with organizing the work on the book. Detailed discussions with Lothar Ratschbacher, Ronald L. Bruhn, Andreas Henk and Joseph N. Moore improved those parts of the book focused on Asian orogens, Alaskan orogens, finite-element modelling and reservoirs, respectively. Joseph N. Moore and Bruce R. Rosendahl helped with editing. Research assistance during the writing of the book was provided by Eva Franců, Eric Cline and Chelsea Christensen. Photographs and figures were provided by John W. Cosgrove, Mark G. Rowan, Keck Geology Consortium, Veronika Vajdová, Atilla Aydin, Piotr Krzywiec, Cathy L. Hanks, Wesley K. Wallace, Brent A. Couzens-Schultz, David V. Wiltschko and Kurt Sternlof. R. Eric Higgins took some of the photographs during our joint work in the field. Juraj Tomana, Chelsea Christensen, Benjamin K. Welker, Douglas G. Jensen, Clay G. Jones, Bree Christensen and Melissa J. Wilkinson all helped with the drafting. Steven P. Clausen, Lea Sýkorová, Chelsea Christensen, Clay G. Jones, Paul D. Jones and Margaret L. Schmidt assisted with references.

PART ONE
Fundamentals of Thrustbelts

1 Introduction to the topic of thrustbelts

For the purposes of this book the term 'thrustbelt' is given a broad meaning to encompass any deformed belt in which contractional or transpressional brittle and brittle/ductile structural styles dominate over other types of structures, including conventional thrustbelts, transpressional ranges, toe thrusts and accretionary prisms (Figs. 1.1–1.6, Tables 1.1–1.6).

Conventional thrustbelts

Conventional thrustbelts evolve out of either passive margin or intracratonic rift systems and their consequent sedimentary basins (Fig. 1.7). Examples of passive margin sediments involved in a thrustbelt are seen in the Appalachians, Andes or Alps. Examples of orogenic belts evolved out of intracratonic rifts are the Atlas Mountains, Palmyrides or the northern Andes. The rift systems, whether of pure extensional or transtensional origin, form the fundamental crustal weaknesses that focus compressional stress and provide the volume of rocks that subsequently become incorporated into the thrustbelt. Nice examples of extensional and transtensional rifts later involved in thrusting come from the

Urals. Their different geometries in relation to the direction of compression determined different structural styles in different parts of the Urals. Passive margin basins, with their broad post-rift sedimentary prisms tapering out onto the nonrifted cratons favour '*thin-skin*' structural styles in which the sedimentary cover strata are detached and deformed independently of the underlying basement (Fig. 1.8). Intracratonic rift systems, on the other hand, tend to produce '*thick-skin*', or basement-involved thrustbelts in which inverted half-grabens or uplifted basement blocks are a dominant feature (Figs. 1.9 and 1.10). However, the distinctions between thin- and thick-skin thrustbelt styles are not rigid. Even in the thin-skin variety, the basal thrust surfaces root within displaced basement elements, many of which can be demonstrated to have been older normal faults (Fig. 1.11). Elements of thin-skin styles are frequently encountered in inverted graben systems, especially where salt or thick shale deposits flank the precursor intracratonic basin. A nice example of the salt thickness controlling thin-skin versus thick-skin structural style comes from the inverted Broad Fourteens basin in the North Sea.

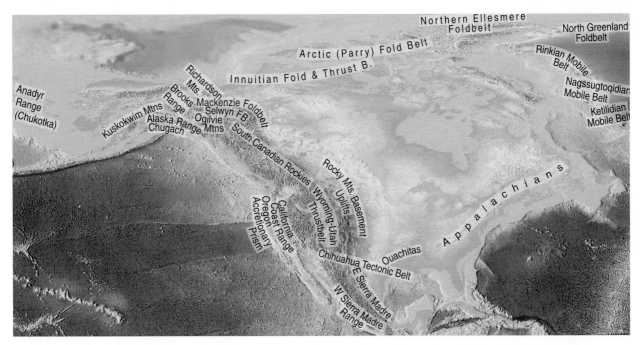

Fig. 1.1. Thrustbelt map of the North American continent. The topographic map is taken from Smith and Sandwell (1997).

Fig. 1.2. Thrustbelt map of the South American continent. The topographic map is taken from Smith and Sandwell (1997).

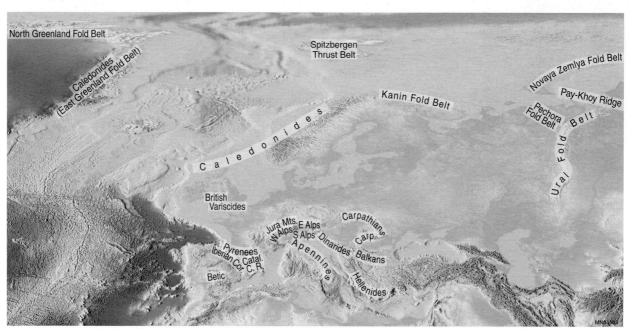

Fig. 1.3. Thrustbelt map of Europe and adjoining North Africa. The topographic map is taken from Smith and Sandwell (1997).

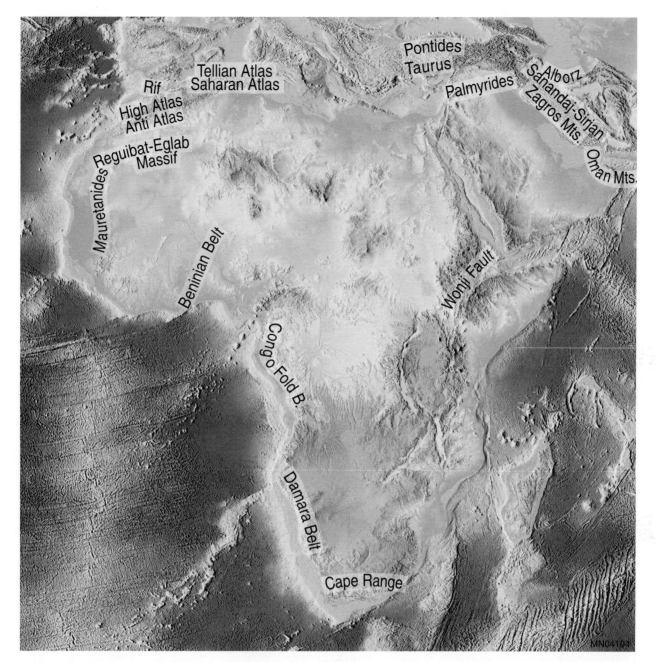

Fig. 1.4. Thrustbelt map of the African continent. The topographic map is taken from Smith and Sandwell (1997).

For a conventional thrustbelt to develop, basement rocks must be shortened somewhere within the width of the belt. In some instances, this involves partial restoration of the extension accompanying rifting, resulting in an inverted rift system (Fig. 1.9). In other cases the shortening results in contractional translation of the original rift elements for very large distances and the generation of thin-skin structural styles where the thrusts cut up into and displace thin slabs of the sedimentary cover (Fig. 1.12). Many thrustbelts exhibit both thin- and thick-skin structural styles in different portions of the belt (Figs. 1.13a and b). Some, such as the Andes, change style along strike (Fig. 1.13a; Allmendinger *et al.*, 1997), whereas others, such as the US Cordillera-Rocky Mountains, exhibit thin-skin styles in their interior and inverted basement styles in their exterior (Fig. 1.13b; Hamilton, 1988), or vice-versa.

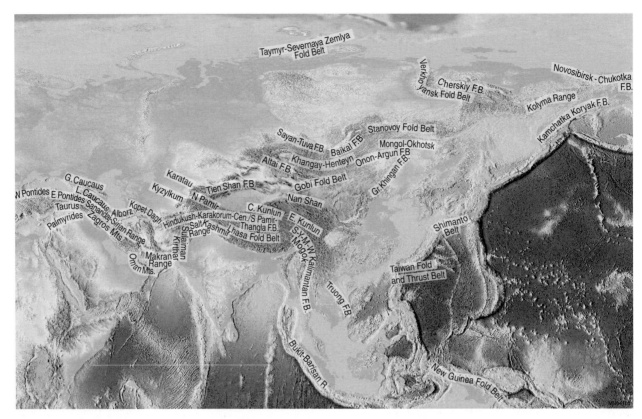

Fig. 1.5. Thrustbelt map of the Asian continent. The topographic map is taken from Smith and Sandwell (1997).

Fig. 1.6. Thrustbelt map of the Australian continent. The topographic map is taken from Smith and Sandwell (1997).

Table 1.1. Names, ages, prevailing structural styles and vergency of thrustbelts on the North American continent.

Thrustbelt name	Thrustbelt age	Structural style	Vergency
Alaska Range	Late Cretaceous–Tertiary (Eisbacher, 1976; Fischer and Byrne, 1990; Dover, 1994; Nokleberg *et al.*, 1994)	Thick-skin	Roughly to N and NW
Anadyr Range (Chukotka)	Late Paleocene–Eocene (Zonenshain *et al.*, 1990; Bocharova *et al.*, 1995)		
Antler thrustbelt	Mississippian	Probably thin-skin	E-vergent (?)
Appalachians	Late Paleozoic (Rodgers, 1995; Mitra, 1987)	Thin-skin	West- to NW-vergent, NE-vergent (Spraggins and Dunne, 2002)
Arctic (Parry) fold belt	Middle Devonian–Early Carboniferous (Harrison, 1995)		
Brooks Range	Two main phases: early Brookian orogeny (Jurassic–Early Cretaceous), late Brookian orogeny (Late Cretaceous–Early Tertiary, Tertiary tectonics predominated in northern Brooks Range, but compressive stresses present until now (Kelley and Foland, 1987; Grantz *et al.*, 1987, 1990, 1994; Dover, 1994; Moore *et al.*, 1994; Plafker and Berg, 1994; De Vera *et al.*, 2001; Wallace, 2003)	Thin-skin	NNE-vergent
California Coast Range	Late Miocene–Quaternary (Medwedeff, 1989)	Transpressional (some structures inverted)	Both W- to SW-vergent and E- to NE-vergent
Chihuahua tectonic belt	Paleocene (Drewes, 1988)		
Chugach–Saint Elias Mts.	30 Ma–present (Plafker and Berg, 1994)	Predominately thin-skin	S-vergent (Plafker and Berg, 1994)
Eastern Sierra Madre Range	Early Tertiary (Davis and Engelder, 1985)		NE-vergent
Innuitian fold and thrustbelt	Probably Late Cretaceous–early Tertiary	Probably thin-skin	SE-vergent
Ketilidian mobile belt	Proterozoic (Stendal and Frei, 2000; Garde *et al.*, 2002)		
Kuskokwim Mountains	Late Cretaceous–Middle Eocene (Oldow *et al.*, 1989; Grantz *et al.*, 1990)		
Mackenzie fold belt	Late Cretaceous–Middle Eocene (Lane, 1996)	Thin-skin	E-vergent
Nagssugtoqidian mobile belt	Proterozoic (Marker *et al.*, 1999; Connelly and Mengel, 2000)		
North Greenland fold belt	Middle Devonian–Early Carboniferous (Harrison, 1995), Late Paleocene–Eocene (Tessensohn and Piepjohn, 1998)		
Northern Ellesmere fold belt	Middle Devonian–Early Carboniferous (Trettin and Balkwill, 1979; Trettin, 1989; Harrison, 1995)	Thin-skin	N-vergent
Ogilvie Mountains (Charley River thrustbelt)	Late Cretaceous–Middle Eocene (Lane, 1996), main phase: Cenomanian–Campanian (Dover, 1994)	Predominately thin-skin	U-shaped belt with SE and NE-vergencies (Dover, 1994)
Oregon accretionary prism	Late Miocene–Quaternary (MacKay *et al.*, 1995)	Thin-skin	W-vergent

Table 1.1 (cont.)

Thrustbelt name	Thrustbelt age	Structural style	Vergency
Ouachitas	Late Mississippian–Early Permian (Arbenz, 1989b; Shumaker, 1992)	Thick- and thin-skin	NW-vergent
Richardson Mountains	Late Cretaceous–Middle Eocene (Lane, 1996)	Thin-skin	E-vergent
Rinkian mobile belt	Proterozoic (Grocott and Pulverlaft, 1990; Kalsbeek *et al.*, 1998)		
Rocky Mountain basement uplifts	Paleocene–Middle Eocene (Oldow *et al.*, 1989)	Thick-skin	Both SW- to W-vergent and NE- to E-vergent
Selwyn fold belt	Late Cretaceous–Middle Eocene (Lane, 1996)		
South Canadian Rockies	Late Jurassic–Eocene (Monger *et al.*, 1982; Brown *et al.*, 1992; Rubin and Saleeby, 1992)	Thin-skin	E-vergent
Western Sierra Madre Range	Cretaceous–Eocene (Weislogel, 1998; Lawton and Giles, 2000)		
Wyoming–Utah thrustbelt (North Sevier thrustbelt)	Late Jurassic–Middle Eocene (Royse *et al.*, 1975; Dixon, 1982; McMechan and Thompson, 1989; Royse, 1993)	Thin-skin	E-vergent

Table 1.2. Names, ages, prevailing structural styles and vergency of thrustbelts on the South American continent.

Thrustbelt name	Thrustbelt age	Structural style	Vergency
Andean Cordillera–northern part	Mesozoic–Tertiary	Thin-skin	S- vergent
Andean Cordillera–N. Peru	Eocene–Quaternary (Lamb *et al.*, 1997; Jaillard *et al.*, 2002)	Thick-skin	NE- to E-vergent
Andean Cordillera–Central and Southern Peru–Bolivia	Eocene–Quaternary	Thin-skin	E-vergent
Austral-Magallenes	Late Cretaceous–Quaternary	Thin-skin	E- and NE-vergent
Barbados accretionary prism	Eocene–Quaternary (Speed *et al.*, 1991; Mascle *et al.*, 2003)	Thin-skin	E-vergent
Brasiliano orogenic belt	Late Proterozoic (Almeida *et al.*, 1981; de Azevedo, 1991)	Thick- and thin-skin	W-vergent
Chilean accretionary prism	Late Miocene–Quaternary (Jaillard *et al.*, 2002)	Thin-skin	W-vergent
Coastal Cordillera	Eocene–Quaternary (Eva *et al.*, 1989; Golonka, 2000)	Thick-skin	S-vergent
Cordillera Central (N. Andes)	Paleogene	Thick-skin	E-vergent
Cordillera Occidental (N. Andes)	Cretaceous–Quaternary (Chigne *et al.*, 1996)		Doubly vergent, both E- and W-ward
Cordillera Oriental (N. Andes)	Late Miocene–Quaternary		Doubly vergent, both E- and W-ward
Espinhaco fold belt	Late Proterozoic (Almeida *et al.*, 1981; de Azevedo, 1991)		
Merida Andes	Miocene (Chigne *et al.*, 1996)	Thick-skin	Doubly vergent, N- and S-ward

Table 1.2. (cont.)

Thrustbelt name	Thrustbelt age	Structural style	Vergency
Peru accretionary prism	Pliocene–Quaternary (Von Huene *et al.*, 1988)	Thin-skin	W-vergent
Sierra de Perija	Middle Miocene–Quaternary	Thin- and thick-skin	S-vergent (doubly vergent?)
Sub-Andean Zone (Central Andes)	Late Oligocene–Quaternary (Baby *et al.*, 1992)	Thin-skin	E-vergent

Table 1.3. Names, ages, prevailing structural styles and vergency of thrustbelts on European continent.

Thrustbelt name	Thrustbelt age	Structural style	Vergency
Apennines	Oligocene–Quaternary (Doglioni, 1993a, b; Pialli and Alvarez, 1997; Albouy *et al.*, 2003)	Thick- and thin-skin	NE- to E- to SE-vergent
Balkans	Middle Jurassic–Oligocene (Sengor and Natalin, 1996; Sinclair *et al.*, 1997; Tari *et al.*, 1997; Vangelov, personal communication, 2003)	Thin-skin in Forebalkan Unit, thin- and thick-skin in Balkan Unit	NNE to NE-vergent
Betic Chain	Oligocene–Miocene (Desegaulx *et al.*, 1991; Vissers *et al.*, 1995)	Thin- and thick-skin	N-vergent
British Variscides	Devonian–Carboniferous (Raoult and Meilliez, 1987; Franke, 1989; Gayer *et al.*, 1993, Gayer and Nemčok, 1994)	Thin-skin	NW-vergent
Caledonides	Silurian–Devonian (Dewey *et al.*, 1993; Harrison, 1995; Torsvik *et al.*, 1996; Milnes *et al.*, 1997)	Thick- and thin-skin	NE-vergent
Carpathians	Cretaceous–Quaternary (Nemčok *et al.*, 1998a; Plašienka, 1999)	Both thick- and thin-skin	NE- to E-vergent
Catalonian Coastal Range	Paleogene (Burbank *et al.*, 1996)	Thick-skin	NW-vergent
Dinarides	Jurassic–Late Early Cretaceous (Yilmaz *et al.*, 1996), Paleocene–Miocene (Doglioni, 1993a, b)	Both thin- and thick-skin	SW-vergent
East Greenland fold belt	Silurian–Devonian (Dewey *et al.*, 1993; Harrison, 1995; Torsvik *et al.*, 1996)		
Eastern Alps	Two main phases: Late Lower Cretaceous–Late Eocene dextral transpression and Miocene sinistral wrenching (Ratschbacher *et al.*, 1991; Linzer *et al.*, 1997)	Thin-skin in Northern Calcareous Alps	NW-vergent and E-vergent during first and second phases
Hellenides	Early Cretaceous (Yilmaz *et al.*, 1996), Oligocene – Quaternary (Davis and Engelder, 1985; Robertson *et al.*, 1991)	Thin-skin	SW-vergent
Jura Mountains	Eocene–Miocene (Butler, 1992; Laubscher, 1986, 1992)	Thin-skin	W-vergent
Iberian Cordillera	Late Paleozoic (Navarro, 1991), Alpine strike-slip faulting (Bergamin *et al.*, 1996), Neogene (Simon Gomez and Cardona, 1988)		
Kanin fold belt	Proterozoic (Valasis and Gornostay, 1989)		
Novaya Zemlya fold belt	Late Proterozoic–Cambrian (Golonka, 2000), Late Carboniferous–Early Jurassic (Zonenshain *et al.*, 1990; Puchkov, 1991, 1997; Nikishin *et al.*, 1996)	Thin-skin	W-vergent

Table 1.3 (cont.)

Thrustbelt name	Thrustbelt age	Structural style	Vergency
Pay-Khoy Ridge	Late Proterozoic–Cambrian (Golonka, 2000), Late Carboniferous–Early Cretaceous (Zonenshain *et al.*, 1990; Puchkov, 1991, 1997; Nikishin *et al.*, 1996)	Thick- and thin-skin	S-vergent
Pechora fold belt	Early Carboniferous–Middle Triassic	Thick-skin	Doubly vergent, both NE- and SW-ward
Pyrenees	Aptian–Oligocene (Roure *et al.*, 1989; Dinares *et al.*, 1992; Munoz, 1992; Puigdefabregas *et al.*, 1992; Verges *et al.*, 1992)	Both thick- and thin-skin, mainly thick-skin	Doubly-vergent, S- and N-ward
Southern Alps	Cretaceous–Pleistocene (Doglioni, 1993a, b; Carminati and Siletto, 1997)	Thick- and thin-skin	S-vergent
Spitzbergen thrustbelt	Paleogene (Bergh *et al.*, 1997)	Thick- and thin-skin	E-vergent
Ural fold belt	Late Devonian–Triassic (Davis and Engelder, 1985; Golonka, 2000)	Thick- and thin-skin	W-vergent
Western Alps	Late Cretaceous–Miocene (Debelmas, 1989; Butler, 1992; Laubscher, 1992; Froitzheim *et al.*, 1996; Kley and Eisbacher, 1999)	Thick-skin	W-vergent

Table 1.4. Names, ages, prevailing structural styles and vergency of thrustbelts on the African continent.

Thrustbelt name	Thrustbelt age	Structural style	Vergency
Anti-Atlas	Cretaceous–Paleogene		
Beninian belt	Late Proterozoic–Ordovician (Clauer *et al.*, 1982)	Thin-skin (?)	W-vergent
Cape Range	Late Carboniferous–Permian (Golonka, 2000)	Thin-skin (?)	N-vergent
Congo fold belt	Proterozoic (Affaton *et al.*, 1995)		NW-vergent
Damara belt	Late Proterozoic–Cambrian (Miller, 1983; Henry *et al.*, 1990; Tankard *et al.*, 1995)		
High Atlas (Morocco)	Cretaceous–Eocene (Beauchamp *et al.*, 1996; Ricou, 1996; Frizon de Lamotte *et al.*, 1998)	Thick-skin	Doubly vergent, N- and S-vergent
Mauretanides	Late Carboniferous–Permian (Clauer *et al.*, 1982; Black and Fabre, 1983; Lecorche *et al.*, 1989)	Thin-skin	E-vergent
Reguibat-Eglab Massif	Archaic–Proterozoic (Feybesse and Milesi, 1994; Potrel *et al.*, 1998)		
Rif	Eocene–Miocene (Morley, 1992, 1993)	Thin-skin	Generally S-vergent
Saharan Atlas	Eocene–Recent (Beauchamp *et al.*, 1996; Ricou, 1996; Frizon de Lamotte *et al.*, 1998)	Thick-skin	S-vergent
Tellian Atlas	Jurassic–Paleogene (Beauchamp *et al.*, 1996)	Thick- and thin-skin	S-vergent

Table 1.5. Names, ages, prevailing structural styles and vergency of thrustbelts on the Asian continent.

Thrustbelt name	Thrustbelt age	Structural style	Vergency
Alborz Range	Late Triassic–Early Jurassic (Golonka, 2000), Eocene–Oligocene (Dercourt *et al.*, 1986)		SW-vergent
Altai fold belt	Late Devonian–Early Carboniferous (Wang *et al.*, 1999; Cunningham *et al.*, 2000; Golonka, 2000)		
Baikal fold belt	Late Proterozoic–Cambrian (Zonenshain *et al.*, 1990), L. Devonian–E. Carboniferous (Cowgill and Kapp, 2001), Late Carboniferous–M Jurassic (S-ward thrusting) (Mattern and Schneider, 2000), M–L. Jurassic (Zorin, 1999), Tertiary–Quaternary (Cunningham *et al.*, 2000)		
Central Kunlun fold belt	Carboniferous–E Jurassic (Delville *et al.*, 2001; Wenjiao *et al.*, 2002), Tertiary–Quaternary (Treloar *et al.*, 1992; Searle, 1996), Paleogene (strike-slips) (Jolivet *et al.*, 2001), Oligocene–Miocene (major compression) (Jolivet *et al.*, 2001)	Thick-skin (Yin and Harrison, 2000)	
Cherskiy fold belt	Late Jurassic–Early Cretaceous (Parfenov *et al.*, 1993; Golonka, 2000)		
East Pontides	Late Carboniferous–Eocene (Ustaomer and Robertson, 1997; Yilmaz *et al.*, 1996; Golonka, 2000)		Roughly N-vergent, later (20 Ma) NE-vergent
Eastern Kunlun fold belt	Triassic–Jurassic (Jolivet *et al.*, 2001), Paleogene (strike-slips) (Jolivet *et al.*, 2001), Oligocene–Miocene (major compression) (Jolivet *et al.*, 2001)		
Gobi fold belt	Late Paleozoic (Zheng *et al.*, 2000), Late Triassic–Jurassic (Davis *et al.*, 2001; Zheng *et al.*, 2000), Tertiary–Quaternary (Cunningham *et al.*, 2000),		
Greater Caucasus	Paleocene–Quaternary (Sobornov, 1996; Golonka, 2000; Schelling *et al.*, 2003)		Doubly vergent, S- to SE-vergent in the south and N-vergent in the north
Great Khingan fold belt	Jurassic accretionary wedge (Gonevchuk *et al.*, 2000), Albian-Cenomanian (Natalin and Chernysh, 1992; Markevich *et al.*, 1996)		
Hindukush–Karakorum–Central and South Pamir-Thangla fold belts	Jurassic–E. Cretaceous (Zanchi *et al.*, 2000), Late Cretaceous–Quaternary (Liu *et al.*, 1996), Tertiary–Quaternary (Treloar *et al.*, 1992; Searle, 1996), Miocene (Hubbard *et al.*, 1999)		Early dextral transpression controlled by SE–NWσ_1, then S- to SE-ward thrusting (Ratschbacher, 2003, personal communication) Cenozoic Central Pamirs: doubly vergent under N–S contraction with sub-vertical and E–W extension, associated with Oligocene–Miocene metamorphism and magmatism South Pamirs: Jurassic to Early Cretaceous emplacement of ophiolite-bearing nappes probably rooting in the Rushan Phart

Table 1.5 (cont.)

Thrustbelt name	Thrustbelt age	Structural style	Vergency
			suture; Miocene transpression related to Karakoram fault Zone (Ratschbacher, 2003, personal communication)
Kamchatka Koryak fold belt	Cretaceous–Miocene (Zonenshain *et al.*, 1990; Golonka, 2000)		
Karatau Range	Proterozoic (Pre-Riphean) (Allen *et al.*, 2001), Silurian-Devonian (Allen *et al.*, 2001), Late Paleozoic (Schelling *et al.*, 1998), Late Triassic–E. Jurassic (inversion) (Schelling *et al.*, 1998), Jurassic–Cretaceous boundary (compression) (Schelling *et al.*, 1998), Neogene (transpression) (Schelling *et al.*, 1998)		
Kashmir		Thin-skin (Davis and Lillie, 1994)	S-vergent (Davis and Lillie, 1994)
Khangay–Henteyn fold belt	Late Paleozoic (Watanabe *et al.*, 1999)		
Kirthar Range	Tertiary–Quaternary (Treloar *et al.*, 1992; Searle, 1996), Pliocene–Quaternary (E-vergent thin-skin) (Schelling, 1999)	Thin-skin (Banks and Warburton, 1986; Shelling, 1999)	E-vergent (Shelling, 1999) E- and W-vergent (Banks and Warburton, 1986)
Kolyma Range	Late Jurassic–Late Cretaceous (Parfenov *et al.*, 1993; Golonka, 2000)		
Kopet Dagh fold belt	Oligocene–Pliocene (Kopp, 1997; Torres, 1997; Lyberis and Manby, 1999; Golonka, 2000)		SE-vergent
Kyzylkum Massif	Silurian–Devonian (W-vergent) (Babarina, 1999), Late Paleozoic (S-vergent) (Babarina, 1999)		
Lesser Caucasus	Paleocene–Pliocene (Golonka, 2000)		NE-ward
Lhasa-Amdo-Shiquenhe fold belt	Cretaceous–Tertiary (Treloar *et al.*, 1992; Searle, 1996), Early Cretaceous (Otto *et al.*, 1997)	Thin-skin, mostly pre-India – Asia collision; *c.* 50% shortening (Murphy *et al.*, 1997; Yin and Harrison, 2000; Kapp *et al.*, 2003)	S-ward (Ratschbacher, 2003, personal communication)
Makran Range	Paleocene–Quaternary (Golonka, 2000)		
Mogok belt	Miocene (Ji *et al.*, 2000)		
Mongol–Okhotsk belt	Late Carboniferous–Permian (Sengor and Natalin, 1996), E–M. Jurassic closure of the Mongol–Okhotsk Ocean (Zorin, 1999), M–L Jurassic (collision) (Zorin, 1999), Jurassic-M Cretaceous (transpression) (Zorin, 1999)		
Nan Shan fold belt	Early Paleozoic (N-ward arc) (Yin *et al.*, 2000), Oligocene–Miocene (Western Nan Shan–thin-skin, Eastern Nan Shan–thick-skin) (Rumelhart *et al.*, 1997; Yin *et al.*, 1999), Tertiary–Quaternary (Treloar *et al.*, 1992; Searle, 1996)		

Table 1.5 (cont.)

Thrustbelt name	Thrustbelt age	Structural style	Vergency
Northern Pamir–Muztagh fold belt	Jurassic–E Cretaceous (Zanchi *et al.*, 2000), Paleogene–Quaternary (Leith and Alvarez, 1985; Hamburger *et al.*, 1992; Treloar *et al.*, 1992; Strecker *et al.*, 1995; Searle, 1996; Arrowsmith and Strecker, 1999; Coutand *et al.*, 2002; Strecker *et al.*, 2003), Late Paleogene–E Neogene (Czassny *et al.*, 1999), Late Oligocene–E Miocene (traps similar to S Andreas system in N Afghan Platform) (Brookfield and Hashmat, 2001), Neogene (Burtman, 2000)	Oligocene–Recent thin-skin thrusting with a dextral strike-slip component; major deformation since the Late Miocene, shortening rates from GPS data up to 2 cm yr^{-1}	
Novosibirsk–Chukotka fold belt	Middle Jurassic–Late Cretaceous (Zonenshain *et al.*, 1990; Bocharova *et al.*, 1995)		
Oman Mountains	Late Cretaceous–Late Miocene (Pillevuit *et al.*, 1997; Golonka, 2000), Late Cretaceous–Paleogene compression, Miocene transpression (own work)	Thick (dominant) and thin-skin (own work)	WNW-vergent on western side (own work)
Onon–Argun fold belt	E–M Jurassic (continent-arc collision) (Zorin *et al.*, 2001), M–L Jurassic (collision) (Zorin, 1999), Jurassic–M Cretaceous (transpression) (Lamb *et al.*, 1999)		
Palmyrides	Late Cretaceous–Paleocene (Golonka, 2000)		
Salt Range	Oligocene–Quaternary (Nizamuddin, 1997), Pliocene-Quaternary (Jaswal *et al.*, 1997)	Thick-skin (Pogue *et al.*, 1999)	S-vergent (Pogue *et al.*, 1999)
Sanandaj–Sirjan Range	Jurassic–Eocene (Guiraud and Bellion, 1996; Ricou, 1996; Sengor and Natalin, 1996; Golonka, 2000)		SW-vergent
Sankiang–Yunnan–Malaya–West Kalimantan fold belt	Late Cretaceous–Paleogene (Williams *et al.*, 1988, 1989)		
Sayan–Tuva fold belt	Late Proterozoic–Ordovician (Zonenshain *et al.*, 1990), Tertiary–Quaternary (Cunningham *et al.*, 2000)		
Shimanto belt (inland of Nankai accretionary prism)	Cretaceous–Tertiary (Ohmori *et al.*, 1997)		
Stanovoy fold belt	Proterozoic (Pre-Riphean) + Late Proterozoic (Vernikovskiy, 1995)		
Sulaiman Range	Cretaceous–Paleocene boundary–Quaternary (Allemann, 1979; Jadoon *et al.*, 1992)	Thick-skin (Davis and Lillie, 1994)	
Taurus	Late Cretaceous–Paleocene (Sengor and Natalin, 1996), Early Miocene (Dercourt *et al.*, 1986)		S- to SW-vergent, young SE-vergent (20 Ma)
Taymyr–Severnaya Zemlya fold belt	Late Proterozoic–Cambrian (Vernikovskiy, 1995, 1996), Late Carboniferous–Middle Jurassic (Zonenshain *et al.*, 1990; Parfenov *et al.*, 1993; Vernikovskiy, 1995, 1996)		

Table 1.5 (cont.)

Thrustbelt name	Thrustbelt age	Structural style	Vergency
Tien Shan fold belt	Early Paleozoic (arc, thrusting) (Shu *et al.*, 1999; Brookfield, 2000), Carboniferous (collision) (Bazhenov *et al.*, 1999; Brookfield, 2000), Late Paleozoic strike-slip faulting (Shu *et al.*, 1999), Late Permian–E Jurassic (transpression) (Bazhenov *et al.*, 1999), Late Paleogene–E Neogene (Czassny *et al.*, 1999), Neogene (Burtman, 2000), Tertiary–Quaternary (Treloar *et al.*, 1992; Cobbold *et al.*, 1993; Searle, 1996; Gao *et al.*, 1998; Allen *et al.*, 1999; Burchfiel *et al.*, 1999)	Paleozoic: thick-skin (Brookfield, 2000)	Paleozoic: N to NE-vergent (Brookfield, 2000), Tertiary–Recent: S-vergent, active since 20 Ma
Verkhoyansk fold belt	Late Jurassic–Late Cretaceous (Parfenov *et al.*, 1993; Golonka, 2000)		
West Pontides	Late Carboniferous–Eocene (Ustaomer and Robertson, 1997, Yilmaz *et al.*, 1996; Golonka, 2000)		Roughly N-vergent
Zagros Mountains	Oligocene–Quaternary (Dercourt *et al.*, 1993; Sinclair, 1997; Vrielynck, 1997), Middle Miocene (Letouzey *et al.*, 2003), Cretaceous–Paleogene (Letouzey *et al.*, 2003)	Thin-skin (Middle Miocene) thick-skin (Cretaceous–Paleogene)	SW-vergent (Middle Miocene)

Table 1.6. Names, ages, prevailing structural styles and vergency of thrustbelts on Australian continent.

Thrustbelt name	Thrustbelt age	Structural style	Vergency
Bukit–Barisan Range	Miocene (Huchon and Le Pichon, 1984; Lee and Lawver, 1994)		
Lachlan fold belt	Late Ordovician–Early Carboniferous (O'Sullivan *et al.*, 1996, 2000; Foster and Gray, 2000)	Thin-skin (Foster and Gray, 2000; Taylor and Cayley, 2000)	N-vergent (Taylor and Cayley, 2000)
New England fold belt	M–L Devonian (Morand, 1993), Carboniferous (subduction accretionary complex) (Skilbeck *et al.*, 1994; Caprarelli and Leitch, 1998)		
New Guinea fold belt	Early Eocene–Oligocene–Quaternary (Medd, 1996; Longley, 1997; Sinclair, 1997), 55 Ma–collision of the New Guinea passive margin with intraoceanic arc, 25 Ma–collision of the New Guinea passive margin with Philippines–Halmahera–New Guinea arc system (Hall, 1997)		N- to NW-vergent
Southern Alps	Pliocene–Quaternary (Beaumont *et al.*, 1992; Barnes, 1996)		
Taiwan fold and thrustbelt	Late Miocene–Quaternary (Suppe, 1981; Davis *et al.*, 1983; Teng, 1990; Sinclair, 1997), evidence concerning Eocene–Oligocene (Hall, 1997; Kong, 1998), after Pliocene (Mou and Su, 2003)	Thin-skin	W-vergent
Truong fold belt	Early Triassic (Lepvrier *et al.*, 1997), E Cretaceous (Thanh, 1998), Aptian–Turonian (Lepvrier *et al.*, 1997), Oligocene–E Miocene (Lepvrier *et al.*, 1997)		

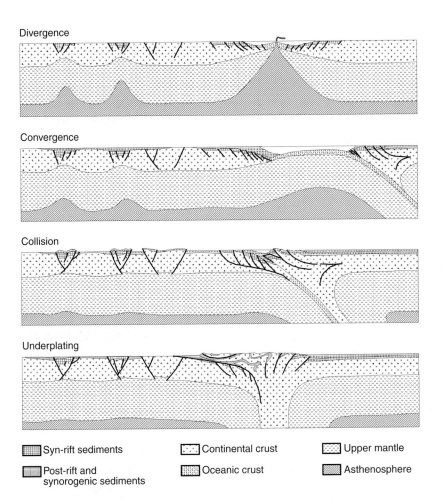

Divergence

Convergence

Collision

Underplating

Syn-rift sediments

Post-rift and
synorogenic sediments

Continental crust

Oceanic crust

Upper mantle

Asthenosphere

Fig. 1.7. Schematic cross sections
through an evolving convergence zone
(Ziegler, 1989).

Transpressional ranges

Transpressional ranges evolve either at oblique plate boundaries where strike-slip movements are added to the overall convergence, or at restraining bends, over-steps and splays of transform or larger transcurrent faults (*sensu* Sylvester, 1988). The South Wales Variscan, the South Trinidad thrustbelt and part of the South Carpathians are all examples of the first category, displaying more contraction than transcurrence in the combined strike-slip faulting and thrusting. The Great Sumatran fault zone, the El Pilar fault system of Northern Venezuela and the Alpine fault in New Zealand are examples of the strike-slip end member in their combined strike-slip faulting and thrusting. The Transverse Ranges of the San Andreas fault system are an example of deformation developed at a restraining bend. Transpressional ranges can have either a thin- or thick-skin character, controlled by the same factors as in conventional thrustbelts. The South Wales Variscan and the Central Sumatra basin are good examples of thin- and thick-skin styles, respectively.

An increasingly significant class of thrustbelt is

expressed as *toe thrusts* at the leading edge of large-scale gravity-driven glide complexes along the outer portions of passive margin sedimentary prisms or in thick, young delta complexes (McClay *et al.*, 2000). Examples of toe thrusts in passive margin prisms come from the Santos basin in Brazil (e.g., Cobbold *et al.*, 1995), Gulf of Mexico (e.g., Worrall and Snelson, 1989; Wu *et al.*, 1990; Rowan *et al.*, 2000), Nova Scotia (e.g., Keen and Potter, 1995), Eastern Mediterranean (e.g., Gaullier *et al.*, 2000), East Africa (Coffin and Rabinowitz, 1988), Northwest Africa (Tari *et al.*, 2002) and Southwest Africa (e.g., Liro and Coen, 1995; Marton *et al.*, 2000). Examples of toe thrusts from young delta complexes are known in the Mahakam or Niger delta (e.g., Ferguson and McClay, 2000; Bilotti and Shaw, 2001). Some of the gliding complexes are detached on salt, like the toe thrusts in the Gulf of Mexico, and others on thick over-pressured shale, such as the Niger delta or northern Brazilian toe thrusts. These thrustbelts neither root into basement, nor do they require lithospheric shortening. They are balanced by the extension in the upper portion of the glide complex and by the down-slope advance of

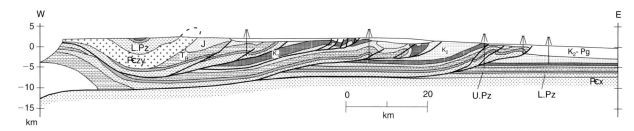

Fig. 1.8. Structural cross section through the Wyoming thrustbelt (after Royse *et al.*, 1975 and Pícha, 1996). PCx, Archean; PCzy, Proterozoic; L.Pz, Lower Paleozoic; U.Pz, Upper Paleozoic; Tr, Triassic; J, Jurassic; K$_1$, Lower Cretaceous; K$_2$, Upper Cretaceous; K$_2$-Pg, Upper Cretaceous–Paleogene.

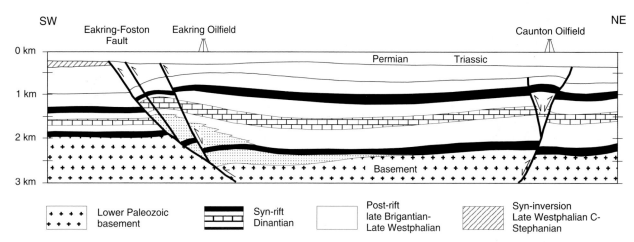

Fig. 1.9. Cross section through the Eakring Oil Field in England (modified from Fraser *et al.*, 1990). A half-graben-bounding normal fault with two synthetic normal faults has been partially inverted. Inversion also caused a pop-up structure of the Caunton Field.

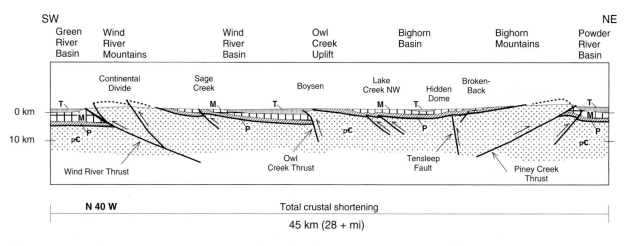

Fig. 1.10. Cross section through the Rocky Mountain foreland uplifts in Wyoming drawn parallel to the direction of shortening (after Brown, 1988). The profile is located in Fig. 1.13(b). The strike of individual basement uplifts controls the amount of shortening. For example compare E–W-striking Owl Creek thrust and Tensleep fault with NW–SE-striking Wind River and Piney Creek Thrusts. pC – Precambrian basement, P – Paleozoic, M – Mesozoic, T – Tertiary.

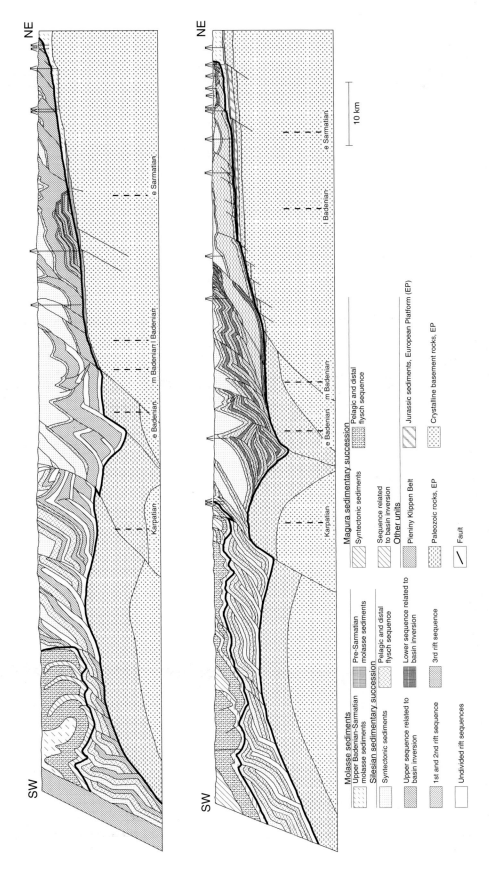

Molasse sediments

Upper Badenian-Sarmatian molasse sediments

Pre-Sarmatian molasse sediments

Silesian sedimentary succession

Syntectonic sediments

Pelagic and distal flysch sequence

Upper sequence related to basin inversion

Lower sequence related to basin inversion

1st and 2nd rift sequence

3rd rift sequence

Undivided rift sequences

Magura sedimentary succession

Syntectonic sediments

Pelagic and distal flysch sequence

Sequence related to basin inversion

Other units

Pieniny Klippen Belt

Jurassic sediments, European Platform (EP)

Paleozoic rocks, EP

Crystalline basement rocks, EP

Fault

10 km

Fig. 1.11. Two profiles through the eastern portion of the West Carpathians, deformed by both thin- and thick-skinned tectonics. The cross-cutting relationships of faults indicate that the thick-skin shortening is very young.

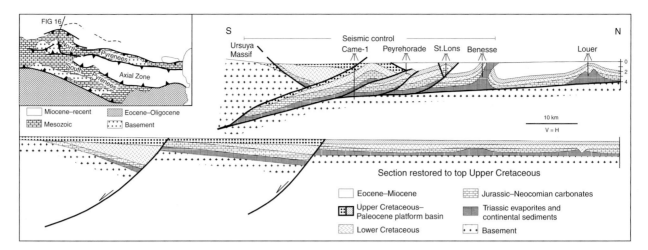

Fig. 1.12. Balanced and restored cross section through the North Pyrenees (Hayward and Graham, 1989). The inset shows the location.

the glide complex. Sediments above the detachment are deformed in a brittle manner. The upper extensional province of the glide complex is located on the inclined basin margin. Extension develops structures such as basinward dipping tear normal faults, salt rollers in footwalls, salt walls along conjugate normal faults, extensional rafts and turtle anticlines (Fig. 1.14). The lower contractional province of the glide complex is located on the flat basin floor. Compressional structures developed here include growth folds, asymmetric salt walls above reverse faults, salt tongues, salt wedges, salt canopies, fold trains and thrust sheets. Both provinces can be quite large and accommodate significant strains. Extensional and contractional domains in the Campos basin accommodate about 100 km of extension and contraction, respectively (Demercian *et al.*, 1993). The total amount of down-slope gliding can also be quite large. The distance travelled by the gliding complex in the Santos basin, determined by comparison of a salt extent map (Ojeda, 1982; Pereira and Macedo, 1990; Demercian *et al.*, 1993) with the location of the continental–oceanic crust boundary (Karner, 2000), is of the order of 60–160 km.

Gravity gliding may not be the only driving force for toe thrusts. Cobbold *et al.* (2001), for example, appealed to Andean compression or possibly ridge push to contribute to the development of the folds and thrusts occurring in the Santos basin in Brazil. The structure of toe thrusts varies considerably (Table 1.7) because it is affected by the interaction of several controlling factors such as: (1) the distribution, thickness and rheology of the layer, which becomes a detachment horizon; (2) the temporal/spatial distribution and the rate of sedimentary loading; (3) the width and dip of the basin margin;

and (4) the presence of gliding prohibitors such as large tilted normal fault footwalls or seamounts. The importance of the detachment material distribution can be documented by comparing the Nova Scotia example with the Southwest African examples. While the syn-rift Upper Triassic–Lower Jurassic evaporite distribution in Nova Scotia forms scattered fault-bounded units and frequently does not allow the development of a larger toe thrust, Aptian post-rift SW African evaporites frequently form continuous and very efficient detachment horizons.

Accretionary prisms
Accretionary prisms develop during the early stages of the plate convergence at continental forearc and island arc margins (Fig. 1.15). They evolve out of sediments detached from the oceanic crust and stacked in thrust sheets in front of the advancing plate overriding the oceanic plate. Along most of these margins, a large portion of the ocean-basin layer consisting of pelagic and hemipelagic sediments is subducted. This layer contains roughly up to 50% oceanic section. The other 50% is composed of material derived from the overriding plate. Examples come from the Peruvian continental margin (Kukowski *et al.*, 1994), the Barbados accretionary wedge (Westbrook *et al.*, 1988; Ferguson *et al.*, 1993) and the Nankai accretionary prism (Morgan and Karig, 1995). The largest difference between accretionary prisms and conventional thrustbelts is a lack of sediment lithification, i.e. a lack of rheological contrast among various sedimentary layers in the deformed section.

There are two extreme types of such continental forearc or island arc margins: accretionary and erosional (*sensu* von Huene and Scholl, 1991, 1993).

(a)

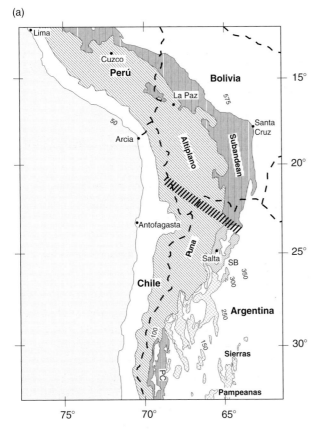

(b)

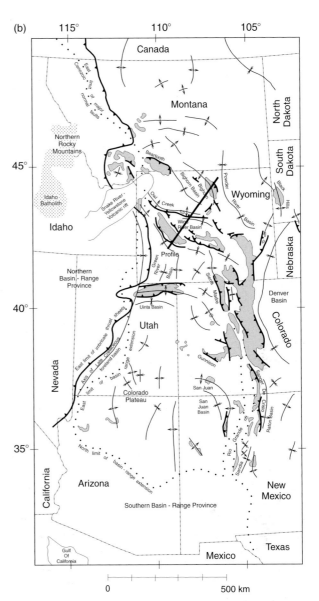

Fig. 1.13. (a) Map showing the difference in structural style along the eastern side of the Peruvian to Argentinean portion of the Andes (modified from Allmendinger *et al.*, 1997). Oblique hachure shows area higher than 3 km above the Sea level. Vertical hachure indicates thin-skin thrustbelts in the sub-Andean ranges of Bolivia and the Precordillera (PC) in Argentina. Stippled pattern shows thick-skin foreland deformation in the Sierras Pampeanas and Santa Barbara System (SB). The NW–SE hachured zone located roughly along the Bolivian-Argentinean border is the boundary between the Altiplano and Puna. (b) Schematic map of the Rocky Mountain region indicating the deformation front of the thin-skin thrustbelt and the location of most prominent basement uplifts in the adjoining foreland (after Hamilton, 1988). Most of the foreland basins and uplifts are of Laramide age, i.e. very late Cretaceous to early Paleogene. The outcrop of Precambrian rocks in front of the thin-skin thrustbelt to the north of the Basin and Range province are shaded.

The accretionary type characterizes margins, which are currently growing by sediment accretion. Typical examples are the Aleutian, Barbados or Oregon accretionary prisms (Gutscher *et al.*, 1996; Lallemand *et al.*, 1995; Table 1.8).

The erosional type characterizes margins, which are either stable or shrinking by the removal of the arc material. Typical examples are the North Chilean, North

Kermadec or Tonga accretionary prism (Lallemand *et al.*, 1994; Lallemand, 1995; Table 1.8).

In both types, the thickness of sediment bypassing the margin core is about the same. Lallemand (1995) describes a thickness of about 1 km. The convergence rate and subducted sedimentary flux along erosional margins is 2.5 times higher than the rate along accretionary margins. Sandbox experiments that simulated both margin types indicate significant differences in kinematics and structural styles of accretionary and erosional wedges (e.g. Lallemand *et al.*, 1994; Gutscher *et al.*, 1996). The main differences are in the fault spacing and the rate of advance of the deformation front. The other important difference is the variation of the frontal slope dip.

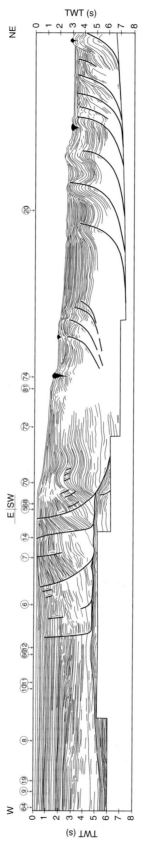

Fig. 1.14. Coupled gravity-collapse and toe-thrust structures in the offshore Eastern Venezuela basin; line tracing of regional seismic profile (redrawn after Di Croce, 1995).

Table 1.7. Characteristics of toe thrusts detached on evaporites.[a]

Toe thrust	Evaporite setting	Basin width (km)	Water depth (km)	Active	Fold train: asymmetry	Fold train: wavelength (km)	Fold train: max amplitude (km)	Fold train: fold envelope
Safi Haute Mer, Morocco	Syn-rift	120	4–5	Yes	—	—	—	—
Ras Tafelney, Morocco	Syn-rift	200	1.5–2.5	No	Yes	3–5	0.5	Subhor
Corisco, Eqatorial Guinea	Post-rift	140	1.5–2	Yes	—	—	—	—
Lower Congo basin, Gabon	Post-rift	250	2.8–3.2	No	Yes	2–4	0.5	Subhor
Lower Congo basin, Angola	Post-rift	350	3.2–4	Yes	Yes	2–5	0.8	Basin-ward dip
Kwanza basin, Angola	Post-rift	300	2.5–3.5	Yes	Yes	2–5	0.6	Basin-ward dip
W Somali basin, Kenya	Syn-rift	200	1.5–2.5	No	Yes	6–8	0.8	Subhor
Majunga basin, Madagascar	Syn-rift	150	1–3	No	No	8–10	1.2	Subhor
Mississippi Fan foldbelt, USA	Syn-rift	800	2.2–3	No	Yes	5–10	2	Subhor
Perdido Fan foldbelt, USA	Syn-rift	800	2.8–3.2	No	No	10–15	2.5	Basin-ward dip

Note:
[a] Tari *et al.* (2002).

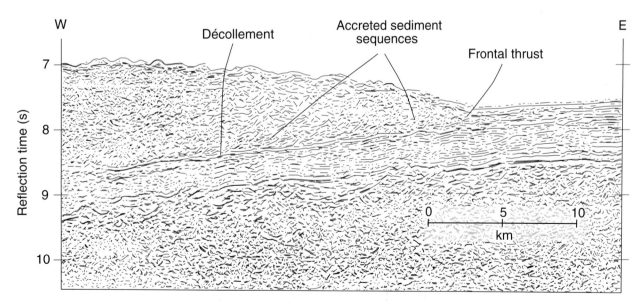

Fig. 1.15. Migrated reflection seismic section through the Barbados accretionary prism (Westbrook *et al.*, 1988). The décollement at the base of the wedge is indicated by high-amplitude seismic reflectors. The décollement is predominantly stratigraphically controlled, but it appears to cut down through section in places and truncates some underlying horizons. Above the décollement, patches of finely layered reflectors may represent bedding of accreted sediments.

Table 1.8. Mean tapers and other important parameters measured for 28 natural transects.[a]

Name	Symbol	Latitude	Longitude	Alpha (deg.)	Beta (deg.)	Trench sediment thickness (km)	Convergence rate (cm yr⁻¹)	Input flux (10 to −1 to 2 km Myr⁻¹)	Vertical offset oceanic faults (km)	Seafloor roughness
Typical accretionary wedges										
S Barbados	A	12°N	57°W	0.9	3.2	8	2	16	0	1
Barbados	A	13°N	58°W	1.1	1.6	4.5	2	9	0	1
Martinique	A	15°40′N	59°W	0.9	3.6	1	2	2	0	1
Guadeloupe	A	16°12′N	59°W	0.6	1.5	0.9	2	1.8	0	1
Barbuda	A	18°N	60°W	1.1	4.1	0.8	2	1.6	0	1
Hikurangi	A	41°S	178°E	0.7	2.6	1.7	4	6.8	0	1
Nankai	A	32°N	134°E	4	5.5	1.7	3.5	6	0	1
Oregon Central	A	47°N	126°W	1.1	2.7	2	3	6	0	1
Aleutian	A	51°N	175°W	3.4	3.9	1	6.5	6.5	0.3	1
Intermediate accretionary wedges										
Manila	I	20°N	120°E	4.5	7.5	1.5	0	0	0	13
Sumba	I	11°S	119°E	1.5	6.7	2	7.5	15	0	13
S Kermadec	I	31°S	177°W	3.7	6.8	0.4	7	2.8	1.5	13
Kashima	I	36°N	144°E	2.5	4.6	0.9	10.5	9.5	0.7	13
Japan 37°	I	37°N	144°E	3.1	4.9	0.9	10.5	9.5	0.5	13
Japan 39°40″	I	39°40′N	144°E	5.3	5.3	0.4	10.5	4.2	0.4	13
Japan 40°10″	I	40°10′N	144°E	6	6.9	1.5	10.5	15.8	0.4	13
Japan 40°40″	I	40°40′N	144°E	4.4	4.1	0.7	10.5	7.4	0.4	13
S Kurile	I	42°N	146°E	1.3	9.7	0.5	7.5	3.8	0.3	13
Peru	I	9°S	84°W	4.5	6.6	1	10	10	0.7	13
New Hebrides	I	16°S	167°E	4.3	2.3	0.3	12	3.6	0.3	13
Nonaccretionary wedges										
New Britain	N	7°S	151°E	4.1	17.3	0.4	12	4.8	0.3	15
Tonga 19°	N	19°S	173°W	5.7	13.6	0.4	17	6.8	0.5	15
Tonga 20°	N	20°S	173°W	9	15.9	0.4	17	6.8	0.5	15
Tonga 23°	N	23°S	174°W	14.3	13.2	0.4	16.5	6.6	1	15
Osbourn	N	26°S	176°W	3.2	16.2	0.4	8	3.2	1	15
N Kermadec	N	28°S	176°W	5.9	12.7	0.4	9	3.6	1	15
N Chile	N	21°S	71°W	8.5	12.3	0.4	9.5	3.8	0.5	15
Costa Rica	N	10°N	86°W	4.5	3.8	0.6	4.5	2.7	0.6	15

Note:
[a] Lallemand *et al.* (1994).

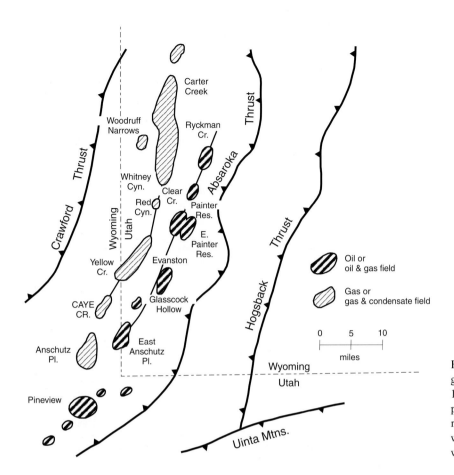

Fig. 1.16. Map of the Absaroka oil and gas fields (modified from Lamerson, 1982 and Warner, 1982). All significant petroleum of the Wyoming thrustbelt is restricted to the Absaroka thrust sheet, with the majority occurring in hanging wall traps.

Petroleum systems in thrustbelts

Petroleum systems in thrustbelts vary. In conventional 'thin-skin' thrustbelts hydrocarbon accumulations are generally limited to the first, second, or less commonly, third thrust sheets. Furthermore, oil and biogenic gas fields tend to be situated in an outer belt near the front of the thrustbelt, with thermogenic gas fields lying within a more internal belt. This pattern is ideally displayed in the southern Wyoming–Utah thrustbelt (Fig. 1.16), but is recognized in many other regions as well (i.e., southern Canadian Cordillera, Eastern Venezuela) where the thrustbelt is developed in an internally thickening passive margin succession. Beyond the first or second thrust sheet, the allochthonous source rocks are generally capable of generating only gas, if they are not already completely 'burned out'. The higher levels of maturity in the interior of these thrustbelts is due in part to the degree of pre-thrust burial in the passive margin prism, and in part to the fact that the source rock commonly is a detachment unit left behind in the footwall of the thrustbelt taper. In contrast, if the frontal thrusts carry broad sheets detached on salt, as in the Potwar Plateau, Zagros and Central Appalachians, the zonation of oil and gas fields may be less obvious due to the greater width of the region of displaced, but little deformed, strata.

Oil and gas fields are usually less predictably distributed in thick-skin thrustbelts, especially where the individual thrust elements are tied to inverted half-grabens. The generation, migration and entrapment of hydrocarbons may be controlled by local conditions preceding basin inversion or by the specific magnitude of the inversion. The one pattern that is predictable, however, is that the inversion structures will be aligned preferentially along the original rifted margins of the graben or pull-apart basin system.

Traps in the thrustbelt can be filled following the cessation of thrusting if the hydrocarbon kitchens continue to operate, but the migration pathways are likely to be complex fracture networks within the volume of the thrust sheets, not the thrust surfaces themselves. In these situations, vertical migration is favoured. Geological events external to the specific portion of the thrustbelt under consideration are generally required to continue burial and hydrocarbon generation after the close of thrusting. Most commonly, it is burial beneath the foredeep debris of another thrust system or beneath a successor basin formed by orogenic collapse that

Table 1.9. Giant oil accumulations in thrustbelt or inversion settings.[a]

Field name	Thrustbelt	Oil-in-place ($\times 10^9$ bbl)	Depth (m)	API (°)	Age of emplacement	Evidence for destruction
Gach Saran	Zagros	53	1036	31	Neogene	Oil seep
Marun	Zagros	52	2865	33	Neogene	Gas seep; large gas cap
Kirkurk	Zagros	38	800	36	Maast–Paleocene	Oil seep
Ahwaz	Zagros	36	2484	32	Neogene	
Agha Jari	Zagros	28	2268	34	Neogene	Gas seep
Rag-e-Safid	Zagros	22	1900	29	Neogene	Gas seep; large gas cap
Paris	Zagros	15	1423	39	Neogene	Gas seep; large gas cap
Bai Hassan	Zagros	15	1400	34	Neogene	
West Qurna	Zagros	14	2100		Neogene	
Naft-e-Safid	Zagros	12	1500	37	Neogene	Oil seep; large gas cap
Carito	E. Venezuela	12	2000	31	Neogene	
Kupal	Zagros	12	3200	33	Neogene	
Furrial	E. Venezuela	11	2000	26	Neogene	
Halfayah	Zagros	11	3800		Neogene	
Abu Ghirab	Zagros	10	2987	24	Maast–Paleocene	
Bibi Hakimeh	Zagros	10	1300	30	Neogene	Gas seep; large gas cap
Sassan	Zagros	10	2500		Neogene	
Haft Kel	Zagros	9	2000	38	Neogene	Oil seep
Uzen	Trans-Caspian	8	1050	35	Maast–Paleocene	
Malgobek	Caucasus	8	550	28	Eo-Oligocene	Large gas cap
Delhluran	Zagros	8	3627	31	Neogene	
Masjid-I-Suleiman	Zagros	7	400	39	Neogene	Oil seep
Karanj	Zagros	7	2300	34	Neogene	Gas seep; large gas cap
Midway-Sunset	CA Coast Range	6	200	18	Neogene	Oil seep
Huntington Beach	CA Coast Range	6	1500	18	Neogene	Large gas cap
Saddam	Zagros	6	2100		Maast–Paleocene	
Pazanan	Zagros	6	2210	38	Neogene	Gas seep; large gas cap
Usa	Pechora	6	1200	36	Mid–Late Permian	
Cusiana	North Andes	6	4200	50	Neogene	Large gas cap
La Paz	North Andes	5	1250	30	Neogene	Oil seep
Kuh-e-Mund	Zagros	5	578	11	Neogene	Oil seep
Coalinga	CA Coast Range	5	600	16	Neogene	Oil seep
Shushufindi	North Andes	4	2979	30	Eo-Oligocene	
Sarkhan	Zagros	4	2300		Neogene	
Jambur	Zagros	4	2400	38	Maast–Paleocene	
Jabal Fauqui	Zagros	4	3000	26	Maast–Paleocene	
Buzurgan	Zagros	4	3000	23	Maast–Paleocene	
Ventura Avenue	CA Coast Range	4	1500	30	Neogene	Oil seeps
Long Beach	CA Coast Range	3	2200	23	Neogene	
Vozey	Pechora	3	2000	36	Mid–Late Permian	
Neft Dashlary	Caucasus	3	900	28	Neogene	Oil seeps
Musipan	E. Venezuela	3	2000	22	Neogene	
Tikrit	Zagros	3	2499	23	Maast–Paleocene	
Khabbaz	Zagros	3	2100		Maast–Paleocene	
Sarvestan	Zagros	3	2300		Maast–Paleocene	
Hamrin	Zagros	3	2400		Maast–Paleocene	

Table 1.9 (cont.)

Field name	Thrustbelt	Oil-in-place ($\times 10^9$ bbl)	Depth (m)	API (°)	Age of emplacement	Evidence for destruction
Karamay	Junggar	3	300	15	Mid–Late Jurassic	Oil seeps
Elk Hills	CA Coast Range	3	2400	35	Neogene	Large gas cap
Buena Vista	CA Coast Range	3	1500	28	Neogene	Large gas cap
Cano Limon	North Andes	3	2316	29	Neogene	
La Cira-Infantas	North Andes	2	400	23	Neogene	Oil seeps
Sacha	North Andes	2	28	28	Eo–Oligocene	
Santa Fe Springs	CA Coast Range	2	1400	35	Neogene	Oil seeps
Stargrozny	Caucasus	2	500	38	Maast–Paleocene	Oil seep; large gas cap
Nur	Zagros	2	2100		Neogene	
Eman Hassan	Zagros	2	2300		Maast–Paleocene	
Jufeyr	Zagros	2	2300		Neogene	
Nargesi	Zagros	2	2300		Neogene	
Siah Makan	Zagros	2	2300		Neogene	
Rangely	Rocky Mountains	2	1800		Eo–Oligocene	Oil seeps
Salt Creek	Rocky Mountains	2	520	37	Late Cretaceous	Large gas cap
Belridge South	CA Coast Range	1	500	13	Neogene	Oil seeps
Yarega	Pechora	1	160	10	Mid–Late Permian	Oils seeps; hydrodynamic effects
Elk basin	Rocky Mountains	1	1100		Late Cretaceous	Oil seeps

Note:
[a] Macgregor (1996).

keeps the hydrocarbon kitchens active for some time after the end of thrusting. For instance, it is burial beneath about 2 km of sediments shed into the Green River basin from the flanking Paleocene–Eocene Rocky Mountain uplifts that was responsible for the post-thrust generation of hydrocarbons from Lower Cretaceous foredeep basin source rocks in the Late Cretaceous Wyoming–Utah thrustbelt. Without the superposed post-thrust burial, this extremely prolific thrustbelt would have been oil- and gas-poor. The timing of the burial of the source rocks before, during and after thrusting is critical.

Although the size and geometries may differ slightly, the types of traps that dominate in both thin- and thick-skin thrustbelts are the same: hanging wall anticlines and fault-sealed footwall traps. Very gentle, doubly plunging frontal or inversion anticlines form the largest traps, and by far are the most desirable exploration targets. In many of the young thrustbelts, the quality of the seal may be very poor, but the active petroleum systems continue to keep the leaky traps filled to spill. However, the presence of gas and condensate in Paleozoic thrustbelts worldwide attests to the possibility of having high-quality shale and diagenetic seals, even in the deformed rocks of a thrustbelt.

Most diagenetic and mechanical processes in thrustbelts have a negative effect on reservoir quality. Only fracturing and dissolution can improve reservoir quality. The critical factors in predicting reservoir quality are the effects of compaction and cementation before deformation, and the effects of pressure solution and fracturing during thrusting. Thrustbelts with the lowest reservoir risk are those that are the least deeply buried (minimal compaction and cementation) and with the least tectonic shortening (minimal internal strain and pressure solution). However, a relatively low risk will also be associated with sites within a thrustbelt that are likely to be intensely fractured. Local high effective stress, fine grain size and low initial (pre-thrust) porosity favour such sites.

Long-term preservation of oil in thrustbelt traps would appear to be difficult. The processes that lead to post-entrapment destruction of oil accumulations include erosion, fault leakage, gas flushing and biodegradation. Of the 350 giant oil fields identified by Macgregor (1996), 64 (18%) were in thrustbelt or inverted foreland basin settings. Half of these thrust-related fields had evidence of post-entrapment destruction (Table 1.9). Not surprisingly 64% of the fields had been charged in the Neogene, and nearly all had been charged since the Mid-Cretaceous.

2 Mechanics of thrust wedges

The kinematics and dynamics of thrustbelts have been investigated through a range of two-dimensional models based on the development of thin-skinned thrust wedges by horizontal shortening of sediments. All of them include the effects of gravity and topography, and describe thrustbelts as wedges detached along a basal décollement, which is either pushed foreland-ward by a rigid buttress in the hinterland (e.g., Chapple, 1978; Cowan and Silling, 1978; Davis *et al.*, 1983; Stockmal, 1983; Dahlen *et al.*, 1984; Lehner, 1986; Platt, 1986; Mandl, 1988; Molnar and Lyon-Caen, 1988; Dahlen and Barr, 1989; Beaumont *et al.*, 1994; Strayer *et al.*, 2001) or is pulled by the subducting plate (Silver and Reed, 1988; Barr and Dahlen, 1989; Willett *et al.*, 1993).

The earliest thrustbelt model described wedge material as being ideally plastic (Chapple, 1978). This material underwent shortening as it advanced in a piggy-back mode, from the back to the front of the wedge. A surface slope, with a critical taper was created, allowing the wedge to experience the plastic yield state and to slide along its décollement towards the foreland.

Subsequent thrustbelt models described deforming material as noncohesive, frictional Coulomb wedges (Davis *et al.*, 1983; Dahlen, 1984) or frictional plastic wedges (Mandl, 1988). Although these models and the previous one differ in the behaviour of the deforming material and the yield conditions, they all describe thrust wedges as tapering toward the foreland. In all of these cases, the wedge would advance if the sum of the basal and topographic slopes reached a critical angle (Fig. 2.1). Other common properties include hinterland-dipping décollement, below which there is little deformation and significant horizontal compression in the wedge material. Finite-element and experimental models show that a proportionately large convergence is necessary for the build-up of a critically tapered wedge compared to the thickness of the deformed material (Davis *et al.*, 1983; Harry *et al.*, 1995; Gutscher *et al.*, 1996). In naturally occurring thrustbelts, after approximately 150–200 km of shortening, topography develops to cause a balance among rock strength, gravitational forces and shear stress along the décollement to dominate the deformation. Roughly, after this amount of shortening pre-existing mechanical heterogeneities exert

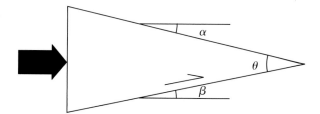

Fig. 2.1. Geometry of an orogenic wedge (Mitra and Sussman, 1997). Wedge taper, θ, consists of the surface slope, α, and basal décollement dip, β.

less influence on the thrustbelt development. A similar but less precise quantification can be implied from the taper and shortening data of the Rhenohercynian thrustbelt of the Central European Variscides, which achieved stable sliding with a taper oscillating around the critical value only during the final stage of contraction (Plesch and Oncken, 1999).

Two-dimensional Coulomb wedge model

In the following section the basic force balance within the thrust wedge based on a noncohesive, frictional Coulomb wedge model is outlined.

The Coulomb wedge deforms until it attains a critical taper. A critically tapered wedge that accretes no new material is the thinnest possible thrustbelt that can advance over its décollement without further internal deformation. The entire volume of such a critically tapered wedge is on the verge of shear failure. A critically tapered wedge that accretes new material deforms internally in order to accommodate the additional material and maintain its critical taper (Davis *et al.*, 1983). Such a wedge thickens with progressing deformation and its advances from the rigid buttress in the hinterland. The rigid buttress is formed commonly by either an island arc or the inner thicker part of the mountain range. The buttress is generally stronger than the wedge material because it is thicker. However, it is never completely rigid in nature. There is field evidence (e.g., Westbrook *et al.*, 1988; Nemčok *et al.*, 1998; Guliyev *et al.*, 2002; see Fig. 2.2; Sperner *et al.*, 2002) and experimental models (Malavieille, 1984; Bonini *et al.*, 1999; Storti *et al.*, 2000; Persson, 2001; Persson and Soukotis, 2002) indicating internal deformation of the

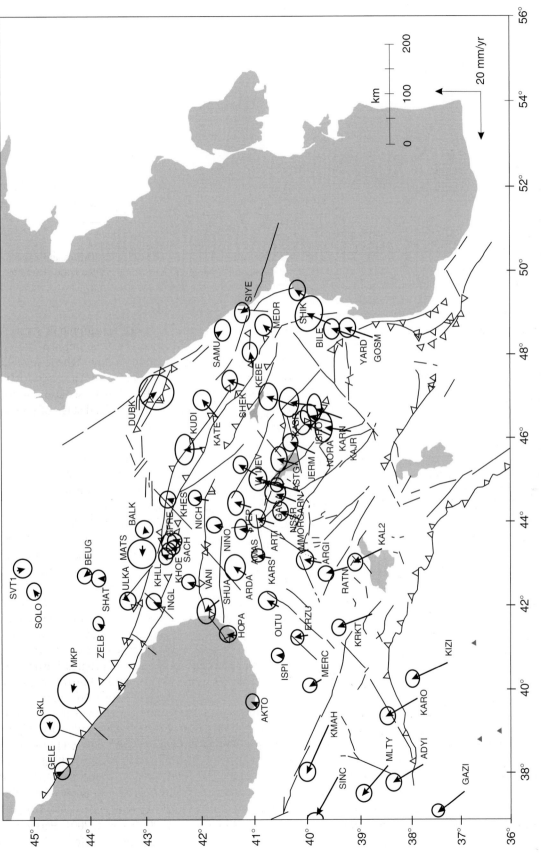

Fig. 2.2. Map of the present-day horizontal motion vectors of the broader Caucasus region determined from Global Positioning System (GPS) data, related to the collision of Eurasian and Arabian Plates (Guliyev *et al.*, 2002). The arcuate thrust in the lower left corner outlines the front of the Arabian indenter. Area to the northwest of the tip of the indenter is the escaping eastern Turkey. Its horizontal motion vectors are larger than motion vectors of the Arabian front, indicating internal deformation of the Arabia in its frontal portion.

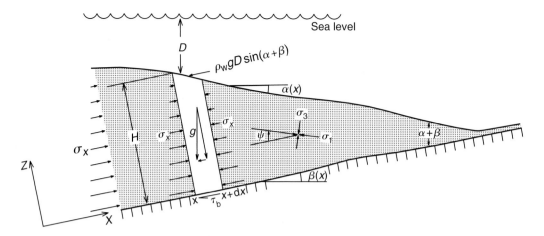

Fig. 2.3. Sketch of the noncohesive, frictional Coulomb wedge subject to horizontal compression at the verge of Coulomb failure throughout (Davis *et al.*, 1983). The figure shows the force balance on an arbitrary column of width dx. σ_x is the orogenic stress caused by the forelandward push of the rigid buttress. σ_1 and σ_3 are the maximum and minimum principal compressional stresses operating inside the wedge. ψ is the angle between the décollement and σ_1. Surface slope, α, and basal décollement dip, β, form the wedge taper. τ_b is the basal shear traction, g is the acceleration of gravity and ρ_w is the water density. H is the thickness of the arbitrary wedge column and D is the water column above it.

buttress. Rock deformation inside the wedge exhibits pressure-dependent, time-independent Coulomb behaviour (e.g., Byerlee, 1978; Paterson, 1978):

$$|\tau| = c + \mu(\sigma_n - p), \tag{2.1}$$

where τ is the shear stress, c is the cohesion, μ is the coefficient of friction, σ_n is the normal stress and p is the pore fluid pressure. Among these factors, cohesion can be relatively unimportant in siliciclastic accretionary wedges and thrustbelts, as indicated by laboratory measurements (e.g. Hoshino, 1972). Pore fluid pressure plays an important role in influencing the effective normal stress acting on the detachment and strength of the wedge material (e.g., Strayer *et al.*, 2001; Saffer and Bekins, 2002). In the case of a subaerial wedge, the water table may be located at the topographic surface. In this case the pore fluid pressure is directly related to the hydrostatic regime, but, below some stratigraphically controlled point, can rise above the hydrostatic value. When the water table is below the surface, fluid pressures are reduced and the wedge material is stronger. The wedge material in submerged wedges is typically weak.

Critical tapers of subaerial and submerged wedges are controlled by the balance of three and four forces, respectively (Davis *et al.*, 1983):

(1)   gravitational body force exerted by the wedge material;
(2)   gravitational body force exerted by the overlying water (in the case of submerged wedges);

(3)   frictional resistance to sliding along the décollement; and
(4)   compressive push (Fig. 2.3).

The simplified critical taper equation (see Davis *et al.*, 1983) is

$$\alpha + \beta = \frac{(1 - \lambda)\mu_b + (1 - \rho_w/\rho)\beta}{(1 - \rho_w/\rho) + (1 - \lambda)K}, \tag{2.2}$$

where α is the dip of the topographic wedge surface, β is the dip of the detachment, λ is the ratio between pore fluid pressure and lithospheric pressure, μ_b is the basal coefficient of friction, ρ_w is the density of water, ρ is the density of the wedge material and K is a dimensionless quantity defined as

$$K \cong \frac{\sin\phi}{1 - \sin\phi} + \frac{\sin^2\phi_b + \cos\phi_b(\sin^2\phi - \sin^2\phi_b)^{1/2}}{\cos^2\phi_b - \cos\phi_b(\sin^2\phi - \sin^2\phi_b)^{1/2}}, \tag{2.3}$$

where ϕ is the angle of internal friction for the wedge material and ϕ_b is the angle of basal friction. Equations (2.2) and (2.3) describe the critical taper $\alpha + \beta$ of the wedge. The variation of K with μ and μ_b is shown in Fig. 2.4. An increase in basal friction increases the critical taper. This relationship has also been shown by experimental modelling (Gutscher *et al.*, 1996). An increased effective internal friction leads to decreased critical taper. Internally stronger wedges are thinner than weaker ones.

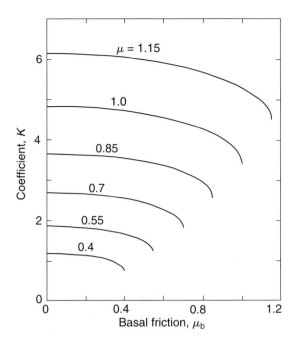

Fig. 2.4. Sensitivity of coefficient K to basal and apparent internal coefficients of friction μ_b and μ (Davis *et al.*, 1983). If μ_b is small in comparison with μ, $K \cong 2 \sin \phi/(1 - \sin \phi)$, but as μ_b approaches μ, K decreases significantly and $\partial K/\partial \mu_b \rightarrow -\infty$.

The equation for the basal shear traction (τ_b) is (Davis *et al.*, 1983)

$$\tau_b = (\rho - \rho_w)gh\alpha + (1 - \lambda)K\rho gh(\alpha + \beta), \qquad (2.4)$$

where g is the acceleration due to gravity and h is the local wedge thickness measured along the z axis. This equation describes the frictional sliding resistance along the décollement counteracted by two deformation-driving forces (the two terms on the right-hand side of the equation). The first term on the right is the gravity acting on the dipping topographic surface of the wedge. The second term on the right, depending on the wedge taper $\alpha + \beta$, is a horizontal push, acting from the rear of the wedge.

The described critical taper theory explains the wedge kinematics based on a simple comparison of its taper with its calculated critical taper. If the wedge taper is below the critical value, the rear of the wedge reacts by shortening and thickening in order to achieve the critical taper value necessary for thrustbelt advance. Such a thickening can be achieved by:

(1) contractional inversion of former basins;
(2) out-of-sequence thrusting along pre-existing thrusts;
(3) back-thrusting; or
(4) duplexing along the décollement.

The regional thrustbelt dynamics reacts by long-term deformation events but short-term deformation events are also common, related to local differential wedge adjustments (e.g., DeCelles and Mitra, 1995; Braathen *et al.*, 1999).

The Spitsbergen thrustbelt provides an example, in which the wedge increases its taper in its hinterland portion by the inversion and uplift of the Carboniferous St Jonsfjorden Trough, following the development of the initial low-taper contractional wedge (Braathen *et al.*, 1999).

Examples of out-of-sequence thrusting or back-thrusting increasing the taper in the orogenic hinterland can be found, for example, from the Alps (Hurford *et al.*, 1989; Sinclair and Allen, 1992) or West Carpathians (Sperner *et al.*, 2002).

The Sevier orogenic wedge provides a wealth of data on wedge taper adjustments reverting back to critical taper after major advance periods (e.g., DeCelles and Mitra, 1995). The overall wedge development can be characterized by forelandward progression of episodic thrusting and related folding (e.g., Royse *et al.*, 1975; Dixon, 1982; Coogan, 1992a, b; Royse, 1993; Fig. 2.5), dated by unconformities and syn-orogenic conglomerates (Fig. 2.6). After each wedge advance, the wedge taper was maintained by shortening in its rear part, which was achieved by growth of the large antiformal duplex in the Wasatch culmination (DeCelles and Mitra, 1995; Fig. 2.5). Similar duplex or antiformal stack adjustments of the wedge taper were described from the Southern Alps (Schonborn, 1992) and from the Appalachians (Mitra and Sussman, 1997).

If the wedge taper is above the critical value, the wedge reacts: by increasing its length with the accretion of new material at its front in order to achieve the critical taper; or by extension; or by collapse near the top of the wedge, where the extension is triggered by gravitational instabilities.

An example of a wedge increasing its length can be documented from the West Carpathians where the thrust-sheet advance was impeded by an elevated basement horst. Internal thrusting in the wedge caused the taper to increase and enabled the wedge to climb over the basement step (Fig. 2.7). It then advanced onto the Middle Badenian evaporitic horizon, which resulted in a taper decrease in the area above the evaporitic horizon and lengthening of thrust sheets (Nemčok *et al.*, 1999). The propagation of the Carpathian thrust system during the Middle Badenian is also indicated by a pronounced increase of the advance rate (Fig. 2.8). Similar examples are known from the Southern Pyrenees where the southwestward advancing portion of the thrustbelt reached

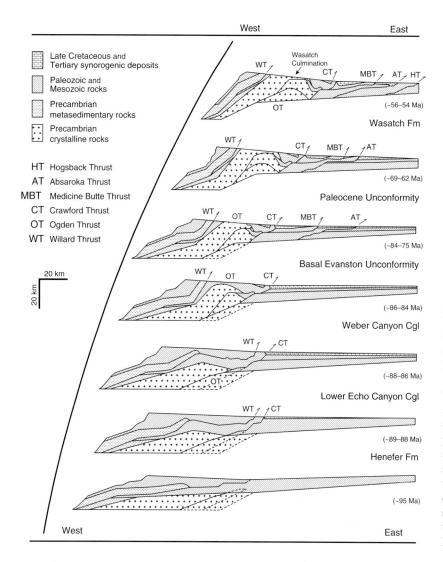

Fig. 2.5. Sequential deformation of the Utah thrustbelt shown by series of balanced cross sections at the approximate latitude of Ogden in Northern Utah (modified from Coogan, 1992; Yonkee, 1992; DeCelles, 1994; Willis, 1999). Restoration is defined by structural relationships, age of syn-orogenic conglomerates, and paleontological and radiometrical dates summarized in DeCelles (1994). Figure shows the eastward advance of the thrustbelt with a piggy-back ramp development, related deposition and the subsequent deformation of the syn-tectonic sediments from Cenomanian to Early Eocene.

small local Lutetian to Early Oligocene evaporitic basins that resulted in taper reduction (Verges *et al.*, 1992).

The effect of decreasing the convergence rate (oro-genic force) below a threshold value has been shown by analogous material modelling (e.g., Rossetti *et al.*, 2000). Below a certain threshold value the topographic load can no longer be sustained by the strength of the wedge material and the wedge reacts by syn-convergence extension and related exhumation of its deeper levels. The Sevier orogenic belt provides a well-constrained natural example of such a collapse (Constenius, 1996). A rapid decrease in North America–Pacific plate convergence resulted in a reduction of the E–W compression in the belt, which caused Middle Eocene–Early Miocene extension after a brief 5–7 Ma period of quiescence. Gravitational collapse of the wedge resulted in westward horizontal spreading that ceased about 20 Ma when the wedge became stable.

Accommodating normal faults had listric geometries, utilizing earlier thrust planes for detachment (Fig. 2.9).

Experimental Coulomb wedge models and natural thrust wedges document a relationship between the dip of the topographic wedge slope and the décollement dip (Davis *et al.*, 1983; Figs. 2.10 and 2.11). The response of wedge taper to changes in various key factors can be illustrated as follows.

Wedges composed of strong material, e.g., those made by thick-bedded, well-cemented siliciclastic sedi-ments, need lower taper values to advance forelandward than wedges with weak material, e.g., those made by thin-bedded carbonate/shale sediments (Fig. 2.12).

The strength of the décollement zone controls the wedge behaviour because the critical wedge taper is pro-portional to the basal coefficient of friction (Fig. 2.13). A weak zone decreases the taper as known from the cases of thrustbelts moving along décollements within

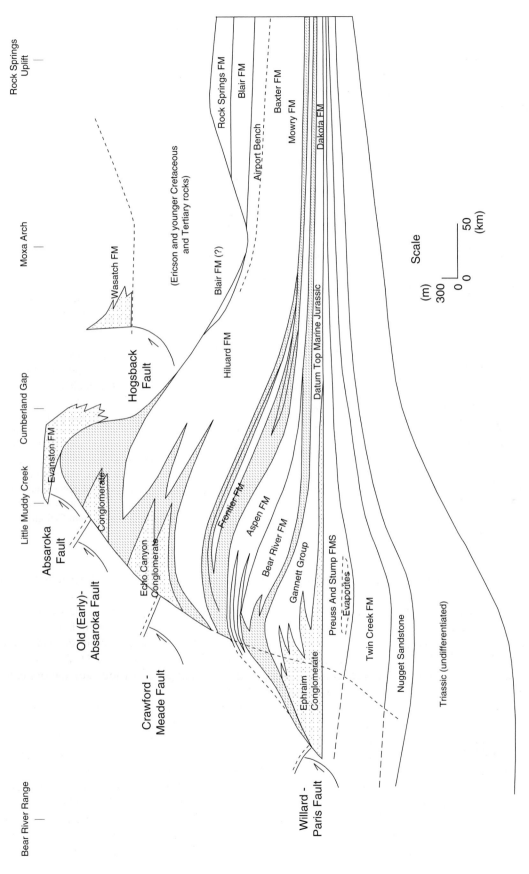

Fig. 2.6. Schematic illustration of syn-orogenic conglomerates deposited in reaction to the development of a specific thrust sheet in the development history of the Wyoming thrustbelt (Lammerson, 1982). Activity of the Willard thrust is dated by the Lower Cretaceous Ephraim conglomerate, activity of the Crawford thrust is recorded by the Upper Cretaceous Echo Canyon conglomerate. The early and late Absaroka thrust phases are dated by the Upper Cretaceous Adaville Formation and the Upper Cretaceous–Paleocene Evanston conglomerate. Hogsback-thrust movements are recorded by the Eocene Wasatch conglomerate.

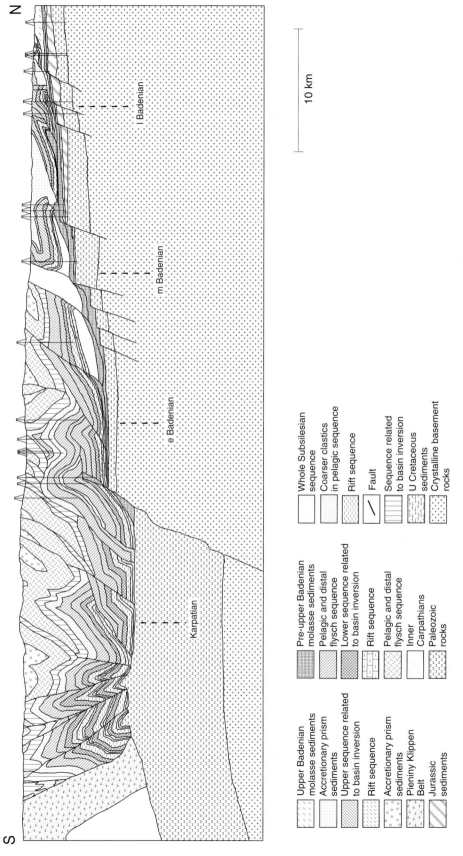

S

N

10 km

I Badenian.

m Badenian.

e Badenian.

Karpatian

Upper Badenian molasse sediments

Accretionary prism sediments

Upper sequence related to basin inversion

Rift sequence

Accretionary prism sediments

Pieniny Klippen Belt

Jurassic sediments

Pre-upper Badenian molasse sediments

Pelagic and distal flysch sequence

Lower sequence related to basin inversion

Rift sequence

Pelagic and distal flysch sequence

Inner Carpathians

Paleozoic rocks

Whole Subsilesian sequence

Coarser clastics in pelagic sequence

Rift sequence

Fault

Sequence related to basin inversion

U Cretaceous sediments

Crystalline basement rocks

Fig. 2.7. Regional balanced cross section through the West Carpathians (Nemčok *et al.*, 2000; Profile 2 in Fig. 2.8). The Slopnice antiformal stack was formed when the wedge increased its taper to overcome a pre-existing normal fault in order to advance further forelandward. This normal fault-related step represented a local increase in the basal friction.

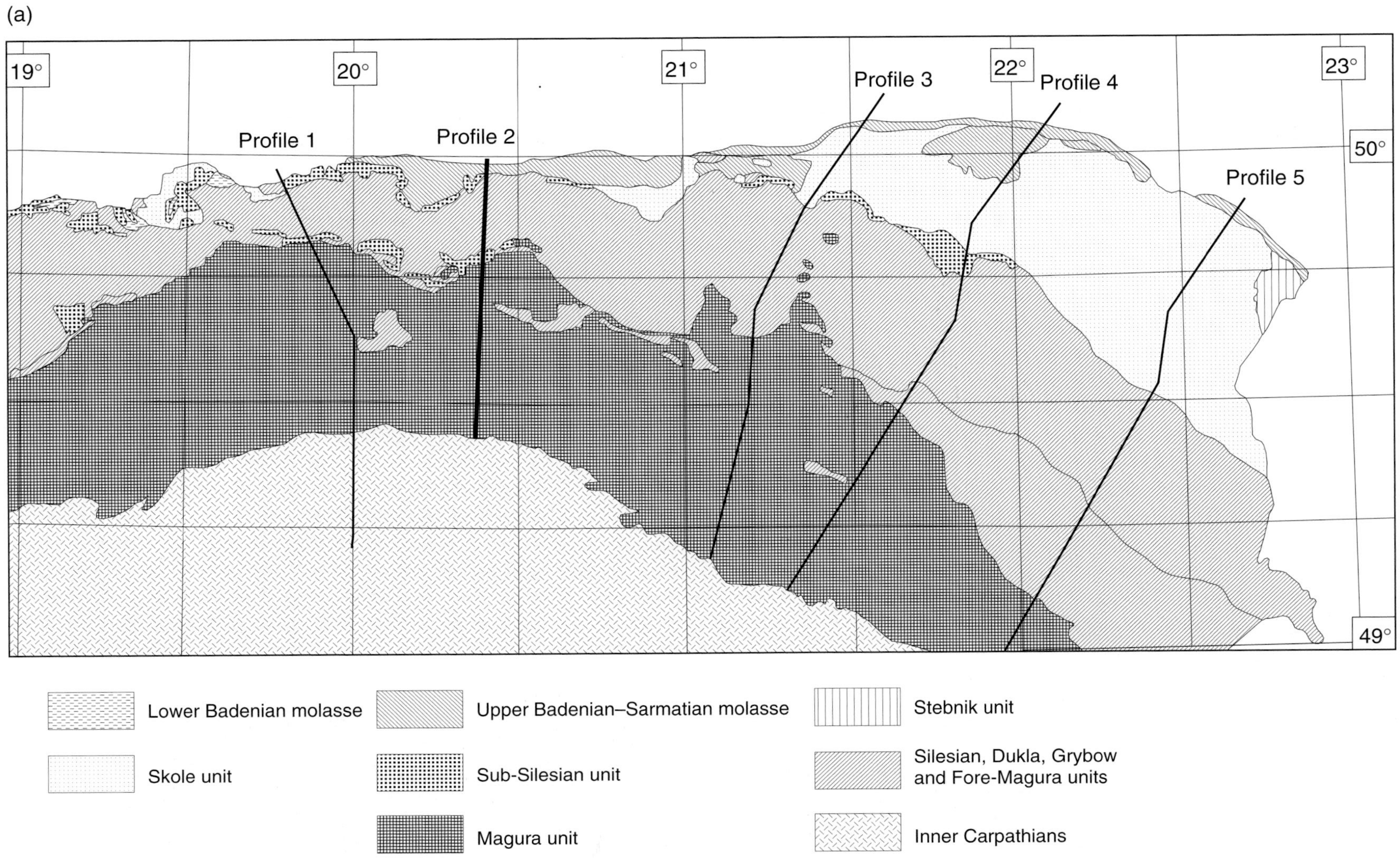

(a)

Fig. 2.8. Wedge response to locally decreased basal friction, West Carpathians (modified from Nemčok *et al*., 1999). Part (a) shows the location of five regional balanced cross sections. Part (b) shows the advance of the thrustbelt front towards the foreland. The travelled distance for different time periods was calculated from the hinterlandward extent of the underthrusted autochthonous sediments located below the wedge. Columns from left to right indicate Karpatian, early Badenian, middle Badenian, late Badenian and Sarmatian. Dramatic early Badenian decrease of the thrustbelt advance is related to the overcoming of the pre-existing normal fault and taper increase by internal deformation shown in Fig. 2.7. Rapid middle Badenian increase of the wedge advance reacts to decreased basal friction where the wedge advanced on top of middle Badenian evaporites after overcoming the basement step.

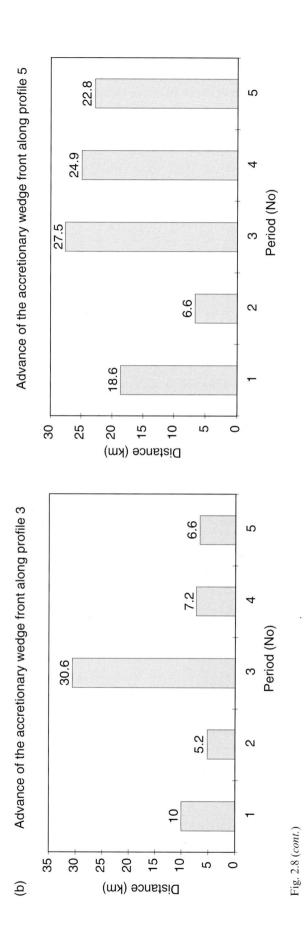

Fig. 2.8 (*cont.*)

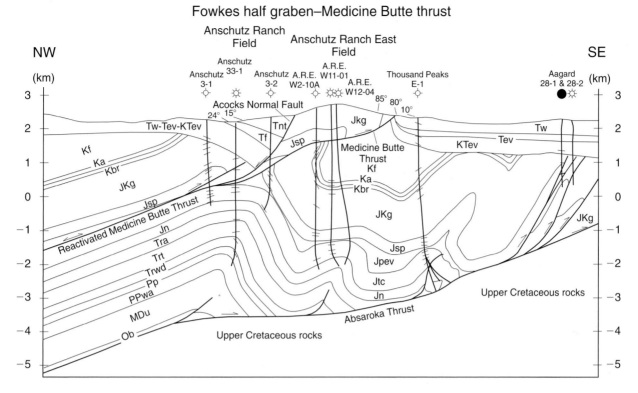

Fig. 2.9. Geological cross section through the Anschutz Ranch and Anschutz Ranch East fields, Wyoming thrustbelt (Constenius, 1996). Figure shows the Acocks normal fault, reactivated Medicine Butte thrust and Fowkes half-graben formed during the orogenic collapse that took part during 49–20 Ma, about 5–7 Ma after the end of orogenic shortening. Ob-Jtc are formations ranging from the Ordovician Big Horn dolomite to Jurassic Twin Creek limestone, Jpev is evaporites of the Jurassic Preuss Formation, Jsp-Kf are formations ranging from the Jurassic Stump-Preuss Formations to Cretaceous Frontier Formation, Kev is the Maastrichtian Evanston Formation, Tev is the Paleocene Evanston Formation, Tw is the late Paleocene–early Eocene Wasatch Formation, Tf is the middle Eocene Fowkes Formation and Tnt is the late Eocene Norwood Tuff.

evaporitic horizons (e.g., Wiltschko and Chapple, 1977; Davis and Engelder, 1985; Harrison, 1995). Reduced basal coupling along evaporitic horizons permitted thrustbelts to maintain very low tapers: as low as 1° or less in the Appalachian Plateau, about 1°–2° in the Spitsbergen thrustbelt or less than 4° in the Pyrenees (Davis and Engelder, 1985; Verges *et al.*, 1992; Braathen *et al.*, 1999). Thrustbelts also lack a consistently dominant thrust vergence and have widely but regularly spaced folds (Dahlen, 1990; Verschuren *et al.*, 1996; Braathen *et al.*, 1999; Nemčok *et al.*, 1999). The low shear stress results in extreme thrustbelt width and sub-horizontal décollement in the Parry Island thrustbelt (Harrison, 1995). The presence of an extremely weak detachment horizon, such as large thicknesses of salt, can be even more important than the influence of the tectonic load, resulting in gravitationally unstable salt-cored anticlines, which continue to grow diapirically (e.g. Davis and Engelder, 1985).

The loss of a weak décollement zone or an increase in its strength due to lateral facies changes results in an

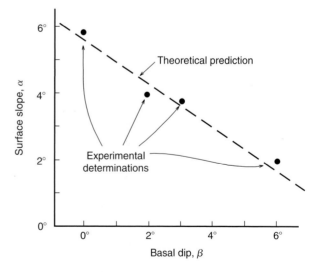

Fig. 2.10. Mean surface slope versus décollement dip from sandbox experiments (Davis *et al.*, 1983). Dots represent the average of eight experimental runs at décollement dip $\beta = 0°$, two at $\beta = 2°$, fourteen at $\beta = 3°$ and nine at $\beta = 6°$. Theoretical prediction curve is calculated from the function $\alpha = 5.9° - 0.66\beta$.

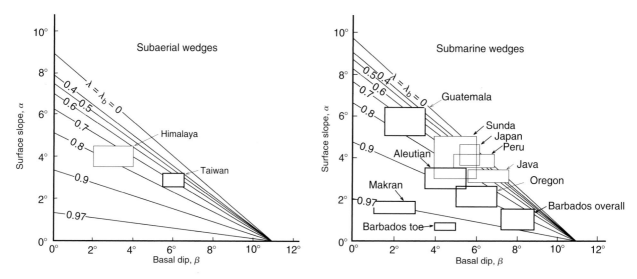

Fig. 2.11. Theoretical linear wedge taper prediction curves $F = \alpha + R\beta$ calculated for various fluid pressure ratios, which are identical for the wedge and its base ($\lambda = \lambda_b$), assuming friction along the base $\mu_b = 0.85$ and wedge friction $\mu = 1.03$ (Davis *et al.*, 1983). Boxes indicate tapers of active wedges, which were used to infer the fluid pressure ratios within them. Heavy outlines indicate wedges with direct fluid pressure data. A wedge sediment density $\rho = 2400\,\mathrm{kg\,m^{-3}}$ was used for submarine wedges. Other density values yield very similar results, because the calculation sensitivity to density is minimal.

increase of the wedge taper. For example, in the Appalachian Plateau where the décollement crosses obliquely the boundary of the Silurian salt, the increase in basal shear stress causes a change in structural style to form a stack of splay faults within an anomalously oriented en-echelon anticline. The influence of the salt horizons in the Pyrenees is even more pronounced. The distribution of Lutetian to Lower Oligocene salt horizons in basins, the depocentres of which shifted successively to the southwest in front of the advancing Pyrenean thrust sheets, control the staircase geometry of the detachment fault. Salt horizon pinch-out at the southern boundary of the Cardona basin resulted in the formation of a passive back-thrust set. Another example comes from the Spitsbergen thrustbelt, where in the western portion of the eastward-propagating thrust system, Paleozoic sediments are coupled with metamorphic basement rocks, and have an increased taper in comparison with the eastern portion, which is underlain by Permian evaporites. Orogen strike-parallel variation in basal shear stress within the wedge is one of the factors that can cause the development of a transfer zone, as documented by sandbox modelling (Calassou *et al.*, 1993).

Accordingly, local fluctuations in décollement friction can result in local taper adjustments (Fig. 2.14). Studies on variations of basal shear stresses either along recently active accretionary wedges or analogue material models not only show such taper adjustments

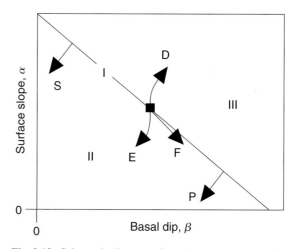

Fig. 2.12. Schematic diagram of a wedge response to various taper modifying factors shown in the surface slope/décollement dip coordinate system (DeCelles and Mitra, 1995). I indicates the wedge development stage when a wedge taper is critical; II shows the stage when a taper is subcritical; and III indicates a supercritical wedge taper stage. The wedge advances in self-similar form by accretion of material across its base in stage I. It stalls in stage II because of insufficient taper. The wedge is stable and capable of sliding forelandward without a material accretion in stage III. Increased erosion of the upper surface, E, increased durability of the upper surface, D, and flexural subsidence, F, modify an ideal critical taper denoted by black square. Increased wedge strength, S, and increased pore pressure along the décollement or decreased basal strength, P, modify the location of the critical taper line in the diagram but its angle remains constant.

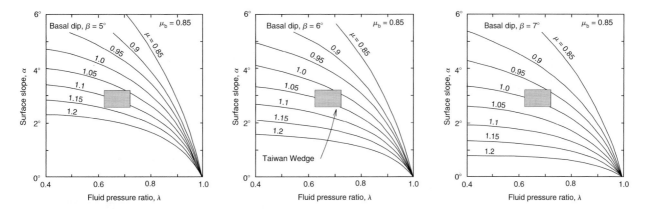

Fig. 2.13. Surface slope α/fluid pressure ratio λ plot for three basal wedge dips β and a value of friction $\mu_b = 0.85$ along the décollement (Davis *et al.*, 1983). The rectangle indicates measured α and $\lambda = \lambda_b$ in western Taiwan. The best fitting effective coefficient of internal friction for the wedge is $\mu = 1.03$, indicating that the basal dip of the wedge is $6° \pm 1°$.

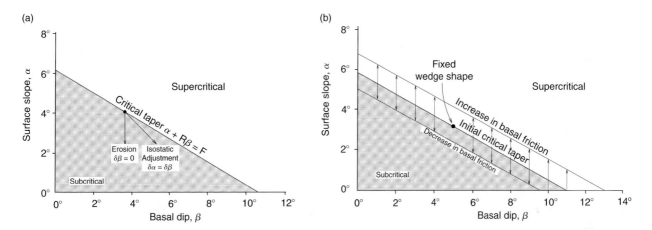

Fig. 2.14. The effect of the change in basal friction, either in λ_b or μ_b on wedge taper (Davis *et al.*, 1983). Increased basal friction, which obeys the condition $(1 - \lambda_b)\mu_b \leq (1 - \lambda)\mu$, triggers the wedge deformation to increase the taper. Its decrease results in a wedge that advances over its foreland without accretion of new material or its internal deformation. The wedge behaviour is very sensitive to these changes. Taper changes indicated in the diagram represent a response of the submarine wedge to only a $\pm 2\%$ change in basal pore pressure, λ_b, from initial $\lambda_b = \lambda = 0.8$.

but also the development of distinct structural styles (e.g., Lallemand *et al.*, 1992, 1994; Gutscher *et al.*, 1996). The latter are controlled by the resulting angular relationship between the maximum compressive stress σ_1 and the basal décollement angle (e.g., Davis and Engelder, 1985; Fig. 2.15), which is also controlled by the angle of the internal friction of the wedge (Davis and Engelder, 1985; Lallemand *et al.*, 1992; Figs. 2.16 and 2.17). Such changes in structural style along the same thrustbelt, which have been attributed to basal shear stress variations, are documented from the Appalachian Plateau (Davis and Engelder, 1985), the Parry Island thrustbelt, Arctic Canada (Harrison, 1995) and the South Pyrenees (Verges *et al.*, 1992).

The influence of pore fluid pressure on the décollement strength–wedge taper relationship is shown in Figs. 2.11 and 2.12, and Table 2.1. Increased pore fluid pressure reduces the strength of the décollement. Yet another strength-reduction mechanism along the décollement is the continued thrustbelt movement and related strain accumulation along the décollement. This mechanism can change a décollement microstructure or composition, causing either strain softening (Schmid, 1982; Mitra, 1984; Wojtal and Mitra, 1986) or reaction softening (Gilotti, 1989). Examples of such strain softening come from the Ogden duplex of the Sevier orogenic belt where the Ogden duplex underwent about 40 km displacement along the 100–500 m thick basal

(a)

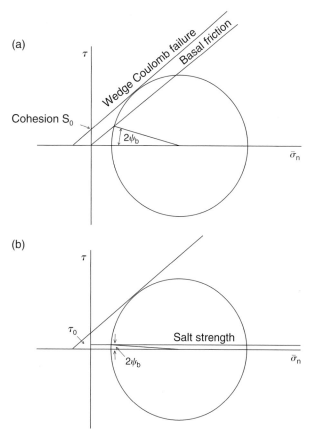

(b)

Fig. 2.15. Coulomb–Mohr diagram showing the contrast in the dip of the σ_1 axis with respect to the basal décollement, ψ_b, for a strong (a) and weak (b) décollement layer (modified after Davis and Engelder, 1985).

décollement and experienced strains as high as γ equal to 80–400 (Yonkee, 1992).

Both topographic slope and décollement dip react to isostasy, erosion and deposition (Fig. 2.12). Erosion at the back of the wedge combined with deposition at the wedge front results in a taper decrease, modifying the topographic slope of the wedge. Erosion of the wedge decreases its topographic slope, which may cause a decrease of the décollement dip due to isostatic rebound. Such an isostatic adjustment in response to erosion can result in some re-initiation of the internal deformation. The exposure of erosionally resistant rocks on the surface reduces the erosion rate, which further affects taper adjustment. Good examples of erosion–taper interactions come from the Sevier orogenic belt (DeCelles and Mitra, 1995). Erosional events progressively reached the wedge uplift, during each major thrust displacement cycle (Fig. 2.5). The wedge reacted by stalling, having less than critical taper and sediments were accumulated on top of the wedge during

the static intervals, allowing the taper to rebuild by shortening and uplift in the wedge rear.

A wedge also responds isostatically to increased overburden. The stacking of thrust sheets in the rear parts of the wedge tends to result in a greater downward isostatic adjustment at the rear than in the front, causing flexural subsidence. This results in an increase of the décollement dip. Because the total wedge taper is conserved in the isostatic process, the dip of the topographic slope is correspondingly reduced. The exchange of topographic slope dip for décollement dip during isostatic adjustment causes the wedge to become subcritical, which renews the deformation until the critical taper is attained (Fig. 2.12).

Sandbox models satisfactorily validate the frictional model (Davis *et al.*, 1983; Dahlen, 1984, 1990; Malavieille, 1984; Zhao *et al.*, 1986; Mulugeta and Koyi, 1987, 1992; Mulugeta, 1988; Colletta *et al.*, 1991; Liu and Dixon, 1991; Dixon and Liu, 1991; Lallemand *et al.*, 1992, 1994; Liu *et al.*, 1992; Marshak *et al.*, 1992; Marshak and Wilkerson, 1992; Calassou *et al.*, 1993; Malavieille *et al.*, 1993; Koyi, 1995; Storti and McClay, 1995; Gutscher *et al.*, 1996; Wang and Davis, 1996; Nieuwland and Saher, 2002), confirming that it can explain the development of thin-skin thrustbelts under plane-strain conditions and which have structural strikes perpendicular and parallel to the controlling stresses. Consequently, the critical taper theory has been successfully used to explain the structural style and stress in thin-skin thrustbelts in external portions of orogens (e.g., Davis *et al.*, 1983; Dahlen, 1984, 1990; DeCelles and Mitra, 1995; Constenius, 1996; Mitra and Sussman, 1997). The theory successfully explains periodic wedge growth, consistent with thrustbelt studies showing significant variations in taper over relatively short time intervals, such as in the Spanish Pyrenees (Meigs *et al.*, 1996; Meigs and Burbank, 1997) and numerous erosional unconformities related to multiple individual tectonic events, as in the Lachlan fold belt, Australia and the Apennines (Foster *et al.*, 1996; Amato and Cinque, 1999). It should also be stated that a good agreement between observed wedge geometries and Coulomb wedge predictions has been achieved most commonly for wedges with lower basal frictions (compare Davis *et al.*, 1983; Dahlen, 1984; Gutscher *et al.*, 1996). As discussed by Bombolakis (1994), there are physical problems associated with relating to mechanical formulations of the critical-wedge model applied to subaerial thrustbelts, where inertia and elastic stiffness are significant. At present, numerical models of the Coulomb wedge suffer from a lack of material memory in the development stages following the initial stage (Strayer *et al.*, 2001), because analogue material models

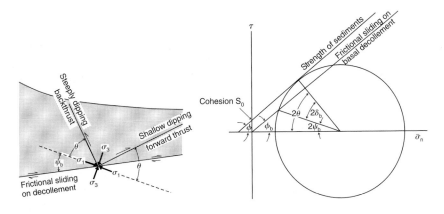

Fig. 2.16. Sketch indicating why a thrust in the thrust wedge dips more shallowly than a back-thrust (modified after Davis and Engelder, 1985).

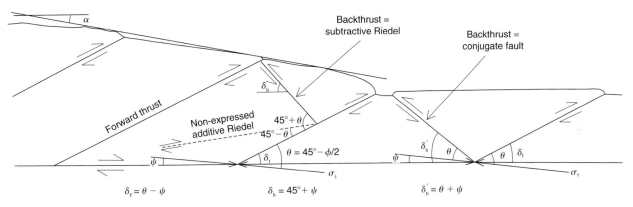

Fig. 2.17. Rheological control over potential thrust faults inside an accretionary wedge (modified after Lallemand *et al.*, 1992). Determination of the angle ψ is shown in Fig. 2.16.

have shown the importance of various structures in later strain located inside a developing wedge (e.g., Mulugeta, 1988; Dixon and Liu, 1991). Deformation also tends to strengthen the wedge during its evolution, which is difficult to simulate in numerical models.

Limitations of Coulomb wedge models

All published studies of Coulomb wedge models assumed the following: that the thrust sheets slide stably; that the internal deformation of the wedge is homogeneous; and that slip occurs over the entire thrust plane. These are assumptions that are not typically found in nature (Wiltschko and Dorr, 1983; Nieuwland and Saher, 2002). There are numerous indications that fault slip is intermittent and faults do not slip everywhere simultaneously (e.g., Price, 1988; Knipe, 1993; Malin, 1994; Gutscher *et al.*, 1996; Mueller and Suppe, 1997; Asanuma *et al.*, 2002). For example, Fig. 2.18 documents an earthquake swarm, reactivating various locations along the Rose Valley fault zone that lasted about 60 days. The swarm cannot be characterized simply as progressing in a specific direction with time. It also

involves some switching back and forth. There are also numerous known cases of heterogeneous distributions of earthquake shear stress drop over much of the rupture surface, determined from tomographic images of earthquake slip (e.g., Bouchon, 1997; Bouchon *et al.*, 1998; Day *et al.*, 1998).

It is also known that different parts of the wedge experience contrasting deformation during orogenic loading (e.g., Geiser, 1974; Wojtal and Mitra, 1986; Mitra, 1987; Koyi, 1995). Loading applied to the undeformed sedimentary cover results in straining, which precedes propagation of large-scale thrust faults and fold development (Mandl, 1988). Figure 2.19 shows such a layer-parallel straining affecting Precambrian conglomerate layers at the base of the Sevier orogenic belt near Salt Lake City, Utah followed by younger small-scale thrusting. The conglomerates underwent both deformation events under ductile conditions at a depth of at least 10 km.

The rheological control over the deformation during such layer-parallel shortening results in different structures in different rocks, as documented in the

Table 2.1. Wedge frictions, décollement frictions, pore fluid pressure ratios and tapers from active accretionary prisms.[a]

Accretionary prism	μ	μ_b	λ	$\alpha + \beta$ (°)
Taiwan			0.675 ± 0.05	*c.* 9
Aleutian	0.45 ± 0.09	0.28 ± 0.07	$\cong 0.87$	7.3
Nankai	0.50 ± 0.10	0.20 ± 0.10		9.5
Guatemala			High	*c.* 8.1
Oregon	0.62 ± 0.10	0.23 ± 0.10	0.85 ± 0.03	3.8
Barbados			$\cong 1$	2.7–4.1
Makran	052 ± 0.10	0.24 ± 0.09	$\cong 1$	3.5

Note:

[a] Modified from Davis *et al.* (1983) and Lallemand *et al.* (1994).

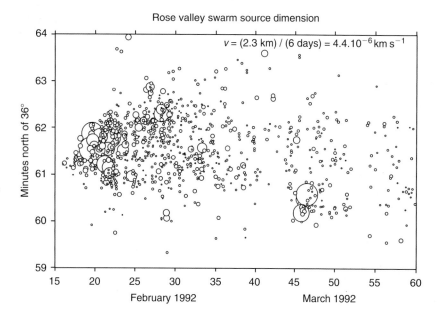

Fig. 2.18. The space–time evolution of the Rose Valley earthquake swarm, California, during its first month (Malin, 1994). The circles have radii proportional to the rupture lengths of the micro-earthquakes that took place at the time and latitude shown. The radii were determined using the moment–stress drop relations of Abercombie and Leary (1993). An important feature is the build up and decay of the sequence, including the larger events south of the initial swarm and almost one month later. Another part of this earthquake swarm took place seven months later.

Appalachians (Mitra, 1987). In units susceptible to pressure solution, such as argillaceous or ferruginous sandstone and limestone, this shortening takes place by the development of bedding-normal stylolites, clay-carbon partings and other seams of insoluble residues. In more argillaceous units, cleavage development by the preferred alignment of micaceous and/or clay minerals takes place. Competent units are shortened by extensive minor faulting.

Rheologically different rocks within a multilayered system can experience various deformation mechanisms during the same time interval, passing the brittle–ductile limit at different times. For example, a study in the Valley and Ridge province (Mitra, 1987) shows that the control of the rock composition on the intensity of pressure solution is more important in determining the mechanism of sliding than temperature and pressure. Typically, lithological types susceptible to pressure solution are characterized by pressure-solution slip in the region where pure orthoquartzites exhibit frictional sliding under the same temperature and pressure conditions.

The amount of layer-parallel shortening varies from those thrustbelts deformed during and after lithification to accretionary wedges shortened before lithification. Shortening determined from deformed Silurian Skolithos burrows from the Appalachian Valley and Ridge varies between 5 and 25% (Geiser, 1974). Internal strains determined in various thrust sheets of the Sevier orogenic belt represent 10–35% of the total shortening (McNaught, 1990; Protzman and Mitra, 1990; Mitra, 1993; Mitra and Sussman, 1997). Shortening, deter-

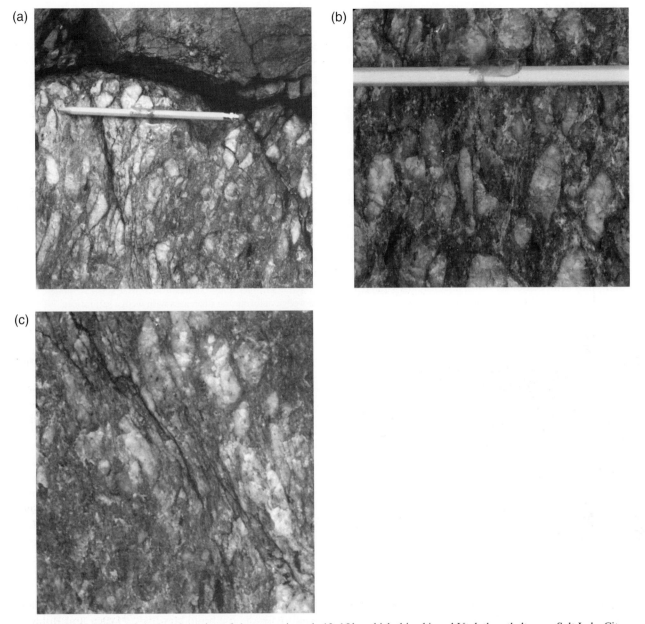

Fig. 2.19. Deformation of the basal portion of the approximately 12–15 km thick thin-skinned Utah thrustbelt, near Salt Lake City, Utah. (a) The deformation of the Precambrian conglomerate layer below the quartzite layer. The first deformation event was ductile layer-parallel shortening. It was followed by the development of ductile shear zones having a dip of 50–60°. Details of layer-parallel shortening and shear zone are shown in (b) and (c), respectively.

mined from the seismic and the borehole data from the Nankai accretionary prism, varies around 68% (Morgan and Karig, 1995).

It should be emphasized, however, as documented by sandbox modelling (e.g., Koyi, 1995, 1997), that the layer-parallel shortening, as discussed above, is not only a function of different rheological properties of layers within the deforming multilayer system. The other strong control is the amount of overburden. Sandbox models indicate that while deep layers experience 40–50% internal layer-parallel shortening, the uppermost layers imbricate and fold after a negligible lateral compaction.

The applicability of critical taper models of frictional wedges to orogenic natural wedges experiences further problems because parameters such as pore fluid pressure, wedge rock and décollement strengths during deformation are practically not quantifiable (DeCelles and Mitra, 1995).

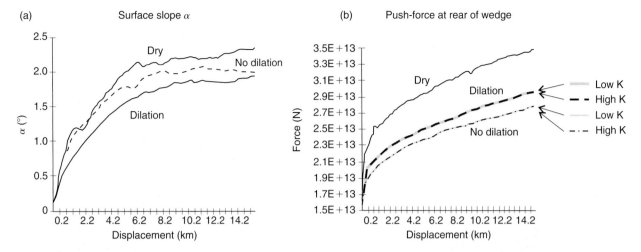

Fig. 2.20. Interpretation of continuum finite-difference models of dry and wet accretionary wedges (Strayer *et al.*, 2001). (a) Comparison of surface slopes of dry, saturated nondilational and saturated dilational wedges indicate that a dilational wedge saturated by fluids is stronger than a nondilational saturated wedge, which is stronger than a nonsaturated wedge. (b) A similar comparison indicates that the orogenic force necessary for the internal wedge deformation decreases with wedge saturation.

The thrust emplacement is accommodated by transient poro-elastic loading and subsequent hydraulic failure of rocks at elevated pore fluid pressures, indicating an important relationship among the volumetric strain rate, the pore fluid pressure and the hydraulic diffusivity of deforming rocks (e.g., Cello and Nur, 1988; Knipe *et al.*, 1991; Brown *et al.*, 1994). This interaction among deformation, fluid flow and fluid pressure is the key element controlling the development of thin thrust wedges (Strayer *et al.*, 2001). Although numerical models of this interaction have been made (Strayer *et al.*, 2001), their results can be summarized only with some caution, because the continuum finite-difference code used for a study does not reproduce the variety of structures, common in thrustbelts, characterized by discontinuous deformation. These models have indicated that fluids decrease the value of the orogenic force necessary for wedge deformation (Fig. 2.20). A dilational wedge saturated by fluids is stronger than a nondilational saturated wedge.

Two-dimensional brittle–ductile wedge model

The application of the previously outlined critical taper models of frictional wedges to thick orogenic wedges requires the addition of temperature-dependent deformation at greater depths, because thermally activated ductile flow becomes dominant in this region (e.g., Pavlis and Bruhn, 1983; Platt, 1986a, b; Jamieson and Beaumont, 1988; Barr and Dahlen, 1989; Willett *et al.*, 1993). The rheological behaviour of most orogenic wedges is controlled by dominant water-rich quartz and feldspar lithologies (Pavlis and Bruhn, 1983; Barr and Dahlen, 1989). These lithologies, for

temperatures above half of their melting temperature, undergo power-law creep:

$$(\sigma_1 - \sigma_3) = \left(\frac{\varepsilon s^p}{A}\right)^{1/n} \exp\left(\frac{E}{nRT}\right) \qquad (2.5)$$

where $(\sigma_1 - \sigma_3)$ is the differential stress, ε is the effective strain rate, s is the grain size, p is the grain-size exponent, A is the generalized viscosity coefficient, n is the stress power-law exponent, E is the activation energy, R is the universal gas constant and T is the temperature. Controlled by geothermal gradients of $20–30\,°C\,km^{-1}$ and a thermal diffusivity of about $10^{-6}\,m^2\,s^{-1}$ (Pavlis and Bruhn, 1983; Barr and Dahlen, 1989; Ord and Hobbs, 1989), the brittle–ductile transition of thicker orogenic wedges lies at depths of 10–20 km. Under typical orogenic strain rates of $10^{-16}–10^{-13}\,s^{-1}$, the wedge strength decreases dramatically with depth below the brittle–ductile boundary (e.g., Pfiffner and Ramsay, 1982).

The presence of contrasting deformation mechanisms in different portions of the wedge controls the wedge shape (Suppe and Connors, 1992; Williams *et al.*, 1994; Carminati and Siletto, 1997; Fig. 2.21). Low-angle surface slopes are typical for the frontal parts of wedges, which deform in a purely brittle manner. The steep slopes behind it are formed above the region where the wedge has already passed the brittle–ductile transition and starts to creep, while the basal décollement, which deforms at higher strain rates, still remains brittle. Further to the rear, where both the wedge base and the décollement are ductile, the surface slope flattens and forms a plateau. Natural examples of brittle–ductile wedges are known from the Peruvian–Bolivian–

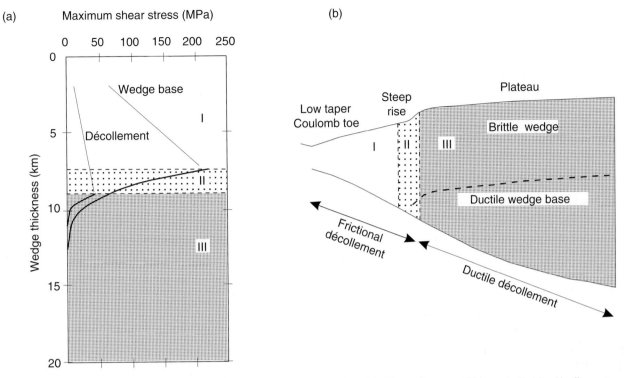

Fig. 2.21. Sketch of the brittle–ductile critical taper model (Williams *et al.*, 1994). The wedge taper (b) is controlled by décollement and wedge base strengths (a). The low taper Coulomb toe in the orogenic front is controlled by both strengths being brittle (zone I). This wedge portion deforms in a frictional Coulomb manner. The steep rise forms where the wedge base is already ductile but décollement is still brittle-frictional (zone II). The plateau develops in the region characterized by both strengths being ductile (zone III). Part (a) indicates the calculated maximum shear stress along the décollement and in the wedge base at different depths. Linear portions of curves are calculated for brittle behaviour, exponential for ductile.

Argentinean Andes (Williams *et al.*, 1994) and Southern Alps (Carminati and Siletto, 1997) on Earth, and Maxwell Montes and Uorsar Rupes on Venus (Williams *et al.*, 1994). The brittle–ductile critical wedge taper model assumes that the décollement position underneath the wedge does not change with time. It is based on the knowledge that grain sizes decrease in localized shear zones such as a basal décollement, resulting in diffusion creep being the dominant ductile deformational mechanism. Therefore, the effective strain rate along the décollement is higher than in the ductile portion of the orogenic wedge, while the ductile strength is lower because of the weakening by dynamic recrystallization. The ductile portion of the wedge deforms by dislocation creep. For a brittle–ductile critically tapered wedge, equation (2.2) is rewritten as (Williams *et al.*, 1994)

$$\alpha + \beta = \frac{(\rho - \rho_w)gh\beta + \tau_b - J}{(\rho - \rho_w)gh + \Delta h}, \qquad (2.6)$$

where α is the dip of the topographic wedge surface, β is the dip of the detachment, ρ is the density of the wedge material, ρ_w is the water density, g is the acceleration due

to gravity, h is the wedge thickness measured along the vertical coordinate, τ_b is the basal shear traction (which is calculated differently for brittle and ductile behaviour) and J is a term that appears only when the wedge portion above the décollement undergoes ductile deformation, since differential stress in the ductile regime depends on changes of both wedge thickness and taper along the x coordinate. Δh also depends on the mode of deformation in the wedge above the décollement.

The boundary conditions of the brittle–ductile taper model are simplifications, because the décollement can be weakened by strain weakening mechanisms other than the transition to diffusion creep (Tullis and Yund, 1985; Evans and Dresen, 1991; Williams *et al.*, 1994), but the model satisfactorily explains several examples of natural wedges mentioned earlier. The values of the model input parameters could also cause errors, because of large uncertainties with the determination of flow parameters and their assignment to specific geological environments (e.g., Kirby, 1983; Carter and Tsenn, 1987; Paterson, 1987; Ranalli and Murphy, 1987; Strehlau and Meissner, 1987; Rutter and Brodie, 1991). The model neglects the advective heat transfer within the wedge but sensitivity analysis suggests that the errors are no larger than 5% (Williams *et al.*, 1994).

Analogue-material experiments provide knowledge of the internal deformation of brittle–ductile orogenic wedges. Experimental viscous wedges indicate a self-similar development if the strain rate remains constant, indicating that the wedge enters a force equilibrium field analogous to Coulomb wedges (Rossetti *et al.*, 1999). The wedge reacts to changes in convergence rate by approaching a new equilibrium state, corresponding to its new strength at the new convergence rate. Paraffin wax models indicate that the wedge deforms mainly by forelandward migration of the wedge front while the rear wedge portion undergoes mainly thickening and vertical movements (Rossetti *et al.*, 2001). Wedge topography and size of the deformed area depends on the convergence rate (Rossetti *et al.*, 2000).

Fast convergence rates control high wedge tapers, high plateau areas and localized deformation zones inside the wedge. Fast convergence scenarios are characterized by material paths towards the wedge rear and material accretion along the base, resulting in homogeneous thickening. The basal décollement does not propagate forelandwards during the wedge development.

Slow convergence rates control low wedge tapers, low plateau areas and distributed deformation inside the wedge. Low convergence scenarios are characterized by progressive deformation propagation towards the foreland. A comparison of fast and slow convergence scenarios shows that underplating in the rear wedge portion is prevalent in orogenic wedges undergoing fast convergence, while accretion at the wedge front is prevalent in wedges controlled by slow convergence rates.

Three-dimensional wedge models

All the models described so far can be characterized by their two-dimensional character. Studies of the three-dimensional behaviour of orogenic wedges indicate that wedge taper mechanics becomes much more complex when a wedge deforms in a true three-dimensional sense (e.g., Byrne *et al.*, 1988; Davy and Cobbold, 1988; Koons, 1990; Marshak *et al.*, 1992; Calassou *et al.*, 1993; Lallemand *et al.*, 1994; Wang and Davis, 1996; Macedo and Marshak, 1999; Marques and Cobbold, 2002). A well-documented example of true three-dimensional evolution is the development of the toe thrust system in the Niger Delta in the Gulf of Guinea, where gliding occurs along shale that has minimum basal friction due to its material properties and increased fluid pressure. The frontal wedge portion can be divided into a whole set of domains with different thrusting directions and arcuate thrusts, indicating different lateral adjustments to the complex sea floor morphology in time and space.

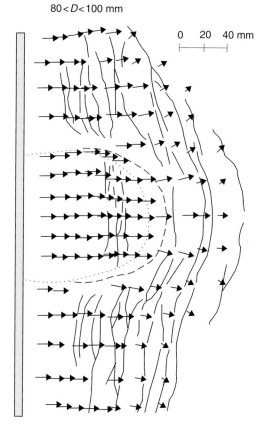

$80 < D < 100$ mm

0   20   40 mm

Fig. 2.22. Displacement vectors and thrust faults sketched from the final stage of the sandbox model simulating the effect of the long plateau on the three-dimensional character of deformation (after Marques and Cobbold, 2002). Initial deformation developed discrete thrusts in both plateau and surrounding regions. Ramp spacing was larger in plateau due to its higher strength. Transfer zones accommodated different thrusting in plateau and its surrounding. New thrusts developed around the plateau front during subsequent development stages, having arcuate shapes. They accreted to the plateau, making it larger. Syn-shortening progressive sideways collapse of the plateau was characteristic for the further thrustbelt development. It was accommodated by dip-slip motions on mentioned arcuate thrusts around the initial plateau and dip-slip motions on plateau thrusts.

Similar complex three-dimensional wedge adjustments have been produced by analogue-material experiments, which concentrated on arcuate thrustbelts and focused on the effect of foreland obstacles, the shape of the rigid indenter, variable detachment depth and along-strike detachment strength variations (e.g., Marshak, 1992; Macedo and Marshak, 1999; Marques and Cobbold, 2002; Fig. 2.22). The results of these models indicate that wedges with low basal friction are very sensitive to variations in surface topography. They undergo shortening along arcuate thrusts, which are convex against high topography areas and concave against low topography areas. Localized erosion in the hinterland could allow local thrusts to remain active, inhibiting the development of new thrusts in the respective frontal portions of the wedge. Stronger orogen portions can become stable and surrounded forelandward by arcuate thrust systems. A natural analogue for the arcuate thrust around a stronger portion is, for example, the Tromen volcanic area in the Argentinean Neuquén basin (Marques and Cobbold, 2002).

Another example of three-dimensional wedge adjustments comes from wedges adjacent to transpressional plate margins (e.g., Gayer and Nemčok, 1994; Gayer *et al.*, 1998; Braathen *et al.*, 1999). Towards the orogenic hinterland, where sediments are better coupled to the basement, the wedge is stronger and undergoes orogen-parallel or orogen-oblique deformation. In the frontal portions of the wedge, where sediments are weaker and decoupled, a broad zone of orogen-perpendicular or orogen-oblique contraction is developed.

3 Mechanics of thrust sheets

Folding/faulting interaction in thrust sheet development

Folding and faulting in thrustbelts are interdependent and/or competing processes that have been described in terms of models of fold–thrust interaction (e.g., Berger and Johnson, 1982; Suppe, 1983; Williams and Chapman, 1983; Suppe and Medwedeff, 1984; Jamison, 1987). Over the past decade, several quantitative geometric models, described later, have been developed for thin-skin thrustbelts providing specific relationships between the fault and fold geometries, such as fault-bend, fault-propagation and detachment folding (Suppe, 1983; Suppe and Medwedeff, 1984; Jamison, 1987; Chester and Chester, 1990; Erslev, 1991; Epard and Groshong, 1995). In general, these geometric models are based on line-length or area balancing and assume kink-type or tri-shear folding, where flexural slip is the dominant deformation mechanism. Some models have been expanded by adding localized thickening or thinning within the fold, translation along imbricate thrust faults or over ramps, and progressive changes in fold geometry with displacement (e.g., Jamison, 1987; Mitra, 1990, 2002; Suppe and Medwedeff, 1990; Erslev, 1991; Erslev and Mayborn, 1997; Hardy and Ford, 1997; Almendinger, 1998). The models provide guidelines for seismic interpretation and construction of balanced cross sections. They have become valuable tools for defining trap geometry in the subsurface (e.g., Suppe and Namson, 1979; Mitra, 1986; Namson and Davis, 1988; Mount et al., 1990). Because of their geometric basis, however, the models do not allow one to determine why, when, and where a particular process, such as imbrication or detachment, occurs. Neither do they allow the determination of the exact geometrical response of the deforming section, which is unique for each mechanical stratigraphy and character of the faults involved. For example, both fault zone drag and folding resistance of the thrust sheet influence the fold geometry and the internal strain distribution, and are important in resisting forelandward motion (e.g., Elliott, 1976; Wiltschko, 1979a, b, 1981; Berger and Johnson, 1980, 1982). The folding resistance of the sheet depends on the rheological properties of each stratigraphic layer and on the degree of mechanical anisotropy of the unit as a whole. Therefore, it is critical to understand the physical aspects of structural trap

development in thrustbelts in order to define the trap geometries correctly.

The way in which the rock package deforms, whether by folding or faulting, is controlled by the mutual relationship between the faulting and buckling instability envelopes (Jamison, 1992; Fig. 3.1a).

Conical faulting instability is defined by the Drucker–Prager failure criteria (Drucker and Prager, 1952):

$$\sigma_f = \beta J_1 + J_2^{1/2}, \tag{3.1}$$

where σ_f is the failure stress, J_1 and J_2, are the first stress invariant and the second invariant of deviatoric stress, respectively, and β is a coefficient (≥ 0 for compression) that may vary as a function of J_1. The axis of the cone is defined by the hydrostatic stress relationship, $\sigma_v = \sigma_{h1} = \sigma_{h2}$, where both horizontal stresses and the vertical stress are equal. For cohesive materials the apex of the cone lies in the tension octant of the working stress space at a distance of $\tau_0 \cot \varphi$ from the origin, where τ_0 and φ are the cohesion and the angle of internal friction, respectively. The half apex angle γ (Fig. 3.1b) is related to the angle of internal friction as follows:

$$\tan \gamma = 3^{1/2} \sin \varphi (3 + \sin^2 \varphi)^{-1/2}. \tag{3.2}$$

When the stress state is represented by a point inside the cone, the material is stable. When the rock follows a burial path, points are inside the cone (Fig. 3.1c). Any addition of tectonic load will result in a stress path moving toward the instability envelope (Fig. 3.1d).

It has to be noted that the cone described does not have an ideal shape. It has a rounded apex, located in the tension octant. The other deviation from an ideal cone occurs in the compression octant, in the area where normal stresses are high. This results from the fact that inter-grain friction is a pressure-sensitive shear resistance of materials. It depends on the normal stresses that act across the particle boundaries, and these tend to increase with progressive burial or tectonic loading. The angle of internal friction is described by the Coulomb–Mohr equation:

$$\tau = c + \tan \varphi (\sigma_n - P_f), \tag{3.3}$$

where c is the cohesion, φ is the angle of internal friction, σ_n is the normal stress, P_f is the fluid pressure and

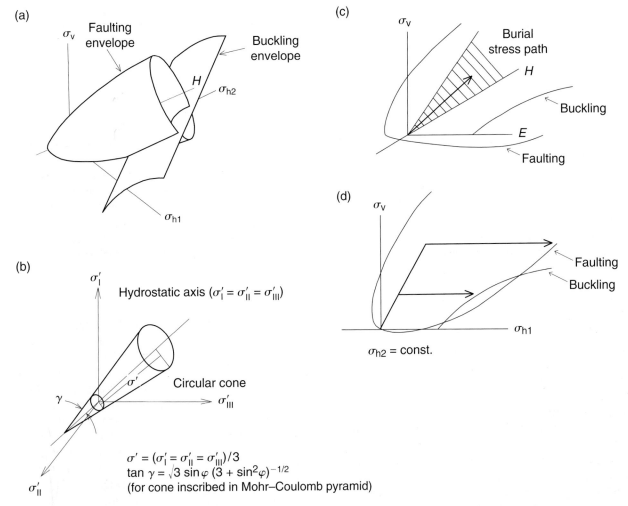

Fig. 3.1. (a) Faulting and buckling instability envelopes viewed together in the working stress space (Jamison, 1992). (b) Drucker–Prager yield cone in three-dimensional stress space (Mandl, 1988). $\sigma' = (\sigma'_I + \sigma'_{II} + \sigma'_{III})/3$. $\tan\gamma = (\sin\varphi\,(3 + \sin^2\varphi)^{-1/2}$. (c) The stress path for burial within a basin with no lateral stresses of tectonic origin (thus, with $\sigma_{h1} = \sigma_{h2}$) will fall in the σ_v–E plane and within the cross-hatched region (Jamison, 1992). (d) The application of regional horizontal stresses will change the horizontal, but not the vertical stresses. The stress path will move towards the intersection with the instability envelopes (Jamison, 1992).

τ is the shear strength of the rock. It has a constant value, as the dependence of the shear strength on the normal stress is linear. However, results from rock mechanic tests show that, depending on rock type, deviations from linearity can become more pronounced under larger confining pressures (Fig. 3.2). Tests on a great variety of rocks indicate that these deviations always imply a decrease in the peak friction angle with increasing normal effective stress, causing an apex decrease of the Drucker–Prager cone. The major controlling factor of this phenomenon is dilatancy suppression under the higher confining stresses. The associated friction angle reduction becomes significant at quite different confining pressures for different materials, because the suppression of dilatancy works under a broad range of confining pressures for different materi-

als. A particularly drastic reduction is to be expected when the confining pressure required to suppress dilatancy causes a transition from brittle microdeformations to intracrystalline gliding.

The convex planar folding instability for layers of uniform geometry, elastic properties and frictionless contacts (Fig. 3.1a) is defined by elastic buckling (Timoshenko, 1936; Johnson and Page, 1976):

$$\sigma_{f0} = \frac{E}{1 - v^2}\left[\frac{t^2 k^2}{12} + \left(\frac{\pi}{n^2 t k}\right)^2\right], \tag{3.4}$$

where E is Young's modulus, v is Poisson's ratio, k is the shape factor, n is the number of layers and t is the individual layer thickness. The shape factor k is given by

$$k = lt, \tag{3.5}$$

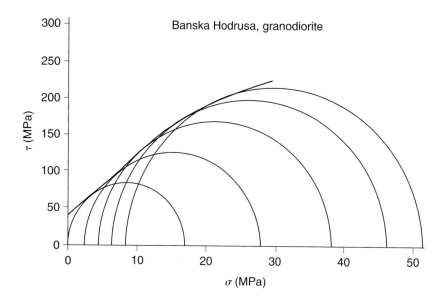

Fig. 3.2. Mohr-circle envelope determined by triaxial testing of the upper Badenian–lower Sarmatian granodiorite from Banska Hodrusa, West Carpathians. The determined cohesion is 38.9 MPa, the angle of internal friction varies from 40° to 20°, depending on the increasing confining pressure.

where l is the wave number of the sinusoid. The wave number l, given in km^{-1}, is calculated from

$$l = \frac{2\pi}{L},\qquad(3.6)$$

where L is the thrust sheet length.

In the working stress space typical for thrustbelts (Fig. 3.3), the burial stress path lies in the $\sigma_v\!-\!E$ plane (Fig. 3.1c), in the stability field for both faulting and folding. An addition of tectonic stress related to thrusting moves the stress path toward the instability envelope (Fig. 3.1d).

In order to discuss the interplay between folding and faulting, we now assume the existence of one homogeneous layer with varying thickness. At low values of the vertical stress σ_v, indicating relatively shallow burial, the stress path will intersect the folding instability envelope first, resulting in detachment folding. At moderate values of vertical stress σ_v, indicating moderate burial, the stress path will intersect the folding and faulting instability envelopes synchronously, resulting in fault-propagation folding. At high values of vertical stress σ_v, indicating relatively deep burial, the stress path will intersect the faulting instability envelope first, resulting in fault-bend folding.

It is important to keep in mind that both competing mechanisms (equations 3.1 and 3.4) are controlled by mechanical stratigraphy of the deforming rock package. The faulting is controlled by cohesion, friction and pore fluid pressure (equations 3.2 and 3.3). The folding is controlled by elastic constants (equation 3.4). Each of them further depends on other factors, as discussed in Chapter 7.

Dynamics and kinematics prior to and during thrust sheet detachment

Before thrust sheets in thin-skinned thrustbelts move along their faults and/or experience folding, they undergo initial straining and detachment (e.g., Geiser, 1974; Koyi, 1995, 1997). The thrust sheet can only detach along a rock layer of lower shear strength. Such a layer may consist, for instance, of weak limestone, the shear strength of which may show little dependence on the effective stresses in the pertinent pressure range (e.g., Riffault, 1969). The reason for this is that the main part of the shear strength in this limestone is formed by the pressure-insensitive cohesional strength, controlled by cementation. Other detachment layers are commonly formed by shale, which has a shear strength that is dependent on the effective stresses. The reason for this is that the main part of the shear strength in this shale is formed by the pressure-sensitive frictional strength, which increases with the amount of overburden and decreases with increasing pore fluid pressure. The detachment layers are also commonly made of evaporites, for which the shear strength is dependent on the viscosity, the layer thickness and the displacement velocity.

The initial straining and detachment of the thrust sheet can be visualized by considering a rectangular sheet composed of competent elastic/frictional plastic material, separated from a rigid basement by a thin layer of incompetent elastic/frictional plastic material (Mandl, 1988; Fig. 3.4). The deformation and stress changes inside the thrust sheet, simulated by finite-element modelling (Mandl, 1988), result from a horizontal displacement of the back end of the competent rectangular sheet. The straining initiates with elastic

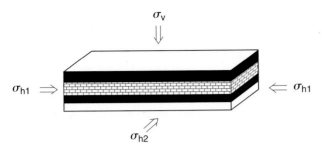

Fig. 3.3. Stress space defined by principal stress axes oriented parallel and perpendicular to the bedding (Jamison, 1992).

straining followed by plastic hardening and softening (Geiser, 1974; Wojtal and Mitra, 1986; Mitra, 1987). A displacement as large as 0.75% of the thrust sheet height, shown in Fig. 3.4, induces different strain patterns in the competent and incompetent layers. Plastic softening penetrates far into the incompetent detachment horizon, while the competent rock layer is either still elastically deforming or in the plastic hardening range. The analysis indicates that when a detachment starts to propagate in the incompetent layer, the shear strength of the detachment layer does not have its peak value, but has a residual value, which is approached in the main part of the softened zone. A completely detached thrust sheet rests on a detachment layer, which is in a shear-softened state. The material of the detachment layer, prior to detachment and thrusting, experiences shear hardening, which includes tectonic compaction, followed by shear softening. The shear softening contains both brittle and ductile components (e.g., Scholz, 1968; Krantz and Scholz, 1977; Lockner *et al.*, 1992). The brittle shear softening includes mechanisms which produce a shear and stress reduction that are only slightly affected by the rate of applied straining. They are shear dilatancy (e.g., Handin *et al.*, 1963; Brace *et al.*, 1966; Scholz, 1968; Knipe *et al.*, 1991), cataclastic diminution and rounding of grains (e.g., Aydin, 1978a, b; Antonellini and Aydin, 1994; Antonellini *et al.*, 1994), and the growth of micro-cracks or macroscopic joints (e.g., Dula, 1981; Mitra, 1987; Srivastava and Engelder, 1990; Chester *et al.*, 1991).The ductile shear softening mechanisms include, for example, pressure solution and dynamic recrystallization (e.g., Elliott, 1976a; Mitra, 1988b; Knipe, 1993; Butler and Bowler, 1995). Both are strongly affected by the rate of applied straining, since they are controlled by some type of diffusion.

This model indicates that if the competent elastic/frictional plastic material is not separated from a rigid basement by a thin layer of incompetent elastic/frictional plastic material, a thrust sheet cannot be detached.

Finite-element analysis (Mandl, 1988; Figs. 3.5a–c) shows that the shear-softening zone in this case extends from the lower rear corner of the competent layer to the surface and no detachment is formed. The zone fully develops when horizontal displacement of the back end of the competent layer is as small as 3.75% of its thickness. At this stage, deformation in all other parts of the competent layer stops. The listricity of the thrust fault is caused by basal shear stresses and related σ_1 stress deflections (Fig. 3.5d).

The plastic detachment layer and the competent thrust sheet interact mechanically. Therefore, stresses and deformation in the thrust sheet depend on those in the detachment layer. The mechanical behaviour of the detachment layer controls the propagation and location of the detachment zone, the propagation and location of the slip plane system of the detachment zone and the development of secondary faults underneath the competent thrust sheet (e.g., Riley *et al.*, 1995; Couzens *et al.*, 1997; Strayer, 1998; Kukowski *et al.*, 2002). It also controls the amount of penetrative deformation of the competent thrust sheet and its ramp spacing (Riley *et al.*, 1995; Gutscher *et al.*, 1996; Couzens *et al.*, 1997; Strayer and Hudleston, 1997).

In order to analyse the behaviour of the plastic detachment layer, Mandl (1988) assumed simple plane strain conditions and an infinitely long detachment layer. The plastic detachment layer has a stress-independent shear strength that helps to obtain an analytical solution. This layer consists of a stack of layers with different shear strengths, located between two rigid plates, representing a rigid basement and the competent thrust sheet. The weak detachment layer is compressed and sheared by a vertical normal stress σ_y and a horizontal shear stress τ, respectively. They satisfy the limit condition of perfect plasticity, together with horizontal normal stress σ_x:

$$(\sigma_x - \sigma_y)^2 + 4\tau^2 = 4K(y)^2 \qquad (3.7)$$

and equilibrium equations:

$$\frac{\partial \sigma_x}{\partial x} + \frac{\partial \tau}{\partial y} = 0, \quad \frac{\partial \tau}{\partial x} + \frac{\partial \sigma_x}{\partial y} = \gamma y, \qquad (3.8)$$

where $K(y)$ is the shear strength which changes with depth (y) and γ is the specific weight of the detachment layer material. The solution is (Mandl and Shippam, 1981)

$$\tau = \frac{\tau_1 + \tau_2}{2} - ay$$
$$\sigma_y = \sigma_y^0 + \gamma(H - y) + ax \qquad (3.9)$$
$$\sigma_x = \sigma_y \pm 2[K(y)^2 - \tau^2]^{1/2}$$

with the positive constant

(a)

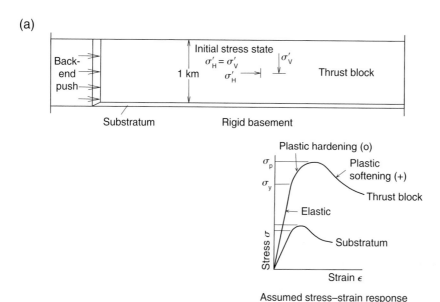

Fig. 3.4. Finite-element model of the thrust sheet lying on the weak substratum (Mandl, 1988). (a) Definition of the boundary conditions and the rheology of a thrust sheet lying on a weak substratum. Elastoplastic behaviour is assumed. The rate of plastic dilation reaches a maximum at the peak stress and decreases to zero in the softening range. (b) Finite-element net of the end part with growing plastic hardening ($\bigcirc$) and softening ($+$) zones in the thrust sheet and weak substratum. (c) Directions and relative magnitudes of σ_1.

(b)

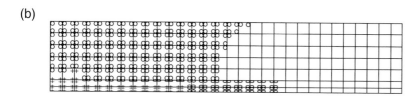

(c)

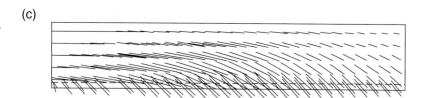

$$a = -\frac{\tau_1 - \tau_2}{2H} > 0, \qquad (3.10)$$

where τ_1 and τ_2 are negative shear stresses along the upper and lower boundaries, respectively, of the plastic detachment layer, H is the half-thickness of this layer and, σ_y^0 is the compressive vertical stress loading the layer, at the point with coordinates $x = 0$, $y = \pm H$. Directions of potential slip are identical to directions of maximum shear stress (e.g., Wallace, 1951; Bott, 1959; Angelier, 1979). They bisect the right angle between principal stress directions. This is valid for the purely cohesive plastic material with angle of internal friction, φ, approaching $0°$, unlike the frictional plastic material where the angle α between the slip plane and σ_1 is

$$\alpha = \frac{\pi}{2} - \varphi. \qquad (3.11)$$

Calculated slip lines inside the weak detachment layer have curved trajectories (Fig. 3.6a). It is assumed that

the shear strength, K, of the weak substratum varies according to (Fig. 3.6b)

$$K = \left(1 - \frac{1}{2}\cos^2 y\right)K_\infty, \qquad (3.12)$$

where K_∞ is the shear strength of the weak substratum at its boundaries and the tangential stress at the base of the thrust sheet, τ_1, equals $0.6 K_\infty$. The location of the principal displacement zone of the detachment is determined by the contact of shear stress and strength curves (Fig. 3.6b). The graph shows that the principal displacement zone lies above the plane with the lowest shear strength.

Location of the detachment is further complicated when the detachment layer is formed by alternating incompetent and competent, i.e. plastic and elastic, beds. The horizontal shear stress τ varies only in incompetent beds and the vertical normal stress remains unchanged with depth throughout the stack of beds in this case (Fig. 3.6c). Since the variation of the horizontal shear stress τ

(a)

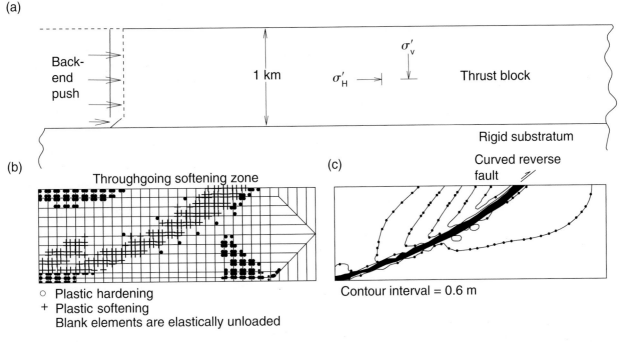

(b)

Throughgoing softening zone

(c)

Curved reverse fault

o Plastic hardening
+ Plastic softening
Blank elements are elastically unloaded

Contour interval = 0.6 m

Fig. 3.5. Horizontal compression inside rectangular sheet of rock attached to a rigid basement: elastic–frictional plastic finite-element model (Mandl, 1988). (a) Scheme of configuration, initial stresses and the boundary displacement. (b) Finite-element mesh of the rear part of the sheet with a throughgoing slightly listric softening zone after 37.5 m of total displacement. Note the second, steeper, plastic zone, which ceased growing and unloaded elastically, when the main shear zone reached the surface. Blank elements of the mesh are elastically unloaded, $\bigcirc$ indicates plastic hardening and + plastic softening. (c) Contours of the magnitudes of the displacement increments that took place during the last increment of boundary displacement. The bundling of contour lines indicates a concentration of simple shearing. (d) Curvature of a thrust fault as a consequence of the perturbation of σ_1 trajectories caused by basal shear stress. Note that the low basal shear stress does not affect σ_1 trajectories, while the progressive increase of the basal shear stress results in a deflection of σ_1 trajectories towards the basement.

(d)

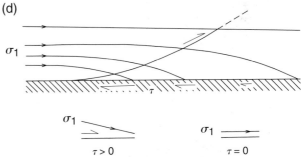

with depth is determined by the horizontal gradient of the overburden stress σ_y and does not depend on the local shear strength, different gradients may, in a given bedded detachment layer, mobilize different basal slip planes. A stronger increase in σ_y in the thrust direction, caused by folding or local thickening, is associated with detachment at a shallower depth of the detachment layer or even ramping up in a stratified detachment layer (Fig. 3.6d). This can even involve a propagation of the detachment into a bed with higher shear strength or a propagation of two detachments, one above another, in two beds.

A change in the overburden, by either facies changes or thickening, is not the only cause of detachment ramping. Gliding on a weak horizon may experience a local increase in shear strength of the horizon. This can be caused, for example, by grain-size decrease or horizon pinch-out. A tectonic cause for such a shear strength increase is documented in the Spitsbergen Tertiary thrustbelt, where a step in the detachment coincides with wedging-out of Lower Carboniferous shale against the pre-existing horst (Bergh *et al.*, 1997). This results in detachment ramping to a higher horizon formed by the evaporitic Gipshuken Formation.

Although ramps and secondary faults in the detachment zone are always P shears (sensu Tchalenko, 1970), they have different origins. Ramps form at local disturbances of the stress field (Wiltschko and Eastman, 1983; Schedl and Wiltschko, 1987), while secondary faults in the detachment zone are formed in an undisturbed stress field (Mandl, 1988; Fig. 3.6a). These secondary faults are mobilized when the base of the competent

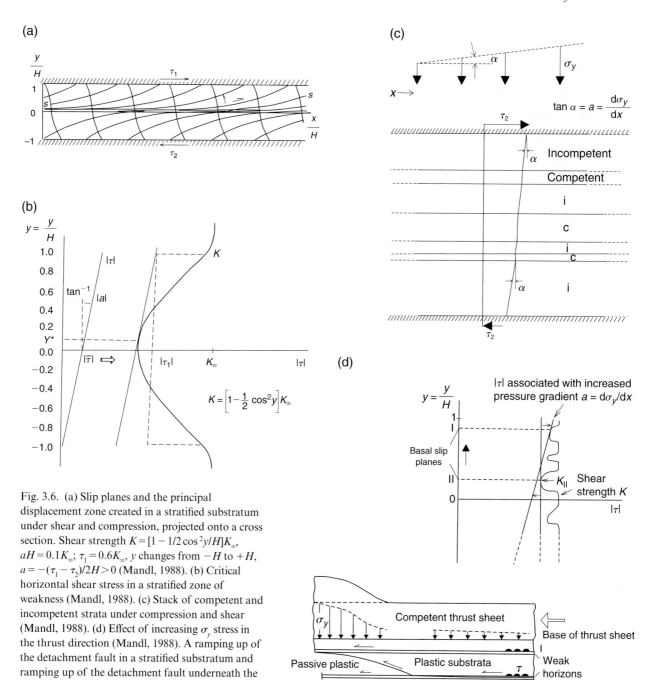

Fig. 3.6. (a) Slip planes and the principal displacement zone created in a stratified substratum under shear and compression, projected onto a cross section. Shear strength $K = [1 - 1/2 \cos^2 y/H] K_\infty$, $aH = 0.1 K_\infty$; $\tau_1 = 0.6 K_\infty$, y changes from $-H$ to $+H$, $a = -(\tau_1 - \tau_2)/2H > 0$ (Mandl, 1988). (b) Critical horizontal shear stress in a stratified zone of weakness (Mandl, 1988). (c) Stack of competent and incompetent strata under compression and shear (Mandl, 1988). (d) Effect of increasing σ_y stress in the thrust direction (Mandl, 1988). A ramping up of the detachment fault in a stratified substratum and ramping up of the detachment fault underneath the thrust sheet are shown.

thrust sheet deforms or when compaction of a weak substratum enables minor slips. They can have very small displacements, but can occur in large sets. A natural case of such a set of secondary faults comes from the South Wales coalfield, at the front of the British Variscides (Fig. 3.7). They are analogous to those calculated in Fig. 3.6(a).

The case of two coexisting detachment faults above each other can lead to the formation of P shears, which

interconnect the two fault planes in a sigmoidal fashion. This could be visualized if we replaced the sinusoidal strength profile in Fig. 3.6(b) by a profile with two sinusoidal segments, both of which make contact with the straight shear stress line, and by duplicating the weak layer shown in Fig. 3.6(a). This deformation results in a special type of duplexing (sensu Boyer and Elliott, 1982).

The cases of weak detachment behaviour discussed above assumed ideally plastic behaviour with stress-inde-

Fig. 3.7. Secondary faults in a detachment zone developed within the White Four coal seam at Nant Helen opencast site in the Variscan foreland of South Wales.

pendent shear strength. Thus they did not describe the effects of fluid pressure. Detachments in thrustbelts are usually formed in pressure-sensitive materials, which would make the analysis more complex. Nor does the plasticity approach described above account for strain rates. Therefore, it cannot provide information on the detachment propagation rate or on the movement rate of the detached thrust sheet. Further, it gives no information about the rate of the sheet imbrication, when its movement stops at the propagation front. In order to add a strain rate dependence to the analysis would require the assumption that the incompetent detachment layer also has a viscous component in its deformational response.

The stresses and deformation in the thrust sheet depend on those in the detachment layer. For example, the ramp spacing can be affected by the shortening of the ductile rock of the detachment layer under compression prior to detachment propagation. As shown by analogue material experiments, this can impose a pervasive shortening upon the overlying thrust sheet, which controls the ramp location (Dixon and Liu, 1991; Liu and Dixon, 1991). Similar results are provided by numerical models (Riley *et al.*, 1995; Couzens *et al.*, 1997).

The detachment layer also influences the stress transmission inside the thrust sheet. When the layer behaves as a viscous material the transmission is significantly enhanced. Unlike the shear resistance of the plastic detachment layer described earlier, the shear resistance of the viscous layer is strain rate dependent. Therefore, stresses inside the thrust sheet may change over time, even when the horizontal displacement of the back end of the competent rectangular sheet remains constant. In this case, the stress transfer is described by a stress-diffusion model (Bott and Dean, 1973; Fig. 3.8). This

(a)

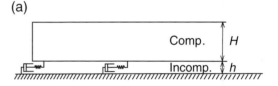

(b)

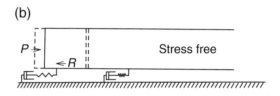

(c)

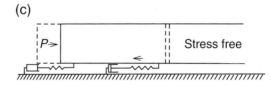

Fig. 3.8. Moving 'stress front' in a competent thrust sheet lying on a viscoelastic substratum (Mandl, 1988). (a) Stage prior to thrusting. The visco-elastic behaviour of the substratum is modelled by an elastic spring and a viscous dashpot in series (Maxwell fluid). (b) Application of a horizontal load P causes the rear segment of the competent sheet to shorten concomitantly with elastic extension and onset of viscous flow in the underlying substratum. Initial resistance R of the deformed part of the substratum balances the driving force. (c) The relaxation of shear resistance, which accompanies the development of creep flow in the substratal segment, causes a transfer of the shear load to a neighbouring substratal segment and extension of the stressed region in the competent sheet.

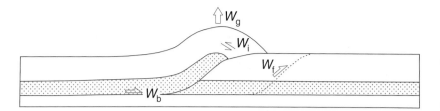

Fig. 3.9. Sketch illustrating work involved in initiation of fracturing (W_f), overcoming basal friction (W_b), opposing gravity (W_g) and internal deformation (W_i) (Zoetemeijer, 1993).

model describes how the push force applied to the rear causes elastic shortening of the rear part of the competent thrust sheet and the onset of viscous flow in the underlying detachment layer. It also describes how the 'stress front' advances from the stressed part of the competent thrust sheet into its unstressed part. This movement has been demonstrated by photoelastic experiments (Blay *et al.*, 1977). The time, T, until the elastic sheet stresses to 90% of its constant end stress is

$$T = \frac{\eta}{E}\left(\frac{L^2}{Hh}\right), \tag{3.13}$$

where h is the thickness and η is the viscosity of the viscous detachment layer. L is the length, E is Young's modulus and H is the thickness of the elastic sheet. Despite the significant simplifications involved in this concept of the moving 'stress front', it serves as a good tool in understanding the stress transfer. For example, the moving 'stress front' 'maps' the rock section and locates weaknesses, which may localize future faults. Or, for example, given the time span of the applied tectonic force, the transfer time can determine which parts of the rock sequence will be deformed.

Dynamics and kinematics after thrust sheet detachment
The energy accumulated by orogenic forces controls deformation and stress changes inside the moving thrust sheet and its foreland. Each episode of the detachment fault rupturation and thrust sheet displacement and folding, accompanied by elastic unloading, is triggered by the preceding stress buildup (Yielding *et al.*, 1981; Rockwell *et al.*, 1988; Philip *et al.*, 1992; Treiman, 1995). Accumulated energy is consumed on (Mitra and Boyer, 1986; Fig. 3.9):

(1)   propagation of new faults;
(2)   reactivation of pre-existing faults;
(3)   opposing gravity;
(4)   internal deformation of the thrust sheet.

When one tries to unravel the sequence of local tectonic events affecting the thrust sheet and its neighbourhood, it should be possible to find a unique solution, following the principles of minimum physical work (Nadai, 1963). Each new tectonic event in a succession of events would

be the one that requires the least amount of energy. Therefore, it should be theoretically possible to choose a unique sequence of events from alternatives suggested by a geometric structural interpretation. Unfortunately, attempts to perform such an energy balance have been successful only in natural cases of simple structures, such as a simple duplex scenario (e.g., Mitra and Boyer, 1986). The reason for this is that it is extremely difficult to quantify the physical work involved in these events. The work involved in fracture initiation is difficult to quantify, because the location of the thrust initiation involves too many factors. It is also difficult to quantify the basal shear stress, because it is not constant, as has been assumed in several balancing studies (Elliott, 1976a; Mitra and Boyer, 1986) and it does not increase linearly with depth, as has been assumed in several studies (McGarr, 1980; Williams, 1987). It usually varies with lateral changes in the controlling parameters that include cohesion, friction, overburden load and fluid pressure, which are themselves controlled by lateral lithological changes. Basal friction and cohesion can also vary in time, because they depend on the amount of internal deformation. The amount of internal deformation and the evolution of the fluid source, both of which change with time, further control the fluid pressure. An estimate of the work done opposing gravity is based on the evolution of the topographic relief. Determining this evolution is rather difficult, because it is modified by the flexural response of the lithosphere to both thrust loading and erosional unloading. Lithospheric flexure and erosion may also affect the amount of internal deformation, influencing the quantity of work involved in flexural folding of the thrust sheet. The study of internal deformation can be affected by problems of precision in determining the strain.

Thus, energy balancing seems to be a rather difficult tool to use in complex local thrust sheet structures. Therefore, in these cases, it may be more realistic to choose one of the following:

(1)   geometric and kinematic models accompanied by a simplified analysis of physical constraints; or
(2)   validation of the structural solution by analogous material modelling.

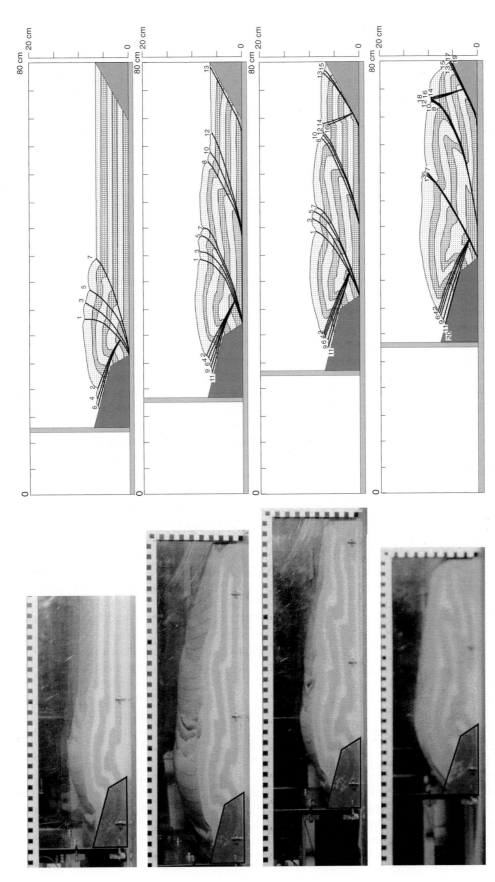

Fig. 3.10. Sandbox model of the energy balance law in the accretion history of a thrust wedge. The numbers indicate the sequence of fracturing and rupturing events. Note that each steepening of the active thrust fault causes its displacement to stop and triggers the initiation or reactivation of the next thrust fault. Thrust faults are seen to develop in sets of splays or sets of sub-parallel fault planes.

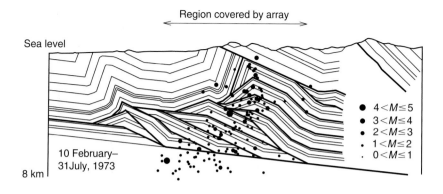

Fig. 3.11. Toe of the active western Taiwan fold-and-thrustbelt, showing hypocentres of micro-earthquakes near the cross section in the foothills of southern Taiwan (data from Suppe, 1980; Wu *et al.*, 1979; see also Davis *et al.*, 1983). Seismicity at the toe is a good indication that the interior of the wedge is at or near its critical taper.

Analogue material modelling seems to be the most suitable tool for detailed validation of several alternative structural solutions derived by geometric balancing, allowing an energy balance control. Detailed video-recorded histories of accretion events in a modelled wedge allow the deformational response to energy balance constraints to be evaluated (Fig. 3.10). The use of miniature stress gauges in the analogous material modelling brings the possibility of determining the energy balance behind sequences of events. Fig. 3.10 documents one sequence of fracturation and rupturation events, controlled by the criterion of minimal physical work. It also shows that the rear part of the accretionary wedge attains a critical taper and becomes stable after a rather large number of repeated rupturation events. This results in a mechanism in which a set of thrusts within a broad zone is likely to be active until they become too difficult to reactivate. A similar energy balance control over a succession of thrust events, switching back and forth between a major back-thrust and the first thrust of the accretionary wedge, was observed in analogue material models by Malavieille *et al.* (1993). There are also natural examples that document the energy balance concept. For example, balanced cross section and earthquake data from Taiwan (Wu *et al.*, 1979; Suppe, 1980; Fig. 3.11) show a broad zone of seismoactive thrusts within the accretionary wedge. Any of them could be used for the energy release after each new energy build-up cycle.

The energy balance determines the ramp spacing, i.e. the widths of individual thrust sheets. A thrust sheet, that has been detached and sheared from its substratum by tectonic force, moves on a detachment fault. It experiences an increase in basal friction as it climbs a progressively larger area of the ramp. It feels an increase in the gravity resistance against its movement as it climbs up. It experiences buttressing forces at its front as the sheet ploughs into its foreland or gets stranded on strata of higher frictional resistance. As the push from the tectonic force continues, the stresses inside the thrust sheet grow until, at some position within the thrust sheet, they exceed the strength of the material and induce faulting, thus, controlling the ramp spacing.

In order to evaluate ramp spacing, let us look again at the rectangular thrust sheet of Mandl (1988). The sheet is defined by boundary thrusts. It has uniform material properties and is sufficiently 'brittle' to reach the faulting instability envelope sooner than the folding instability envelope (see Fig. 3.1a). The assumed boundary conditions are either prescribed by the horizontal normal stress from the rear or by displacement along the base. Regardless of the prescription, the major features of the principal stress trajectories are the same in all models. σ_1 is horizontal in the rear of the sheet. Its magnitude decreases in the thrust direction and its direction deflects towards the detachment, caused by basal shear stress (see Fig. 3.5d). In order to study variations of the shear stress, and to indicate zones of failure, the thrust sheet has an angle of internal friction of 30°, which is uniform inside the sheet. The cohesive part of the shear strength, τ_0, then equals

$$\tau_0 = (\sigma'_\mathrm{I} - \sigma'_\mathrm{III})2\sqrt{3}, \qquad (3.14)$$

where σ'_I and σ'_III are principal stresses. It defines the minimum cohesive strength of the thrust sheet material that prevents the sheet from faulting. It is the strength of this material, the thrust sheet thickness, the basal friction and the pore pressure in both thrust sheet and detachment layer, which determine the ramp spacing and ramp trajectories under the same stress regime. After the ramp has formed, the initial downward deflected σ_1 trajectories, known from the pre-failure model (Fig. 3.5d), change. σ_1 trajectories in the rear basal part become straight and horizontal, which is caused by elastic unloading during ramp formation. The straightening of the earlier deflected σ_1 trajectories by the ramp is much more interesting, because it is a mechanism which 'channels' the push from the rear further into the thrust sheet. This may enable the next ramp to develop under very similar conditions to those accompanying the first faulting. This requires the first ramp to lock up, after the first thrust sheet has climbed sufficiently, and the rear parts have gained sufficient strength by thickening to withstand further deformation. This mechanism then cyclically repeats itself towards the foreland.

The principal difference between the model and nature is that natural thrust sheets do not have rectangular cross sections. They are tapered, which is a result of the primary deposition or later shortening. A slight taper shifts the first ramp towards the foreland, because of magnification of the sub-horizontal σ_1 stress in the thrust direction while the sub-vertical σ_3 stress decreases only slightly.

4 Thin-skin thrustbelt structures

There are various natural types of thrust sheets in thin-skin thrustbelts. Their development, driving mechanisms and internal deformation differ. However, there are various natural kinematic transitions between these thrust sheet types, driven by continuous shortening, changes in controlling rheologies or potential internal deformation of various types of structures.

Fault-propagation fold

When the stress path of the rock section loaded by burial and tectonic stress intersects the faulting instability envelope synchronously with the folding instability envelope (as discussed in the previous chapter), the deformation results in fault-propagation folding. Fault-propagation fold development should be rather rare, considering the statistical chance of the stress path intersecting both instability envelopes at the same time. However, there are numerous case studies in the literature of fault-propagation folds from thrustbelts around the world. Fairly recently Mitra (2002b), following earlier suggestions by Morley (1994), Mitra (1997) and Storti et al. (1997), provided the most likely explanation for this anomaly. Faulted detachment folds roughly resemble fault-propagation folds, which in regions with poorer quality data could lead to their misinterpretation and interchange, as he documents in several cases from the Albanide thrustbelt, the Wyoming thrustbelt, the Papua New Guinea thrustbelt and the Mississippi Fan thrustbelt in the Gulf of Mexico.

Fault-propagation folds develop by a mechanism described by the flexural-slip model (Suppe and Medwedeff, 1984; Suppe, 1985; Jamison, 1987), which is shown in Fig. 4.1. Natural examples have been recognized in the Carpathians (Roure et al., 1993; Krzywiec, 1997; Fig. 4.2; Nemčok et al., 1999), the Alberta Foothills (Dahlstrom, 1969; Lebel et al., 1996), the Appalachians (Bally, 1983; Perry, 1978; Mitra, 2002b), the Atlas mountains (Saint Bezar et al., 1998), the Eastern Himalayas (Kent et al., 2002), the Greater Caucasus (Sobornov, 1996), the Andes (Beer et al., 1990), the European Southern Alps (Tissot et al., 1990; Doglioni, 1993b), the Apennines (Pieri, 1989; Zoetemeijer, 1993), the Wyoming thrustbelt (Royse et al., 1975; Lamerson, 1982), Nankai and Barbados accretionary prisms (Westbrook et al., 1988; Morgan and Karig,

1995), New Zealand (Nicol et al., 1994; Barnes, 1996), western Taiwan (Suppe, 1985) and the South Wales Variscan (Fig. 4.3). Because of their complex structural geometry and steep limb dips, fault-propagation folds are characterized normally by poor seismic expression, and their detailed geometries are difficult to resolve (Mitra, 1990), as can be seen in our Carpathian example (Fig. 4.2). Our experience with seismic reflection profiles through fault-propagation folds in the Carpathians and other thrustbelts indicates that seismic data tend to fail in imaging steeply dipping beds in the footwall and forelimb of the folds. The poor seismic expression can lead to a mistaken interpretation as a fault-propagation fold instead of a faulted detachment fold.

It should be noted that there is no alternative model for fold-propagation fold development. Although tempting, neither the original trishear model (Erslev, 1991) nor modified trishear models (Hardy and Ford, 1997; Erslev and Mayborn, 1997; Allmendinger, 1998; Zehnder and Allmendinger, 2000), described later, apply. The reason for their failure to describe fold-propagation fold development is as follows. Mechanical models of trishear-like folds (Johnson and Johnson, 2002a, b) indicate that the triangular zone of deformation assumed by trishear kinematic models develops only in the case of the forced fold mechanism and only under certain conditions. The term forced fold mechanism means a single or multilayer cover sequence that deforms more or less passively over rigid basement blocks that are displaced along planar or listric faults (Stearns, 1978). The only conditions when the triangular zone develops are when the cover is strongly coupled to the basement. In any other circumstances it does not develop. Furthermore, a triangular deformation zone develops only when the reactivated fault is planar and not listric. The mechanical models of Johnson and Johnson (2002a, b) also imply that fault-propagation folds produce a backlimb deformation to accommodate a fault arrest instead of the triangular region of concentrated deformation around the tip of the ramp as assumed by the trishear model.

The flexural-slip model for fault-propagation fold development creates a distinct fold asymmetry (see Fig. 4.1). Its front limb is much steeper than the rear limb. If we assume no complex deformation of the

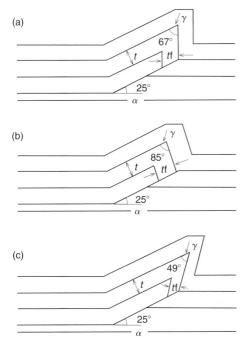

(a)

(b)

(c)

Fig. 4.1. Fault-propagation fold (Jamison, 1987). γ is the interlimb angle, α is the ramp angle, t is the layer thickness and t_f is the layer thickness in the front limb.

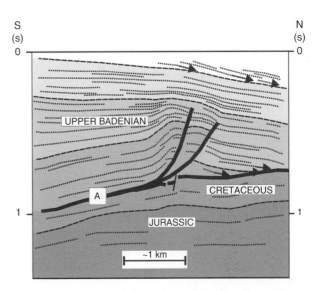

Fig. 4.2. The first frontal thrust sheet of the West Carpathians imaged by reflection seismic section (after Krzywiec, 1997). The overall anticline asymmetry with steeper front limb and a trend of truncated reflectors along the ramp that dies out roughly at a depth level of 0.5 s indicates that a thrust sheet was developed by fault-propagation folding.

Fig. 4.3. Fault-propagation fold in Carboniferous limestone, Black Rock Quarry, South Wales Coalfield. The photograph indicates that ramp propagation was synchronous with folding, because folding is present only in the hanging wall. The fold represents a relatively early stage of its development. Note how ramp tip is located at the focal point of the overlying syncline.

fault-propagation fold, we can recognize several geometric constraints inherent to this fold. The fold develops as a reaction to the thrust in depth. The displacement along this thrust decreases to zero at its propagating tip where the deformation is accommodated only by folding (Fig. 4.3). If we project the axial plane through the frontal syncline, it transects the thrust tip. The syncline in the rear part of the structure indicates a point where the footwall flat changes into the footwall ramp. The axial plane of the fault-propagation anticline intersects the line where the hanging wall flat changes into the hanging wall ramp. Bedding above the hanging wall flat is parallel to the footwall flat and the ramp. The portion of the hanging wall, which is cut by the thrust fault, is folded into a single anticline. Beds, which are not cut by the ramp, are folded by two anticlines. The transition from one deep anticline to two shallow anticlines, together with the axial plane of the frontal syncline, helps to locate the ramp tip line.

The interlimb angle γ, determined from the deeper single anticline, is a function of the ramp angle α and the amount of thickening or thinning in the front fold limb (Figs. 4.1 and 4.4a). The ramp angle is controlled by rheology and pre-existing anisotropies. Figure 4.4(b) shows that a fault-propagation fold with a ramp angle α of 25° can have an interlimb angle of between 21° and 105°. When the thickness of the folded layers remains constant the interlimb angle equals 67°. If the front limb of the fold thickens, the interlimb angle increases. If the front limb thins, the interlimb angle decreases.

Fault-propagation folds frequently change along strike from open to tight folds with increasing fault displacement (Mitra, 1990). This phenomenon is best illustrated by the Turner Valley anticline in the Alberta Foothills (Gallup, 1954; Fig. 4.5). The fault-propagation fold in section AA′, in which the top of the Mississippian is displaced by about 2450 m along the Turner Valley sole fault has an interlimb angle of 86°. In section BB′, which has a greater displacement of about 3350 m, the fold has an interlimb angle of 57°. Assuming that variations in geometry along strike commonly reflect variations with time, this suggests that many fault-propagation folds initiate as open folds and become tighter with increasing fault displacement (Mitra, 1990). This kinematic model, involving a progressive change in interlimb angle with time, is also supported by the result of experimental rock modelling by Chester (1988). Such complex models involve transmission of the interbed shear through the entire structure and lead to area-balanced models, which are characterized by differential thickness changes of units (Mitra, 1990).

The mechanism of fault-propagation folding requires a continuous folding of beds at the tip of the fault.

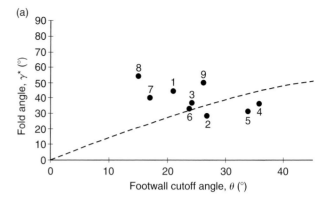

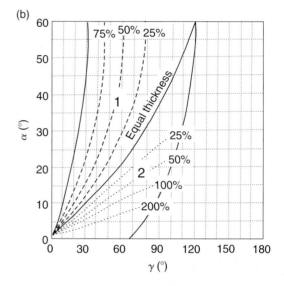

Fig. 4.4. (a) Plots of the fold interlimb half-angle for faulted strata (γ^*) against the footwall cutoff angle for a number of fault-propagation folds and translated fault-propagation folds compared with the theoretical values predicted by Suppe's (1985) model. 1, A part of the Turner Valley anticline, Alberta Foothills with a small displacement (Gallup, 1954); 2, another part of the Turner Valley anticline, with a large displacement (Gallup, 1954); 3, the Meilin anticline, Taiwan (Suppe, 1985); 4, the Digboi anticline, Naga Hills thrustbelt (Mathur and Evans, 1964); 5, the Taipei anticline, Taiwan (Suppe, 1985); 6, the Wills Mountain anticline, West Virginia Appalachians (Perry, 1975); 7, a minor anticline, near Dunlap, Tennessee, Appalachians (Serra, 1977); 8, a minor anticline, Hudson Valley fold belt, New York. (b) Relationship between the interlimb angle γ and the ramp angle α (Jamison, 1987). The graph indicates the case where the thickness in both limbs is the same. Area 1 shows the case where folding results in front limb thinning. Area 2 shows the case where folding results in front limb thickening.

Because of the variation in mechanical properties of layered sedimentary units, this mechanism may operate only within some units of a multilayered sequence. A fault may initiate within a fault-propagation fold and

(a)

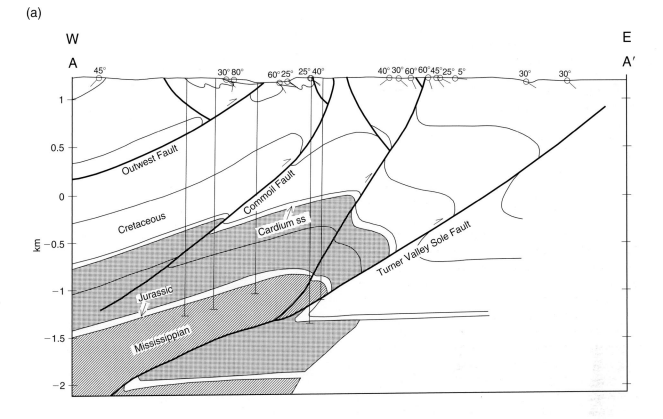

(b)

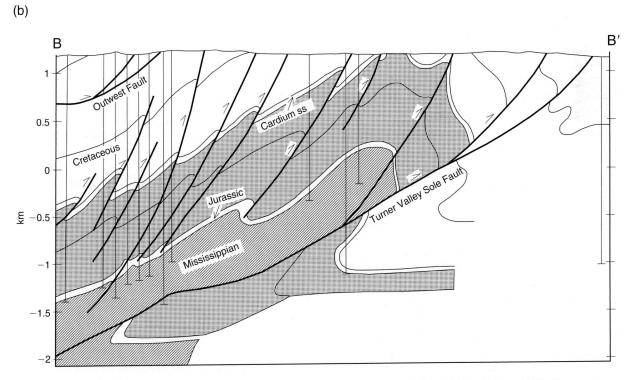

Fig. 4.5. Cross sections through the Turner Valley anticline, Alberta Foothills (after Gallup, 1954 and Mitra, 1990). Cross section BB′ (b), which has a larger thrust displacement than cross section AA′ (a) also has a smaller fold angle (γ^*), suggesting that many fault-propagation folds tighten with increasing fault displacement. No vertical exaggeration. Vertical solid lines show well control.

(a) Propagation through undeformed section

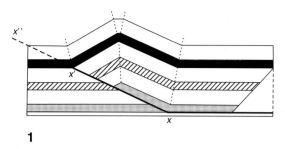

1

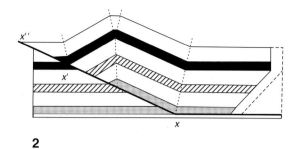

2

(b) Propagation through axial plane of syncline

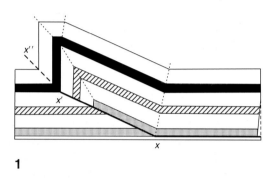

1

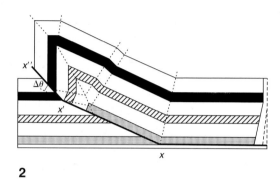

2

(c) Propagation through anticlinal forelimb

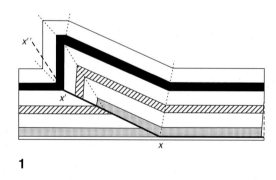

1

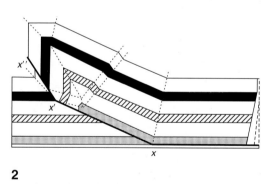

2

Fig. 4.6. Translation of fault-propagation folds along thrusts through: (a) the undeformed section; (b) the synclinal axial plane; and (c) the anticlinal forelimb (Mitra, 1990). Models are constructed with equal line lengths in the deformed state. Shortening profiles show relative shear between units, which must be transmitted through the structure or consumed by relative thickness changes.

subsequently propagate through additional units (Suppe, 1985; Mitra, 1990; Figs. 4.6–4.8), branch into a number of imbricate splays (Lamerson, 1982; Mitra, 1990; Fig. 4.9), or flatten into a detachment within an incompetent unit (Perry, 1975; Mitra, 1990; 4.10). In all of these cases, the geometry of the earlier formed fault-propagation fold may be modified during subsequent translation through fault bends.

If the fold is open, the synclinal axial plane dips considerably more steeply than the thrust, so that propagation through the synclinal axial plane or forelimb requires a significant change in the fault dip. In this case the fault is most likely to propagate through the undeformed section with little or no dip change, and carry the synclinal hinge in the hanging wall (Mitra, 1990; Fig. 4.6a). An example of this structure is a meso-scale

NW   SE

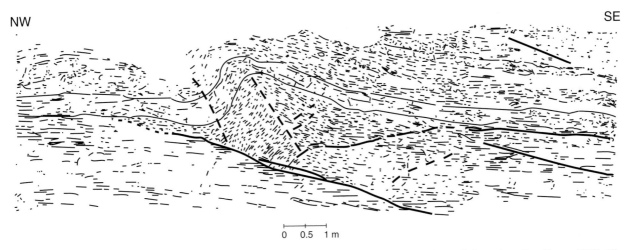

0   0.5   1 m

Fig. 4.7. Translated fault-propagation fold in which the thrust continues to propagate into the undeformed section (Serra, 1977). The fault-propagation fold is located at the tip of a fault-bend fold. The fold is within sandstones and siltstones of the Pennsylvanian Gizzard Group is the southern Appalachian thrustbelt near Dunlap, Tennessee. The segment of the thrust related to fold propagation terminates at the point of intersection between the thrust and the synclinal axial plane.

fault-propagation fold within sandstones and siltstones of the Pennsylvanian Gizzard Group from the Dunlap road cut in the Tennessee Valley and Ridge province of the Appalachians (Harris and Milici, 1977; Serra, 1977; Suppe, 1985; Boyer, 1986; Fig. 4.7). If the fold is tight, the fault may propagate along the synclinal axial plane because this case requires a relatively small synclinal bend in the thrust fault (Mitra, 1990; Fig. 4.6b). Translation of the fold through the synclinal bend results in the development of an additional backlimb panel parallel to the new fault geometry and some additional folding of the front limb. An example case comes from the Naga thrust, the frontal boundary of the Naga Hills thrustbelt in India (Mathur and Evans, 1964; Fig. 4.8).

Alternatively, the fault may propagate through the forelimb of the fold (Fig. 4.6c), leaving the synclinal axis in the footwall of the thrust, as seen in the Taipei thrust, western Taiwan (Suppe, 1985).

The main thrust may branch into a number of forelimb imbricates during its propagation or additional imbricates may develop on the back limb of the anticline (Mitra, 1990). The former case is illustrated by one of the folds within the hanging wall of the Absaroka thrust (Fig. 4.9) and the latter case is illustrated by the Turner Valley anticline (Fig. 4.5). The development of one or more forelimb imbricate thrusts and the progressive abandonment of slip on deeper thrusts may be responsible for the observed listric geometry of many thrust faults.

If the fault tip within a fault-propagation fold reaches an incompetent bed in the sedimentary section, the fault may bend into a detachment within this incompetent unit (Fig. 4.10a). Many fault-bend folds with steeply

dipping front limbs may be fault-propagation folds that have been translated over ramps onto upper detachments. The Wills Mountain anticline (Fig. 4.10b), which marks the western boundary of the Valley and Ridge province in Pennsylvania, Maryland and West Virginia, is a good example of a fold that has been translated onto an upper detachment. The lower detachment of this structure is in the Cambrian Rome Formation, and the upper detachment is in the Upper Ordovician Martinsburg Formation.

Fault-propagation folds, which have not undergone translation, are characterized by a lack of slip transfer into or out of the structure. Therefore, their typical location is in thrustbelt areas where the structures can extend to a greater stratigraphic depth, where shortening is relatively small, and where the slip cannot be transferred to adjacent structures. Fault-propagation folds commonly occur in the frontal parts of fold and thrustbelts. Some good examples come from the Wills Mountain anticline (Fig. 4.10b), which marks the western edge of the Valley and Ridge province (Perry, 1975), the Jaipur anticline, which marks the boundary of the Naga Hills thrustbelt (Mathur and Evans, 1964), and the Turner Valley anticline (Fig. 4.5), which constitutes the frontalmost major structure in the Paleozoic carbonate sequence in the southern part of the Alberta Foothills (Gallup, 1954). Fault-propagation folds can be related to secondary imbricates of major thrust sheets, such as those in the Absaroka thrust sheet (Fig. 4.9).

Simple and translated fault-propagation folds form important structural traps in thrustbelts (Mitra, 1990). The most important traps in fault-propagation folds are in the crests of major anticlines. Example are the Turner

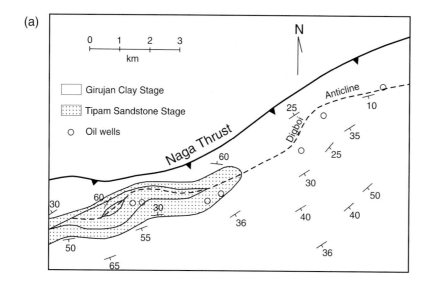

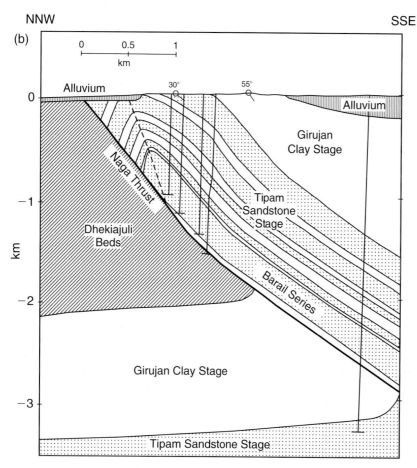

Fig. 4.8. Map (a) and cross section (b) through the Digboi oil field in the Naga Hills thrustbelt (Mathur and Evans, 1964) showing a fault-propagation fold with a fault bend in the footwall, suggesting thrust propagation through the synclinal axis.

Valley anticline, a major oil and gas producer in the Alberta Foothills (Evers and Thorpe, 1975; Fig. 4.5), the Pineview Field in the Absaroka thrust sheet of the Wyoming thrustbelt or the Jaipur anticline, a major oil producer in India (Mathur and Evans, 1964). Fault traps may be present along backlimb thrusts (Fig. 4.5), between imbricates on the forelimbs, and in upturned beds in the footwall. Secondary traps may also be present within major thrust sheets, particularly at the leading edge of the thrust sheet and above footwall ramps (Fig. 4.9). Both types of trap are known to exist in the Wyoming thrustbelt (Mitra, 1990). An example of

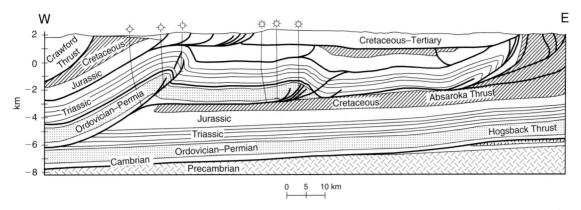

Fig. 4.9. Structural positions of secondary fault-propagation folds in a thrustbelt, the Absaroka thrust sheet case (Lamerson, 1982; Mitra, 1990). Major thrust fault has a staircase geometry and slip transfer to the surface. Secondary fault-propagation folds are located at the leading edge of this thrust fault and above the footwall ramp dissipating fault slip. Yet another fold-propagation fold between those two folds shows the translation by imbricate thrusts.

the former trap type is the Lodgepole oil and gas field at the leading edge of the Absaroka thrust. An example of the latter trap type is the Cave Creek field in the hanging wall of the Absaroka thrust (Lamerson, 1982).

Detachment fold

When the stress path of the rock section loaded by burial and tectonic stress intersects the folding instability envelope earlier than the faulting instability envelope, the deformation results in detachment folding (Figs. 4.11–4.14).

A simplified detachment fold is shown in Fig. 4.11 (Jamison, 1987). The structure is dominated by folding. Propagation of the detachment fault is a consequence of the folding, accommodating discontinuities in strain patterns associated with folding. Detachment folds are common in rock packages characterized by significant variations in competency and thickness of individual horizons (Fig. 4.15). The detachment horizon is commonly an incompetent layer, which can be more or less mobile. Examples of the more mobile layers forming detachments include evaporites or shale. Examples of less mobile horizons consist of carbonates. The detachment horizon is typically overlain by a thick competent layer, typically formed by carbonates or competent siliciclastics. The best examples of detachment folds come from the Jura Mountains (Buxtorf, 1916), the Zagros fold belt (Stocklin, 1968), the Wyoming thrustbelt (Webel, 1987; Coogan and Royse, 1990), the Canadian Rocky Mountains (Jamison, 1997), the Appalachian Plateau (Wiltschko and Chapple, 1977; Davis and Engelder, 1985; Mitra, 1986), the South Pyrenees (Holl and Anastasio, 1993; Burbank and Verges, 1994; Hardy *et al.*, 1996), the Culm Measures of the SW British Variscan (Fig. 4.13), the Brooks Range, Alaska (Homza and Wallace, 1997; Fig. 4.16) and the Mississippi Fan

fold belt (Fig. 4.14). Detachment folds are characterized by a vertical distance between the base of the competent layer and the detachment, which is related to the fold amplitude, measured as the height difference between the fold crest and the surface of the undeformed sedimentary section. The initial wavelength is affected by the thickness of the competent layers that undergo folding (Currie *et al.*, 1962). Figure 4.11(b) shows the dependence between the interlimb angle γ and the back limb angle α. In contrast to fault-propagation folds, the angle α changes during the whole period of growth of a detachment fold, with changing amplitude. Detachment folds (see Fig. 4.11a) develop most typically above a highly mobile detachment horizon. As the fold grows in amplitude, the detachment has to rise in order to maintain constant bed length (Fig. 4.12), which could be kinematically impossible above a stratigraphically fixed detachment horizon. Examples of detachment folds formed above a highly mobile detachment horizon come from the southern Pyrenees (Hardy *et al.*, 1996), the Appalachians (Wiltschko and Chapple, 1977), and the Catalan Coastal Ranges (Burbank *et al.*, 1996). Detachment folds also develop above less mobile detachment horizons. These cases are less common and require a deformation in the fold core in order to keep the growing void underneath the anticline filled. Examples of detachment folds with less mobile detachment horizons come from the Tip Top field in the Wyoming thrustbelt (Webel, 1987; Groshong and Epard, 1994; Fig. 4.17) and from the Deer Park anticline in the Appalachians (Epard and Groshong, 1995; Fig. 4.18). For the growing anticline to remain balanced above a less mobile detachment horizon, appropriate amounts of shortening must accumulate simultaneously at all levels within the structure. For example, all the second-order folds and faults within the Tip Top

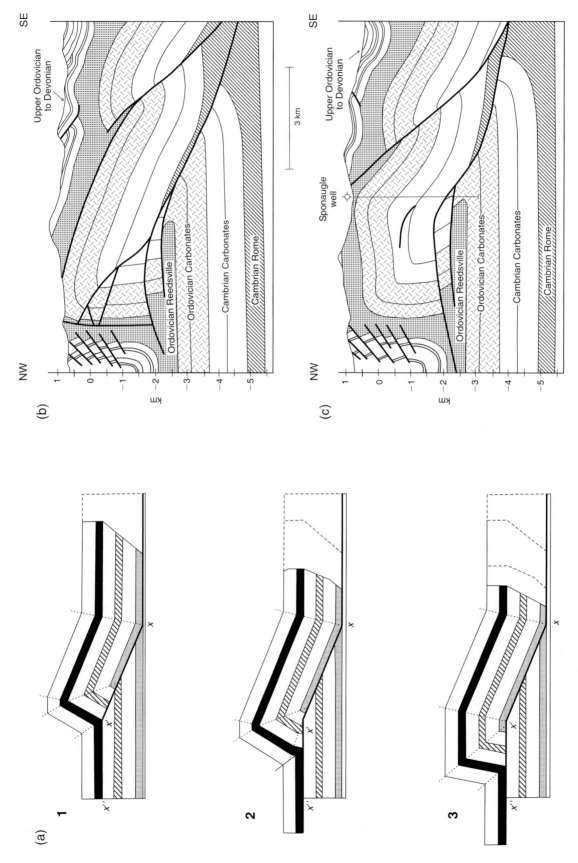

Fig. 4.10. (a) Translation of a fault-propagation fold over a thrust ramp to form a ramp-related anticline (Mitra, 1990). Models are constructed with equal line lengths in the deformed state. Deformed profiles show relative shear between units that must be transmitted through the structure or consumed by relative thickness changes. (b), (c) Cross sections through the Wills Mountain anticline in the West Virginia Appalachians (Perry, 1975). The northern section (b) resembles model a2, whereas the southern section resembles model a3. Note the development of a secondary hanging-wall thrust through the small axial plane predicted by the model in the hanging wall.

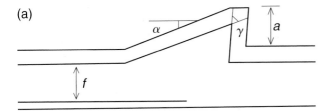

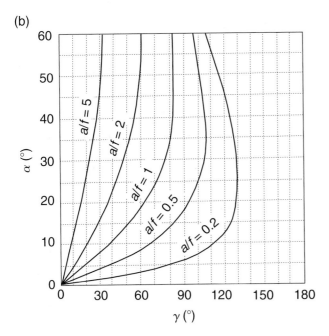

Fig. 4.11. (a) Detachment fold (Jamison, 1987). γ is the interlimb angle, α is the back-limb angle, f is the vertical distance between the base of the competent layer and detachment, and a is the fold amplitude. (b) Relationship between the interlimb and back-limb angle (Jamison, 1987). An explanation is given in the text.

fold hinge migration. Thrustbelts with detachment folds, such as the Appalachians, Zagros or Alps (Goguel, 1962; Hull and Warman, 1970; Gwin, 1964), indicate that both mechanisms are equally important (Mitra, 2002b). An increasing amount of data from thrustbelts such as the Albanide thrustbelt, the Wyoming thrustbelt, the Papua New Guinea thrustbelt and the Mississippi Fan thrustbelt in the Gulf of Mexico suggests that two additional mechanisms are involved in the growth (Mitra, 2002b; Figs. 4.12g and h). The first one is a rotation of the frontal limb segment towards a steeper position without any internal deformation (Fig. 4.12g). It affects rocks prone to flexural slip and has a tendency to operate in early growth stages. The second mechanism causes such a segment rotation by shear (Fig. 4.12h). It affects rocks that are rather resistive to flexural slip and has a tendency to operate in late growth stages. The second mechanism can develop strained forelimbs (Jamison, 1987; Suppe and Medwedeff, 1990), which makes detachment folds prone to subsequent translation during ongoing shortening (Mitra, 1990). Detachment folds may be subsequently translated along forelimb thrusts, similar to the break thrusts described by Willis and Willis (1934) in the Appalachians (Figs. 4.19 and 4.20) and also along backlimb back-thrusts, similar to those described by Rowan (1997) in the Mississippi Fan thrustbelt in the Gulf of Mexico (Figs. 4.21 and 4.22). Some of the resulting structures (Fig. 4.23) resemble translated fault-propagation folds. This resemblance is most probably the main reason behind a large number of published fault-propagation case studies worldwide despite the unique dynamics needed for their growth. They require a stress path that simultaneously meets the faulting and folding instability of the deforming rock package, which makes it very unlikely that they would occur in large numbers worldwide. As indicated in several cases from the Albanide thrustbelt, the Wyoming thrustbelt, the Papua New Guinea thrustbelt and the Mississippi Fan thrustbelt in the Gulf of Mexico (Mitra, 2002b), many of earlier-interpreted fault-propagation folds are in fact translated detachment folds. It is quite possible that the classic detachment folds from the Zagros belt also have been translated along forelimb thrusts (Hull and Warman, 1970; Mitra, 1990).

Some characteristic features, shown in Fig. 4.23 distinguish translated detachment folds from translated fault-propagation folds. The best indicator is the down-warping of synclines below the regional level (Wiltschko and Chapple, 1977; Fig. 4.19), which is caused by the transfer of material of the detachment horizon from synclines into anticlines. This phenomenon is controlled by the mobility of the detachment horizon. Other diagnostic features depend on a detailed knowledge of the

anticline in the Wyoming thrustbelt (Fig. 4.17) grew simultaneously with the fold growth, helping to fill the growing void (Groshong and Epard, 1994). The presence of these layer-parallel strains allowed a detachment fold to form above a stratigraphically fixed detachment horizon. In the case of the Tip Top fold, most of the required shortening in the anticlinal core occurred by the formation of second-order folds and faults. The remaining strain probably occurred as meso-scale structures or homogeneous strain at a micro-scale.

Detachment folds were suggested initially to grow by the two mechanisms shown in Figs. 4.12(a–f) (Epard and Groshong, 1995), which compete. The first mechanism is characterized by a fixed distance between the pair of hinges during the fold growth (Figs. 4.12a–c). In this mechanism the fold grows by limb migration. In the second mechanism the distance between the fold hinges increases as the fold grows (Figs. 4.12d–f) and involves

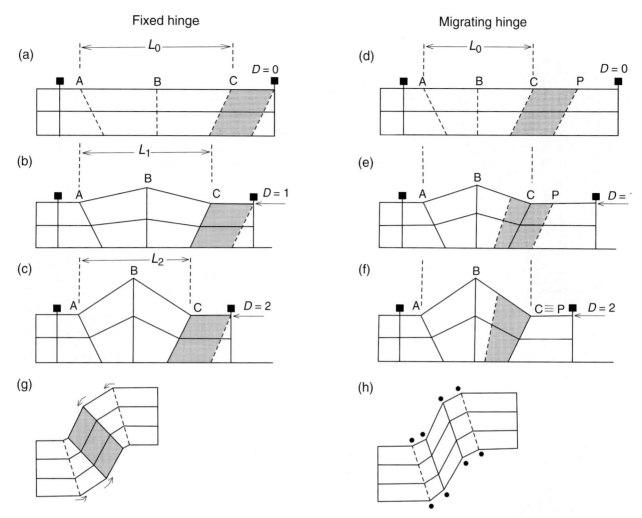

Fig. 4.12. Four mechanisms involved in development of the detachment fold (modified from Epard and Groshong, 1995; Mitra, 2002b). (a)–(c) Fold-hinge folding. (d)–(f) Partially migrating fold-hinge folding. The stippled regions are points fixed with respect to the layers. (g) Rotation of the limb segment into a steeper geometry without internal deformation. It is achieved by bed migration through the outer hinges indicated by dashed lines. (h) Limb segment rotation with internal deformation. It is achieved by rotation along a pair of fixed hinges. It introduces an internal shear inside a rotating segment. It may or may not require migration through outer hinges, as indicated by dashed lines.

growth of detachment folds. Figure 4.23 shows the results of the analysis of numerous cases of translated detachment folds. Each of them is controlled by the specific rheology of the deformed rock package and the specific combination of growth mechanisms described in Fig. 4.12. It seems that it is the capability of the cover units to undergo flexural slip that determines the extent of straining in the rotated segments of the forelimb and backlimb, which then controls break-thrust propagation during later folding stages.

Simple and translated detachment folds form by far the largest structural traps in thrustbelts, as documented by Table 1.9, where the largest fields come from the detachment folds of the Zagros fold belt. The most important traps in detachment folds are major anticlines. Examples

of other detachment fold forming structural traps come from the hydrocarbon fields of the Fort Norman area in the Northwest Territories of Canada, the fields of the Ionian Zone of the Albanide thrustbelt, the Painter and East Painter fields of the Wyoming thrustbelt, and the Puri anticline of the Papua New Guinea fold belt (McLean and Cook, 1999; Mitra, 2002b; Fig. 4.24).

Fault-bend fold
When the stress path of the rock section loaded by burial and tectonic stress intersects the faulting instability envelope before the folding instability envelope, the deformation results in fault-bend folding.

A simple fault-bend fold (Suppe, 1983) is shown in Fig. 4.25. Case examples come from the Appalachians

Fig. 4.13. Circular detachment fold and kink detachment folds in the Carboniferous Crackington Formation in Hartland Quay, North Devon, UK (photograph courtesy of John Cosgrove). Detachment horizon is made of shale.

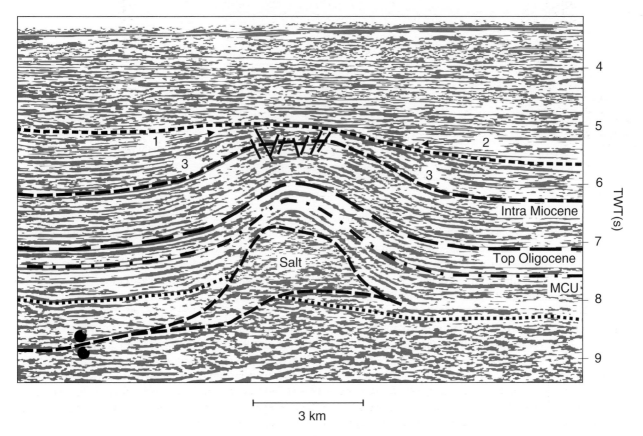

Fig. 4.14. Detachment fold in the Mississippi Fan fold belt, Gulf of Mexico (Rowan *et al.*, 2004).The detachment horizon is formed by the Upper Triassic–Lower Jurassic Louann Salt Formation, which fills the fold core.

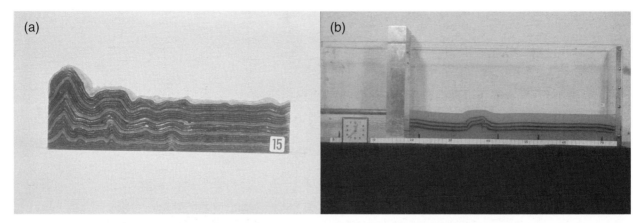

Fig. 4.15. (a) The formation of chevron folds and box folds above a basal detachment in a paraffin wax multilayer compressed by a piston moving in from the left-hand side (photograph courtesy of John Cosgrove). (b) The fold formation in a multilayer made up of five gelatine layers sandwiched between two thick weaker gelatine layers (photograph courtesy of John Cosgrove). In this experiment both boundaries of the multilayer are folded, i.e. the folds do not form above an undeflected detachment horizon as in (a).

Fig. 4.16. Detachment fold train in the backlimb of a larger detachment fold in the Fourth Range in the northern part of the Arctic National Wildlife Refuge, Brooks Range, Alaska (photograph courtesy of Wesley K. Wallace). The fold is formed in carbonates of the Mississippian and Pennsylvanian Lisburne Group (grey) above incompetent shale of the Mississippian Kayak Shale (black). Increasing parasitic folding is evident downward in the Lisburne Group toward the Kayak Shale.

(Suppe, 1980; Kulander and Dean, 1986; Woodward, 1985; Mitra, 1986; Srivastava and Engelder, 1990), the Wyoming thrustbelt (Royse *et al.*, 1975; Williams and Dixon, 1985), the South Californian Ranges (Namson and Davis, 1990), the South Wales Variscides, the Spitsbergen thrustbelt (Bergh *et al.*, 1997), Papua New Guinea thrustbelt (Medd, 1996), the Alberta Foothills (Liu *et al.*, 1996; Spratt and Lawton, 1996), the Spanish Variscides (Alvarez-Marron, 1995), the Sierra de Perijá (Gallango *et al.*, 2002; Fig. 4.26), the Western Alps (Charollais *et al.*, 1977), the European Southern Alps (Schonborn, 1992), the Kirthar and Sulaiman ranges in Pakistan (Banks and Warburton, 1986), the Apennines (Bonini *et al.*, 2002; Fig. 4.27), the Andes (Baby *et al.*, 1992; Shaw *et al.*, 1999), the Mackenzie and Franklin

Mountains, Canada (Cook and MacLean, 1999), Taiwan (Suppe, 1985) or the Pyrenees (Williams, 1985; Burbank *et al.*, 1992).

A simple fault-bend fold forms in the following way. In this model it is assumed that all movement is by slip of the hanging wall, with a passive footwall. A flat thrust propagates within an incompetent layer, ramps up through the overlying competent layer and becomes flat along a stratigraphically higher incompetent layer (Fig. 4.25a). A nice example of flats and ramps formed in various lithologies comes from the restored portion of the Absaroka thrust, which forms the Anschutz Ranch gas and East Anschutz Ranch oil and gas fields (West and Lewis, 1982; Fig. 4.28). The figure shows that flats are formed within the Cambrian shale unit,

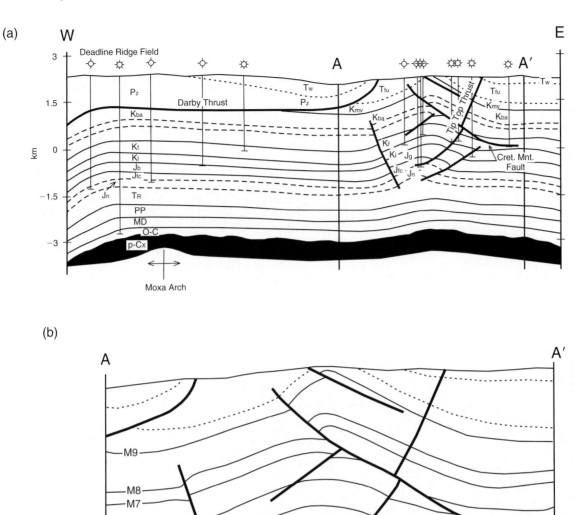

Fig. 4.17. Tip Top field, Wyoming (Webel, 1987; Epard and Groshong, 1995). (a) Cross section showing the well control. (b) Enlargement of the Tip Top area, with wells deleted for clarity. Dotted horizons are unconformities, the top of crystalline basement is black.

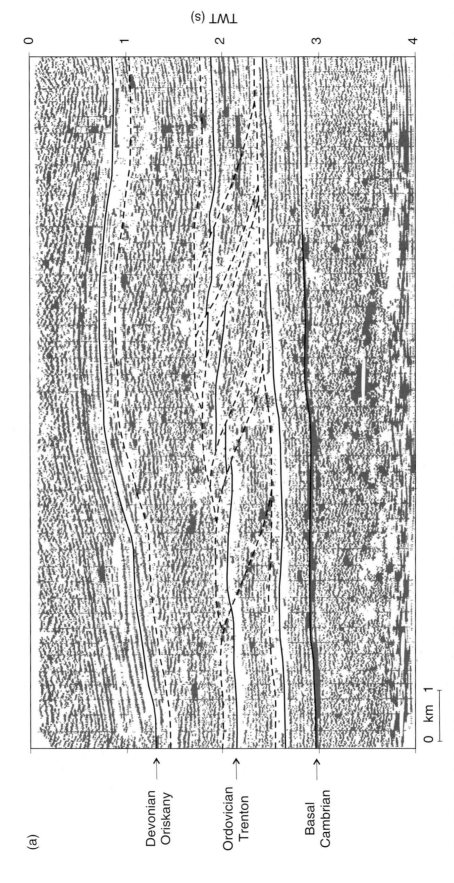

Fig. 4.18. Deer Park anticline, Appalachian Plateau, Pennsylvania. (a) Interpreted seismic section (Mitra, 1986). (b) The depth section is based on seismics from (a) (Epard and Groshong, 1995). Level 1: base of the middle Ordovician (detachment horizon); T: Trenton limestone; level 2: base of the upper Ordovician; level 3: base of the Devonian.

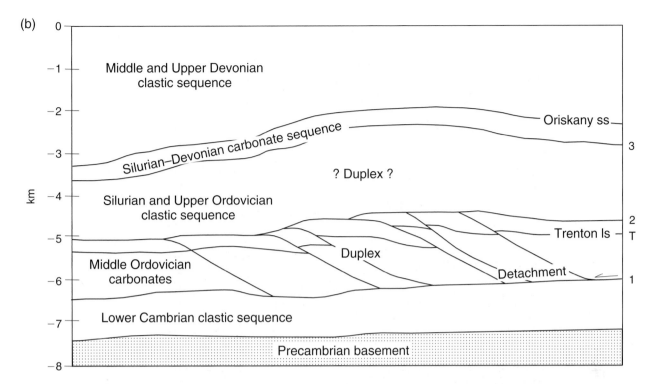

Fig. 4.18 (*cont.*)

Fig. 4.19. Detachment fold translated by a forelimb thrust. Coastal section in the Variscan-deformed early Westphalian interbedded sandstones and shales, Broadhaven, Pembrokeshire, South Wales, UK. A hammer placed in the lower part of the frontal syncline indicates the scale (the photographed part of the outcrop is about 15 m high).

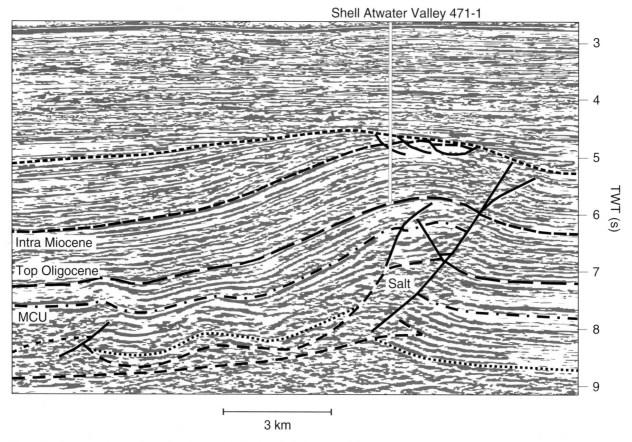

Fig. 4.20. Asymmetric translated detachment fold in the Mississippi Fan fold belt (Rowan *et al.*, 2004).

Fig. 4.21. Symmetric translated detachment fold (photograph courtesy of the Keck Geology Consortium; photograph taken by Kevin Pogue). It is developed in the Pennsylvanian Calico Bluff Formation in Alaska. The cliff is approximately 250 m high.

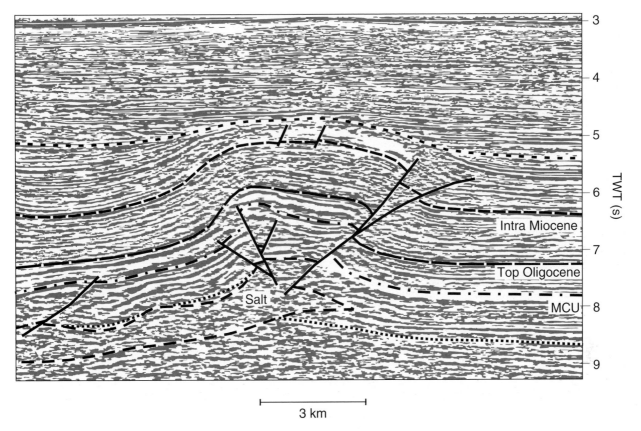

Fig. 4.22. Symmetric translated detachment fold in the Mississippi Fan fold belt (Rowan *et al.*, 2004).

the Permian Phosphoria shaly Formation and the Triassic Ankareh shaly Formation. Ramps are cut through the Mississippian Madison carbonate unit, the limestone of the Triassic Thaynes Formation, the Jurassic Twin Creek limestone unit and the Jurassic Nugget sandstone unit.

Propagation of the controlling fault causes forward movement of the hanging wall generating a syncline–anticline pair above the upper flat. These are syncline A and anticline B in Fig. 4.25(b). Further displacement causes straightening of anticline B, which disappears (Fig. 4.25c). New migrating anticlinal pairs are generated. Continuous displacement brings syncline A and newly formed anticline B_1 forward from the ramp area. The distance between syncline A and anticline B_1 progressively grows until a constant distance is reached and A and B_1 continue migrating as a pair, called the migrating fold pair of the system (Dahlstrom, 1969; Fig. 4.25d). In this situation syncline A indicates the transition from the hanging wall ramp into the upper hanging wall flat and anticline B_1 indicates the transition from the hanging wall ramp into the lower hanging wall flat. An estimate of the thickness of the competent layer is possible if the thrust angle, the fold limb length and the distance between the migrating fold pair are known.

In the rear part of the thrust, a fixed fold pair is formed synchronously with the migrating pair (Fig. 4.25d). Syncline D forms above the line of intersection between the lower footwall flat and the footwall ramp. The position of this syncline is fixed from the initial growth stage of the structure. The initial rear anticline C is formed close to syncline D and migrates progressively forward, until it reaches the position above the footwall transition into the upper footwall flat. From this stage onwards both folds remain fixed and help with determining the thickness of the competent layer.

Although the positions of syncline D and anticline C become fixed with respect to the footwall, material in the hanging wall continuously migrates through both structures with progressive slip along the thrust, so that rock layers in the hanging wall are successively folded and unfolded as they pass through the fold structures. Any permanent strain markers, such as en-echelon vein sets or cleavage, would be expected to record this strain history in rock layers of the lower hanging wall flat that now overlie the upper footwall flat. This deformation requires an interlayer slip during the migration of layers

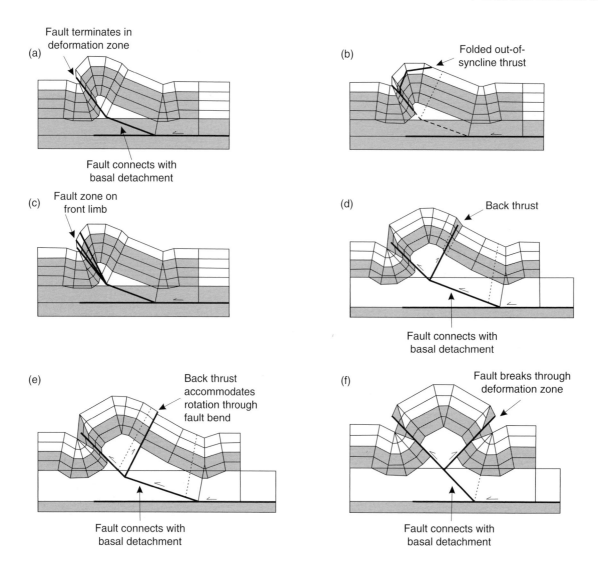

Fig. 4.23. Asymmetric and symmetric translated detachment folds (modified from Mitra, 2002b). Figure (a) shows the asymmetric detachment fold developed in a following rock package, characterized by high competency contrast between detachment and cover units. It consists of a thin incompetent detachment unit, thick competent unit prone to flexural slip and the uppermost thin moderately competent unit prone to some internal strain. Fold growth involves limb rotations and hinge migrations in early stages, followed by fault propagation through the competent folded unit. Fault temporarily ends within deformation zone the upper competent units in the forelimb and within upper levels of the detachment unit. This fault eventually breaks through the whole forelimb and connects with basal detachment. This development model can have two modifications. The first one, shown in (b) involves out-of-syncline back-thrusts, which deform the anticline and become folded during its later development. The second one shown in (c) involves fault branching into imbricate splays. Figure (d) shows the asymmetric detachment fold developed in a following rock package, characterized by moderate competency contrast between detachment and cover units. It consists of a thick incompetent detachment unit, moderately competent unit highly prone to flexural slip and the uppermost incompetent unit prone to internal strain. Fold growth starts with much less asymmetric fold and more pronounced flow inside detachment unit than in the case (a). Flexural slip is more dominant than in the case (a). It allows segment rotations without any larger internal strains. The fault zone initially propagates inside competent unit between locked hinges in the forelimb. It is narrow in the moderately competent unit and progressively widens outward in the uppermost incompetent unit. This fault eventually connect with basal detachment and breaks through the forelimb. Backlimb is also deformed by a thrust fault. This back-thrust either develops independently and terminates against a forelimb thrust when this one connects with detachment (d) or it capitalizes on the backlimb shear caused by bed migration through the bend in the main fault (e). Figure (f) shows the development of the symmetric detachment fold, which is almost identical to the case (d). It is the sub-horizontal dip of the detachment horizon and its extremely low friction that causes fold symmetry. The other consequence is that both thrust in the forelimb and back-thrust in backlimb develop as a conjugate pair and one of them eventually takes over and connects with detachment.

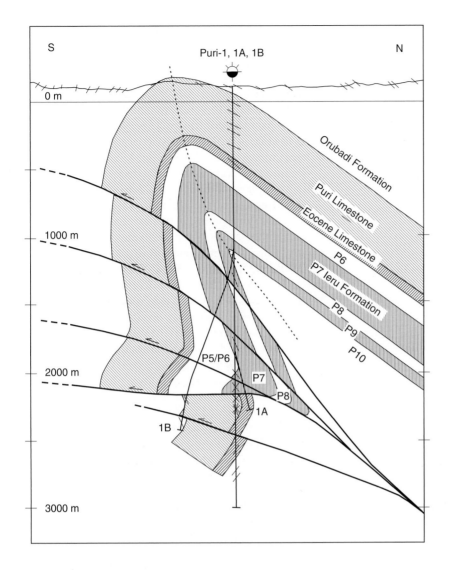

Fig. 4.24. Cross section through the Puri anticline, Papua New Guinea thrustbelt (Mitra, 2002b). Thrusts dissecting the frontal limb developed during detachment fold translation.

from syncline to anticline. The curved interrupted lines in Figs. 4.25(c, d) indicate the migrating axial surfaces, which where folded and straightened during the development. It could be argued that the upper part of the section migrates forward somewhat more than the lower parts of the hanging wall. It is possible that thrust angles and their lengths will be slightly modified during thrusting. These modifications depend on a number of factors, such as the initial lithostratigraphy, the angle of the ramp through the competent layer, the thickness of incompetent layers, the amount of shear and folding parallel to the main thrust, the migration distances of the thrust layers above and below the competent layer, development of a ductile layer, and others.

Fault bend folds most likely develop in response to episodic, earthquake-related slip along the controlling fault. They can be formed by a number of deformation mechanisms.

For example, the growth of the Wheeler Ridge anticline, in the San Joaquin basin, California, which is underlain by blind faults, is dominated by kink-band migration (Mueller and Suppe, 1997). Evidence for the development of a narrow kink-band formed during thrusting comes from the Northridge thrust-type earthquake (Treiman, 1995). Data from this earthquake indicate that fault-bend folds, which grow by movement of the hanging wall strata from flatter fault surfaces upward onto the base of thrust ramps, or above wedge tips, can exhibit bed thickening. The bed thickening is accommodated by small thrust faults, which crosscut the bedding. These small-scale structures were observed in a synclinal bend formed above a wedge tip that was uplifted during the magnitude 6.8 Northridge earthquake (Treiman, 1995).

Flexural slip faulting, or bedding-parallel shear, represents another mechanism that controls the

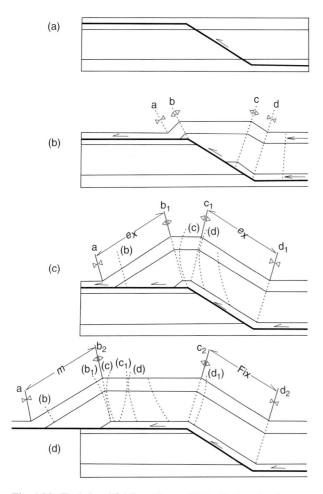

Fig. 4.25. Fault-bend folding (Crane, 1987). Explanation in text.

development of fault-bend folds. This mechanism occurred during the Superstition Hills earthquake (Klinger and Rockwell, 1989), and the Northridge earthquake (Treiman, 1995), in the cores of the Wheeler Ridge anticline (Medwedeff, 1992) and the Ventura Avenue anticline (Rockwell *et al.*, 1988).

Although the growth of a fault-bend fold is driven by compression, extension in local areas is also common, represented by normal faults and tensile fractures. Extensional faulting is common either in fold limbs, which form by the collapse of the hanging wall strata onto the top of ramps, or in upward-convex fault bends, as known from the Apennines (Bonini *et al.*, 2002; Fig. 4.27). Examples of such extensional faulting associated with historic thrust-type earthquakes are known from Algeria (Yielding *et al.*, 1981), Armenia (Philip *et al.*, 1992), and in the eastern United States along the Reelfoot Scarp fault-bend fold, which was uplifted during the 1812 earthquake (Kelson *et al.*, 1996), and serve as examples of such extensional faulting.

Unlike fault-propagation and detachment folds, fault-bend folds can accommodate a significant shortening (compare Figs. 4.1, 4.11 and 4.25). They require a slip transfer along the upper detachment to adjacent structures (Fig. 4.29). This slip can be dissipated in other structures (Fig. 4.29a), transferred backwards along a passive roof thrust (Fig. 4.29c), transferred forward (Fig. 4.29b), or transferred to the topographic surface along a major thrust (Fig. 4.29d).

Common natural cases of slip dissipation structures in fault-propagation and faulted detachment folds were described earlier in the chapter. Natural examples of their role in dissipating slip have been shown in the Wyoming thrustbelt (Mitra, 1990) and the Ebro basin in front of the Catalan Coastal Ranges (Burbank *et al.*, 1992, 1996). Both of these dissipation structures develop inside the thrustbelt (e.g., Lamerson, 1982; West and Lewis, 1982) or in the frontal terminations of the thrust system, where they occur together with penetrative strains, broad zones of thrust faults above a detachment (Fig. 4.30a), and frontal monoclines (Fig. 4.30b; Morley, 1986).

Simple fault-bend folds that form producing traps in thrustbelts are not easy to find, because most of the fault-bend fold-related fields are always combined with duplexes or the fault-bend fold forms an individual horse of a duplex. Therefore, most of the examples will be discussed later in the section dealing with duplexes. The only possible exceptions are the Madrejones oil field in the Argentinean Sub-Andes (Echavarria *et al.*, 2003), the Wheeler Ridge anticline in the San Joaquin basin, California (Medwedeff, 1992; Mueller and Suppe, 1997), fault-bend fold structures in the southeastern Pyrenees (Martinez *et al.*, 1997) and the Perija thrustbelt (Gallango *et al.*, 2002), although the latter two examples are at present exploration areas with oil indications in wells and seeps. The whole Wheeler Ridge structure, being a triangle zone with a buried fault-bend fold, also does not qualify as a simple fault-bend fold.

Most typical traps are anticlinal crests of major folds. Potential secondary traps are faulted blocks controlled by back-thrusts, thrusts and normal faults affecting crestal regions (see Fig. 4.27; Bonini *et al.*, 2002), controlled by the shear strength of the main fault and the mechanical stratigraphy of the deformed sediments. Combinations of thrust and syn-tectonic and post-tectonic normal fault-related traps is typical for the Northern Apennines. Such a syn-contractional extension was observed in action during the El Asnam earthquake in Algeria, which controlled SE-dipping normal faults in the hanging wall anticline coeval with SE-vergent thrusting (King and Vita-Finzi, 1981; Meyer *et al.*, 1990). More complex traps including a fault-bend

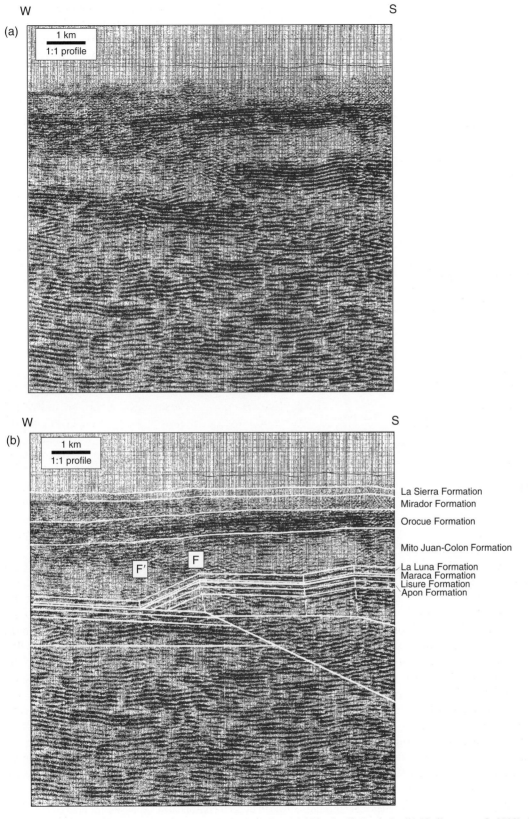

Fig. 4.26. (a) Reflection seismic profile through the fault-bend fold in the Sierra de Perijá (Gallango *et al.*, 2002). (b) Geological cross section interpreted from (a).

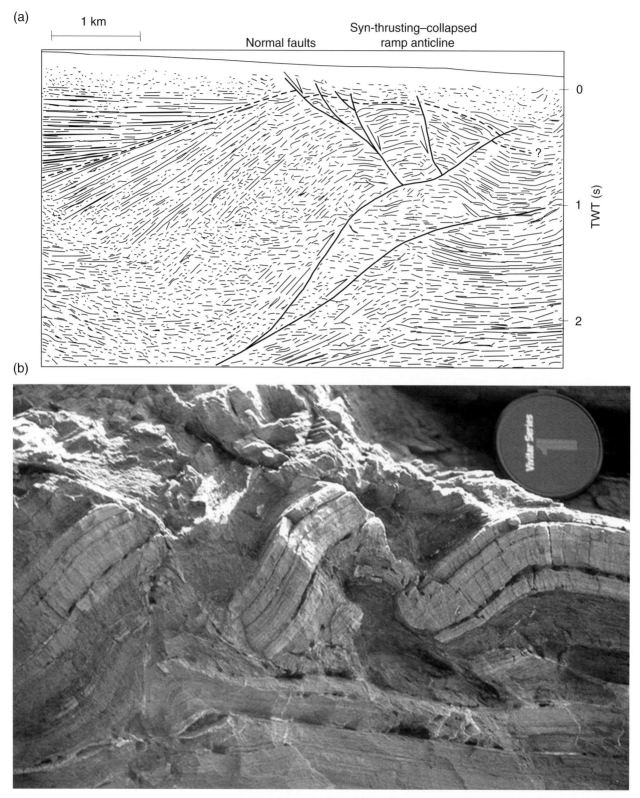

Fig. 4.27. (a) Reflection seismic profile through the fault-bend fold in the Apennines, affected by normal faulting in its crest (Bonini *et al.*, 1999). The crestal collapse is caused by weak lateral constraints provided by semi-consolidated sediments. (b) Fold crest deformed by a set of normal faults, Portovery, Italy (photograph courtesy of the Keck Geology Consortium; photograph taken by David Bice). Deformation is an analogue for the large-scale anticline collapses in the Northern Apennines.

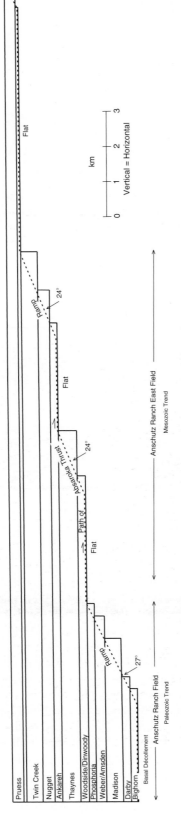

Fig. 4.28. Restored balanced cross section through the Anschutz Ranch gas and East Anschutz Ranch oil and gas fields, Wyoming thrustbelt, showing the effect of mechanical stratigraphy on the location of ramps and flats (West and Lewis, 1982). Explanation in text.

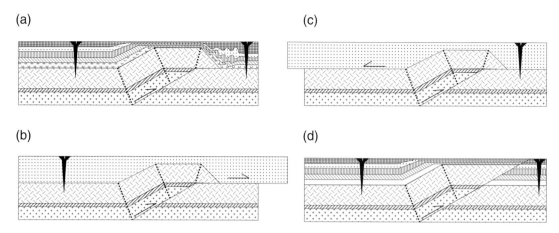

Fig. 4.29. Slip transfer required by fault-bend folds (modified after Zoetemeijer, 1993). (a) The thrust propagates into a basin. Syn-tectonic sediments are coupled with pre-tectonic sediments and deform. (b) Shortening along the fault is transmitted to the foreland and an active roof thrust is formed. (c) Shortening along the fault is transmitted to the hinterland and a passive roof thrust is formed. (d) The thrust propagates into syn-tectonic sediments and cuts through them.

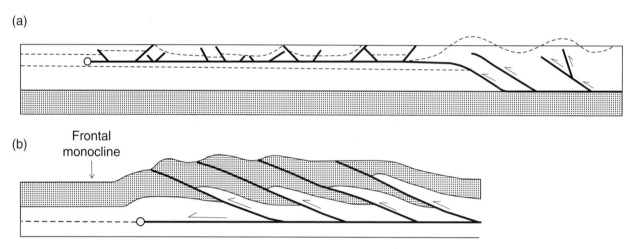

Fig. 4.30. Slip dissipation structures at frontal terminations of thrust systems (from Morley, 1986). (a) Broad zone of penetrative strain or discrete thrust faults above detachment. (b) Frontal monocline.

fold with an underlying duplex structure, such as an example documented from the Appalachians (e.g., Suppe, 1980; Fig. 4.31; Mitra, 1986), are discussed later.

Duplex

When the stress path of the rock section loaded by burial and tectonic stress intersects the faulting instability envelope earlier than the folding instability envelope, the deformation can result in development of a duplex. A duplex is formed by horses stacked between two thrusts or detachment faults (Figs. 4.32–4.34).

Duplexes can occur in either the footwall or hanging wall of a dominant detachment fault, the frontal zone of a ramp anticline, or in an anticlinal core (Mitra, 1986). In each of these positions, the explanation for the deve-

lopment of the duplex is different. Footwall duplexes, developed in the ramp region, help to transfer slip to the foreland (e.g., Boyer and Elliot, 1982). Each new horse forms in the footwall when the physical work necessary for its separation is less than the physical work necessary to move the existing thrust sheet. Typical examples come from oil-producing structures of the Pine Mountain thrust system in the Southern Appalachians (Fig. 4.31; Suppe, 1980) and from the gas-producing Waterton field in the Canadian Foothills (Gordy *et al.*, 1977, 1982). Hanging wall duplexes transfer slip to higher stratigraphic levels as a consequence of increased friction along the existing detachment (e.g., Hedlund *et al.*, 1994; Shaw *et al.*, 1999). Examples are known from the Sawtooth Ranges, Montana (Mitra, 1986), the

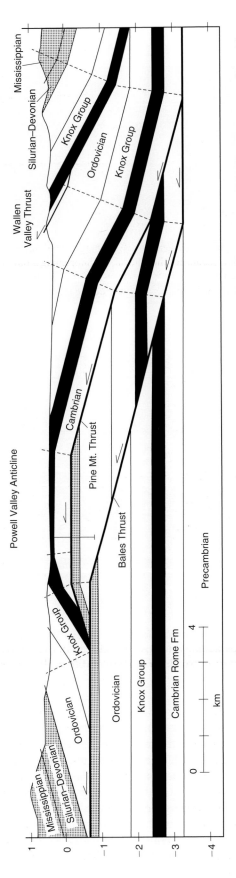

Fig. 4.31. Balanced cross section through the surroundings of the Rose Hill oil field in the Southern Appalachians (modified from Suppe, 1980). The structure was formed by a complete overlap of the lower anticline by the Powell Valley anticline along the Pine Mountain thrust.

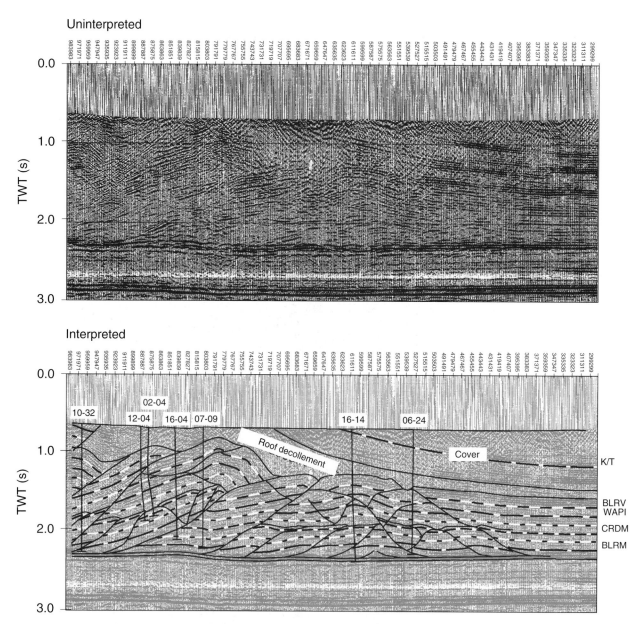

Fig. 4.32. Passive roof duplex from the Canadian Foothills (Couzens-Schultz and Wiltschko, 2000). (a) Uninterpreted seismics. (b) Interpreted seismics. The line is located near the Red Deer River near Sundre. K/T is the Cretaceous/Tertiary boundary, BLRV is top of the Belly River Formation, WAPI is top of the Wapiabi Formation, CRDM is the Cardium sandstone and BLRM is top of the Blairmore Group. Solid lines are thrust faults.

Canadian Rocky Mountains (Fermor and Price, 1976), the Idaho thrustbelt (Hedlund *et al.*, 1994) and the Southern Appalachians (Harris and Milici, 1977). The increase of friction can be caused by lateral facies or pore fluid pressure changes. Duplexes located in frontal zones of ramp anticlines help to dissipate the large-scale displacement accommodated by a fault-bend fold. Typical examples are duplexes in front of the Wills Mountain anticline in the Central Appalachians, such as

the Keyser gas field area (Bagnall *et al.*, 1979; Mitra, 1986). Duplexes in anticlinal cores help to accommodate the required volume changes. A typical example is the Deer Park anticline in the Appalachian Plateau, Pennsylvania, where duplexes fill the core of the detachment fold, which lacks a viscous detachment horizon (Mitra, 1986; Epard and Groshong, 1995; Fig. 4.18).

Duplex geometries vary considerably, covering a range from hinterland sloping duplexes (Boyer and

(a)

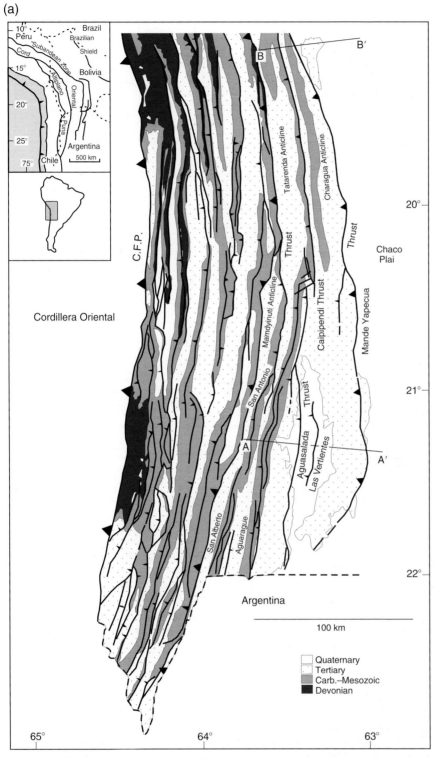

Fig. 4.33. Passive roof duplexes from the Bolivian sub-Andes (modified from Baby *et al.*, 1992). (a) Geological map of the Southern Bolivian sub-Andes with the location of balanced cross sections. Hachure 1 is documenting Quaternary, 2 Tertiary, 3 Carboniferous-Mesozoic and 4 Devonian rocks. CFP is the Cabalgamiento Frontal Thrust. (b) Balanced cross section A–A′. 1 is Tertiary sediments, 2 – Carboniferous–Mesozoic, 3 – Iquiri Formation, 4 – Los Monos Formation, 5 – Icla – Huamampampa Formation, 6 – Kirusillas – Tarabuco – Santa Rosa Formations, 7 – Ordovician rocks. Kir. Det. L.M. Det. and Icla Det. are Kirusillas, Los Monos and Icla detachments. T is thrust and A is anticline. (c) Balanced cross section B–B′.

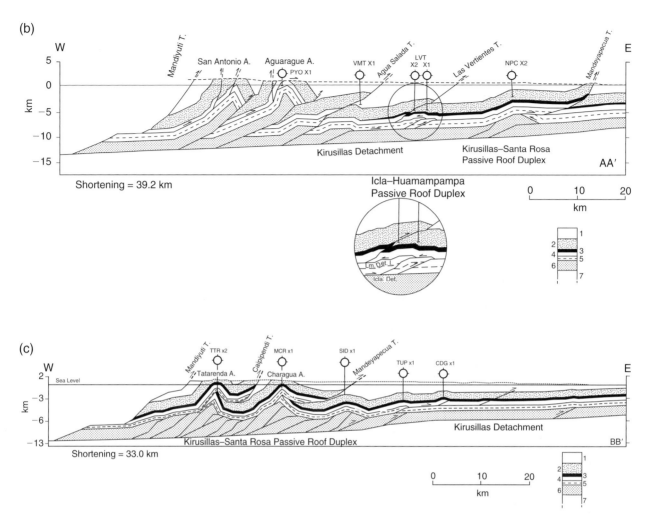

Fig. 4.33 (*cont.*)

Elliott, 1982; Mitra, 1986; Fig. 4.35), the roof thrusts of which dip towards the hinterland through true duplexes (Fig. 4.36), with roof and floor thrusts parallel (Boyer and Elliott, 1982; Mitra and Boyer, 1986; Contreras and Suter, 1997), to foreland sloping duplexes (Boyer and Elliott, 1982; Mitra, 1986) in which the roof thrust dips forelandward (Fig. 4.37), and to fully overlapping ramp anticlines, termed anticlinal stacks.

Natural examples of hinterland sloping duplexes come from the Grandfather Mountain Window in the Southern Appalachians (Boyer and Elliott, 1982), part of the Walker Brook duplex in the Devonian flysch of the Acadian orogen, Maine (Bradley and Bradley, 1988, 1994) and the Sulphur Springs Window in the Appalachians (Mitra, 1986).

Natural examples of true duplexes are known from the Haig Brook, Foinaven, Lighthouse and Eribol area duplexes of the Moine thrustbelt, NW Scotland (Elliott and Johnson, 1980; Coward, 1984; Bowler, 1987;

Fermor and Price, 1987), the lower Basse Normandie duplex in the Boulonnais Variscan thrust front, France (Cooper *et al.*, 1983; Ramsay and Huber, 1987), the Tombstone and other duplexes of the Lewis thrust in the Canadian Rocky Mountains (McClay and Insley, 1986; Fermor and Price, 1987), the Appalachian Plateau and foreland (Bosworth, 1984; Nickelsen, 1986), the Makran accretionary prism (Platt and Leggett, 1986), the Ouachita Mountains (Shanmugam *et al.*, 1988), the Queen Charlotte Islands, British Columbia (Lewis and Ross, 1991), the Boso Peninsula, Japan (Hirono and Ogawa, 1998), the Doublespring duplex from the Idaho thrustbelt (Hedlund *et al.*, 1994) and part of the Walker Brook duplex in the Devonian flysch of the Acadian orogen, Maine (Bradley and Bradley, 1988, 1994).

Natural examples of foreland sloping duplexes occur in part of the Walker Brook duplex in the Devonian flysch of the Acadian orogen, Maine (Bradley and Bradley, 1988, 1994), the Mountain City Window and

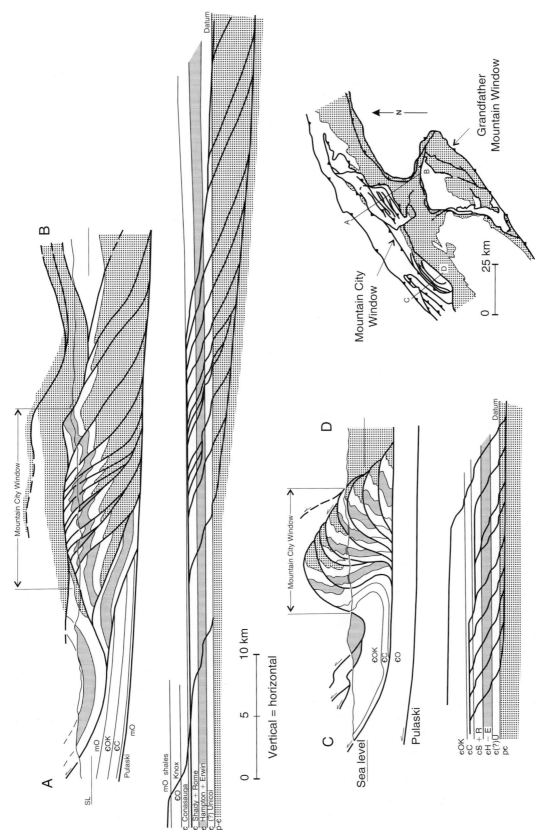

Fig. 4.34. Mountain City Window duplex, Appalachians (Mitra and Boyer, 1986).

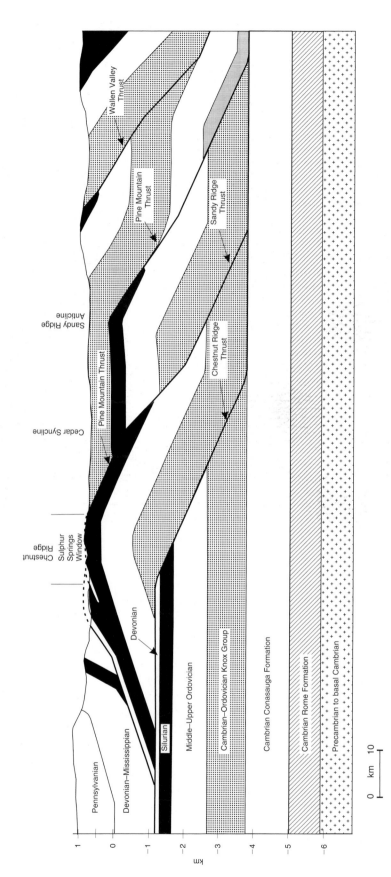

Fig. 4.35. Hinterland-sloping duplex at Sulphur Springs Window, Appalachians (Mitra, 1986).

Fig. 4.36. A small-scale duplex formed in a shale décollement horizon in Widemouth Bay, North Cornwall, UK (photograph courtesy of John Cosgrove).

oil-producing Chestnut Ridge and Sandy Ridge structures in the Southern Appalachians (Boyer and Elliott, 1982; Mitra, 1986), the Anderson Ridge structure and oil-producing Martin Creek structure in the Central Appalachians (Mitra, 1986), the Jumping Pound triangle zone in the Alberta Foothills (Slotboom *et al.*, 1996; Fig. 4.37) and some duplexes above the Lewis Thrust in the Canadian Rocky Mountains (Boyer and Elliott, 1982), including the Waterton gas field (Harris and Milici, 1977; Mitra, 1986).

The geometry of a duplex is controlled by the mechanical interaction of its lower detachment layer, the duplex, and the roof thrust. The stresses and deformation of each depend on this interaction, which was discussed theoretically in a previous chapter.

The combination of a weak horizon forming the future floor thrust fault and a normally competent duplex horizon produces greater footwall movement (Riley *et al.*, 1995). A stronger horizon for the future floor thrust fault slows down the development of the new horse, i.e. a larger displacement occurs on the existing ramp before a new horse is formed (Riley *et al.*, 1995). This is because stress generated by push from behind is more 'funnelled' to an existing horse.

A floor thrust fault with low strength results in large displacements of the structures above it, producing overlapping ramp anticlines with a tendency to form antiformal stacks (Couzens *et al.*, 1997). A weak floor thrust fault causes fewer horses within the duplex and a wider ramp spacing (Strayer, 1998). Each horse accommodates more of the total shortening. A low strength floor thrust fault allows large penetrative strains to affect the duplex sequence before the thrust ramps are generated. Again, this is due to the 'stress

(a)

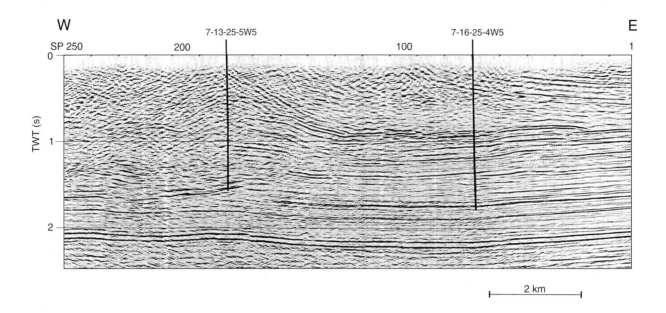

(b)

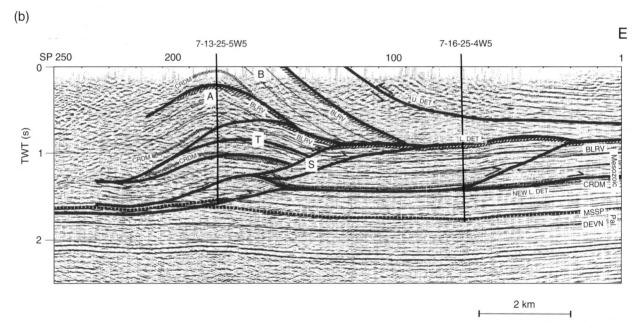

Fig. 4.37. Jumping Pound triangle zone in the Alberta Foothills (Slotboom *et al.*, 1996). (a) Uninterpreted seismic section. (b) Interpreted seismic section. PAL – Paleozoic rocks, MSSP – Mississippian rocks, DEVN – Devonian Rocks, CRDM – Cardium Formation, BLRV – Belly River Formation, U.DET. – upper detachment, L.DET. – lower detachment. A, B, T and S are thrust sheet labels.

funneling effect'. Several horses can move simultaneously in this scenario. A strong floor thrust fault results in horses with a small displacement, each forming independent ramp anticlines, more horses and a closer ramp spacing (Couzens *et al.*, 1997; Strayer, 1998). Each

horse accommodates less of the total shortening. A high strength floor thrust fault allows only small penetrative strains to affect the duplex sequence before the thrust ramps are generated. Horse development would be strictly sequential in this scenario. This scenario is also more likely to form passive roof duplexes, which are described later.

For the same floor thrust fault strength and varying duplex layer strengths, the ramp spacing inside the duplex widens with increasing strength. A stronger floor thrust fault causes a greater footwall deformation than a weaker floor thrust (Gutscher *et al.*, 1996).

Sandbox experiments (Kukowski *et al.*, 2002) suggest that the roof thrust fault of a duplex develops along a horizon with greater strength than the horizon of the floor thrust fault. Its strength, together with mechanical stratigraphy of the roof thrust determines the deformation style of the roof thrust. It also determines whether the roof thrust moves forelandward with the duplex, whether it locally deforms above the duplex or whether it moves hinterlandward (Banks and Warburton, 1986; Dunne and Ferrill, 1988; Geiser, 1988a, b; Groshong and Epard, 1992). The behaviour of the roof thrust defines an allochthonous roof duplex, a passive roof duplex and a duplex with local roof compensation (Fig. 4.38; Boyer and Elliott, 1982; Smart *et al.*, 1997; Averbuch and Mansy, 1998).

Natural examples of allochthonous roof duplexes come from the Valley and Ridge and Plateau Provinces of the Central and Southern Appalachians (Perry, 1978; Geiser and Engelder, 1983; Kulander and Dean, 1986; Mitra, 1986; Geiser, 1988a, b; Ferrill and Dunne, 1989; Smart *et al.*, 1997), the Central Apennine front (Averbuch *et al.*, 1995), the eastern MacKenzie Mountains, Canada (Vann *et al.*, 1986), the Gaissa Nappe, Finland (Towsend *et al.*, 1986), the southern Central Pyrenees (Williams, 1985) and the French Alps (Charollais *et al.*, 1977; Butler, 1985).

Natural examples of duplexes with at least partial local compensation are known from the lower Basse Normandie duplex in the Boulonnais Variscan thrust front, France (Cooper *et al.*, 1983), the Central Appalachians (Ferrill and Dunne, 1989), the Appalachian Plateau (Mitra, 1986; Groshong and Epard, 1992, 1994; Fig. 4.18), the northeastern Brooks Range, Alaska (Wallace and Hanks, 1990a, b), the Osen-Roa thrust, Norway (Morley, 1987) and the French Alps (Charollais *et al.*, 1977; Butler, 1985). Roof thrust deformation studies such as the study of Smart *et al.* (1997) suggest that meso- and micro-scale processes could take up to 75% of such a local compensation and could be easily missed by seismic study.

Passive roof duplexes together with case studies are described in the following section on triangle zones. A passive roof duplex requires the roof sequence to be coupled with its base forelandward of the duplex but able to decouple above the duplex (Banks and Warburton, 1986; Dunne and Ferrill, 1988; Geiser,

1988a, b). Examples are known from the Bolivian Sub-Andes (Baby *et al.*, 1992; Fig. 4.33), the Canadian Rocky Mountains (Thompson, 1979, 1982; Price, 1981; Jones, 1982; McMechan, 1985; Couzens-Schultz and Wiltschko, 2000; Fig. 4.32), the Kirthar and Sulaiman Mountain Ranges (Banks and Warburton, 1986; Humayon *et al.*, 1991; Jadoon *et al.*, 1994) and the Taiwan thrustbelt (Suppe, 1983).

Duplex-related traps in thrustbelts are quite common. Examples, except for the passive roof duplexes mentioned in the following section, include the Sulphur Springs, Big Fleenorton, Martin Creek, Chestnut Ridge and Possum Hollow oil-producing areas and the gas-producing Broad Top duplex and Keyser field of the Southern and Central Appalachians (Rowlands and Kanes, 1972; Jacobeen and Kanes, 1975; Harris and Milici, 1977; Mitra, 1986), Savanna Creek, Jumping Pound, and Waterton gas fields in the Canadian Rocky Mountains (Davidson, 1975; Hennessey, 1975; Gordy *et al.*, 1977, 1982).

Triangle zone

Examples from the Bolivian Sub-Andes (Baby *et al.*, 1992), the Peruvian Ucayali basin (Shaw *et al.*, 1999) and the Alberta Foothills (Liu *et al.*, 1996; Slotboom *et al.*, 1996; Fig. 4.37) show that fault-bend folds, sometimes in combination with other fold types, frequently form complex structures, which originated either above one detachment or a number of detachments. Their stress control is difficult to generalize because not all of the involved faults have to initiate earlier than folding and in the cases of other associated fold types the faults are coeval or younger than the folds, sometimes being formed as accommodation faults in complex fold structures. There are basically two end members of the triangle zone structures (Couzens and Wiltschko, 1996; Tables 4.1–4.3).

The first end member is a fold-dominated structure with opposed thrusts off the same detachment (Fig. 4.39), predominately formed above a weak viscous horizon (Table 4.1). Typical examples come from the Pyrenees (Burbank *et al.*, 1992; Verges *et al.*, 1992; Fig. 4.40) and the Western Carpathians.

The second end member comprises two or more detachments, propagated inside weak viscous or plastic layers (Table 4.1), as in, for example, the Sulaiman Range in Pakistan (Banks and Warburton, 1986). This category also comprises a passive roof duplex (Fig. 4.38). The mechanical stratigraphy survey (Table 4.1) of Couzens and Wiltschko (1996) shows that most passive roof duplexes develop as a result of an incompetent roof thrust, a large strength difference between the roof

(a)

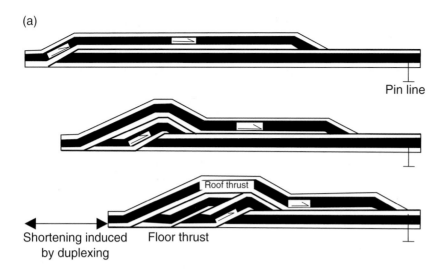

Pin line

Roof thrust

Shortening induced
by duplexing   Floor thrust

(b)

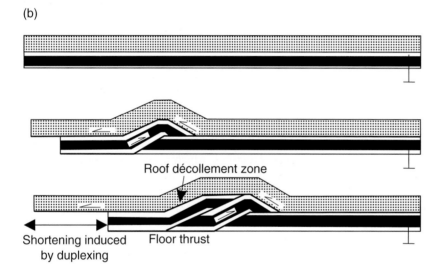

Roof décollement zone

Shortening induced
by duplexing   Floor thrust

(c)

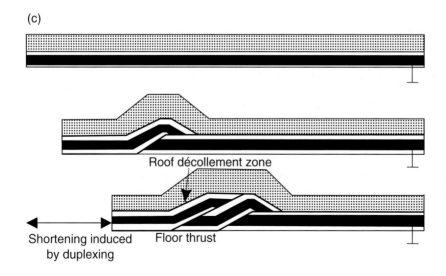

Roof décollement zone

Shortening induced
by duplexing   Floor thrust

Fig. 4.38. Roof thrust responses to the
emplacement of the duplex (modified
from Smart *et al.*, 1997; Averbuch and
Mansy, 1998). (a) Allochthonous roof
duplex. (b) Passive roof duplex.
(c) Duplex with local roof
compensation.

Table 4.1. Example triangle zones in thrustbelts (Couzens and Wiltschko, 1996). The type I geometry is a fold-dominated structure with opposed thrusts off a single detachment and the type II geometry is a thrust-dominated structure with two (or more) detachments, known as a passive-roof duplex.

Orogen	Location	Geometry	Detachment		Cover lithology	Duplex lithology	Ref.
			Floor	Roof			
South Can. Rockies	Frontal	II	sh	sh	sh/ss/co(syn)	ca/ss/sh	1
Central Can. Rockies	Frontal	II	?	sh	ss/sh/ca(syn)	ca/ss/sh	2, 3
West McKenzie Mts.	Frontal	II	?	ev	ca/ss/sh	cryst	4
Brooks Range, Alaska	Frontal	II	?	sh	sh/ss(syn)	ca	5
Wyoming thrustbelt	Frontal	II	sh	sh	sh/ss(syn)	ca/ss/sh	6
Madison Range, Montana	Frontal	II	?	sh	sh/ss	?	7
East Ouachitas	Frontal	II	sh	sh	sh/ss	ss/sh	8, 9
West Ouachitas	Frontal	II	sh	sh	sh/ss	ss/sh	10
Great Valley, California	Frontal	II	?	sh	sh/ss	cryst	11, 12
S. Coast Ranges, California	Frontal	II	?	sh	sh/ss	sh/ss	13
Barbados	Frontal	II	?	sh	sh/ss	sh/ss	14, 15
Ellesmerian Front, Greenland	Frontal	II	?	sh	sh/ca/ss	ca/ss	16, 17
N. Argentina Precordillera	Frontal	II	sh	sh	sh/ss	ca	18
S. Argentina Precordillera	Frontal	II	sh	sh	sh/ss/volc	ss/volc/sh	19
Papua New Guinea	Not frontal*	II	sh	sh	sh/ss/ca	cryst/ss/sh	20, 21
Venetian Alps	Frontal	II	?	sh	sh/ss(syn)	ca	22
German Alps	Frontal	II	sh	sh	sh/ss	sh/ss	23
North Urals	Frontal	II	sh	sh/ev	sh/ss(syn)	ca/ss/sh	24
Dagestan belt, Caucasus	Frontal	II	?	sh	sh/ss(syn)	ca/sh	25
Kohat Plateau, Pakistan	Frontal	II	salt	salt	sh/ss(syn)	ca/ss/sh	26, 27
Kirthar and Sulaiman Ranges	Frontal	II	salt	sh	sh/ss(syn)	ca/sh	28
Potwar Plateau, Pakistan	Frontal*	II	salt	sh	sh/ss(syn)	ca/sh/ss	29
Parry Island thrustbelt	Front–internal	II	salt	sh	sh/ss	ca	30, 31
Central Pyrenees	Internal	II	?	ev	ca/ss/sh	cryst	32, 33
Potwar Plateau, Pakistan	Front Plateau	I	salt		ca/sh/ss		29
Parry Island thrustbelt	Front Plateau	I	salt		ca/sh/ss		30, 31
Central Pyrenees	Front Plateau	I	ev		ss/sh		32, 33
Central Appalachians	Front Plateau	I	salt		ss/sh/coal		34
East McKenzie Mts.	Front region	I	salt		ca/ss/sh		35
South Appalachians	Internal*	I	sh		ca/ss/sh		34
South Urals	Frontal	I	salt		ss/sh		36

References:
1, Price (1986); 2, Thompson (1981); 3, McMechan (1985); 4, Vann *et al.* (1986); 5, Hanks (1993); 6, Williams and Dixon (1985); 7, Tysdal (1986); 8, Blythe *et al.* (1988); 9, Arbenz (1989b); 10, Hale-Erlich and Coleman (1993); 11, Unruh and Moores (1992); 12, Glen (1990); 13, Namson and Davis (1990); 14, Unruh *et al.* (1991); 15, Torrini and Speed (1989); 16, Soper and Higgins (1987); 17, Soper and Higgins (1990); 18, VonGosen (1992); 19, Ramos (1989); 20, Hill (1991); 21, Hobson (1986); 22, Doglioni (1992); 23, Muller *et al.* (1977); 24, Sobornov (1992); 25, Sobornov (1994); 26, McDougall and Hussain (1991); 27, Abbasi and McElroy (1991); 28, Banks and Warburton (1986); 29, Pennock *et al.* (1989); 30, Harrison *et al.* (1991); 31, Harrison (1993); 32, Verges *et al.* (1992); 33, Burbank *et al.* (1991); 34, Woodward (1985); 35, Cook (1992); 36, Kazantsev and Kamaletdinov (1977).

Notes: sh, shale; ev, evaporites; ss, sandstone; ca, carbonate; volc, volcanic rocks; cryst, crystalline rocks; co, coal; (syn), syntectonic sediments; ?, unknown.

* Papua New Guinea, one thrust to foreland of the triangle zone; Potwar Plateau, Pakistan, type II triangle zone behind broad plateau underlain by salt; Southern Appalachians, possible influence of basement topography, forward-verging structures in front.

Table 4.2. Lithology of duplexes in thrustbelts with either back-thrust or forethrust motion on roof detachment (modified after Couzens and Wiltschko, 1996). Italic indicates duplexes involving over 50% of relatively strong material and bold indicates duplexes involving 50% of relatively strong material.

No	Location	Cover response	%coal	%sh	%ss	%ca	Ref.
1	South Can. Rockies	Back-thrust	2	48	**33**	**17**	1–5
2	Central Can. Rockies	Back-thrust	1	49	**29**	**21**	1–5
3	S. Wyoming thrustbelt	Back-thrust	0	32	*26*	*42*	6,7
4	Sulaiman Range, Pakistan	Back-thrust	0	40	*18*	*42*	8
5	Quachitas (west Arkansas)	Back-thrust	0	56	44	0	9
6	Potwar and Kohat Plateau	Back-thrust	0	15	*48*	*37*	8
7	Taiwan	Back-thrust	0	46	*54*	*0*	10
8	Northern Urals	Back-thrust	0	11	*5*	*84*	11
9	Dagestan belt, Caucasus	Back-thrust	3	19	*46*	*32*	12
10	Southern Alps, Italy	Forethrust	0	8	*8*	*84*	13
11	Central Appalachians, PA	Forethrust	0	3	*1*	*96*	14–16
12	Appalachians (PA) Devonian detachment	Forethrust	0	21	*16*	*63*	16
13	Central Appalachians (VA/WV)	Forethrust	0	28	1	*71*	17–19
14	Appalachians (VA/WV) Devonian detachment	Forethrust	0	43	6	*51*	18, 19
15	French Alps	Forethrust	0	12	4	*84*	20

References:
1, Webb and Hertlein (1934); 2, Mellon (1967); 3, Gibson (1985); 4, Frebold (1957); 5, Brown (1952); 6, Anderson (1956); 7, Hunter (1988); 8, Ibraham (1977); 9, Merewether (1971); 10, Yeh and Yang (1990); 11, Sobornov (1992); 12, Sobornov (1994); 13, Schonborn (1992); 14, Inners (1987); 15, Folk (1960); 16, Faill (1987); 17, Diecchio (1986); 18, Kulander *et al.* (1986); 19, Dennison (1971); 20, Charollais *et al.* (1977).

Table 4.3. Lithology of cover sequences in thrustbelts with either back-thrust or forethrust motion on roof detachment (modified after Couzens and Wiltschko, 1996). Italic indicates cover sequences involving over 50% of relatively strong material and bold indicates cover sequences involving 50% of relatively strong material.

No	Location	Cover resp.	%coal	%sh	%ss	%ca	Ref.
1	South Can. Rockies	Back-thrust	1	63	36	0	1
2	Central Can. Rockies	Back-thrust	15	37	48	0	1
3	S. Wyoming thrustbelt	Back-thrust	0	72	28	0	2
4	Sulaiman Range, Pakistan	Back-thrust	0	54	43	3	3
5	Quachitas (west Arkansas)	Back-thrust	0	85	15	0	4
6	Potwar and Kohat Plateau	Back-thrust	0	55	45	0	3
7	Taiwan	Back-thrust	0	56	44	0	7
8	Northern Urals	Back-thrust	0	50	**50**	0	8
9	Dagestan belt, Caucasus	Back-thrust	0	48	*52*	0	9
	Ouachitas (west Arkansas 2)	Back-thrust	0	68	32	0	5
	Venetian Alps, Italy	Back-thrust	0	57	31	12	6
10	Southern Alps, Italy	Forethrust	0	7	0	*93*	10
11	Central Appalachians, PA	Forethrust	1	44	*54*	*1*	11–13
12	Appalachians (PA) Devonian detachment	Forethrust	5	21	*72*	*2*	11–13
13	Central Appalachians (VA/WV)	Forethrust	1	41	*48*	*10*	14–17
14	Appalachians (VA/WV) Devonian detachment	Forethrust	3	32	*58*	*7*	14–17
15	French Alps	Forethrust	0	40	*60*	*0*	18
	Eastern Pyrenees	Forethrust	0	14	*8*	*78*	19

References:
1, Mack and Jerzykewicz (1989); 2, Donnell (1961); 3, Ibraham (1977); 4, Merewether (1971); 5, Merewether and Haley (1969); 6, Massari *et al.* (1986); 7, Yeh and Yang (1990); 8, Sobornov (1992); 9, Sobornov (1994); 10, Schonborn (1992); 11, Edmund and Eggleston (1989); 12, Inners (1987); 13, Folk (1960); 14, Englund *et al.* (1986); 15, Diecchio (1986); 16, Kulander *et al.* (1986); 17, Dennison (1971); 18, Charollais *et al.* (1977); 19, Betzler (1989).

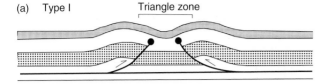

(a) Type I

Triangle zone

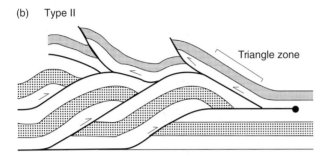

(b) Type II

Triangle zone

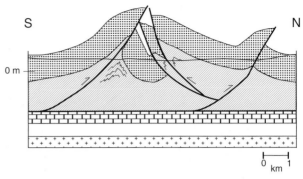

S N

0 m

0 km 1

Fig. 4.40. Type I triangle zone from the South Pyrenees (Verges *et al.*, 1992). A N–S cross section runs across the culmination zones between the Sanauja and Vilanova anticlines.

Fig. 4.39. Two end member types of triangle zones (Couzens and Wiltschko, 1996). (a) The type I end member is a fold-dominated structure with opposing thrusts branched off a single detachment. (b) The type II end member is a passive roof duplex.

(1) frictional resistance of the lower detachment;
(2) work required to flex and deform the duplex as it moves through the ramp towards the foreland;
(3) frictional resistance of the upper detachment; and
(4) work required to deform the roof thrust so that it accommodates the geometry of the advancing duplex (Jamison, 1996).

thrust and the duplex material, and a weak horizon at the base of the future roof thrust (Tables 4.2 and 4.3). Smectite–illite transition studies (Couzens-Schultz and Wiltschko, 2000) indicate that the presence of a weaker rheology along the future roof thrust is not necessary if such a boundary coincides with the smectite–illite transition within a thick sequence. This phenomenon has been described in the Alberta Foothills, where it was compared with the smectite–illite transition in the Gulf of Mexico, which was correlated with an overpressured zone also indicated by increased porosity and decreased density.

Passive roof duplexes themselves can be divided into two types, comprising a simple wedge or a stack of sheets beneath the roof thrust (Jamison, 1996). The former is represented by the Turner Valley triangle zone in the Canadian Cordillera, Alberta (MacKay *et al.*, 1994), the Negro Mountain and Deer Park anticlines in the Appalachians (Woodward, 1985), the triangle zones in the Ebro basin, Spain (Verges *et al.*, 1992) and triangle zones of the Potwar Plateau (Pennock *et al.*, 1989). The second type comprises the Wildcat Hills triangle zone in the Alberta Foothills (Lawton *et al.*, 1994), the triangle zones of the Bolivian (Fig. 4.41) and Argentinean sub-Andes (Echavarria *et al.*, 2003).

The energy balance in the case of a passive roof duplex is very different from that of fold-bend folding. The forward advance of the duplex in a passive roof thrust is opposed by a number of forces, including:

The friction along the lower detachment controls how much stress is transferred into the footwall, controlling how much the footwall is affected by deformation, or eventually when and where a new ramp is developed (Jamison, 1996). The existence of footwall faulting in nature indicates either a weaker rheology for the footwall than for the hanging wall or a more effective stress transfer through the lower detachment.

The mechanical stratigraphy of the duplex itself controls its deformation. Physical models of thrust movement through a ramp region (Morse, 1977; Chester *et al.*, 1991) indicate that this deformation can vary from fracture patterns accommodating structural thickening in the case of a thick homogeneous roof thrust to primarily bedding parallel slip with minor thickening. The core region of a triangle zone is a region of complicated structure that comprises faulting and folding.

A database of data from triangle zones in thrust-belts worldwide (Couzens and Wiltschko, 1996; Tables 4.1–4.3) indicates that triangle zones most frequently occur in frontal parts of thrustbelts (see also Vann *et al.*, 1986). Typical examples of hydrocarbon fields in triangle zones therefore come from frontal portions of the thrust-belt and include thrustbelts such as the Canadian Cordillera, comprising fields like Bullmoose, Quirk, Shaw, Sukunka, Turner Valley and others (MacKay *et al.*, 1996 and references therein). Other examples are the Argentinean sub-Andean triangle zones with fields such as, for example, the Duran field (Echavarria *et al.*, 2003).

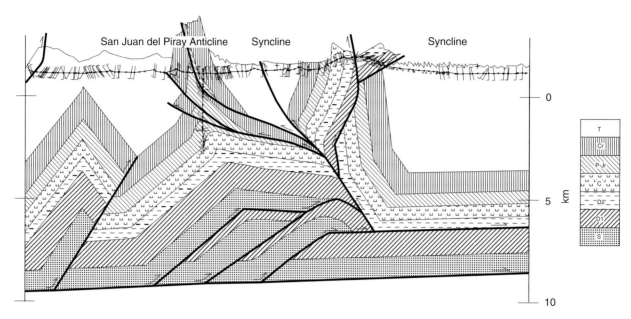

Fig. 4.41. Passive roof duplex from the Bolivian sub-Andes.

5 Thick-skin thrustbelt structures

Inverted graben

When the stress path of the rock section, loaded by burial and tectonic stress, intersects the faulting instability envelope earlier than the folding instability envelope due to decreased cohesion and friction along pre-existing rifting-related normal faults, the deformation results in their inversion (Figs. 5.1 and 5.2). Owing to the amount of stress transfer through the pre-existing normal fault into the footwall, inverted grabens can form either by the inversion of normal faults combined with footwall-edge shortcut faults, in efficient transfer cases or by the inversion of normal faults alone, in less efficient transfer cases. An efficient transfer case occurs when the friction along a pre-existing normal fault is higher. Less efficient transfer cases occur when this friction is low. In these cases the stress is funnelled mostly through the hanging wall. A softer pre-existing graben fill would also have a tendency to avoid buttressing against the footwall edge, which further enhances the chance of avoiding the propagation of short-cut reverse faults through the footwall edge.

Examples of inverted grabens and half-grabens come from the North Sea (e.g. Badley et al., 1989; Figs. 5.3 and 5.4; Nalpas et al., 1995), the north Pyrenees (Hayward and Graham, 1989; Fig. 5.5), the Walls basin, British Isles (Coward et al., 1989; Fig. 5.6), Salta province in northern Argentina (Lowell, 1995), the Gippsland basin, Australia (Davis, 1983), southern Altiplano, Bolivia (Lowell, 1995), and Dneper-Donetsk, Russia (Ulmishek et al., 1994).

Inverted structures (Fig. 5.1) are characterized by different structural features recorded by syn-rift, post-rift and syn-inversion sedimentary sections. The syn-rift sedimentary sequence usually thickens towards the controlling normal fault. The upper portion of the syn-rift section along the inverted normal fault documents a thrust reactivation of the fault. The lower portion preserves the initial normal fault character of the fault. The boundary between these two portions is relatively sharp, as indicated by a natural example from the Bristol Channel, UK (Fig. 5.7). The transition from normal to reverse fault geometry in the syn-rift sequence migrates down the inverted normal fault with increasing inversion and its position defines the amount of inversion. The post-rift section usually has a constant thickness

and does not have an extensional record. It can be affected by subsequent contractional structures. The syn-inversion sedimentation over reactivated normal faults produces a growth anticline that is characteristic of a positive structural inversion (Fig. 5.1). This growth fold displays thinning over the fold crest and thickening away from the reactivated basin margin fault. In cases where listric extensional faults had been reactivated during inversion, hanging wall anticlines are developed and pop-up structures are characteristic (Fig. 5.8).

Normal fault reactivation frequently results in a propagation of the main fault into the post-extension and syn-inversion strata (Fig. 5.2). In the cases where crestal collapse grabens and strong antithetic faulting are developed, inversion compresses these features such that thrust faults propagate from this region producing pop-up structures. A polyphase set of extensional and contractional structures in this case resembles a flower structure, similar to those in a strike-slip setting (Harding, 1985a; Buchanan and McClay, 1991; Mitra, 1993). Normal fault inversion characterized by effective stress transfer through the pre-existing fault results in the development of shortcut thrusts in the footwall edge (Fig. 5.8a). These are typically convex upwards and form small fault-propagation folds at their tips. At high contraction values, the footwall-shortcut thrusts may generate a lower-angle smooth thrust plane and control the incorporation of exotic basement slices within the inversion structure.

The inversion is also controlled by the geometries of pre-existing normal fault patterns. Examples of this control are shown schematically in Fig. 5.9 (McClay, 1995). They illustrate the differences both in extension and inversion between the simple listric and simple planar faults and the fundamental architectures found in ramp/flat and domino fault arrays.

Typical locations of inverted grabens are either in the orogenic foreland (e.g., Badley et al., 1989; Nalpas et al., 1995) or inside the orogen itself (Hayward and Graham, 1989; Roure et al., 1993; Nemčok et al., 2001). Inverted structures in the orogenic foreland can be located very far from the colliding continental plate boundaries, which generate the controlling compressional stress, as documented by the latest Cretaceous and Cenozoic compressional–transpressional inverted structures in

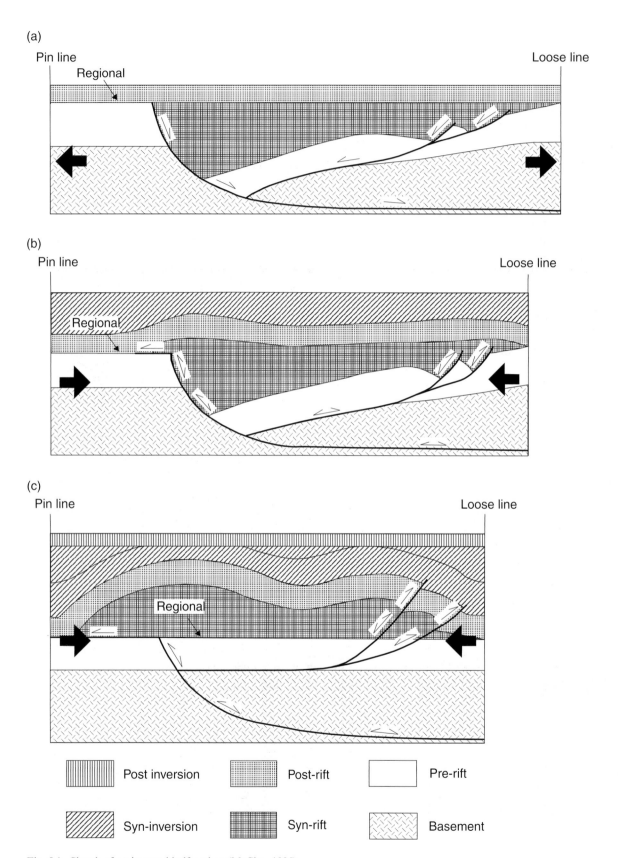

Fig. 5.1. Sketch of an inverted half-graben (McClay, 1995).

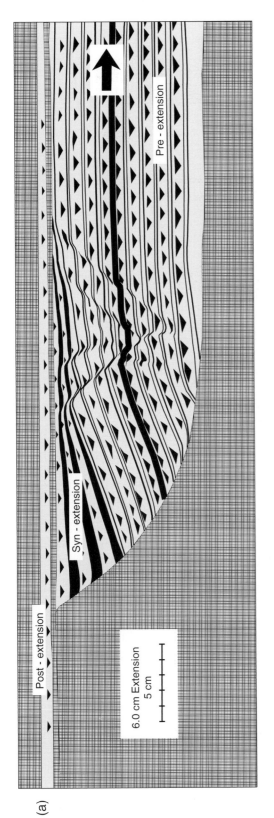

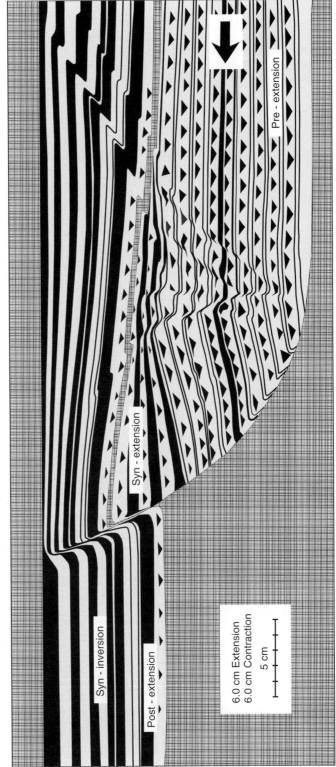

Fig. 5.2. Analogue material model of the inversion of a simple 60° dipping listric normal fault (McClay, 1995).

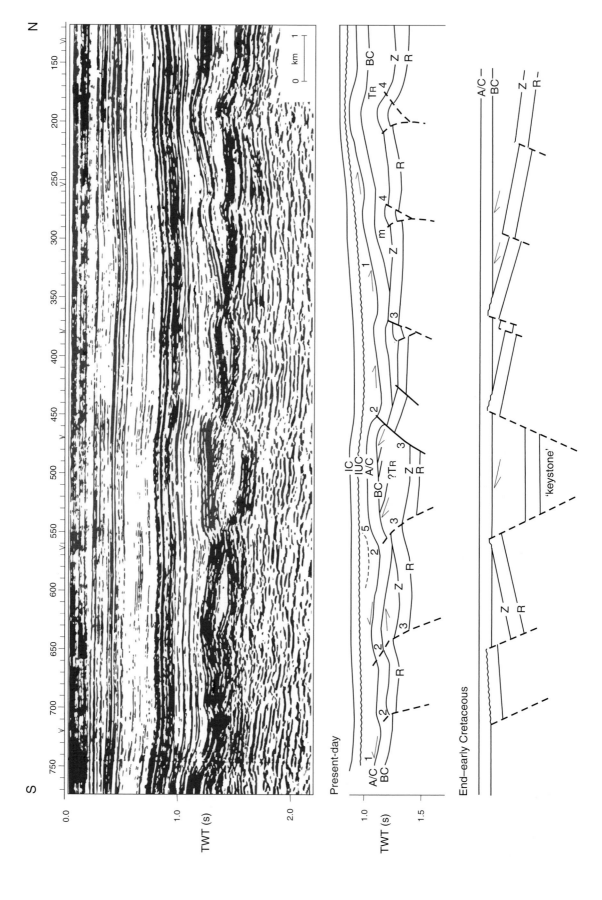

Fig. 5.3. A seismic section across the axis of the now partially subsided Late Jurassic uplift NE of the South Hewett fault, 53/54 quadrants of the UKCS, North Sea (Badley et al., 1989). The line drawings show an interpretation of the end Early Cretaceous and present-day configurations. The end Early Cretaceous reconstruction shows that the most prominent features are Jurassic extensional faults and fault blocks. These contain no syn-extensional sediments. Reflectors showing apparent baselap are in multiples. The base Cretaceous unconformity everywhere erodes into the pre-extension sequence. Fault blocks are parasitic on a more regional uplift centred on the 'keystone' block. The elevation of fault block crests increases towards the 'keystone' from both NE and SW as a result of this uplift. Faults are interpreted as outer-arc extensional structures around a large wavelength anticline. The anticline is interpreted as the product of a possible underlying Jurassic thermal anomaly. The Albian sequence shows relatively uniform thickness across the periclinal axis, indicating a period of uniform subsidence across the area. The present-day configuration is interpreted to be primarily the result of a relatively short-lived post-Cenomanian to pre-Campanian compressional event. Before the compressional event occurred, however, differential subsidence had begun to affect the area following deposition of the Albian sequence. Subsidence was largest above the 'keystone' marking the previous periclinal axis. Onlap to the SW and NE, away from the periclinal axis, onto the top Albian is clearly evident (1). The subsidence was interrupted by the compressional event. The compression caused reactivation of, and reverse movement on, most pre-existing faults. The overlying cover rocks, including the Albian sequence and overlying Chalk, were thrown into forced asymmetric folds above reversed faults. In some cases faults propagated up into the folded cover (2). Reverse movement on many of the faults was insufficient to cancel the original normal throw (3). In other cases it produced net reverse movement and flower structures that are particularly evident at top Zechstein level (4). The burial depth of the pre-existing normal faults immediately prior to the compressional event is unlikely to have been more than several hundred metres. Relatively low values of the deviatoric stress (probably in the range of 10–30 MPa) required for fault reactivation at such shallow burial depths are in accordance with the observation that a majority of pre-existing faults were reactivated. The compressional event was short-lived and no compressional features affect the intraupper Cretaceous (IUC) and younger reflectors. Terminations of reflectors beneath IUC reflector at (5) date the cessation of uplift and the return to subsidence. The early return of subsidence to this part of the area is shown by the conformable nature of reflector IC. Further to the south, the reflector IC is known to represent a major truncational unconformity.

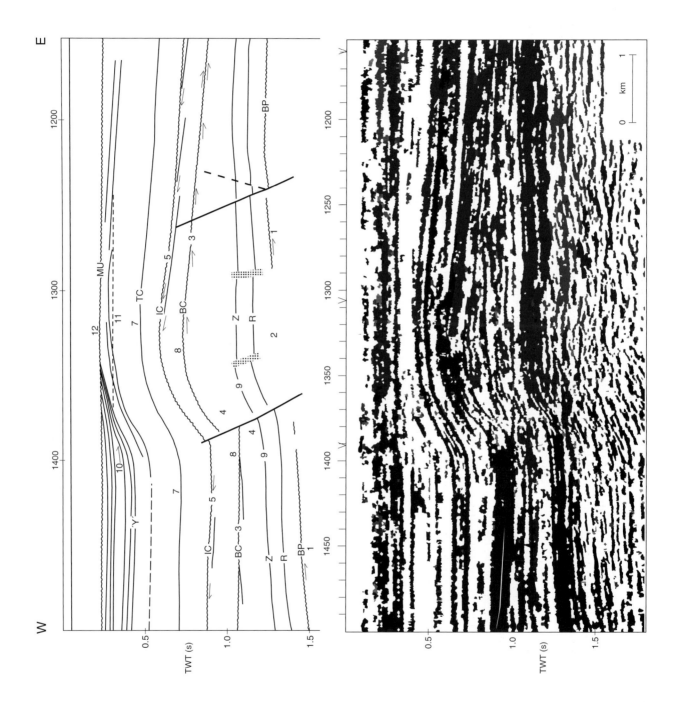

Fig. 5.4. A seismic section and line drawing interpretations of a profile across the South Hewett fault and associated hanging wall fold, 53/54 quadrants of the UKCS, North Sea (Badley et al., 1989). The profile is oriented at approximately 50° to the regional dip direction. The basal Permian unconformity in this area is not characterized by a strong or consistent reflection coefficient. The unconformity can only be located in seismic sections where angularity between Carboniferous and truncating Rotliegend reflectors can be demonstrated. Terminations of Carboniferous reflectors mark the unconformity. Angular discordance between Carboniferous reflectors and the Rotliegend can be seen at (1). Poor data quality in other areas precludes any interpretation (2). The prominent unconformity at the base of the Cretaceous sequences, here truncating SW-tilted Triassic beds, is readily located by both reflector terminations and a reflector at the unconformity surface, characterized here by a positive reflection coefficient (3). The original (pre-Cretaceous) NW downthrow of the South Hewett fault can be recognized by the larger thickness of Triassic beds in the hanging wall than in the footwall (4). The truncational nature of the intraCampanian unconformity is obvious (5). Lower Cretaceous sequences have NE dips imparted by the tilting associated with differential subsidence towards the 'graben axial' area and later folding related to the Upper Cretaceous compressional event. The unconformity in the area imaged by the seismic section juxtaposed Campanian chalk and Lower Cretaceous sand and shale. The resulting negative reflection coefficient produces a clear trough/peak (white above black) response in the seismic section. Note how truncated reflectors terminate beneath the peak (the black of the unconformity reflector, at least 20 ms beneath the actual subsurface truncation position). The South Hewett fault was reactivated during the Miocene compressional event, resulting in a second phase of reverse movement. The amplitude of the fold affecting the top Chalk provides a measure of the magnitude of the Miocene reverse movement at this location (7). The offset of the base Cretaceous unconformity (8) and top Zechstein reflector represents the minimum aggregate value of the Upper Cretaceous and Tertiary reversals. The smaller offset of the top Zechstein reflector (9) gives a measure of the pre-Cretaceous downthrow to the NE. The growth history of the fold developed above the South Hewett fault is recorded in the complex sequence beneath the Miocene unconformity (10). Reflectors beneath Y are sub-parallel. Those above indicate sequences thinning towards the fold. Reflector Y within the late Eocene–Oligocene marks the onset of folding. Reverberations (11) within the first 350 m of seismic section make interpretation of the shallow sequences difficult. The regional truncational intraMiocene unconformity (12) can, however, be followed throughout the area.

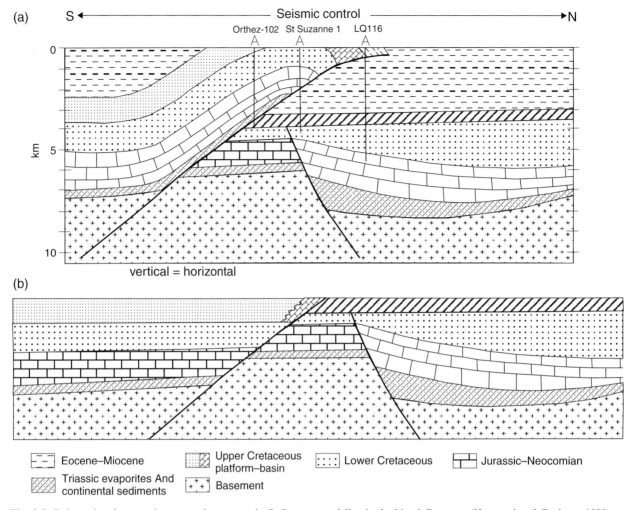

Fig. 5.5. Balanced and restored cross section across the St Suzanne anticline in the North Pyrenees (Hayward and Graham, 1989). The section restored on top of the Upper Cretaceous sediments shows a control on ramp location by a pre-existing normal fault.

Western and Central Europe, related to the Alpine orogeny (e.g., Ziegler, 1988). Stress transmission over such a long distance was documented by strain data from the Cumberland Plateau in front of the Appalachians (Engelder and Engelder, 1977; Engelder, 1979) and the North American interior 2100 km beyond the Ouachita and Appalachian orogens (Craddock *et al.*, 1993).

Inversion of pre-existing grabens in the orogenic foreland requires efficient stress transfer between the foreland plate and advancing orogen and potential decoupling horizons inside the foreland plate that would accommodate the crustal shortening associated with the inversion of grabens. Such decoupling was documented from the intracrustal level in the Rockall Trough, British Isles (Roberts, 1989) or the crust–mantle boundary in the Western Alps, the Walls basin, British Isles or Sumatra (Butler, 1989; Coward *et al.*, 1989; Lowell, 1995).

Although essentially syn-orogenic, inversion movements affect wide areas, sometimes at considerable distances from the collision front, and the detailed timing varies. There are examples documenting an inversion progradation in front of the orogen, such as the West Netherlands Broad Fourteens basin–Sole Pit basin in front of the Alps (Ziegler, 1989) or inverted grabens of the Bohemian Massif and West European Platform in front of the Eastern Alps and the West Carpathians (Nemčok *et al.*, 2000). The inversion progradation scenario requires the orogenic foreland to be progressively more attenuated towards the orogen. There is an example documenting an inversion retrogradation in front of the orogen, such as the Western Shelf area in front of the Alps, where the inversion of the distal Western Approaches basin postdates the inversion of the proximal Fastnet–Celtic Sea basin (Ziegler, 1989). This scenario implies that pre-existing grabens further

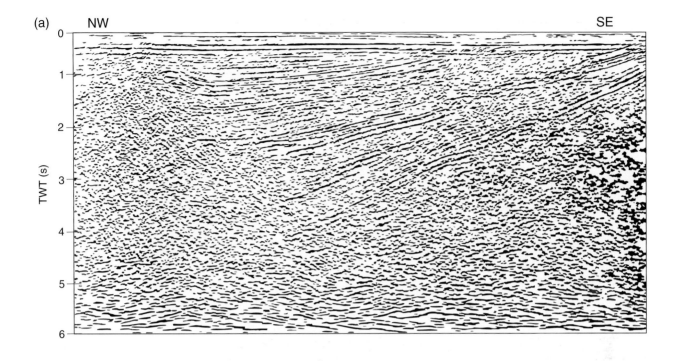

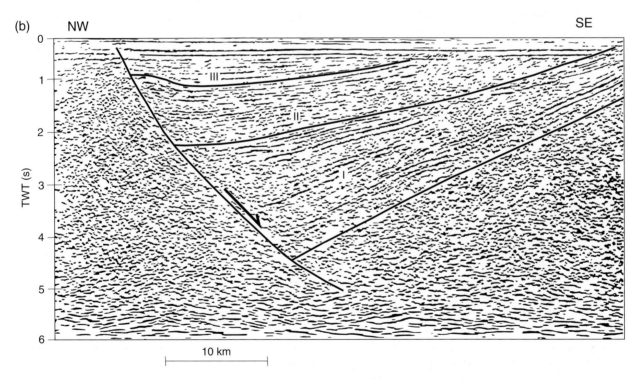

Fig. 5.6. Uninterpreted and interpreted seismic profile through the northwest bounding normal fault of the West Orkney basin that underwent a slight inversion (Coward *et al.*, 1989).

Fig. 5.7. Photograph of the boundary between the upper portion of the syn-rift section indicating contraction and the lower portion indicating extension from the Bristol Channel, UK. The outcrop is in the Lower Jurassic limestone and shale sequence. Note that the boundary is relatively sharp. The upper levels record a footwall drag related to the thrusting. The lower levels record a footwall drag related to normal faulting and tensile fractures in dragged limestone beds.

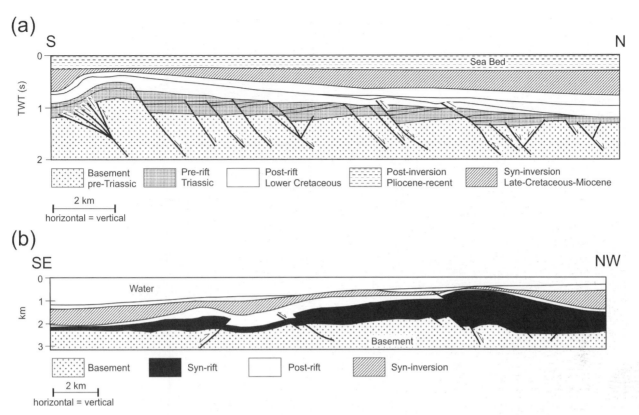

Fig. 5.8. (a) Interpreted seismic section across the South Hewett fault region, southern North Sea (McClay, 1995). The inverted domino fault array shows a major growth anticline on the left-hand side. High-angle thrust faults produced 'arrowhead' or 'harpoon' structures on the right side. Numerous normal faults between them still preserve their original normal fault displacement. (b) Cross section through part of the Sunda Arc (after Letouzey, 1990 modified by McClay, 1995) showing inversion of half-graben fault systems to produce characteristic 'arrowhead' or 'harpoon' geometries.

from the orogen were weaker and that stronger pre-existing grabens occurred closer to the orogen.

Not all of the pre-existing normal faults are necessarily inverted at the same time and, as alluded to in the previous text, not all of the pre-existing normal faults in the same basin are required to be inverted. In rift basins formed by pure shear, coincident thinning of the lithosphere creates areas of weakness that are later selectively shortened and inverted. Thinning in simple shear also occurs, but the locus of thinning is offset from the rift and it is probable that a mechanical detachment controls the rift inversion (Gibbs, 1987). In both pure and simple shear, listric normal faults are ideally suited to being reactivated as thrusts, along the lower, flatter portion, with associated uplift in the upper, steeper part (Letouzey *et al.*, 1990). Inverted extensional fault systems can vary from reactivation of isolated faults to reactivation of all of them. Reactivation of isolated faults is typical for normal fault systems, which have undergone relatively small amounts of positive inversion, such as faults in the

South Hewett fault region in the North Sea (McClay, 1995; Fig. 5.8a). Here inversion reactivated the main basin bounding fault and developed a fault-propagation growth anticline together with a number of small footwall shortcut thrusts. There is only a localized reactivation of extensional faults in the main domino array within the basin.

At greater values of contraction, reactivated normal faults propagate and flatten upwards into the post-rift and syn-inversion sequences (Fig. 5.8b). Complex 'arrowhead' or 'harpoon' geometries result from the wedge of syn-rift sediments being elevated above the regional level (Koopman *et al.*, 1987; Buchanan and McClay, 1991, 1992; McClay, 1995). Their resultant shape depends upon the geometry of the underlying normal fault.

The inversion of the half-graben or graben is further affected by the direction of compression in relation to the graben geometry and the rheology of the base of the post-rift sequence. Compression perpendicular to the graben trend is prone to developing structures affected

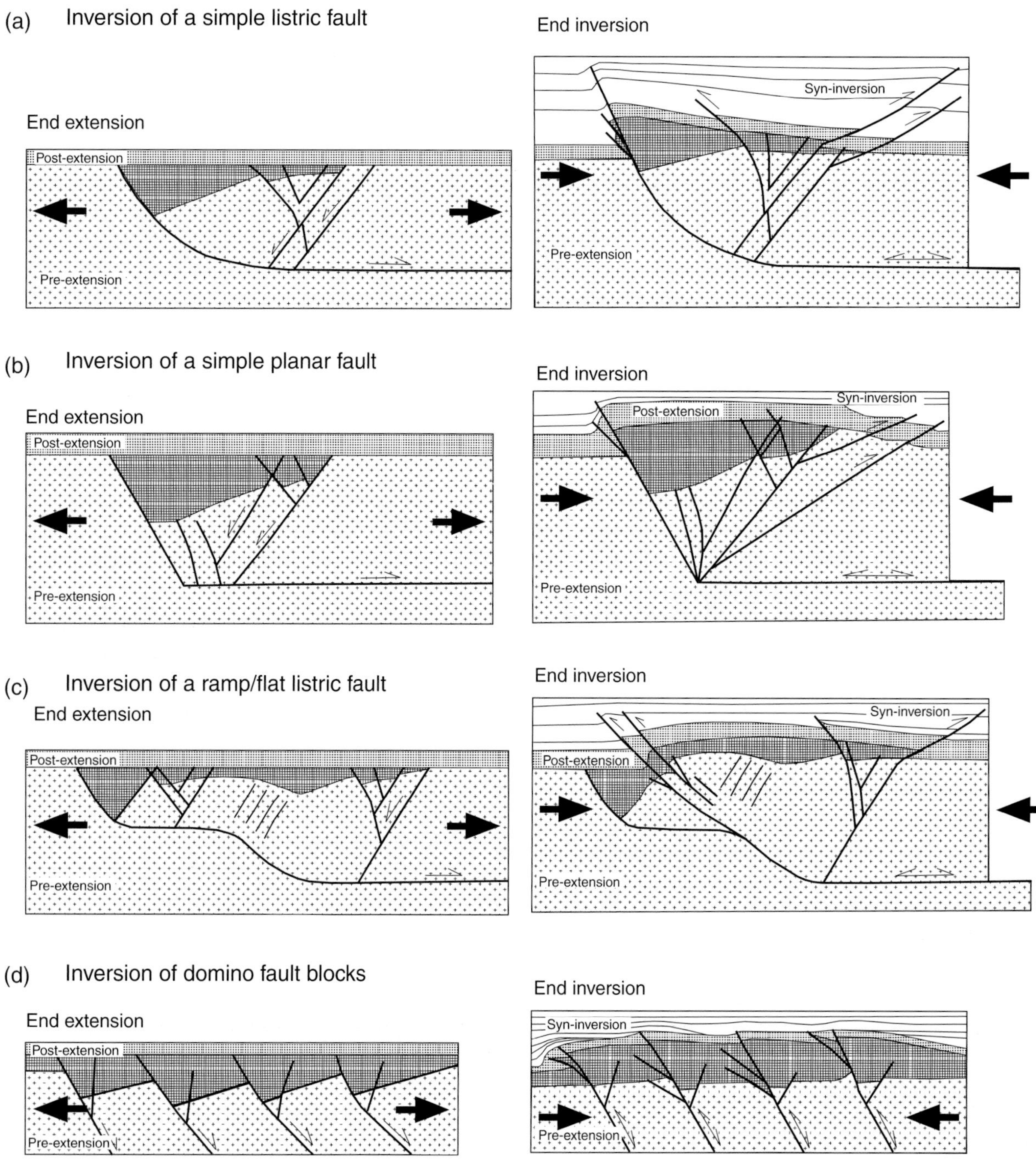

Fig. 5.9. Sketch of inversion scenarios of simple extensional fault systems (McClay, 1995). (a) Inversion of a simple listric fault; (b) inversion of a simple planar fault; (c) inversion of a ramp–flat listric fault; and (d) inversion of a domino fault array.

by buttressing while compression at a low angle to the graben trend commonly develops less energy-related structures. The role of basement anisotropy is discussed later in a Chapter 8.

Inversional structures form important structural traps, such as the Eakring and Caunton oil fields, East Midlands, British Isles (Roberts, 1989; Fraser *et al.*, 1990), the Tandun field in central Sumatra (Mertosono, 1975), the Kimmeridge Bay, Wytch Farm and Wareham oil fields, British Isles (Selley and Stoneley, 1987), the Rhourde Nouss in Algeria (Boudjema, 1987), the Daquing in Songliao, China (Yang, 1985; Li, 1991), the Tupungato and Piedras Coloradas fields in the Cuyo basin, Argentina (Uliana *et al.*, 1995) and the

Table 5.1. Principal geometric elements of basement uplifts in the Rocky Mountain foreland.[a]

| Structure | Shape | Maximum of (evidence) rotation basement | | Fault dip regional relative to dip (type) | | Upper termination of main fault | Deformation in cover (dip refers to bedding) |
		Hanging wall	Footwall	Main fault	Shallow-level fault		
Casper Mountain	Monocline	25° (cover)	35°–50° (cover)	41° (thrust)	88° (normal)	Disappears in cover (décollement likely)	Dip-parallel shortening in syncline, drag folds with same shear as shallow fault
Big Thompson anticline	Monocline	None (foliation)	?	35° (thrust)	70° (reverse)	Disappears in cover	Dip-parallel shortening (lift-off fold in anticline). Dip parallel extension in steep limb
Willow Creek anticline	Monocline	35° (dipmeter, cover strata)	?	41° (thrust)	79° (normal)	Décollement (structural wedge?)	None?
Five Springs	Monocline	45°–55° (pegmatite dikes)	?	24° (thrust)	82°? (reverse?)	Surface break	Dip-parallel shortening (lift-off fold in steep limb). Out-of-syncline bedding plane slip
Banner Mountain	Monocline with 8° forward dip	26° (cover) 15°–20° (foliation)	?	40°–55° (thrust)	77° (normal)	Surface break	Dip-parallel extension in steep limb
Rattlesnake Mountain	Monocline with 15° back dip	None (pegmatite dikes)	?	?	80° (normal)	Disappears in cover (minor surface break)	Dip-parallel extension near cover/basement contact and near anticline axial surface. Shortening in steep limb

[a] Narr and Suppe (1994).

Ijsselmonde in the west Netherlands (Bodenhausen and Ott, 1981; De Jager *et al.*, 1993).

Basement uplift
When the stress path of the rock section loaded by burial and tectonic stress intersects the faulting instability envelope earlier than the folding instability envelope due to decreased cohesion and friction along pre-existing faults in the basement beneath the sedimentary cover, the deformation results in basement uplift. The basic control of this thrust structure is the role of displacements imposed at high angles to the layering. The patterns of layer elongation and layer shortening associated with these major flexures, as revealed by both macroscopic examinations and petrographic studies (e.g., Friedman *et al.*, 1976b; Stearns, 1978; Couples *et al.*, 1994; Narr and Suppe, 1994; Table 5.1), imply that

basement uplifts create significant bending moments. These localized large moments result in the forced folding of the layered cover sequence over the uplifted and rotated basement blocks beneath (Reches and Johnson, 1978; Stearns, 1978). They are responsible for the propagation and decay of a wavetrain of small-scale flexural deformations to locations away from the uplift centre (Couples and Lewis, 1998). The evidence for relatively undeformed basement blocks that moved during the folding of the cover sequence comes from several folds in the Rocky Mountain foreland (Prucha *et al.*, 1965; Stearns, 1971; Mathews, 1986; Erslev *et al.*, 1988; Erslev and Rogers, 1993). More complex folding mechanisms than simple forced folding would need to be assumed in cases where even basement blocks are folded (e.g., Berg, 1962; Blackstone, 1983; Brown, 1984a, b; Narr, 1993; Narr and Suppe, 1994; Table 5.2).

Table 5.2. Characteristic features of basement uplifts.[a]

Basement structures are commonly monoclines
The structures commonly form above a contractional fault in the basement
The basement behaves as a rigid block in some structures, but it is folded in others
The main fault can disappear as it proceeds up into the cover sequence
The steep limbs of folds commonly form by the cover draping over a faulted edge of basement as the hanging wall is uplifted
Deformation in the cover is concentrated in the steep limb and can involve both layer-parallel shortening and layer-parallel
 extension

[a] Narr and Suppe (1994).

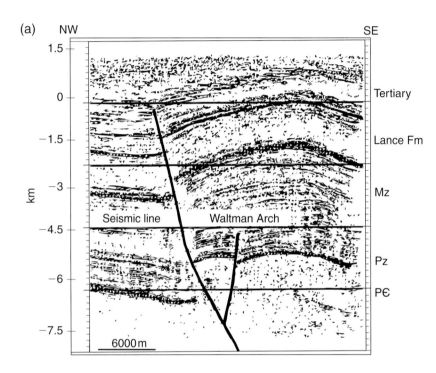

Fig. 5.10. (a) NW–SE seismic profile through the Waltman Arch in the vicinity of the Cave Gulch field in the Wind River basin, Rocky Mountain foreland, Wyoming (Montgomery *et al.*, 2001). (b) Simplified NE–SW cross section interpreted along available seismic data, displaying the structural setting of the field in relation to the Owl Creek thrust and the Waltman arch. The seismic interpretation is modified from Natali *et al.* (2000).

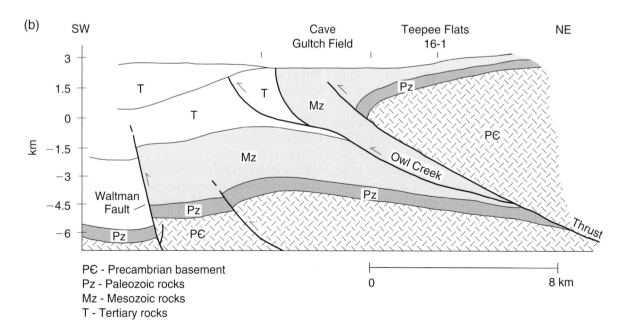

PЄ - Precambrian basement
Pz - Paleozoic rocks
Mz - Mesozoic rocks
T - Tertiary rocks

Table 5.3. Orogenic belts, which are associated with low-temperature basement uplifts.[a]

Orogenic belt	Deformation age	Reference
North American plate		
Boothia–Cornwallis	Devonian	Kerr, 1977; Okulitch *et al.*, 1986
Ancestral Rocky Mountains (including Arbuckle and Wichita Mts.)	Pennsylvanian	Blythe *et al.*, 1988; Mallory, 1972
Eastern Rocky Mountains	Paleogene	Berg, 1962; Cross, 1986
Transverse Ranges	Neogene	Sylvester and Smith, 1979, Dibblee, 1982
South American plate		
Venezuelan Andes, Sierra de Perija, Santa Marta Massif	Neogene	Kellogg and Bonini, 1982; Meier *et al.*, 1987
Central Cordillera (Colombia)	Neogene	Butler and Schamel, 1988
Eastern Cordillera (Colombia)	Neogene	Julivert, 1970; Dengo and Covey, 1993
Eastern Cordillera and sub-Andean Lowland (Peru)	Neogene	Sebrier *et al.*, 1988; Barazangi and Isacks, 1979
Sierras Pampeanas (Argentina)	Neogene	Allmendinger *et al.*, 1983
Carribean plate		
Trinidad and Tobago	Neogene	Robertson and Burke, 1989
African plate		
Western High Atlas	Jurassic to recent	Froitzheim *et al.*, 1988
Central High Atlas	Tertiary	Fraissinet *et al.*, 1988
Cape Ranges	Late Paleozoic to Triassic	Rodgers, 1987
Witwatersrand	Archean	Myers *et al.*, 1989
Eurasian plate		
Harz Mountains	Late Cretaceous	Rodgers, 1987
Pyrenees Mountains	Paleogene	Williams and Fischer, 1984; Choukroune, 1989
Zagros Mountains	Neogene	Jackson and Fitch, 1981; Molnar and Chen, 1983
Tien Shan	Neogene	Tapponnier and Molnar, 1979; Molnar and Chen, 1983
Central Sumatra basin	Neogene	Eubank and Makki, 1981
Indian–Australian plate		
New Guinea fold-and-thrustbelt	Neogene	Abers and McCaffrey, 1988
Ngalia basin, Australia	Carboniferous	Copper *et al.*, 1971

[a] Narr and Suppe (1994).

Natural examples of basement-cored uplifts come from the Rocky Mountains foreland in Utah, Colorado, Wyoming, and Montana (e.g., Stearns, 1971, 1978; Rodgers, 1987; Narr and Suppe, 1994; Table 5.1; Stone, 2002; Fig. 5.10), from the Tertiary structures of the southeast Turkey–north Syria foreland belts (Mitra, 1990), the external Northern Apennines (De Donatis *et al.*, 2001), the upper Cook Inlet basin in Alaska (Haeussler *et al.*, 2000), the Argentinean Salta province in front of the Andes (Allmendinger *et al.*, 1997), the Ghadames basin in the northern Algeria (Mitra and Leslie, 2003), and many other regions (Table 5.3).

Basement uplifts are most commonly associated with orogens formed by continental collision, such as the Variscan (Ziegler, 1989), Uralian (Ulmishek, 1982; Matviyevskaya *et al.*, 1986), Alpine (Butler, 1989) or Pen-Guan Massif next to the Da-Qin Long belt (Rodgers, 1995). They are also known to be associated with orogens that formed in response to high convergence rates between continental and oceanic plates, such as parts of the Andean foreland in Peru and Argentina (Rodgers, 1995; Allmendinger *et al.*, 1997) or the Rocky Mountain foreland (e.g., Brewer *et al.*, 1980; Narr and Suppe, 1994). The lack of basement uplifts or any kind of compressional intraplate deformation in an orogenic foreland, as in the Canadian cordillera of Alberta, can be related either to the absence of major intraplate discontinuities in the foreland or to the lack of mechanical coupling between the foreland and orogen (Ziegler, 1989).

Natural examples of basement uplifts with different kinematics of the cover sequence have been simulated

Cover welded to basement   Cover detached from basement

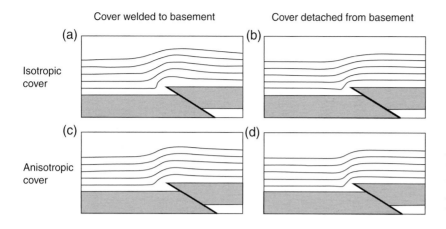

Fig. 5.11. Four kinematic end members
of a basement uplift utilizing a 30°
dipping thrust fault (Johnson and
Johnson, 2002b). End members are
controlled by various combinations of
isotropic cover, anisotropic cover and
the degree of coupling with its
basement. Forced folds in the case of a
coupled cover with basement display
bulging of the anticlinal hinge and slight
backlimb rotation. Folds in the case of a
decoupled cover do not experience any
backlimb rotation.

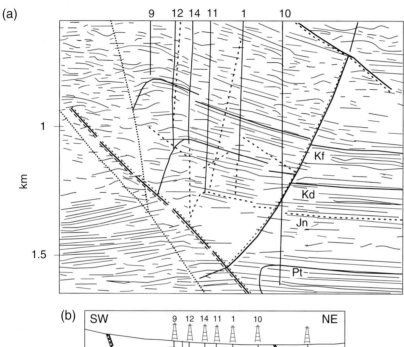

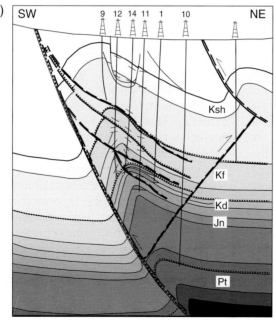

Fig. 5.12. Structural cross section and
seismic profile through the Crooks Gap
oil field in the Great Divide basin,
Wyoming (Wellborn, 2000). The seismic
section is a 2D section through a 3D
volume but still fails to image the frontal
limb completely and obscures the steeper
portion of the backlimb. The structural
cross section is constrained by dipmeter
data from projected wells. Note the
difference between the geologic cross
section and interpreted seismic data. Kf
is the Cretaceous Frontier Formation,
Kd is the Cretaceous Dakota
Formation, Jn is the Jurassic Nugget
Formation and Pt is the Permian
Tensleep Formation.

W   E

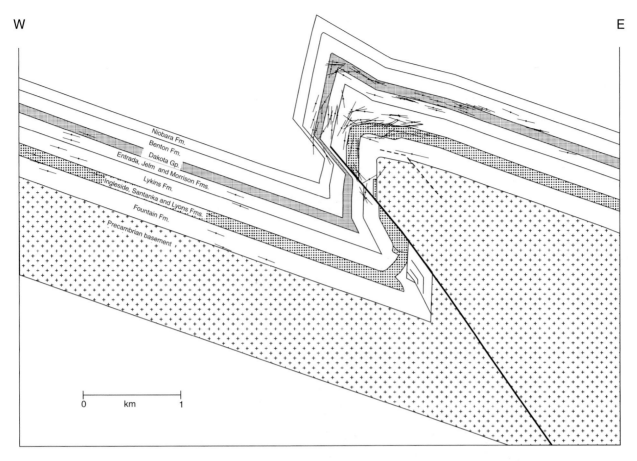

Niobara Fm.
Benton Fm.
Dakota Gp.
Entrada, Jelm and Morrison Fms.
Lykins Fm.
Ingleside, Santanka and Lyons Fms.
Fountain Fm.
Precambrian basement

0   km   1

Fig. 5.13. Balanced cross section through the Big Thompson anticline in the Rocky Mountains foreland, Colorado (Narr and Suppe, 1994).

by physical transverse-load models. Some of them reproduced shortening of the cover layers (Friedman *et al.*, 1976a, b, 1980; Logan *et al.*, 1978). Others reproduced a cover layer affected by neither shortening nor extension (Stearns and Weinberg, 1975; Weinberg, 1979; Couples *et al.*, 1994; Couples and Lewis, 1998) and yet another group reproduced extended cover layering (Patton *et al.*, 1998).

The control behind various kinematic responses of the cover sequence to basement uplifts includes the degree of anisotropy of the cover sequence and the friction between the cover and its basement (Johnson and Johnson, 2002a, b; Fig. 5.11).

The degree with which specific layers of the cover are welded together, i.e. the degree of the cover anisotropy, controls the forced fold geometry. Isotropic cover reacts to folding by widening its forelimb upward from the basement fault tip (Figs. 5.11a and b). Forelimb bedding also has a progressively shallower dip upwards. Roughly similar conclusions can be derived from the physical experiments of Withjack *et al.* (1990), who used a homogeneous clay layer for the cover. Natural examples

of a roughly isotropic cover are perhaps the Middle Ground Shoal fold in the Cook Inlet, Alaska, described by Haeussler *et al.* (2000) and the Crooks Gap fold in Wyoming described by Wellborn (2000) (Fig. 5.12). An anisotropic cover maintains both a constant forelimb width and bedding dip (Figs. 5.11c and d), resembling a kink fold with rounded hinges. Roughly similar conclusions can be derived from the physical experiments of Friedman *et al.* (1980), who used lubricated interlayers of sandstone and limestone forming the cover. Typical natural examples of more anisotropic cover are the Big Thompson anticline or the Willow Creek anticline in the Rocky Mountain foreland in Colorado described by Narr and Suppe (1994) (Fig. 5.13).

The degree of coupling between the basement and cover controls the direction of displacement in the cover over the hanging wall basement block. During the reverse movement of the basement fault, the decoupled cover moves vertically (Figs. 5.11b and d). A natural example for a decoupled cover comes from the Rattlesnake Mountain anticline in the Rocky Mountain foreland, Wyoming (Stearns, 1978). A partially coupled

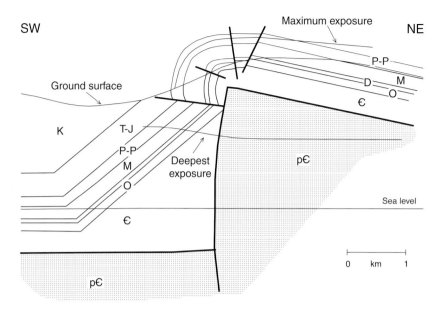

Fig. 5.14. Cross section through the Rattlesnake Mountain uplift (after Stearns, 1971, 1978; Couples, 1986). Small faults developed in the cover sequence have a minor effect on the layer lengths, but more significantly represent hinge zones.

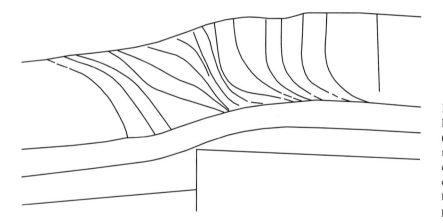

Fig. 5.15. Stress trajectories in the limestone layer of a physical model (Couples *et al.*, 1994). The heavy lines represent the maximum principal compressive stress σ_1. Layers of dolostone and lead, and the top of the forcing steel assembly are indicated for positional reference.

cover moves at an acute angle to the basement fault movement, whereas a coupled cover mimics the fault movement (Figs. 5.11a and c). A natural example of an approximately coupled cover comes from the Uncompahgre uplift in Colorado (Stearns, 1978).

The cover–basement detachment in the case of the Rattlesnake Mountain anticline propagated along a Cambrian shale horizon (Stearns, 1971, 1978; Stone, 1984; Couples, 1986; Brown, 1988; Fig. 5.14). Physical models simulating decoupled cover combined with essentially isotropic cover indicate that, during folding, both detachment and layer-parallel slip occur between the cover and the forcing basement block (Weinberg, 1978; Couples *et al.*, 1994; Couples and Lewis, 1998). At a certain distance from the fold, there is a pattern of movement in which the layered sequence initially moves away from the uplift. Significant stress perturbations start right from the beginning of the uplift (Mandl, 1988; Couples *et al.*, 1994; Fig. 5.15). Later, when a greater structural relief is built, motions of the layered sequence reverse their sense to become layer-parallel translations towards the uplift. These layer-parallel translations are not driven by any layer-parallel load from far-field sources, but rather by mechanical processes local to the fold itself. During this stage, a ductile horizon at the base of the cover sequence flows laterally, especially across the crest of the uplift to the footwall block (Fig. 5.16) in response to pressure gradients, which are inherent to asymmetric uplifts (Jamison, 1979). Numerical models (Jamison, 1979) demonstrate that areas of high and low mean stress develop as a response either to offset at the base of the cover sequence or to asymmetric folding, but not to simple bending.

(a)

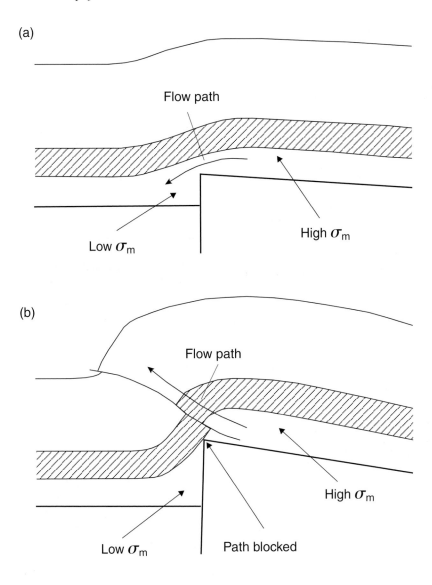

(c)

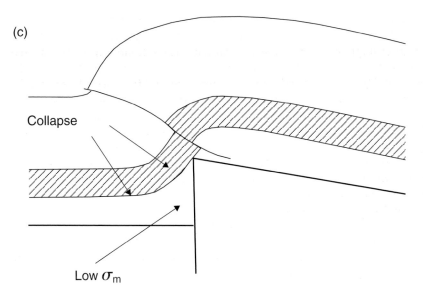

Fig. 5.16. Pressures due to folding and movement resulting from flow of lead in a physical model of forced folding (Couples *et al.*, 1994). (a) Early stage of folding with high pressure σ_m in the hanging wall and low pressure in the footwall. The average flow of lead is shown by the thicker line with an arrow. (b) The later stage of folding with the continuation of pressure zones from (a), but with the previous lead flow path disrupted. The new flow path within the hanging wall is from the high-pressure zone across the forelimb to the lower pressure inside higher parts of the cover sequence. (c) The same relationships as for (b), but with collapse of layers into the synclinal area. Note the change in shape of the triangular area of lead in the footwall from that of (a).

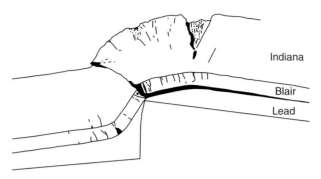

Fig. 5.17. Deformed physical model of forced folding (Couples and Lewis, 1998). Indiana and Blair indicate limestone and dolostone, respectively. The top of the steel block forcing assembly is shown by the lowermost line. The dolostone layer is 1.55 mm thick, the lead is 2 mm thick and the limestone is 7.35 mm thick. The setup simulates an essentially isotropic cover sequence, which is decoupled from the basement. The short lines indicate microfractures visible on a slabbed medial surface or in thin section. The large black areas are voids that opened following release of the confining pressure. Note particularly the prominent graben in the antiform crest, the reverse fault through the synform hinge, the rigid pieces of dolostone in the fold, and the striking thickness changes of the lead.

Cover–basement coupling combined with cover that can be characterized as roughly isotropic can be found on the eastern side of the Uncompahgre uplift in Colorado (Stearns, 1978; Couples *et al.*, 1994). There, Lower Paleozoic strata are missing and the cover sequence has Triassic clastic sediments at its base. Sandstones above the base–cover boundary underwent a ductile deformation during the uplift and folding in the cover. Despite the rather small uplift of 350 m, forced folding resulted in sandstone thinning from 300 to 50 m (Jamison and Stearns, 1982; Heyman, 1983).

The cover sequence deformation controlled by various combinations of degree of anisotropy, degree of coupling and controlling fault geometry plays a critical role in hydrocarbon entrapment in basement uplifts. There are two most critical deformational consequences controlling this entrapment. They are the internal deformation of the cover layer sequence during the uplift

(Fig. 5.17) and the progressive propagation of large-scale faults from the basement to the surface through the cover sequence with continuing uplift (Fig. 5.18). Both factors are controlled by the rheology of the cover sequence and the displacement along basement faults. Internal deformation of the cover sequence can also be controlled by variations in layer-parallel translation, resulting in lateral variations in layer-parallel strain within the layered packages (Couples and Lewis, 1998; Johnson and Johnson, 2002a, b). The alternation in longitudinal strain 'highs' and 'lows' resembles the stretching and bunching of an inchworm (Price, 1988; Means, 1989). These strains have a different expression in different rheologies. For example, limestone with a porosity of about 15% composed of bioclasts and ooids deforms by fracturing only in the area of the basement uplift. Other areas experience deformation by calcite twin gliding and grain–boundary movements (Couples and Lewis, 1998). Brittle dolostone with no significant porosity and very small crystal size primarily undergoes fracturing, which is prominent in the uplift area (Couples and Lewis, 1998). More distant areas are characterized by fracture clusters, apparently controlled by bedding-parallel slip domains (Cooke and Pollard, 1994, 1997; Cooke *et al.*, 1998; Fig. 5.19). Strain variations mentioned earlier also affect the reservoir properties of the cover sequence, causing permeability variations (Couples and Lewis, 1998).

Basement uplifts form important structural traps in fold and thrustbelts, both in anticlinal crests and upturned beds in footwalls. Examples come from the Garzan-Germik oil field, located in a Miocene basement-involved compressional structure in the foreland of southeastern Turkey (Sanlav *et al.*, 1963) or from the Rhourde el Baguel oil field in northern Algeria (Figs. 5.20 and 5.21). In both cases productions come from anticlinal crests. Other examples of basement-uplift fields come from the Crooks Gap, Cave Gulch and Casper Mountains uplift fields, and other fields of the Rocky Mountain foreland (e.g., Narr and Suppe, 1994; Wellborn, 2000; Montgomery *et al.*, 2001; Stone, 2002) and the Middle Ground Shoal and North Cook Inlet fields in the Cook Inlet, Alaska (e.g., Haeussler *et al.*, 2000).

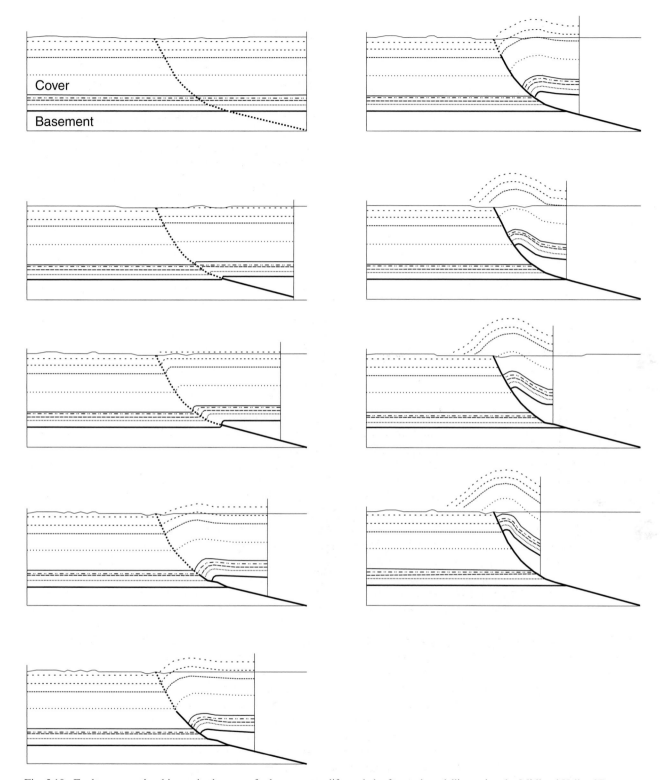

Fig. 5.18. Fault-propagation history in the case of a basement uplift made by forward modelling using the Midland Valley 2Dmove. The dotted and solid lines indicate future and actual fault planes, respectively. The first line from the bottom indicates the top of the basement.

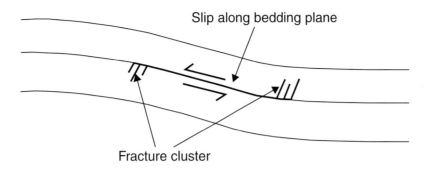

Fig. 5.19. Sketch showing the spatial arrangement of fracture clusters associated with localized bedding-plane slip (Cooke and Pollard, 1994, 1997).

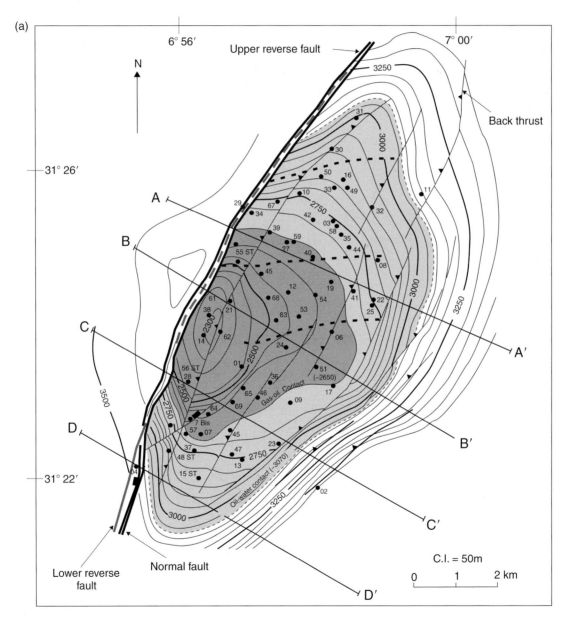

Fig. 5.20. (a) Structural map for the top of the Cambrian sandstone unit in the Rhourde el Baguel field in the Ghadames basin, Algeria (Mitra and Leslie, 2003). The map is based on well data, seismic interpretation and geological cross sections AA′, BB′, CC′ and DD′.

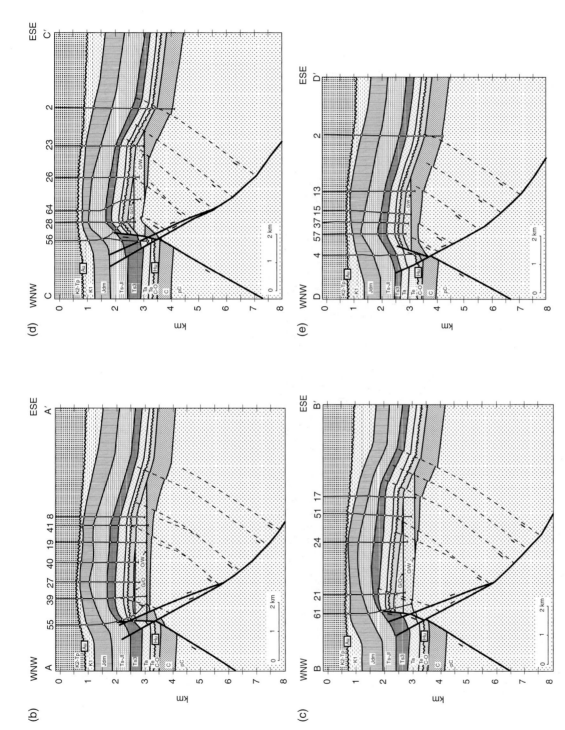

Fig. 5.20. (*cont.*) (b) Structural cross section AA' incorporating data from wells 8, 19, 27, 39, 40, 41 and 55. The structure forms a basement uplift related to reverse fault splays formed during the Albian. Reverse faults cut through a pre-existing Triassic normal fault. (c) Cross section BB' based on data from wells 17, 21, 24, 51 and 61. The uplift is tighter than in AA' and the normal fault has a larger throw. (d) Cross section CC' based on data from wells 2, 23, 26, 28, 56 and 64. The tightening is further increased. Liassic salt units above the normal fault are thickened. (e) Cross section DD' based on data from wells 2, 4, 13, 15, 37 and 57. This section documents the southwestern termination of the field and shows the lowest relief. K2-Tp is the Albian–Pliocene section, Au is the Albian Austrian unconformity, K1 is the Neocomian–Aptian section, Jdm is the Doggerian–Malmian section, Ts-Jl is the Triassic Salifere 1 to Liassic section, Ts3 is the Triassic Salifere 3 section, Ta is the Triassic Argileux section, Te is the Triassic Eruptif section, Hu is the Late Paleozoic Hercynian unconformity, C-O is the Cambrian–Ordovician section, C is the Cambrian section, pC is the Precambrian section, O/W is the oil/water contact and G/O is the gas/oil contact.

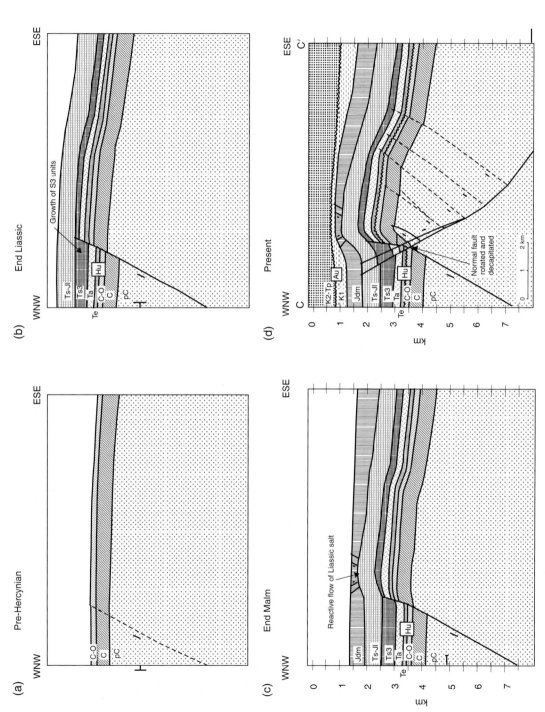

Fig. 5.21. Development of the Rhourde el Baguel basement uplift based on a stepwise restoration of cross section CC' from Fig. 5.20(d) (Mitra and Leslie, 2003). (a) Restoration to the time predating the formation of the Late Paleozoic Hercynian unconformity. (b) Restoration to end Liassic, showing normal faulting and development of tilted blocks. (c) Restoration to end Malmian, indicating the development of low-amplitude salt structure in the Triassic S3 and Liassic salt units during the deposition of the Malmian section. The Malmian section is thinned and faulted over the crest of the salt structure. (d) The present deformed cross section, indicating Albian Austrian contractional deformation forming a basement uplift and reverse faults that cut and rotate pre-existing normal fault. The Albian Austrian unconformity is gently warped by younger deformational events. See Fig. 5.20 for further explanation.

6 Determination of timing of thrusting and deformation rates

Timing of thrusting

A wide range of methods have been used to date thrust sheet events and this section aims to outline the most important of these, using specific case studies to illustrate the principles of each method.

The timing of thrusting is best determined when pre-tectonic, syn-tectonic and post-tectonic depositional sequences (Fig. 6.1) are all preserved. Syn-tectonic sediments also provide information concerning the sequence of thrusting, even in cases where the geometry of pre-tectonic sediments does not allow a distinction to be made between a piggyback sequence of thrusting and an overstep sequence (Fig. 6.2). Since several structures can be active during a specific time interval, piggyback basins can develop between two growing anticlines or two groups of growing anticlines in piggyback-thrusting scenarios. Natural examples are, for instance, the Paleocene–Middle Eocene piggyback basin in the Balkan Unit of the Balkans (Vangelov, personal communication 2003), Eocene–Miocene Caroni and Erin-Ortoire piggyback basins in Trinidad or the Miocene–Pliocene Iglesia piggyback basin in the Argentinean Precordillera (Beer *et al.*, 1990). Development of a piggyback basin starts when the basin is flanked at the foreland side by the thrust tip, which moves upward and cuts the basin. At the hinterland side the basin is flanked by a more mature growth structure. Later, the displacement is transferred to a ramp located further towards the foreland. As the piggyback basin develops in the new structural situation, the depocentre migrates towards the foreland and the basin undergoes internal shortening and uplift. A piggyback basin can also develop in the case of an overstep thrusting sequence. The basin reacts to a hinterlandward migration of shortening by migration of its depocentre toward the hinterland as documented by numerical modelling (Zoetemeijer, 1993).

The syn-tectonic sedimentary record helps to determine the kinematic evolution of thrust sheets (Figs. 6.3 and 6.4). When the deposition rate decreases in relation to the thrusting rate, syn-orogenic deposits are deformed, and important local angular unconformities and erosion surfaces occur (Burbank and Verges, 1994). When the thrusting rate decreases in relation to the deposition rate, important local onlaps form. When syn-tectonic sediments are well recorded, which is rarely the

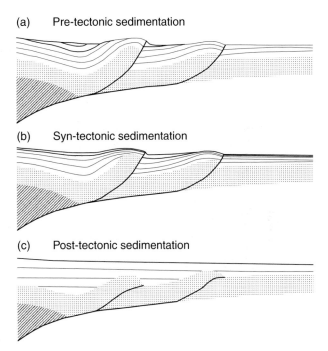

(a) Pre-tectonic sedimentation

(b) Syn-tectonic sedimentation

(c) Post-tectonic sedimentation

Fig. 6.1. Schematic representation of possible basin configurations with sediment deposition (a) before, (b) during and (c) after the thrusting (Zoetemeijer, 1993).

case in outcrop data, both the timing and the detailed folding mechanism can be recognized. This has been demonstrated by numeric modelling of fault-bend, fault propagation and detachment folds in four scenarios with different growth versus background deposition rates (Hardy *et al.*, 1996) or detachment folds developed by two different mechanisms at high and low syn-tectonic deposition rates (Poblet *et al.*, 1997). Different fold mechanisms create distinctly different syn-tectonic sediment records. Each mechanism has its characteristic constraints imposed by the specific geometries of the axial surfaces (Fig. 6.4) and the presence or lack of limb rotation. As material passes through the axial surfaces, the syn-tectonic stratal boundaries are parallel to the pre-tectonic strata in the fold limbs (Fig. 6.4). When limb rotation occurs, the syn-tectonic stratal boundaries are fanning outwards, forming beds progressively thinning towards the fold crest and decreasing in dip upwards (Fig. 6.5).

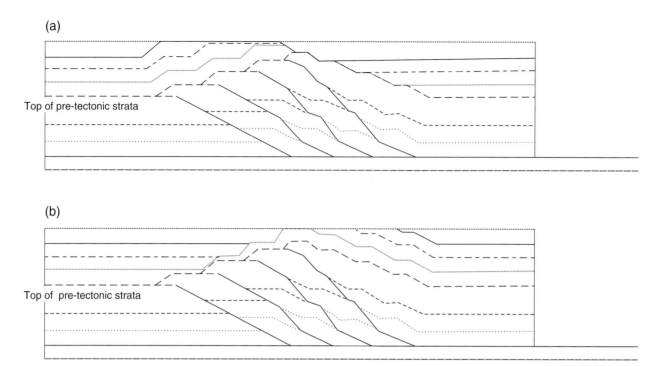

Fig. 6.2. Synthetic model of imbricate fault-bend folding (Zoetemeijer, 1993). The order of the thrust activation is not obvious from the duplex geometry, but from the overlying sediments, which indicate (a) a piggyback sequence and (b) an overstep sequence.

However, a certain degree of natural variability in the syn-tectonic sediment record for the same folding mechanism is introduced by differences in syn-tectonic lithology (Poblet *et al.*, 1997). When the anisotropy in growth strata is weak, the strata can deform by shear parallel to the pre-growth units. When syn-tectonic sediments are weak, they have a tendency to deform by vertical simple shear.

Provided a complete syn-tectonic sedimentary record can be obtained, the relative timing of the growth structure can be established. More commonly, when a complete syn-tectonic sedimentary record is not available due to erosion, the two following approaches are useful for determination of the timing.

The first approach, used in the case of a slightly eroded syn-tectonic sedimentary record, is a projection of the missing sediments. An example is provided by the combined cross section balancing and sedimentological study of the Penagalera detachment fold in the Catalan Coastal Ranges, Spain (Burbank *et al.*, 1996; Fig. 6.6).

The second approach, used in the case of a strongly eroded syn-tectonic sedimentary record, is a provenance study. As an example, a case study is that of the Beartooth Range, Wyoming and Montana (DeCelles *et al.*, 1991). The basic assumption for provenance modelling is that the changes in areal exposure of resistant rocks in an eroding growth structure are represented by the vertical sequence of sediments deposited basinward of the structure. The modelling quantitatively matches sediment and source-growth structure compositions in order to determine the unroofing history of a growth structure. The source stratigraphic section is divided into intervals of arbitrary thickness (Fig. 6.7, 1). The array of these intervals is turned into a thickness matrix reflecting percentages of rock types in different parts of the source region (Fig. 6.7, 2, 3). Then a recalculation is made, treating detritus yields as direct proportions of both erosional resistance and transport durability of different lithologies (Fig. 6.7, 4). The initial thickness matrix thus becomes converted into a matrix with corrected frequency distributions of clast types that would be produced by erosion of different parts of the source stratigraphic section. The resulting different parts of the source section are called exposure gates.

At the end of modelling, the initial exposure gate and unroofing model are specified. The calculated sequence of the generated sedimentary record model is compared with the real one. The timing of thrusting is found by trial and error matches of modelled and real sedimentary records located basinward of the growth structure (Fig. 6.8). A particular danger in the successful application of this method is the potential recycling of older clasts. Other possible dangers are the development of specific scenarios, which prevent the formation of a sedimentary record from directly representing the areal

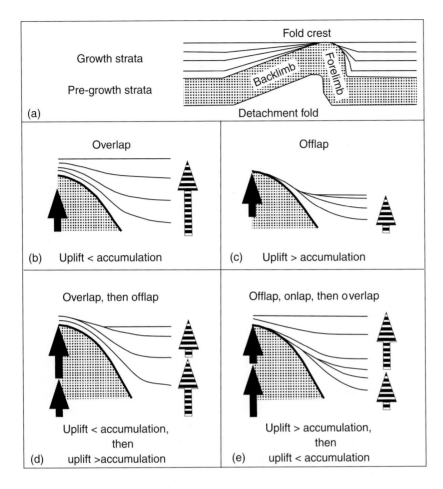

Fig. 6.3. Relationship between uplift and deposition rates indicated by the sedimentary record (Burbank and Verges, 1994). (a) Nomenclature for detachment fold and growth strata deposited during folding. (b)–(e) Models of overlap, offlap and onlap. Predictable geometries of syn-tectonic strata result from contrasts in the relative rates of crestal uplift versus coeval rates of accumulation. Crestal uplift is measured with respect to either (1) the base of syn-tectonic strata adjacent to the fold or (2) the position of correlative marker beds found in both the anticline and the adjacent syncline. When rates of accumulation are consistently greater than the rate of crestal uplift, overlap will occur (b), whereas lower rates of accumulation versus uplift lead to offlap (c). Reversals in the relative magnitude of these rates cause a switch in the bedding geometry (d, e). Onlap (e) occurs following offlap and a change to more rapid accumulation rates.

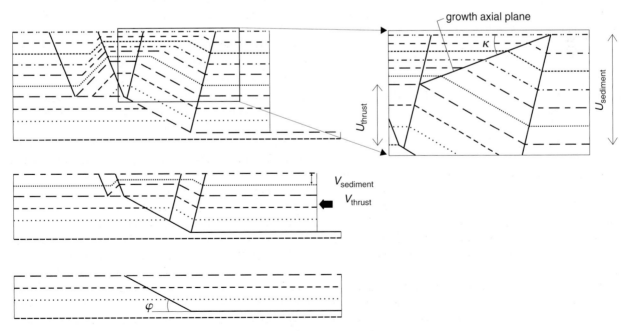

Fig. 6.4. Deposition coeval with thrusting develops angular unconformities in stratigraphy, the geometry of which depends on the thrust velocity, the ramp angle and the deposition rate (Zoetemeijer, 1993). Typical triangular shape of the growth structure is formed by back-limb of the thrust sheet, autochthonous kink-plane and growth axial plane. κ is angle of the growth axial plane with horizontal plane.

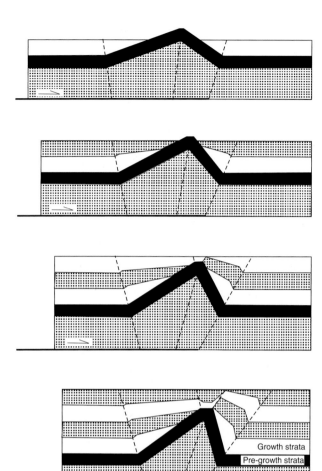

Fig. 6.5. Model of a detachment fold growth by a mechanism with conserved limb length and changing limb angle at high syn-tectonic deposition rate (Poblet *et al.*, 1997). The figure shows four successive stages of shortening with the uplift rate being lower than the sedimentation rate except for the first two stages. Growth strata are deformed by flexural slip-flexural flow.

exposure of the source area. An example would be a scenario where very fine-grained sediment eroded from nonresistant rocks of the growth anticline is allowed to bypass the zone proximal to the eventual trace of the fault. A second example is a scenario where the exposure of resistant, gravel-yielding rocks generates abundant detritus, which accumulates on top of the overturned and eroded fold limb as well as on top of the fine-grained early syn-orogenic sediment. The accuracy of the method is also affected by the subsidence of the basin in front of the structure, which affects sediment dispersal.

Similar rules and success constraints apply to the method based on the correlation of thrust movement with the arrival of a distinctive heavy mineral assemblage in the sedimentary record (Sinclair and Allen, 1992).

Once the relative timing of the growth structure is established by any of the above methods, sediments coeval with its growth need to be dated in order to determine the exact timing. While paleontological dating can be relatively straightforward in the case of marine sediments, it is commonly difficult in the case of continental sediments. In this and other cases where paleontological dating is limited, other methods such as magnetic and isotope stratigraphy, detailed well log correlation, determinations from vitrinite reflectance, illite crystallinity, T_{max}, and K–Ar, $^{40}Ar/^{39}Ar$ and fission-track dating must be used.

Magnetic polarity stratigraphy and isotopic methods were used for age determination of fluvial sediments in the Argentinean Precordillera (Johnson *et al.*, 1986), and were combined with fission-track data on air-fall zircon, allowing a determination of the depositional rate of syn-tectonic sediments. This dating was later also combined with $^{40}Ar/^{39}Ar$ analysis of biotite (Beer *et al.*, 1990). The $^{40}Ar/^{39}Ar$ ages were mostly based on disturbed spectra with partial plateaux developed only at high-temperature heating steps. $^{40}Ar/^{39}Ar$ and fission-track ages differed by about 5 Ma. Beer *et al.* (1990) allowed dated outcrops to be correlated. These outcrops contained growth structures defined in reflection seismic profiles by the kink-band method (Fig. 6.9) described later.

A combination of magnetic polarity stratigraphy and biostratigraphy was used in the south-central Pyrenees (Burbank *et al.*, 1992), where fine-grained syn-tectonic sediments facilitated the successful application of the method, and allowed a detailed determination of both fault-slip and deposition rates. Detailed mapping of the Oliana anticline (Fig. 6.10) delineated four conglomerate units, which were interbedded with occasional siltstones adjacent to the growth fold, but replaced by sandstone and siltstone away from the fold. The studied Oliana magnetic section included the top 400 m of the Igualada marine marl, conglomerate units 2 and 3, and the basal part of the conglomerate unit 4, shown in Fig. 6.10. Twelve magnetozones were defined in this section (Fig. 6.11). Given the Late Bartonian–Early Priabonian age of the Igualada marl, magnetozones N1–N3 correlate with some portion of chron 17 and possibly chron 18 (Burbank *et al.*, 1992). Given the reversal pattern of the Eocene–Oligocene magnetic polarity time scale (Berggren *et al.*, 1985) and the Priabonian–Early Rupelian age of the terrestrial conglomerates (Saez, 1987), magnetozones N3–R6 are correlated with chrons 16N to 13R.

A similar detailed magnetic stratigraphy study was

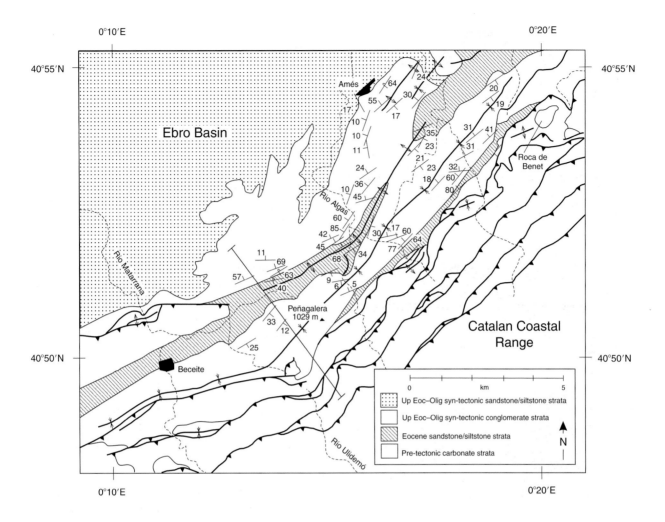

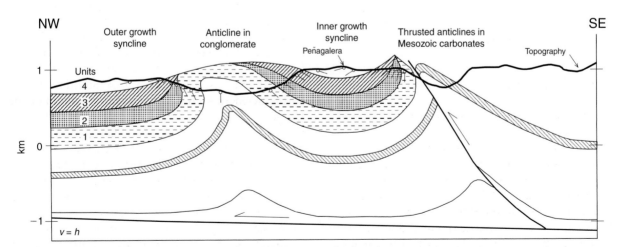

Fig. 6.6. Simplified cross section near Penagalera in the Catalan Coastal Ranges showing a detachment growth fold (Burbank *et al.*, 1996). Note strong overturning of the forelimb of the Penagalera detachment fold and the constant thickness of conglomerate unit 1 across the structure. This indicates that growth of the detachment fold began after deposition of unit 1.

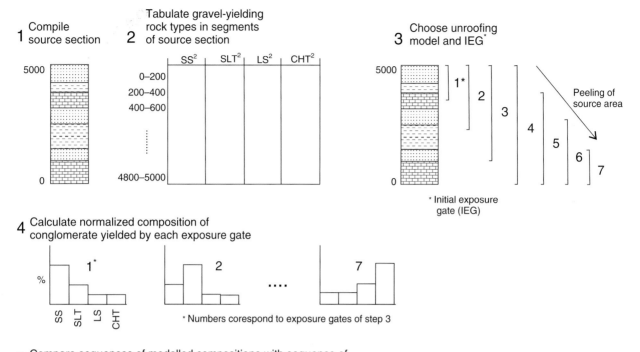

Fig. 6.7. Technique for lithologic provenance modelling (DeCelles *et al.*, 1991). Further explanation is given in the text.

made around the Mediano detachment fold in the southern Pyrenees (Holl and Anastasio, 1993), where the recurrence interval of polarity reversals for the Eocene allowed the history of deformation and sedimentation to be determined with a resolution of 0.5 Ma. Syn-tectonic sedimentation around the anticline, including siliciclastic turbidite and mudstone, allowed calibration by biostratigraphic data, which included nannoplankton and mammal fauna. The correctness of the magnetostratigraphic determination was checked by examining the correlation of the profiles made in each fold limb.

High-resolution magnetopolarity-based chronologies were also established on highly fossiliferous sections of the Napf fan delta 30 km from the studied cross section, which allowed verification of possible hiatuses in the Alpine foreland basin, Switzerland (Schlunegger *et al.*, 1997).

An example of the use of seismic stratigraphy combined with biostratigraphy and global oxygen isotope stratigraphy (Fig. 6.12) for the timing of events in the marine part of an accretionary wedge comes from the North Canterbury region of New Zealand (Barnes, 1996). Seismic stratigraphic analysis resulted in the identification of 12 seismostratigraphic units within

the framework of the sequence stratigraphic model. The youngest 11 units represent nine sea-level-cycle sequences labelled I to IX in Fig. 6.12, which developed during high-amplitude glacio-eustatic fluctuations during the Quaternary. Sequences were correlated with oxygen isotope stages 1–19, spanning from about 0.75 Ma to the present. The identified unconformity surfaces became useful strain markers for the analysis of folding rates and regional tilting across the deformation front. The exposure of most of the units along the coast by uplift and erosion in the crests of fault-propagation folds allowed sampling and nannofloral biostratigraphy in units 1, 3, 4, 6, 7, 9 and 12 shown in Fig. 6.12. The ages of units 1 and 2 were constrained by the radiocarbon dates of shells from cores (Herzer, 1981).

A detailed well log correlation can be used for timing tectonic events in an analogous manner to those of magnetic and isotope stratigraphy. An example is that of the Zayantes thrust fault in the Loma Prieta region of the south Californian part of the San Andreas fault system (Bischke and Suppe, 1990). The use of electric logs allowed movement on the fault, buried beneath the Pliocene–Pleistocene alluvial sediments, to be dated and also the start of the thrust movements near the base of the Pleistocene strata to be determined.

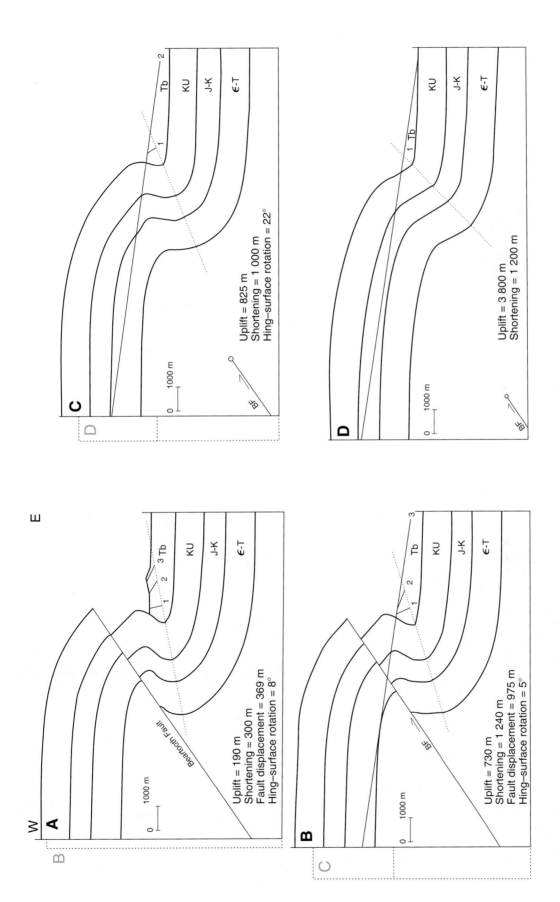

Fig. 6.8. Four-step retro-deformation of the Clark's Fork fan (DeCelles *et al.*, 1991). Symbols for rock units are: Tb – Beartooth Conglomerate, KU – Upper Cretaceous formations, J-K – Jurassic through Lower Cretaceous formations, C-T – Cambrian through Triassic formations. Numbered dip symbols (1, 2, 3) in the Beartooth Conglomerate are orientations of conglomerate beds above basal and intraformational angular unconformities. Dot-dashed lines are traces of hinge surface in the footwall syncline.

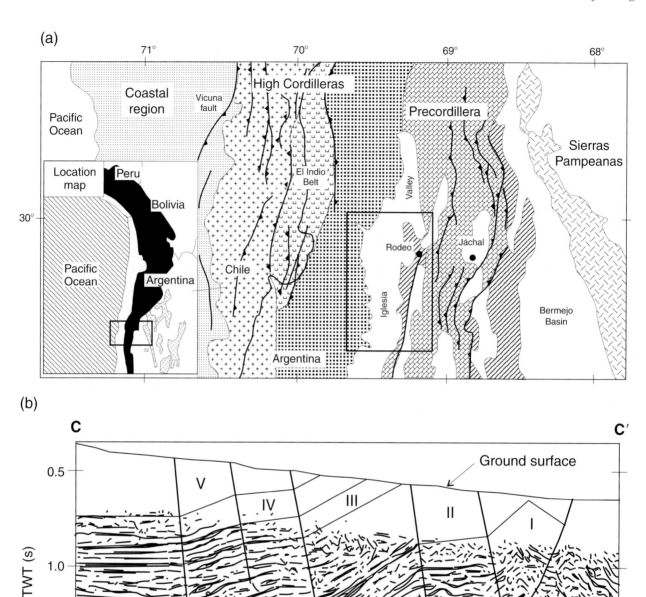

Fig. 6.9. (a) Kink-band method used to extrapolate seismic reflectors to outcrops with dated sediments in the Iglesia basin of the Argentinean Precordillera (Beer *et al.*, 1990). The profile is composed of panels of homogeneous dip, called dip domains. They are located between faults and axial planes of kinks. Five layers of the layer-cake stratigraphy are defined seismostratigraphically, projected to the surface and compared to outcrop sections. The projection assumes layer thickness preservation in the zone with missing seismic data and honours each axial plane. Inset shows the main geological provinces of the Andes at the latitude of the Iglesia basin. Rectangle indicates rough location of the profile. Stipple pattern shows foreland basin strata in the Bermejo basin and piggyback strata in the Iglesia Valley. V pattern indicates Miocene–Pliocene rocks of the El Indio belt. All other patterns indicate the extent of labelled geological provinces. (c) Kink-folds in a well-foliated schist in Luarca, Northern Spain (photograph courtesy of John Cosgrove).

(c)

Fig. 6.9. (*cont.*) (c).

A case study on dating movements using a systematic analysis of source rock maturity data was made in the Pine Mountain thrust sheet, Appalachians, Kentucky (O'Hara *et al.*, 1990). Here measurements of maximum, R_{max}, and random mean, R_{mean}, vitrinite reflectance were made on a single stratigraphic horizon, the mid-Pennsylvanian Fire Clay coal. The coal was dated at around 310 Ma by $^{40}Ar/^{39}Ar$ on sanidine, which is a constituent of a volcanic ash fall within the coal. The cooling age of the sanidine corresponds to the depositional age of the coal-forming peat. The general pattern of determined R_{max} values (Fig. 6.13) indicates pre-thrusting maturation followed by maturation affected by the Pine Mountain overthrust. Detailed vitrinite reflectance data allowed determination of the uplift history of the hanging wall, which in combination with forward balancing allows the detailed slip rate to be calculated.

It must be stressed at this point that organic matter is sensitive to thermal and stress effects along local faults. This process may lead to anomalous maturation levels within fault cores and damage zones around them (e.g. Bustin, 1983). The input of frictional heating along several local faults was demonstrated using a dense pattern of samples from sheared and unsheared sediments (O'Hara *et al.*, 1990). The absence of chemical alteration of coals along the faults suggests that the higher values in sheared samples are due to frictional heating and not due to hydrothermal fluid circulation along fault planes.

It also should be noted that the reflectance fabric also behaves as a strain gauge. Studies of faults deformed approximately by simple shear indicate that the reflectance fabric shows an increase in R_{max}, a decrease in R_{min} and no change in $R_{intermediate}$ (Levine and Davis, 1983). Although a three-dimensional reflectance analysis was not carried out in the Pine Mountain thrust study, the observation that R_{mean} as well as R_{max} display higher values along local faults led O'Hara *et al.* (1990) to the conclusion that the combined effects of differential stress and frictional heat have increased the reflectance values.

Another example of a systematic vitrinite reflectance study in a thrustbelt comes from the Shimanto accretionary prism of Japan (Ohmori *et al.*, 1997), where it

(a)

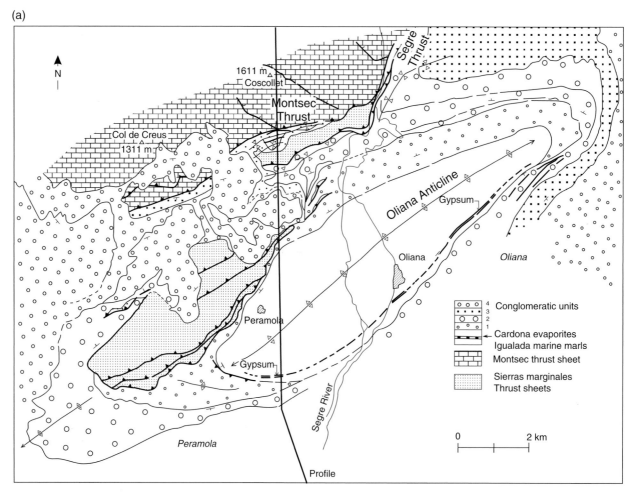

Fig. 6.10. Oliana fault-bend growth fold, the Ebro basin (Burbank *et al.*, 1992). (a) Location of cross section. Conglomerate formations 1–4 are directly related to intervals of thrusting and show that the Sierra Marginales thrusts and the Montsec thrust formed in an overstep sequence.

helped to date an out-of-sequence fault, which did not affect the maturation inside the prism. The vitrinite reflectance data allowed a detailed reconstruction of the thrust sheet geometries (Fig. 6.14) and zircon fission-track data constrained the deformation age, which in combination resulted in an exact displacement rate calculation.

Isotope and fission-track analyses were used for cooling calculations in the thrust wedge of the North Alpine Foreland basin (Sinclair and Allen, 1992). Isotopic ages were determined from biotite and phengite micas. Fission-track data were gathered from apatite and zircon. Calculated cooling rates were then translated into exhumation rates. If the geometry and development mechanism of the growth structure were known one would be able to determine slip rates along the associated fault.

The K–Ar dating method has been used in the

Rhenish Massif to determine the differences in ages of anchimetamorphic sediments by Ahrendt *et al.* (1978). Systematic differences in ages allowed a determination of the syn-metamorphic wave that passed through the Rhenish Massif for about 30 Ma at a calculated deformation progress rate of 0.5 mm yr^{-1}.

The isotope work in the Swiss Alps (Hunziker *et al.*, 1986) provided the evidence of the age of underplating. It indicated that the allochthonous cover to the Aar Massif was incorporated at later stages of the wedge development into the Helvetic nappes. The initial overthrusting is dated as being active between 30 and 35 Ma using the K–Ar and Rb–Sr methods, whereas K–Ar dates for deformation of the underlying autochthon to the Aar Massif occurred between 25 and 15 Ma. This reflects the accretion of the uppermost part of the foreland plate onto the base of the growing thrust wedge approximately 10 Ma after the initial overthrusting of

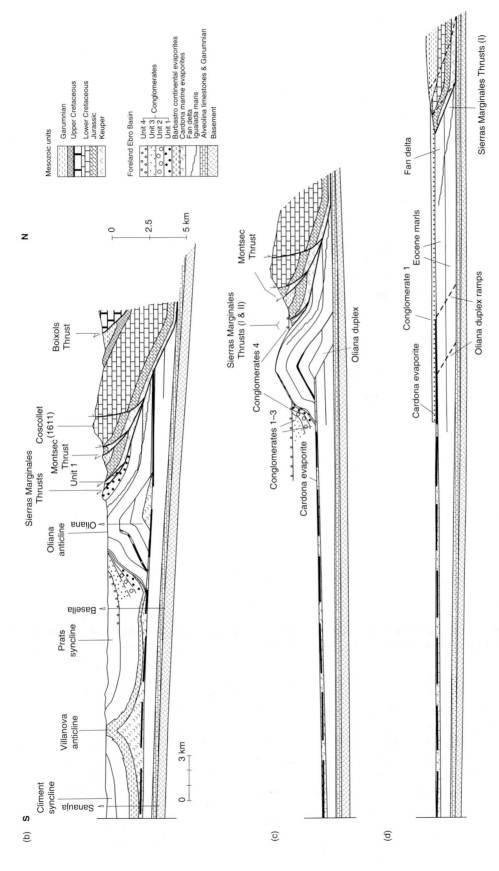

Fig. 6.10. (cont.) (b) Balanced cross section based on bore hole, seismic and surface data. (c) Restored cross section at the time of initial deposition of conglomerate 4. The position of the Oliana duplex with respect to its translation along the footwall ramp and the height of the ramp itself are determined from altitudinal and deformational relationships presently visible across the base of conglomerate formation 4. Conglomerate formations 1–3 have been deformed by both the Oliana duplex and the imbricate thrusts, whereas conglomerate formation 4 remains undeformed. (d) Restored cross section at the time of deposition of conglomerate formation 1, which is clearly involved in the imbricate host system. Whereas the several future ramps through the marls are specified here, the cutoff for the Mesozoic strata of the thrust sheet is unknown. Future thrusts are marked by dashed lines.

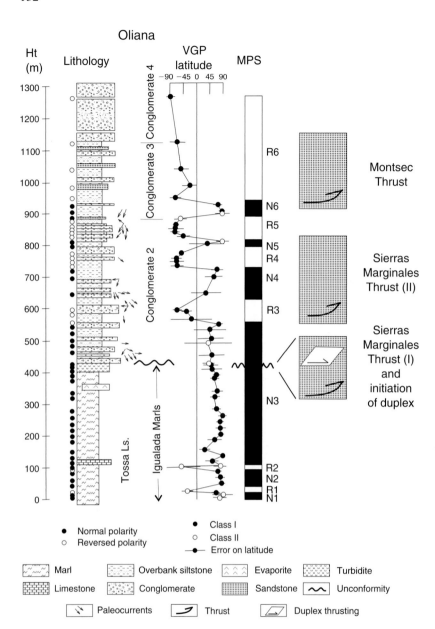

Fig. 6.11. Magnetic polarity stratigraphy and simplified lithostratigraphy for the Oliana growth fold in the south-central Pyrenees (Burbank *et al.*, 1992). Rectangles enclosing thrust symbols indicate the interval, during which a specific thrust was active.

the area. It also correlates with a period of rapid uplift within the wedge that took place between 22 and 18 Ma (Sinclair *et al.*, 1991).

The Fatric Nappe of the West Carpathians provides an example of the K–Ar dating of syn-kinematic sericite from a thrust fault zone (Nemčok and Kantor, 1990), where an Albian age determined by isotopic dating is in accordance with the biostratigraphic age of the youngest sediments present in local thrust sheets.

^{40}Ar/^{39}Ar dating is perhaps most commonly used for structural timing. Dating structural events that took part at temperatures between 150 and 450 °C by ^{40}Ar/^{39}Ar geochronology was made possible by recent advances in the understanding of the kinetics of argon diffusion in potas-

sium feldspar, the differential resetting of syn-kinematic and relict white micas by chemical exchange or neocrystallization (Dunlap, 1993). By combining the thermal history information from multi-diffusion-domain modelling of potassium feldspars with isotopic, geochemical and microstructural data from syn-kinematic and relict white micas it is possible, in certain cases, to date specific structural events as opposed to dating cooling due to exhumation. The case study discussed here was made in the Carboniferous Ruby Gap duplex in central Australia (Dunlap, 1993). The duplex comprises five thrust sheets. Each of them contains white mica-bearing quartzite mylonite. The dated samples comprise micas from entirely recrystallized quartzite mylonite, unrecrystallized mica

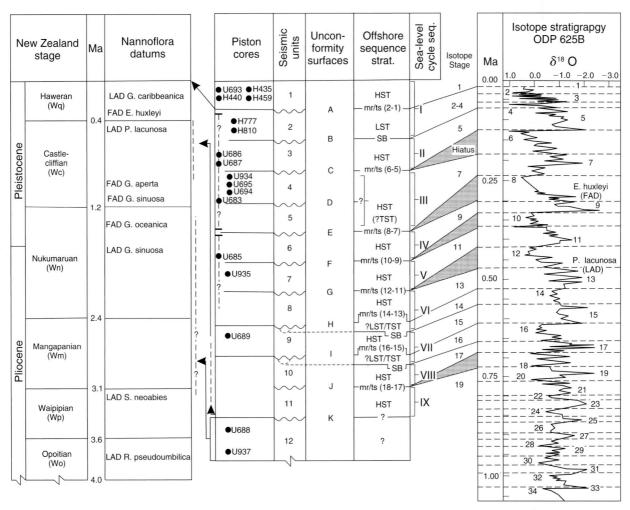

Fig. 6.12. Stratigraphic summary and interpretation of the Pliocene–Pleistocene succession beneath the North Canterbury inner to middle shelf (Barnes, 1996). Seismic stratigraphy, key biostratigraphic data, and sediment cores are tied to New Zealand stage designations. Nannofossil abbreviations include: G – Gephyrocapsa, E – Emiliania, P – Pseudoemiliania, S – Sphenolithus, R – Reticulofenestra. FAD and LAD are first and last appearance data, respectively. Seismic units 1–11 are late Pleistocene–Recent in age, and unit 12 is late Pliocene to early Pleistocene. Dotted lines within units 9 and 10 indicate localized buried channels below unconformities H and I. Vertical dashed lines from cores indicate the uncertainty in the seismic unit sampled by some cores. Sequence stratigraphic abbreviations include: HST – highstand systems tract, LST – lowstand systems tract, TST – transgressive systems tract, mr/ts (2–1) – marine ravinement surface, transgressive surface and inferred correlation to oxygen isotope stage in parentheses, SB – Type 1 sequence boundary. Inferred sea-level-cycle sequences are labelled I to IX and tentatively correlated with the global oxygen isotope sea-level proxy. Oxygen isotope stratigraphy is the orbitally tuned record from Ocean Drilling Program borehole 625B in the Gulf of Mexico (Joyce *et al.*, 1990).

porphyroclasts and potassium feldspars from mylonitic gneiss. Microstructural analysis indicates that the recrystallized white micas from quartzite mylonite grew during the deformation related to the duplex development, over a range of differential stresses. The plateau-like segments of the ^{40}Ar/^{39}Ar age spectra of the recrystallized white micas range from 311 to 325 Ma, and vary systematically with position in the duplex. Unrecrystallized white mica porphyroclasts yield older ages, because they escaped

recrystallization. Thermal modelling of potassium feldspars from the uppermost thrust sheets indicates that the duplex cooled through a temperature range of 320–280 °C, while white micas were recrystallizing 311–325 Ma. The combined results suggest that recrystallized white micas record the cessation of deformation within the duplex and migration of the deformation front toward the hotter internal zones of the thrust system over time.

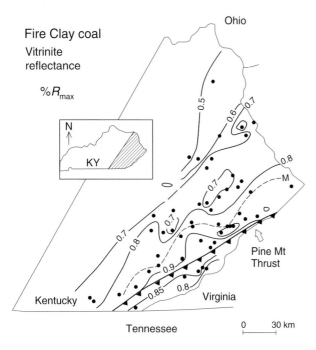

Fig. 6.13. Contoured vitrinite maximum reflectance map for the Pennsylvanian Fire Coal in eastern Kentucky deformed by the Pine Mountain thrust (O'Hara *et al.*, 1990). Dashed line represents the 0.85% contour. Values increase in footwall from 0.5% in the northwest to approximately 1.0% to the southeast. Reflectance values are 0.8–0.85% in the hanging wall. The concordance of contours within approximately 10 km of the trace of the fault and the lower reflectance values in the hanging wall at the same stratigraphic level indicate that a proportion of the footwall maturation was caused by the thrust. Note that cases, where thrusting post-dates thermal maturation in the Appalachians or Shimanto accretionary prism in Japan, are indicated by higher-grade rocks in the hanging wall (e.g. Epstein *et al.*, 1976; Harris and Milici, 1977; Ohmori *et al.*, 1997).

Alluvial chronology (e.g., Zepeda *et al.*, 1998) is the final method for structural timing discussed here. The example comes from alluvial deposits at Wheeler Ridge and the timing is based on the degree of soil development in them. It is a technique used to define the age of deformed geomorphic surfaces (Birkenland, 1984). Age constraints for the studied alluvial sediments included ^{14}C analyses of charcoal and uranium series dating of pedogenic carbonate (Zepeda *et al.*, 1998).

Deformation rates

Deformation rates can be determined from balanced and restored cross sections (e.g., Dahlstrom, 1969; Elliott, 1983; Woodward *et al.*, 1985; De Paor, 1988; Geiser, 1988b; Mitra, 1992), provided that we know the time interval during which different growth structures developed. The calculation involves the distance that a

thrust sheet travelled from initial to deformed state along a specific fault during a known time interval (Crane, 1987). The deformed state is constructed by balancing techniques and is represented in a deformed balanced profile or volume. The initial state is obtained by restoration of the deformed balanced profile or volume.

The simplest balancing techniques are based on an assumption that finite strains within typical structures of the brittle part of the thrustbelt, such as fault-bend and fault-propagation folds, should not be very high, and generally arise by localized or diffuse layer-parallel simple shear. In such a deformation, the bedding planes coincide with surfaces of no finite longitudinal strain. Cross sections through brittle parts of thrustbelts therefore do not show a change of bed length from that of the original bed length. Because simple shear is a plane strain deformation, the area of cross sections of the original layers will remain unchanged, provided the cross section is constructed parallel to the plane containing the maximum and minimum longitudinal strains. This direction will normally lie parallel to the thrust translation direction, when there is no movement of material into or out of the section during thrust deformation. Changes in length or longitudinal strains would be recorded either by extension, defined as the change in length divided by the original length (e.g. Ramsay and Huber, 1983):

$$e = \frac{l_2 - l_1}{l_1}, \tag{6.1}$$

where l_1 and l_2 are the original and new length, respectively, and the extension e is negative for shortening or by shortening:

$$s = l_1 - l_2, \tag{6.2}$$

or by the ratio r_1 of the deformed length to its original dimension:

$$r_1 = \frac{l_2}{l_1}. \tag{6.3}$$

Before the constraints of line length balancing are used in section construction, it is strongly advised that a field check is made to confirm that there is little or no shape change in the bedding surfaces (Fig. 6.15a). The technique is only correct for concentric and kink folds where the finite strains are built up by heterogeneous simple shear parallel to the layering (Fig. 6.15b). If line balancing can be applied, the most typical methods of constructing structures use kink-band types of folding associated with the thrusting (e.g., Coates, 1945; Gill, 1953; Coward and Kim, 1981; Sanderson, 1982; Suppe, 1983, 1985; Jamison, 1987) and assume layer-parallel shear (Kligfield *et al.*, 1986; Groshong

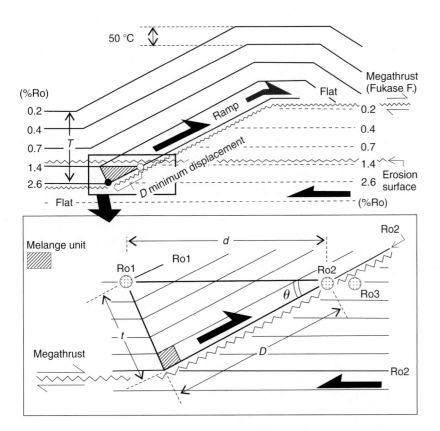

Fig. 6.14. Model for formation of the paleothermal structure around the Fukase thrust in the Shimanto accretionary prism, Japan (Ohmori *et al.*, 1997). Based on a present-day distance *d* of 9.2 km between Ro1 and Ro2 points, projected parts of the Fukase thrust sheet geometry, original burial depth difference *t* of 1.7 km, ramp angle θ equal to 11° and minimum displacement *D* equal to 9 km were calculated. Projection resulted in the thrust sheet thickness *T* of 6.7 km.

and Usdansky, 1986) or vertical shear along ramps (Jones and Linsser, 1986). They are governed by material conservation (Chamberlin, 1910; Dahlstrom, 1969; Hossack, 1979) and constrained by the fold geometries of fault-bend folds (Suppe, 1983), fault-propagation folds (Suppe and Medwedeff, 1984; Chester and Chester, 1990; Mitra, 1990; Mosar and Suppe, 1992), detachment folds (Jamison, 1987; Epard and Groshong, 1995; Mitra, 2002b), duplexes (Boyer and Elliott, 1982; Banks and Warburton, 1986; Mitra, 1986; Dunne and Ferrill, 1988; Smart *et al.*, 1997; Averbuch and Mansy, 1998), triangle zones (Erickson, 1995; Couzens and Wiltschko, 1996; Jamison, 1996) and basement uplifts (Friedman *et al.*, 1976b; Stearns, 1978; Couples *et al.*, 1994; Narr and Suppe, 1994), described in previous chapters. Valid balanced sections are characterized by the compatibility of bedding cutoff geometries (Crane, 1987; De Paor, 1988), map section compatibility (Hossack, 1983; Diegel, 1986) and include the effect of lithostatic flexure caused by thrust loading (Price, 1973; Elliott, 1977; Mugnier and Vialon, 1986).

Whether it is made using bedding readings from wells and surface exposures (Fig. 6.16a) or seismic reflectors (Fig. 6.9), the kink-band method (Coates, 1945; Gill, 1953; Suppe, 1983, 1985; Kligfield *et al.*, 1986) starts with the determination of panels characterized by homogeneous slip and sub-parallel surfaces. These panels are called dip domains. Their boundaries are interpreted as being formed either by faults or by the axial planes of kink folds (Fig. 6.9). The next stage is to include information concerning detachments, faults and stratigraphic boundaries, and axial planes of kinks are extended for each whole thrust sheet (e.g., De Paor, 1988). The deformed balanced cross section is then constructed starting from the undeformed part of the cross section and working towards the deformed part and from the surface downwards (e.g., Dahlstrom, 1969; Woodward *et al.*, 1985) as shown in Fig. 6.16(b). The undeformed cross section is achieved by restoration of the deformed balanced cross section, again commencing at the undeformed end of the deformed cross section and at the surface (Geiser, 1988b; Fig. 6.16c).

The kink-band construction gives a true representation for very brittle lithologies but produces very angular geometries for other lithologies. Therefore it is frequently combined with the round constructions of Hewitt (1920) and Busk (1929). These round constructions determine the thickness of strata between neighbouring nonparallel bedding readings by constructing their normals and connecting their intersection with bedding by circular arcs.

Deformed sediments often depart from the rules of

(a)

(b)

Fig. 6.15. (a) Photograph of a similar fold formed in Cambrian slate in Sardinia with apparent thickening in the hinge area (photograph courtesy of John Cosgrove). Note the well-developed axial cleavage. (b) Two indications of bedding parallel slip: flexural duplex developed in a shale horizon in the fold limb and slickenside on the bedding plane, Balkans.

line balancing in nature (e.g., Hrouda, 1982; Mitra, 1987, 1994; De Paor, 1988; Geiser, 1988b; Mitra *et al.*, 1988; Chester *et al.*, 1991; Chester, 1992; Tarling and Hrouda, 1993; Smart *et al.*, 1997; Wibberley, 1997). Because rock layers have different rheological properties, each layer has its own characteristic strain distribution, which departs from that of the layer-parallel simple shear model. For example, folded competent layers often show layer-parallel stretching around the outer arcs of folds and layer-parallel shortening around the inner arcs. Although related variations in bed length might approximately compensate for one another, it also might be important to make corrections in the case of competent–incompetent rock interfaces. Another case when simple line length balancing is inappropriate occurs when early layer-parallel shortening has taken place prior to and during the early stages of folding. Also in the case where competent layers are widely spaced within incompetent material, the layer lengths of the incompetent material in the contact strain zones are significantly shorter than their original lengths.

If line balancing cannot be applied, area balancing or key bed balancing needs to be applied (e.g., De Paor and Bradley, 1988; Geiser, 1988b; Mitra and Namson, 1989) or data on the finite strains are needed in order to establish correction factors in order to restore the bed lengths to their original lengths (Hossack, 1979; Woodward *et al.*, 1986; Ramsay and Huber, 1987; Geiser, 1988b).

The use of correction factors can be demonstrated using the example of the deformed and restored fold in Fig. 6.17. The equation that relates the quadratic extension to the values of the principal reciprocal quadratic elongations λ_1' and λ_2', and the angle ϕ' between the long axis of the maximum extension direction and the bedding trace is (Ramsay and Huber, 1987)

$$\lambda' = \lambda_1' \cos^2\phi' + \lambda_2' \sin^2\phi', \qquad (6.4)$$

where $\lambda = 1 + e$.

This equation may be rewritten using the multiplying correction factor $F_1 = \lambda'^{1/2}$, necessary to restore a bed length to its undeformed state, the ellipticity of the strain ellipse $R = (\lambda_1\lambda_2)^{1/2}$ and the area change that has taken place in the section $\Delta_A = (\lambda_1\lambda_2)^{1/2} - 1$ (Ramsay and Huber, 1987):

$$F_1 = (1 + \Delta_A)^{1/2}[R^{-1} + (R^2 - 1)R^{-1}\sin^2\phi']^{1/2}. \qquad (6.5)$$

The area change Δ_A is typically the only unknown parameter after the field check. However, because balancing is performed only in terrains where volume is preserved, plane strain is assumed in the cross section. Thus $\Delta_{volume} = 0$, $e_2 = 0$ and $\Delta_A = 0$.

A correction factor F_1 greater than 1 requires an increase of the measured bed length, where the long axis

of the strain ellipse makes a high angle with the bedding trace. A correction factor F_1 of less than 1 requires a decrease in the measured bed length, where the long axis of the strain ellipse makes a low angle with the bedding trace. A correction factor $F_1 = 1$ indicates simple folding by bedding-parallel simple shear.

The next step in the restoration of the bed length throughout the folded layer requires putting together correction factor data from different positions in the fold. If the field strain determinations are scattered (Table 6.1), it is assumed that the correction factor changes systematically from outcrop to outcrop, in accordance with the concept of strain compatibility. Data for the correction factor F_1 and the measured length of a bed from some pin line l' are plotted and points are connected by a smooth curve (Fig. 6.17c). This multiplying correction factor is then used to calculate the restored original bed length from the deformed bed length (Fig. 6.17d). Its inverse form $F_d = 1/F_1$, relates the restored original thickness d and deformed thickness d'.

Available balancing techniques can be divided into commercial software packages based on geometrical constraints (e.g., Moretti and Larere, 1989; Paradigm Geophysical, 1999; Midland Valley, 2000), some containing a third dimension (Paradigm Geophysical, 1999; Midland Valley, 2000; Sassi *et al.*, 2003), and routines that either allow for utilizing various strain-correction factors or are characterized as algorithms, which are not guided solely by geometrical constraints (e.g., Erslev, 1991; Suppe and Hardy, 1997; Allmendinger, 1998; Johnson and Johnson, 2002a, b and references therein; Sassi *et al.*, 2003).

When balanced and restored cross sections and volumes are successfully constructed, they can be used for strain, slip and wedge advance rate calculations. The strain rate ε for a regional balanced cross section is determined from the ratio of the final length to original length divided by the time in seconds. Typical strain rates for orogenic belts are $10^{-12}\,\mathrm{s}^{-1}$ to $10^{-16}\,\mathrm{s}^{-1}$ with a mean value around $3 \times 10^{-14}\,\mathrm{s}^{-1}$ (Kukal, 1990). The slip rate is determined from the shortening in millimetres divided by the time in seconds. The typical slip rates of thrust sheets are listed in the chapter on syn-orogenic deposition and erosion. Typical slip rates for nappes are listed in Table 6.2. The orogenic wedge advance rate is determined as the rate of the frontal thrust advance towards the foreland. The orogenic advance can be quantified, for example, according to a relationship between autochthonous foreland sediments and the accretionary wedge. The northeastward shift of the molasse basin in front of the northeastward-advancing West Carpathians, which is well documented by the shift of its southwestern margin, provides a good example

(a)

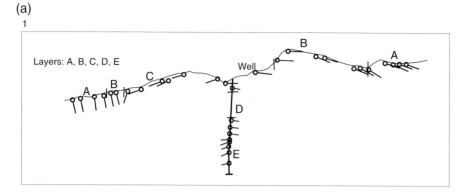

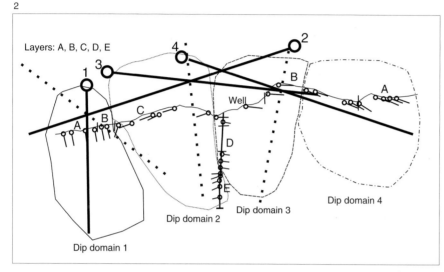

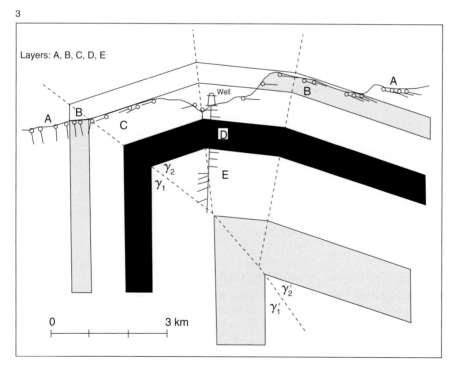

Fig. 6.16. (a) Bedding plane readings, dip domains and deformed balanced cross section through the Meilin anticline in western Taiwan (modified from Suppe, 1985).

(b)

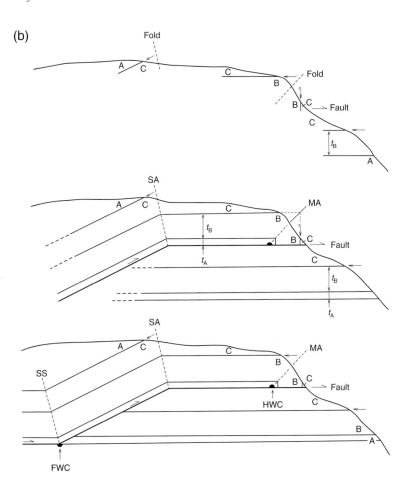

Fig. 6.16. (b) Construction of a deformed balanced cross section by the kink-band technique (De Paor, 1988). (c) Restoration of an undeformed cross section from a deformed balanced cross section (Geiser, 1988).

(c)

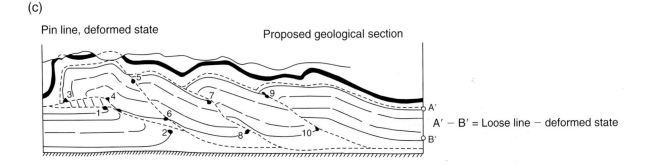

Pin line, deformed state

Proposed geological section

A′ − B′ = Loose line − deformed state

$1 + e_{\text{stiff layer}} = 0.72$

$1 + e_{\text{cover}} = 0.89$

·—·—· Uncorrected fracture array

——— Corrected fracture array

A − C = Displacement profile

Stiff layer

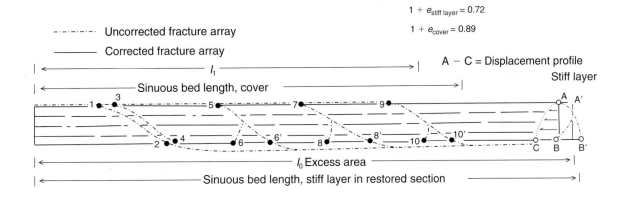

(a)

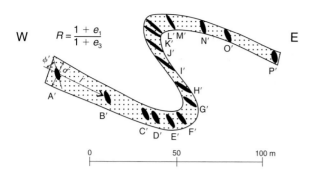

Fig. 6.17. Strain-corrected restoration of a cross section (modified from Ramsay and Huber, 1987). (a) Strained layer in a fold. (b) Variations of measured orthogonal thickness with position in a fold. (c) Correction factor graph constructed for a cross section. (d) Reconstruction of the original length and thickness of a folded layer.

(b)

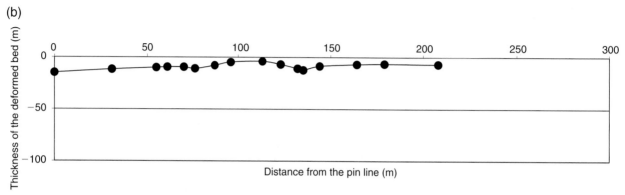

(c)

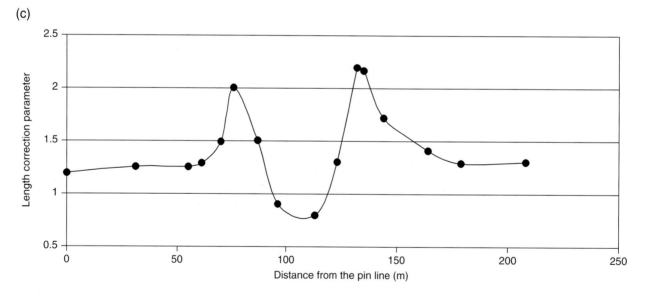

(d)

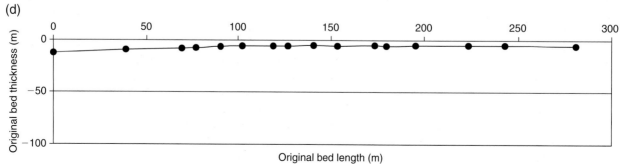

Table 6.1. Field and calculated data for a strain-corrected cross section from Fig. 6.17.[a]

Location	Deformed bed length (m)	Strain ratio R	Angle φ between long axis of the strain ellipse and bedding trace (deg)	Thickness of the deformed bed (m)
A	0	2.3	−47	14.9
B	31	2.8	−45	11.8
C	55	3	−43	10
D	61	3	−45	9.5
E	70	3.1	−56	9.5
F	76	4.3	−75	11
G	87	4.8	42	8
H	96	5	21	4.8
I	113	13	12	4
J	123	6.3	30	7
K	132	0.04	64	11
L	135	0.03	68	12.5
M	144	4.3	−55	8.6
N	164	3	−52	7
O	179	2.5	−51	6.5
P	208	2.5	−52	7

Note:
[a] Modified from Ramsay and Huber (1987).

Table 6.2. Rates of nappe movement in some orogens.[a]

Orogen	Advance length (m)	Movement duration (Myr)	Displacement rate (mm yr⁻¹)
Bettics	>70	?	Several
Southern Apennines	150–200	18	9–14
Central Alps	50	25	2
	140	15	9
	160	16	10
Himalayas	100	26–37	2.2–4
Banda Arc	>55	8–15	2.7–5.6
Western Canada	30	10	3
Venezuela	35	35	1

Note:
[a] Kukal (1990).

(Nemčok *et al.* 1998c). This margin was in contact with the advancing accretionary wedge. The calculated shift rates are roughly 18 mm yr⁻¹ for the Early Miocene and 6 mm yr⁻¹ for the Middle Miocene.

It is always useful to compare thrust sheet or nappe slip rates with plate movement or plate convergence rates (Table 6.3) in related convergent zones. It indicates the proportion of the plate movement rate accommodated by both internal wedge deformation and orogenic advance, since the advance of the rear side of the thrustbelt is transferred into both internal deformation and orogenic advance (e.g., Boyer, 1995). Various scenarios controlling the proportion of these two components are discussed in the chapter on syn-orogenic deposits and erosion. Studies in the Eastern Alpine–Carpathian–Pannonian region (e.g., Balla, 1984; Csontos *et al.*, 1992; Nemčok *et al.*, 1998c) indicate further complexities in the transfer of the convergence rate into the wedge deformation and its advance, caused by movements of microplates included in the system. Separate movements of these microplates, including both translation and rotations (Tari, 1991; Csontos *et al.*, 1992; Márton and Fodor, 1995) can cause local convergence rates to be slower or faster than the large-scale convergence rate.

As discussed earlier, if local thrust sheets are well defined in balanced cross sections and if time constraints are available, one can determine the slip rate along their thrust faults. When strata are deposited across a growing fold, these syn-growth strata commonly record limb angles and limb length through time. Because the upper surfaces of the growth strata are usually sub-horizontal at the time of their deposition, post-depositional rotations are easily defined. These rotations are often associated with growth strata that are tabular in more distal locations, but which either taper or remain tabular as they are incorporated into the forelimb of the fold (e.g., Burbank *et al.*, 1996). If tapered, the angle of taper can also be used to define the magnitude of differential uplift and forelimb rotation during deposition (Burbank *et al.*, 1996; Fig. 6.18). If untapered, but incorporated in the fold forelimb, such strata provide the evidence for lateral growth of the fold and migration of axial surfaces (Suppe *et al.*, 1992). Similar taper and bedding geometries can also be used to define rotation of the backlimb.

When growth strata are well dated, providing a set of isochrons, unfolding of these strata reveals changes in rates of uplift, shortening, forelimb rotation and aggradation (Fig. 6.18). Depositional sequences in a piggyback basin, related to thrust building processes, provide such a pattern of isochrons. A new horizontal isochron is available after each incremental step in thrust deformation, simulating the modification of relief by combined erosion and deposition (Zoetemeijer, 1993; Fig. 6.4). Distances between isochrons differ because of the continuous thrusting. The maximum distance between isochrons is (Zoetemeijer, 1993)

$$t = v_{\text{sediment}}. \tag{6.6}$$

Table 6.3a. Rates of lithospheric plate movements.[a] Methods used include: GPS, satellite laser ranging method, very long baseline interferometry method, paleomagnetic methods, reference to hot spots method and reconstruction based on stratigraphic results of deep-sea drilling. Today, some plates move quickly (50–100 mm yr^{-1}) and others slowly (about 20 mm yr^{-1}), but such differences are not indicated for the Tertiary.[b] However, there is a similarity in that the equatorial plates display a tendency to move faster than the plates closer to the poles.

Plate	Method	Time	Velocity (mm yr^{-1})	Reference
Eurasian plate	Paleomagnetic		92 or 113	Grette and Coney (1974)
Pacific plate	Paleomagnetic, hot spot		80	Kukal (1990)
Pacific plate	Deep-sea drilling	Last 80 Myr	44 followed by 20	Heezen *et al.* (1973)
Pacific plate	Hot spot		41–106.2	Pollitz (1986)
Indian plate		Late Cretaceous–Early Oligocene	100–180	Kukal (1990)
Indian plate		80 Ma	175	Johnson *et al.* (1976)
Indian plate	Paleomagnetic	70–40 Ma	149 ± 45	Patriat *et al.* (1982)
Indian plate	Paleomagnetic	Present	52 ± 8	Patriat *et al.* (1982)

Notes:
[a] Kukal (1990).
[b] Solomon *et al.* (1977).

Table 6.3b. Rates of subduction with brief characteristics of subduction zones.[a]

Region	Subduction of lithospheric plates	Length of zone (km)	Rate of subduction (mm yr^{-1})
Kuriles, Kamchatka, Honshu	Pacific below Eurasian	2800	75
Tonga and Kermadec Islands, New Zealand	Pacific below Indian	3000	82
Middle America and Mexico	Cocos below North American	2800	95
Aleutians	Pacific below North American	3800	35
Java, Sumatra, Burma	Indian below Eurasian	5700	67
South Sandwich Islands	South American below Scotian	650	19
Caribbean	South American below Caribbean	1350	5
Aegean Sea	African below European	1550	27
Solomon Islands, New Hebrides	Indian below Pacific	2750	87
Bonin Island, Mariana Island	Pacific below Philippine	4450	12
Iran	Arabian below Eurasian	2250	45
India	Indian below Eurasian (?)	?	55

Note:
[a] Toksoz (1977), Kukal (1990).

The decrease of this distance above the growing anticline is a function of the thrust sheet velocity and ramp angle:

$$\Delta t = v_{thrust} \sin \varphi, \tag{6.7}$$

where v_{thrust} is the thrust displacement during the time-step Δt and φ is the ramp angle (Fig. 6.4).

A typical triangular feature develops within syntectonic sediments, above the ramp of the growing structure (Fig. 6.4). The lower side of the triangle is formed by the back limb of the thrust sheet, the right-hand side is formed by the autochthonous kink-plane and the left-hand side is formed by the growth axial surface (*sensu* Suppe *et al.*, 1991; Fig. 6.4). The angle κ relative to the horizon is given by

$$\tan \kappa = \frac{u_{sediment} - u_{thrust}}{(u_{thrust}/\tan \phi + u_{sediment}/\tan \gamma)}, \tag{6.8}$$

where $\gamma = \pi - \varphi/2$, $u_{sediment}$ is the maximum height buildup between the thrust sheet and the uppermost isochron, u_{thrust} is the maximum vertical component of the thrust sheet displacement (Fig. 6.4). When u_{thrust} converges to 0, κ converges to γ. When $u_{sediment}$ converges to 0, κ converges to $-\varphi$.

The slip rate can also be determined along thrust

Table 6.3c. The rate of subduction in some subduction zones.[a] O indicates oceanic and C continental subduction, referring to the overriding plate. The age of subduction is in million years before present as determined by magnetic lineations. Dips are measured at depths where slabs show constant inclination.

Subduction zone	Subduction rate (mm yr^{-1})	Type	Age (Ma)	Dip
New Hebrides	98	O	50	68
Kermadec	77	O	100	66
Tonga	95	O	100	58
Izu-Bonin	98	O	130	68
Western Aleutians	67	O	65	62
Middle America	85	O	30	70
New Zealand	58	C	100	60
Japan	127	C	110	40
Alaska	71	C	45	60
Central Aleutians	89	C	65	70
Southern Kurile	107	C	115	44
Northern Kurile–Kamchatka	63	C	115	45

Note:
[a] Furlong *et al.* (1982).

faults, which actually cut the depositional surface, a feature which is more readily observed in modern examples (e.g. Rockwell *et al.*, 1984; Meghraoui *et al.*, 1988). Identification of these thrusts rests largely on the interpretation of syn-tectonic strata. A narrow zone of brecciation is commonly associated both with thrusts that cut older, previously deposited strata, and thrusts that extended to the active depositional surface. Rapid facies changes away from a fault and the presence of syn-tectonic depositional breccias or talus derived from the hanging wall indicate faults that breached the surface (Burbank *et al.*, 1996).

It has to be noted that average slip rates calculated from balanced cross sections comprise episodes of rapid seismic displacements alternating with long periods of relative quiescence. In some cases aseismic slow movements can also be observed in the interval between two earthquakes. A definite indication of the relationship between the seismic and aseismic slip could be the proportion of plate motions accommodated by earthquakes (Fig. 6.19). However, this would be obscured by an unknown amount of seismic slip related to small undetected earthquakes. Selected slip rates from recently active seismic areas show that seismic displacements are very fast in comparison with long-term tectonic movements along seismically active faults (Kukal, 1990). As a result of random summing of seismic and interseismic slip during an observed period of time,

rates of horizontal movement vary in space and time (Kukal, 1990). Various branches of the same fault system move at different rates. Various points of the same fault display different rates of movement, with the largest displacements being towards the centre of the fault zone ellipsoid and displacement progressively decreasing towards its margins (Walsh and Watterson, 1990). A comparison of the global correlation of fault plane area versus earthquake magnitude (Wells and Coppersmith, 1994) with the correlation of fault slip versus earthquake magnitude (Dolan *et al.*, 1995) indicate a linear relationship between the size of displacement and the fault plane area.

The growth of the Wheeler Ridge anticline in the San Joaquin basin, California, provides a chance to study the interaction of folding and faulting during periods of both interseismic and seismic slip (Mueller and Suppe, 1997). Syn-tectonic alluvial sediments, described in the chapter on syn-orogenic deposition and erosion, recorded folding events. Their age was determined by alluvial chronology (Zepeda *et al.*, 1998), described earlier.

The front limb of the Wheeler Ridge anticline contains a number of geomorphic landforms, including 2–4 m wide terraces separated by 9–12 m wide risers that have been produced by active kink-bend migration (Mueller and Suppe, 1997). Alluvial sediment ridges on the front limb acted as a finite strain marker, which recorded the sense of limb widening in relationship to the slip direction on the underlying blind thrust. The front limb was formed by kink-band migration above a northward-migrating thrust wedge, which lies 2 km below the surface. Folding occurred as coseismic events. These events were caused by sudden uplift during episodic earthquakes on the blind thrust that underlies the fold, similar to known historic thrust-type earthquakes, which were accompanied by folding (Yielding *et al.*, 1981; Philip *et al.*, 1992; Treiman, 1995). Upward propagation of the active axial surface through onlapped alluvial sediments, deposited during interseismic periods, created a prism of uplifted, but otherwise undeformed strata. This mechanism produces terrace-like features, the outer edges of which mark the former position of the active axial surface, prior to the folding event. The terrace forms the upper surface of the syn-tectonic sediment onlapped above material already bent into the front limb of the Wheeler Ridge anticline. The distance between the outer edges of adjacent terraces measured parallel to bedding is equivalent to the amount of fault slip released during the earthquake on the underlying blind thrust. While the average fault slip rate was determined at 10 mm yr^{-1} during the last 65 ka, individual seismic slip events had an average displacement of 10 m (Mueller and Suppe, 1997).

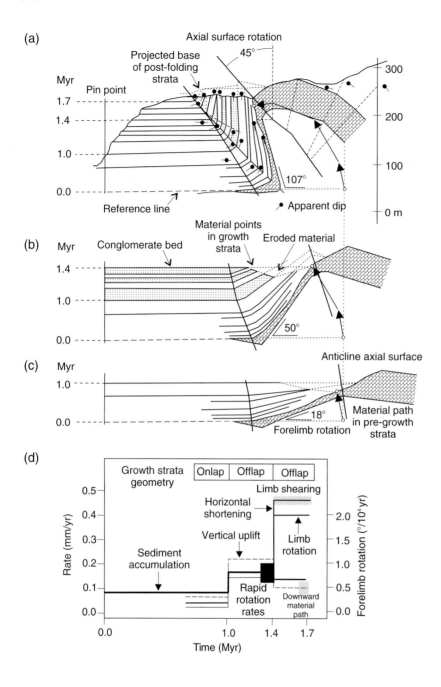

Fig. 6.18. Reconstruction of a growth syncline development at Can Juncas, showing the position of magnetostratigraphy (Verges *et al.*, 1996; Burbank *et al.*, 1996). Time periods of the deformation are related to the time interval ⟨0, 1.7⟩, where 0 and 1.7 Ma are the start and end of the deformation. Based on a process of unfolding, the upper surface of each syn-tectonic stratal unit is restored to horizontal, and underlying units are unfolded or back-rotated by an amount equivalent to the dip removed from the topmost stratum. Stratal areas are preserved during unfolding. Motion of material points, axial surfaces and fold hinges can be reconstructed in each step.
(a) Deformed state cross section at the end of the folding, at 1.7 Myr after the start, depicting observed bedding geometries in a growth syncline and adjacent anticline. Note the uniform thickness of the strata at the pin line and bed tapering in the fold forelimb. The forelimb is overturned by 17°.
(b) Restoration at 1.4 Myr after removal of 67° of the forelimb dip. Note that material points in the growth strata are uplifted with respect to the anticlinal crest in the next youngest step following the 1.7 Ma step. This causes a pronounced offlap. (c) Restoration at 1.0 Myr after a removal of 89° of the forelimb dip. Note that the initial path of a material point in the anticlinal crest is upward, and that the anticlinal hinge is almost vertical. In younger steps, the anticlinal hinge rolls forward by 45°, and the material point in the pre-growth strata moves upward, forward and finally downward. (d) Calculated rates of the crestal uplift, shortening, forelimb rotation and aggradation based on unfolding.

As an alternative to stratigraphic data, deformation microstructures may be used to determine strain rates for small rock volumes, by comparing deformation mechanisms inferred from observed deformation products and comparing these with experimentally defined deformation mechanism maps (e.g., Groshong, 1988; Knipe, 1989; Sibson, 1989). In a case study from the sub-Alpine thrustbelt of southeastern France (Butler and Bowler, 1995) microstructural observations are linked to a time-averaged displacement rate, obtained from balanced cross sections, of 6 mm yr⁻¹ during a period of 5–10 Ma.

Thrust sheets developed by piggyback fault-propagation folding, which changed into fault-bend folding during continuous shortening. Bulk displacement rates across local fault zones were calculated by integrating the inferred strain rate for deformed rocks within the fault zone across the finite width of the fault zone (Butler and Bowler, 1995; Fig. 6.20). The results were in the range 10^{-2}–10^{-6} mm yr⁻¹. The remaining time-averaged displacement rate obtained from balanced cross sections is formed by displacement along large-scale thrusts. Each time, several of them were coevally active.

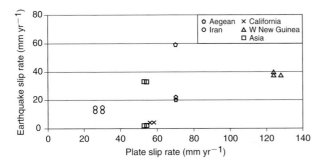

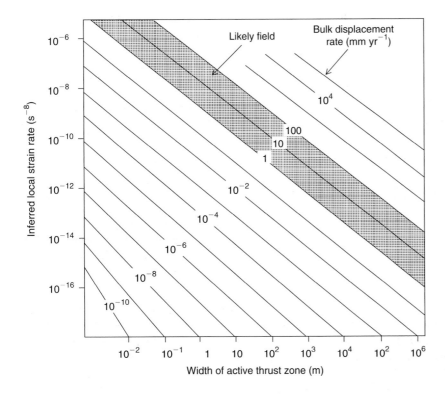

Fig. 6.19. Proportion of plate convergence accommodated by medium and large earthquakes (calculated from data by Ekstroem and England, 1989). These medium and large earthquakes account for only 25–70% of the deformation that accommodates plate convergence. The missing 75–30% of this deformation is accommodated by both aseismic slip and undetected small earthquakes. Note that only earthquakes in the Aegean are a dip-slip accommodation of the dip-slip plate boundary movements, earthquakes and plate boundary movements in Iran and Central Asia have both dip-slip and strike-slip components. Earthquakes in Western New Guinea and California are strike-slip accommodation of the strike-slip plate boundary movements.

Fig. 6.20. Relationship between the local strain rate inferred from deformational mechanisms, the width of active fault zones, across which the rocks deform at the local strain rate and the bulk displacement that is achieved by integration (modified after Butler and Bowler, 1995). The likely bulk displacement field of 1–10 mm yr^{-1} is determined from the time-averaged displacement rate estimates obtained by linking stratigraphic data to shortening values obtained from cross section restoration. The diffusion mass-transport-dominated deformation operates at strain rates of 10^{-15}–10^{-11} for the Chartreuse section, which achieves only a time-averaged displacement rate of 10^{-2}–10^{-6} mm yr^{-1}. The remaining, dominant, proportion of the regional shortening occurred primarily on several large-scale thrusts, with active widths of several metres, within which a relatively high time-averaged strain of about 10^{-6} s^{-1} operated.

PART TWO

Evolving Structural Architecture and Fluid Flow

7 Role of mechanical stratigraphy in evolving architectural elements and structural style

The lithostratigraphy in this chapter will be interpreted as mechanical stratigraphy, which determines the nature of the deformational response of the shortened rock package to applied stresses. Each mechanical stratigraphy is characterized by particular faulting and folding strengths. When the rock section is loaded by orogenic stress, strengths control whether the section deforms by:

(1) fold-first deformation (e.g. Heim, 1919; Wiltschko and Eastman, 1983; Fisher et al., 1992; Woodward, 1992; Fischer and Anastasio, 1994; Morley, 1994);

(2) fault-first deformation (Rich, 1934; Fox, 1959; Royse et al., 1975; Suppe, 1983; Medwedeff, 1989; Hedlund et al., 1994; Jamison and Pope, 1996); or

(3) contemporaneous fold–fault deformation (e.g. Dahlstrom, 1970; Brown and Spang, 1978; Williams and Chapman, 1983; Chester and Chester, 1990; Suppe and Medwedeff, 1990; Alonso and Teixell, 1992; Couzens and Dunne, 1994; Tavarnelli, 1994).

The mechanical stratigraphy controls the energy balance, described in Chapter 3, because it affects the propagation of new faults, the reactivation of pre-existing faults, the gravity forces of the uplifting part of the thrust sheet and the internal deformation of the thrust sheet.

Role of mechanical stratigraphy in thrust sheet initiation

As described in Chapter 3, the mechanism of thrust sheet detachment is controlled by the competition of faulting and folding strengths during the orogenic stress buildup. The strength limit, which is reached first along the stress path of the autochthonous rock section, determines the structural type of the future thrust sheet.

If the critical strength for folding is the lower of the two critical strengths, detachment folds will form. The critical strength for folding is controlled by Young's modulus and Poisson's ratio, for the thicknesses of the rocks and the number of layers in the rock sequence, the friction at their contacts and the individual layer geometries.

An increase in Young's modulus produces a linear increase in critical strength (Fig. 7.1a), showing progressively greater folding resistance, starting with weak rocks such as siltstone and shale, followed by limestone, and ending with rigid rocks that are difficult to fold, such as sandstone with siliceous cement or igneous rocks. Young's modulus is affected by a number of rock properties and thermodynamic conditions. The study of deformation impact at Sieniawa Dam, Poland (Theil, 1995) illustrates the nonlinear sensitivity of Young's modulus to internal deformation of the rock (Fig. 7.2a). Young's modulus is further affected by rock density variations (Fig. 7.2b), the effective normal stress (Fig. 7.2c), the rock anisotropy, cracks, the loading rate, the temperature and the confining pressure (Lama and Vutukuri, 1978). Increasing crack density lowers the modulus. A higher modulus is recorded when rock is stressed parallel to cracks or layers. The rate of loading is effective in nonlinearly elastic rocks, where the strain rate varies. A higher loading rate results in a higher Young's modulus, for example, in sandstone. Increasing the temperature decreases the modulus. The influence of the confining pressure varies depending on the rock type. Its influence on strong low-porosity rocks is small, but is high on partially cracked, high porosity or weathered rocks. An increased confining pressure increases Young's modulus.

The relationship between the critical stress for folding and Poisson's ratio is not as simple as in the case of Young's modulus (Fig. 7.1b). There is a possible linear relationship, but the data scatter is significant. The scatter may, in part, be due to experimental error in the determination of Poisson's ratio, but is probably also due to rock anisotropy and crack or porosity effects (see Lama and Vutukuri, 1978). Poisson's ratio is affected by rock properties and thermodynamic conditions. The possible temperature effect on Poisson's ratio of rocks can be implied from a documented nonlinear increase with increasing temperature in the case of silica glass (Clark, 1966; Fig. 7.3a). An increase in confining pressures might lower Poisson's ratio in the case of a weak rock, but for stronger rocks, it may have no obvious influence (Fig. 7.3b). An increase in Poisson's ratio with stress in some rocks may be associated with plastic deformation, resulting in a greater rise of lateral strain with depth

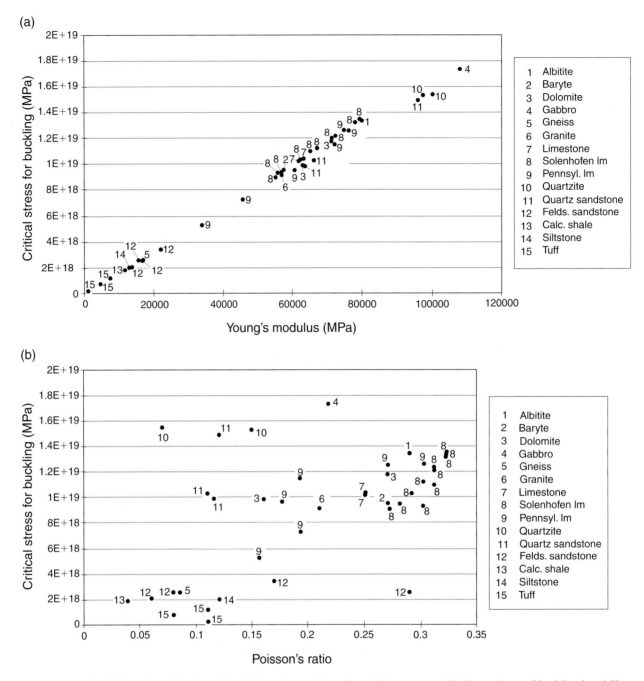

Fig. 7.1. (a) Dependence of buckling instability on Young's modulus of a multilayer system. (b) Dependence of buckling instability on Poisson's ratio of a multilayer system. Both elastic constants are from Clark (1966).

(Lama and Vutukuri, 1978). Other controlling factors are the presence or absence of cracks, porosity and the loading rate. The presence of cracks oriented parallel to the direction of applied stress may open and, thus, increase Poisson's ratio (Lama and Vutukuri, 1978).

An increase in both the number of layers and the individual layer thickness in the autochthonous rock section increases the resistance of the multilayer system

to folding. It can be implied from the simplified case of a multilayer system with frictionless layer contacts (Figs. 7.4a and b). Neither of these two relationships is linear. The increase in individual layer thickness has a larger impact on the increase of the critical stress, needed for folding, than the number of layers.

Friction at layer contacts determines whether a multi-layer system behaves as a set of individual layers or as

(a)

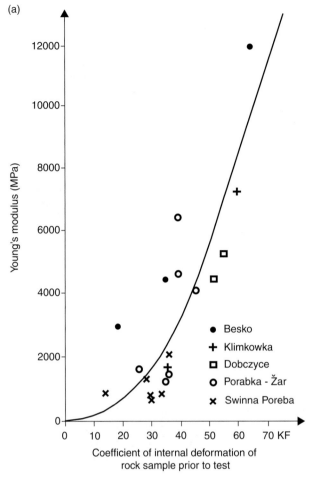

(b)

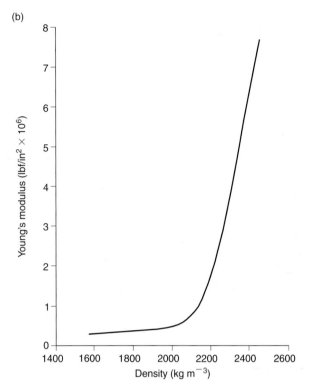

Fig. 7.2. (a) Dependence of Young's modulus on deformation for a flysch rock, Sieniawa Dam, Polish Carpathians (modified after Theil, 1995). The coefficient of deformation grows toward the intact rock. (b) Variation of Young's modulus with density for marlstone (modified after Lama and Vutukuri, 1978). Young's modulus is determined experimentally. (c) Variations of Young's modulus with effective normal stress for Westerly granite (modified after Lama and Vatokuti, 1978).

(c)

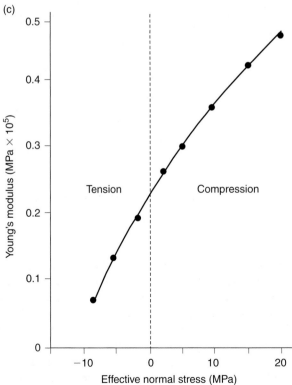

some form of coupled packages. Layer contacts are not completely frictionless in natural cases, as assumed for the multilayer system studied in Fig. 7.4. Friction, controlled by the vertical normal stress, can vary from frictionless to completely locked in the extreme case. Increasing the vertical stress controls this friction to the extent that it can completely inhibit layer-parallel slip. In this case, layers do not behave as individual units, but rather as fewer beams with larger individual thicknesses. Effectively, the thickness of individual units becomes greater and the number of units becomes smaller, causing an increase in the critical stress necessary for folding. This would be the case when the rheology of the material within each such beam affects its total rheology. This case is documented by a study of Young's modulus and Poisson's ratio of a flysch beam from the Krosno Unit of the West Carpathians, Poland (Theil, 1995). Young's modulus increases and Poisson's ratio decreases with the sandstone content in the sandstone–shale flysch (Table 7.1).

(a)

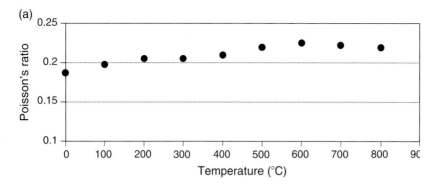

(b)

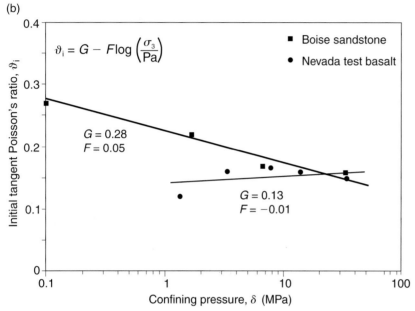

Fig. 7.3. (a) Dependence of Poisson's ratio on temperature. Data for silica glass are taken from Clark (1966). (b) Variation of initial tangent Poisson's ratio with confining pressure for Boise sandstone and Nevada test basalt (modified after Lama and Vutukuri, 1978).

If the layer contacts are close to frictionless, but the multilayered system includes several rock types, their individual rheologies become important. This could cause special scenarios, such as that where the weaker rocks in a limestone–shale or sandstone–shale system have reached folding instability, but the stronger rocks have not, thus preventing the whole system from folding.

Based on comparisons of natural folds with experimental and numerical models in multilayer systems (e.g., Ramsay, 1967; Hudleston, 1973a, b; Ramsay and Huber, 1987, Cruikshank and Johnson, 1993; Hudleston and Lan, 1993), buckle folds can develop in isolated stiff layers surrounded by a less stiff matrix, or even in homogeneous systems, but with anisotropic materials (Cobbold *et al.*, 1971; Ridley and Casey, 1989). The main control on the fold shape is the stress exponent of the stiff layer surrounded by a less stiff matrix and the viscosity ratio between the stiff layer and its matrix (Fletcher, 1974; Smith, 1975, 1977). Apart from a rheological control, any initial irregularities,

which could be up to a 10% variation in the layer thickness, will to a lesser extent influence the fold growth. The influence decreases with increased shortening. Initial wavelengths can be important for the early fold growth stages when they are not too different from the preferred wavelength (Fletcher and Sherwin, 1978) and when the stiff layer has the character of a Newtonian material, surrounded by a less weak matrix (Hudleston and Lan, 1994). The numerical models of Hudleston and Lan (1994) document the fact that for a strongly nonlinear stiff layer in a weak matrix, any initial irregularities lose their influence on fold growth after a shortening of 10%. The folded multilayer system, thus, indicates a rapid convergence to a preferred fold shape. Some of the primary wavelengths could just end up being weak parasitic folds located in the fold limb. In contrast, for a Newtonian stiff layer in a less weak matrix, initial irregularities influence the fold growth up to a shortening of 60%, indicating a slow convergence to a preferred fold shape. These models show that the folding strength of multilayer systems increases with the stress exponent of

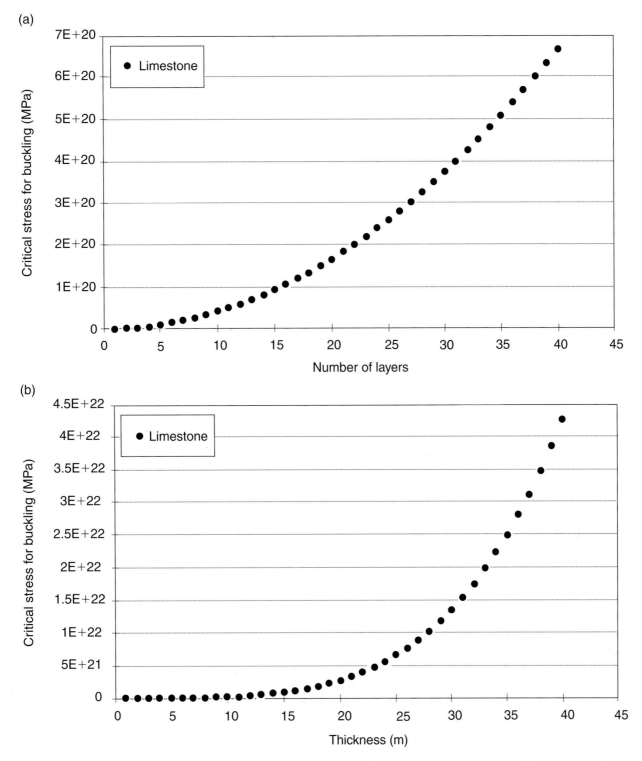

Fig. 7.4. (a) Dependence of buckling instability on the number of equally thick layers within a rock section. All layers are 5 m thick, have Young's modulus of 63 500 MPa, and Poisson's ratio of 0.252. Elastic constants are typical of Montreal limestone (Clark, 1966). (b) Dependence of buckling instability on the individual layer thickness of a multilayer system. The thickness of all layers is the same. Elastic constants are given in (a). Both calculations are made using equations from Chapter 3.

Table 7.1. Relationship of Young's modulus (*E*) and Poisson's ratio (*ν*) to the sandstone/shale ratio in a multilayered flysch system.[a]

Sandstone content (%)	Young's modulus *E* (MPa)	Poisson's ratio *ν*
>85	>8200	0.25–0.3
50–80	4300–8200	0.25–0.3
15–50	1500–4300	0.3–0.35
<15	800–1500	0.3–0.35
0	<1000	0.3–0.4

Note:
[a] Theil (1995).

the stiff layer provided the stiff layer/matrix viscosity ratio remains the same. With an increase of the stress exponent, fold limbs become longer and straighter, and fold hinges are more localized and sharper. An increase of the viscosity ratio between the stiff layer and the matrix increases the tendency for fold hinge thickening. Typical sediments with higher stress exponents include stiffer limestones or fine-grained sandstone or siltstone. These rock types would have a tendency to form sharp chevron folds with long limbs, when surrounded by a shale matrix, as documented from the Valley and Ridge province of the Central Appalachians (Hudleston and Lan, 1994). A comparison of such natural fold cases with experiments on deformation controlled by crystal–plastic mechanisms (Kirby and Kronenberg, 1987) indicates that the stress component might be only moderately high and chevron folds with long straight limbs can be further enhanced by strain softening (Neurath and Smith, 1982) and anisotropy (Cobbold, 1976; Ridley and Casey, 1989).

The multilayer system becomes anisotropic when bedding planes allow for layer-parallel slip but the slip only occurs along certain bed boundaries. Packets of welded beds bounded by movement horizons could slip independently of one another during folding (Tanner, 1989). Initially isotropic rocks such as massive limestone can develop flexural slip horizons inside an otherwise homogeneous rock unit and become anisotropic. Typical multilayer systems, which are initially highly anisotropic and prone to flexural slip are distal turbidites or other planar-bedded sequences that have a tendency to form flexural chevron folds (Tanner, 1992). A somewhat more complex flexural slip system develops in more proximal and irregularly bedded sedimentary sequences. Here flexural duplexes develop, driven by local bed thickness variations, lateral facies changes and the presence of large-scale soft-sediment structures. These higher friction areas trigger a development of

flexural duplexes facilitating a ramping-up of flexural slip surfaces to higher 'easy-glide' horizons by footwall collapse (Tanner, 1992).

If the critical strength for folding is equal to the critical strength for faulting, fault-propagation folds form. The strength controls for folding have been discussed above and the controls for faulting will be discussed together with fault-first folds below.

If the faulting strength is the lower of the two strengths, fault-bend folds, duplexes, fault-bend fold-based triangle zones, inverted grabens or basement uplifts form. Each of these structures initiates when the rock section fails in faulting first and specific structures are determined by the unique combination of factors controlling the affected mechanic stratigraphy.

The critical strength of the rock section for faulting is controlled by the cohesion, friction and the fluid pressure of individual layers.

Together with the angle of internal friction, the cohesion is characteristic for a particular lithostratigraphy. Cohesion is a pressure-insensitive property, thought to result from the strength imparted to an aggregate by electrostatic bonding between grains. Therefore, the shear strength of a weak limestone may show a slight pressure dependence, formed mainly by cohesion controlled by cementation, close to the shear strength of the ideal plastic material. Cohesion of several basic rock types, for comparison, is shown in Fig. 7.5(a). This figure demonstrates that igneous rocks and calcite or silica cemented rocks have the largest cohesion. Cementation becomes important, for example, when sediments in an accretionary prism become cemented before the formation of faults and folds, such as in the Alaskan accretionary complex, where lithification by calcite and ankerite cementation preceded the imbrication (Sample, 1990). Cementation changes the rheology of sandstone by filling the pore space and increasing the peak strength (Fig. 7.6) and cohesion so that particulate flow is no longer a viable deformation mechanism. Cohesion of heavily consolidated intact rocks, being outside the crystal–plastic deformation range, can be reduced by Griffith's microcracks (e.g. Brace, 1961). Briggs (1980), Singh (1976) and Byerlee (1968) have shown that cohesion in siltstone and mudstone increases linearly with increased quartz grain content, and decreases hyperbolically with increased illite content.

Friction has a pressure-dependent character. It is commonly visualized as arising directly from the physical friction between grains, although it is a strictly mathematical function. Therefore, the shear strength of a sandstone controlled by cohesion and pressure-dependent friction, characterized as an elastic/frictional plastic material with strain hardening and softening, shows a

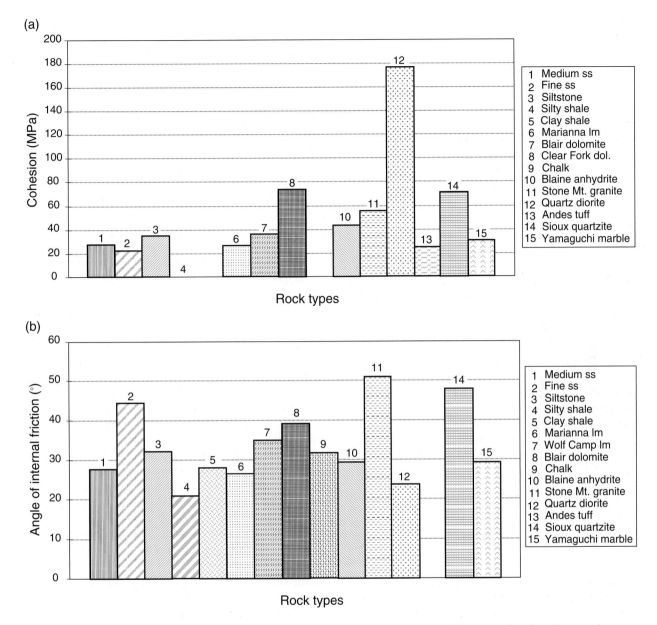

Fig. 7.5. (a) Cohesion of basic rock types. (b) Angle of internal friction of basic rock types. Values are taken from Lama and Vutukuri (1978).

pressure dependence. Friction for several basic rock types is shown in Fig. 7.5(b). This figure illustrates several features, such as an increase of friction with a decrease in grain size, which has been documented in numerous experimental studies, implying a dependence on porosity and particle shape (e.g. Karig, 1986; Maltman, 1994). The material dependence of the angle of internal friction is shown in Fig. 7.7(a), where it increases with the sandstone content in flysch rock. The dependence of the angle of internal friction on the mineralogical composition has been documented for

numerous rocks (e.g. Briggs, 1980; Zoback *et al.*, 1987; Radney and Byerlee, 1988; Hardcastle and Hills, 1991). For example, it increases with increased quartz grain content and decreases with illite content (Briggs, 1980; Singh, 1976; Byerlee, 1968). A similar level of control can be implied from the shear stress studies of Theil (1995) (Fig. 7.7b), where the decreasing control of clay and the increasing control of quartz have been shown.

It should be noted that thrustbelt structures may be formed by mechanical stratigraphies other than those already discussed. These include rocks roughly similar to

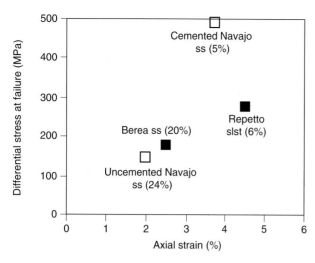

Fig. 7.6. Stress–strain values for experimentally deformed sandstone at peak strength (modified after Sample, 1990). Among sandstones, the primary control on ultimate strength appears to be porosity regardless of sandstone composition. Furthermore, cemented sandstone has a higher peak strength than siltstone, but data with constant pore fluid pressure are not available. Porosity of starting materials is shown in parentheses. Berea sandstone is a moderately well cemented graywacke, Navajo sandstone samples are quartz arenite, one with abundant void space and the other cemented by calcite, Repetto siltstone is shown to approximate the strength of shale. Direction of maximum compressive stress during the experiment at room temperature was perpendicular to bedding. Samples were jacketed to maintain pore fluid pressure. Confining pressure was 200 MPa for Berea and Repetto, 100 MPa for Navajo. Effective confining pressure was 50 MPa.

ideal plastic materials or pressure-dependent elastic/frictional plastic materials. A special category is rocks, such as evaporites, which have a viscosity-, thickness- and displacement velocity-dependent shear strength. These rocks can be characterized as being viscous material. Their importance for thrustbelt structures is apparent when one compares the proportion of thin-skin structures detached along shale horizons with those detached along evaporitic horizons (Tables 7.2 and 4.1).

Fluid content is an important control on the strength of all three types of materials. The fluid content decreases the shear strength of the rock as shown by tests on wetted clay (Arch et al., 1988; Theil, 1995; Fig. 7.8). The decrease shown in Fig. 7.8 is not only achieved by the subtraction of the fluid pressure from the total normal stress (the effective stress), but is also affected by the reaction of clay to water saturation, as shown by the nonlinear dependence. The same experimental data indicate an exponential increase of the deformation as a consequence of wetting, with a significant increase

after about 20% water saturation. The results of experiments, which test the reaction cohesion and friction of the mudstone and siltstone on wetting, also indicate their reduction with saturation (e.g. Briggs, 1980; Singh, 1976; Byerlee, 1968; Table 7.3). The saturation effect on the viscous material, salt, is much more complex, as it is controlled by the deformation and the confining pressure (Urai, 1985). Long-term creep experiments indicate that trace brine is always present in natural salt, causing distinct weakening at low strain rates (Urai et al., 1986).

Given that the faulting strength (controlled by cohesion, friction and fluid pressure) of the rock section is weaker than the folding strength, a specific fault-first structure is determined by the presence of low-strength pre-existing faults and by the mechanical stratigraphy. Weak pre-existing faults might favour the development of inverted grabens or basement uplifts. The occurrence of a low-strength potential detachment horizon might favour the development of triangle zones, comprising thrusts branching from the same detachment. The presence of a higher-strength potential detachment horizon might favour the formation of fault-bend folds with vergency towards the foreland. The presence of a higher-level detachment could favour duplexing, while the addition of a rather thick but relatively soft overburden might change an allochthonous duplex into a passive roof duplex. Note that specific mechanical stratigraphies might favour a particular structural style, but it is the combination of mechanical stratigraphy and the orogenic stress distribution that determines the outcome, as discussed in Chapter 3.

The important consequence of the autochthonous rock section being strained before thrust sheet development is that it is not the mechanical stratigraphy of the intact rock that controls the structural style but the mechanical stratigraphy of the deformed material. The initial straining can change the mechanical stratigraphy of the autochthonous rock section by changing its density, degree of interlocking or the anisotropy of the rock fabrics.

The impact of the degree of interlocking and rock anisotropy is well documented in rock mechanics handbooks (e.g. Clark, 1966; Dreyer, 1972; Vutukuri et al., 1974; Lama and Vutukuri, 1978), but is usually related to the compressive strength. These analyses, however, provide some understanding of the influence of these particular factors on the faulting strength.

The determination of the interlocking index for cohesive materials is an important parameter, because it controls the frictional part of the shear strength of these materials. The interlocking index (G) is defined (Niggli, 1948) by

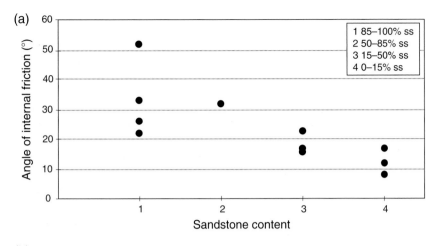

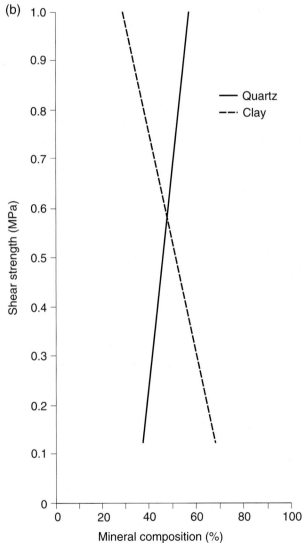

Fig. 7.7. (a) Dependence of the angle of internal friction on the sandstone content of a flysch rock. Data are taken from the Krosno Formation, Polish Carpathians (modified after Theil, 1995). Sandstone content: 1 – 85–100%, 2 – 50–85%, 3 – 15–50%, 4 – 0–15%. (b) Dependence of the shear strength on the quartz and clay proportion in a flysch rock, Sieniawa Dam, Polish Carpathians (modified after Theil, 1995). Normal stress was kept at 0. 2 MPa.

Table 7.2. Examples of thin-skinned thrustbelts developed at least in part atop an evaporite layer.[a]

Fold belt	Location	Peak deformation	Evaporite layer age	Reference
Franklin Mts.	NW Canada	L. Cret – E Ter	Cambrian	1, 2
Parry Island belt	Canadian Arctic	L. Dev – E Miss	Ordovician–Permian	3, 4
Appalachian Plateau	NE United states	L. Penn – Perm	Silurian	5, 6
Sierra Madre Oriental	NE Mexico	E. Tertiary	U. Jurassic	7, 8
Cordillera Oriental	Colombia	L. Miocene – present	L. Jurassic	9–11
Atlas Mts.	Morocco, Algeria	Eocene	Triassic	12
Pyrenees	France, Spain	Eocene	Triassic	13, 14
Jura Mts.	Switzerland	Eocene	Triassic	15, 16
Carpathians	Rumania	Pliocene	Miocene	17
Zagros Mts.	Iran	Active	L. Precam., Miocene	18
Salt Range	Pakistan	Active	Cambrian	19
Tadjik belt	Tadjik Rep.	Active	Jurassic	20, 21
Southern Urals	Russian Rep.	L. Paleozoic	L. Paleozoic	22
Amadeus basin	Central Australia	Devonian	Precambrian	23, 24
Verkhoyansk belt	Siberia	Cretaceous	Devonian (?)	25
Hellenic Arc	Aegean Sea	Active	U Miocene	26

Note:

[a] Davis and Engelder (1985).

References:

1, Cook and Aitken (1976); 2, Cook and Bally (1975); 3, Davies (1977); 4, Balkwill (1978); 5, Rodgers (1963); 6, Rodgers (1970); 7, De Cserna (1971); 8, Rodgers *et al.* (1962); 9, Ujeta (1969); 10, Campbell and Burgl (1965); 11, McLaughlin (1972); 12, Tortochaux (1978); 13, Brinkman and Logters (1968); 14, Liechti (1968); 15, Spicher (1980); 16, Laubscher (1972); 17, Paraschiv and Olteanu (1970); 18, Farhoudi (1978); 19, Yeats and Lawrence (1985); 20, Leith (1984); 21, Keith *et al.* (1982); 22, Kazantsev and Kamaletdinov (1977); 23, Stewart (1979); 24, McNaughton *et al.* (1968); 25, Nalivkin (1973); 26, Le Pichon *et al.* (1982).

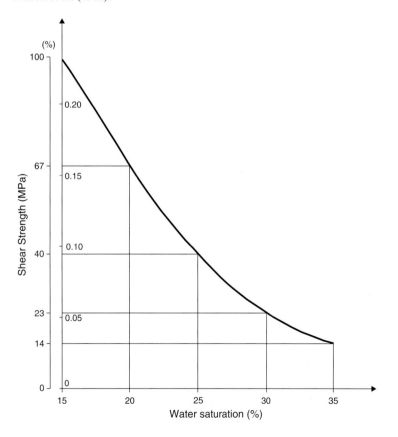

Fig. 7.8. Dependence of the shear strength on the water saturation of clay (modified after Theil, 1995). Clay samples are from Lipowica, Kotelnica and Dobczyce in the Polish Carpathians. Nonlinear relationship indicates no simple relation of shear strength to wetting, which involves a reaction of clay to water saturation.

Table 7.3. Cohesion and friction data determined experimentally from mudstone and silty mudstone.

	Rock cohesion (MPa)		Angle of internal friction (degrees)				S
			peak		residual		
	dry	wet	dry	wet	dry	wet	
slt/mdst	38.00	22.00	21.5	20.5	25.0	21.5	1
	30.00		25		24.5	24.0	1
Mdst	32.00	5.90	23	0–12	26.6	0	1
	47.30	12.42	12	10	0		2
	9.08	8.50	24	6	3		2
	0.6	4.90	03	5	8		2
	36.20		22		29		2
	3.66		6		26		2
	18.20		17		21		2

Abbreviation:

mdst = mudstone, slt/mdst = silty mudstone.
(S, sources of data: 1, Briggs (1980); 2, Singh (1976).)

$$G = \frac{1}{n} \sum_{i=1}^{n} \frac{O_i^{1/2}}{V_i^{1/3}}, \tag{7.1}$$

where n is the number of grains, V_i is the grain volume and O_i is the part of the grain surface that is in contact with neighbouring grains. The modification for a study in thin sections is

$$G = \frac{1}{n} \sum_{i=1}^{n} \frac{U_i^{1/2}}{A_i^{1/3}}, \tag{7.2}$$

where A_i is the grain area and U_i is the portion of the grain perimeter that is contact with neighbouring grains.

Experimental results show that the compressive strength of a rock structure increases with increased interlocking (e.g., Dreyer, 1972) and the anisotropy of the rock sample (e.g., Dreyer, 1972). The influence of the degree of interlocking is linear and the influence of the rock anisotropy is nonlinear.

The control of the anisotropy of the rock fabric over the rock strength requires a detailed discussion. The rock anisotropy prior to orogenic straining is given by the primary rock fabric. The anisotropy at the time of thrust sheet initiation varies for the same rock type, because it was modified by inelastic processes during the straining, which accompanies the buildup of the stress state to the limit necessary for thrust sheet initiation (e.g. Knipe, 1985; Mandl, 1988). These strains consist mainly of loosening of grain aggregates. Such load-induced anisotropies in sand and clay, although having a minimal effect on the orientation of slip planes, may greatly affect the amount of continuous straining that precedes the actual fracturing.

Analogue-material experiments indicate than once the choice of the initiated structure is made by a combination of the orogenic stress with competing faulting and folding strengths, it is the thickness of the autochthonous rock section that determines the size of this structure. The simplest possible rock section representation is the one-layer case. When we reduce the rock section into one homogeneous sediment layer, and the only parameter allowed to vary is the thickness, the role of thickness on the thrust sheet width can be observed without a need to specify the rheological properties, keeping them all fixed. As proved by sandbox modelling (e.g. Colletta *et al.*, 1991; Marshak and Wilkerson, 1992) and study of natural examples (Boyer, 1995; Goff and Wiltschko, 1992), the width of the thrust sheet is a linear function of the sediment layer thickness (Fig. 7.9; Marshak and Wilkerson, 1992).

Role of mechanical stratigraphy in thrust sheet development

After an initial discussion of the impact of mechanical stratigraphy on thrust sheet initiation, this section analyses the impact of mechanical stratigraphy on the energy balance, which controls the development following thrust sheet initiation.

Assuming one homogeneous sediment layer for simplicity, a stress equilibrium calculation can be made to demonstrate the impact of mechanical stratigraphy on the energy balance during the thrust sheet development (Fig. 7.10). The balance of horizontal tectonic forces, following and modifying Mandl (1988) requires

$$F = F' + \tau_b \Delta L + F_p, \tag{7.3}$$

where F and F' are tectonic forces in the rear and front of the thrust sheet, ΔL is the length of the thrust flat, τ_b is the basal shear stress along the flat and F_p is the force causing a plastic deformation in the flat portion of the thrust sheet. The force F' is balanced by counteracting toe forces:

$$F' \cos\beta = T' + W' \sin\beta + F_p', \tag{7.4}$$

where β is the ramp angle, T' is the frictional force along the ramp, W' is the weight of the toe and F_p' is the force causing a plastic deformation in the thrust sheet toe. In order to move the shortened thrust sheet along the new potential ramp behind the existing ramp

$$F \cos\beta = T + W \sin\beta + F_p'', \tag{7.5}$$

where T is the resistance against propagation of a new ramp, W is the weight of the potential new toe and F_p'' is the force causing plastic deformation of the sediments. In the case of initiating the new thrust sheet in front of

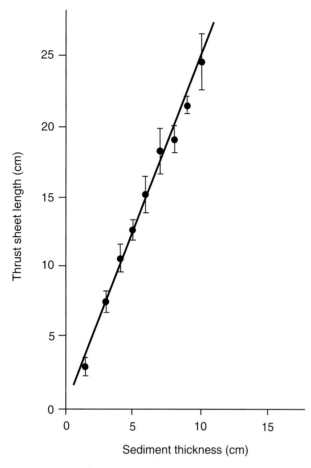

Fig. 7.9. A graph of the distance from the backstop to the first-formed thrust plotted against the initial layer thickness (Marshak and Wilkerson, 1992). The dry sand layer has well-sorted, sub-rounded grains, that are 0.3–0.6 mm in diameter, with a density of 1570 kg m^{-3} and an angle of internal friction of 30°. In order to enhance the development of discrete ramps, the sand was saturated with water at a ratio of about 1 cm^3 of water per 2000 cm^3 of sand. Note that a range of results from repeated runs is obtained for a given initial layer thickness but the mean values define a straight line. Bar lengths indicate a standard deviation.

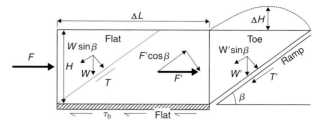

Fig. 7.10. Stress equilibrium during thrust sheet development (modified after Mandl, 1988). See the text for an explanation.

release mechanisms is affected by the mechanical stratigraphy of the deforming sedimentary section.

The term $T + W \sin \beta + F_p''$ describes the control over whether a new thrust sheet will develop, separated by a newly propagated fault. The term $\tau_b \Delta L + T'$ documents the factors that determine whether a pre-existing thrust fault is reactivated. The term $W' \sin \beta$ describes the control over whether an existing thrust sheet is moved further against gravity. The term $F_p + F_p'$ documents the factors that determine whether the existing thrust sheet is further deformed.

The resistance against propagation of a new ramp, the amount of plastic deformation within a rock before it fails in faulting and the weight of the potential new toe, controlling propagation of the new fault, are all affected by the mechanical stratigraphy. The resistance against propagation of the ramp is controlled by the cohesion, friction and pore fluid pressure of the partially strained rock section. The fluid pressure acts against the total vertical stress, resulting in a decrease of the effective vertical stress.

Layer-parallel strain prior to any larger-scale fault propagation changes the strength of the affected rock section, decreasing cohesion and friction. At levels of differential stress significantly lower than those necessary for the main fault propagation, randomly oriented microcracks develop and increase in number, forming clusters, as demonstrated in rock mechanics tests (e.g., Scholz, 1968; Krantz and Scholz, 1977; Lockner *et al.*, 1992). The amount, size and distribution of these microcracks affects the rock strength. For example, the compressional strength of the rock C_0 is inversely proportional to the square root of the crack half-length of the longest Griffith's crack in the rock sample:

$$C_0 = 8 \left(\frac{2E\gamma}{\pi a} \right)^{1/2}, \tag{7.6}$$

where E is the Young's modulus and γ is the specific surface energy.

Chapter 3 indicated that the development of a new

the existing ramp, the term F should be replaced by the portion of the tectonic force F' that continues through the existing ramp, having lost the portion that supports the existing thrust sheet.

Each new orogenic stress buildup (e.g., Yielding *et al.*, 1981; Rockwell *et al.*, 1988) generates stresses that are consumed on propagation of the new fault, reactivation of the pre-existing one, work against opposing gravity or internal deformation (Mitra and Boyer, 1986). Each new energy consumption step will be determined by choosing the minimum physical work (Nadai, 1963). The set of equations (7.3)–(7.5), helps to illustrate in a very simplified form, how each of these four energy

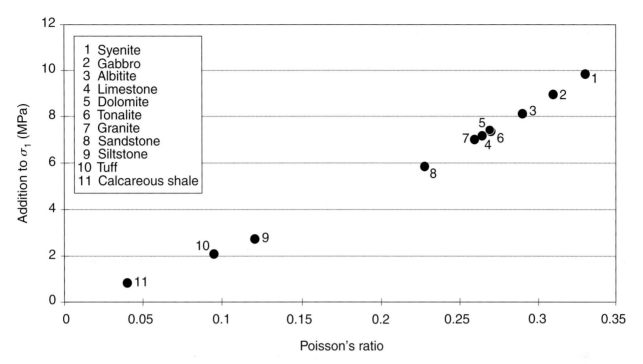

Fig. 7.11. Calculation of the horizontal stress contribution of the overburden following poro-elasticity stress–strain relationships (e.g. Jaeger and Cook, 1976) for isotropic and linearly elastic material. The overburden load is kept at 20 MPa, Poisson's ratios of tested rocks are taken from Clark (1966).

detachment fault forming a future thrust sheet occurs within a horizon with reasonably low shear stress associated with a sufficiently large shear stress that decreases downward in the rock package. Therefore, in addition to the strength of various horizons, it is also important how the mechanical stratigraphy affects the stress transmission. The stress transmission in various rocks can be illustrated by assuming an isotropic layer and investigating how changes in vertical stress generate changes in horizontal stress, using Poisson's function (e.g. Jaeger and Cook, 1976; Fig. 7.11):

$$\Delta\sigma_h = \Delta\sigma_v \frac{v}{1-v}, \tag{7.7}$$

where v is Poisson's ratio. Figure 7.11 shows that this transmission increases with rock 'competence'.

In order to generate large differential stresses required for fault propagation within a weak detachment layer by stress transmission, it is necessary for the potential detachment horizon to be juxtaposed with competent layers so that large competence contrasts exist. This principle is illustrated, for example, by the finite-element models of Schedl and Wiltschko (1987) and by the presence of combinations of layers with such a high contrast along each thrust flat in the Anschutz and Aschutz East fields of the Absaroka thrust sheet (West and Lewis, 1982; Fig. 4.28). Finite-element models document how

thrust faults propagate along the basement–sediment interface (the surface along which increased differential stresses are generated; Fig. 7.12). The differential stress increase is dependent on the competence contrast between the basement and the sediment layer. The effect of the basement on the sedimentary package is more pronounced in the case of an overlying soft carbonate or shale, where the ratio of Young's moduli for basement versus sediment ranges from 12:1 to 2:1. The effect is less pronounced in the case of fine to coarse indurated clastics, where the range of this ratio is limited to 2.5:1 to 2:1 (Schedl and Wiltschko, 1987).

The shear stress transmitted to a potential detachment horizon varies according to the different rocks of the affected rock section.

The shear strength of a rock close to an ideal plastic material, such as some weak limestones, shows a negligible pressure dependence. The behaviour of such a layer is not affected by slight changes in the overburden. When the tectonic force is raised to a certain level, shearing starts. When the force drops below this critical value, shearing 'freezes' immediately. Experiments with less ideal plastic materials, such as clay, show little variation in the stress–strain relationships with the strain rate (Arch *et al.*, 1988).

The shear strength of an elastic/frictional plastic material, such as sandstone, behaves differently. It is

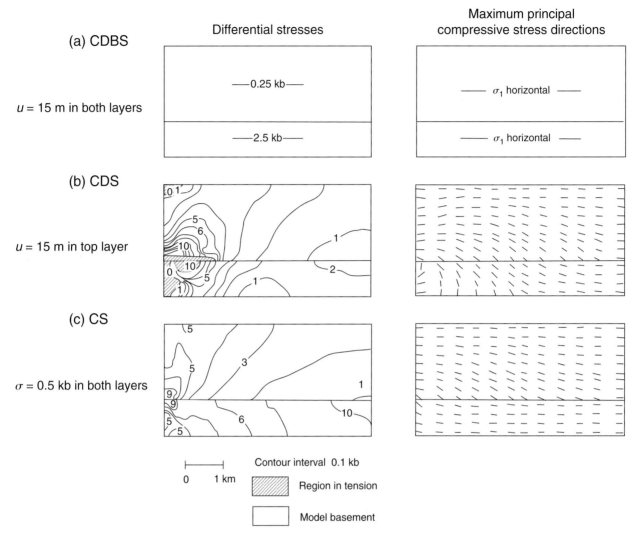

Fig. 7.12. Differential stress calculation for the finite-element model of a sediment layer juxtaposed with basement (modified after Schedl and Wiltschko, 1987). Poisson's ratio is 0.499, CDBS is a model with constant 15 m horizontal displacement in both basement and sediment layer, CDS is a model with constant 15 m displacement in sediment layer alone and CS is a model with a nearly constant stress of 0.5 kb applied to both layers.

pressure-dependent and therefore highly sensitive to overburden changes.

The shear strength of viscous materials, such as rock salt, belongs to a third category of rock strengths. It is dependent on the displacement velocity, the thickness and the viscosity (Figs. 7.13a and b). The relationship between the shear strength and the thrust displacement velocity is linear (Fig. 7.13a). The relationship between the shear strength and the thickness of the viscous layer is hyperbolic; the greater the thickness is the smaller the strength (Fig. 7.13b). Figure 7.13 illustrates how the viscosity of rock salt and Solenhofen limestone, from different parts of the viscous spectrum, changes the shear strength over several orders of magnitude and

demonstrates the control of the material itself. Viscosity is further dependent on the temperature (Fig. 7.13c). A natural laboratory for studying the influence of viscous layer thickness is the Broad Fourteens basin in the North Sea discussed in Chapter 8. The basin is a NW–SE-trending graben formed by Middle Triassic–Early Cretaceous rifting and inverted during the Turonian–Maastrichtian (Nalpas *et al.*, 1995; Brun and Nalpas, 1996). The structural style is different in the southeastern, central and northwestern parts of the Broad Fourteens basin, being affected by the thickness of the Zechstein salt horizon. Salt is missing in the southeastern part of the basin and this results in a complex system of normal faults, some of which indicate

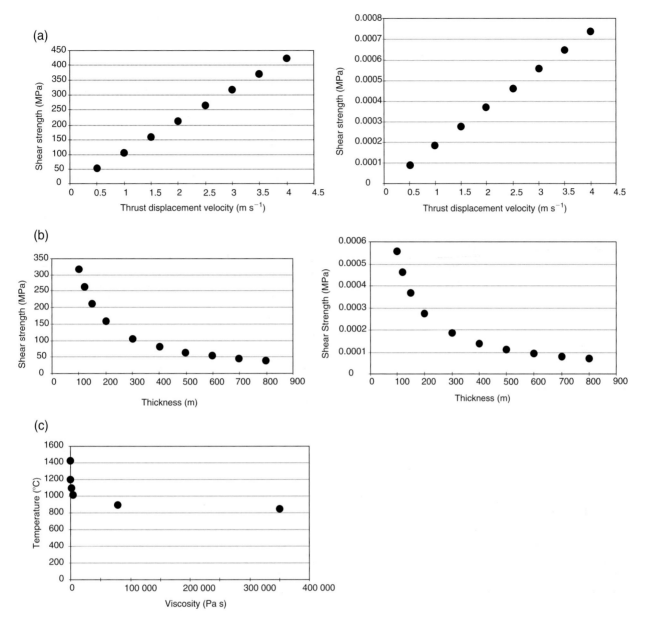

Fig. 7.13. (a) Shear strength dependence of a viscous layer on the thrust displacement velocity. Viscosities of Solenhofen limestone (left plot) and rock salt (right plot) are taken from Clark (1966), velocity from active thrusts (Davis *et al.*, 1989; Medwedeff, 1989, 1992; Mount *et al.*, 1990; DeCelles *et al.*, 1991; Suppe *et al.*, 1992; Burbank *et al.*, 1992; Avouac *et al.*, 1993; Jordan *et al.*, 1993; Zoetemeijer *et al.*, 1993; Shaw and Suppe, 1994). (b) Shear strength dependence of a viscous layer on the layer thickness. (c) Viscosity dependence on temperature. Data are taken from Clark (1966).

inversion. Salt between 0–400 m thick in the central part of the basin results in the development of normal and reverse faults in the 'cover' sequence. These faults are not directly connected to basement faults. The salt horizon with this thickness was sufficiently weak to become deformed by detachment faulting, so that the cover sequence became decoupled from the basement. Salt between 400–600 m thick or greater in the northwestern

part of the basin resulted in a decoupled cover and the development of frequent diapirs, which localized reverse faults (Nalpas *et al.*, 1995; Brun and Nalpas, 1996).

As discussed in Chapter 3, the stress transfer through a multilayer system results in straining which precedes failure, both in incompetent horizons, where flats are developed, and in competent horizons, where ramps are developed. This straining generates plastic hardening

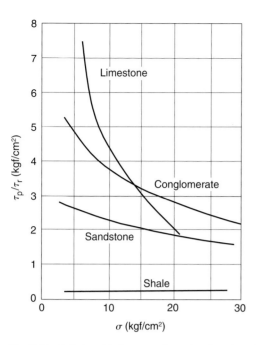

Fig. 7.14. Relationship between the ratio of peak and residual shear strength (τ_p/τ_r) and the normal stress (σ) for various rocks (modified after Vutukuri *et al.*, 1974). The shale is represented by roughly ideal plastic material, other materials are elastic/frictional plastic.

followed by plastic softening and finally failure but has a different history and character in the different layers of a deforming mechanical stratigraphy. Such plastic deformation of limestone from the Southern Appalachians (Wojtal and Mitra, 1986) and of quartz-ite and sandstone from the Central Appalachians (Mitra, 1987) involves strain hardening caused by the development of dense dislocation patterns and 'grain boundary' sliding, followed by grain fracturing which causes strain softening. The failure develops in the sof-tened zone, which already has a strength value changed from a peak to a residual value, as discussed earlier. Therefore it is the residual shear strength that co-determines which horizon will become a detachment horizon. Experiments on mudstone and silty mudstone (Table 7.3) illustrate the reduction of the cohesion and the angle of internal friction from a peak to a residual value. Figure 7.14 shows that the difference between peak and residual shear stress values can vary signifi-cantly in natural rocks and depends on the normal stress in elastic/frictional plastic materials.

Straining resulting from stress transfer is well docu-mented in sandbox models (e.g. Koyi, 1995, 1997). They show that layer-parallel shortening, which results from this straining, is not only a function of the different rhe-ological properties of the layers within the deforming

multilayer system. The other strong control is the amount of overburden. Sandbox models indicate that while deep layers experience a layer-parallel shortening of 40–50%, the uppermost layers imbricate and fold after a negligible lateral compaction.

The rheological control over the deformation during the layer-parallel shortening, which precedes the onset of folding or faulting, results in different structures in different rocks, as documented in the Appalachians (Mitra, 1987; Fig. 7.15). In units susceptible to pressure solution, this shortening takes place by the development of bedding-normal stylolites, clay-carbon partings and other seams of insoluble residues. In more argillaceous units, cleavage development by the preferred alignment of micaceous minerals takes place. Competent units are shortened by extensive minor faulting.

The amount of layer-parallel shortening in thrust-belts varies from those shortened after lithification to accretionary wedges shortened in the unlithified stage. The amount of shortening, determined from Silurian Skolithos burrows from the Appalachian Valley and Ridge, varies between 5 and 25% (Geiser, 1974). Shortening, determined from seismic and borehole data from the Nankai accretionary prism, varies around 68% (Morgan and Karig, 1995).

Straining associated with stress buildup can have an important consequence when the deforming potential thrust sheet contains layers that behave in a ductile manner. These layers experience pervasive shortening prior to the formation of a detachment fault (e.g. Mandl, 1988). This can lead to the development of small folds, which concentrate differential stress and influence the location of potential thrust ramps (e.g. Mandl, 1988; Schedl and Wiltschko, 1987; Goff *et al.*, 1996; Fig. 7.16).

Stress transfer facilitated by straining can be visual-ised using the concept of the moving 'stress front' dis-cussed in Chapter 3. This concept suggests that the stress transfer acts as a 'search' mechanism, which locates weaknesses in the autochthonous rock section, and which may localize future ramps and flats. The transfer time determines which parts of the multilayer system deform, given a certain time interval for the activity of the applied tectonic load. Figure 7.17 dem-onstrates how the thrust sheet section, simplified to one layer, and the detachment layer interact in influencing the time for the stress transfer. The transfer time is con-trolled by the thickness of the thrust sheet layer (Fig. 7.17a), the thickness of the detachment layer (Fig. 7.17b), the viscosity of the detachment layer (Fig. 7.17d) and Young's modulus of the thrust sheet layer (Fig. 7.17c). The viscosity control is linear and has the largest impact on the stress transfer time (Fig. 7.17d).

Fig. 7.15. Contrasting behaviour of limestone and dolostone in a sequence of deformed Ordovician sedimentary rocks below the Sudbury thrust at Lemon Fair Bluffs, near Sudbury, Vermont (photograph courtesy of the Keck Geology Consortium; photograph taken by Barbara Tewksbury). Dolostone forms boudins with calcite veins. Limestone is ductilely deformed.

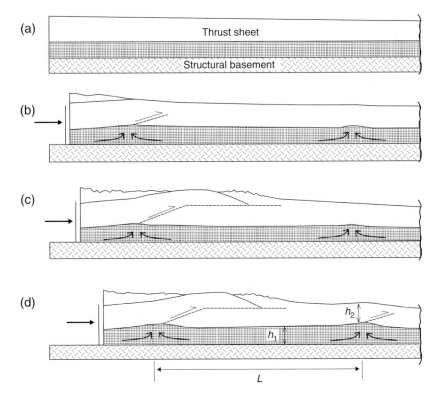

Fig. 7.16. Development of a thrust fault as a consequence of folding and material flow within a weak décollement layer (modified after Goff *et al.*, 1996). (a) Initial configuration before deformation. (b) Folding occurs during straining of the sedimentary sequence prior to thrusting. If the material flow cannot keep pace with shortening, a ramp will develop, located by the fold. (c) Thrust sheet emplacement is accompanied by erosion and syn-tectonic deposition. (d) Additional flow occurs in the foreland and the mechanism repeats itself.

The other controls are hyperbolic. The stress transfer time dependence on the 'brittleness' of the thrust sheet material (Fig. 7.17c) indicates, in a somewhat similar manner to Fig. 7.11, that competent materials are the most efficient stress transmitters. The control of the thrust sheet thickness (Fig. 7.17a) is about an order of magnitude more important than the control of the detachment layer thickness (Fig. 7.17b).

The importance of the material of the detachment horizon can be further documented by its influence on the σ_1 stress trajectories. The formation of a detachment fault in horizontally shortened strata is preceded by the development of basal shear stresses, formed by frictional reaction. A very small basal shear stress does not deflect σ_1 stress trajectories, but with higher basal shear stress, σ_1 trajectories are deflected towards the base. This

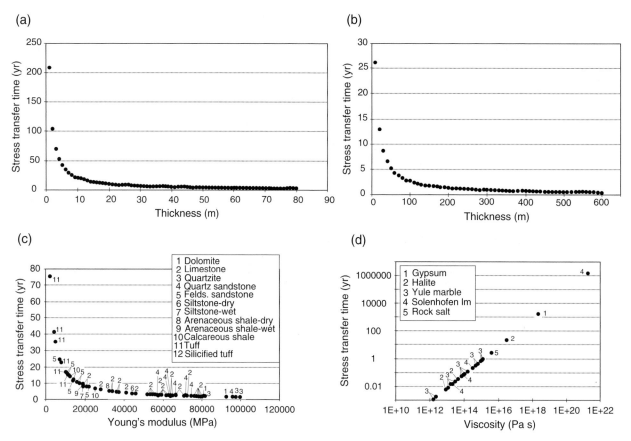

Fig. 7.17. Stress transfer dependence: (a) on the thrust sheet thickness. (b) on thickness of the viscous detachment layer; (c) on the thrust sheet material (d) on viscosity of the detachment layer. Calculations are explained in the text. Physical constants are taken from Clark (1966) and Lama and Vutukuri (1978).

deflection influences the trajectory of the ramp propagation. The maximum deflection is developed when sediments are coupled with their basement. A more rigid basement 'holds back' the overlying sediments, because of the large resistance along the basement–sediment interface (Schedl and Wiltschko, 1987). Various basal shear stresses along recently active accretionary wedges or analogue material models result in distinct structural styles (Gutscher *et al.*, 1996; Lallemand *et al.*, 1992, 1994), controlled by the angle between the basal décollement and σ_1 (Davis and Engelder, 1985), and the angle of internal friction (Davis and Engelder, 1985; Lallemand *et al.*, 1992). Longitudinal variation in the basal shear stress within a wedge results in distinct changes of structural style, as documented, for example, from the Appalachian Plateau (Davis and Engelder, 1985), the Parry Island thrustbelt, Arctic Canada (Harrison, 1995) and the South Pyrenees (Verges *et al.*, 1992). The reduced basal coupling in the presence of salt permits the thrustbelt to maintain a very narrow cross-

sectional taper of 1° or less in the Appalachian Plateau and less than 4° in the Pyrenees. Thrustbelts also lack a consistently dominant thrust vergence and they have widely but regularly spaced folds. A low shear stress results in extreme thrustbelt width and sub-horizontal décollement, as in the Parry Island thrustbelt. The condition of a low basal shear stress breaks down at the oblique boundary of the Silurian salt in the Appalachian Plateau and the structural style changes into a pile of splay faults within en echelon, anomalously oriented anticlines. The influence of salt horizons in the Pyrenees is even more pronounced. The distribution of Lutetian to Lower Oligocene salt horizons in basins, the depocentres of which shifted successively to the southwest in front of the advancing Pyrenean thrust sheets, controls the staircase geometry of the detachment fault. Salt horizon pinch-out at the southern boundary of the Cardona basin resulted in the formation of a set of passive back-thrusts. Lateral variation in basal shear stress within the wedge is one of the factors that can cause transfer zone development, as documented by sandbox modelling (Calassou *et al.*, 1993). The presence of an extremely weak detachment horizon, such as a thick salt layer, can even predominate over the influence of the tectonic load, resulting in gravitationally unstable salt-cored anticlines, which continue to grow diapirically (e.g., Davis and Engelder, 1985).

The above discussion of the propagation of new ramps in thin-skin settings leads in to a consideration of basement uplifts where this propagation is preceded by forced folding in the sedimentary cover. As in the case of buckling driven by lateral shortening, forced folding develops more or less passively over rigid basement blocks and is driven by displacement of the blocks along planar or listric faults (Stearns, 1978). The rheology of the multilayer system and the layer contacts also represent important controls (Johnson and Johnson, 2002a, b). The anisotropy of the system can be described by the ratio of the viscosity for shortening or lengthening, η_n, to the viscosity for shearing parallel to the layers, η_s (Johnson and Johnson, 2002b). Both viscosities are recalculated for the multilayer system (Johnson and Johnson, 2002b):

$$\eta_s = \frac{t\eta_1\eta_2}{t_2\eta_1 + t_1\eta_2} \tag{7.8}$$

$$\eta_n = \frac{t_1\eta_1 + t_2\eta_2}{t}, \tag{7.9}$$

where the total thickness t is calculated from the thicknesses of individual layers t_1 and t_2.

If the ratio equals 1, the sedimentary cover is isotropic. Such a cover can be imagined as being massively bedded isotropic material, which is very resistant to folding. If the ratio is less than 1, the cover is anisotropic and layer-parallel soft to deformation. This material makes folding even more difficult, since the beds have a tendency to thin or thicken rather than fold. If the ratio is greater than 1, the cover is anisotropic and layer-parallel stiff to deformation. It is this type of cover that is the easiest to fold. This cover can be visualized as being formed by many thin layers for which the contacts have a low resistance to sliding. Even the presence of several stiff layers among soft ones transforms the cover as a whole to a layer-parallel stiff package.

The deformation preceding fault propagation in an anisotropic layer-parallel stiff cover has a tendency to be localized, although it widens upwards into the isotropic cover. Isotropic and layer-parallel soft covers develop forced folds with greater thickening in the syncline than a layer-parallel stiff cover. A layer-parallel stiff cover has a tendency to undergo greater thinning in the forelimb than the other two cover types.

Shear resistances along the flat and ramp, which control the reactivation of a pre-existing fault, are affected by the mechanical stratigraphy of the host rock and the rheology of the fault zone itself. Both the shear resistances are dependent on cohesion, friction and pore fluid pressure. As documented by shear tests of cohesionless materials, such as sand or glass beads, friction along the fault zone can be less than the friction of an intact material (Table 7.4). The effect of the basal friction is to produce an overall increase in the basal shear stress, following the equation

$$\tau_b = c + \tan\varphi(\sigma_n - P_f), \tag{7.10}$$

where c is the cohesion, φ is the angle of internal friction, σ_n is the normal stress and P_f is the fluid pressure.

Contact between rheologically different rocks can result in significant local stress perturbations, as documented by Mandl (1988) in the case of an isotropic horizontal overburden above a basement fault. In the overburden, movements along the basement fault extend the fault into a plane of maximum shear strain and stress, which forces the σ_1 trajectories to deflect and results in a change of the fault orientation. The stress deflection differs if the overburden is interlayered with some weak layers. Before the onset of faulting, the layers become flexed and the stress state limit is attained in those flexed regions in which the extension is greatest. This results in a different slip trajectory pattern.

The driving mechanism of the existing thrust sheet against gravity is controlled by the density of the deformed rock section. This resistance increases with the progressive climbing of the sheet over the ramp.

The internal deformation of the rock section incorporated in the existing thrust sheet is controlled by its mechanical stratigraphy. The straining related to the thrust sheet emplacement could have an important consequence in maintaining the geometric style of the related fold. Typical examples are detachment folds in the Appalachian Plateau, which grew in relation to the amount of salt available to flow into their core (Figs. 7.18a and b; Wiltschko and Chapple, 1977). Another example is the distinct thickness change of incompetent layers inside fault-propagation chevron folds in the Andes (Baby *et al.*, 1992; Belotti *et al.*, 1995b; Dunn *et al.*, 1995) or the Carpathians, caused either by shale duplexing or plastic deformation.

Rheologically different rocks within a multilayered system can experience simultaneously various deformation mechanisms and reach the brittle–ductile limit at different times (Fig. 7.15). For example, the study in the Valley and Ridge province of the Appalachians (Mitra, 1987) shows that the control by the rock composition on

Table 7.4. Angle of internal friction of intact and pre-fractured aeolian sand, coloured sand and glass beads. The grain-size is roughly the same for all three materials and both sands have well-rounded clasts.

Material	Angle of internal friction (deg)	Error in determination
Sand, intact	42.1	0.036 93
prefractured	39.1	0.038 31
Coloured sand, intact	32.7	0.023 63
prefractured	22	0.027 74
Glass beads, intact	27.3	0.018 82
prefractured	21.7	0.026 45

the intensity of pressure solution is more important in determining the mechanism of sliding than temperature and pressure. Typically, lithological types susceptible to pressure solution such as argillaceous or ferruginous sandstone and limestone are characterized by pressure-solution slip in the region where pure orthoquartzites exhibit frictional sliding under the same temperature and pressure conditions. The rheology further affects the fold shape, varying from an angular geometry in competent units like sandstone to a more concentric and sinusoidal geometry in less competent units, as documented by natural (Mitra, 1987) or experimental examples (Chester *et al.*, 1991).

The choice of further deforming an existing thrust sheet can become quite important in the energy balance history of a specific structure, as documented by the Wills Mountains duplex in the Appalachian Plateau, West Virginia (Smart *et al.*, 1997), where over 65% of the roof thrust shortening was produced by meso- and micro-scale deformation. This deformation (11.5 km) consists of 1.2 km shortening by meso-scale folding and faulting, 3.3 km by cleavage, 6.3 km by shortening of crinoids and 0.7 km by calcite twinning. This compares with 17.5 km shortening of the underlying duplex. Only the remaining 6.0 km of shortening along the roof thrust was by macro-folding directly above the duplex or by thrusting in front of it.

Evidence of such a duplex undergoing extensive internal deformation comes from the lower of the Basse Normandie duplexes in the Variscan front of France. Here the movement on each subsidiary duplex thrust is less than 50% of the total shortening. The rest of the shortening was taken up by layer-parallel shortening of incorporated beds. A significant layer-parallel shortening inside the duplex is also documented from the Walker Brook duplex in the Devonian flysch of the Acadian orogen, Maine (Bradley and Bradley, 1994).

Commonly when internal deformation occurs, an existing structure is weak and becomes strengthened by the deformation, as is the case in the upper of the Basse Normandie duplexes (Averbuch and Mansy, 1998). Here increased friction in front resulted in out-of-sequence duplexing, followed by the tightening of the frontal fold and later by thrusts cross-cutting the fold due to increasing buttressing effects in front. The out-of-sequence shortening of the already formed Walker Brook duplex (Bradley and Bradley, 1994), bringing the former floor thrust over the former roof thrust, indicates a similar additional adjustment to increased friction. Similar out-of-sequence duplex shortening comes from the Eriboll duplex from the Moine thrust zone, NW Scotland (Bowler, 1987).

Internal deformation sometimes changes the rheology of the structure, as documented from the Doublespring duplex, developed in Mississippian massive limestone in Idaho (Hedlund *et al.*, 1994), which is rheologically not well suited for folding. The material suited for faulting that started each new horse of the duplex by extending the floor thrust and propagating a new ramp was very resistant to folding. As a result, the massive limestone was changed by a series of parallel shear zones to a system of relatively undeformed layers. This deformation is believed to be localized by heterogeneities in otherwise isotropic material (Hedlund *et al.*, 1994). These shear zones abruptly terminate near fold hinges and show high strain gradients at their ends. The evidence thus indicates that folding occurred about a fixed hinge for each horse of the duplex, as a response to the resistance to initial translation of the thrust sheet up the ramp. This resistance also triggered leading-edge antithetic faulting in the horses. After the initial folding resistance was lowered by the development of these shear zones, the horses were translated to form the duplex. This translation left a minimal internal strain record in the horses, indicating lower energy conditions than for the initial buckling.

The internal straining varies with the growth stage of the specific structure, changing its internal deformation style with the development stage of the structure. Typical examples are secondary fold-accommodation faults that react to different events in the fold growth history (e.g., Mitra, 2002a; Fig. 7.19). The first two structures are out-of-syncline and into-anticline thrusts, which react to fold tightening. Other examples are internal wedge thrusts developed in competent units, accommodating variations in penetrative layer-parallel strains between adjacent units. Forelimb thrusts are also important, reacting to rotation and layer-parallel extension in forelimbs during the late stages of the fold growth.

(a)

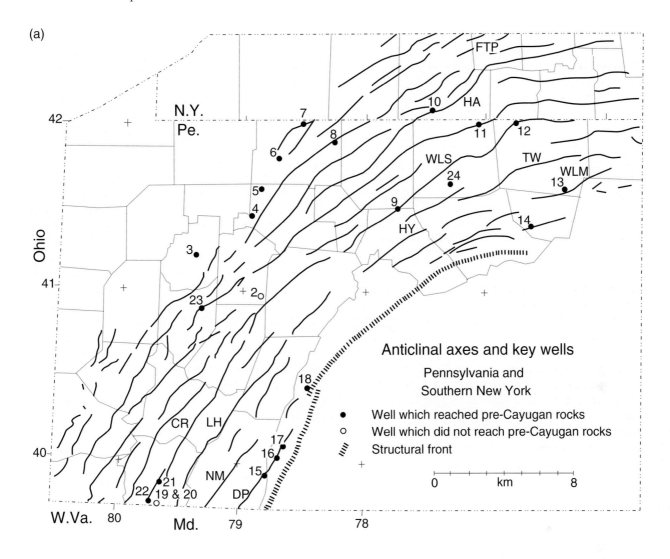

(b)

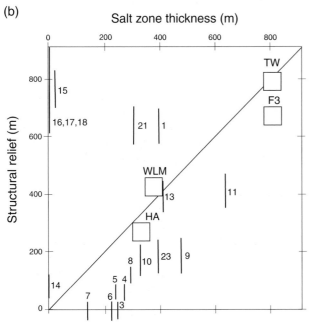

Fig. 7.18. (a) Detachment fold anticlines of the Appalachian Plateau used for structural relief/weak layer thickness study in (b) (modified after Wiltschko and Chapple, 1977). Anticlines: DP – Dear Park, NM – Negro Mountain, CR – Chestnut Ridge, LH – Laurel Hill, HY – Hyner, WLS – Wellsbor, HA – Harrison, FTP – Fir Tree Point, TW – Towanda, WLM – Wilmot. Numbers and letters indicate location of the data used in (b). (b) Structural relief versus crestal salt-zone thickness from locations in (a) (modified after Wiltschko and Chapple, 1977). Error bars and boxes indicate the data precision.

Fig. 7.19. Complex accommodation thrusts in a fold hinge in Coal Measures sandstone, Saundersfoot, South Wales.

8 Role of pre-contractional tectonics and anisotropy in evolving structural style

Large-scale scenarios

Large-scale anisotropies influence the evolving structural style of a thrustbelt by affecting the lithospheric deformation, which includes both elastic and inelastic components (Karner *et al.*, 1993). The elastic deformation is represented by lithospheric flexure. The inelastic deformation is represented by brittle deformation and ductile creep. The lithospheric deformation at a specific depth depends on the local rock strength. The rock strength varies with depth in the lithosphere in response to changes in ambient temperature, stress, strain and compositional variations (e.g. Goetze and Evans, 1979; Brace and Kohlstedt, 1980; Kirby, 1983).

At shallow depths in the lithosphere, the rock strength depends mainly on the confining pressure and it yields by frictional sliding on fracture surfaces (Brace and Kohlstedt, 1980). Such a brittle deformation of the upper lithosphere is influenced by the distribution of pre-existing faults and fractures in the crust and their orientation with respect to the applied tectonic force. Confirmation of such an influence is provided by rock mechanic tests (e.g., Donath, 1961; Turner and Weiss, 1963). The critical calculations based on hand specimens (e.g. Jaeger, 1960), on the oceanic lithosphere of the Central Indian Ocean (Karner *et al.*, 1993), the continental lithosphere of the Rocky Mountain foreland (Allmendinger *et al.*, 1982) and on the Pyrenean foreland (Desegaulx *et al.*, 1991) also demonstrate the influence.

Deeper within the lithosphere, the rock strength would be expected to decrease with depth as increasing temperature promotes thermally activated creep processes (Goetze and Evans, 1979; Goetze, 1978). Even relatively small applied forces would be expected to cause the top and bottom of the lithosphere to yield, thereby reducing the thickness of the strong core region of the lithosphere. This strong core controls the wavelength of the flexural deformation due to applied forces, which include plate driving forces and smaller-scale loads.

The smaller-scale loads on the lithosphere can be positive, such as mountains, seamounts and river deltas, causing downward deflections. They can also be negative, related to tectonic or erosional denudation of the lithosphere and buoyancy, causing upward deflections (Karner *et al.*, 1993; Fig. 8.1). The flexural deformation

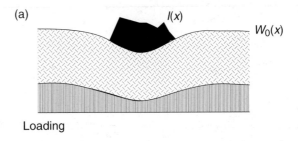

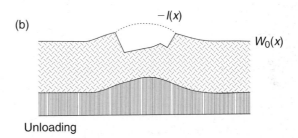

Fig. 8.1. Deflection $w_0(x)$ generated by lithospheric loading and unloading (Karner *et al.*, 1993). (a) Deflected lithosphere due to loading. (b) In the case of unloading, the final basin geometry and the shape of its floor are the results of the basement morphology produced by extension and the isostatic response of the lithosphere to this extension.

caused by these loads is sensitive to lithospheric anisotropy. The anisotropy includes any pre-existing lithospheric deflection and lateral variations in flexural strength of the lithosphere at the time when the force was applied (Waschbusch and Royden, 1992; Karner *et al.*, 1993).

An example of lateral strength variations that controlled the younger lithospheric deformation comes from the Central Indian Ocean. This case study, using reflection seismic data, shows that the upper oceanic crust and its sedimentary cover are broken into blocks bounded by high-angle reverse faults (Fig. 8.2; Karner *et al.*, 1993), which deform the underlying oceanic lithosphere at least down to a depth equivalent to 10 s two-way travel time. The steep dip of these basement faults and their strikes parallel to marine magnetic lineations suggest that they represent a reactivated crustal fabric imparted during the formation of the oceanic crust. The reverse sense of the fault reactivation is consistent with

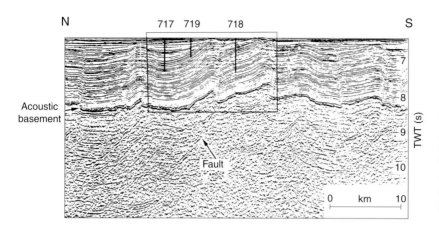

Fig. 8.2. Multichannel seismic reflection profile across Leg 116 drill sites 717–719 imaging high-angle reverse faults within the crust (Karner *et al.*, 1993). They have dips of about 35–40°, spacing of 5–20 km and penetrate into overlying sediments. Some penetrate down to at least 10 s two-way travel time.

the north–south compression determined from earth-quake focal mechanisms in the area (Wiens and Stein, 1983; Lambeck *et al.*, 1984; Fredrich *et al.*, 1988; Petroy and Wiens, 1989).

Quite distinct effects of lithospheric deflection and lateral variations in the flexural lithospheric strength on the structural style of a thrustbelt can be observed in the case of the oceanic lithosphere and the overlying toe thrust. Gravity-driven toe thrusts of gravity glides are apparently more responsive to flexural changes beneath them than classical thrustbelts due to their lower internal strength. Flexural behaviour of the lithosphere underlying large-scale gravity glides, such as those in the Amazon, Niger, Nile or Mahakam Deltas, interacts with the tectonics of overlying gravity glides.

The Niger Delta provides a good example case. Here the different buoyancy of oceanic crustal blocks separated by oceanic fracture zones affects the structural style of gravity gliding. Both gravity gliding and deposition in turn affect the flexural deformation of the underlying lithosphere. Figure 8.3 shows two oceanic blocks juxtaposed along an oceanic fracture zone. The first of these is overlain by an extensional half-graben, while a set of thrust sheets overthrust the second. The former has subsided, and the latter has been uplifted, generating a major basement step for the overlying gravity glide to overcome. Seismic reflection profiles indicate that deposition in the half-graben creates sufficient energy in the extensional region of the gravity glide to drive a new cycle of contraction in the toe thrust. Figure 8.3 indicates that the boundary between the extensional and contractional domains of the gravity glide became locked above the oceanic fracture zone for a relatively long time. The figure also indicates that the stress buildup triggering new contractional events in the toe thrust is cyclic.

The effects of lithospheric deflection and lateral variations in the flexural lithospheric strength on the struc-

tural style of thrustbelts developed within the continental lithosphere have also been described. A case study from the Rocky Mountain foreland shows two different crustal responses to the Laramide deformation. Here, the north–south variation in tectonic style is controlled by the Archean–Proterozoic crustal boundary in southeastern Wyoming (Allmendinger *et al.*, 1982). The Archean basement terrain, northwest of this boundary, which has a crustal thickness of about 38 km, is deformed by thick-skin structures, varying in trend between W and NW. The related basins are broad. Laramide structures in this terrain are strongly controlled by Precambrian structures. The Proterozoic basement terrain, southeast of this boundary, with a crustal thickness of about 48 km, is deformed by thick-skin structures generally parallel to the ancestral Rocky Mountains. The related basins are narrow. Laramide structures in this terrain are controlled by both Precambrian and Paleozoic structures. The thinner Archean crust controlled the greater diversity of Laramide structural trends. It may also have facilitated a Laramide deformation further east from the Rocky Mountains front in the Archean terrain (Allmendinger *et al.*, 1982).

This case study demonstrates the role of lateral strength variations in the continental lithosphere. The following example describes the role of variations in the direction of orogenic advance. The Aquitaine basin in front of the Pyrenees provides a suitable case study and shows the consequences of foreland basin development on thinned continental lithosphere, inherited from pre-orogenic extensional phases (Desegaulx *et al.*, 1991). The bathymetry at the transition from the pre-orogenic extensional basin to the foreland basin and the compaction of pre-orogenic sediments both contribute to the accommodation space for foreland basin sediments and thrust loads. The extension-induced transient thermal state of the lithosphere results in ongoing thermal subsidence

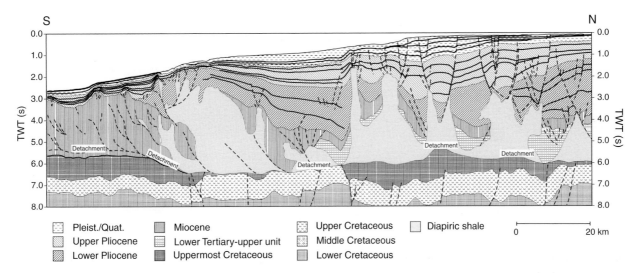

Fig. 8.3. Geological cross section through an extensional–contractional domain boundary in a gravity glide in the Niger Delta, located above a step in the oceanic crust. The step is formed by different flexural responses of the two blocks juxtaposed along an oceanic fracture zone. Further explanation is given in the text.

and changes of the flexural rigidity through time. Numerical modelling of the Aquitaine foreland basin development shows that the inherited transient thermal state of the lithosphere contributes to the total basin depth and width, the post-compressional subsidence history and the forelandward sediment onlap pattern (Fig. 8.4; Desegaulx *et al.*, 1991). Thermomechanical effects of the pre-orogenic extension significantly reduce the flexural rigidity value to 30–43% and the required topographic or thrust load to 40%.

The juxtaposition of different crustal types, variations in thickness of the continental crust and variations in thickness of the sedimentary section have been systematically studied by finite-element modelling to determine their roles in orogen development (Harry *et al.*, 1995). Models focused on these factors simulate 25–500 km of shortening in orogens that developed during contractional deformation at previously rifted continental margins. Natural examples of these models include the Alaskan Brooks Range, the Ouachita Mountains, the Southern Appalachians, the Mesozoic western Cordillera of North America, the West Carpathians, the Betic and Balearic chain, the Iberic Cordillera and the Pyrenees (Guimera, 1983; Choukroune *et al.*, 1989; Oldow *et al.*, 1989; Rankin *et al.*, 1989; Roca and Desegaulx, 1989; Viele and Thomas, 1989; Roure *et al.*, 1993). Mechanical heterogeneities that influenced the style of contractional deformation in these orogenic belts result from seaward thinning of the continental crust at the rifted margin. Other heterogeneities include the boundary between weak continental lithosphere and strong oceanic lithosphere, and the

presence of thick sedimentary accumulations on the continental shelf.

Strength distributions can vary significantly for different crustal scenarios, as shown in Fig. 8.5 (Harry *et al.*, 1995). Figures 8.5(a, b) show that regions with thick crust are weaker than adjacent regions because a portion of the strong uppermost mantle has been replaced by weaker crustal material. Figure 8.5(c) demonstrates that thick sedimentary sections, with sediment thickness up to 15 km found at many rifted continental margins, weaken the middle crust by replacing relatively strong crystalline rocks with weaker sedimentary rocks. Figure 8.5(d) shows that a juxtaposition of different crustal types is most pronounced at the ocean–continent boundary, where the strong oceanic lithosphere abuts the weaker continental crust.

Contraction rates of 1–5 km Myr^{-1} have been used to simulate orogens. The lower mechanical lithosphere is allowed to become warmer during contraction, because the duration of this orogenic contraction is set at the characteristic thermal time constant for the lithosphere of about 60 Myr.

The models indicate that orogenic décollements develop at middle and lower crustal depths, partitioning strain into the upper crustal, lower crustal and mantle strain domains. The magnitude and spatial distribution of strain within each domain are nearly uniform but differ significantly between domains. Modelled deep-seated heterogeneities produced by crustal thickness variations influence the strength of the lithosphere most strongly in the uppermost mantle, focusing strain into a narrow region of the lower crust. Associated surface

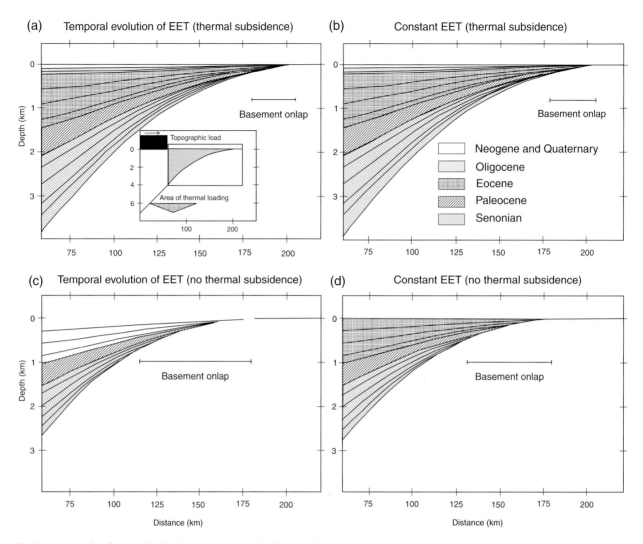

Fig. 8.4. Aquitaine foreland basin development models (Desegaulx *et al.*, 1991). (a) Stratigraphic predictions for a prograding thrust load model. Arrow indicates the direction in which the thrust load increases through time. Chronostratigraphic horizons are given for the Senonian (90, 84, 80, 75, 70 and 66 Ma), Paleocene (58 Ma), Eocene (55, 50, 40 and 36 Ma), Oligocene (24 Ma) and Neogene–Quaternary (0 Ma). a) Model incorporating both temporal evolution of effective elastic thickness EET and ongoing thermal subsidence. (b) Constant effective elastic thickness (EET) and ongoing thermal subsidence. (c) Temporal evolution of EET and zero thermal subsidence. (d) Constant EET and zero thermal subsidence.

deformation and uplift are broadly distributed, with the intensity of the surface strain decreasing monotonically on either side of the heterogeneity. No well-developed foreland basins form.

Modelled weaknesses in the upper crust produced by thick sedimentary sections reduce the strength of the middle and upper crust. They result in strongly localized deformation at the surface, more diffuse deformation at the base of the crust and the formation of well-developed foreland basins.

In modelled areas in which the strong crust abuts weaker crust, the deformation is strongly localized near the surface and is most intense in the transitional region

separating the two types of crust. Strong oceanic lithosphere behaves as a relatively rigid indenter into weaker continental lithosphere. A basal detachment develops on the seaward side of the orogen, in the weak layer of the lower crust. A second shear zone develops within the orogen at the base of the weak sedimentary section in the middle crust, partially decoupling strain in the lower crust from the strain within the upper crust.

Basement anisotropies controlling the structural style of the orogen can be formed by a ridge (Fig. 8.6), seamount or an active basement thrust slice (Fig. 8.7) in the subducting oceanic plate. Their effects on the accretionary wedge development have been studied by sandbox

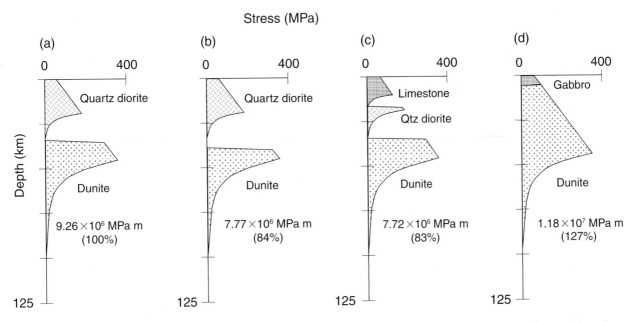

Fig. 8.5. Flow stress in the lithosphere (Harry *et al.*, 1995). Parameters: surface heat flux (continent) – 55 mW m^{-2}, mantle heat flux (continent) – 29 mW m^{-2}, surface heat production (continent) – 3 × 10^{-6} W m^{-3}, thermal decay depth (continent) – 10 km, thermal expansion coefficient – 3.1 × 10^{-5} K^{-1}, crust conductivity – 2.5 W m^{-1} K^{-1}, mantle conductivity – 3.4 W m^{-1} K^{-1}, sediment specific heat – 800 J kg^{-1} K^{-1}, continental crust specific heat – 875 J kg^{-1} K^{-1}, oceanic crust specific heat – 875 J kg^{-1} K^{-1}, mantle specific heat – 1250 J kg^{-1} K^{-1}, initial constant (log$_{10}$ *A*) (limestone) – 4.3 MPa^{-n} s^{-1}, initial constant (log$_{10}$ *A*) (quartz diorite) – 2.9 MPa^{-n} s^{-1}, initial constant (log$_{10}$ *A*) (gabbro) – 9.7 MPa^{-n} s^{-1}, initial constant (log$_{10}$ *A*) (dunite) – 2.6 MPa^{-n} s^{-1}, activation energy (limestone) – 213 kJ mol^{-1}, activation energy (quartz diorite) – 219 kJ mol^{-1}, activation energy (gabbro) – 497 kJ mol^{-1}, activation energy (dunite) – 498 kJ mol^{-1}, exponent *n* (limestone) – 1.7, exponent n (quartz diorite) – 2.4, exponent *n* (gabbro) – 3.4, exponent *n* (dunite) – 4.5, density (limestone) – 2650 kg m^{-3}, density (quartz diorite) – 2850 kg m^{-3}, density (gabbro) – 2900 kg m^{-3}, density (dunite) – 3300 kg m^{-3}. The net lithospheric strength is obtained by integrating the flow stress over the thickness of the lithosphere. It is indicated at the bottom of each figure, with comparison to the initial continental lithosphere. (a) Continental lithosphere with 35 km thick crust. (b) Continental lithosphere with 40 km thick crust. (c) Continental lithosphere with 35 km thick crust, including a 17.5 km thick sedimentary section. (d) Oceanic lithosphere.

modelling (Lallemand *et al.*, 1992; Domingues *et al.*, 1994), and deep ocean drilling data and reflection seismics from the Tonga and Nankai convergent margins (Lallemand *et al.*, 1992).

Figure 8.6 shows the evolving structural style when a ridge is subducted. When the ridge reaches the wedge front, a new thrust sheet begins to override it. The detachment surface coincides with the ridge surface, because the slope of the ridge surface approximates to the dip of the newly formed ramp. The taper of the wedge changes significantly. The topographic slope angle of the wedge hinterlandwards of the ridge decreases from 7° to horizontal, due to an increase in the detachment dip and a halt in the wedge advance. All pre-existing thrust sheets are deformed and elongated and pre-existing thrusts and back-thrusts are reactivated. The front of the wedge recedes, accompanied by a strong uplift as the ridge is subducted. The topographic slope angle of the wedge forelandwards of the ridge steepens until the slope angle reaches the repose angle of sand (30°). This slope is maintained until a new accretionary prism forms in the front of the ridge. Sediments forelandwards of the ridge, lying in the wake of the ridge, are subducted rather than accreted.

Figure 8.7 shows the evolving structural style when an active basement thrust slice is subducted. The slice develops a pop-up structure in front of the developed accretionary wedge. A secondary pop-up structure also develops at the toe of the wedge. This results from the hinterlandward tilting of the σ_1 stress axis, when the floor is tilted by the activity of the basement thrust slice. The hinterlandward tilting of the hanging wall of the basement thrust slice increases the dip of the basal detachment of the accretionary wedge. It delays the forward migration of the wedge. This delay in migration results in the wedge thickening, which causes an increase of the frontal topographic slope dip and a decrease of the slope toward the hinterland. The frontal

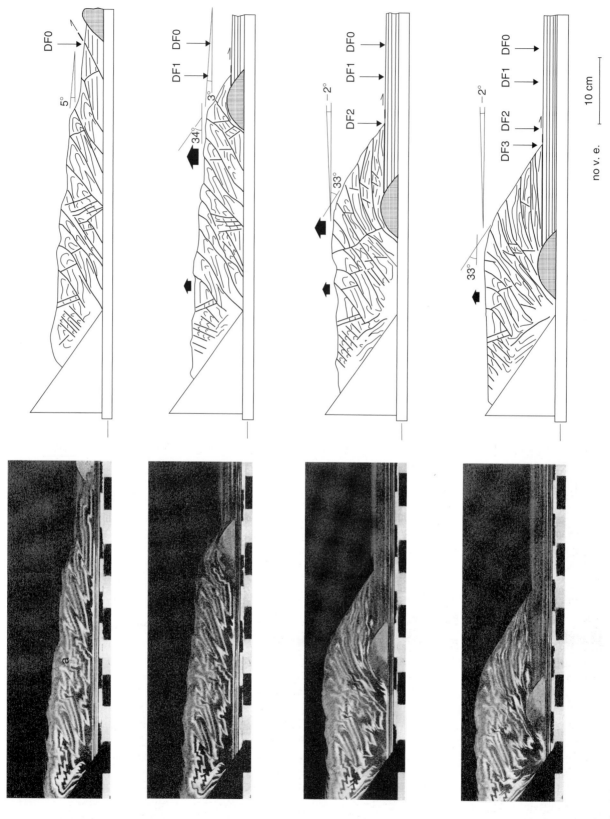

Fig. 8.6. Four stages of massive ridge subduction with related line drawings (Lallemand *et al.*, 1992). DF, deformation front; thick arrow, uplift.

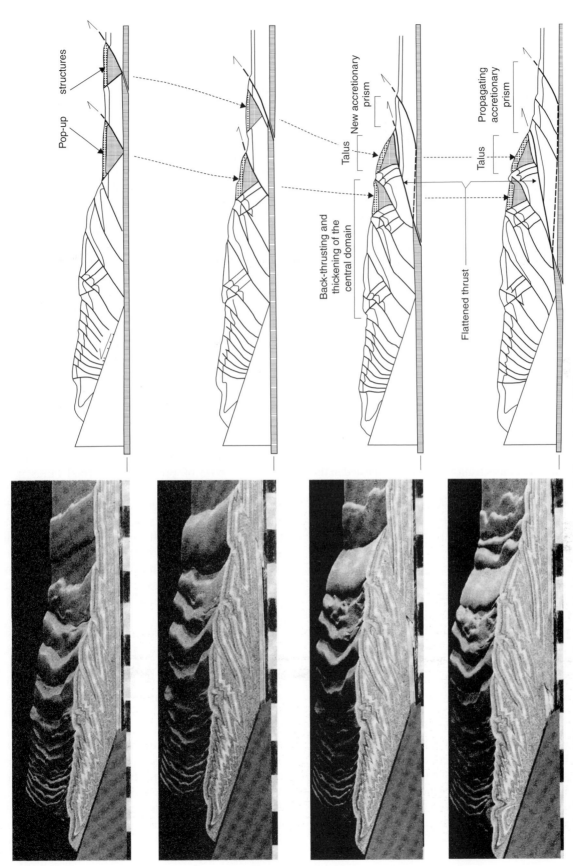

Fig. 8.7. Four stages of basement slice subduction with related line drawings (Lallemand *et al.*, 1992).

ramp of the wedge merges with the thrust, which is an extension of the basement thrust. Both pop-up structures are accreted to the front of the wedge. Internal deformation of the wedge creates a new critical taper and the wedge again starts to propagate forwards. In this example, a complex structure that includes a plateau, bordered by two uplifted pop-up structures and a talus, both lying above a flattened thrust, are preserved during later stages of the wedge advance.

There are numerous natural examples that document the effect of basement morphology on the developing toe thrust systems in front of gravity glides, which are particularly sensitive to the underlying morphology. The Eratosthenes seamount in front of the Nile Delta, for example, redirected the advance patterns of different parts of the gliding system (Gaullier *et al.*, 2003). It acted as a distal buttress, which was felt in a wide frontal portion of the system. It caused strike-slip reactivation of earlier thrusts, necessary for adjustments to the advance. Another example comes from the southeastern lobe of the Niger Delta gravity gliding system, which reacts to steps under the detachment controlled by oceanic fracture zones. The system advances stepwise through these steps, building sufficient energy for the next move after each step. It also reacts to positive floor morphology associated with the Cameroon volcanic line, which forms a constraint for its advance by changing its advance vector from southeastward to south-southwestward. Yet another example comes from a gravity glide located in the north Brazilian offshore, which reacts to the higher basal friction upon arriving into the deep basin by an overstep thrusting in its contractional domain. This domain became shingled by a system of thrusts, which became younger hinterlandwards.

Meso-scale scenarios
Meso-scale anisotropies influence the evolving structural style of the thrustbelt by affecting the local deformation, which includes either direct reactivation of the pre-existing structure or the impact of these earlier structures on the location of new structures.

A systematic study of both direct reactivation and the impact of earlier structures was made by Brun and Nalpas (1996) for the case of pre-existing grabens affected by younger compression. Two basic configurations were studied. The first included a brittle basement and a sedimentary cover, the second involved a basement separated from a sedimentary cover by a weak viscous layer consisting of evaporites, which allowed the development of a detachment. Both configurations were subjected to compression applied at varying angles α to the graben trend, ranging from 0° to 90° with steps of 15°.

The results of the pure brittle experiment are summarized in Fig. 8.8. For α equal to 90°, pre-existing normal faults show no reactivation, the graben fill is slightly uplifted and new reverse faults develop outside the graben. For α equal to 45°, pre-existing normal faults are barely reactivated, the graben fill is uplifted and new reverse faults develop at a slightly steeper dip than those formed with α equal to 90°. For α equal to 15°, pre-existing normal faults are reactivated and pass upward into reverse faults giving an upward convex shape to the graben borders, analogous to the natural example from the Sole Pit basin (Badley *et al.*, 1989). The fact that the dip of the reverse faults does not significantly vary for α values below 45° suggests that the oblique displacement is partitioned as nearly pure dip-slip on newly formed reverse faults and nearly pure strike-slip along pre-existing normal faults. At α equal to 45° and smaller, newly formed reverse faults show an increased dip and a decreased throw, and totally disappear when α is smaller than 15°.

The results of the experiment with a brittle basement separated from a sedimentary cover by a weak viscous horizon are summarized in Fig. 8.9. For α equal to 75°, pre-existing normal faults in both basement and cover show no signs of reactivation, while the graben fill is slightly uplifted. Newly formed reverse faults develop in the basement outside the graben, similar to those shown in Fig. 8.8, as well as in the cover. In all experiments with α values greater than 45°, reverse faults in the cover initiate at the base of pre-existing normal faults in the cover, where the cover sediments are thinnest, but the normal faults are not reactivated. For α equal to 30°, pre-existing normal faults in the cover are reactivated and the graben fill is uplifted. In all experiments with α less than 45°, basement normal faults show no evidence of reactivation, but early normal faults within the cover are reactivated. At α less than 60°, strike-slip faults develop inside the graben at an angle of 45° to the graben boundary faults.

The existence of a décollement in the weak viscous layer below the cover sediments causes the development of faults, which are very different from the faults developed in the pure brittle system. Normal faults in the basement are steeper than normal faults in the cover, both formed by pre-existing extension (Fig. 8.10). Newly formed reverse faults in the cover are steeper than the newly formed reverse faults in the basement (Fig. 8.10). Only the dips of both normal and reverse faults formed in the basement are comparable to those developed in the pure brittle system. The main difference from the brittle model occurs in decoupled cover sediments, where the dip of normal faults ranges from 55° to 65° and the dip of reverse faults ranges from 35° to 40°. The reactivation of normal faults starts when

(a)

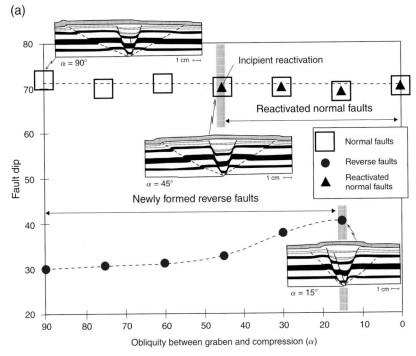

(b)

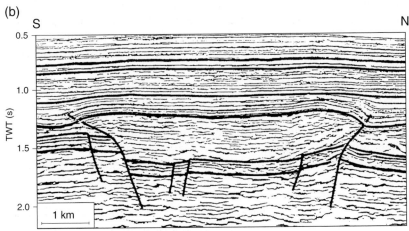

Fig. 8.8. Diagram showing the mean dip of normal and reverse faults as a function of the obliquity α between the compression and graben trend, within an entirely brittle system (Brun and Nalpas, 1996). (a) Cross sections of models for $\alpha = 90°$, 45° and 15°. (b) Example of an inverted graben comparable to the bottom inset of (a) from the Sole Pit basin (after Badley *et al.*, 1989).

α is between 45° and 60°. In reactivated normal fault zones, the presence of strike-slip faults that trend obliquely to the graben borders indicates that the deformation has been partitioned, which has not been observed when the cover sediments are coupled to the basement. The reactivation of pre-existing normal faults and the development of new reverse faults is further modified by the presence of salt diapirs, formed during the older extension and located either along the graben borders or on the graben platforms. These diapirs become preferential locations for thrust-fault development during the younger contraction. This is particularly the case for diapirs and salt walls located in cover sediments along graben borders.

All the modelled structural styles are analogues for natural examples. The model analogue for the Sole Pit basin (Badley *et al.*, 1989) indicates the role of the pre-existing anisotropy in a pure brittle system. Analogues for various parts of the Broad Fourteens basin (Nalpas *et al.*, 1995) show scenarios, with a decoupled brittle system with diapirs, a decoupled brittle system and a pure brittle system, all controlled by the distribution of Zechstein salt. This distribution is characterized by at least 400–600 m thickness of salt in the north, which reduces to a thickness of 0–400 m in the basin centre, and disappears in the south (Fig. 8.11).

The influence of pre-existing faults can also be seen in classic thrustbelt regions, where the amount of shortening is much greater than is present in inverted rifts in the orogenic foreland. Examples from the eastern Jura

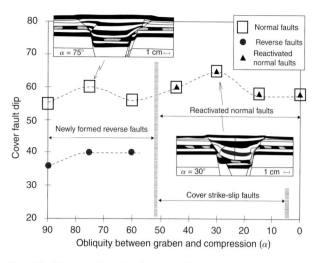

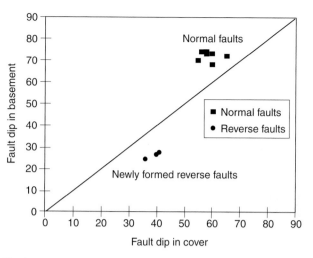

Fig. 8.9. Diagram showing the mean dip of normal and reverse faults plotted as a function of the obliquity α between the compression and graben trend, within a system involving décollement of the cover (Brun and Nalpas, 1996). Insets show cross sections of models for $\alpha = 75°$ and $30°$.

Fig. 8.10. Diagram showing the mean dip of normal and reverse faults in the cover plotted as a function of the mean dip of normal and reverse faults in the basement in models involving a décollement of the cover without diapirs (Brun and Nalpas, 1996).

Mountains (Laubscher, 1986; Fig. 8.12) show how pre-existing normal faults bounding the Late Paleozoic graben influenced the Neogene shortening. The Neogene foreland thrustbelt is detached here along Middle Triassic evaporites. This detachment displays warps, gentle slope changes and steps above pre-existing normal faults. These features acted as stress concentrators during the later contraction, which resulted in the development of several thrust ramps such as the Mandach thrust. Apparently, large fault steps were not necessary to localize these ramps. Slight breaks of slope or slight changes in curvature were sufficient for the stress concentration required to generate a ramp (Laubscher, 1986).

Outcrop examples from the Dauphinois zone of the French external Alps (Davies, 1982) document a similar ramp location mechanism. In this area, pre-existing normal faults were developed during Mesozoic extension related to the opening of the Tethys Ocean. They cut both the basement and the sedimentary cover, including weak Triassic horizons, which were later used by the thrustbelt detachment. During the Alpine shortening, the basal detachment propagated up the steps created by these normal faults. It ramped upwards and either rejoined the weak Triassic sequence in the footwall or climbed higher in the stratigraphic section where it propagated within the weak Liassic sequence (Davies, 1982). Thrusts that ramp off this detachment are frequently localized above pre-existing normal faults, which initiated duplexing above them. The pre-existing normal faults themselves were also sometimes reactivated (Davies, 1982).

Ramps located above pre-existing normal faults are common in thrustbelts. Apart from the above examples, they can be observed in seismic profiles from the Zagros Mountains (Jackson, 1980) and also the Ouachita Mountains (Lillie, 1984). They have been described in profiles from the Appalachians, the US Western Cordillera and the Carpathians (Jacobeen and Kanes, 1974; Seguin, 1982; Thomas, 1983; St.-Julien *et al.*, 1983; Skipp, 1987; Nemčok *et al.*, 1998b; Schmidt *et al.*, 1988).

The effect of a pre-existing basement topography on thrust ramping has been systematically modelled by both finite-element (Schedl and Wiltschko, 1987) and photoelastic (Wiltschko and Eastman, 1983) methods in order to understand the controlling stress concentration. The finite-element approach was used in the case of a basement coupled with cover sediments. The photoelastic approach has been used in the case of cover sediments decoupled from the basement.

Models of the basement coupled with overlying sediments have simulated three distinct tectonic scenarios (Fig. 8.13). In the first both the basement and cover underwent a constant displacement, simulating the action of a rigid indenter affecting the system. The second scenario experienced a constant horizontal displacement in the cover sediments, simulating thin-skin shortening. The third scenario was characterized by a constant stress applied to both basement and sediments, simulating deformation near a compressional plate boundary. Each of these scenarios tested the influence of two forms of basement step on the contractional structural style. One of the basement steps had a height

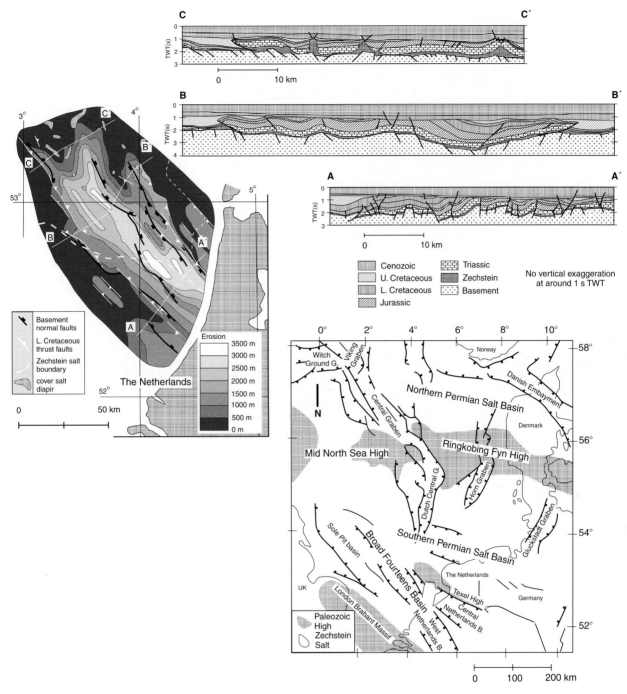

Fig. 8.11. Structural map and line drawings of seismic sections showing a variation of the structural style in the Broad Fourteens basin as a consequence of the development of the Zechstein salt horizon (modified from Nalpas *et al.*, 1995). AA′: Profile through a pure brittle system. BB′: Profile through a decoupled brittle system. CC′: Profile through a decoupled brittle system with diapirs.

equal to one-ninth of the sedimentary cover thickness (Fig. 8.13). In the other the height was half the cover thickness (Fig. 8.14).

The modelling indicates that the greater the difference between Young's moduli for the basement and cover, the stronger the effect of the basement on the stress regime in cover sediments. The highest differential stresses develop around basement steps (Fig. 8.13). The differential stresses are only perturbed over a region one to two times the height of the basement step. Models

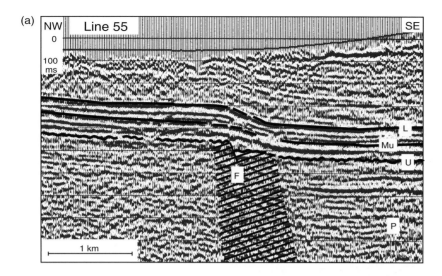

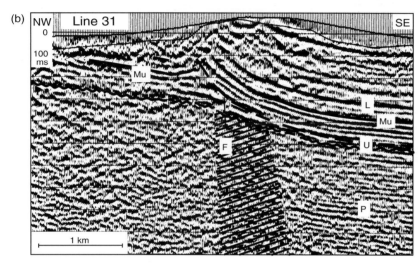

Fig. 8.12. Two interpreted reflection seismic profiles through the Neogene Mandach detachment in the Jura Mountains (Laubscher, 1986). (a) Reflection seismic profile 55, located to the east of the Aare River (stack). (b) Reflection seismic profile 31, located to the west and south of the Aare River (migrated). P – Paleozoic, U – unconformity at the base of the Mesozoic sediments, Mu – top of the Muschelkalk, L – top of the Liassic sediments, F – Neogene thrust, dot-dash line – Mandach thrust branching off the detachment, cross-hatched zone – normal fault bounding the late Paleozoic trough.

with a 45°-dipping step show that both top and bottom corners of the step are stress concentrators and a prominent stress shadow develops forelandward of the step (Fig. 8.14). The upper corner of a 90°-dipping step concentrates stress more than the upper corner of a 45°-dipping step. It also has a more pronounced stress shadow forelandwards. This comparison demonstrates that 90°-dipping steps are stronger ramp locators.

Models indicate that the location of ramps in thrust-belts can be affected by basement topography not only in the case of rigid indentation but also with thin-skinned shortening and deformation near compressional plate boundaries. Photoelastic models (Wiltschko and Eastman, 1983) indicate that decoupled systems produce even higher stress concentrations. They display regions of perturbed stresses over much larger areas than do coupled models. This is because basement steps are essentially rigid in decoupled models, but in coupled models basement corners are deformable,

which results in the corners having a lower stress concentration.

The above coupled and decoupled models are end-member analogues of natural cases. They indicate that thrust faults affected by basement steps initially propagate from basement corners as well as along the basement–cover sediment interface, because in these locations differential stresses are higher (Fig. 8.15). If a critical stress is required for faulting, the low-stress region behind a basement step is unlikely to contain faults. Using this principle, faults that propagate past a basement step would be expected to ramp upwards. The higher the contrast between Young's moduli for the cover sediments and the basement, the stronger the stress perturbationis. A stronger stress perturbation represents a greater chance for a ramp location.

The stress perturbation is a function of the basement step height and the magnitude of the stress concentration is proportional to the acuteness of the step corner.

(a)

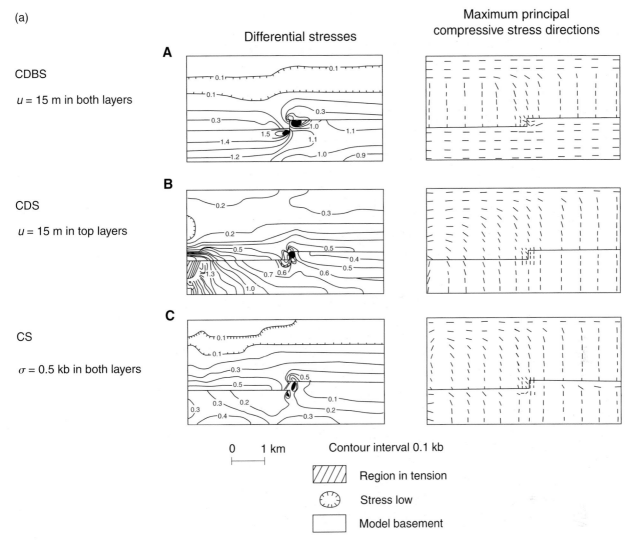

Fig. 8.13. Finite-element model of the stress perturbation at a small basement step in a coupled basement–sedimentary cover system (Schedl and Wiltschko, 1987). Models in (a) differ from models in (b) by Poisson's ratio of the basement. $\nu = 0.499$ in (a) and $\nu = 0.25$ in (b). Shared values include: angle of the basement corner $= 90°$, sediment density $= 2300\,kg\,m^{-3}$, basement density $= 2700\,kg\,m^{-3}$, Young's modulus of sediments $E = 1.1 \times 10^{11}$ dyne cm^{-2}, Young's modulus of the basement $E = 1.1 \times 10^{12}$ dyne cm^{-2}. Differential stresses are contoured on the left-hand side of the figure and corresponding σ_1 stress directions are shown on the right. CDBS – constant displacement in basement and sedimentary cover (15 m), CDS – constant displacement in cover sediments (15 m), CS – constant stress applied to basement and cover sediments (0.5 kbar).

If a basement relief is to localize a thrust ramp, then a critical basement/cover combination of contrasting Young's moduli, the height of the basement topography and the angularity of the basement topography is needed. There is no direct relationship between the basement relief and the ramp location because of the influence of other controlling factors such as sedimentary heterogeneities and pre-existing tectonic anisotropies in the sedimentary cover (Wiltschko and Eastman, 1983). However, there are several natural examples where basement control apparently dominated. One of these is in

the southernmost Appalachians (Drahovzal *et al.*, 1984). Here the pre-existing normal and reverse faults in the basement have controlled to varying degrees the magnitude of duplexes and the location of ramps and associated anticlines, depending upon the local structural relief of the basement. Large buttresses, formed by large-throw basement normal faults of the Birmingham trough, triggered the most distinct deformational response, resulting in polyphase, break-back-thrusting.

Another example of direct basement control comes from the Umbria–Marches Apennines (Montanari *et*

(b)

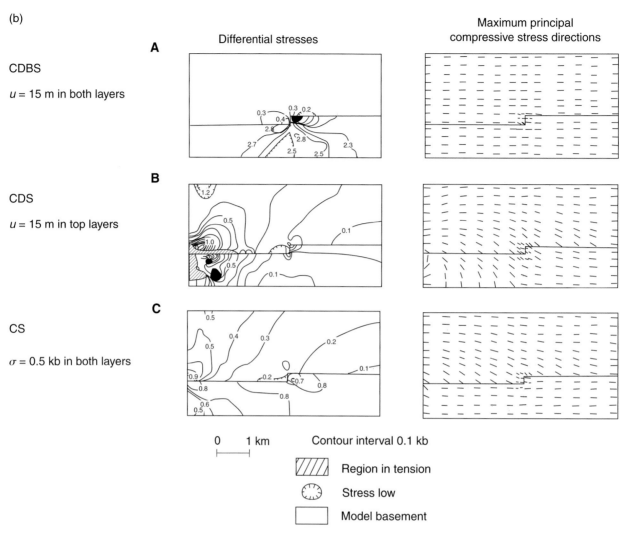

Fig. 8.13. (*cont.*).

al., 1983). Here sets of Late Hettangian listric normal faults influenced the development of the Late Tertiary thrustbelt. The normal faults dip both west and east, and their original spacing was apparently controlled by their depth to the detachment (Montanari *et al.*, 1983), resulting in a 5–10 km wide spacing in the northern and southern part of the thrustbelt and a 1 km wide spacing in the central part. Ramp development in the Late Tertiary eastward advance of the thrustbelt was controlled by west-dipping pre-existing listric normal faults. The east-dipping normal faults apparently impeded thrusting and caused buckling or development of large box folds. Wide spacing of west-dipping normal faults controlled the development of long and continuous thrust sheets. A closer spacing of west-dipping normal faults resulted in the development of short thrust sheets with complex fault mosaics in their anticlines.

An interesting insight into factors that controlled the transition between thin- and thick-skin tectonics comes from the Tertiary thrustbelt of central Spitsbergen (Bergh *et al.*, 1997). This east to northeast-vergent thrustbelt consists of three major zones with distinct structural styles. It comprises a western basement-involved thrust complex, a central zone of thin-skinned thrust units and an eastern zone in which a frontal duplex system is developed, bounded in the east by steep basement-involved thrusts (Fig. 8.16).

In each of these zones a combination of mechanical stratigraphy and pre-existing structures controlled the Tertiary thrustbelt development, as indicated by balanced cross sections (Bergh *et al.*, 1997; Fig. 8.16b). The restored sections show that the top of the basement has a ridge–basin configuration that reflects the shapes of Devonian and Carboniferous basins bounded by normal

(a) Stress distribution for a 45° slope

(b) Principal stress directions for a 45° slope

$U_y = 0$

(c) Stress distribution for a 90° slope

0   1 km

Contour interval 0.1 kb

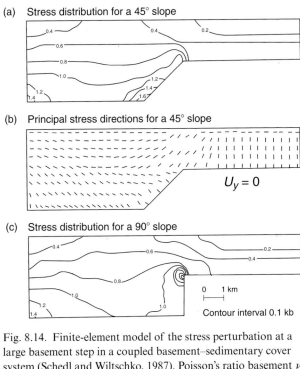

Fig. 8.14. Finite-element model of the stress perturbation at a large basement step in a coupled basement–sedimentary cover system (Schedl and Wiltschko, 1987). Poisson's ratio basement ν = 0.25, angle of the basement corner = 45° and 90°, sediment density = 2300 kg m^{-3}, Young's modulus of sediments E = 1.1×10^{11} dyne cm^{-2}. Constant stress is applied to basement and cover sediments. (a) Contours of the σ_1 stress for the model with a 45°-dipping basement slope. (b) σ_1 stress direction for the model with a 45°-dipping basement slope. (c) Contours of the σ_1 stress for the model with a 90°-dipping basement slope.

(a) CS and CDS boundary conditions

(b) CS and CDS boundary conditions

(c) CDBS boundary conditions

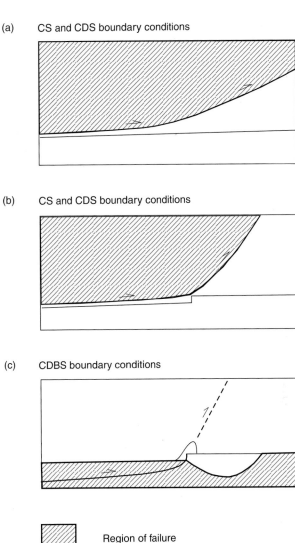

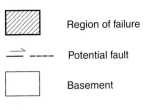

Region of failure

Potential fault

Basement

Fig. 8.15. Summary diagram of failure regions and potential ramps for geometries and boundary conditions from Figs. 8.13 and 8.14 (Schedl and Wiltschko, 1987). (a) A basement step localizes thrust faults in sediments under both CS and thin-skin, CDS, boundary conditions from Fig. 8.13. (c) A basement step localizes thrust faults in the basement under the thick-skin, CD BS, boundary conditions from Fig. 8.13.

and strike-slip faults. The northeast wedging-out of the Lower Carboniferous strata against the central basement ridge coincides with a step in the frontal thrust to higher stratigraphic levels near the western–central zone transition. Some of the northeast-striking normal faults apparently acted as oblique ramps, accommodating out-of-plane movements in the thrustbelt. The thin-skinned nature of the central zone is caused by the presence of sub-horizontal strata above relatively flat basement and evaporite lithologies favourable for detachment transfer. A block in the eastern zone, to the east of the Lappdalen thrust ramp formed a buttress against strain transfer further east, causing a step in the Permian detachment. The profiles also indicate that local buttressing effects controlled the location of the late out-of-sequence thrusts both in the western zone and along the boundary between the central and eastern zones. The pre-existing basement anisotropy located the basement-involved faulting in the eastern zone.

All of the above relationships deduced from balanced cross sections (Bergh *et al.*, 1997) indicate a time-trans-

gressive horizontal shortening and an in-sequence, thick-skinned–thin-skinned translation in an eastward direction. They were followed by thick-skinned basement uplift in the east and subsequent out-of-sequence thrusting in the west. The thick-skinned event in the east is coeval with, or follows the time-transgressive shortening

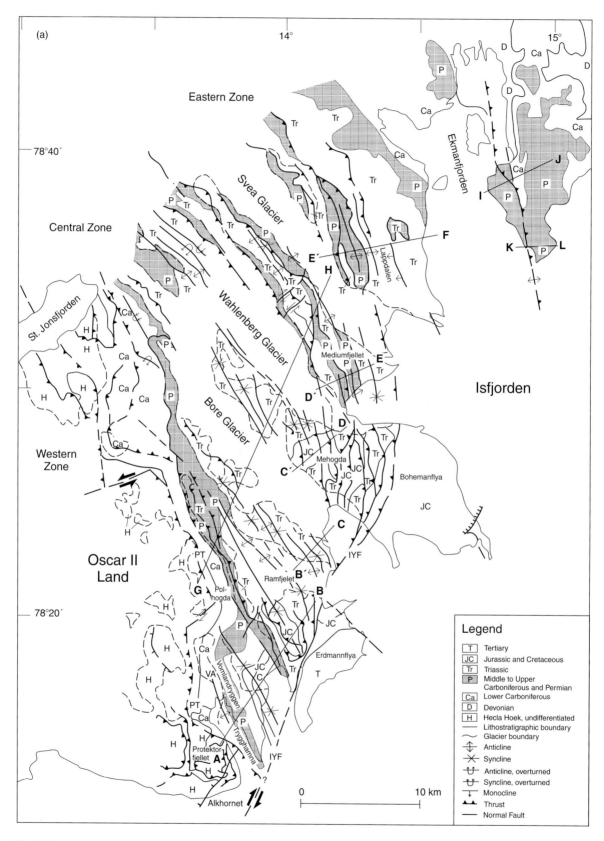

Fig. 8.16. Geological cross sections through central Spitsbergen (Bergh *et al.*, 1997).

(b)

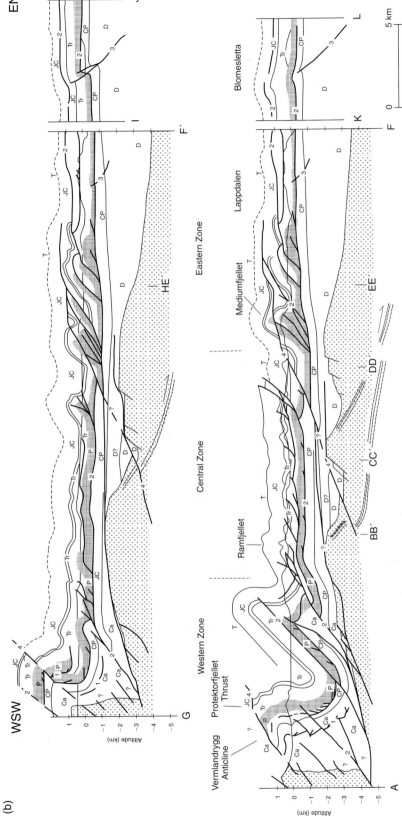

Fig. 8.16 (cont.). Geological cross sections through central Spitsbergen (Bergh et al., 1997). Sections are located in (a). The basement is indicated by a cross pattern. T – Tertiary, Jc – Jurassic and Cretaceous, Tr - Triassic, Pks – upper Permian, Cp - Middle-Upper Carboniferous and lower Permian, Ca – Lower Carboniferous, and D – Devonian strata. The deeper parts of sections are based on line drawings of Statoil's seismic profile ST8815-227 by extrapolating structures and using the regional structural grain as projection axis. Numbers on different faults refer to the relative thrust stages. Stage 1 was a sub-horizontal north-northeast shortening that produced internal detachment folds and thrusts in the western zone. Stage 2 resulted in major northeast-southwest uplift and folding of the western zone and in-sequence foreland-propagating fold–thrust development. It also resulted in folding of stage 1 detachments and propagation of successively higher stage 2 detachment levels eastward. Stage 3 was characterized by basement-involved thrust uplift and folding of stage 2 detachments in eastern areas. Stage 4 was an out-of-sequence thrusting and decapitation of earlier structures in both western and eastern zones. Stage 5 is represented by late extensional overprint on the western basement high. A mid-crustal detachment, linking the western hinterland domains and the deep-seated thrusts in the foreland, is proposed to explain interaction of thin- and thick-skinned structures (Anderson et al., 1994).

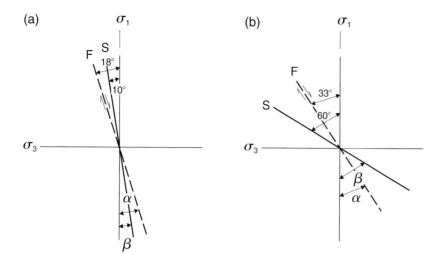

Fig. 8.17. Schematic diagram of the failure F in an anisotropic rock system, in which the shear stress τ deviates from its minimum value along pre-existing S-planes (Turner and Weiss, 1963). Further explanation is given in the text.

in the west, whereas the late out-of-sequence thrusting is contemporaneous in both regions.

A similar overlap of tectonic events and styles of kinematic interaction have been demonstrated in the Rocky Mountain thrustbelt and its foreland (e.g., Hamilton, 1978; Schmidt and Garihan, 1983; Kulik and Schmidt, 1988). The foreland responded to the thin-skinned tectonics of the thrustbelt by bilaterally symmetric uplifts of the basement and by inversion of structures located along pre-existing anisotropies. Thrusts later propagated through already uplifted foreland blocks, reacting to buttressing effects or adjusting the wedge taper by out-of-sequence thrusting. Detachments were then folded by the late basement uplift faulting.

Small-scale scenarios

Small-scale anisotropies affect the evolving structural style of a thrustbelt by influencing local fault and fracture development in brittle and transient environments. They include inhomogeneities in the original rock structure such as bedding and foliation planes and pre-existing fractures. Their role in the propagation of new faults can be either calculated or determined by rock mechanic testing of relatively small cylindrical specimens.

An example of such a calculation is the study of the role of a single plane of weakness, such as schistosity or a similar group of S-planes, performed by Jaeger (1960). The shear strength in the studied rock system varies continuously from a maximum on planes perpendicular to S-planes to a minimum parallel with S-planes. The calculation indicates that the rock mass would not be disrupted by movement along a single group of fault planes F, obliquely cutting S-planes, and having an orientation similar to that in the isotropic rock. It also shows no movement along S-planes themselves. The newly propagated faults F are always formed between

S-planes and the nearest potential fault planes predicted by the Coulomb–Navier theory for isotropic rock (Fig. 8.17). If α and β are the angles between the maximum principal compressive stress σ_1 and the S and F planes, respectively, then low values of the angle β will result in an angle α relatively smaller than α in a corresponding isotropic system (Fig. 8.17a). High values of the angle β will be accompanied by an angle α somewhat greater than α in a corresponding isotropic system (Fig. 8.17b).

This calculation was later confirmed and refined by rock mechanic testing (Donath, 1961). The experiments tested several kinds of anisotropy, such as bedding, cleavage and schistosity, at room temperature and under gradually increasing confining pressure. Anisotropy effects were simulated in brittle, brittle–ductile and ductile deformation environments. Compressive tests were carried on jacketed cylinder-shaped specimens submerged in kerosene to simulate the confining pressure.

The results indicate that the anisotropy orientation distinctively affects the differential stress at failure (Fig. 8.18a). Pre-existing planes located close to potential fault planes predicted by the Coulomb–Navier theory for isotropic rock reduce the differential stress to a value about five to six times smaller than its value in the case of unfavourably oriented anisotropies, parallel or perpendicular to the maximum principal compressive stress σ_1. Figure 8.18a shows that differential stress is increased by increased confining pressure.

The Coulomb–Navier theory for isotropic rock predicts an angle of $\alpha = \pm(45° - \varphi/2)$, where α is the angle between the fracture and the maximum principal compressive stress axis σ_1, and φ is the angle of internal friction. In an anisotropic rock, as shown by the above tests, the angle α will depend on the attitude of the anisotropy relative to σ_1. The rock mechanics tests (Donath, 1961),

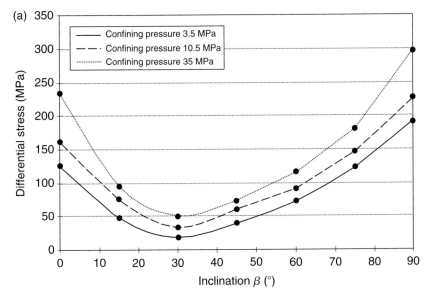

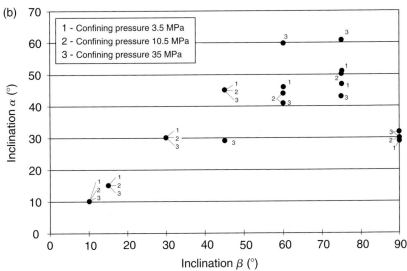

Fig. 8.18. (a) Differential stress necessary for failure plotted as function of the inclination β of the cleavage from the maximum principal compressive stress σ_1 and the confining pressure in the Martinsburg slate (Donath, 1961). (b) Influence of the cleavage on the angle α of the shear plane from the maximum principal compressive stress σ_1 in the Martinsburg slate (Donath, 1961). Note that the inclination of the failure α only approaches zero because the shear stress τ does not exist on cleavage planes parallel to σ_1 and therefore a shear failure along cleavage with an inclination β of 0° cannot take place.

however, show that in certain specific cases newly formed faults can be parallel to the pre-existing anisotropy and can have the orientation as predicted by Coulomb–Navier theory. This phenomenon happens when the angle β between anisotropy and σ_1 is above 0° and below 30°, and, in some cases, at an angle of $\beta = 45°$ (Fig. 8.18b). Some influence of the anisotropy orientation on the failure plane can be noted even at angles of $\beta = 60°$ and 75°. This latter phenomenon occurs when the deviation angle β attains 90° and shears are formed according to Coulomb–Navier theory.

The increasing confining pressure tends to reduce the effects of anisotropy on shear fractures. Ductility, being a function of confining pressure and rock type, also reduces the effect of anisotropy. The intensity of the influence depends on the type of anisotropy. Cleavage and schistosity were shown to be more favourable than irregular bedding.

Normal and shear stresses acting on a newly formed shear plane are also influenced by anisotropy (Donath, 1961). Experimental results indicate a linear relationship between the normal stress σ_n and the shear stress τ on the planes of anisotropy, governed by the empirical formula $\tau = 89.8 + 0.329\sigma_n$. The formula yields an angle of internal friction φ of 18° and an average cohesive strength c of 89.8 bar. The curve governed by the formula and failure stresses calculated from experimental data are shown in Fig. 8.19. The circles mark the state of stress during failure for an expected cleavage orientation, with individual points on the circles indicating τ and σ_n for these orientations. In cases of cleavage inclinations with β being 30°, 45°, 15° and 60°, these

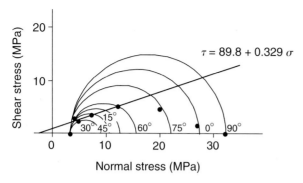

Fig. 8.19. Mohr circles of failure in the Martinsburg slate at a confining pressure of 350 bar with failure criterion for the cleavage plane (Donath, 1961). Further explanation is given in the text.

points lie very close to the line representing the failure along cleavage. If the cleavage inclination β amounts to 60°, a shear plane is formed at an angle α of 49° in relation to the maximum principal compressive stress σ_1 (Fig. 8.19). The shear stress τ is zero on cleavage planes parallel and perpendicular to σ_1 and therefore a shear failure along cleavage cannot take place.

The maximum principal compressive stress σ_1 at failure and shear resistance τ_0 also vary with the orientation of the planar anisotropy in the rock (Donath, 1961; Fig. 8.20). All of the described results (Figs. 8.17–8.20) indicate that anisotropy strongly affects the strain geometry of rocks displaying at least moderately brittle behaviour. Rocks with a clear planar anisotropy and having at least a moderately brittle behaviour can be cut by shear fractures at any angle α of up to 60°

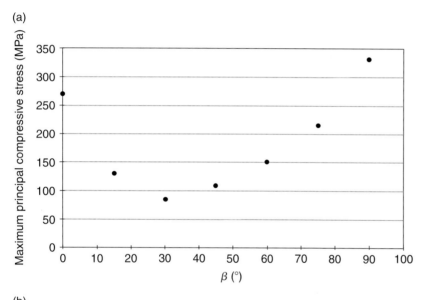

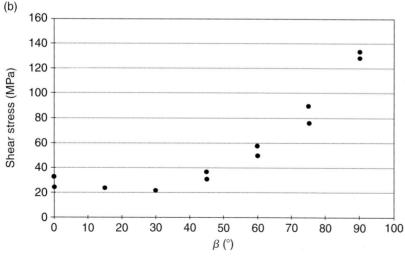

Fig. 8.20. (a) Maximum compressive stress σ_1 at a confining pressure of 350 bar during the failure of the Martinsburg slate plotted as function of the inclination β of the anisotropy from the maximum principal compressive stress σ_1 (Donath, 1961). (b) Shear resistance plotted as function of the inclination β of the anisotropy from the maximum principal compressive stress σ_1 (Donath, 1961).

relative to the maximum principal compressive stress σ_1. The described evidence, however, does not contradict Coulomb's law. Although a planar anisotropy can cause a large deviation of a single shear fracture from the σ_1 direction, it does not affect newly developed conjugate shears equally. In a brittle environment, the development of folds, for which the limbs lie at an angle θ of 45° in relation to the maximum principal stress σ_1, starts to be controlled by slip along bedding planes. At higher angles of θ, the deformation is controlled by different mechanisms, such as flow within beds or flow across bedding planes.

The above conclusions have been documented by numerous natural examples of shear deformation structures and kink bands in slates indicating that such rocks are dissected by faults oblique with respect to the cleavage (e.g., Turner and Weiss, 1963). Shear along S-planes has a tendency to take place only in a fractured rock system that was pre-weakened in the direction parallel to the S-planes, but is more or less equally strong in all other directions (Jaeger, 1960). Even here, S-planes will be preferentially used rather than other anisotropies only in a particular range of orientations; i.e. when their deviation β from the σ_1 direction varies from 10° to 40°.

9 Role of syn-orogenic erosion and deposition in evolving structural style

In addition to tectonic forces, pre-existing structures, fluids and mechanical stratigraphy, the evolution of a thrustbelt is also controlled by syn-orogenic erosion and deposition. Their influence modifies the geometry of both local thrust structures and also the whole thrust wedge. The following sections are devoted to these two aspects.

Role of syn-orogenic erosion and deposition in an evolving thrustbelt

Syn-orogenic erosion and deposition influence both the evolving thrustbelt and basin development at its front, the fill of which is progressively accreted into the thrustbelt and thus controls its structural style.

Direct influence of syn-orogenic erosion and deposition in an evolving thrustbelt

Syn-orogenic deposition can influence the developing thrustbelt by modifying the wedge due to the addition of sediments both between and in front of growing structures, whilst syn-orogenic erosion can modify the wedge by changing the shapes of growing structures.

The addition of new material during syn-tectonic deposition can take place at various stages of thrust development and at various rates (Table 9.1). This material becomes involved in subsequent thrustbelt deformation (Fig. 9.1). Local control of the deposition rate is especially likely in piggyback basins, i.e. those basins that form above, and are transported by, actively deforming thrust sheets. Natural examples in marine settings are present in the Po basin in front of the Apennines (Ori and Friend, 1984; Fig. 9.2), the syn-orogenic Čelovce Formation in the West Carpathians (Nemčok et al., 1996) and the Paleocene–Middle Eocene Formations of the Balkan Unit in the Balkans. Examples in continental settings have been documented in the Pyrenees (Ori and Friend, 1984; Fig. 9.3; Bentham et al., 1992). Connections between the main basin and a local piggyback basin are controlled by the interaction of many factors that include: climate, sediment supply, the subsidence rate of the basin, the sediment compaction rate, isostasy, river grade and uplift rates of bounding thrust structures (e.g., Johnson et al., 1986; Beer et al., 1990). Climate changes can affect nonmarine deposition by changing the stream discharge, or the ratio of chemical to mechanical detritus eroded from the source area, or the lake levels, as documented by studies in the Iglesia piggyback basin in the Argentinean Andes (Beer et al., 1990) and in the Beartooth Range in the northern Rocky Mountains (DeCelles et al., 1991). The sediment supply depends on the resistance to erosion and the relief in the source area (e.g., Beer et al., 1990; DeCelles et al., 1991). When the rate of progressive uplift of bounding structures is larger than the rate of erosion, the relief increases and space is created in the internally drained basin, as known from the Argentinean Andes, the Transverse Ranges, California and the Brooks Range, Alaska (Beer et al., 1990; Burbank et al., 1996). When the rate of progressive uplift of bounding structures is only episodically larger, the piggyback basin undergoes periodic closing and opening, as documented in the Argentinean Andes (Beer et al., 1990). Accelerated uplift in the source area results in an increase in the mean sediment accumulation rate in the piggyback basin, as known from the Siwalik system in the Himalayas and the Huaco system in the Andes (Johnson et al., 1986).

Detailed studies of the interaction of a river system with such growing thrust structures were made in the Wheeler Ridge, the Transverse Ranges, California and the Brooks Range, Alaska (Burbank et al., 1996), the Oliana anticline in the Ebro basin, Southern Pyrenees (Burbank et al., 1992; Burbank and Verges, 1994). These studies highlight the influence of two ratios on the ability of rivers to maintain their course across a growing structure. The first of these is the aggradation/uplift ratio. As long as aggradation keeps pace with or exceeds the uplift, a transverse river can maintain an antecedent course across a growing structure. In such cases, no topographic relief may develop, despite ongoing folding of the subsurface. If uplift exceeds aggradation, a positive topography develops above the depositional plain. In this case the second ratio, that of stream power/erosional resistance, comes into play. In this case both ratios decide whether an antecedent river can maintain its course across the fold. When uplift exceeds aggradation, the river must be able to erode the crest of the fold at a rate equal to the difference between the uplift and aggradation rates in order to sustain a forelandward gradient across the uplifting back limb of the fold.

Table 9.1. Sedimentation rates in different environments.

Sedimentation rate (mm yr^{-1})	Setting	Reference
0.06–0.13	Scala Dei Group, Ebro foreland basin	Colombo and Verges (1992)
0.07–0.2	Mediano anticline, South Pyrenean foreland basin	Holl and Anastasio (1993)
0.14–0.23	Perdido fold belt, Gulf of Mexico	Mount *et al.* (1990)
0.06–0.25	Syn-orogenic alluvial sediments, South-Central Unit, Pyrenees	Burbank and Verges (1994)
0.1–0.5	Siwalik Molasse, Pakistan	Johnson *et al.* (1986)
0.01–0.5	North Apennine foreland basins, Italy	Ricci Lucchi (1986)
0.5–1.3	North Apennine foreland basins, Italy	Zoetemeijer *et al.* (1993)
0.25–0.5	Escanilla Formation, Ainsa basin, Spain	Bentham *et al.* (1993)
0.1–0.9	Continental sediments, Precordilleran foreland basins, Argentina	Jordan *et al.* (1993)
1.8	Wheeler Ridge, San Joaquin Valley, California	Medwedeff (1992)
0.24–0.88	Bermejo Basin, Precordilleran foreland basin, Argentina	Jordan (1995)
0.1	Marine Bartonian–Early Priabonian Igualada marls, Pyrenean foreland	Caus (1973), Powers (1989)
0.2–0.25	Upper Eocene–Lower Oligocene continental conglomerate, fluvial fan-delta deposition, Central Pyrenean foreland	Burbank *et al.* (1992)
>0.35	Igualada Bartonian–Early Priabonian marine marls, during episodes of thrust loading (accelerated deposition), Central Pyrenean foreland	Burbank *et al.* (1992)
>0.6	Upper Eocene–Lower Oligocene continental fan-delta deposition immediate next to thrusts, Central Pyrenean foreland	Burbank *et al.* (1992)
0.17–0.92	14–2 Ma, Huaco sequence, Precordillera of western Argentina, arid climate, alluvial fan, playa deposits	Johnson *et al.* (1986)
0.4	Trough axis, Timor trough, minor amounts of siliciclastic detritus from Timor island	Charlton (1988)
0.1	Lower slope, Timor trough, minor amounts of siliciclastic detritus from Timor island	Charlton (1988)
0.0014–0.002	Late Miocene–Pleistocene, frontal part of the Barbados accretionary prism in abyssal plain	Mascle *et al.* (1988)
0.0019	Last 460 000, undeformed region immediately in front of the Barbados accretionary prism in the abyssal plain	Henry *et al.* (1990)

A direct relationship between thrusting, which changes local geometries and syn-orogenic sediments located on the wedge, is commonly indicated by episodes of paleocurrent reorganization and facial changes. A classical example is the occurrence of conglomerates associated with the displacement of each major thrust of the Wyoming–Idaho–Utah thrustbelt (e.g., Lamerson, 1982). Their origin is attributed to an increased slope in the sediment catchment area in response to the thrust activity (Wiltschko and Dorr, 1983). Other examples indicate more complicated relationships between tectonics and deposition, such as when a thrust tip warps up a sediment trap, which results in a lack of sediment supply associated with thrust activity (Blair and Bilodeau, 1988). A more complex relationship occurs with an oblique boundary such as that of the central unit in the southern Pyrenees (Verges and Munoz, 1990). Here, each thrust tip is covered by a conglomerate wedge, which in turn is cut by the younger thrust formed hinterlandwards of the older one.

The reason for different depositional reactions to thrusting is the sedimentary response to episodic tectonism, which varies among different depositional environments. For example, low gradient fluvial, lacustrine and marine environments respond faster to tectonic subsidence than do fans or braided plain environments (Blair and Bilodeau, 1988). The slow response time of fan deposition is caused by its character. Fan deposition is controlled by extremely intense low-frequency rainfall,

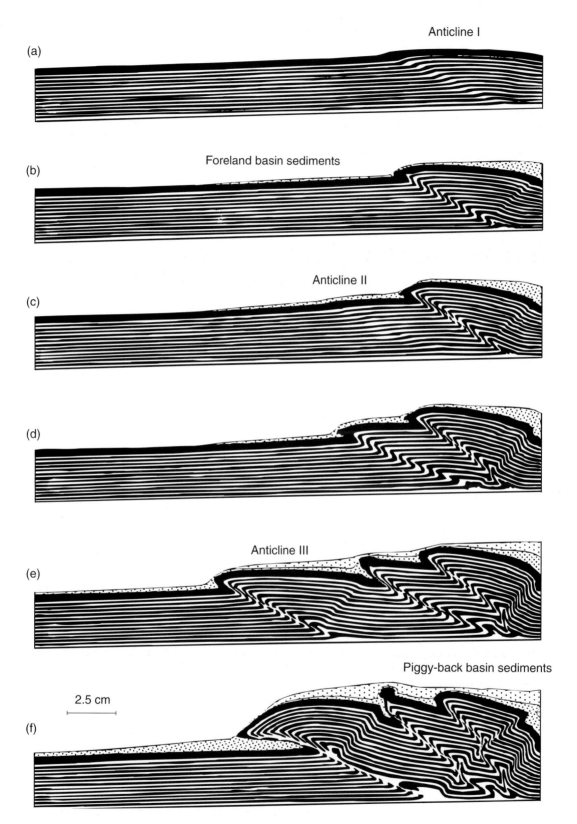

Fig. 9.1. Line drawings from time-lapse photographs of a sandbox model with syn-tectonic deposition, after contraction of (a) 6 mm, (b) 14 mm, (c) 22 mm, (d) 30 mm, (e) 54 mm and (f) 90 mm (Storti *et al.*, 1997). The black layer indicates the top of the pre-tectonic sediments.

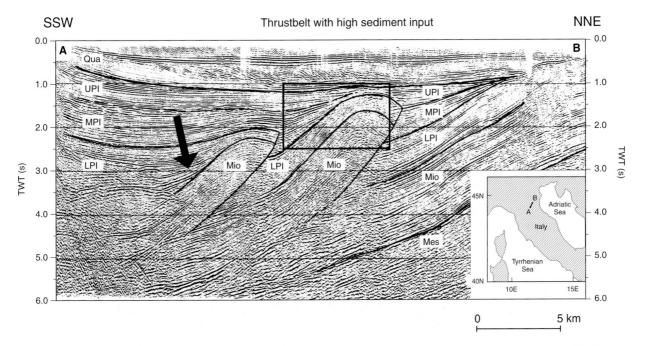

Fig. 9.2. A migrated reflection seismic profile through the Ferrara–Romagna thrust arc, Northern Apennines, Italy (modified from Pierri, 1989; Zoetemeijer, 1993). The inset shows the location of the profile. Mes is Mesozoic, Mio – Miocene, EPl – Early Pliocene, MPl – Middle Pliocene, LPl – Late Pliocene and Qua – Quaternary. The Pliocene–Quaternary sediments show angular unconformities formed by tectonic movement. The arrow points to a truncation that indicates new thrust activation. The box shows a significant wedging of horizons, recording deep-seated thrust activation.

which supplies the coarse sediment to small drainage basins. These conditions require a long time for their development, even in the case of high topographic relief. Fluvial systems longitudinal to thrustbelt structures react more rapidly to subsidence, because their deposition is characterized by more effective water and sediment discharge from drainage basins, which are thousands to tens of thousands times larger than those of the fans. If the local depocentre of a thrustbelt advancing into the sea subsides below sea level, and there is a connection between the depocentre and the sea, the development of a marine embayment would be virtually instantaneous (Blair and Bilodeau, 1988).

Syn-tectonic deposition can trigger various responses from the angles α and β of a thrustbelt wedge taper, described in Chapter 2. These responses depend on whether the topographic slope is decreased or whether the angle of the basal slope is increased by the isostatic response to the added overburden. An increase in basal slope is usually the more important effect because it shifts the wedge taper towards the supercritical field, resulting in a decrease of internal deformation and an increase in the thrust front advance.

As discussed in Chapter 2, erosion has a tendency to alter the balance of forces within a thrust wedge towards the triggering of internal deformation (e.g., Chapple, 1978; Davis *et al.*, 1983). Numerous orogens have responded to erosion by a perturbation of the distribution of internal velocities that acts to restore the critical taper (e.g., Jamieson and Beaumont, 1988, 1989). The types of deformation initiated by increased erosion in the internal parts of a wedge include processes such as out-of-sequence thrusting, back-thrusting and underplatting. Erosion can even cause renewed activity in the most proximal dominant thrust sheets, which have experienced little deformation since the onset of their emplacement (e.g., Platt, 1986a; Sinclair *et al.*, 1991; Sinclair and Allen, 1992; Boyer, 1995; Schlunegger *et al.*, 1997). Erosion destroys the critical taper of the deforming wedge by reducing the stress tensor component within the wedge that is related to the weight of the overburden. Internal deformation follows because the yield strength of the noncohesive frictional Coulomb wedge material decreases as overburden is removed and deformation makes the wedge stronger.

The nature of deposition in an adjacent foreland basin can determine whether the primary cause of orogen uplift is thrusting, resultant thickening or enhanced erosion (e.g., Burbank, 1992). Crustal thickening causes asymmetric subsidence within the adjacent

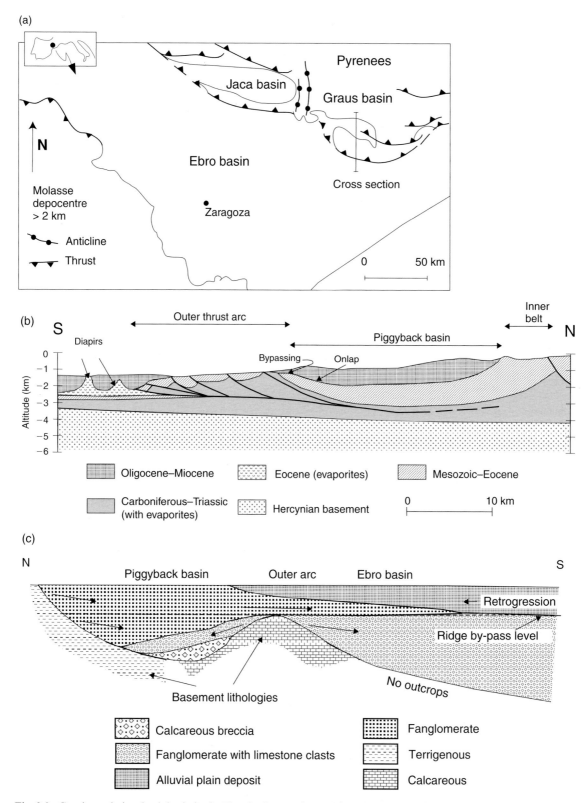

Fig. 9.3. Continental piggyback basin in the Ebro basin complex, southern Pyrenees (Ori and Friend, 1984). (a) Location of the cross section. (b) Cross section through the Graus piggyback basin and the marginal area of the Ebro basin. Onlap and bypassing have been observed in outcrops. (c) Section based on field observations showing the sedimentary infill of the western tip of the Graus piggyback basin. Arrows indicate paleocurrents.

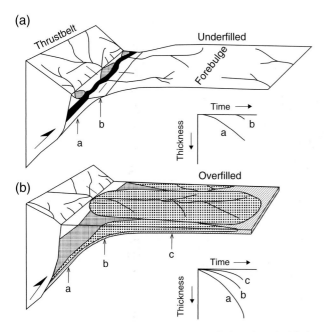

Fig. 9.4. Comparison of the geometry and facies of underfilled and overfilled foreland basins (Jordan, 1995). (a) Underfilled basins form a valley, and receive sediment from both the thrustbelt and forebulge. (b) In overfilled basins, the forebulge receives sediment from the thrustbelt and is expressed only by a slower accumulation rate and thinner strata than in the proximal zone. Insets show the accumulation history at increasing distances from the thrust front, revealing that the accumulation rate is slower in distal sites compared to proximal sites. Although the tectonic-subsidence histories may be identical for these two basin types, the accumulation histories are quite different.

foreland (Fig. 9.4). Considerably more subsidence and deposition occurs in the proximal parts of the foreland and this subsidence inhibits the progradation of transverse fans across the basin. Transverse rivers are therefore restricted to proximal areas, and longitudinal or axial rivers flow along the medial part of the foreland near the fan toes. In cross section, the basin appears underfilled, because the rate of subsidence exceeds the rate of sediment supply. Enhanced erosion drives isostatic recovery that reduces the crustal root and flexural uplift occurs across the foreland in response to the reduced mountain load (Fig. 9.4). Enhanced erosion also produces an increased sediment flux into the basin. This distributed sedimentary load, if preserved within the basin, both shifts the loading centre away from the hinterland and also increases the width of the foreland basin. Along with possible uplift of the proximal foreland basin, subsidence across the foreland is less asymmetric, because of the distributed sedimentary load. Reduced subsidence in the proximal foreland and

enhanced sediment flux into the basin cause transverse fans to expand and longitudinal rivers to be displaced toward the distal basin.

A linkage between erosion and syn-tectonic deformation has been studied using viscous models (Beaumont *et al.*, 1992), which are more appropriate for orogenic mechanics at large scale. These models were applied to the Southern Alps of South Island, New Zealand, formed by collision of the Pacific and Indian plates at a convergence rate of $2\,\mathrm{cm\,yr^{-1}}$ along the Alpine fault during the last 5 Myr. They show the importance of the ratio of gravitational to tectonic forces in controlling the velocity and the strain-rate distribution. Gravitational forces limit orogenic thickening. When erosion removes mass, gravitational stabilization is reduced and the strain rates increase correspondingly. Vertical strain, coupled with isostatic adjustment, induces an enhanced uplift as a response to erosion and causes modification of the syn-tectonic deformation of an orogen. The simulated erosion includes both short- and long-range transports. Soil creep, land sliding and hill wash represent the processes of short-range transport, which transfer material from elevated areas to valleys. The processes of long-range transport include fluvial transport in bed load, suspended load and dissolved load. These models also indicate that time scales of 0.5–5 Myr are sufficient for profound climatic effects in collisional orogens.

The modelling of the Southern Alps suggests that the recorded shortening would produce a much larger orogen than has actually occurred, indicating that significant erosion has taken place. It has to be noted, though, that the model does not include the effect of isostasy, and, therefore, provides only basic development trends. The model, which applies erosion controlled by a westerly wind, shows a distinct orogenic geometry (Fig. 9.5a; Beaumont *et al.*, 1992). The geometry is significantly affected by dominant erosion on the western orogenic flank because precipitation is greatest on that side, and because the boundary condition for surface transport requires the Tasman Sea to be an infinite catchment area for the eroded sediment. The dotted line in Fig. 9.5(a) indicates the initial surface projected to a position using its cumulative vertical displacement. The distance between the model surface and this line thus provides a measure of the total erosion in the spatially fixed model coordinate system. This measure would equal the exhumation experienced by rocks currently at the surface were the uplift vertical. However, the horizontal component of the tectonic movement carries the uplifting crust progressively to the west and rocks reach the surface to the west of their position indicated by the dotted line. Consequently, once a steady state between

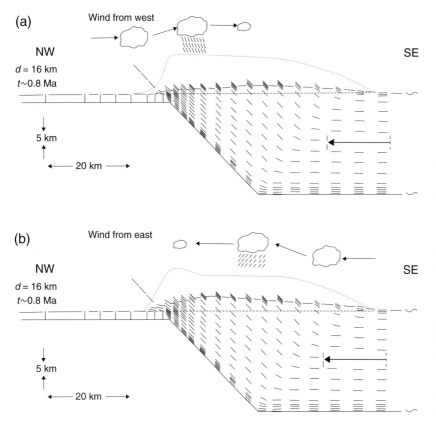

Fig. 9.5. Plane-strain deformation model of the Southern Alps, New Zealand (Beaumont *et al.*, 1992). (a) Model with orographic precipitation derived from a westerly source. Velocity vectors are relative to a fixed Indian plate. The dotted lines show the projected position of the initial surface. The westerly vergent tectonic wedge has been suppressed by erosion. The western sink for mass also stunts the growth of the orogen as a whole and a near steady state is reached with only a small easterly vergent tectonic wedge. It is clear that erosion has modified structural and metamorphic styles throughout the model orogen, not just locally. Bold horizontal arrows show the total amount of shortening, *d*. (b) Model with orographic precipitation derived from an easterly source. The source has the same water vapour flux at its eastern boundary as model in (a).

erosion and uplift has been achieved, the rocks exhumed from the greatest depth reach the surface of the model adjacent to the Alpine fault. It can also be implied that rocks, which have undergone the largest exhumation, do not reach the surface where the erosion rate is greatest, even in a steady-state orogen. Modelling also shows that rivers on western and eastern slopes of the orogen are different. The eastward-draining rivers have smaller equilibrium sediment-carrying capacities for equivalent sized watersheds. Therefore, only rivers that acquire a large discharge by developing larger watersheds can achieve the same degree of incision as their western equivalents.

If the erosion distribution is affected by an easterly wind, the orogen geometry differs from geometries affected by no erosion or by erosion dominated by a westerly wind (Fig. 9.5b; Beaumont *et al.*, 1992). As the orogenic surface uplifts and orographic rainfall develops, the eastern flank of the orogen is subject to the largest precipitation and erosion. The efficiency in extracting precipitation from vapour controls the orogenic profile. As it increases (Beaumont *et al.*, 1992), the belt of exhumed deepest rocks migrates eastward, away from its position adjacent to the Alpine fault. It may even cross the drainage divide. With declining precipitation on the western orogenic flank, neither erosion nor

horizontal advection is sufficient to expose deeply buried rocks against the Alpine fault.

The erosion is affected by numerous factors such as climate, relief and geological composition, each of them having several other constraints (e.g., Saunders and Young, 1983; Kukal, 1990; Jordan, 1995).

An example of climatic control comes from ODP Sites 717C and 718C located in the Bengal fan in front of the Himalayas, showing a decrease in sediment accumulation since 7 Ma (Burbank *et al.*, 1993). It is a response to waning mechanical denudation as a result of increasing slope stabilization from dense plant cover under intensified monsoonal climatic conditions (Harrison *et al.*, 1993 and references therein). Seasonality plays an important role (e.g., Sinclair, 1997); the larger the seasonal changes the faster the erosion occurs (e.g., Ahnert, 1970; Kukal, 1990). A long dry season produced in a monsoonal, a Mediterranean, a tropical savanna or a steppe climate leaves the ground bare and vulnerable to erosion by the rains of the wet season. The rapid removal of waste by the rare, but intensive runoff over the little-protected soil in the dry Rocky Mountain region, USA, is just as effective as the slower, but more continuous, waste removal processes in the humid Appalachian region of the USA, Western Appalachians or the Papua New Guinea foldbelt. The

Table 9.2. Rate of erosion of the South American Andes in various climatic zones.[a]

Climatic zone	Rate of erosion (mm yr^{-1})
Tropical arid	0.05
Temperate humid	0.5
Polar	1.1–1.65

Note:
[a] Scholl *et al.* (1970).

difference lies in the amount removed rather than in the mode of removal. The rivers of the Rocky Mountain region carry only a small percentage of their total load in solution, whereas the dissolved load of the rivers in humid areas is considerably larger.

A comparison of the climatic effects on erosion in the same orogen was made in the Andes (Table 9.2; Scholl *et al.*, 1970). Table 9.2 shows the effect of accelerated erosion of glaciers and glacial melt waters. Especially in places with substantial morphological differences, the rate of erosion increases towards glaciated regions. Similar conclusions can be drawn from the large world-wide data set of Saunders and Young (1983) or from the Himalayan data (Einsele *et al.*, 1996).

An example of the control of relief comes from the Argentinean sub-Andes where the study of Flemings and Jordan (1989) determined a linear relationship between the orogenic slope and sediment transport. The implication is that uplift tends to increase the relief, exerting an influence on the erosion rate (Ahnert, 1970). During the initial orogen development, uplift usually exceeds erosion, which leads to an increase of relief. Erosion progressively accelerates until it equals the uplift. Following this stage there is no further relief change, as long as all the involved rates remain constant. The orogen, thus, reaches a steady-state dynamic equilibrium between erosion and uplift (Hack, 1960). Examples of steady-state orogens are known from the specific development stages of the Argentinean Precordillera (Johnson *et al.*, 1986), Taiwan thrustbelt (Covey, 1986; Dahlen and Suppe, 1988), the Himalayas (Johnson *et al.*, 1985; Cerveny *et al.*, 1988) and the Southern Alps of New Zealand (Adams, 1985). Case studies showing periods of enhanced erosion postdating periods of dominant uplift include the Swiss and French Alps (Schlunegger *et al.*, 1997; Sinclair, 1997).

A classic example of a mountain belt, where denudation was adjusted to uplift, and where orogenic precipitation and climate changed due to relief buildup, is that of the Himalayas (Johnson *et al.*, 1985; Cerveny *et al.*, 1988). The present-day topography of the Himalayas is largely determined by erosion and northward retreat of the steep southern slope of the mountain range (Masek *et al.*, 1994). Particularly high rates of orographically induced precipitation, higher than 2000 mm yr^{-1}, on the lower slope and the resulting removal of rock detritus have maintained a steep frontal slope. High peaks at the plateau edge and deep canyon incision across the plateau margin mainly reflect isostatic uplift as a result of rapid erosional unloading (e.g., England and Molnar, 1990; Montgomery, 1994). These processes were promoted by highly effective glacial erosion (Einsele *et al.*, 1996).

During the early stages of the development of the Himalayas, erosion was exceeded by the tectonic mass influx along the Gangdese thrust system active since about 27 Ma and the Main Central thrust active since about 22 Ma. It caused the southward advance and increased elevation of the earlier Gangdese mountain front and the younger High Himalayan mountain front (Einsele *et al.*, 1996). Later in the Neogene, following the end of slip along the main central thrust and the north Himalayan normal fault zone 18 Ma, the erosion-controlled retreat of the mountain front had a greater influence than tectonic mass influx (Masek *et al.*, 1994). This retreat was enhanced by a double-sided climate with a moist monsoonal climate in the south characterized by high erosion rates and an arid climate in the north characterized by limited erosion. This climatic contrast resulted in a retreat of the southern slopes by 3.5 mm yr^{-1}.

Even if the orogen never obtains a steady-state relief, such as, for example, in cases of progressively increasing and then abruptly ending uplift, the relation between relief and uplift and between erosion and relief acts as a causal linkage between uplift and erosion, via relief. So that a tendency always exists towards establishing a dynamic equilibrium between the rate at which the land is raised and the rate at which it is worn down, with the latter striving to match the former. The adjustment of erosion to uplift requires time, and this time lag varies among different erosional processes (Ahnert, 1970). As a result of headward erosion, the adjustment is transmitted upstream and into tributaries. Erosion of the streams steepens the lower parts of the valley side slopes and thus accelerates the transport of rock waste from the slopes to the streams. With steepening of the valley sides and related acceleration of waste movement, the waste transport extends progressively further upslope until it reaches the interfluves and summits. Faster removal causes the thinning of the waste cover. It also causes more rapid denudation of the bedrock. Eventually the rate of waste production by denudation equals the new higher rate of waste removal. This is the stage when summits, slopes and streambeds are being lowered at the same rate at which the land is uplifted. The total lag time from the beginning of the uplift to the eventual adjustment of the rate of summit lowering is likely to be longer for large uplifted areas, with large drainage basins, than for small ones. This is because of the time needed to transmit the adjustment of stream courses throughout the area.

The relationship between erosion and uplift also has an impact on orogenic advance toward the foreland (e.g., Boyer, 1995). If the erosion rate is less than the uplift rate, the orogen becomes supercritical, which leads to the accretion of additional frontal thrust sheets. The advance will thus be accompanied by growth in the thrustbelt width. If the erosion rate equals the uplift, the orogen advances at a steady-state relief. If the erosion rate exceeds the uplift, the internal deformation increases and the thrustbelt might actually retreat or decrease in width.

When the range of relief values in an orogen is great, the morphologically effective climate factors themselves vary as functions of the relief and thus implicitly enter into the denudation–relief relationship (e.g., Ahnert, 1970). Areas of high orogenic relief encompass a great range of altitudes and hence reach through several altitudinal climatic zones. Above the timberline, processes of denudation tend to be more intensive than below it.

An example of the control of geological composition on erosion comes from parts of the Himalayas and the Rocky Mountains, where erosion-resistant rocks have slowed down the erosion (DeCelles *et al.*, 1991; France-Lanord *et al.*, 1993). The presence of rocks with varying resistance to weathering causes variations in the erosion rate. This is particularly important in the case of small drainage basins that lie on only one or two types of rock. With increasing size, the basins usually contain a great variety of rock types for which the different resistances are likely to balance out so that the effect of lithological differences upon the total sediment load may become neutralized. The geological composition further controls the subsequent transport towards the basin. Transport areas located further away from the source terrane record transport-related progressive clast reduction controlled by rock durabilities, as indicated by systematic petrologic analysis of various clast-size categories in eastern Alpine rivers (Kováč *et al.*, 1992).

Indirect influence of syn-orogenic erosion and deposition on an evolving thrustbelt via the influence on foreland basins

The influence of syn-orogenic erosion and deposition on an evolving foreland basin is dynamic and includes the reaction to episodic thrustbelt development. However,

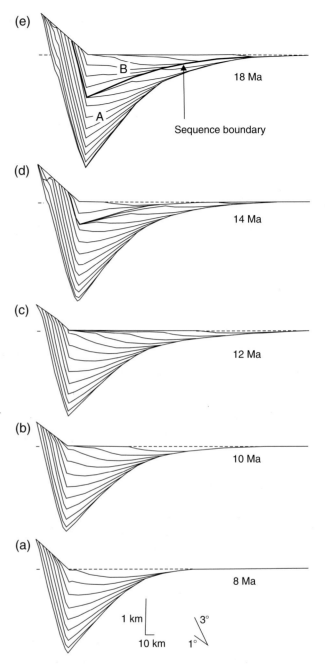

Fig. 9.6. Dynamic foreland basin model showing the stratigraphic response to cyclical thrust activity (Jordan and Flemings, 1991). From 4 to 8 Ma and 12 to 16 Ma thrusting was steady; but from 8 to 12 and after 16 Ma no thrusting occurred. Parts (a)–(d) show sequential steps through the thrust cycle and (e) shows the final stratigraphy with sequences mapped. Time lines show 1 Ma intervals. The uppermost line is also the depositional topography. The dashed line is the height of the hinge at the beginning of the model and sea level throughout this model. Inclined lines indicate 1° and 3° slopes at the vertical exaggeration of the cross section.

(f)

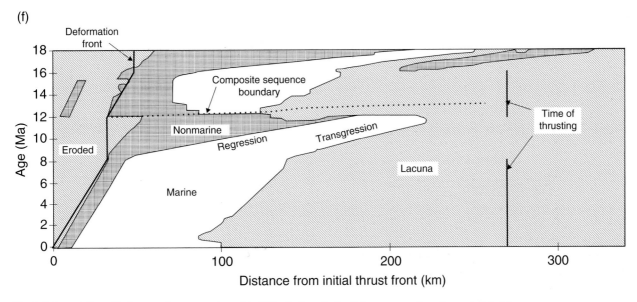

Fig. 9.6. (*cont.*) Part (f) shows a chronostratigraphic (Wheeler) analysis of the cross section shown in (a)–(e).

the stratigraphy of a foreland basin results from an inter-relationship between episodic tectonics, eustasy and sediment supply, the latter being controlled by climate, relief and the erosional resistance of the rock (e.g., Blair and Bilodeau, 1988; Flemings and Jordan, 1989, 1990; Jordan and Flemings, 1991; Beaumont *et al.*, 1992). Tectonic subsidence rates are generally an order of magnitude greater than eustasy, making the structural input of greater significance in controlling syn-orogenic deposition (Jordan and Flemings, 1991).

During the episodic development of a thrustbelt, the stratigraphy of a foreland basin records the various episodes. The stratigraphic record includes unconformities, which form on both proximal and distal margins of the basin and which are more pronounced on the forebulge.

The cessation of thrusting is recorded by an immediate shift of the available sediment accommodation space from the region proximal to the thrust front (Fig. 9.6a) to the space filled by water and left over from previous time intervals. Subaerial erosion cuts into the most proximal part of the basin in the last half of the tectonic quiescence interval (Fig. 9.6c). A small amount of sediment is eroded from the forebulge and supplied to the distal side of the basin during the quiescence, because the forebulge profile flattens due to broadening of the load on the proximal side of the basin.

Renewed thrusting instantly shifts the balance of accommodation space back to a proximal zone (Fig. 9.6d). The transition from quiescence to renewed thrusting is marked by an abrupt shift from broad lenticular stratal packages (Fig. 9.7b) to wedge-shaped

stratal packages that build outward from the thrust front (Fig. 9.7c) (Flemings and Jordan, 1990). As subsidence outstrips sediment supply, the underfilled part of the basin is re-established a short distance from the thrust front.

The above cyclic thrust activity results in two distinct unconformities. On the proximal side of the basin, an unconformity forms during periods of quiescence (Fig. 9.6), as strata offlap and the most proximal parts of the basin are bypassed and eroded. This unconformity is rapidly onlapped at the next initiation of thrusting (Fig. 9.6e). On the distal side of the basin, an unconformity is cut at the beginning of thrusting as the forebulge migrates toward the thrustbelt. The proximal and distal unconformities merge. The progradational strata of a single tectonic cycle underlie the unconformity in both locations, but the timing of the erosion is entirely different. Transgressions on the distal side correlate to regressions on the proximal side and distal regressions correlate to proximal transgressions (Fig. 9.6f). The described sedimentary record is characterized by major progradation during tectonic quiescences.

The influence of syn-orogenic erosion and deposition on an evolving foreland basin in front of the thrustbelt front interacts with several other factors. A list of the most significant factors, discussed briefly later, includes (Zoetemeijer *et al.*, 1993, Fig. 9.8):

(1)   the flux of sediment discharged into the basin, characterized by the transport coefficient of the mountain belt, κ_m;

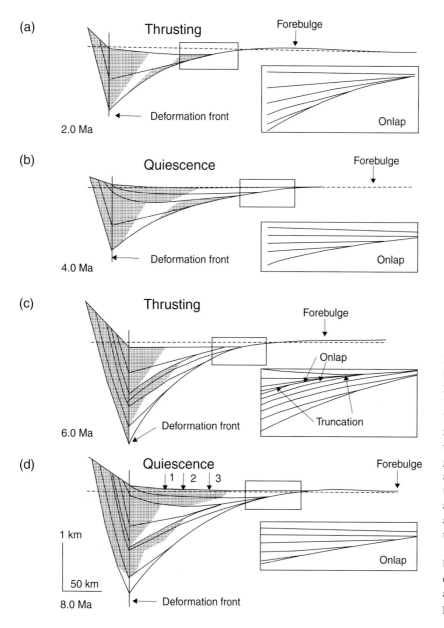

Fig. 9.7. Stratigraphic evolution of a foreland basin over 8 Myr, shown at 2 Myr intervals (Flemings and Jordan, 1990). Time lines show 1 Ma intervals. Vertical exaggeration is 40 times. Shading indicates basin margin facies, which were deposited at a gradient greater than 0.0005. Basin-axis facies, accumulated at a smaller gradient are unshaded. Right margin of the basin is amplified in boxes and 200 000 year lines are used to illustrate onlap and truncation at vertical exaggeration of 100 times. Bedding on the left side has been uplifted and deformed by leading edge of the thrustbelt. Thrustbelt advance was 20 km during the second phase of thrusting.

(2)   the efficiency with which sediments are distributed in the basin, characterized by the transport
coefficient of the basin, κ_b;
(3)   the rigidity of the lithosphere, expressed by
means of its effective elastic thickness, T_e; and
(4)   the growth of the thrust wedge, expressed by the
topographic slope angle, α, and the thrust front
advance rate, v_{thrust}.

The last two factors control the shape of the foreland
basin by their control on the lithospheric flexure of the
overridden plate. However, the total orogenic load on
the overridden plate results from the growth of the
thrust wedge combined with the progressive filling of

the foreland basin by sediment eroded from the same
wedge. The amount of sediment is controlled by the
amount of eroded material, which can be represented by
diffusion models (e.g., Flemings and Jordan, 1989;
Sinclair *et al.*, 1991; Peper, 1993) and controlled by the
interaction of weathering and transport processes (e.g.,
Beaumont *et al.*, 1992). The amount of sediment is also
controlled by the sediment distribution within the basin
and serves to illustrate how complexly linked the controlling factors are.

A critically tapered thrust wedge incorporates the
basin sediments into the thrust deformation as it
advances towards the foreland. While the sediments
become part of the wedge, a new sediment flux enters

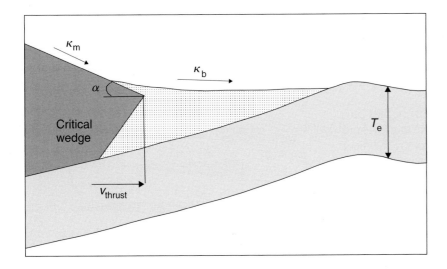

Fig. 9.8. Main factors influencing an evolving foreland basin in a thrustbelt front (Zoetemeijer, 1993). Factors are: the thrust front advance rate, v_{thrust}, the angle of critical taper, α, the efficiency coefficient of the sediment distribution, κ_{b}, and the erosion products of the same belt, κ_{m}, and the effective elastic thickness, T_{e}, of the lithosphere.

the basin, arising from erosion of the accretionary wedge. This results in the frequent recycling of sediment, documented from accretionary wedges (e.g., Ori *et al.*, 1986; Steidtmann and Schmitt, 1988). The situation in which the amount of sediment incorporated in the wedge is balanced by the amount of erosional products entering the basin is described by Covey (1986) as steady-state foreland basin evolution. It corresponds to a stage when the geometry of the thrustbelt and foreland basin is constant and the system only migrates onto the foreland. However, the relationship between the thrusting rate and the efficiency of erosion and deposition results in a whole variety of basins with synorogenic sediments. High thrust front advance rates combined with low sediment flux into the basin or inefficiency of sediment distribution in the basin lead to underfilled basins, whereas overfilled basins are characterized by low thrust rates combined with efficient erosion and deposition (Flemings and Jordan, 1989; Peper, 1993).

Sediment flux and distribution
The processes of sediment supply to the foreland basin include both erosion and deposition understood as diffusive processes, in which sediment transport is proportional to the surface gradient. Both marine and nonmarine transport can be expressed by the diffusion equation (Jordan and Flemings, 1991):

$$\frac{\partial h}{\partial t} = \kappa \frac{\partial^2 h}{\partial x^2}, \qquad (9.1)$$

where h is the elevation of the Earth's surface, t is the time, x is the horizontal distance and κ is a proportionality constant, with units of length^2 time^{-1}, referred to as the transport coefficient. In erosional regimes, the diffusion equation is used to describe the denudation of

fault and shoreline scarps (e.g., Nash, 1980; Hanks *et al.*, 1984), whereas in a depositional setting it describes both fluvial and deltaic processes (Begin *et al.*, 1981; Kenyon and Turcotte, 1985). The sediment flux into a basin represents a denudation rate multiplied by the width of the mountain belt (e.g., Flemings and Jordan, 1990). The denudation rate in actively deforming mountain belts is much higher than in old inactive mountain belts (Pinet and Souriau, 1988), because the rate is proportional to relief and it falls rapidly when no uplift is occurring. The relief of an actively evolving orogenic belt further influences the erosion rate by influencing the local precipitation rates. The erosion rate is further controlled by factors such as climate, rock erosional resistance and orogenic belt double-sided climate.

On a thrustbelt scale, the sediment can originate from several different source areas, each possibly having different activity and depositional mechanisms. Several source areas per thrustbelt are known from the Venetian basin located between the Dinarides, Alps and Apennines (Massari *et al.*, 1986) or the Alpine Molasse Zone located in front of the eastern and western Alps (Schlunegger *et al.*, 1997; Frisch *et al.*, 1998). The existence of several source areas can result in an interfingering system of clastic wedges, characterized by different orientations within the foreland basin stratigraphy.

The role of the erosion and sediment transport coefficients, κ_{m} and κ_{b}, on an evolving foreland basin stratigraphy can be studied when the thrusting rate of the advancing thrustbelt, v_{thrust}, and the flexural rigidity of the plate underlying the basin, T_{e}, are kept fixed (Flemings and Jordan, 1989). The effect of a distinct increase in the erosional transport coefficient, κ_{m} is to increase the volume of the foreland basin and to increase the elevation at which sediments onlap the mountain belt. The basin is larger, because it must store the material supplied by an increased flux. Sediments onlap onto the thrust front to achieve a higher gradient, in order to accommodate the increased flux of sediment from the thrustbelt. Thus, higher-energy facies are needed to transport the increased flux of supplied sediment.

The efficiency of the sediment transport, κ_b has profound effects on the final geometry of the foreland basin. As the ability of the basin to transport sediment is increased, more of the strata are deposited further from the thrustbelt. Instead of a narrow basin, where much of the strata are deposited at a steep gradient proximal to the thrust, the sediment is spread out across the basin at lower gradients.

Lithospheric flexure
The role of flexural rigidity of the underlying plate beneath the foreland basin, T_e, on the evolving basin stratigraphy can be studied when the erosional and sediment transport coefficients, κ_m and κ_b, and the thrusting rate of the thrustbelt, v_{thrust}, are kept constant (Flemings and Jordan, 1989). Low rigidity results in a basin that is wide and shallow, forming a small volume of sediments. Increased flexural rigidity causes a narrower and deeper basin resulting in a larger volume of sediments.

Thrustbelt load
The role of the thrusting rate on an evolving foreland basin stratigraphy can be studied when the erosional and sediment transport coefficients, κ_m and κ_b, and the flexural rigidity of the plate beneath the basin, T_e, are kept constant (Flemings and Jordan, 1989). At a slow thrusting rate the new sediment accommodation space created due to flexure is less than the rate at which sediment is supplied. Consequently, the basin is nearly overfilled and becomes quite deep. At a high thrusting rate, the basin is rapidly consumed and creates more space at each time-step.

Foreland basin control over thrustbelt style
An important control over syn-tectonic sediments on the evolving structural style of a thrustbelt is in their ability to modify the angle of the critical taper of the advancing wedge (Fig. 9.8). As described in chapter 2, the oscillations of the wedge taper around the critical taper control the internal deformation and advance of the thrustbelt. When the advancing thrustbelt incorporates the sedimentary fill of the foreland basin into its structure, the taper of the sedimentary fill controls the subsequent thrustbelt development. The effect of this basin taper on the thrustbelt development was systematically analysed by Boyer (Fig. 9.9; 1995). Figure 9.9

(a)
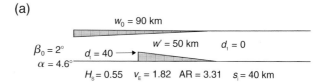

(b)
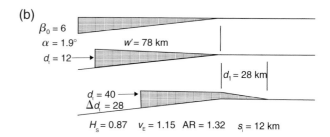

(c)
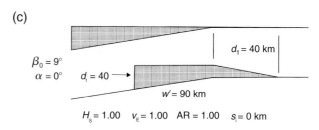

Fig. 9.9. Thrust-front advance for basin tapers of $\beta_0 = 2°, 6°$ and $9°$ (Boyer, 1995). The translation at the back of the thrust wedge d_t is absorbed forelandward by (1) internal horizontal shortening, S_i, of the wedge necessary to build taper and by (2) the advance of the thrust front, $d_1 \cdot H_s$ is the amount of internal shortening expressed as a ratio of original thrustbelt width, w_0, divided by the deformed width, w'. V_E is the vertical stretch, given by present height at the back of the thrust wedge divided by the original height prior to deformation. AR is the axial ratio of strain, given by V_E/H_s. (a) For a basement dip of $2°$ all of the translation of the back of the thrust wedge is absorbed in building a critical taper. There is no frontal advance of the thrust front relative to the basin margin. (b) For a basement dip of $6°$ more of the translation of the back of the thrust wedge is available for frontal advance. 12 km is absorbed in internal shortening and the remaining 28 km results in frontal advance. (c) For a basement dip of $9°$ no internal deformation is required to build a taper. The prism is stripped from its basement as a single thrust sheet. All translation of the back of the thrust wedge translates into frontal advance of the thrustbelt.

shows that natural thrust wedges vary between two extremes; those with a small original taper, which absorb the rear translation by building a critical taper and those with a large original taper, which absorb the rear translation by forelandward advance of the wedge.

The deformation required to build a critical thrustbelt taper is a function of the taper of the pre-orogenic and syn-orogenic sediment fill in front of the propagating basal décollement (Boyer, 1995). Since the distribution of syn-orogenic sediments varies in space and time during the wedge development, as does the basement dip, these variations can produce a corresponding spectrum of structural styles within the wedge. Most thrust-belt-related sedimentary basins not only have a wedge shape, they are also convex upward. Their basal slope increases toward the deeper parts of the basin. According to sandbox experiments this affects the thrust spacing (e.g., Marshak and Wilkerson, 1992), which typically decreases in most thrustbelts towards the foreland (Goff and Wiltschko, 1992).

Further complexity is induced by strike-parallel variations of basement dip, which can be either smooth or, at fault boundaries, abrupt. The area balance considerations introduced in Fig. 9.9 can predict changes in internal shortening and the magnitude of thrust advance related to these along-strike changes in the basin taper. Lateral variations in internal shortening may be reflected not only by the intensity of the thrust imbrication and folding, but also by the intensity of penetrative mechanisms, such as pressure-solution and tectonic compaction.

Smooth lateral variations in basement dip produce an interweaving pattern of imbricate faults (Calassou *et al.*, 1993), which transfer displacement between the large, continuous thrust sheets in regions of high basin dip and more closely spaced and intensely shortened imbricates in adjacent regions of low initial basin dip (Boyer, 1995). Alternatively, very rapid lateral changes in basement taper may lead to the formation of cross-strike partitioning faults (e.g., Thomas, 1990; Lawton *et al.*, 1994).

The described sedimentary basin taper exerts control over many aspects of the growth and advance of thrust-belts (Boyer, 1995):

(1) the initiation of dominant thrust sheets;
(2) the magnitude of shortening within the thrust wedge;
(3) the spatial and temporal variation in the widths of thrustbelts and thus the the rates at which thrustbelts grow in width; and
(4) the rates of thrustbelt advance.

The greater the shortening required for building and maintaining the critical taper, the narrower the thrustbelt will be in its final stage. Throughout its development, the thrust wedge cyclically increases and decreases in width as convergence alternately translates into propagation of the frontal thrust and advance of the thrust wedge accompanied by internal shortening to maintain critical taper (Mulugeta and Koyi, 1992).

An example of basin taper influence on thrustbelt

style is the occurrence of dominant thrust sheets. A dominant thrust sheet develops in the hinterland parts of the thrustbelt, where the basin fill reaches critical or supercritical taper values (Boyer, 1995). Their tectonic load depresses the foreland crust, increasing the dip of the basement and causing the foredeep to migrate fore-landward (e.g., Price, 1973; Beaumont, 1981). The dip of the foreland crust continues to increase as a function of tectonic load, until it reaches a critical value. Then the next thrust initiates. Thus, thrust advance can be a self-perpetuating process. The advance of the initial, dominant thrust sheet increases basin dip, which favours the initiation of the next thrust (Boyer, 1995). The feedback loop between basement dip and thrust advance leads to rapid thrust advance in regions of maximum basement dip and produces prominent thrust salients, such as, for example, the Provo thrust salient of the Wyoming thrustbelt (Boyer, 1995).

Role of syn-orogenic erosion and deposition in evolving local structure

Syn-tectonic deposition influences the trajectories of thrust planes, which bound newly accreted thrust sheets, as shown by sandbox models (Storti and McClay, 1995; Storti *et al.*, 1997). Syn-tectonic deposition in these models is simulated by the addition of growth strata onto the developing thrust wedge after each increment of deformation. Varying depositional rates are simulated by varying the amount of growth strata added, representing no deposition, low, intermediate and high depositional environments. It should be noted that fore-landward migration, shortening and wedge taper are strongly affected in these models by the fact that the base of the sandbox is not allowed to react to the over-burden load. Therefore, only the number of thrust faults and their geometries can realistically be assessed by the modelling.

Models ranging from no deposition to intermediate deposition are characterized by a stable Coulomb wedge of regularly spaced foreland-vergent thrust sheets, each with minor back-thrusts at the leading edge (Storti and McClay, 1995). With an increase in the syn-tectonic deposition rate, the number of thrust sheets and the slope angle α decrease, the final thrust-ramp geometries become markedly steeper and the wedges achieve their critical state faster. The models with no deposition have staircase ramp-flat geometries. The fault trajectories in the models with small to intermediate deposition have smoothly varying sigmoidal trajectories. In the no-deposition models, the thrusts rapidly climb through the pre-tectonic layers and then flatten out parallel to the layering. In the low to intermediate deposition models, the thrusts flatten inside the syn-tectonic sediments

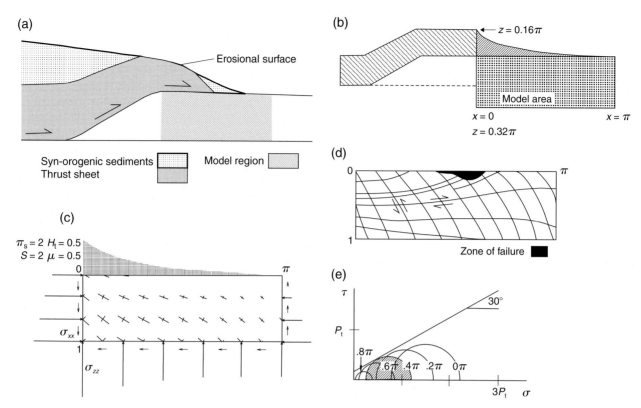

Fig. 9.10. Summary of models for exponentially decreasing surface load (modified after Goff and Wiltschko, 1992). (a) Idealized geometry of a thrust fault system, showing overlap of a hanging wall and a footwall and the development of a syn-tectonic sedimentary cover. The model area shown is the area of interest in the footwall for which stresses are calculated. The effect of more hinterlandward piggy-back thrusts is ignored. (b) Geometric representation of the surface load for the model, simulating point source sedimentation. (c) The configuration of the model showing the surface boundary conditions and the calculated stress state. Stress bar lengths external to the model are scaled to half the maximum height of overburden at $x = 0$. Internal stress bars show the relative magnitudes of the principal stresses at each point. (d) Fracture trajectories determined using a Coulomb criterion and an internal angle of friction of 30°. The shaded area represents the first area that will be in failure. The corresponding thrust fault is dashed. (e) Mohr circle representation of stress along the surface. $z = 0$ shows how the combination of changing mean stress and differential stress interact and result in failure occurring near $x = 0$, $z = 0.5\pi$. Each circle is labelled to correspond to its position along the surface.

(Storti *et al.*, 1997). Natural cases with low syn-tectonic rates of deposition exist in the Barbados Ridge (Westbrook *et al.*, 1988), where the thrusts have typical staircase trajectories and commonly flatten close to the surface of the wedge.

The high deposition model has a different structural style from the other models. It includes several much steeper small-displacement thrusts, which develop close to the backstop. The syn-tectonic deposition rate strongly affects thrust plane trajectories in these models. Thrust trajectories are planar and are propagated through syn-tectonic sediments at a constant angle. There is no change in the kinematic mechanisms from thrust-tip folding to thrust ramp folding and upper ramp-flat fault trajectories do not form. Natural examples with high syn-tectonic rates of deposition have been

described, for example, from the Argentinean Andes (Jordan *et al.*, 1993) and the Northern Apennines (Pieri, 1989; Zoetemeijer, 1993; Fig. 9.2).

Syn-tectonic deposition influences the resolved normal stress onto the basal décollement surface. An increased normal stress can alter the shear strengths of potential detachment horizons that can in turn lead to the propagation of the detachment in another horizon, as discussed in Chapter 3.

The presence of a weak horizon within syn-orogenic sediments can have an impact on the location of the basal detachment. It is very common to observe sequence boundaries and maximum flooding surfaces acting as the main décollement horizons, or the evaporitic layers of transgressive system tracts as detachment horizons. A typical example occurs in the southern

Pyrenees (Vergés *et al.*, 1992), where the staircase propagation of the basal décollement during the Late Eocene–Late Oligocene was controlled by the location of syn-orogenic evaporitic basins developed in front of advancing thrust sheets. This example shows the importance of lateral facies changes in the syn-orogenic section subsequently accreted into the thrustbelt. Such lateral facies changes and also thickness changes introduce lateral rheological changes. For example, a variation in deposition from siltstone to mudstone may increase cohesion and the angle of internal friction. The distance of the deposition from the source area controls both the degree of the grain reworking and the grain size, which affects the angle of internal friction (e.g., Brace, 1961). Variations of the sand/shale ratio, typical for flysch depositional systems, affect Young's modulus and Poisson's ratio as well as the cohesion and the angle of internal friction. Typical examples of variations in cohesion within syn-orogenic sediments are provided by studies of active accretionary wedges, controlled by burial and tectonic compaction (e.g., Karig, 1986; Maltman, 1994; Morgan and Karig, 1995). Studies of historical accretionary wedges (e.g., Sample, 1990) have shown that cohesion is further increased by cementation. Variations in cohesion and friction are also affected by fluid pressure. The extent to which the pore fluids control compaction depends on whether or not the fluid is allowed to escape during deformation. Apart from normal consolidation, where pore fluid loss has adjusted to the compaction, overconsolidated sediments have the mechanical properties of having once been more compacted. On the other hand, the underconsolidated sediments have not yet reached a compaction state in equilibrium with the surrounding environment. An overpressure buildup by localized rapid sedimentation in low-permeability syn-tectonic sediments is another control on the strength of potential detachment horizons.

Localized syn-tectonic deposition plays a role in the development of ramps, as shown by finite-element modelling (Goff and Wiltschko, 1992). The stresses associated with the overriding thrust sheet and syn-orogenic sedimentary cover result from the overburden weight and the shear stress along the interface between the overriding thrust sheet and the footwall. This interface is assumed to separate mechanically the footwall from the overlying thrust sheet and the syn-tectonic sedimentary cover. The syn-orogenic sedimentary cover thins exponentially towards the foreland (Fig. 9.10). In this case, the principal stresses slowly decrease in magnitude forelandward and away from the maximum compressive stress and the surface load imposed by the climbing thrust sheet (Fig. 9.10c). The interplay between a decreasing horizontal regional compression and a decreasing effective mean stress, the latter resulting from the decrease in overburden, results in a change in orientation of the near-surface principal stress trajectories. This controls the propagation of curved fault trajectories in the foreland portion of the model (Fig. 9.10d). The failure propagates as a thrust flat beneath the thickest overburden and curves upward as the overburden, formed by both the thrust sheet nose and the syn-tectonic sedimentary wedge, thins. This footwall failure develops only when the regional compression builds up to a level significantly greater that the combined weight of the overriding thrust sheet and syn-tectonic sediments. This combined weight has a tendency to stabilize the footwall due to the large increase of the effective mean stress. Therefore, the length and mass of the overriding thrust sheet together with its syn-orogenic cover directly control the location of new thrust faults forming in the footwall. An increase of the pore pressure in the footwall shifts the location of the failure towards the hinterland. Based on force balance arguments alone (e.g., Marshak and Wilkerson, 1992), the length of each new thrust sheet decreases as the thrust sheet nose, together with its syn-tectonic sedimentary wedge, thins out.

An important aspect of the role of deposition and erosion on thrustbelts is their interaction with growing thrustbelt structures. This interaction results in the development of stratigraphically complex units that form by the erosion and resedimentation of rocks at the crests of actively growing structures on the basin floor. Just as in the case of an entire thrustbelt wedge, the process of erosion and deposition can be approximated by a diffusion law whereby the transport rate is proportional to the topographic gradient (Begin *et al.*, 1981; Kenyon and Turcotte, 1985; Moretti and Turcotte, 1985; Flemings and Jordan, 1989; Goff and Wiltschko, 1992; Fig. 9.10). In erosional regimes, the diffusion equation is used to describe the denudation of fault and shoreline scarps (Nash, 1980; Hanks *et al.*, 1984), whereas in depositional settings it describes both fluvial and deltaic processes (Begin *et al.*, 1981; Kenyon and Turcotte, 1985). An exponential decrease in the thickness of syn-tectonic sediments away from the thrust sheet (Goff and Wiltschko, 1992) is an end member of a range of possible geometries. It is consistent with the observation that clastic fan deposits can be approximated as point sources in which the volume of sediment decreases exponentially from the source (Wright and Coleman, 1974; Lawrence *et al.*, 1990).

Most natural examples of the interaction of the uplift of growing structures with erosion and deposition are from continental environments, because they allow

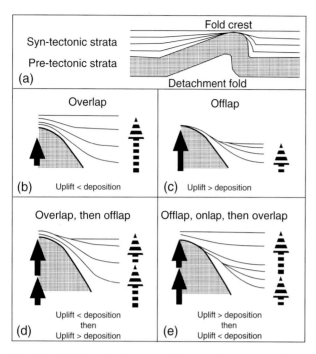

Fig. 9.11. Relationship between uplift and deposition rates indicated by the sedimentary record (modified from Burbank and Verges, 1994). (a) Nomenclature for the detachment fold and the syn-tectonic strata deposited during folding. (b)–(e) Models of overlap, offlap and onlap. Predictable geometries of syn-tectonic strata result from contrasts in the relative rates of crestal uplift versus coeval rates of accumulation. Crestal uplift is measured with respect to either (1) the base of syn-tectonic strata adjacent to the fold or (2) the position of correlative marker beds found in both the anticline and the adjacent syncline. When rates of deposition are consistently greater than the rate of crestal uplift, overlap will occur (b), whereas lower rates of deposition versus uplift lead to offlap (c). Reversals in the relative magnitude of these rates cause a switch in the bedding geometry (d, e). Onlap (e) occurs following offlap and a change to more rapid deposition rates.

detailed studies of outcrops. These studies have demonstrated that four variables act as first-order controls over fluvial interactions with growing folds (Burbank *et al.*, 1996):

(1)   the characteristics of the fluvial system;
(2)   the geometry and rates of growth of individual folds and faults;
(3)   the three-dimensional development of multiple structures; and
(4)   the pre-existing inherited topography.

The response of a fluvial system to the growth of a structure is strongly influenced by the stream power, the rates of aggradation and erosion, and the pre-existing topography. The surface expression of growing folds is a function of the rates of crestal uplift, the substrate

resistance and the fold geometry. A positive topography above the depositional plain develops when the rate of structural uplift exceeds the rate of aggradation. The amount of erosion of the growing structure in this case is determined by the ratio between stream power and the erosional resistance of the emergent rocks.

The interaction between uplift, erosion and deposition rates leaves a distinct record on local structures, which can be observed in both geological and seismic profiles (Fig. 9.11; Jordan, 1995). The resulting typical local structures, including fault-bend folds, fault-propagation folds and detachment folds, can be divided into four categories (Hardy *et al.* (1996)):

(1)   the background deposition keeps pace with the growing structure;
(2)   the base level rise is faster than the background deposition, resulting in a limited amount of local deposition;
(3)   the background deposition is low so that the growing structure can produce local deposition; and
(4)   the growing structure is the only source of deposition.

These four cases were simulated by numerical models (Hardy *et al.*, 1996), which used a 2 km thick pretectonic sediment package, a slip rate of 1.5 mm yr^{-1} and a time interval of 1 Myr. Such a slip rate is known from compressional structures (Table 9.3), as are the chosen background deposition rates of 2, 1.3, 0.5 and 0 mm yr^{-1} for each of the four cases simulated (Hardy *et al.*, 1996). Local erosion, transport and deposition were simulated by a diffusion model with a diffusion coefficient of 3 m^2 yr^{-1}. The resulting structures are described in the following subsection.

Fold structures when the background deposition keeps pace with the growing structure
The case when background deposition is as fast as the growth of a fault-bend fold is characterized by several distinct features (Fig. 9.12a). As the base level rise prevails over the creation of structural relief, growth strata are continuous, but thin, across the growing structure. This results from decreasing accommodation space being generated across the growing fault-bend fold. Kink bands appear as narrowing upward triangles within the growth strata. The apexes of these triangles (Fig. 9.12a) mark the end of deformation. The triangles record the locus of material which has passed through the active axial surfaces. The flat-lying strata on the fold crest represent material that has passed through neither of the active axial surfaces.

Similar features are recorded by growing fault-propagation folds in this scenario (Fig. 9.12b). Growth strata

Table 9.3a. Selected slip rates from compressional settings.

Slip rate (mm yr^{-1})	Structure	Reference
0.33	Beartooth fault, Wyoming and Montana	DeCelles *et al.* (1991)
0.82	Lost Hills thrust, San Joaquin Valley, California	Medwedeff (1989)
0.1–0.9	North Canterbury blind thrusts, New Zealand	Barnes (1996)
1.25	Tugulu-Dushanzi thrust, Northern Tien Shan	Avouac *et al.* (1993)
1.3	Channel Islands thrust, Santa Barbara Channel, California	Shaw and Suppe (1994)
1.3	Pitas Point thrust, Santa Barbara Channel, California	Shaw and Suppe (1994)
0.0–2.0	Precordillera thrust belt, West-Central Argentina	Jordan *et al.* (1993)
1.0–2.0	Santa Barbara Channel and Los Angeles basin, California	Suppe *et al.* (1991)
2.1; 2.5; 2.5–4; 4.6	Sierras Marginales thrust system, South-Central Pyrenees, various time periods	Burbank *et al.* (1992)
0.23; 1.4–2.3	Perdido fold belt, Gulf of Mexico	Mount *et al.* (1990)
4.3	Wheeler Ridge thrust, California	Medwedeff (1992)
1.9–5.2	Pliocene thrusts, Los Angeles area	Davis *et al.* (1989)
Max 2.5	North Apennines foreland thrusts	Zoetemeijer *et al.* (1993)
0.21–0.26	Fault-propagation fold-related faults, offshore, North Canterbury, New Zealand	Barnes (1995)
0.5	Mediano detachment fold, South Pyrenees	Holl and Anastasio (1993)
c. 10	Wheeler Ridge thrusts, California	Mueller and Suppe (1997)
7	Blind thrusts in Swiss Molasse basin	Homewood *et al.* (1986)

associated with the fault-propagation folds are continuous and thin across the growing structure.

A distinct fanning of the growth strata is recorded in the case of detachment folds with beds thickening away from the rotating fold limb (Fig. 9.12c). As a result of the decreasing uplift rate with continued slip and the constant background deposition rate, the growth strata first onlap the fold limb and eventually overlap the crest. This syn-tectonic sedimentary record records the very rapid initial uplift and emergence, which occurs at the start of limb rotation.

Fold structures when the base level rise is faster than background deposition and growing structures give rise to local deposition

When the base level rise is faster than the background deposition rate, the syn-tectonic strata are prevented from filling every available accommodation space as occurred in the previous case. This scenario results in sediments draping a growing structure. It also results in common slumping and down-slope material transport, caused by the steep limbs of the structure.

A growing fault-bend fold in this scenario (Fig. 9.13a) is much broader than the fault-bend fold in the previous case (Fig. 9.12a). Diffusion mechanisms acting in the area of the growing structure result in the thickening of the syn-tectonic strata towards the structure and thinning over it.

A fault-propagation fold (Fig. 9.13b) would also show a broader structure than in the previous case (Fig. 9.12b). The syn-tectonic strata can be seen to thicken towards the structure, but thin over it. In particular, growth strata are noticeably thinned over the uplifting crest of the structure. Natural examples of fault-propagation folds showing these relationships have been recognized in the northern Po basin (Fig. 9.2; Pieri, 1989; Zoetemeijer, 1993). Here, the rate of uplift at the beginning of the middle–late Pliocene was faster than burial by syn-tectonic marine sediments (Pieri and Groppi, 1981; Ori and Friend, 1984). Thrust sheets produced emergent ridges during much of the period of syn-tectonic deposition. Reflection seismic profiles and geological cross sections show (Fig. 9.2) that younger syn-tectonic sediments onlap and later overlap against the thrust sheets as deposition progressively caught up with uplift at later growth stages of structures. Another example is known from the frontal Polish West Carpathians (Fig. 9.14).

In the case of detachment folding (Fig. 9.13c), diffusion processes result in the erosion of both limbs and the crest of structures and in downdip syn-tectonic deposition of the eroded material. The oldest syn-tectonic

Table 9.3b. Selected depositional rates from compressional settings.

Sedimentation rate (mm yr⁻¹)	Setting	Reference
0.06–0.13	Scala Dei Group, Ebro foreland basin	Colombo and Verges (1992)
0.07–0.2	Mediano anticline, South Pyrenean foreland basin	Holl and Anastasio (1993)
0.14–0.23	Perdido fold belt, Gulf of Mexico	Mount *et al.* (1990)
0.06–0.25	Syn-orogenic alluvial sediments, South-Central Unit, Pyrenees	Burbank and Verges (1994)
0.1–0.5	Siwalik Molasse, Pakistan	Johnson *et al.* (1986)
0.01–0.5	North Apennine foreland basins, Italy	Ricci Lucci (1986)
0.5–1.3	North Apennine foreland basins, Italy	Zoetemeijer *et al.* (1993)
0.25–0.5	Escanilla Formation, Ainsa basin, Spain	Bentham *et al.* (1993)
0.1–0.9	Continental sediments, Precordilleran foreland basins, Argentina	Jordan *et al.* (1993)
1.8	Wheeler Ridge, San Joaquin Valley, California	Medwedeff (1992)
0.24–0.88	Bermejo Basin, Precordilleran foreland basin, Argentina	Jordan (1995)
0.1	Marine Bartonian–Early Priabonian Igualada marls, Pyrenean foreland	Caus (1973), Powers (1989)
0.2–0.25	Upper Eocene–Lower Oligocene continental conglomerate, fluvial fan-delta deposition, Central Pyrenean foreland	Burbank *et al.* (1992)
>0.35	Igualada Bartonian–Early Priabonian marine marls, during episodes of thrust loading (accelerated deposition), Central Pyrenean foreland	Burbank *et al.* (1992)
>0.6	Upper Eocene–Lower Oligocene continental fan-delta deposition immediate next to thrusts, Central Pyrenean foreland	Burbank *et al.* (1992)
0.17–0.92	14–2 Ma, Huaco sequence, Precordillera of western Argentina, arid climate, alluvial fan, playa deposits	Johnson *et al.* (1986)
0.4	Trough axis, Timor Trough, minor amounts of siliciclastic detritus from Timor Island	Charlton (1988)
1.7	Convay Trough, thrust belt on the Australia–Pacific plate boundary, offshore North Canterbury, New Zealand	Carter *et al.* (1982)
0.1	Lower slope, Timor Trough, minor amounts of siliciclastic detritus from Timor Island	Charlton (1988)

Table 9.3c. Time-averaged slip rates determined from fossil thrusts for longer periods of time.

Slip rate (mm yr⁻¹)	Period of activity	Structure	Reference
1.4	100 Ma	Whole Sevier belt, Western USA	Royse *et al.* (1975)
6.0	12 Ma	Meade thrust, Sevier belt	Royse *et al.* (1975)
5–10	1–2 Ma	Prospect thrust, Sevier belt	Wiltschko and Dorr (1983)
3–7	Past 10.5 Ma	Whole Precordillera belt, Argentina	Sarewitz (1988)
0.7	3 Ma	Basement uplift, Beartooth, Wyoming/Montana (note enormous uplift for this minor slip)	DeCelles *et al.* (1991)
3–4	Eocene–Oligocene	Eastern Swiss Alps, frontal thrust	Pfiffner (1986)
10–14	Eocene–Oligocene	Western Swiss Alps, frontal thrust	Pfiffner (1986)
2–4	33 Ma–Seravalian	Western Swiss Alps, frontal thrust	Pfiffner (1986)
0.6	30–25 Ma	Back-thrusting event, Insubric fault zone, Alps	Sinclair and Allen (1992)
2	30–25 Ma	Alpine thrust front, Switzerland	Sinclair and Allen (1992)
6	10–5 Ma	Chartreuse Massif, French Subalpine thrust belt	Butler and Bowler (1995)
6–12	110–35 Ma	Alpine thrust front, Switzerland	Sinclair and Allen (1992)
2	25–2 Ma	Alpine thrust front, Switzerland	Sinclair and Allen (1992)

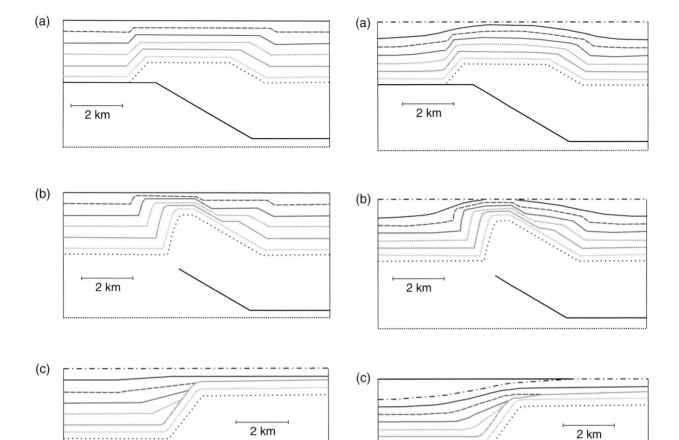

Fig. 9.12. Fold structures when the background deposition keeps pace with the growing structure (Hardy *et al.*, 1996). Included syn-tectonic strata are recorded at 200 ka intervals. Total run time is 1 Ma. Base level rise during model run is 2 mm yr^{-1} and background deposition rate is 2 mm yr^{-1}. (a) Fault-bend model. Slip is constant from right to left during the model run at 1.5 mm yr^{-1}. (b) Fault-propagation model. Slip is constant from right to left during the model run at 1.5 mm yr^{-1}. (c) Detachment fold model. Figure shows the frontal part of the structure showing progressive limb rotation. Limb length is 2 km. A constant slip rate of 0.75 mm yr^{-1} (equivalent to 1.5 mm yr^{-1} for a symmetric detachment fold) was applied.

Fig. 9.13. Fold structures with base level rise faster than background deposition resulting in a growing structure with local deposition (Hardy *et al.*, 1996). Models include a diffusion model for erosion, transport and deposition. Total run time is 1 Ma. Base level rise during model run is 2 mm yr^{-1} and background deposition rate is 1.3 mm yr^{-1}. The diffusion model uses a diffusion coefficient of 3 m^2 y^{-1}. Growth strata are recorded at 200 ka intervals. (a) Fault-bend fold model. Slip is constant from right to left during the model run at 1.5 mm yr^{-1}. (b) Fault-propagation fold model. Slip is constant from right to left during the model run at 1.5 mm yr^{-1}. (c) Detachment fold model. Figure shows the frontal part of the structure showing progressive limb rotation. Limb length is 2 km. A constant slip rate of 0.75 mm yr^{-1} (equivalent to 1.5 mm yr^{-1} for a symmetric detachment fold) was applied.

strata onlap the uneroded fold limb, whereas upsection the onlap surface is also an erosional surface. This is a result of the progressive development of the detachment fold, which produces steeper slopes, and therefore more erosion, with time.

Fold structures when the background deposition is low and growing structures result in faster local deposition
With low background deposition and faster local deposition, derived from growing folds, dramatic effects on fault-bend folding can be observed (Fig. 9.15a). Background deposition across the uplifting fold becomes significantly reduced, which leads to both the

onlap of strata on the forelimb and a thinned and condensed sedimentary sequence over the fold crest. The onlap surface on the forelimb can be seen to climb through the stratigraphy and to become the topmost surface on the backlimb of the fold. Such unconformities have been observed in the Lost Hills anticline from the Temblor fold belt in California (Medwedeff, 1989, 1992; Fig. 9.16). This anticline started to grow during the Late Miocene and continued to grow during

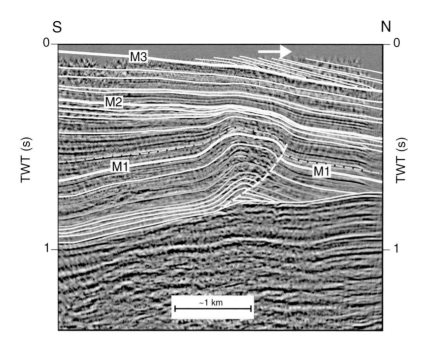

Fig. 9.14. Reflection seismic profile through the frontal growth fold in the Polish part of the West Carpathian accretionary wedge and its interpretation (modified after Krzywiec, 1998, personal communication).

the late Pleistocene or Holocene. Syn-tectonic strata of the Etchegoin and San Joaquin formations are represented by shallow marine to brackish water sediments, while the youngest unit, the Tulare Formation, consists of fluvial and lacustrine siltstone, sandstone and conglomerate.

Another fault-bend fold example in this scenario is the Oliana fold in the Ebro basin, Spain. Here, the uplift rate during the Late Eocene–Oligocene was faster than burial by syn-tectonic alluvial conglomerates (Burbank *et al.*, 1992; Fig. 9.17). The growth history of the structure was determined by reconstructing the structure for the time period of the deposition of an early syn-tectonic conglomerate 1 (Fig. 9.17d) and for the time period of the deposition of a late syn-tectonic conglomerate 4 (Fig. 9.17c). Restoration of the initial growth stage shows the position of the southward-prograding fan delta that crops out extensively in the northern flank of the Oliana anticline. The end of the fan-delta deposition corresponds to the closure of the marine basin and deposition of the early Priabonian Cardona evaporite, the top of which formed an extensive, subhorizontal surface. Restoration of the late growth stage corresponds in time with the base of conglomerate 4, which was deposited unconformably across the growing structure. The crestal uplift rate of the structure was very slightly greater than the sediment accumulation rate. It resulted in syn-tectonic strata with gradually tapering beds that are incorporated and recognizable in the forelimb of the Oliana fold (Burbank and Verges, 1994).

A distinctly different syn-tectonic record develops in the case of fault-propagation folding (Fig. 9.15b). The background deposition across a growing fold is significantly reduced. This leads to the onlap of syn-tectonic strata on the backlimb of the fold and to the subsequent thinning of these strata. This also results in a truncation of the syn-tectonic strata on the forelimb. There is no growth sequence present over the fold crest, due to a greater uplift than was the case in the fault-bend folding case for the same amount of displacement. Pre-tectonic strata in the crestal area are quite deeply eroded. The onlap surface on the backlimb climbs through the syn-tectonic strata and becomes a truncation surface on the forelimb.

In the case of detachment folding (Fig. 9.15c), the low background deposition rate produces dramatic thinning and erosion on the frontal limb and the crest of the structure. The syn-tectonic record also includes a downdip re-deposition of the eroded material.

Fold structures when the growing structures are the only source of deposition
Fold structures in this scenario are quite different from the folds of the previous scenarios developed in marine environments. In the case of fault-bend folding (Fig. 9.18a), the pre-tectonic strata are deeply eroded, especially on the forelimb of the structure. As the forelimb always has a steeper dip than the backlimb in fault-bend folds, differential erosion rates, brought about by the contrasting limb dips, result in asymmetric erosion. Two wedges of syn-tectonic strata are located on the limbs of

(a)

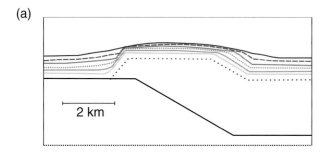

(b)

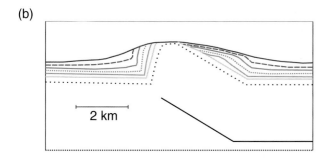

(c)

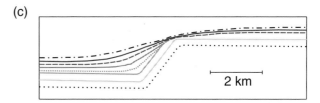

Fig. 9.15. Fold structures with low background deposition resulting in a growing structure producing faster local deposition (Hardy *et al.*, 1996). The models include a diffusion model for erosion, transport and deposition. Growth strata are recorded at 200 ka intervals. Total run time is 1 Ma. Base level rise during model run is 2 mm yr^{-1} and background deposition rate is 0.5 mm yr^{-1}. The diffusion model uses a diffusion coefficient of 3 m^{2}yr^{-1}. (a) Fault-bend fold model. Slip is constant from right to left during the model run at 1.5 mm yr^{-1}. (b) Fault-propagation fold model. Slip is constant from right to left during the model run at 1.5 mm yr^{-1}. (c) Detachment fold model. Figure shows the frontal part of the structure showing progressive limb rotation. Limb length is 2 km. A constant slip rate of 0.75 mm yr^{-1} (equivalent to 1.5 mm yr^{-1} for a symmetric detachment fold) was applied.

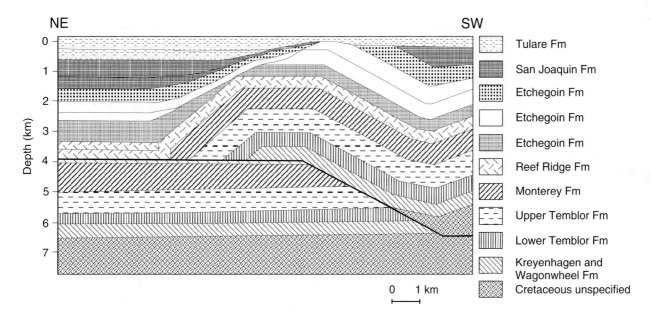

Fig. 9.16. The Lost Hills structure (Medwedeff, 1989).

the structure (Fig. 9.18a). Although both wedges were derived from the same fold, their stratal architectures are different. Strata on the forelimb consist of a series of foreland dipping and thinning beds, which onlap the structure. The backlimb strata consist of hinterland dipping and thinning beds, which offlap the structure and are eroded and truncated up-dip.

The Wheeler Ridge fault-bend anticline offers an ana-

logue for a sub-aerial fault-propagation fold in this scenario (Fig. 9.19a). Its uplift rate has been more rapid than the burial by syn-tectonic alluvial fan deposits during the last 125 ka (Medwedeff, 1992; Burbank *et al.*, 1996; Mueller and Suppe, 1997). Its syn-tectonic sediments were deposited as relatively small alluvial fans fed by drainage on the front limb of the fold and by much larger drainage basins, funnelling sediment through

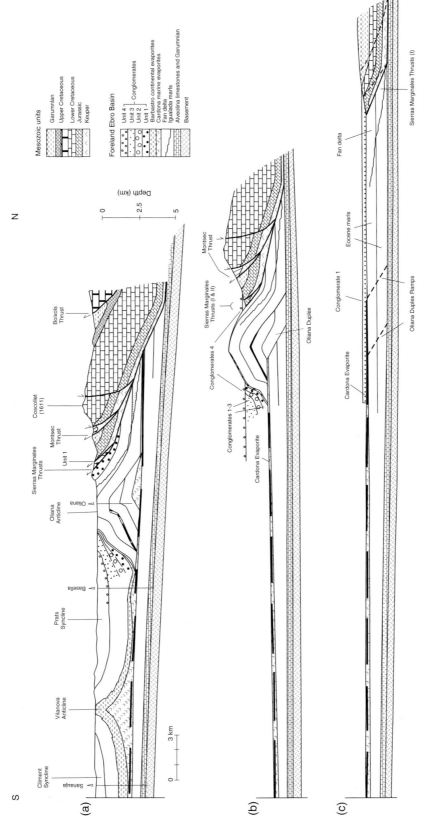

Fig. 9.17. Oliana fault–bend growth fold, the Ebro basin, southern Pyrenees (Burbank *et al.*, 1992). (a) Balanced cross section based on bore hole, seismic and surface data. (b) Restored cross section at the time of initial deposition of syn-tectonic conglomerate 4. The position of the Oliana duplex with respect to its translation across the footwall ramp and the height of the ramp itself are determined from altitudinal and deformational relationships presently visible along the base of conglomerate formation 4. Syn-tectonic conglomerate formations 1–3 have been deformed by both the Oliana duplex and the imbricate thrusts, whereas conglomerate formation 4 remains undeformed. (c) Restored cross section at the time of deposition of conglomerate formation 1, which is clearly involved in the imbricate host system. Whereas the several future ramps through the marls are specified here, the cutoff for the Mesozoic strata of the thrust sheet is unknown. Future thrusts are marked by dashed lines.

(a)

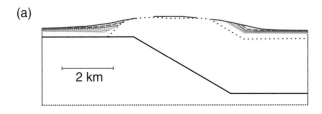

(b)

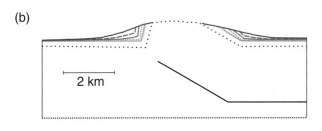

(c)

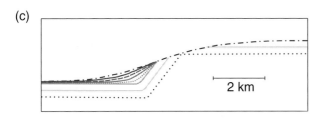

Fig. 9.18. Fold structures where the growing structure is the only source of deposition (Hardy *et al.*, 1996). Models are developed in a sub-aerial setting. Growth strata are recorded at 200 ka intervals. Total run time is 1 Ma. A diffusion model of erosion, transport and deposition is included, and there is no background deposition or base level rise. The diffusion model uses a diffusion coefficient of $3\,m^2\,yr^{-1}$. (a) Fault-bend fold model. Slip is constant from right to left during the model run at $1.5\,mm\,yr^{-1}$. (b) Fault-propagation fold model. Slip is constant from right to left during the model run at $1.5\,mm\,yr^{-1}$. (c) Detachment fold model. The figure shows the frontal part of the structure showing progressive limb rotation. Limb length is 2 km. A constant slip rate of $0.75\,mm\,yr^{-1}$ (equivalent to $1.5\,mm\,yr^{-1}$ for a symmetric detachment fold) was applied.

water and wind gaps, which cut the fold. There are regularly spaced terraces located on the front limb, typically 2–4 m wide and separated by 9–12 wide risers. Terraces at the base of the limb generally have sharp outer edges, which become increasingly more diffuse and eroded at higher positions on the limb. Other landforms formed on the front limb include elongated ridges of the alluvial sediment, which form distinctive corrugations. The crests of these features are elevated 6–8 m above adjacent areas on the fold limb (Fig. 9.19b).

The determined rate of crestal uplift is $3.2\,mm\,yr^{-1}$. It is considerably greater than the deposition rate of

$1.8\,mm\,yr^{-1}$ determined at locations close to the anticline (Medwedeff, 1992). The result of this rate combination is to form syn-tectonic strata, which pinch-out abruptly against the forelimb and display a strong offlap (Burbank and Verges, 1994), similar to that shown in Fig. 9.11(c).

In the case of fault-propagation folding, the pretectonic strata are also deeply eroded, especially on the crest (Fig. 9.18b). Two wedges of syn-tectonic strata are located on the limbs. Their beds offlap the structure and are truncated up-dip.

Typical cases of fault-propagation folds are known from the western part of the Ebro basin in the southern Pyrenees. Here, the uplift rates during the earlier part of the latest Eocene to Miocene were faster than burial by syn-tectonic continental sediments (Ori and Friend, 1984). The emergence of a group of fault-propagation folds (Fig. 9.3) is recorded by the coarse sedimentary breccias, which contain local material. They are located exclusively in the immediate vicinity of folds, as seen in outcrops.

Deep erosion on the front limb and crest is also seen in the case of detachment folding. The syn-tectonic strata form a wedge, which firstly thickens and then thins in the direction away from the rotating limb (Fig. 9.18c). A natural example of this type of detachment folding is the Penagalera detachment fold in the Catalan Coastal Range, Spain. The Paleogene uplift rate was more rapid than burial by syn-tectonic alluvial conglomerates (Burbank *et al.*, 1996; Fig. 9.20). Projections of differentially tilted conglomerate horizons across the anticlinal crest indicate syn-tectonic erosion of about 400–500 m in the crestal region. Carbonate breccias derived from local carbonate bedrock are preserved along the onlap contact. Whereas the contact between the growing fold and syn-tectonic conglomerates dips steeply, the conglomerate beds dip gently toward the contact and clearly indicate that the wall of carbonates represents a paleorelief progressively onlapped by the aggrading syn-tectonic conglomerates.

It should be noted that the ratio of the uplift to background deposition can change during the development history of a thrustbelt structure. An example of this phenomenon is the Mediano detachment fold from the southern Pyrenees (Fig. 9.21). Its growth is recorded by angular and progressive unconformities with syn-tectonic Eocene sediments (Holl and Anastasio, 1993). Figure 9.21 shows the western fold limb. Here, the syn-tectonic strata display an early growth stage when uplift was faster than background deposition followed by a later stage in which uplift was slower than deposition. The Mediano detachment fold does not necessarily indicate that the rate of the background deposition

(a)

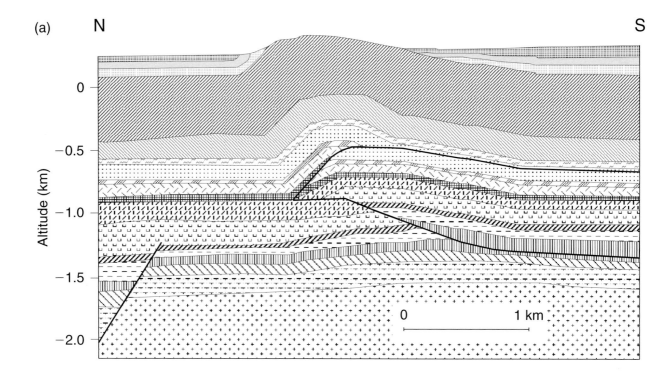

(b)

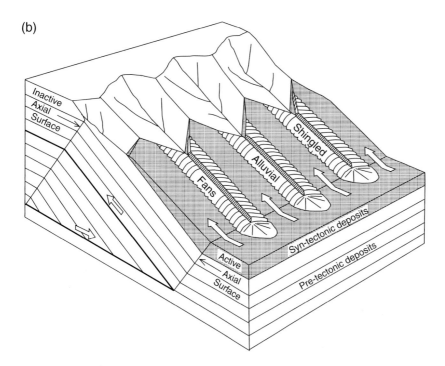

Fig. 9.19. (a) The Wheeler Ridge anticline, California (Medwedeff, 1992). The cross section shows the inferred structural and stratigraphic relationships using a wedge–thrust solution. The Wheeler ridge appears to be structurally complex and while Medwedeff (1992) interprets this structure as a wedge thrust, he acknowledges that the density of structural and stratigraphic data available is insufficient to uniquely identify the nature of the structure, the interpretation of which can have several alternatives. (b) A block diagram illustrating the architecture of alluvial sediments on the front limb of the Wheeler Ridge anticline (Mueller and Suppe, 1997). Note the shallow wedge thrust at depth that migrates to the north, continually incorporating alluvial fan deposits into the front limb. The unique aspect of the fluvial ridges is the incised drainages located along their crests.

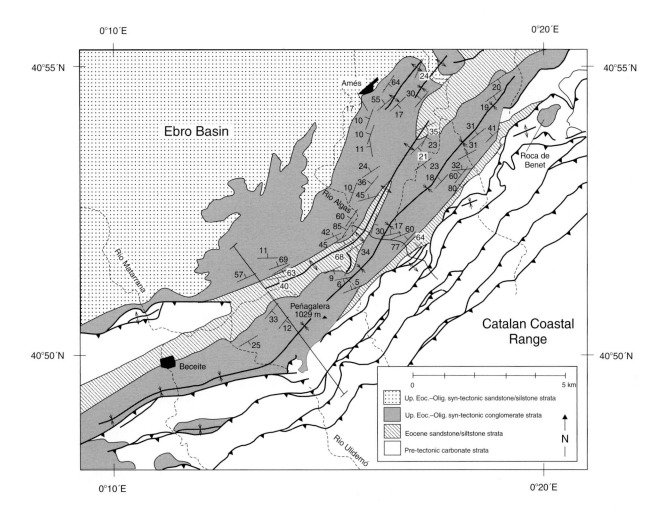

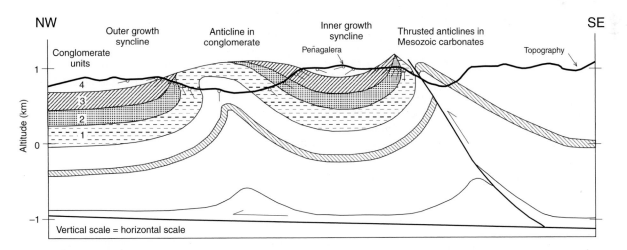

Fig. 9.20. Simplified cross section near Peñagalera in the Catalan Coastal Ranges, Spain, showing a detachment growth fold (Burbank *et al.*, 1996). Note the strong overturning of the forelimb of the Peñagalera detachment fold and the constant thickness of the conglomerate unit 1 across the structure. This indicates that growth of the detachment fold began after deposition of unit 1.

WSW   ENE

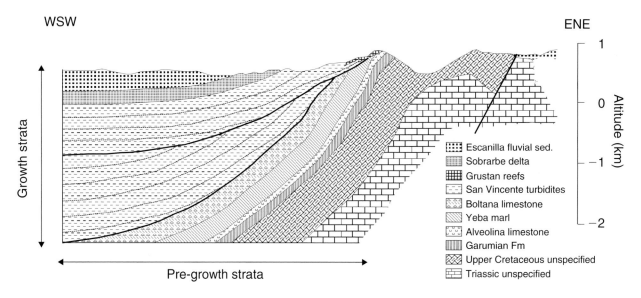

Fig. 9.21. Cross section of the western limb of the Mediano detachment fold, southern Pyrenees (Hardy *et al.*, 1996).

changed during the fold growth. The deposition rate could have remained constant, while the uplift rate of the detachment fold has changed, as indicated by the results of the numerical modelling of Hardy *et al.* (1996).

As pointed out earlier, the erosion rate is not a direct response to the uplift, but a response to the steepness of slopes of the relief created by the uplift. As indicated by the four fold growth scenarios described above, if the fault-bend, fault-propagation and detachment folds undergo the same uplift, they are affected by erosion differently. A fault-propagation fold with a sharp crest and steep limbs will be eroded more efficiently than either a fault-bend or a detachment fold with wider flat crests and shallower limbs (Figs. 9.12, 9.13, 9.15 and 9.18).

Just as the syn-tectonic erosion is affected by the inherent fold geometry of the thrustbelt structure, the amount of uplift is also affected by the inherent kinematics of the specific thrustbelt structure. It is a necessary feature of thrustbelt structures that the same amount of displacement along its detachment fault creates a different amount of uplift for different types of fault-related folds (Fig. 9.22a). The uplift response of various thrustbelt structures to shortening (Fig. 9.22a) can be divided into three categories. The first one includes those with a small uplift response to shortening, typical for basement uplifts, which have the smallest uplift response to the shortening. The displacement/ uplift ratio for this category is greater than 1. Basement uplifts with steep thrust faults have the greatest uplift response in this category. Uplifts controlled by listric

faults can have a higher uplift response during the later stages of the shortening.

The second category is characterized by a displacement/uplift ratio roughly equal to 1. This category includes fault-bend folding and various inversion structures associated with pre-existing normal faults. In the case of detachment folding with conservation of limb angles but with changing limb lengths, the ratio also oscillates around unity. Full inversion of normal faults causes larger uplift during the later stages of shortening compared with inversion, in which a short-cut thrust fault is developed through the footwall edge.

The third category contains fault-propagation folding and detachment folding with limb lengths conserved and limb angles changing during the fold development. These structures are characterized by a displacement/uplift ratio smaller than 1.

Uplift paths in relation to the shortening can also be grouped according to the shape of their curves in Fig. 9.22(a). Folds, where faulting precedes folding, and detachment folds with conserved limb angles and changing limb lengths, show uplift increasing proportionally with shortening from the origin, giving concave upwards curves. In structures, where faulting is contemporaneous with folding, the initial displacement causes an abrupt uplift. The later increments of displacement result in decreasing increments of uplift, so that uplift to displacement curves are convex upwards. The response of uplift to displacement along a detachment (Fig. 9.22a) is controlled by the hanging wall thickness (Figs. 9.22d and e) and the ramp angle (Figs. 9.22b and c). These two factors work differently in structures

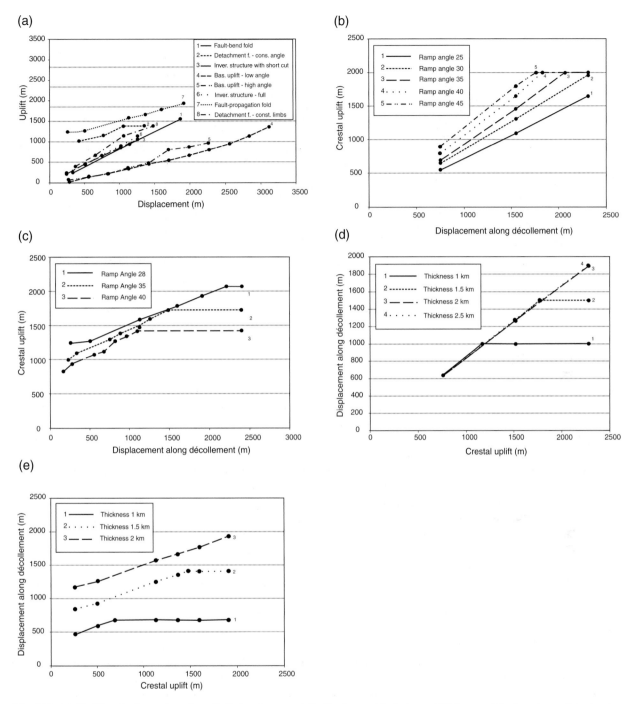

Fig. 9.22. (a) Uplift paths for various thrustbelt structures modelled by forward techniques. Thickness of the hanging wall is about 2 km. Fault-bend and fault-propagation ramp angles are 28°. Note that the uplift path for the detachment fold is higher for the case with varying limb lengths but conserved limb angles than in the case with constant limb lengths and varying limb angles. Note that basement uplift with steep listric thrust fault accelerates the uplift at later stages of the shortening in contrast to a basement uplift with a shallow thrust fault. (b) Uplift paths related to displacement along the décollement fault of a fault-bend fold controlled by increasing ramp angle. Each model is run for a 2 km thick sediment package. (c) Uplift paths related to displacement along the décollement fault of a fault-propagation fold controlled by increasing ramp angle. Each model is run for a 2 km thick sediment package. (d) Uplift paths related to displacement along the décollement fault of a fault-bend fold controlled by sediment package thickness. Each model is run for a ramp angle of 30°. (e) Uplift paths related to displacement along the décollement fault of a fault-propagation fold controlled by sediment package thickness. Each model is run for the ramp angle of 28°.

where faulting is contemporaneous with folding and structures where faulting predates folding.

In the case of fault-bend folding, the ramp angle affects the uplift until the hanging wall flat climbs on the footwall flat (Fig. 9.22b). The uplift value then remains the same during subsequent shortening. This maximum uplift value is different for different thrust sheet thicknesses. Before the maximum uplift is achieved, a linear relationship is developed, the higher the ramp angle is the steeper the displacement–uplift curve.

The ramp angle–uplift relationship for fault-propagation folding is different from the relationship for a fault-bend fold (Fig. 9.22c). The largest possible uplift for a shallow-dipping ramp in a fold-propagation fold is greater than the largest possible uplift for a steep ramp in a fault-bend fold. Each displacement–uplift curve for a fault-propagation fold starts with an abrupt increase of uplift with shortening, giving a convex upwards curve, whereas with a fault-bend fold, the initial relationship is linear (Figs. 9.22b and c). The only similarity between these two types of folding is the existence of a maximum possible uplift, beyond which further shortening cannot be accommodated. In the case of a fault-propagation fold, this is the point when the fault-propagation fold converts to a fault-bend fold and further shortening does not create further uplift (Fig. 9.22c).

The effect of thickness on fault-bend and fault-propagation folding also varies (Fig. 9.22d). The uplift–shortening curve for fault-bend folding is identical for all thicknesses, until the curves branch off at points when the hanging wall flats climb over footwall flats. At these points, for each thickness, each fold reaches its maximum uplift, proportional to the thickness of the sedimentary package. Uplift–shortening curves for fault-propagation folds are different. Each uplift increase with displacement reaches a point where fault-propagation folding stops or changes into fault-bend folding. Beyond this point further shortening does not produce further uplift. The above curves show greater amounts of uplift with greater hanging wall thicknesses.

10 Fluid flow in thrustbelts during and after deformation

The basic equation for porous flow under the effects of external loading, aquathermal pressuring, time and dehydration or precipitation, developed by Biot (1941), Bear (1972), Cooper (1966), Sharp (1983) and Shi and Wang (1986), in simplified form is

$$
\left(\frac{\alpha_n}{1-n} + n\alpha_s + n\beta\right)\frac{dp}{dt} = \frac{k}{\eta}\nabla^2(p - \rho_f\phi)
$$
$$
+ \left(\frac{\alpha_n}{1-n} + n\alpha_s\right)\frac{d\sigma_m}{dt} + [n(\gamma_f - \gamma_s) + (\gamma_s - \gamma_b)]\frac{T}{dt}
$$
$$
+ \frac{1}{1-n}\left(-\frac{\partial n}{\partial t}\right) + \frac{h}{\rho_f}, \tag{10.1}
$$

where p is the pore fluid pressure, ρ_f is the fluid density, ρ_s is the density of solid grains, $\alpha_n = -\partial n/\sigma_e$ is the porosity compressibility, $\alpha_s = 1/\rho_s(\partial\rho_s/\sigma_e)$ is the compressibility of solid grains, β is the fluid compressibility, n is the porosity, γ_f is the thermal expansion coefficient of the fluid, γ_s is the thermal expansion coefficient of the solid, γ_b is the bulk thermal expansion coefficient, σ_m is the total mean stress due to overloading and tectonic compression, σ_e is the effective mean stress ($\sigma_m - p$), $\phi = gd$ is the gravitational potential, g is the acceleration due to gravity, d is the depth below the surface, T is the temperature, k is the absolute permeability, η is the fluid viscosity and h is the fluid source or sink due to dehydration of clays and other causes. $d(\ )/dt$ denotes the derivative of a quantity with respect to time.

The term on the left-hand side of the equation is formed by the product of the rate of change in pore pressure with specific fluid storage. The first term on the right-hand side is the pore fluid diffusion term and the other terms on the right are various source terms. The relative magnitude of diffusion and the source terms determine the generation, maintenance or dissipation of pore fluid pressure. If the source terms are dominant, the pore pressure will increase. If the fluid supply and dissipation are equal, the pore pressure will be in a steady state. If the diffusion is dominant, the pore pressure will drop.

Depending on which flow driving factors dominate thrustbelts host various kinds of fluid flow mechanisms. Some of them are responsible for migration of large fluid volumes, some of them are capable of driving

only minor volumes. Their sources differ. Apart from rock permeability, fluid flow mechanisms drive fluids through fracture systems.

A discussion of fluid sources, fluid flow mechanisms and the competition between porous and fracture flow forms the focus of this chapter.

Fluid sources

With the exception of the first term, the terms on the right-hand side of equation (10.1) describe various fluid sources. The source described by the second term is controlled by porosity compressibility and porosity changes driven by rate of change of mean stress. The coupling between the pore fluid pressure, the mean stress and the porosity compressibility is defined by the equation $\alpha_n = -\partial n/\sigma_e$, which shows the porosity compressibility expressed as a porosity change driven by the mean stress reduced by the pore fluid pressure (Table 10.1, Fig. 10.1). The change in the mean stress is controlled by changes in the overburden and the tectonic stress.

The source described by the third term is controlled by the thermal expansion coefficients of the fluid and the rock, the bulk thermal expansion coefficient and the rate of change in temperature.

The fourth term describes the source controlled by porosity changes with time, which result either from a creep of the solid sediment skeleton or from chemical processes. Although this fluid source may be dependent on the mean stress and temperature, it is distinct from either the mechanical or the aquathermal mechanisms.

The fifth term describes either a sink or a source of fluid due to clay dehydration, hydrocarbon generation, cracking to gas, gypsum-to-anhydrite transformation or mineral diagenesis without dehydration (e.g., Osborne and Swarbrick, 1997). The importance of each of these sources is different.

The fluid source resulting from dehydration is typical for smectite, a common detrital component of shale, which contains abundant interlayer water in its crystal structure. The water released during simple dehydration expands in volume, because some of the interlayer water molecules are packed more closely than those of ordinary water (Osborne and Swarbrick, 1997). In highly overpressured rocks, smectite is stable as two or three water-layer complexes at temperatures below 200 °C

Table 10.1. Laboratory measurements of rock compressibilities of sedimentary rocks from uniaxial and triaxial experiments.[a] Compressibility can be also calculated from the bulk modulus: $\alpha = 1/K_b$ or from Young's modulus, E, and Poisson's ratio, ν: $\alpha = [3(1 - 2\nu)]/E$. *In situ* measurements for shale are different from laboratory measurements. It is because of inelastic behaviour, fracture porosity and volume scale effects. Therefore, listed values have to be taken as minimum estimates for large-scale geological applications. Water compressibility can be assumed to be nearly constant and equal to $4.4 \times 10^{-10}\,\text{Pa}^{-1}$.[b]

Rock	Confining stress (MPa)	Equivalent depth (m)	Poisson's ratio	Porosity (fraction)	Bulk grain (Pa^{-1})	Compressibility matrix (Pa^{-1})
Cretaceous Bearpaw shale	0.3	15	0.25	0.25	2.1×10^{-11e}	$4.0 \times 10^{-9*}$
Cretaceous Pierre shale	9.0	400	0.25	0.32	2.1×10^{-11}	5.7×10^{-9}
	24.0	1070	0.25	0.31	2.1×10^{-11}	1.1×10^{-9}
Pennsylvanian Caseyville shale	1.7	75	0.25	0.18	2.1×10^{-11}	6.0×10^{-9}
Devonian Chattanooga shale	0.0	0	0.11	0.016	2.1×10^{-11}	1.9×10^{-10}
	56.0	2470	0.27	0.016	2.1×10^{-11}	8.8×10^{-11}
	112.0	4875	0.42	0.016	2.1×10^{-11}	3.0×10^{-11}
Pennsylvanian Allegheny siltstone	0.0	0	0.08	0.018	2.1×10^{-11}	1.8×10^{-10}
	57.0	2500	0.13	0.018	2.1×10^{-11}	7.2×10^{-11}
	113.0	4875	0.16	0.018	2.1×10^{-11}	6.3×10^{-11}
Alberta Cretaceous sandstone	0.3	15	0.25	0.3	2.6×10^{-11}	2.2×10^{-9}
California Sespe graywacke	3.5	150	0.25	0.24	2.1×10^{-11}	3.2×10^{-10}
	28.0	1220	0.25	0.23	2.1×10^{-11}	1.2×10^{-10}
	90.0	3960	0.25	0.22	2.1×10^{-11}	4.9×10^{-11}
California Sespe arcose	3.5	150	0.25	0.18	2.1×10^{-11}	1.7×10^{-10}
	28.0	1220	0.25	0.18	2.1×10^{-11}	4.9×10^{-11}
	90.0	3960	0.25	0.17	2.1×10^{-11}	3.4×10^{-11}
Mississippi Berea sandstone	10.0	770	0.25	0.22	2.1×10^{-11}	1.76×10^{-10}
	20.0	1540	0.25	0.22	2.1×10^{-11}	9.9×10^{-11}
	30.0	2310	0.25	0.22	2.1×10^{-11}	7.7×10^{-11}
Colorado Weber sandstone	10.0	770	0.25	0.06	2.1×10^{-11}	5.0×10^{-11}
Colorado Kayenta sandstone	—	—	0.25	0.20	2.6×10^{-11}	5.2×10^{-11}
South Dakota Cretaceous Niobara chalk	0.0	0	0.15	0.083	1.4×10^{-11}	1.8×10^{-11}
	15.0	670	0.12	0.083	1.4×10^{-11}	2.6×10^{-11}
	29.0	1280	0.25	0.083	1.4×10^{-11}	2.7×10^{-11}
California sandy limestone	3.5	150	0.25	0.24	2.1×10^{-11}	6.2×10^{-11}
	28.0	1220	0.25	0.23	2.1×10^{-11}	2.6×10^{-11}
Mississippian Maxville limestone	0.0	0	0.23	0.009	1.4×10^{-11}	2.4×10^{-11}
	73.0	3350	0.25	0.009	1.4×10^{-11}	2.2×10^{-11}
	147.0	6400	0.27	0.009	1.4×10^{-11}	2.1×10^{-11}
Missouri Cambrian Bonneterre limestone	0.0	0	0.05	0.023	1.4×10^{-11}	1.4×10^{-11}
	98.0	4270	0.14	0.023	1.4×10^{-11}	6.5×10^{-11}
	196.0	8535		0.023	1.4×10^{-11}	4.2×10^{-11}
Pennsylvania Cambrian dolomite	—	—	0.30	—	1.4×10^{-11}	1.2×10^{-11}
West Virginia Blair dolomite	0.0	0	0.25	0.0	1.4×10^{-11}	1.2×10^{-11}
	100.0	4570	0.25	0.0	1.4×10^{-11}	1.2×10^{-11}

Note:

[a] Ge and Garven (1992).

[b] Freeze and Cherry (1979).

[*] *In situ* measurement.

[e] estimated.

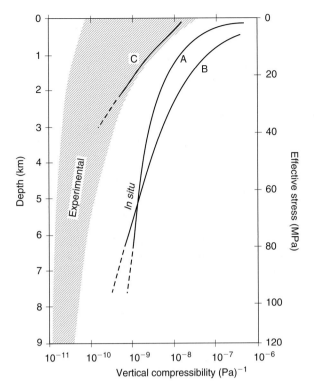

Fig. 10.1. Experimental and *in situ* compressibility data as a
function of depth or confining pressure (Ge and Garven, 1992).
The labelled curves are derived from observed porosity–depth
trends in sedimentary basins. Curve A is for Cenozoic mud and
sand in the Gulf Coast, curve B is for Cenozoic mud and sand
in the North Sea and curve C is for Paleozoic shale and
sandstone in north central Oklahoma. Laboratory
measurements of bulk compressibility can be orders of
magnitude smaller than effective long-term values derived from
in situ porosity–depth relations.

(Colten-Bradley, 1987). Under high effective stresses,
the first water layer is expelled at temperatures below
60 °C, the second one at a temperature of 67–81 °C and
the third one at a temperature of 172–192 °C (Colten-
Bradley, 1987). Because this fluid source requires high
effective stresses and is inhibited by overpressure development in shale, it has only a minor importance compared with other fluid sources.

The release of structurally bound water together with
Na, Ca, Mg, Fe and Si ions from smectite also occurs
during its transformation to illite by the addition of Al
and K ions. The problem with the evaluation of the
importance of this fluid source, however, is that this
chemical reaction and its volumetrics are not fully
understood (Osborne and Swarbrick, 1997). This transformation also changes the sediment rheology, because
the smectite clay framework collapses and the water
release increases the sediment compressibility. The smec-

tite-to-illite transition may be a more important fluid
pressure source than clay dehydration, because ions
released in the reaction may precipitate as cements and
enhance the pressure buildup (e.g., Bekins *et al.*, 1994;
Osborne and Swarbrick, 1997). For example, in the Gulf
Coast of the USA (Bruce, 1984), or in the Devonian
shale in the Appalachian basin (O'Hara *et al.*, 1990),
there is a close regional relationship between the onset of
overpressure and the smectite-to-illite transformation.

Pore pressure controlled by volume changes during
kerogen maturation varies widely depending on
assumptions made about gases generated and the
density of the residual kerogen and pyrobitumen. There
are results indicating both volume increase (e.g.,
Meissner, 1978) and decrease (e.g., Ungerer *et al.*, 1983).
The effect of maturation on the fluid pressure is most
pronounced when large volumes of liquid hydrocarbons
are generated within a short time period in low-
permeability rocks, where primary migration is difficult.
This is relevant for any of the conversions from solid to
liquid mentioned earlier, such as clay dehydration. If the
fluid does not escape, the change in rheology of the
mudrock during conversion of solid kerogen into liquid
hydrocarbons produces disequilibrium compaction in
the source rock. The amount of compaction depends on
the effective permeability. Therefore, any related pore
pressure buildup is the result of disequilibrium compaction rather than volume expansion (Osborne and
Swarbrick, 1997).

However, the maturation reaction rate is controlled
by pore pressure buildup. Increased pore pressure slows
down the organic metamorphism, as indicated by laboratory experiments (Price and Wenger, 1992). However,
the additional factor in this slowdown is the source rock
permeability. Shales with permeability of the order of a
microdarcy do not hinder petroleum migration (Burrus
et al., 1994). Laboratory experiments further indicate
that hydrocarbons, water and minerals are all in rever-
sible redox equilibrium and can exchange oxygen and
hydrogen with each other when reaction conditions
change (Seewald, 1994). For example, overpressured
basins in China have source rocks that are anomalously
immature considering their temperature history (Fang
et al., 1995). All the evidence indicates that maturation
itself is a pore pressure source of uncertain importance.
Only disequilibrium compaction triggered by solid-to-
liquid conversions can be an important source.

Methane gas is generated biogenically within sediments during shallow burial at temperatures below
80 °C (Barker, 1987). In addition, gas hydrates can form
in relatively shallow reservoirs at low temperatures and
high pressures (e.g., Shipley *et al.*, 1979; Kulm and
Suess, 1990; Moore *et al.*, 1990c; Yamano *et al.*, 1992;

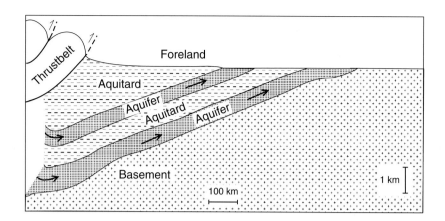

Fig. 10.2. Topographically driven groundwater flow in a foreland basin.

Ferguson *et al.*, 1993). As burial increases and the temperature rises to 21–27 °C, hydrates become unstable and gas is released, which results in a volume increase (e.g., Davidson *et al.*, 1978; Hunt, 1979; Kvenvolden and McMeanamin, 1980; Martin *et al.*, 1996). At much higher temperatures of 120–140 °C, oil converts to lighter hydrocarbons and ultimately methane by thermal cracking, which also results in a volume increase. Almost complete cracking to gaseous hydrocarbons occurs at temperatures in excess of 180 °C (Hunt, 1979; Tissot and Welte, 1984). There are several basins where the distribution of overpressure is correlated with assumed zones of oil cracking, including the Triassic and Jurassic reservoirs in the northern and central North Sea or the Jurassic Smackover reservoir of Mississippi and Alabama (Osborne and Swarbrick, 1997). The major uncertainty in determining the importance of this pore fluid pressure source is that the increased pore fluid pressure most likely retards the gas generation/oil cracking (Osborne and Swarbrick, 1997).

The transformation of gypsum to anhydrite occurs at a temperature of 40–60 °C and results in the loss of 39% of the bound water (Jowett *et al.*, 1993). It can be a potentially strong source of pore pressure buildup, but occurs during shallow burial.

Diagenesis itself can decrease or increase sediment porosity. Growth of cement in the rock pores reduces the pore volume and can increase the pore fluid pressure. For example, up to 30% by volume of a deeply buried reservoir rock in the North Sea can consist of diagenetic minerals (Giles *et al.*, 1992). Dissolution of minerals in a rock increases the pore volume and can result in a decrease of pore pressure. The ability of diagenesis to control the pore fluid pressure depends on whether cementation or dissolution occurs in a closed or an open setting (Bjorlykke, 1984; Gluyas and Coleman, 1992). Certain diagenetic reactions are controlled by pore fluid pressure, which makes the problem more complex. For example, overpressure tends to

inhibit quartz cementation where the major source of silica is from pressure solution. When the pore pressure increases, it tends to inhibit mineral precipitation (Byerlee, 1993), and it decreases the effective stress (Osborne and Swarbrick, 1997).

Major fluid flow mechanisms

Fluid flow driven by topography or gravity
If precipitation and infiltration of water in regions of high thrustbelt elevation are sufficient to recharge the water table, a continuous supply of groundwater is available to maintain flow. The time constant t_{ch}, which describes the diffusion of transients in regional flow systems, is (Lachenbruch and Sass, 1977)

$$t_{ch} = \frac{L^2}{4D},$$ (10.2)

where L is the diffusion distance and the hydraulic diffusivity, D, is given by

$$D = \frac{K}{S_s},$$ (10.3)

where K is the hydraulic conductivity and S_s is the sediment specific storage. The time constant t_{ch} is usually several orders of magnitude lower than the characteristic time scale of orogenic processes, which build topographic gradients. Therefore, the flow system can be characterized as a quasi-steady-state system (Deming, 1994b). Topographically driven fluid flow in such a system occurs nearly everywhere where there is topographic relief. Typical examples are foreland basins (Fig. 10.2; Garven and Freeze, 1984b). In such a foreland basin setting, the recharge takes place in the mountain range and its foothills. The fluid moves downward and out into the neighbouring foreland basin in the direction of decreasing hydraulic head. The moving fluid seeks a path of least resistance or a combination of the highest hydraulic conductivity with the highest head gradient. This may be the lowest sedimentary sequence in a basin. Total transport distances of several hundred to a thousand kilometres are possible, characterizing such a path. For example, isotopic and trace-element analyses of saline water discharging from natural springs and artesian wells in central Missouri show that fluids most likely originated as meteoric recharge in the Rocky Mountains, located about 1000 km to the west (Banner *et al.*, 1989). The existence of regional, topographically driven groundwater flow systems is also documented in several basins in western North America, including the Kennedy basin (Bredehoeft *et al.*, 1983), the Denver basin (Belitz and Bredehoeft, 1988) and the Big Horn basin (Bredehoeft *et al.*, 1992). Flow velocities in the topographically driven system are directly proportional to the head gradient and the rock permeability, according to Darcy's law described by equation (10.22). Deming *et al.* (1992) inferred average Darcy velocities in the North Slope basin, Alaska, as being of the order of 10 cm yr^{-1} by linking near-surface heat loss and gain to the mass flux through the basin. Average Darcy velocities in the Great Artesian basin, Australia, are similar, as interpreted from dating of radioactive chlorine isotopes (Cathles, 1990). The relatively high fluid velocities associated with topographically driven flow systems indicate that they are efficient transport mechanisms for both heat and mass.

Fluid flow driven by sediment compaction

Sediment compaction described by the second right term in equation (10.1) causes a porosity decrease. Porosity decreases under increasing effective stress σ_e. An empirical relation between the porosity n of shale, unaffected by tectonic stress, and the burial depth, given by Athy (1930) is

$$n = n_0 \exp(-b\sigma_e), \tag{10.4}$$

where $b = 1/(\rho - \rho_f)gH$, n_0 is the porosity at the surface, ρ is the bulk density, ρ_f is the fluid density, g is the acceleration due to gravity and H is a material parameter. The bulk density of a porous medium can be calculated from (Shi and Wang, 1986)

$$\rho = \rho_f n + \rho_s(1 - n), \tag{10.5}$$

where ρ_f and ρ_s are the fluid and the sediment densities, respectively. The exponential relation (10.4) is valid for sandstone (e.g., Hoholick *et al.*, 1984) and carbonates (e.g., Schmoker and Halley, 1982). The value b typically ranges from 10^{-8} to 10^{-7} Pa^{-1} for shale and from 5×10^{-9} to 5×10^{-8} Pa^{-1} for sandstone (Athy, 1930; Hoholick *et al.*, 1984; Fig. 10.3). Figure 10.3 shows that data trends do not correlate with lithology, indicating that some factor other than lithologic characteristics,

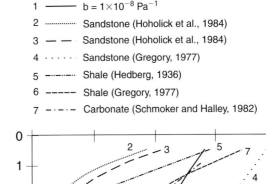

1 ——— $b = 1 \times 10^{-8}$ Pa^{-1}
2 ·········· Sandstone (Hoholick et al., 1984)
3 — — Sandstone (Hoholick et al., 1984)
4 · · · · · Sandstone (Gregory, 1977)
5 —·—·— Shale (Hedberg, 1936)
6 — — — Shale (Gregory, 1977)
7 —· —· — Carbonate (Schmoker and Halley, 1982)

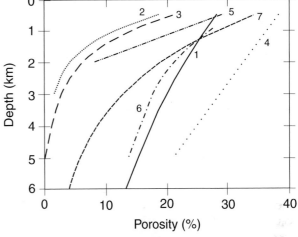

Fig. 10.3. Porosity compaction as a function of depth for six rocks in tectonically undisturbed basins and porosity as a function of depth, assuming hydrostatic pore pressure and using the equation (10.4) (Smith and Wiltschko, 1996).

such as the rock age or the degree of lithification may control the quantity. This is frequently the case when the rocks involved in thrustbelts are lithified prior to their shortening. Therefore, they have a low porosity reduction rate with increasing effective stress, while accretionary prisms are characterized by a high rate.

Studies of modern accretionary prisms show that tectonic compaction plays a significant role in total compaction (Fig. 10.4; Morgan and Karig, 1995) and should modify b values calculated only from burial. Values of layer-parallel shortening, which is a result of straining, increase with depth and can reach values as high as 40–50%, as shown by sandbox experiments (e.g., Koyi, 1995, 1997). The relationship between the porosity n and the effective stress σ_e is also dependent on the loading path (e.g., Lambe and Whitman, 1969). If the active effective stress is smaller than the pre-existing maximum effective stress, the porosity n will change much more slowly with effective stress than during normal consolidation or 'virgin compression'. Thus in the case of burial and no tectonic stress Athy's function (10.4) is modified to

$$n = n_{max} \exp[-b'(\sigma_e - \sigma_{max})]. \tag{10.6}$$

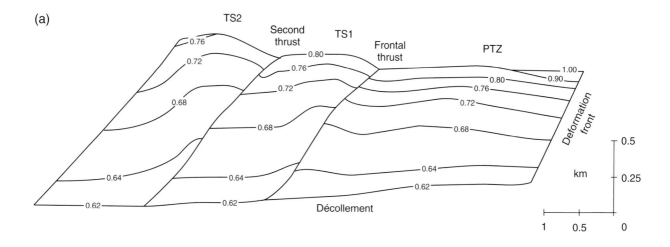

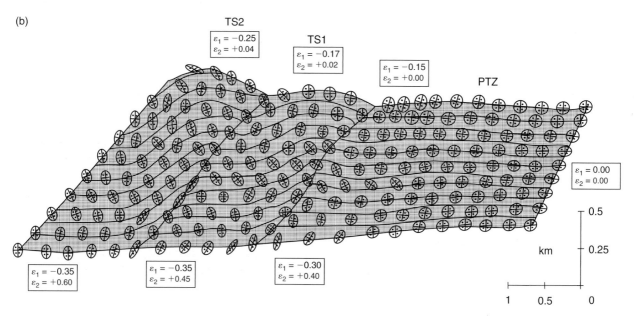

Fig. 10.4. Profile through the Nankai accretionary prism (Morgan and Karig, 1995). (a) Volume ratios determined along the profile. The final volume ratio distribution was determined iteratively as the diffuse strain field was calculated by a technique which integrates porosities measured in the drill cores and estimated from seismic interval velocities, and distortional strains indicated by changes in bed thickness on seismic depth profiles. Derived ratios are corrected according to assuming no volume change along the décollement. (b) Distribution of strains along the profile. Calculated strain ellipses are plotted with element streaklines, or sediment burial paths. Superimposed on the strain ellipses are line elements that were initially horizontal and vertical in the undeformed configuration. Average horizontal and vertical tectonic strains ε_1 and ε_2 are also shown near the top and base of the accretionary prism. Near-surface sediments show little vertical extension. Deeper sediments show significant extension, particularly in trailing edges of the protothrust zone and thrust sheets. Shear strains are not well resolved by this method, introducing irregular rotations of strain ellipses.

The contribution of compaction in pore pressure generation can be resolved from equation (10.1):

$$dp = \frac{\alpha_n}{\alpha_n + n\beta(1-n)} d\sigma_m. \qquad (10.7)$$

When the porosity of sediments decreases due to compaction, pore fluids, which were originally present in the near-surface sediments, are expelled. Sandstones, for example, compact from about 39–49% porosity at deposition (Lundegard, 1992) to about 15–25% porosity at depths of 2–3 km due to grain rearrangement and some chemical dissolution at the grain contacts (Sclater and Christie, 1980). At greater depths there is little potential for a reduction in their porosity due to

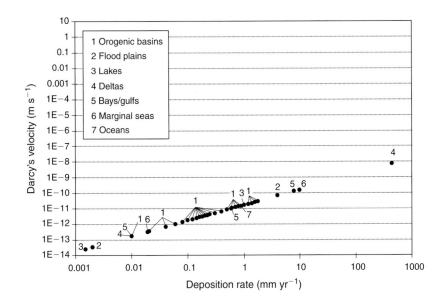

Fig. 10.5. Calculation of Darcy velocity from sediment compaction by the 'waste basket' calculation of Deming (1994b). The surface porosity is chosen arbitrarily as 50% and deposition rates are taken from Kukal (1990).

mechanical compaction. However, a distinct porosity reduction due to diagenetic cementation can occur. The depositional porosities of clays, for example, are 65–80% (Sclater and Christie, 1980). Clays compact by grain rearrangement and ductility at great depths (4–6 km), where the porosity can be reduced down to 5–10%.

An idea about average flow rates of expelled fluids from compacting sediments can be obtained from a simple 'wastebasket' calculation (e.g., Deming, 1994b; Fig. 10.5). Unfortunately, it assumes compaction only by burial, omitting tectonic compaction. Therefore, the calculated flow rates in thrustbelts are underestimated. Similar calculations for accretionary prisms have been made by Bray and Karig (1985), Foucher *et al.* (1990a, b), Langseth *et al.* (1990), Le Pichon *et al.* (1990b) and Screaton *et al.* (1990). The calculation uses an imaginary column or wastebasket with cross-sectional area, A, which is constructed from the surface to the bottom of the sedimentary basin. Sediment enters the wastebasket with a velocity, R, equal to its deposition rate. The total mass accumulation rate is RA and the total fluid accumulation rate is RAn_0, where n_0 is the porosity at the surface. The maximum possible fluid velocity due to sediment compaction is equal to the total fluid accumulation rate RAn_0 divided by the area A (Fig. 10.5). For a typical range of deposition rates in foreland basins of 0.1–1 mm yr^{-1} (Table 9.1), Darcy velocities reach a maximum value of 0.05 cm yr^{-1}. These velocities are more than two orders of magnitude slower than those for topographically driven flow. The decrease of porosity with depth also causes the quantity of pore fluid available for deep circulation to become progres-

sively limited. However, even in the case of a depositional rate as low as 0.005 mm yr^{-1}, the pore compaction is approximately two orders of magnitude more important for driving fluid flow than aquathermal pressuring (Deming, 1994). It should be stressed that compaction-driven fluid flow in thrustbelts is always associated with compression-driven and topographically driven flows, resulting in a fluid flow of the order of centimetres to metres per year.

Permeability anisotropy in sedimentary basins has important consequences for compaction-driven fluid flow, when sediment layers have contrasting permeability. Flow across layers is controlled by the lowest permeability in the sedimentary sequence, while flow parallel to bedding planes is controlled by the highest permeability. Pore fluids expelled from compacting shales are likely to travel vertically until they encounter a reservoir rock (Bethke, 1985). Where reservoirs are isolated within shale sequences, flow is predominantly vertical (Mann and Mackenzie, 1990). Where reservoirs are areally extensive and interconnected, fluid and fluid pressure may be transferred laterally (e.g., Magara, 1976; Mann and Mackenzie, 1990). Such a lateral flow is known from the South Caspian basin, where field data document the regional-scale compaction-driven flow (Bredehoeft *et al.*, 1988). The basin has a Cenozoic deposition rate as high as 1.3 mm yr^{-1}. It has a large sediment thickness and evidence for overpressure decreasing towards the basin margin (Figs. 10.6a and b). Sandstone beds in its fill act as drains for both the underlying and overlying compacting shale layers (Fig. 10.6c).

Disequilibrium compaction is determined as the mechanism generating overpressure in a number of basins, involving thick clay, mud, marl and shale successions during rapid burial. Examples come from the Gulf Coast (Dickinson, 1953), the South Caspian basin (Bredehoeft *et al.*, 1988), the Mahakam delta (Burrus *et al.*, 1994), the Williston basin (Burrus *et al.*, 1994) and the North Sea (Mann and Mackenzie, 1990; Audet and McConell, 1992). Overpressuring acts against compaction and if cementation is negligible, it can preserve higher sediment porosities than expected (Osborne and Swarbrick, 1997).

Fluid flow driven by buoyancy forces resulting from density gradients due to temperature or salinity gradients

By rewriting equation (10.19) (derived later in this chapter) to give (Deming, 1994b)

$$h = z + \frac{p}{\varrho_{f0} g}, \qquad (10.8)$$

an equivalent hydraulic head can be defined with reference density ρ_{f0} such as, for example, an equivalent fresh water hydraulic head. The first and second terms on the right-hand side of this equation are elevation and pressure terms for the hydraulic head h, p is the pore fluid pressure and g is the acceleration due to gravity.

Equation (10.20) (derived later in this chapter) for the volumetric flow rate q, also can be rewritten, yielding (Deming, 1994b)

$$q = -\left(\frac{k\rho_{f0} g}{\eta}\right)\left[\nabla h + \left(\frac{\rho_f - \rho_{f0}}{\rho_{f0}}\right)\nabla z\right], \qquad (10.9)$$

where k is the permeability tensor, ρ_f is the fluid density and η is the fluid dynamic viscosity. Thus equation (10.9) shows that fluid flow results not only from a head gradient ∇h, but also from fluid density gradients, which give rise to buoyancy forces. These density gradients result largely from thermal expansion or salinity gradients (e.g., Hanor, 1979; Frape and Fritz, 1987; Ranganathan and Hanor, 1987).

For the simplified case of a one-dimensional homogeneous porous medium confined between two impermeable boundaries, free convection due to thermal expansion occurs if the Rayleigh number, R, exceeds a critical value of 27.1 (Turcotte and Schubert, 1982)

$$R = \frac{\alpha g \rho_f^2 C k \Delta z^2 \gamma}{\eta \lambda}, \qquad (10.10)$$

where α is the fluid coefficient of thermal expansion, g is the acceleration due to gravity, ρ_f is the fluid density, C is the heat capacity of the fluid, k is the permeability of the porous medium, Δz is the height or thickness of the convecting cell, γ is the thermal gradient, η is the

(a)

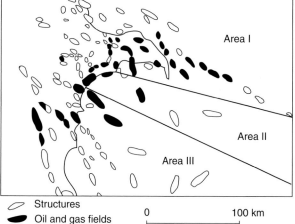

Structures
Oil and gas fields

0   100 km

Fig. 10.6. Regional pore pressure study in the South Caspian basin (Bredehoeft *et al.*, 1988). (a) Area of the investigation near the Baku Archipelago. Areas I–III show differing sand/shale ratios (1:1, 3:7 and 1:9) and pore-pressure regimes.

fluid dynamic viscosity and λ is the thermal conductivity of the porous medium. Thermal expansion, α, is pressure and temperature dependent, but its variation is small. For example, for water in the upper 4 km of a thrustbelt, an average value of $5 \times 10^{-4}\,°C^{-1}$ can be used. The thermal conductivity, λ, of sedimentary rocks ranges from 1 to $7\,W\,m^{-1}\,°C^{-1}$ as described later in Chapter 12.

However, one can consider the increase of fluid density with depth that occurs as the salinity increases (e.g., Hanor, 1979; Frape and Fritz, 1987; Ranganathan and Hanor, 1987). It is documented from sedimentary basins where data show that the salinity increases with increasing depth, although the rate of increase tends to decrease as the depth increases (Hanor, 1979). In the presence of a flat salinity gradient it is possible that the geothermal gradient may be sufficiently high to lead to a density inversion and convective overturn if the average crustal permeability is high enough (Deming, 1994b).

Rayleigh-type analyses assume convection between two plates, which are exactly parallel to each other and perpendicular to the gravity vector, i.e. an assumption that is violated in thrustbelts. In porous media that are tilted with respect to the gravity vector by more than about 5°, convection always occurs. Thrustbelts are examples where circulation may be common, although its presence would not be indicated by a Rayleigh-type analysis (Criss and Hofmeister, 1991).

The full form of Darcy's law from equation (10.9)

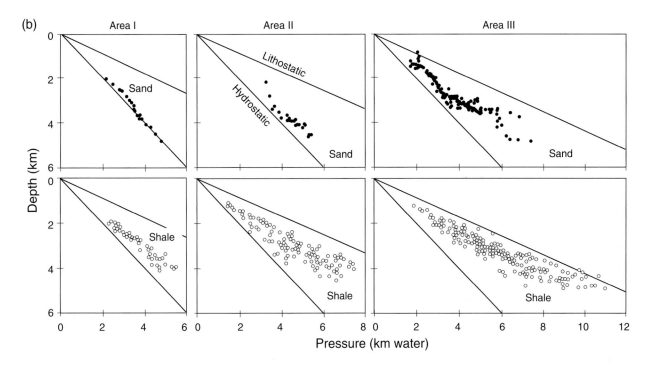

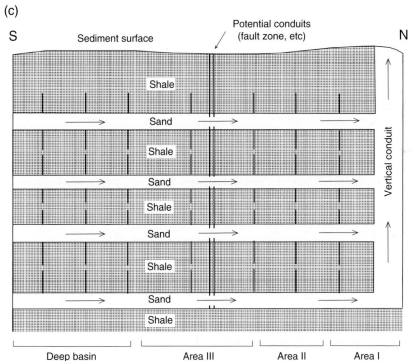

Fig. 10.6. (*cont.*) (b) Pore pressure versus depth for both shale and sandstone in areas I–III. (c) Diagrammatic cross section of a compartmentalized fluid flow system.

implies that vertical and lateral flow must happen any time a nonzero lateral fluid-density gradient exists. It is evident that density-driven flow is probably both pervasive and perpetual through the upper crust, although its magnitude may be small in comparison to topographically driven flow.

There are unique thrustbelt settings in which density gradients dominate over hydraulic gradients to drive the fluid, such as salt bodies in toe thrusts. In this setting the density gradient arises from the high thermal conductivity of salt relative to the surrounding sediments and from changes in salinity due to dissolution of the salt

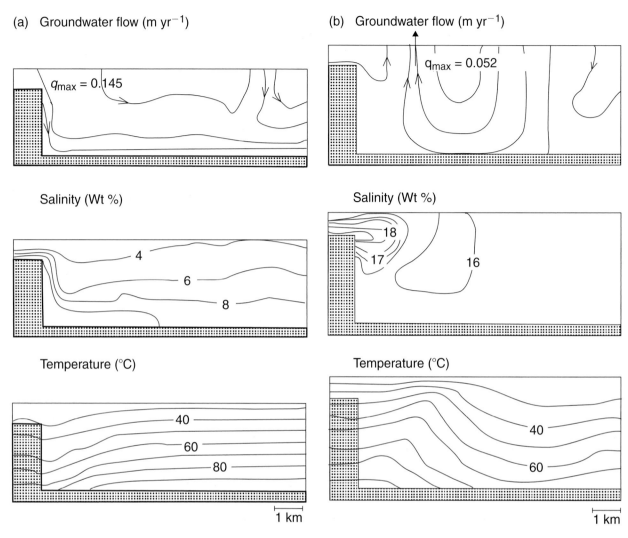

Fig. 10.7. (a) Streamlines, salinity contours and isotherms after 10 Myr for a geological scenario with a salt dome, in which the salt (shaded) builds into a homogeneous medium and ambient groundwater in surrounding sediments has a low salinity (Evans *et al.*, 1991). Downward buoyancy forces resulting from high salinities in the crestal area control the groundwater flow. Isotherms are depressed near the salt dome leading to low surface heat flow above the dome. The upper boundary condition is the specified head, which drops 4 m off the flank of the dome. Modifying head or recharge on the upper boundary did not affect the general flow patterns. (b) Similar simulation of results after 14 Myr, but the diapir builds into sediments containing groundwater with high background salinity. Model parameters and hydraulic boundary conditions are the same as in (a). When the background salinity is high, thermal effects control groundwater density leading to free convection near the salt dome.

itself (Evans *et al.*, 1991). Examples of preferential dissolution of salt dome crests are known from the Gulf of Mexico, caused by meteoric groundwater flow and the fact that dome flanks are usually protected by shale sheets (Atwood and Forman, 1959; Bodenlos, 1970). The dissolution creates a salinity plume above the less dense and gravitationally unstable deeper groundwater.

In order to evaluate the relative importance of different mechanisms driving groundwater near salt domes, Evans *et al.* (1991) numerically modelled geological scenarios including situations when the salinity of ambient groundwater was either low (Fig. 10.7a) or high

(Fig. 10.7b). Their models coupled groundwater flow, heat transport and transport of dissolved salt. The results indicate that a large salinity gradient is established as a result of salt dissolution in the case of low background salinity (Fig. 10.7a). The density gradient drives groundwater flow laterally into the basin, which in turn carries dissolved salt and serves to increase the salinity rapidly throughout the radial basin. Groundwater transports the heat downward, which leads to a low surface heat flow above and adjacent to the salt column. When the salt dome intrudes into sediments containing high salinity groundwater, then the thermal effects near the salt dome become relatively more important (Evans and Nunn, 1989). In order to investigate the trade-off between thermal and salinity effects, modelling at varying initial background groundwater salinity indicates that when the salinity exceeds 15 weight %, thermal effects become dominant and result in upward groundwater flow near the salt column (Evans *et al.*, 1991; Fig. 10.7b).

Minor fluid flow mechanisms

Minor fluid flow mechanisms have lower flow rates than the major mechanisms. This difference is sometimes difficult to evaluate as in the case of compression-driven flow, which can be hard to separate from topographically and compaction-driven flows. Sometimes it is possible to find specific examples of temporarily or spatially restricted situations in which the minor fluid flow mechanism can play a significant role.

Fluid flow driven by compression

A fluid flow mechanism driven by compression is in practice a more complete version of the sediment compaction-driven flow, including components of both vertical and horizontal compaction. This hydrodynamic model is based on a concept in which thrust sheets act like 'squeegees' expelling pore fluids out of foreland basins and into the platform margin (Oliver, 1986). Stable isotope data from the Canadian Cordillera and southern Pyrenees indicate that related fluid expulsion occurs consistently from hinterland to foreland and each thrust sheet forms a separate hydrodynamic unit (Bradbury and Woodwell, 1987). Fluid inclusion data from the Central Appalachians indicate a temperature perturbation (Dorobek, 1989) that may imply rapid fluid expulsion driven by this mechanism. A two-dimensional poroelastic model, which couples fluid flow and tectonic stress as sedimentary strata are compressed and loaded by advancing thrust sheets (Ge and Garven, 1992), indicates that tectonic loading can generate an excess pressure within a foreland basin. It determines time periods needed for this excess pressure dissipation and flow patterns. In addition to setting up both stress-induced and topographically driven fluid migration systems, compression of the foreland results in the development of fracture networks that increase regional permeability of basin strata.

The fluid flow and pressure models of Ge and Garven (1992) simulating compression-driven flow use a simplified foreland basin scenario, which includes only a basal aquifer and an overlying aquitard (Figs. 10.8a and b). The initial fluid flow and pressure, prior to loading, are controlled solely by topography (Figs. 10.8c and d). Hydraulic head contours have elevation values representing the point at which they intersect the water table

surface. The presence of a basal aquifer clearly focuses fluid flow between the elevated recharge region and the discharge region at the platform margin (Fig. 10.8d). After the first 5 years following the onset of orogenic loading, the hydraulic head contours become highly distorted near the thrustbelt front (Fig. 10.8e). A high pore pressure zone is developed in this region and large flow velocities, an order higher than the initial ones, are concentrated in the basal aquifer (Figs. 10.8e and f). After about 1000 years, the high-pressure zone is still visible, although hydraulic head contours become much smoother as the overpressure dissipates (Fig. 10.8g). The fluid velocity pattern shows that the discharge area is much larger than in the initial steady state (Fig. 10.8h), indicating that tectonic compression can induce substantial changes in regional flow patterns and rates. Most of the excess fluid pressure dissipates after about 5000 years and the fluid regime gradually approaches a steady state similar to the initial one (Figs. 10.8i and j). Darcy velocities of the fluid in both the basal aquifer and in the rest of the basin show a relatively rapid buildup after the orogenic loading and a slow decrease with time.

The fluid pressure and flow regimes are affected by hydraulic conductivity, which controls the excess pore pressure dissipation as described by equations (10.1), (10.23) and (10.24) (derived later in this chapter). Except for the hydraulic conductivity, K, the fluid diffusion time, t_d, is controlled by the length of the strata, L, and the sediment specific storage, S_s (Ge and Garven, 1992)

$$t_d = \frac{L^2 S_s}{K}. \tag{10.11}$$

The fluid pressure and flow regimes are affected by compressibility, which, if large, allows for greater pressure buildup at the onset of orogenic loading because of increased deformation. The relationship between the excess pore fluid pressure, p, and the mean stress, σ_m, is (Ge and Garven, 1992)

$$p = \rho g h = \frac{\sigma_m \alpha_n}{\alpha_n + n\beta}, \tag{10.12}$$

where α_n and β are the porosity and fluid compressibilities, n is the porosity, ρ is the bulk density of the porous medium, g is the acceleration due to gravity and h is the hydraulic head.

With a larger effective compressibility, hydraulic diffusivity decreases for the same hydraulic conductivity, which results in longer transient flow periods. A more compressible rock accommodates a greater deformation for the same level of loading. The fact that geological processes have long time scales results in an increase of deformation, with elastic deformation being

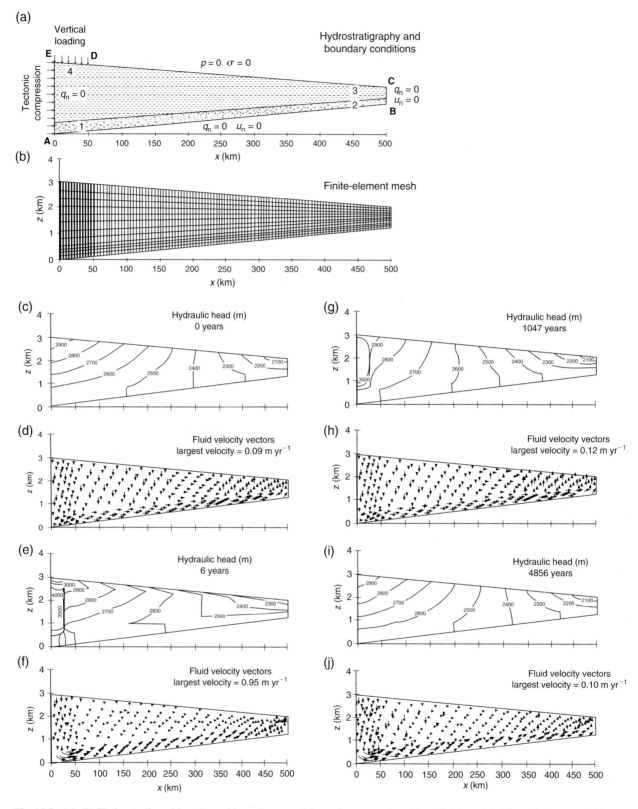

Fig. 10.8. (a), (b) Hydrostratigraphic units and boundary conditions of compression-driven fluid models (Ge and Garven, 1992). (c–j) Numerical models of hydraulic head distribution and fluid velocity fields at several time periods after orogenic loading for the basin characterized by basal aquifer and overlying aquitard from (a).

accompanied by inelastic processes such as creep. The effect of equal loading on a more compressible and less diffusive sedimentary system generates greater excess pore pressures and longer transient flow periods (Figs. 10.9a and b).

Fluid pressure and flow regimes are also affected by heterogeneity, including lithological changes, layering, cementation patterns and fracture patterns (Fig. 10.9c). The perturbation of both regimes is documented by an example with an additional shallow aquifer, which increases the hydraulic gradient above it and already draws more fluid recharge in the initial steady state (Figs. 10.9d and e). This aquifer is also capable of focusing some fluids toward the centre of the basin (Fig. 10.9e). After the onset of orogenic loading, two high pore fluid pressure zones are generated, separated by a shallow aquifer (Fig. 10.9f). The high-pressure gradient below and above the shallow aquifer produces fluid velocities within the shallow aquifer, which are larger than the velocities in the basal aquifer (Fig. 10.9g). An increase in magnitude of the tectonic loading increases the fluid velocity in the basal aquifer.

A disadvantage of the described simulations is the assumption that the emplacement of both vertical and horizontal loads is nearly instantaneous, which exaggerates transient pressure effects, thereby producing high estimates of initial flow rates (Ge and Garven, 1992).

An attempt to model the impact of a progressively emplaced tectonic load on the fluid regime was made by Smith and Wiltschko (1996). They, however, considered only its vertical component and ignored pore pressure changes due to shear stresses, which results in an underestimation of the pore pressure driven by tectonic loading. The other disadvantage is a lack of a boundary condition, which would simulate fracture formation and cause pore pressure dissipation. Therefore, the modelled high pore pressure zones indicate areas of likely rock failure.

The modelled values of excess pressure, p', and the fluid pressure ratio, λ, are based on the following relations:

$$p' = p - \rho_f g d \qquad (10.13)$$

and

$$\lambda = \frac{p}{\rho_b g d}, \qquad (10.14)$$

where ρ_b is the bulk density, ρ_f is the fluid density, g is the acceleration due to gravity and d is the depth. The thrust velocity is estimated to be $0.5 \, \mathrm{cm \, yr^{-1}}$ and the permeability is homogeneous and isotropic.

Prior to thrusting in this model, the pore fluid pressure increased uniformly. With thrust displacement, the orientation of the pore pressure gradient became disturbed beneath the toe of the thrust sheet and within the thrust sheet near the base of the thrust ramp (Fig. 10.10b). This disturbance remained throughout the whole thrust deformation (Figs. 10.10c and d). The upward bulge of pore pressure beneath the thrust ramp developed due to thrust loading. Because the fluid flow is driven by excess pore pressure gradients and the direction of flow is normal to the excess pressure contours during thrusting, fluid flow is generally upward within the hanging wall (Figs. 10.1b–d). The area immediately above the thrust ramp receives components of flow from both the hinterland and the footwall. Beneath the thrust toe, excess pore pressure indicates that fluid in this region is flowing into the foreland. Values of the pore fluid/lithostatic pressure ratio, λ, are high at the end of deposition (Fig. 10.11a), which indicates that conditions favourable to thrust motion were developed prior to thrusting. During thrusting, the highest values of λ are concentrated within and immediately beneath the thrust toe as well as in the hinterland portion of the thrustbelt (Figs. 10.11b–d). The value of λ decreases forelandwards as the pore pressure decays in the absence of loading. Further experiments, changing the homogeneous isotropic permeability, indicate that the pore pressure buildup is enhanced by decreased permeability (Fig. 10.12). A permeability of $10^{-16} \, \mathrm{m^2}$ and larger is high enough to dissipate fluids and prevent pore pressure buildup. Siltstone, mudstone, shale and certain types of limestone are, therefore, likely zones of overpressuring during thrust loading. Sandstone and permeable carbonates are likely zones of low excess pore pressure. Pore pressure buildup is also enhanced by an increased value of porosity compressibility, which raises the excess pressure throughout the model, but most noticeably beneath the thrust sheet and hinterland syn-tectonic sediments. Values of λ are accordingly increased.

Sensitivity analyses by Smith and Wiltschko (1996) allow a study of the influence of a sequence consisting of layers with different permeabilities. The different model studies progressively depart from three equally permeable layers towards a sequence with the least permeable layer at the bottom, a permeable layer in the middle and a second least permeable layer at the top. The models vary the permeability from high to low. The analyses indicate that the pore pressure buildup is highly influenced by these changes. The low permeability layer at the bottom develops the largest excess pore fluid pressure gradients. The high permeability layer in the middle has the lowest gradients. Since the fluid flow in the middle layer is faster than the flow in the top layer, fluid in the middle layer tends to accumulate beneath the top

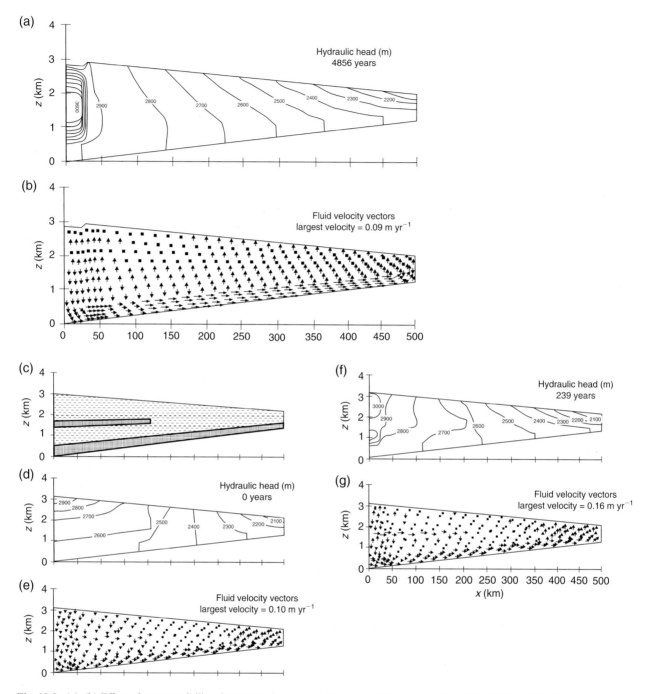

Fig. 10.9. (a), (b) Effect of compressibility, documented on a model for the last time period from Fig. 10.8 in which the bulk compressibility of the overlying aquitard unit has been increased ten times (Ge and Garven, 1992). (c)–(g) Effect of the second, pinching-out aquifer in the basin.

layer. It generates zones of high pore pressure within the hanging wall anticline and beneath the thrust toe. The model with layers of higher permeability develops lower pore pressures, mostly in the lowest shale layer (Fig. 10.13c). It indicates that a low permeability layer at the bottom needs the presence of a lower permeability layer at the top to enhance pore pressure buildup by restrict-

ing vertical flow. The sensitivity analyses also indicated that the buildup of the pore fluid pressure is dependent on the mean stress increase.

It should be stressed that Smith and Wiltschko (1996) did not consider the possibility of changing the thrust fault permeability. Therefore, it should be expected that a fault conduit would tend to reduce pore pressures

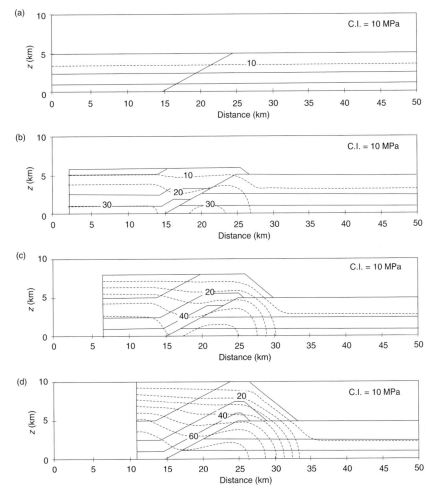

Fig. 10.10. Evolution of excess pore pressure within a thrust sheet with homogeneous and isotropic permeability of $10^{-18}\,m^2$ (Smith and Wiltschko, 1996). The various stages are: (a) the situation prior to thrusting, (b) the situation after 2 km displacement or 18% of the total displacement, (c) the situation after 6.5 km (60%) displacement and (d) the situation at the end of thrusting, where the section at the top of the ramp is doubled. C.I. is the contour interval.

within the surrounding sediments and a fault barrier would increase pore pressures beneath the fault.

All the studies of fluid flow mechanisms driven by compression discussed above, imply the operation of compaction-driven flow, which also includes a horizontal compaction. This mechanism has only rarely been determined as the mechanism generating overpressure. The only case studies are the San Andreas fault (Sleep and Blanpied, 1992), and the Barbados accretionary prism (Fisher and Zwart, 1996). This is partly because the mechanism is not yet fully understood, as indicated by the fact that each of the described studies of this mechanism ignored some of the basic constraints for this mechanism, such as:

(1) changing permeability along the thrust fault plane;
(2) considering both the vertical and horizontal components of the orogenic load;
(3) incorporating pore pressure changes due to shear stresses; or
(4) using realistic emplacement rates for both horizontal and vertical orogenic loads.

Fluid flow driven by seismic pumping

A seismic pumping, or fault-valve mechanism, becomes possible when active faulting occurs in a sedimentary sequence with overpressured layers and a suprahydrostatic vertical gradient in fluid pressure (Sibson, 1981). The valving action depends on the ability of faults to behave as seals when they are tectonically quiescent and as conduits during and immediately following rupturing. The explanation for fault-sealing during interseismic periods is the presence of clays, cataclastic gouge or mineral precipitation within the fault core. Breaching of these seals that form barriers to overpressured zones located within the sedimentary sequence by rupturing leads to fluid flow between regions of fluid pressure contrast connected by the conductive fault. Such a seismic pumping mechanism can occur in any tectonic setting where overpressuring has developed, but extreme fault-valve action tends to be associated with steep reverse faults (Sibson, 1990a).

Figure 10.14 shows the stress and fluid pressure cycling that is associated with valve action on such a steep reverse fault. Prior to rupturing, the stress condition

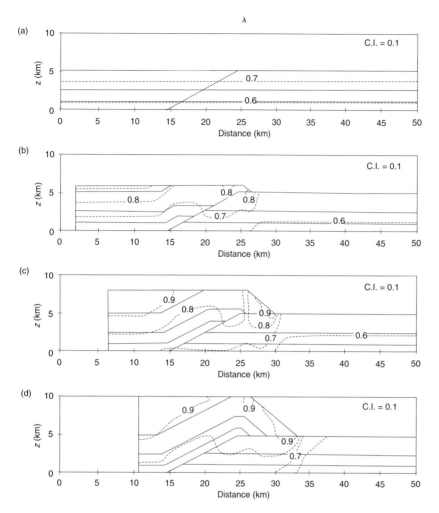

Fig. 10.11. Evolution of the ratio of pore pressure to lithostatic pressure within a thrust sheet in a model shown in Smith and Wiltschko (1996).

$\sigma_1 > p > \sigma_3$ characterizes the overpressured layer, which will become the future rupture nucleation site. Sub-horizontal hydrofractures start to form in this layer during this time period. The rupturing results in shear stress release along the fault, which, in turn, results in the stress condition changing to $p > \sigma_1 \approx \sigma_3$. Under this stress condition, fluids previously trapped in the array of sub-horizontal hydrofractures and in any grain-scale fracture porosity are expelled into the fault zone, which becomes the main conduit. Fluid discharge along the main conduit continues until the pressure gradient changes to hydrostatic or the fault zone, opened by the rupturing, becomes sealed by mineral precipitation. This results in the stress condition changing to $\sigma_1 \approx \sigma_3 > p$. This stage is followed by both the differential stress and the pore fluid pressure gradually rebuilding towards the next rupturing episode.

It is apparent that the seismic pumping mechanism described above depends on the competition between permeability creation and destruction in fault zones during their active and quiescent stages (Sibson, 1992a).

Closure of the fault zone fractures reduces the bulk permeability of the fault zone by various orders of magnitude (Caine, 1999). This might produce a fault zone with significantly lower permeability than the surrounding host rock. As a consequence, the predominant fluid flow would occur in a host rock in a direction parallel to fault. The flow across the fault would be restricted by the reduced permeability of the fault.

A simplified model of fracture aperture changes in the fault zone during pre-failure, failure and post-failure was made by Caine (1999). The impact of fracture opening and closure during fault zone evolution was simulated by uniformly varying fracture apertures in each fault zone from 1000 to 1 μm. The fault zone was located in an impermeable host rock. Cases with a uniform aperture of 100 μm were assumed to represent an intermediate stage in the evolution of a fault zone that undergoes an idealized mechanical deformation cycle from pre-failure, to failure, to post-failure. Stepwise fracture closure from 100 to 10 μm and then to 1 μm represented a series of post-failure to pre-failure

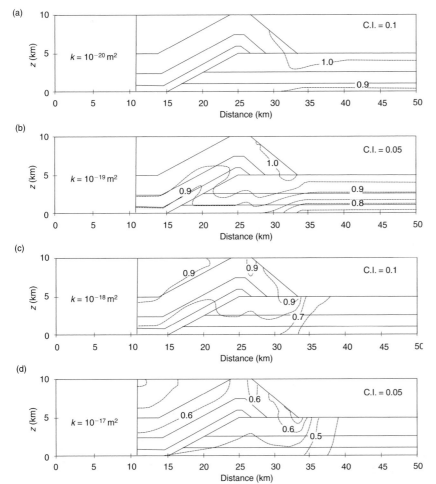

Fig. 10.12. Ratio of the pore pressure to the lithostatic pressure for the final stage of thrust loading for four thrust sheets with different homogeneous and isotropic permeabilities (Smith and Wiltschko, 1996).

stages with progressively more extensive fracture sealing. Uniformly increasing fracture apertures to 1000 μm in a fault zone represented the increase in permeability that probably occurs during failure. Fault zones were classified into four end members, shown in Fig. 10.15. Each of the end members behaves differently during the pre-failure, failure and post-failure stages.

Closure of the fault zone fractures from 100 to 10 μm in both the broad distributed deformation zone and in localized deformation zones reduces the bulk permeability of the fault zones by three orders of magnitude. This results in a fault zone with significantly lower permeability than the surrounding host rock. Consequently, hydraulic gradients oriented sub-parallel to the fault zone would cause most fluid to flow through the adjacent host rock while flow across the fault zone would be restricted.

Reducing fracture apertures from 100 to 10 μm in the core of the composite deformation zone reduces the bulk permeability of the core while leaving the broader damage zone as a primary pathway for the fluid flow.

The resulting permeability structure is a combined conduit–barrier flow system. In this system, fluid movement across the fault zone is restricted and redirected parallel to the fault zone.

During failure, a net increase in fracture aperture is to be expected within the fault zone, even though some fractures are open and some are closed. In the localized, distributed and composite deformation zones bulk permeability normal to the fault zone is slightly enhanced by a factor of 2 or 3. Increased aperture in a single fracture has little impact on bulk permeability normal to the fault because the host rock network controls the fluid flow normal to the fault. In all types of fracture zones, failure, increasing the aperture of the fault zone to 1000 μm, results in a large increase in permeability parallel to the fault zone. The increase can be two to three orders of magnitude. Fluid flow will be focused into the fault zone whenever the hydraulic gradient is not precisely perpendicular to the fault. Thus, the failure stage could be associated with a significant fault-related fluid flow (e.g., Newman and Mitra, 1994;

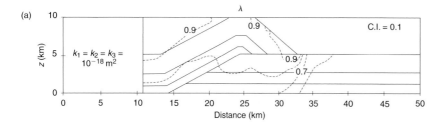

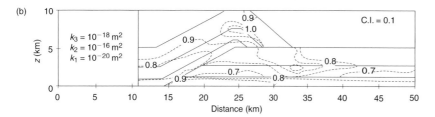

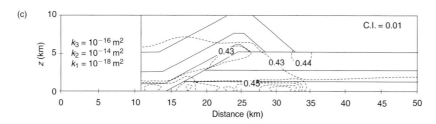

Fig. 10.13. Ratio of the pore pressure to the lithostatic pressure for the final stage of thrust loading for three thrust sheets (Smith and Wiltschko, 1996). (a) A homogeneous and isotropic thrust sheet. (b) and (c) An inhomogeneous, isotropic thrust sheet with permeabilities simulating an upper layer composed of mixed shale, siltstone and sandstone, a middle layer composed of limestone and dolomite and a lower layer mainly composed of shale and siltstone with different permeabilities.

Sibson, 1994a, b; Goddard and Evans, 1995, Davidson *et al.*, 1998; Caine, 1999).

During the post-failure stage, mineral precipitation and lowered stress progressively seal fault zones. Fracture apertures in each type of fault zone can be modelled by reducing the aperture from 100 to 1 μm (Caine, 1999), resulting in a permeability reduction. Significant reductions in bulk permeability occur normal to the fault zone. This reduction is by one to two orders of magnitude for the 10 μm case and by four to five orders of magnitude for the 1 μm case. Permeability parallel to the fault zone is little affected by the sealing. The fluid directed toward the fault zone is primarily transmitted either through the adjacent host rock in the cases consisting of a single fracture, localized and distributed deformation zones or through the damage zone of a composite deformation zone.

However, it should be stressed that natural flow scenarios are different from those modelled by Caine *et al.* (1996) and Caine (1999), because these authors considered only uniform changes of fracture apertures in fault zones. This is not the case in nature where fault zones are composed of fractures of various orientations, in which fracture apertures are adjusted according to their attitude in relation to the local stress field (e.g., Carlsson and Olsson, 1979; Hancock, 1985; Barton *et al.*, 1995; Finkbeiner *et al.*, 1997; Fig. 10.16). Faults and fractures perpendicular to the maximum principal compressive

stress are preferentially closed, thereby reducing the permeability perpendicular to the maximum principal stress. Permeability perpendicular to the minimum principal compressive stress direction is relatively enhanced because the lower resolved normal stress results in less fracture aperture reduction (e.g., Carlsson and Olsson, 1979). Determinations of fault zone permeability are further complicated by increased small-scale fracturing in the vicinity of faults, which are on the verge of shear failure (Barton *et al.*, 1995). The result of such fault zone complexity is an anisotropic transmissivity in such a zone. A large-scale field example of anisotropic fluid transmissivity through fault/fracture systems comes from the Yucca Mountain, Nevada. Here, the σ_1 value of the modern stress field is parallel to the long axis of the hydraulic transmissivity ellipse determined by long-term aquifer pumping tests (Ferrill *et al.*, 1999).

Fluid flow driven by thermal expansion or aquathermal pressuring

The equation for heat transfer is (Shi and Wang, 1986)

$$\frac{c\rho_b DT}{Dt} = -c_f \rho_f q^* \nabla(T) + K\nabla^2 T + Q, \qquad (10.15)$$

where c is the bulk specific heat, ρ_b is the bulk density, T is the temperature, t is the time, c_f is the specific heat of the fluid, ρ_f is the density of the fluid, K is the bulk

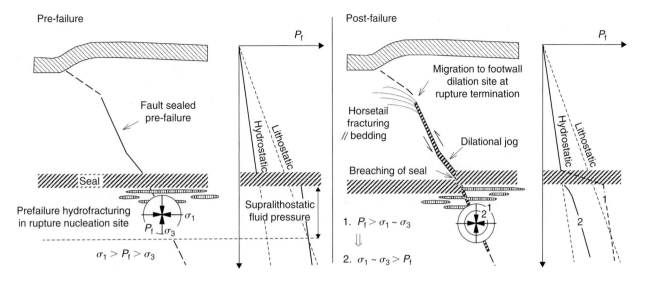

Fig. 10.14. Sketch illustrating the stress and fluid pressure states during both pre-failure and post-failure associated with extreme fault-valve action on a steep reverse fault (Sibson, 1995). The sketch shows the simplest situation where one seal separates hydrostatic and overpressured fluid regimes.

thermal conductivity, q is the fluid discharge and Q is the heat source in unit volume. $D(\ )/Dt$ is the material differential in the deforming coordinates. The specific heat of a porous medium, c, is calculated from (Smith, 1973)

$$c = c_f n + c_s(1 - n), \tag{10.16}$$

where c_f is the specific heat of the fluid, n is the porosity and c_s is the specific heat of the solid. A similar relationship holds for the heat source. The bulk thermal conductivity, K, can be calculated from (Lewis and Rose, 1970)

$$K = K_s\left(\frac{K_f}{K_s}\right)^n, \tag{10.17}$$

where K_s and K_f are the thermal conductivities of the solid and fluid, respectively.

The first term on the right-hand side of equation (10.15) describes the advective heat flow due to the flow of the fluid. The advective heat transfer due to the motion of the solid matrix is handled implicitly by the material derivative term on the left-hand side of equation (10.15). The second term on the right-hand side represents the conductive heat transfer through the solid and the fluid. The third term represents heat sources such as radioactive elements.

Equations (10.15)–(10.17) describe how the thermal transfer is coupled with porous flow and sediment compaction. Mechanical loading of sediments and temperature changes induce porous flow. Drainage of fluid in turn enables the sediment to compact and introduces advective heat transfer. Furthermore, the thermal and

hydraulic conductivity, bulk specific heat and water storage are material properties, which depend on the temperature and the pressure.

The contribution of aquapressuring to pore pressure generation, dp, can be resolved from equation (10.1):

$$dp = \left[\frac{\gamma_f}{\alpha_n(1 - n) + \beta}\right]dT, \tag{10.18}$$

where the term $n(n\gamma_f - \gamma_s) + (\gamma_s - \gamma_b)$ from equation (10.1) is approximated as $n\gamma_f$. This is possible because the thermal expansion of the porous rock, γ_b, is roughly the same as the bulk thermal expansion for the solid grains, γ_s, and both are two orders of magnitude smaller than the thermal expansion of water (e.g., Shi and Wang, 1986).

A comparison of the pore fluid pressure generated by aquapressuring and by sediment compaction (Fig. 10.17; Shi and Wang, 1986) clearly indicates that for the generation of pore fluid pressure in a sedimentary basin under normal geological conditions, the effect of sediment compaction is much greater than aquathermal pressuring. On the other hand, during abnormal, rapid increases of temperature such as those occurring near an intruding igneous body or during unloading due to uplift and erosion, the aquathermal effect may not be negligible in comparison with the mechanical effect.

In general, aquathermal expansion is not a typical overpressuring mechanism, because it requires an extremely low permeability seal, lower than the usual permeability range of shale (Osborne and Swarbrick, 1997). The mechanism could potentially function in

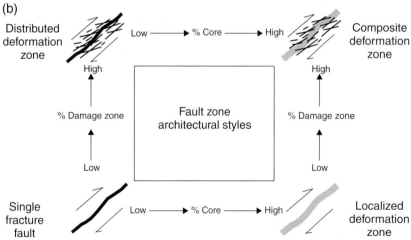

Fig. 10.15. (a) Photograph of a fault zone architecture, example from the Wyoming–Utah thrustbelt, Salt Lake City, Utah. Various types of fault core contain tectonic gouge, cataclasite and mylonite. Damage zones can be formed by small faults, fractures, cleavage, veins and folds. (b) Fault zone architectural styles (Caine, 1999). A localized conduit is a feature of a fault zone, which has slip localized along a single curviplanar surface or along discretely segmented planes. Both fault core and damage zone are absent or poorly developed. A distributed conduit is a feature of a fault zone, which has slip accommodated along distributed surfaces. The fault core is absent or poorly developed. The damage zone has well developed discrete slip surfaces and fracture networks. This fault zone architecture, known also as fault-fracture mesh (Sibson, 1996) is typical for modern accretionary prisms (Moore and Vrolijk, 1992). A localized barrier is a feature of a fault zone, which has localized slip accommodated within a cataclastic zone. The fault core is well developed and the damage zone absent or poorly developed. The fault core behaves like an aquitard within the somewhat higher permeability aquifer of the host rock. Similar fault zones are reported as deformation bands in sandstones (Antonellini and Aydin, 1994). A combined conduit-barrier is a feature of a fault zone, which has its deformation accommodated within a localized cataclastic zone and a distributed zone of subsidiary structures. The fault core behaves like an aquitard sandwiched between two aquifers in the damage zones.

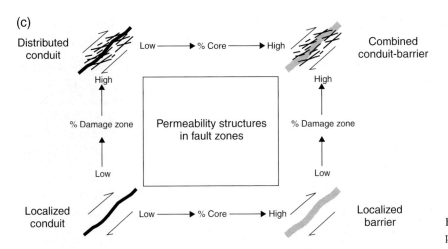

Fig. 10.15. (*cont.*) (c) Fault zone permeability structures (Caine, 1999).

evaporite-rich sedimentary sequences or during the deep burial of very impermeable shale.

Fluid diffusion

Fluid movement in porous media accreted in thrustbelts occurs in response to head gradients and buoyancy forces. The hydraulic head, h, is the mechanical energy per unit mass of fluid, equivalent to the height above an arbitrary datum to which fluid will rise in a tube (Hubbert, 1940):

$$h = z + \frac{p}{\rho_f g}, \qquad (10.19)$$

where z is the elevation term. The pressure term includes the pore fluid pressure, p, the fluid density, ρ_f, and the acceleration due to gravity, g.

The contribution of the laminar flow of an inhomogeneous fluid through an anisotropic porous medium in the pore pressure balance, neglecting inertial forces, is (Bear, 1972)

$$q = -\frac{k}{\eta}(\nabla p + \rho_f \nabla z), \qquad (10.20)$$

where q is the Darcy velocity or specific discharge, k is the permeability tensor, η is the fluid dynamic viscosity, ∇p is the pore fluid pressure gradient, z is the elevation and g is the acceleration due to gravity. The specific discharge is understood as the volumetric flow rate per unit time divided by the cross-sectional area perpendicular to the flow direction. The Darcy velocity, q, is related to the linear velocity, v, by

$$v = \frac{q}{n}, \qquad (10.21)$$

where n is the porosity.

Equation (10.9) shows that fluid flow results not only from a head gradient, ∇h, but also from fluid density gradients, which give rise to buoyancy forces. With the constant density assumption, equation (10.9) reduces to the form of Darcy's law (Darcy, 1856; Hubbert, 1969):

$$q = -\frac{k\rho_{f0}g}{\eta}\nabla h, \qquad (10.22)$$

or

$$q = -K\nabla h, \qquad (10.23)$$

where K is the hydraulic conductivity tensor. Through the implicit assumption of a constant-density fluid, the hydraulic head can be approximated as a hydraulic potential for the purpose of obtaining insight into thrustbelt problems for which an exact answer is unobtainable (Deming, 1994b). In this case, there are no buoyancy forces. The transient movement of fluids in the thrustbelt is described by the diffusion equation. In complex and heterogeneous thrustbelt environments, the assumption of a constant-density fluid is nearly always violated. However, the diffusion equation is nevertheless a useful approximation that can be used to obtain insight into the physics of fluid flow in a thrustbelt. For a porous medium that is both homogeneous and isotropic, the diffusion equation is (Deming, 1994b)

$$\frac{\partial h}{\partial t} = \frac{K}{S_s}\nabla^2 h, \qquad (10.24)$$

where h is the hydraulic head, t is the time, K is the hydraulic conductivity and $\nabla^2 h$ is the curvature of the head field. S_s is the specific storage that can be expressed as the left-hand term from equation (10.1):

$$S_s = \left(\frac{\alpha_n}{1-n} + n\alpha_s + n\beta\right)\frac{dp}{dt}, \qquad (10.25)$$

where α_n is the porosity compressibility, n is the porosity, α_s is the compressibility of solid grains, β is the fluid compressibility and p is the pore fluid pressure. The specific storage refers to the physical ability of a porous

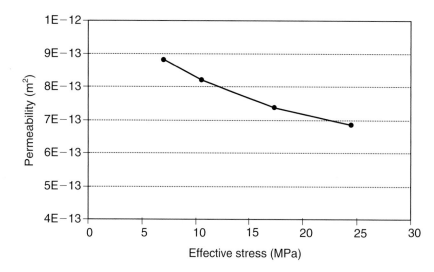

Fig. 10.16. Measured data for fracture permeability dependence on effective stress for volcanic rocks (Sigal, 1998, personal communication).

medium to store fluid. It is largely determined by the compressibility of the fluid and various sediments (Table 10.1; Ge and Garven, 1992).

Fluid diffusion is also controlled by hydraulic conductivity, which is proportional to the intrinsic or absolute permeability of the sediments, proportional to the fluid density and inversely proportional to the viscosity of the fluid. Some of the factors that determine hydraulic conductivity (see equations 10.22 and 10.23) may vary over orders of magnitude in different thrust-belt settings while others are relatively constant. Permeability is such a highly variable factor, capable of variations over more than 11 orders of magnitude.

The fluid density is further determined by temperature, salinity and pressure. Of these three factors, temperature and salinity are more important. For example, the density of water decreases by about 8% when the temperature increases from 25 to 150 °C at constant pressure. Assuming a related pressure increase of 50 MPa at hydrostatic conditions and a temperature gradient of 25 °C km^{-1}, it results in only a 2% increase of water density.

The viscosity of the fluid further depends on both temperature and pressure. The viscosity variation with temperature is much more important than with pressure, as demonstrated with water as an example (Clark, 1966).

Transient fluid flow along thrust faults

As indicated earlier, thrust faults undergo huge permeability changes during passage through the pre-failure, failure and post-failure stages of their development cycles. Studies of these cycles from modern accretionary prisms with anomalously high fluid outflow (Carson *et al.*, 1990; Davis *et al.*, 1990; Fisher and Hounslow, 1990;

Foucher *et al.*, 1990a, b; Henry *et al.*, 1990; Langseth *et al.*, 1990; Le Pichon *et al.*, 1990a, b; Yamano *et al.*, 1990, 1992) and studies of transient and dynamic dewatering of modern and ancient accretionary prisms (Fischer and Hounslow, 1990; Henry and Wang, 1990; Foucher *et al.*, 1992; Fig. 10.18; Le Pichon *et al.*, 1993; Bekins *et al.*, 1994; Fisher *et al.*, 1996; Larroque *et al.*, 1996), led to the discovery of transient fluid flow in thrustbelts.

Among the thrust faults investigated in accretionary prisms, décollement thrusts experience the largest changes. They constitute zones where wedge regimes are juxtaposed against regimes of subthrust deformation. An example of a wedge deformation regime comes from the Barbados accretionary prism. Here the anisotropy of magnetic susceptibility data indicates an abrupt change from lateral compression inside the wedge to vertical compaction at the upper levels of the detachment zone (Housen *et al.*, 1996). The transition is characterized by a high pore fluid pressure (Shipley *et al.*, 1995). Similar high fluid pressures along décollements and other thrusts have been indicated by high-amplitude, reversed polarity reflections from the Oregon accretionary prism (Tobin *et al.*, 1994), the Nankai accretionary prism (Moore *et al.*, 1990a), the Barbados accretionary prism (Westbrook *et al.*, 1988), the Costa Rica accretionary prism (Shipley *et al.*, 1990) and the Cascadia accretionary prism (MacKay *et al.*, 1994). Porosity curves from accretionary prisms such as the Barbados prism, recording abrupt porosity increases in the footwall below thrust faults, have also indicated distinct changes in deformation regimes between footwalls and hanging walls of both décollement and other thrusts of the prism (Fig. 10.19; Morgan and Karig, 1995). These data indicate that a décollement is capable of controlling porosity changes during two stages of its development. First, during the post-failure

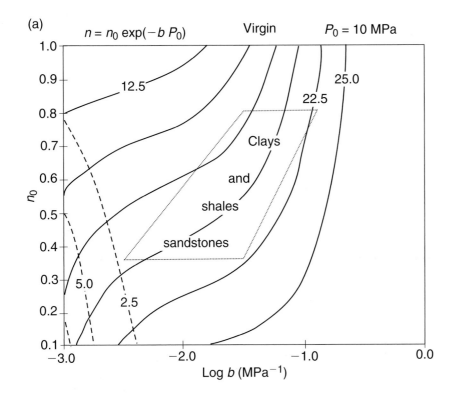

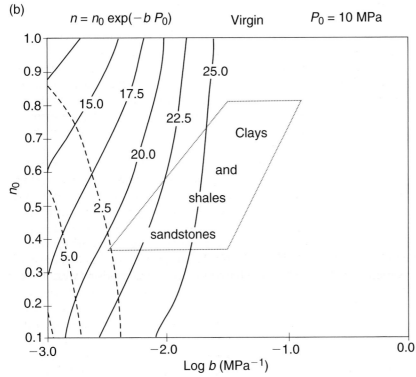

Fig. 10.17. Contours for the coefficients of pore pressure generation due to mechanical loading and aquathermal pressuring, as functions of material with different initial porosity n_0 and $b = 1/(\rho - \rho_f)gH$ (Shi and Wang, 1986). ρ is the bulk density, ρ_f is the fluid density, g is the acceleration due to gravity and H is a material parameter. Geothermal gradient is $30\,°C\,km^{-1}$. Density of solid grains is $2700\,kg\,m^{-3}$. Bulk density is calculated from $(1 - n)\,\rho_s + n\rho_f$. Pore pressure generated by both mechanical loading (solid lines) and aquathermal pressuring (dashed lines) is in kPa. (a) Diagram showing loading at an effective pressure of 10 MPa. (b) Diagram showing loading at an effective pressure of 100 MPa.

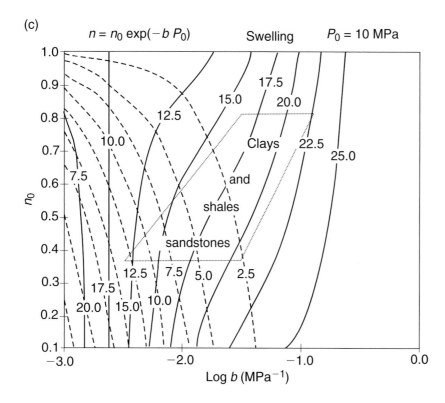

Fig. 10.17. (*cont.*) (c) Diagram showing unloading after the effective pressure is loaded to 100 MPa.

permeability collapse when the décollement seals the subthrust sediments allowing an increase in pore fluid pressure to maintain the higher porosity. Secondly, immediately post-failure when the décollement is permeable, the fluid flow is directed along the progressively compacting décollement, further reducing the weight of the accretionary prism, which has already been reduced by the pore pressure within it (see Mandl, 1988; Karig and Morgan, 1990; Maltman *et al.*, 1992). Hydrogeological studies along the Barbados décollement zone (Screaton *et al.*, 1990) indicate that permeabilities here are three to five orders of magnitude greater that in the surrounding sediments. This occurs despite the fact that the décollement layer consists of 80% clays (Tribble, 1990) with intrinsically very low permeabilities (Freeze and Cherry, 1979), and despite the fact that shear deformation generally reduces permeability along faults, as shown by ring-shear experiments (Brown *et al.*, 1994). However, the Barbados décollement zone contains common veins that occur in the form of syn-tectonic extensional shears (Etheridge, 1983) and hydrofractures (Brown and Behrmann, 1990; Vrolijk and Sheppard, 1991), which have also been observed in other accretionary prisms (e.g., Byrne and Fisher, 1990). These extensional shear and hydrofracture networks in the basal décollement zone occur at lithostatic fluid pressures in order to remain open (e.g., Behrmann, 1991). The low-angle orientation of the maximum compressive principal

stress in the overlying wedge (see Fig. 2.17) restricts vertical hydrofracture development and thus vertical fluid loss. Hydrofractures grow laterally under lithostatic fluid-pressure conditions (Fig. 10.20; Brown *et al.*, 1994). Near the toe of the accretionary prism, elastic swelling of the fault-wall sediments responds to a local pore-fluid pressure increase by faulting. This preferentially intensifies the horizontal compressive stress and prevents development of the vertical hydrofractures (Brown *et al.*, 1994). In the case of a sediment in a laterally confined system undergoing unloading by an increase in pore-fluid pressure, Δp, the change in horizontal effective stress is controlled by $\Delta\sigma'_h = \Delta p[\nu(1-\nu)]$, where ν is Poisson's ratio (Lorenz *et al.*, 1991). The change in vertical stress is given by $\Delta\sigma'_v = \Delta p$ because the top of the wedge is a free bounding surface. As values of ν in the wedge vary between 0.1 and 0.3, the horizontal stress is reduced and falls below the rate of vertical stress reduction as pore-fluid pressures rise. This again constrains any open system of hydrofractures to develop sub-parallel to the free surface at the top of the wedge (Brown *et al.*, 1994). The lower shear strength and cohesion and the active nature of the low-angle fault zones should promote a coupled response between the propagation of strain along the fault and the development of extensional shears and hydrofractures. In locally more indurated lithologies, faulting, with effective normal stress close to zero, also allows the formation of permeable dilatant

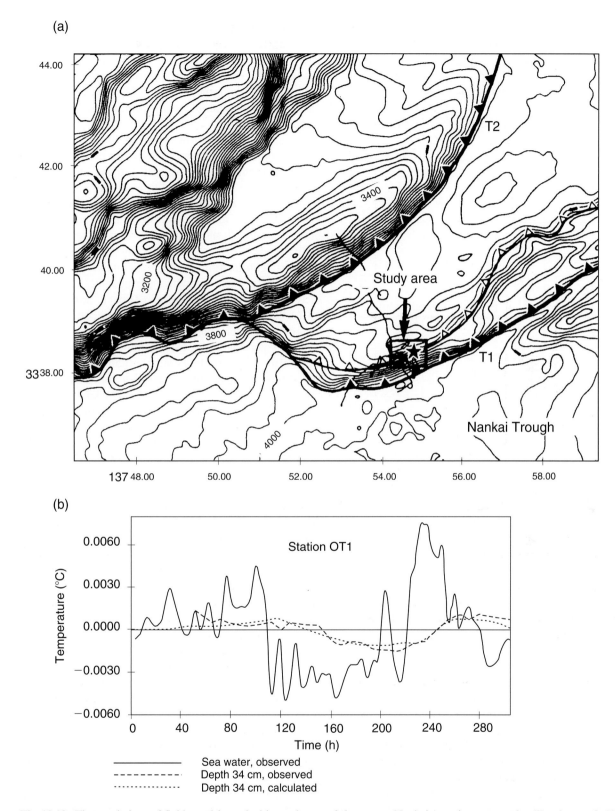

(a)

(b)

Fig. 10.18. Time variations of fluid expulsion velocities at the toe of the eastern Nankai accretionary complex (Foucher *et al.*, 1992). (a) Location map of the study area. T1 and T2 are the frontal and inner thrusts. The intermediate thrust is indicated by open barbs. (b) Filtered temperature fluctuations about mean values in seawater and 34 cm into the sediment are indicated by a continuous and a dashed curve, respectively. The calculated temperature at 34 cm, indicated by dotted curve, is computed from temperature variations in seawater with a thermal diffusivity of $3 \times 10^{-7} \, \text{m}^2 \, \text{s}^{-1}$. There is no correlation between the calculated and observed curves.

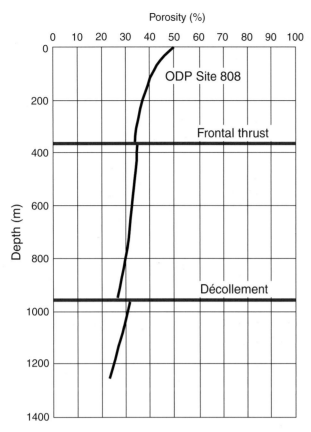

Fig. 10.19. In situ porosities for ODP Site 808 from the Barbados accretionary prism, calculated from shipboard porosities and corrected for the sand component and porosity rebound (Morgan and Karig, 1995).

breccia. Examples of such breccias are seen in the basal décollement of the Nankai accretionary wedge (Maltman *et al.*, 1993).

Character of basic permeable systems in thrustbelts and their interaction

Rock permeability

Rock permeability varies over more than 11 orders of magnitude, from values as high as $10^{-9}\,m^2$ for karst limestone, to values lower than $10^{-20}\,m^2$ for some shales and unfractured crystalline rocks (Deming, 1994a; Smith and Wiltschko, 1996). Presumably the average permeability of lithologies such as salt may be even lower than the permeability of shale.

Rock permeability is a function of the lithological characteristics and the stress state. Grain size and sorting are the primary lithological characteristics that control this permeability. Rocks with a large grain size tend to have higher permeabilities than rocks with a smaller grain size (Prince, 1999). Similarly, well-sorted rocks may have higher permeabilities than poorly sorted rocks due to the reduction of pore space by fine-grained particles occupying the space between larger grains. Chemical and diagenetic processes further affect permeability by changing the lithology. They include dissolution, recrystallization, cementation and mineral transformations.

The stress state controls the mechanical deformation of the rock. Deformation can influence permeability by both the formation of fractures and the reduction of porosity due to compaction. For example, experimental results on sandstones (e.g., Wilhelm and Somerton, 1967; Mordecai and Morris, 1971; Daw *et al.*, 1974) indicate that with increasing differential stress, permeability tends to decrease by generally less than 50% due to matrix compaction. At approximately two-thirds of the ultimate rock strength, permeability begins to increase as fractures develop (Zoback and Byerlee, 1976). The change in permeability due to compaction is insignificant when compared with the potential range of sediment permeabilities (Smith and Wiltschko, 1996). Furthermore, permeability is potentially more sensitive to fracture development than compaction, because fracture permeability can be up to three orders of magnitude larger than the unfractured permeability, as demonstrated by fracture permeability studies (e.g., Caine, 1999; Nemčok *et al.*, 2002).

Rock permeability in thrustbelts has distinctively different vertical and horizontal components. Vertical permeability is likely to be inhomogeneous due to stratification and shallowly dipping fault systems. Horizontal permeability can be also inhomogeneous due to stratigraphic pinch-outs, facies changes and steeply dipping faults. Rock permeability is a highly variable quantity even within seemingly the same facies. For example, permeability measurements on cores from sedimentary basins that sample geological horizons with uniform lithologies commonly vary over five orders of magnitude (Deming, 1994c).

Rock permeability also changes with scale (e.g., Davis, 1969; Brace, 1980). Estimates made from pumping tests in wells commonly yield values one to two orders of magnitude greater than those derived from measurements on core samples. Furthermore, pumping tests can fail when sampling fracture networks that can increase large-scale permeability.

As mentioned earlier, rock permeability is dependent on both the total stress and the fluid pressure (e.g., Wilhelm and Somerton, 1967; Mordecai and Morris, 1971; Daw *et al.*, 1974). The rock permeability–effective stress (total stress minus pore fluid pressure) relationship (Fig. 10.21) is distinctively different from the fracture permeability – effective stress relationship (Fig. 10.16), which can be used as an indicative tool for the

(a)

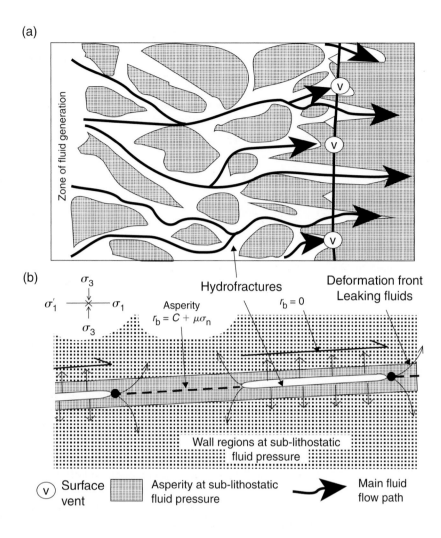

(b)

Fig. 10.20. Heterogeneous fracture development along an accretionary wedge décollement (Brown *et al.*, 1994). (a) Map view of the décollement zone. Asperities are formed by sediments between open fractures, which are at sublithostatic fluid pressures. Consolidation is retarded in lithostatically pressured hydrofractured regions, but still occurs in sublithostatically-pressured asperities. Asperities provide a basal traction component. A small frictional resistance is developed in lithostatically pressured hydrofractured regions that consist of various combinations of extensional shears and hydrofractures during different periods of their dilational history. (b) Cross section of the décollement zone. Near the toe, hydrofractures are at higher pressures than the surrounding regions, to which fluids will be leaked. Internal fluid pressures in asperities are buffered by pore pressures in wall sediments. Therefore the effective normal stress σ'_n is greater than zero. Asperities provide basal frictional resistance r_b that is controlled by mud's internal angle of friction μ and its limited cohesion C.

distinguishing between these two end member permeabilities in a reservoir. In order to derive a rock permeability–effective stress relationship we need to combine permeability–porosity and porosity–effective stress relationships.

The change in rock permeability with porosity can be expressed as (Thompson *et al.*, 1987; Sigal, 1998, personal communication)

$$k = A r_{\text{eff}}^2 n^m, \tag{10.26}$$

where k is the permeability, A is the pore shape factor, r_{eff} is the characteristic pore throat size, which controls access to the primary flow network, n is the porosity and m is the cementation factor. The pore shape factor, A, is calculated as the average pore area from a thin section. Its most common value for sandstone is 1.16, and can be as high as 1.39 for sandstone with calcite cement (Sigal, 1998, personal communication). A good estimate of r_{eff} is the pore throat size on a high-pressure mercury injection curve at which the maximum injection rate occurs (Sigal, 1998, personal communication).

The cementation factor, m, for well-connected pore spaces generally ranges from about 1.7 for poorly consolidated sediments to about 2.1 for consolidated sediments (Sigal, 1998, personal communication).

The change in rock porosity with effective stress, involving reasonable simplifications, can be obtained using a bulk modulus in the following equation (Sigal, 1998, personal communication):

$$n(\sigma) = 1 - (1 - n_0) e^{(\sigma - \sigma_0)/K}, \tag{10.27}$$

where σ is the effective stress, K is the static bulk modulus, n_0 is the initial porosity and σ_0 is the initial effective stress.

By combining equations (10.26) and (10.27) the rock permeability–effective stress relationship can be determined. The pore shape factor, A, and the cementation factor, m, can be treated as constants with effective stress changes (Sigal, 1998, personal communication). However, it is necessary to include the calculation of the pore throat size, r_{eff}, reduction with increasing effective stress (Zimmerman *et al.*, 1986):

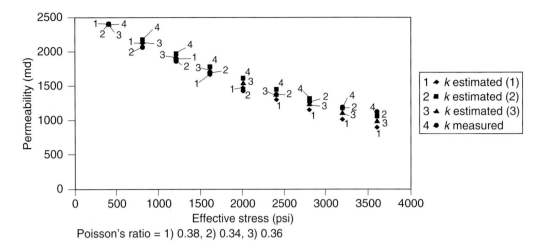

Fig. 10.21. Measured and calculated data of rock permeability dependence on effective stress for hard consolidated sandstone (Sigal, 1998, personal communication).

$$r_{\mathrm{eff}}^2(\sigma) = r(\sigma_0)^2 \mathrm{e}^{-c_{\mathrm{pc}}(\sigma - \sigma_0)}, \qquad (10.28)$$

where c_{pc} is the pore compressibility and r is the initial characteristic pore throat size.

An example of the rock permeability–effective stress relationship for a hard consolidated sandstone is shown in Fig. 10.21. This figure documents a broad range of continuous rock permeability changes with effective stress. The following subsection shows how a similar permeability curve for fracture permeability undergoes rather limited changes between the stage with fractures fully open and the stage when the fracture apertures have collapsed due to fluid pressure dissipation (Fig. 10.16).

Fracture permeability
Fracture permeability includes: (1) permeability related to fault zones, which bound thrust sheets, and (2) permeability related to fracture and fault patterns located inside thrust sheets. The latter is described in Chapter 21. The fluid flow properties of bounding fault zones can be divided into four basic categories (Fig. 10.15c) based on the relationship of the fault zone components such as fault core, damage zone and host rock (sensu Chester and Logan, 1986; Smith *et al.*, 1990; Fig. 10.15a). Four combinations of different fault cores and damage zones (Fig. 10.15b; Caine, 1999) result in four basic architectural styles of the fault zone, where the fault cores may include:

(1)   single-slip surfaces;
(2)   unconsolidated clay-rich gouge zones;
(3)   brecciated and geochemically altered zones; or
(4)   highly indurated, cataclasite zones

and damage zones may include:

(1)   small faults;
(2)   veins;
(3)   fractures;
(4)   cleavage; and
(5)   folds.

A fault core is a part of the fault zone where comminution, fluid flow, geochemical reaction and other fault-related processes can alter the original lithology and its permeability. For example, progressive grain reduction, dissolution, reaction and mineral precipitation during fault zone evolution typically cause the core to have reduced permeability, relative to the adjacent damage zone and the host rock. A damage zone is a part of the fault zone where the host rock permeability can also be changed. The fault core and damage zones are surrounded by relatively undeformed host rock (10.15a).

The four architectural styles of the fault zone include (Caine, 1999):

(1)   single fracture fault;
(2)   distributed deformation zone;
(3)   composite deformation zone; and
(4)   localized deformation zone.

Architectural styles (Fig. 10.15b) determine the fluid properties of a fault zone, comprising localized conduit, distributed conduit, combined conduit–barrier and localized barrier (Caine, 1999; Fig. 10.15c).

Factors controlling the permeability of the fault zone include the lithology, the fault scale, the fault type, the deformation style, the deformation history, the fluid chemistry, the pressure–temperature history, the

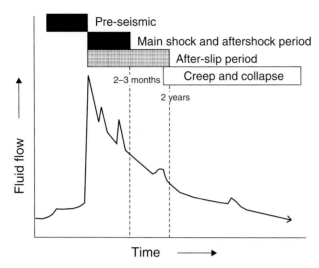

Fig. 10.22. Three stages of permeability evolution, each linked to the earthquake cycle (Knipe, 1993).

component percentage and the component anisotropy. For example, granitic rocks deformed by mature fault zones frequently have feldspars affected by fluid–mineral interactions. This results in fault core lithologies rich in clay gouge materials (e.g., Blanpied *et al.*, 1995; Hickman *et al.*, 1995; Goddard and Evans, 1995), which can significantly reduce the fault core permeability. Quartz-rich sandstones deformed by distributed deformation zones with small displacements have a low permeability fault core developed by grain-size reduction and the formation of deformation bands (e.g., Antonellini and Aydin, 1994). The faulting in sandstone/shale sequences or granitic rocks with a sedimentary cover with combined mechanical and chemical processes leads to the formation of low-permeability fault cores (e.g., Chester and Logan, 1986; Caine, 1999).

As mentioned earlier, the permeability–effective stress relationship for the host rock and fracture system are quite different. The fracture permeability, which is also dependent on both the total stress (e.g., Carlsson and Olsson, 1979; Fisher *et al.*, 1996) and the fluid pressure (e.g., Knipe, 1993; Fisher *et al.*, 1996), which are coupled (e.g., Oliver, 1986; Sibson, 1990; Fisher *et al.*, 1996; Connolly and Cosgrove, 1999), can collapse abruptly when certain stress conditions are met (see Fig. 10.22). The fracture permeability changes with effective stress within a relatively narrow interval of values (Fig. 10.16), from fully opened apertures to collapsed fracture space. This narrow range can be used as an indicative tool to distinguish fracture and rock permeabilities, the latter changing with effective stress within a relatively broad interval of values (Fig. 10.21).

The relationship between the fracture permeability

and stress has been determined by (Walsh, 1981; Sigal, 1998, personal communication)

$$k^{1/3} = A + B\log_e(\sigma),\qquad(10.29)$$

where A and B are constants and σ is the effective confining pressure. A and B can be determined from the data distribution (Fig. 10.16). The fracture permeability can also be calculated from (Sigal, 1998, personal communication)

$$k^{1/3} = k_0^{1/3}\left(1 - \frac{2^{1/2}h}{a_0}\log_e\frac{\sigma}{\sigma_0}\right),\qquad(10.30)$$

where k_0 and a_0 are the permeability and half-aperture of the fracture at a reference effective stress σ_0 and h is the fracture surface roughness factor. Figure 10.16 indicates that there are relatively small permeability changes over a broad range of effective stresses, before the fracture permeability collapses due to pore fluid pressure dissipation below the value of the strength of the surrounding rock. Another example of the permeability–effective stress relationship comes from borehole 948D, which penetrated the décollement of the Barbados accretionary prism (Fisher *et al.*, 1996). At its upper extreme, the bulk permeability approaches $10^{-12}\,\mathrm{m}^2$ when the fluid pressure approaches being lithostatic. At the lower extreme projected beyond the range of the data set, a bulk permeability of about $10^{-18}\,\mathrm{m}^2$ is predicted when the fluid pressure is hydrostatic. This lower extreme is consistent with laboratory tests of fine-grained material from the same area (Taylor and Leonard, 1990). The permeability along the décollement varies with the pore pressure along a continuum, from intergranular flow when the fluid pressure is low, to 'fracture' flow when the pressure approaches being lithostatic (Fisher *et al.*, 1996).

As the fluid pressure drops and sediments in the fault zone compact, the width of fractures within the fault zone decreases. Although compaction is a plastic response to increasing mean stress, the assumption that the fault zone materials respond elastically for the relatively small change in stress required to close the fracture allows the linear relationship between the fluid pressure and the fracture width to be represented as (Roberts *et al.*, 1996)

$$\frac{w - w_0}{l} = -m(p_c - p),\qquad(10.31)$$

where w_0 is the initial fracture width, w is the current fracture width, l is a unit length (1 m), p is the current fluid pressure, p_c is the pressure at which the fracture opened and m is the constant describing the compressibility of fault zone materials.

Interaction of rock and fracture permeability

The fluid flow properties of a fault zone that bounds a thrust sheet influence the fluid flow in and around the fault zone differently for each of the four fault zone categories mentioned earlier (Fig. 10.15). The character of the fluid flow depends on the permeability contrasts between the fault core, the fault damage zone and the host rock.

Caine *et al.* (1996) and Caine (1999) modelled the fluid flow in a host rock deformed by the different types of fault zone. The models were focused on how the fracture networks contribute to the fault zone permeability structure. Flow simulations, using a $20 \times 20 \times 20\,\mathrm{m}$ cubic model, yielded relative magnitudes and anisotropies of the bulk equivalent permeability in three mutually perpendicular directions, parallel to the fault strike, parallel to the fault dip and parallel to the fault normal. All fractures were assumed to act as parallel smooth-walled conduits with rectangular cross sections, which allow each element of the fault-fracture mesh to be assigned a fracture transmissivity:

$$T = \frac{a^3 \rho_\mathrm{f} g}{12 \eta}, \tag{10.32}$$

where T is the transmissivity, a is the fracture aperture, ρ_f is the fluid density, g is the acceleration due to gravity and η is the dynamic fluid viscosity. In each model, a uniform hydraulic gradient consistent with regional scale flow system gradients was applied. The total volumetric fluid flux computed between two opposite faces of the cube was used to compute the equivalent bulk permeability:

$$k = \frac{\eta}{\rho_\mathrm{f} g} \frac{q}{IA}, \tag{10.33}$$

where k is the equivalent bulk permeability, q is the volumetric flow rate, I is the dimensionless hydraulic gradient and A is the cross-sectional area, across which the discharge q flows.

The various models of Caine (1999) give interesting permeability results. They provide a simplified visualization of permeability changes during the fault zone development from the pre-seismic, through the seismic to the post-seismic periods for all four architectural styles of the fault zone. A host rock undergoing different stages of deformation in the development of a specific architectural style does not seem to have a consistent anisotropy of permeability, i.e. none of the three flow directions always coincides with maximum, minimum or intermediate permeability. In contrast to the host rock, the fault zones themselves give rise to a permeability anisotropy with the minimum permeability always being normal to the fault zone, the intermediate permeability always parallel to the dip and the maximum permeability always parallel to the fault strike. The exception is the case of a single fracture fault (see Fig. 10.15b). This anisotropy is not particularly pronounced, except where the deformation zone is distributed. A sensitivity analysis indicates that the permeability anisotropy is affected by the fracture length and the orientation more than by the fracture density.

An interesting comparison is between a host rock with randomly located fractures in a localized continuous fault zone and one with a single fracture fault. The permeability contrast between the host rock and the fault zone for the same size of fracture apertures during the seismic period differs by five orders of magnitude in the two cases. The fault permeability is still three orders of magnitude greater when the fault zone aperture is an order of magnitude smaller than the apertures of the fractures within the host rock during the post-seismic period. The remaining scenarios indicate a permeability that is on average one to two orders of magnitude higher in the fault zone than the permeability of the host rock when the apertures of the fractures in the fault zone and host rock have the same size during the post-seismic period. These permeability differences of several orders of magnitude can have a significant impact on flow patterns near fault zones bounding thrust sheets. Studies of modern accretionary prisms indicate that, where there is a difference in permeability of two to five orders of magnitude between the sediments and the faults, fluid flow is always captured by the faults (e.g., Moore *et al.*, 1990c). Geochemical evidence from the Apache Leap tuff, Arizona, also indicates such a preferential flow along faults (Davidson *et al.*, 1998).

The bulk permeability anisotropy of a fault zone itself is influenced by the anisotropy of individual fault zone components combined with the integrated effect of permeability contrasts between components. These relatively large contrasts can cause significant changes in the patterns and rates of fluid flow in porous or fractured media that would not otherwise occur in homogeneous isotropic permeability structures (Caine, 1999). Each local fracture controls its equivalent permeability by fracture aperture and width (Fig. 10.23):

$$q = \frac{a^3}{12} \frac{\rho_\mathrm{f} g}{\eta} IW, \tag{10.34}$$

where a is the fracture aperture, W is the fracture width, ρ_f is the fluid density, g is the acceleration due to gravity, η is the dynamic fluid viscosity and I is a dimensionless hydraulic gradient. The impact of changes in fracture aperture (Fig. 10.23b) is much greater than the linear impact of the fracture width (Fig. 10.23c), because the total volumetric flux through a fracture is related to the cube of its aperture.

(a)

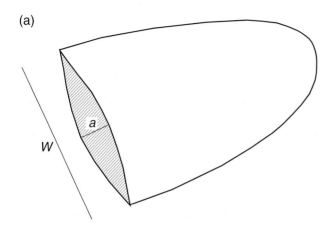

Fig. 10.23. (a) Fracture aperture and width. Note that the fracture width is always measured normal to the direction of flow within the fracture. (b) The impact of the fracture aperture change on the equivalent bulk permeability. (c) The impact of the fracture width change on the equivalent bulk permeability.

(b)

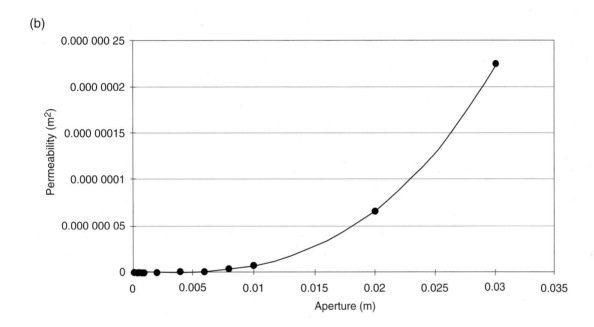

(c)

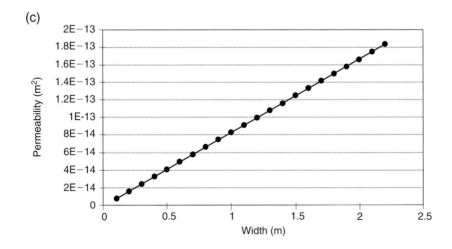

As discussed earlier, thrustbelt structures, which undergo multiple seismic cycles during the thrustbelt development, also undergo multiple adjustments of host rock and fracture permeability. Three stages in the fault zone development characterize each such cycle (e.g., Moore *et al.*, 1990b; Knipe, 1993; Brown, 1995; Roberts *et al.*, 1996):

stage 1   when the fault permeability is several orders of magnitude larger than the permeability of the host rock and the zone is capable of capturing fluid flow;

stage 2   when the permeable system of the fault zone may collapse and entrap fluids undergoing increased fluid pressure, surrounded by a less permeable host rock; and

stage 3   when the permeability system of the fault zone competes with the permeability system of the host rock.

Apart from these general developmental stages, local combinations of a fault zone permeability system with the permeability systems of reservoirs, source rocks and seals introduces spatial complexities. For example, stage 2 can be enhanced by the location of the detachment inside a low-permeability detachment horizon, or the permeability balance between the fault and the host rock in stage 3 can be modified by the presence of a highly permeable reservoir horizon.

When one tries to evaluate the importance of fracture permeability relative to rock permeability in thrustbelts, it is important to treat semi-lithified accretionary prisms and lithified accretionary wedges separately because of their different permeability systems.

Accretionary prisms are characterized by a broad range of burial, compaction and deformation histories, because sediments experience large and rapid changes in their physical and mechanical properties (e.g., Karig, 1986, 1990). The general sequence of deformation progresses from pervasive ductile deformation accomplished by grain boundary sliding, through localized ductile failure and formation of deformation bands, and finally to brittle faulting, as known from the Nankai accretionary prism, Japan (Morgan and Karig, 1995). Fluid flow paths in accretionary prisms include bulk flow through the grain framework of sediments, flow through fracture networks and flow along fault arrays.

The pathway for bulk flow is provided by pores in the grain framework of the sediment undergoing compaction and lithification (see Fig. 10.4). Migration characteristics change with time, controlled by porosity and permeability changes (e.g., Bray and Karig, 1985; Karig, 1990). Microfabrics formed in fine-grained sediments are very heterogeneous, composed of a relatively rigid framework of grain aggregates connected by weak links, each with different strengths (Knipe *et al.*, 1991). Alignment domains are collapse zones, which result from heterogeneous compaction driven by repeated, localized and transient deformation events, which control the porosity and permeability evolution. It thus follows that alignment zones are zones of reduced porosity, which induce permeability anisotropy in sediments.

Flow through the fracture networks utilizes several fracture types. One of these is mud-filled veins, which are formed at shallow depths in fine-grained slope sediments, while anastomosing networks of low-displacement cataclastic shear zones are formed in silts and sands (Knipe *et al.*, 1991 and references therein). Carbonate-filled vein arrays and open fracture networks have also been found in some active margins (e.g., Brown and Behrmann, 1990).

Mud-filled veins are driven by fluid pressure exceeding the cohesion of the sediment during the compaction. Their extensional geometry, evidence of disaggregation, dilation and mineral growth, indicate fluid flow into them (Knipe *et al.*, 1991). They act as small fluid reservoirs during deformation.

Anastomosing cataclastic shear zones commonly develop in partly cemented siltstones/sandstones. They are characterized by grain-size reduction, fracture and brecciation. Their origin is associated with repeated strain hardening events linked to pore pressure fluctuations within the fractured material (Knipe *et al.*, 1991), indicating important episodic fluid flow. Natural examples come from the Cascadia accretionary prism, Oregon (Knipe *et al.*, 1991).

Flow along fault arrays in accretionary prisms has been determined by:

(1)   the identification of biological communities and vents near faults (e.g., Dron *et al.*, 1986; Le Pichon *et al.*, 1986, 1990a; Suess *et al.*, 1987; Griboulard *et al.*, 1989; Han and Suess, 1989; Carson *et al.*, 1990; Lewis and Cochrane, 1990; Moore *et al.*, 1990b; Sibuet *et al.*, 1990; Moore and Vrolijk, 1992);

(2)   the recognition of geochemical and thermal anomalies in fault zone fluids (e.g., Boulegue *et al.*, 1986; Kinoshita and Yamano, 1986; Blanc *et al.*, 1988; Davis *et al.*, 1990; Fisher and Hounslow, 1990; Foucher *et al.*, 1990a, b; Gieskes *et al.*, 1990; Henry *et al.*, 1990; Lallemant *et al.*, 1990a, b; Langseth *et al.*, 1990; Yamano *et al.*, 1990, 1992; Shipley *et al.*, 1995; Larroque *et al.*, 1996; Martin *et al.*, 1996);

(3)   the logging-while-drilling of thrust zones and core analysis (e.g., Moore *et al.*, 1995);

(4) the identification of syn-tectonic extensional shears and hydrofractures (e.g., Etheridge, 1983; Von Huene, 1984; Lundberg and Moore, 1986; Vrolijk, 1987; Brown and Behrmann, 1990; Byrne and Fisher, 1990; Fisher and Byrne, 1990; Kemp, 1990; Mascle and Moore, 1990; Bebout, 1991; Vrolijk and Sheppard, 1991; Maltman *et al.*, 1992; Butler and Bowler, 1995; Larroque *et al.*, 1996);

(5) the preferential growth of new mineral phases along fault zones (e.g., Knipe, 1986; Bebout, 1991); and

(6) seismic anomalies (e.g., Bangs and Westbrook, 1991; Shipley *et al.*, 1993).

The mesoscale geometry of these faults range from about 1 mm to over 20 m thick zones of scaly fabrics, defined by a mesoscopic set of anastomosing, polished and slickensided surfaces, breccias and stratal disruption/melange (e.g., Brown and Behrmann, 1990). Deformation in these zones comprises disaggregation and particulate flow both with and without fracturing (Knipe *et al.*, 1991). Microscale structures, observed in modern fault zones from accretionary prisms, contain micromovement domains of finite porosity collapse, which enclose domains of less aligned material (e.g., Knipe, 1986), representing the end of each rupture event. Also present are randomly arranged aggregates of less than 6 μm in size (e.g., Prior and Behrmann, 1990), which represent the main rupture stage. A series of microfabrics develops during the deformation event and is modified during its late stages and between deformation events (Knipe, 1986). The fluid flow depends on the volume, dilation and migration velocity of strain waves related to each deformation event (Knipe *et al.*, 1991), resulting in transient fluid flow. Evidence for transient fluid flow comes from both direct measurements (e.g., Carson *et al.*, 1990) and experiments (e.g., Brown *et al.*, 1994). Pervasive straining by disaggregation and particulate flow, involving deformation along certain portions of the fault zone during such a deformation event, produces random fabrics that change the deformed part of the fault zone into a conduit. Decrease and collapse of fracture apertures occur at the end of the deformation event when fluid is either expelled or a fluid overpressure is dissipated.

Lithified accretionary wedges are characterized by a less broad range of burial, compaction and deformation histories than semi-lithified accretionary prisms, because their sediments experience fewer changes in their physical and mechanical properties. A competition between sediments and faults in their ability to focus fluid flow largely depends on the type of faulting. Fault-slip events range from earthquakes of large magnitudes, through earthquake swarms, down to micro-earthquakes and aseismic creep waves, which migrate at slow rates and produce small displacements (Knipe, 1993). Large earthquakes associated with large rupture areas have a greater possibility of rapidly linking compartments with different fluid pressures, which results in a seismic pumping flow mechanism, described in detail above. In addition to the modification of pre-faulting fluid pressure distribution, earthquakes also create their own fluid pressure patterns, for example high fluid pressure gradients at the tip zones of the rupture area, which decay with time (e.g., Nur and Booker, 1972; Booker, 1974). The smaller magnitude displacement events associated with aseismic creep (e.g., Scholz, 1990) have the potential to transfer slowly smaller packets of fluid along the fault.

The conduit potential of a fault is critically dependent not only on the decay of fluid pressure gradients affected by the permeability evolution along the network of fractures and faults connecting pressure gradients, but also on the pattern of dilation produced during the faulting event and dilation decay in and around the fault zone (Knipe, 1993). The factors that control the modification of the initial pattern of dilation induced by a large faulting event include:

(1) the distribution and magnitude of aftershocks;

(2) the time-dependent processes that operate in the fault zone volume during and after the aftershock activity; and

(3) the time interval before repetition of the cycle.

The aftershock period is that period during which communication is maintained between isolated and different fluid pressure regimes, which originated by rupturing in the pre-seismic stage. It represents a minimum time period when enhanced fluid flow paths are present in the fault zone. Because aftershocks are known to cluster either in the fault tip zones or at fault bends (e.g., Sibson, 1989; Scholz, 1990), they can modify fluid flow in the fault zone by creating fluid storage at tips and dilatational bridges (e.g., Connolly and Cosgrove, 1999). Thus, both fluid flow and aftershock activity are influenced by the interaction between fluid pressure gradients set up by the faulting itself and pre-existing fluid pressure patterns modified by the propagating fault. During and following the aftershock period, aseismic creep also characterizes some fault zones. It represents the time period when fluid packets may still be able to migrate with creep waves and when dilation is likely (Knipe, 1993). The permeability evolution as related to the seismic cycle can thus be divided into three stages (Fig. 10.22):

(1) a pre-seismic period of fluid redistribution;
(2) a co-seismic period when the main shock and aftershocks maintain the possibility of enhanced flow; and
(3) a post-seismic period when the porosity and permeability of the fault zone can decay or collapse.

The fluid regimes before, during and after failure are well documented from artificially triggered slope failures in soil by Harp *et al.* (1990), who measured pore fluid pressure during the whole experiment using electric piezometers. Despite the fact that each piezometer recorded a specific local pore pressure curve, the composite information shows a pore pressure increase during the pre-failure stage. The first pore pressure drop is due to dilatancy fracturing that started shortly before failure. A major fluid pressure drop occurs during failure and progressive fluid pressure dissipation continues post-failure.

The ability of a fault to act as a focus for fluid flow or as a barrier is determined by the evolution of the porosity and permeability of the deformed rocks in the fault zone and the adjacent volume of damaged rock during the earthquake cycle.

The pre-seismic stage of the cycle is associated with the development of microfractures and fracture arrays, which link during the period before the main event (e.g., Bekins *et al.*, 1995; Roberts and Nunn, 1995). Such structures may form some of the damage zones (Fig. 10.15a).

The main seismic stage results in rupture propagation and displacement (e.g., Sibson, 1990b). It leads to the generation of frictional wear gouges and breccias, where the cataclastic flow is characterized by grain rolling, frictional sliding and comminution by brittle fracture (House and Gray, 1982; Aydin and Johnson, 1983; Blenkinsop and Rutter, 1986; Rutter *et al.*, 1986; Sammis *et al.*, 1986). Dilation associated with the main phase is instrumental in the development of implosion breccias associated with collapse into dilation sites (Sibson, 1986). This dilation can occasionally result in a rapid fluid pressure dissipation, cement precipitation and formation of cement-supported breccias (Sibson, 1986; Sibson *et al.*, 1988).

The aftershock stage associated with the post-seismic phase involves a range of processes similar to those involved in the main event. The post-seismic phase is the most critical for potential fault permeability collapse. This phase comprises potential permeability reduction by fracture network collapse due to a drop in fluid pressure, thermal expansion of the wall rock adjacent to fractures and/or precipitation of mineral cements along the fault zone (e.g., Lowell, 1990; Knipe, 1993; Brown, 1995; Roberts and Nunn, 1995) or potential retention of the residual permeability due to the roughness of fracture or fault walls (e.g., Jiang *et al.* 1997). Chemical healing or the dissolution of minerals in high-energy environments adjacent to the fracture and reprecipitation in lower-energy environments within the fracture efficiently closes microfractures or smaller fractures but leaves larger fractures through which fluid can flow (Hickman and Evans, 1987; Brantley *et al.*, 1990; Evans *et al.*, 1996; Jiang *et al.*, 1997). Thermodynamic arguments suggest that quartz and calcite can be deposited along a flow path when pressure decreases (Bruton and Helgeson, 1983; Capuano, 1990). These minerals are never deposited as fracture filling in overpressured zones of basin sediments (Roberts *et al.*, 1996).

The final result combining all the deformation processes related to faulting discussed above, determines the next state of the fluid flow pattern. The other controlling factor is the timing of fault activity relative to the diagenetic sequence in the area, which controls the interaction of host rock and fault permeabilities affecting the resulting fluid flow pattern. Therefore, the controls on the pore fluid flow and fluid pressure regimes are the timing of the faulting cycle stages together with the type of lithologies affected by the individual stages. Only when the right combination of fault event magnitude, fracture geometry and fracture linking occurs, allowing the creation of a pathway between the fluid source and sink volumes and production and preservation of permeable fault zone fabrics, does extensive fluid flow between the host rock and the fault zone occur. The situation can be even more complicated when hydrocarbon maturation factors and the role of buoyancy forces that aid fluid migration and leakage within and from the fault zone are considered. For example, the influx of hydrocarbons into a fault zone can prevent diagenesis within the fault zone or stop the growth of cement. This provides a mechanism for prolonging the permeability window of the fault (Fig. 10.22), i.e. maintaining a fluid migration pathway, which can be eventually, but more slowly, sealed by tar or asphalt formation.

The elliptical shape of an isolated fault, with displacement decreasing from its centre to its propagating tip (Walsh and Watterson, 1990) has an important consequence for host rock–fault permeability interactions. This is additional to the pressure gradients induced by a faulting event (Figs. 10.8–10.14), the rate of decay of that gradient, and the permeability of fault zones discussed earlier, because the fluid flow is also dependent on the location and size of fluid source and sink areas, where fluids enter and escape from the fault zone.

PART THREE
Thermal Regime

11 Introduction to the thermal regimes of thrustbelts

In this brief introduction the various factors that control the evolving thermal regime of a thrustbelt will be discussed. The thermal regime is important for both the structural style and petroleum systems of a thrustbelt.

Structural style is affected by the thermal regime on both thrust wedge and thrust sheet scales. The structure of a thrust wedge is dependent on the lithospheric strength, which is controlled by the thermal conditions (Fig. 11.1). This has been shown by laboratory experiments where thermal conditions affect the strength of rocks undergoing orogenesis (e.g., Goetze and Evans, 1979; Fig. 11.2). The crust and uppermost lithospheric

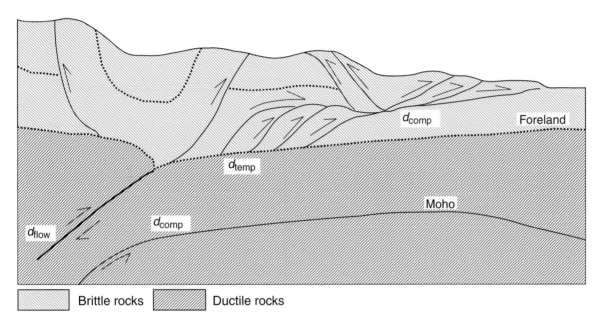

Brittle rocks   Ductile rocks

Fig. 11.1. Concept for a macroscopic deformation mechanism involving brittle and ductile crystalline systems, separated by the brittle–ductile transition (Hatcher and Hooper, 1992). d_{comp} is compositional detachment, d_{temp} is the thermally induced detachment, d_{flow} is the detachment caused by difference in flow or strain rate.

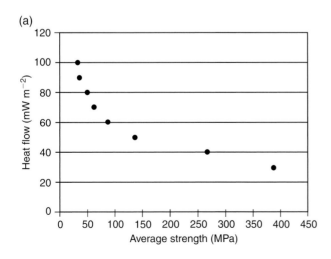

Fig. 11.2. (a) Average strength as a function of heat flow. The strength profile used for the calculation of the average strength was modelled for a 16 km thick quartzite layer for heat-flow values of 30, 40, 50, 60, 70, 80, 90 and 100 mW m^{-2}. Other parameters were kept unchanged (β value for thrusting 3.0 (from Ranalli and Murphy, 1987), pore fluid factor λ for hydrostatic gradient 0.36, surface temperature 273.16 K, strain rate ε 10^{-15} s^{-1}).

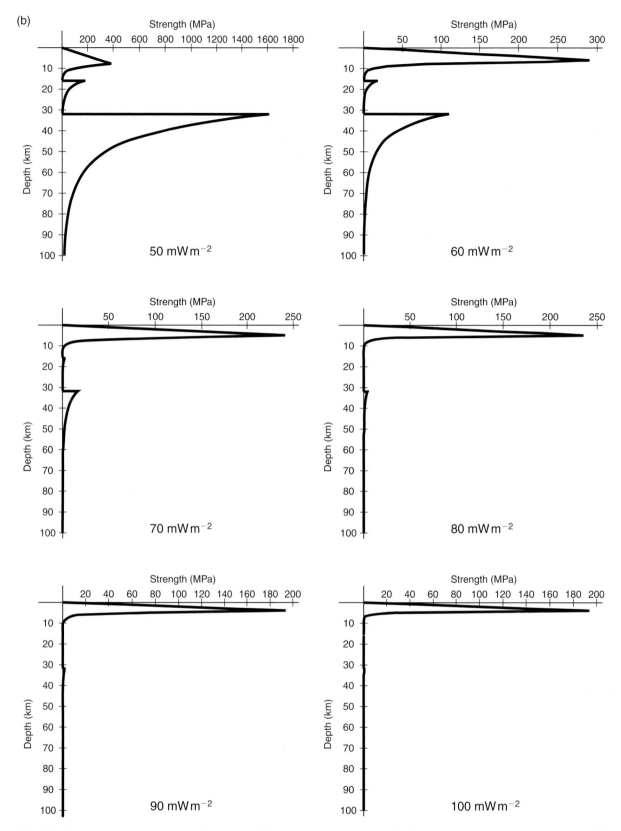

Fig. 11.2. (b) Lithospheric strength profiles modelled for a three-layer quartzite–diorite–dunite system for heat-flow values of 50, 60, 70, 80, 90 and 100 mW m^{-2}. Other parameters are as in (a). Note how the increased temperature regime raises the brittle–ductile boundary and decreases the strength.

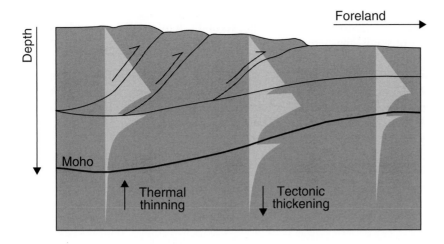

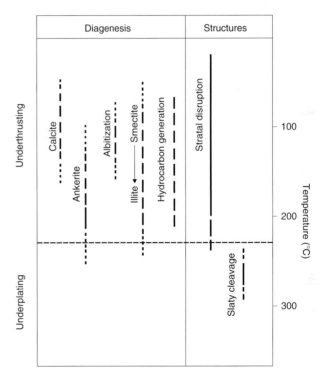

Fig. 11.3. A schematic illustrating different parameters controlling the lithospheric strength in a compressional region (after Zoetemeijer, 1993). The lithosphere is simulated as a three-layer system.

mantle may behave as an elastic entity at low geothermal gradients around $10\,°C\,km^{-1}$ (Ord and Hobbs, 1989). With increasing geothermal gradients, the likelihood for a detachment developing along the boundaries between regions of contrasting strength increases (Fig. 11.3). The rheology of the crustal layers determines both the position of the basal décollement and whether parts of the thrustbelt deform in a brittle or ductile manner, as has been demonstrated in the Appalachians (e.g., Mitra, 1987). The precise location of the detachment depends on the geothermal gradient and the mineralogical composition of the crust. In general, for high geothermal gradients around $50\,°C\,km^{-1}$ detachments are no longer predicted to be present and the crust and mantle below 5 km are uniformly weak. The lithospheric strength controls the lithospheric flexural response to thrust mass loading that, in turn, controls the thrust wedge, together with the other factors described in Chapters 2 and 9. The thermal regime of an orogenic belt evolves with its structural development. An advancing thrustbelt, in turn, controls the properties of the continental lithosphere. Changes in geothermal gradients can result in changes in the lithospheric strength, referred to as tectonic thickening and thermal thinning (Fig. 11.3). Immediately after thrusting the lithosphere becomes stronger than a nonthickened lithosphere. The developing thrustbelt elevates geothermal conditions within the thickened lithosphere by burial of radiogenic-enriched crustal rocks. This results in weakening of the thickened lithosphere after 10–20 million years following the initiation of thrusting. Typical examples of thermal weakening come from the Aegean and Carpathian–Pannonian regions.

At a thrust sheet scale, the strength of a rock sequence responding to initial shortening is controlled by the thermal conditions, which determine whether a rock deforms in a ductile or brittle manner (e.g., Mitra, 1987). In addition, the thermal regime affects the distribution of cements (Fig. 11.4), which precipitate in different tem-

Fig. 11.4. Model for relative timing of cementation of the Alaskan accretionary complex (modified after Sample, 1990). Increasing length of dashes corresponds to increasing confidence in the temperature estimate. Solid lines represent temperatures of stratal disruption and slaty cleavage formation that are constrained by fluid inclusion data and other indicators of metamorphism. The horizontal dashed line is the approximate temperature boundary between deformation dominated by stratal disruption, related to underthrusting, and deformation dominated by cleavage formation, related to underplating.

perature intervals. These cements control the cohesive element of the rock strength. Cementation can dramatically change the strength of an imbricated sedimentary section, as documented in the Alaskan accretionary complex, where lithification by calcite and ankerite cementation preceded the imbrication (Sample, 1990). Such cementation can change the rheology of sandstone by filling pore space, which increases the peak strength and cohesion so that particulate flow is no longer a viable deformation mechanism. Such a peak strength increase due to cementation is discussed in Chapter 7.

The petroleum systems of a thrustbelt are greatly affected by the thermal regime. First, the thermal regime controls source rock maturation and hydrocarbon expulsion and, secondly, it affects the seal and reservoir rocks by controlling their deformational environment and cementation.

The source rock maturation depends on the thermal history of the source rock section (e.g., Waples, 1992a, b). In the case of source rocks deposited prior to thrusting, the thermal history of the precursor basin also becomes important.

12 Role of pre-orogenic heat flow in subsequent thermal regimes

The initial heat flow, present in an area prior to shortening, plays an important role in establishing the later thermal regime of the local thrust sheets. It sets the stage for the syn-orogenic thermal regime by determining its initial character, although its influence progressively diminishes during the subsequent development stages. The flexural study of the Aquitaine foreland basin on the French side of the Pyrenees (Desegaulx et al., 1991) serves as an example of the influence of the heat flow that existed prior to the onset of shortening. This foreland basin developed on thinned continental lithosphere, inherited from pre-orogenic extensional phases. The extension-induced transient thermal state of the lithosphere resulted in ongoing thermal subsidence and changes of the flexural rigidity through time. Numerical modelling shows that the inherited transient thermal state of the lithosphere contributed to the total basin depth and width, the post-compressional subsidence history and the forelandward sediment onlap pattern. The thermomechanical effects of pre-orogenic extension significantly reduced the magnitudes of the flexural rigidity (30–43%) and the topographic or thrust load (40%) required to form this basin.

The variety of basins capable of being later incorporated in thrustbelts can be reviewed according to their tectonic setting or geothermal signatures (Fig. 12.1, Table 12.1; e.g., Allen and Allen, 1990; Platte River Associates, 1995).

The coolest thermal regimes are present in regions formed by old continental shield or old oceanic crust (Figs. 12.1 and 12.2, Table 12.1). Figure 12.2 shows the surface heat-flow contours for the broader Equatorial Atlantic region and allows a comparison of the thermal regimes of numerous settings. The coolest regions are clearly represented by the West African Shield, the Guyana Shield and the São Francisco Craton. The oldest portions of the South Atlantic, the Equatorial Atlantic and the Central Atlantic oceanic crust include areas that form the second coolest region.

Old passive margins, such as the Atlantic margins (Fig. 12.2), can be described as having near-normal geothermal gradients (sensu Robert, 1988). They have present-day geothermal gradients of about 25–30 °C km^{-1}. Typical values are; for example, the Congo with 27 °C km^{-1}, the Gabon with 25 °C km^{-1} and

the Gulf Coast at the Terrebone Parish bore hole with 25 °C km^{-1} (Fig. 12.3). Typical vitrinite reflectance profiles for these geothermal gradients show that the reflectance R_0 is about 0.5% at depths of about 3 km (Fig. 12.3). These profiles have sublinear curves.

Basins, which are cooler than normal (Robert, 1988), include oceanic trenches (Fig. 12.4), outer forearcs (Fig. 12.5) and foreland basins (Fig. 12.1). Ocean trenches are cold, with surface heat flows commonly around 40 mW m^{-2} (Allen and Allen, 1990). Such a case is illustrated by the surface heat flows in the Andean trench (Fig. 12.4), although a continuous zone of low surface heat flow is not present due to local complexities in the thermal regime. The example of Sumatra and Java (Fig. 12.5), however, illustrates more clearly the location of an oceanic trench and outer forearc by a more or less continuous zone of low surface heat-flow values. This cold zone sharply contrasts with the hot thermal regimes of the volcanic arc and backarc. A further example of a cold oceanic trench and forearc can be found in the Eocene–Miocene coals of the Japanese archipelago (Robert, 1988). They occur in two different regions and in two different settings. The first is in Hokkaido, in the north along a branch of the present-day Japan trench. The second is in Kyushu, in the south in the position of a volcanic arc relative to the Ryu-Kyu trench. These two regions have different vitrinite reflectance profiles (Fig. 12.6). The Hokkaido region is cold with poorly evolved coals. They are sub-bituminous with a R_0 value of 0.5% occurring at a present depth of 5 km. The volcanic arc in Kyushu is hot and contains anthracites, which have R_0 values of over 2%. The Mariana trench is a southward continuation of the Japan trench. Its forearc region is also cold, with surface heat flows of less than 42 mW m^{-2} (Allen and Allen, 1990).

An example of a foreland basin is the North Alpine Foreland basin in southern Germany with typical present-day geothermal gradients of 22–24 °C km^{-1} (Teichmuller and Teichmuller, 1975; Jacob and Kuckelhorn, 1977). The Anzing 3 bore hole near Munich, which penetrates the autochthonous molasse filling the basin, has a vitrinite reflectance of only 0.51% at the base of the Tertiary section, a depth of 2630 m. The Miesbach 1 bore hole cuts through about 2 km of

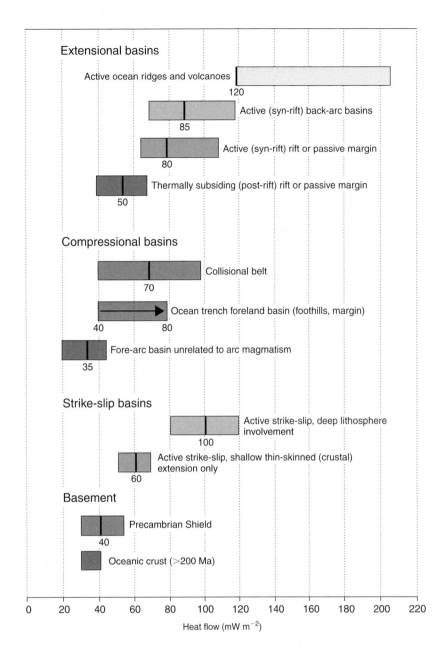

Fig. 12.1. Summary of typical heat-flow values associated with sedimentary basins of various types (Allen and Allen, 1990).

thrust sheets of the frontal Alpine thrust zone (sub-Alpine zone) before bottoming in autochthonous sediments at a depth of 5738 m (Fig. 12.7). Even at its total depth, the vitrinite reflectance is still only 0.6%. It indicates an abnormally low geothermal gradient during the Tertiary, which was caused by rapid subsidence.

Before discussing basins that are hotter than normal it is necessary to mention subduction- and collision-related orogenic belts. If the belts are rather long-lasting such as the Andes or the Greater Caucasus, their syn-tectonic sediments are deposited under their own thermal regimes and subsequently incorporated into the orogenic belt. The thermal regime of the Andes is shown in Fig. 12.4. Although they are rather complex, the

surface heat-flow contours indicate that some orogenic zones, such as the Altiplano Plateau, have significantly hot thermal regimes. The thermal regime of the collisional system in front of the Arabian plate (Fig. 12.8) shows a moderate warming in the contractional zone directly in front of the indenting Arabia, reaching surface heat-flow values of 80–90 mW m^{-2}, and a moderate cooling along the orogenic zone towards the east, characterized by a decrease from values of 80–90 mW m^{-2} to values of 50 mW m^{-2}. This pattern is in accordance with expectations that the geothermal conditions should be most elevated directly in front of the indenter because here the crust undergoes the greatest thickening by burial of radiogenic-enriched crustal rocks.

Table 12.1. Typical values of surface heat flow for geological provinces (Platte River Associates, 1995). Some generalizations can be made within a given province, but there are numerous local anomalies. Therefore, listed values serve only as guidelines in determining appropriate values for further studies. Platte River Associates compilation is based on data from Gretener (1981a), Foucher and Tisseau (1984), Lucazeau and Le Douaran (1984), Von Herzen and Helwig (1984), Čermák and Bodri (1986), Lucazeau (1986), Royden (1986), Allen and Allen (1990).

Geological province/specific example	Heat flow	
	(heat-flow units)	(mW m^{-2})
Average oceanic	1.6	67
Average continental	1.4	58
Eastern USA	1.36	57
England/Wales	1.41	59
Great Plains, Western USA	1.5–2.5	62–104
Young or active extensional basins	>1.9	>80
Pannonian basin, Central Europe	2.0–2.5	85–105
Pattani Trough, Gulf of Thailand	1.9–2.4	80–100
Gulf of Lion margin, Western Mediterranean	1.6–2.1	65–90
Gulf of Suez, Egypt	1.5–2.0	60–84
Viking Graben, North Sea	1.3–2.0	56–82
Uinta basin, Utah	1.0–1.6	40–65
North Biscay margin, East Atlantic	0.8–1.0	36–43
Old shield areas	<1.0	<40
Balkan–Ukrainian Shield	0.7–1.0	30–40
Canadian Shield	0.8	34
West Australia	0.9	39

Note:
Platte River Associates (1995).

Basins which are hotter than normal (Robert, 1988) include those located in regions of lithospheric extension such as backarc basins (Fig. 12.9), oceanic spreading systems (Fig. 12.2), continental rift systems (Fig. 12.10), some strike-slip basins (Fig. 12.11) and the internal arcs of zones of B-type subduction (Fig. 12.5).

Oceanic rifts are zones of very high heat flows, as documented by the data in Fig. 12.2. 126–168 mW m^{-2} are typical values and they can occasionally reach 210–252 mW m^{-2} (Allen and Allen, 1990; Deming and Nunn, 1991).

Some Californian strike-slip basins have very high geothermal gradients, for example, 200 °C km^{-1} in Imperial Valley. This means that even very young sediments in these basins can be highly mature. Figure 12.11 shows a prominent system of high surface heat-flow anomalies associated with a system of pull-apart basins occurring between the Salton Sea and Baja, California. These pull-apart basins are developed inside releasing oversteps between the San Andreas, Brawley, Imperial and San Jacinto dextral strike-slip faults (Fuis and Kohler, 1984; Moore and Adams, 1988; Larsen and Reilinger, 1991). The pull-apart basins are progressively closer to developing oceanic crust by Sea-floor spreading in the direction from Salton Sea to Baja, California and all are associated with tholeitic basalt and rhyolite volcanism.

Continental rifts have high present-day geothermal gradients, e.g. over 50 °C km^{-1} in the Red Sea or up to 100 °C km^{-1} in the Upper Rhine Graben (Allen and Allen, 1990). Typical heat-flow values for young rifts cluster around 60.5–90.5 mW m^{-2} (Morgan, 1984).

Oceanic measurements and deep bore holes in the Red Sea (Girdler, 1970) indicate that high surface heat flows, generally over 126 mW m^{-2}, occur in a broad band at least 300 km wide (Figs. 12.10a and b). This

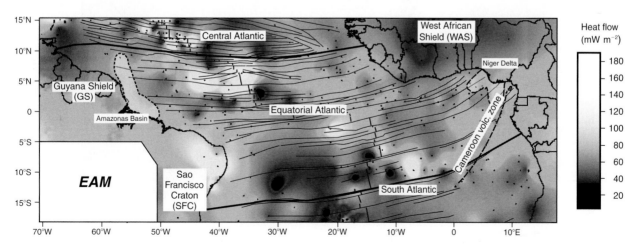

Fig. 12.2. Surface heat-flow contour map of the broader equatorial Atlantic region. Data are taken from the database maintained by the International Heat-flow Commission.

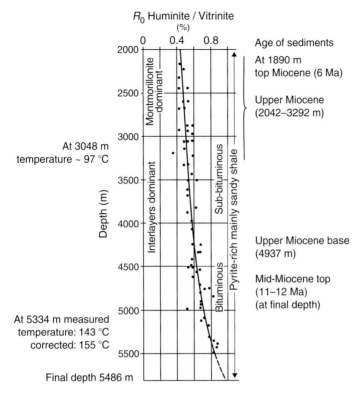

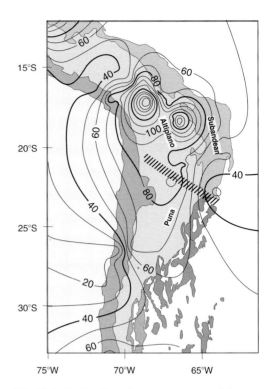

Fig. 12.3. Vitrinite reflectance profile for the Terrebone Parish, Point au Fer bore hole in Louisiana (Heling and Teichmuller, 1974). The profile is sub-linear and continuous, indicating a near constant geothermal gradient through time.

Fig. 12.4. Surface heat-flow contour map of the Andes between 10° S and 35° S longitudes. Data are taken from the database maintained by the International Heat-flow Commission.

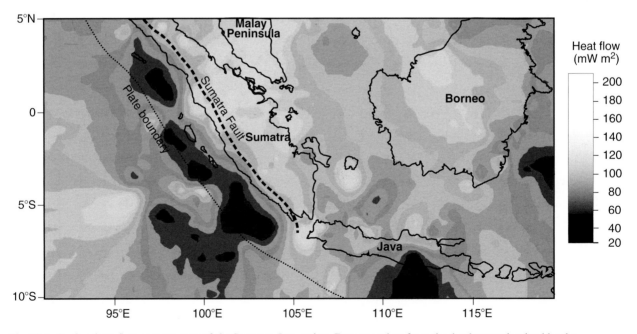

Fig. 12.5. Surface heat-flow contour map of the Sumatra–Java region. Data are taken from the database maintained by the International Heat-flow Commission.

(a)

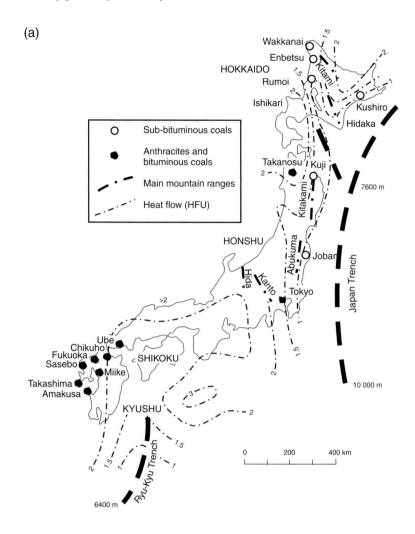

(b)

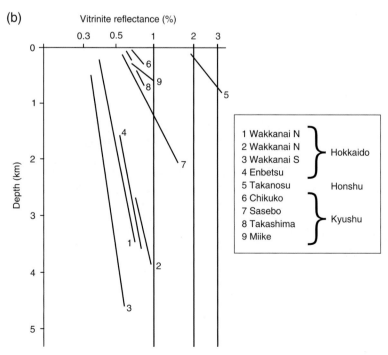

Fig. 12.6. (a) The deposits of Tertiary coals in Japan and the surface heat flow in HFU (1 heat-flow unit $\cong$ 42 mW m^{-2}) (Robert, 1988). Anthracites and bituminous coals are found in Kyushu and the extreme SW of Honshu whereas sub-bituminous coals are found in the north of Honshu and on Hokkaido. (b) Vitrinite reflectance profiles for the locations shown in (a). The first set of locations 1–4 belongs to a branch of the oceanic Japan trench. They are characterized by low reflectances and are associated with low present-day heat flows. The second group of locations 6–9 corresponds to the internal arc relative to the Ryu-Kyu trench. They are highly evolved at shallow depths of burial and are associated with high present-day heat flows. The single location 5, from the northwest of Honshu, occurs in the current volcanic area where the heat flow exceeds 64 mW m^{-2}, and the vitrinite reflectance is very high.

(a)

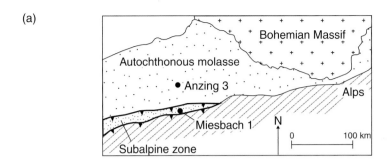

(b)

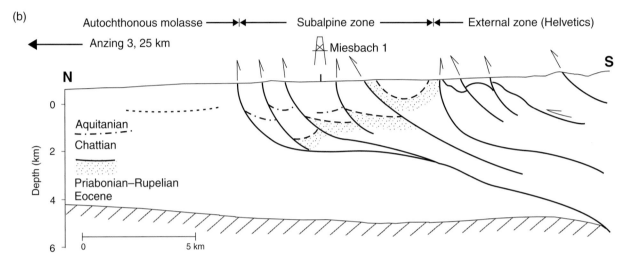

(c)

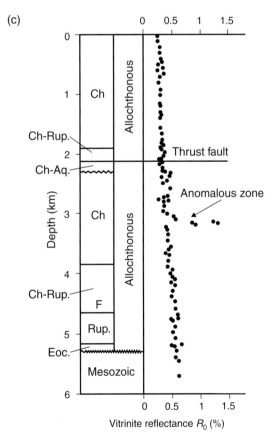

Fig. 12.7. (a) Location of the Bavarian part of the North Alpine Foreland basin in southern Germany (Teichmuller and Teichmuller, 1975; Jacob and Kuckelhorn, 1977). Anzing 3, near Munich, and Miesbach 1 are bore holes discussed in the text. (b) Cross section of the southernmost part of the Bavarian section of the North Alpine foreland basin, showing the location of bore hole Miesbach 1 in the shortened sub-Alpine Zone (Teichmuller and Teichmuller, 1975). (c) The vitrinite reflectance profile at Miesbach (Jacob and Kuckelhorn, 1977) shows that the autochthonous molasse under the basal sub-Alpine thrust is poorly evolved. It does not exceed 0.6% even at 5738 m depth. This is indicative of a very low geothermal gradient during the period of rapid Oligocene deposition.

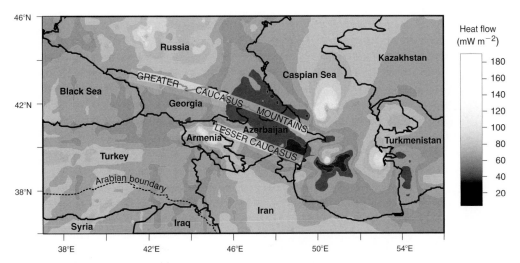

Fig. 12.8. Surface heat-flow contour map of the broader region around the colliding Eurasian and Arabian plates. Data are taken from the database maintained by the International Heat-flow Commission.

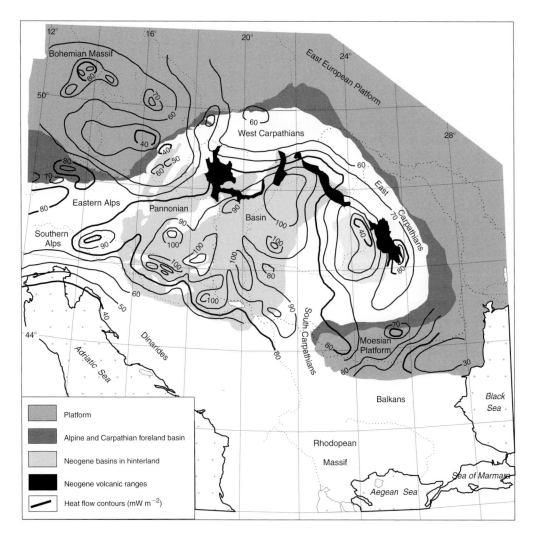

Fig. 12.9. Heat-flow map of the Carpathian–Pannonian region (Nemčok *et al.*, 1998c). The number of boreholes used for the heat-flow calculations by Čermák and Hurtig (1977) and Krus and Šutora (1986) varies in different countries. The largest numbers come from Hungary (100), Bohemia (89) and Slovakia (59). Bore holes are concentrated mainly in basins with industrial activity.

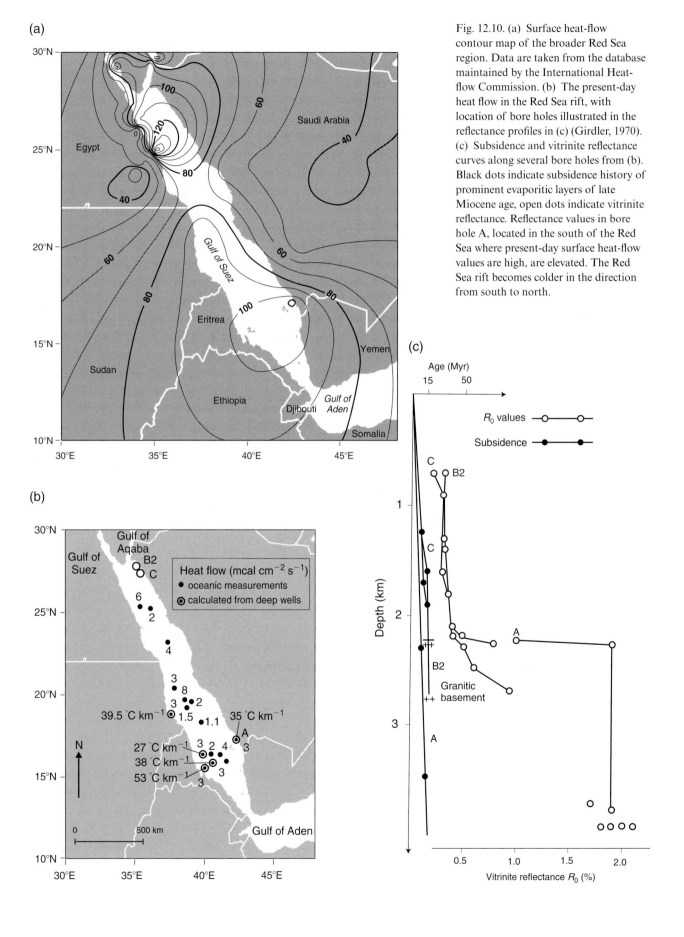

Fig. 12.10. (a) Surface heat-flow contour map of the broader Red Sea region. Data are taken from the database maintained by the International Heat-flow Commission. (b) The present-day heat flow in the Red Sea rift, with location of bore holes illustrated in the reflectance profiles in (c) (Girdler, 1970). (c) Subsidence and vitrinite reflectance curves along several bore holes from (b). Black dots indicate subsidence history of prominent evaporitic layers of late Miocene age, open dots indicate vitrinite reflectance. Reflectance values in bore hole A, located in the south of the Red Sea where present-day surface heat-flow values are high, are elevated. The Red Sea rift becomes colder in the direction from south to north.

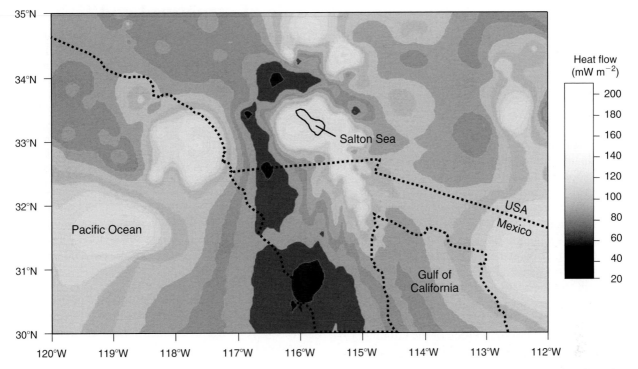

Fig. 12.11. Surface heat-flow contour map of the broader Salton Sea and northern Baja Californian region. Data are taken from the database maintained by the International Heat-flow Commission.

band is centred along the rift axis. The organic maturation shown by vitrinite reflectance profiles, oil, gas and condensate occurrences indicates that the highest maturity is present in the south of the Red Sea. Intermediate values are located in the centre of the Red Sea and the lowest values occur in the Gulf of Suez (Fig. 12.10c). This cooling trend can be correlated with different amounts of extension, which is largest in the south of the Suez–Red Sea rift system (Allen and Allen, 1990). Thus, the Oligocene–Miocene elevated heat flows of the Gulf of Suez have now decreased to near normal values, while high heat-flow values are present in the still actively rifting southern Red Sea.

Internal arc heat flows are elevated because of magmatic activity (Allen and Allen, 1990; Fig. 12.5). The Tertiary anthracites of Honshu, Japan, with a vitrinite reflectance of 2–3%, mentioned above, serve as an example. Similar patterns are found in ocean–continental collision zones such as the Andean Cordillera, and are also found, but less commonly, in parts of continent–continent collision zones such as the Alps.

A case area for lithospheric extension with the development of backarc basins is the orogenic hinterland of the Carpathians, the Pannonian basin system, where surface heat-flow values exceed $100\,\mathrm{mW\,m^{-2}}$ in several basins (Fig. 12.9). The extension started during the Eggenburgian (20 Ma). The region experienced several local basin inversion events during the Late Miocene and Quaternary. The subduction that drove the extension continues to the present day only along a small segment located at the junction of the Eastern and Southern Carpathians (e.g., Nemčok *et al.*, 1998c).

13 Role of structural and stratigraphic architecture in thermal regimes

In this chapter it will be shown how stratigraphic architecture controls the thermal regime of a thrustbelt through lithostratigraphy-dependent thermal conductivity, specific heat, the heat production rate and thermal diffusivity. Because thrusting modifies the stratigraphic architecture in thrustbelts, by causing a repetition of portions of initial rock sections and juxtaposing lateral facies changes, the structural architecture also dictates the thermal regime. Lateral changes of lithofacies can cause temperature disturbances, as has been shown in the Brooks Range and its Foothills, Alaska, and in the western Canada sedimentary basin. Thrusting also modifies the temperature, pressure, porosity, density and composition of the saturant of the original lithostratigraphy and thus further influences the thermal regime.

Before discussing the role of thermal conductivity, specific heat, heat production rate and thermal diffusivity on the thermal regime of a thrustbelt, the theory of heat conduction will be outlined and the above controlling factors defined. Heat convection and its interaction with the conductive thermal regime are described in Chapter 16.

Heat conduction

The temperature change ΔT in a rock body is accompanied by a change in the heat content ΔQ. The change of heat content is

$$\Delta Q = cm\Delta T, \tag{13.1}$$

where c is the specific heat capacity and m is the body mass.

Fourier's first equation of heat conduction shows that the quantity of heat Q flowing through a plate equals

$$Q = -\lambda \frac{T_2 - T_1}{h} At = -\lambda \frac{\Delta T}{h} At, \tag{13.2}$$

where λ is the thermal conductivity, $T_2 - T_1$ is the temperature difference between the two boundary surfaces, h is the plate thickness, A is the area of the plate and t is the time during which the heat flows.

For unit area and unit time, the heat-flow density q is given by (e.g., Buntebarth, 1984; Haenel et al., 1988)

$$q = -\lambda \left(\frac{dT}{dh}\right). \tag{13.3}$$

The heat-flow density is expressed in three dimensions as

$$q = -\lambda \operatorname{grad} T = -\lambda \nabla T, \tag{13.4}$$

where $q\, (= (q_x, q_y, q_z))$ and $\nabla T\, (= (\partial T/\partial x, \partial T/\partial y, \partial T/\partial z))$ are vectors. ∇T is the temperature gradient and λ is the thermal conductivity tensor:

$$\lambda = \begin{pmatrix} \lambda_{xx} & \lambda_{xy} & \lambda_{xz} \\ \lambda_{yx} & \lambda_{yy} & \lambda_{yz} \\ \lambda_{zx} & \lambda_{zy} & \lambda_{zz} \end{pmatrix}, \tag{13.5}$$

which has a negative sign for anisotropic media. If it is assumed initially that heat flows only in a vertical direction, the resulting one-dimensional equation for the heat flow is

$$q = \lambda_{zz}\left(\frac{dT}{dz}\right). \tag{13.6}$$

Because rocks incorporated in thrustbelts, especially sediments, are frequently anisotropic, the thermal conductivity λ has to be measured in a vertical direction for one-dimensional calculations. Also the usual vertical heat-flow density can be distorted by local heat sources or by geometrical patterns of rocks with various thermal conductivities (Tables 13.1 and 13.2). Such patterns in thrustbelts include the sequence of various lithostratigraphic units, the interfingering of facies and stratigraphic repetitions in thrust sheets. The thermal conductivity depends on temperature (Čermák and Haenel, 1988):

$$\lambda(T) = \frac{\lambda_0}{1 + CT}, \tag{13.7}$$

where C is of the order of 10^{-3} K. It is a constant, which is different for each rock and needs to be determined experimentally.

Fourier's second equation expresses the behaviour of the rock body, through which the heat flows. The difference between inflowing and outflowing heat equals the thermal energy remaining in the body volume dV. This difference, calculated along the z axis equals

$$dq_z = \left(\frac{\partial q_z}{\partial z}\right)dz. \tag{13.8}$$

The thermal energy stored in the rock volume element during unit time is

$$\rho c \, dV \left(\frac{\partial T}{\partial t} \right), \qquad (13.9)$$

where ρ is the density, c is the specific heat capacity, dV is the rock body volume and $(\partial T/\partial t)$ is the temperature change during unit time. Fourier's second equation of heat conduction is

$$-\left(\frac{\partial q_x}{\partial x} + \frac{\partial q_y}{\partial y} + \frac{\partial q_z}{\partial z} \right) = \rho c \frac{\partial T}{\partial t}. \qquad (13.10)$$

Substituting for q with the corresponding equations from equation (13.4) gives

$$\left\{ \frac{\partial[\lambda_x(\partial T/\partial x)]}{\partial x} + \frac{\partial[\lambda_y(\partial T/\partial y)]}{\partial y} + \frac{\partial[\lambda_z(\partial T/\partial z)]}{\partial z} \right\} = \rho c \frac{\partial T}{\partial t}. \qquad (13.11)$$

When the rock material is isotropic and homogeneous and the thermal conductivity is equal in all directions, equation (13.11) reduces to

$$\lambda \left(\frac{\partial^2 T}{\partial x^2} + \frac{\partial^2 T}{\partial y^2} + \frac{\partial^2 T}{\partial z^2} \right) = \rho c \frac{\partial T}{\partial t} \qquad (13.12)$$

or $\partial T/\partial t = (\lambda/\rho c)\Delta T = \kappa \Delta T = \kappa \, \text{div}(\text{grad } T) = \kappa \cdot \nabla^2 T$, where $\kappa = \lambda/\rho c$ is the thermal diffusivity, $\Delta = \nabla^* \nabla = (\partial^2/\partial x^2 + \partial^2/\partial y^2 + \partial^2/\partial z^2)$ is the Laplace operator and $\nabla = (\partial/\partial x + \partial/\partial y + \partial/\partial z)$ is the nabla or del operator.

It is important to extend Fourier's second equation to the version that takes into account the heat source or sink within the volume dV of the assumed rock body (e.g., Haenel *et al.*, 1988):

$$\frac{\partial T}{\partial t} = \kappa \Delta T + \frac{H}{\rho c}, \qquad (13.13)$$

where H is the heat production rate, which is the heat generated in a volume element of the rock of unit mass during unit time. Under steady-state conditions, where the temperature change with time equals zero, equation (13.13) reduces to

$$\Delta T = -\frac{H}{\lambda}. \qquad (13.14)$$

In order to consider a temperature-dependent thermal conductivity $\lambda(T)$ and a depth-dependent heat production rate $H(z)$, equation (13.14) can be rewritten (Čermák and Haenel, 1988):

$$H(z) + \frac{d}{dz}\left[\lambda(T) \frac{dT}{dz} \right] = 0. \qquad (13.15)$$

Typical analytical solutions, after defining the boundary conditions as $T_0 = T_{\text{surface}}$ and $q_0 = \lambda (dT/dz)_{z=0}$, vary according to the thermal parameters of the case lithostratigraphy (Čermák and Haenel, 1988). For n sediment layers with a negligible heat production rate $(H(z) = 0)$ and with thermal conductivity λ_i constant

across each layer Δz_i thick, the temperature at depth z equals

$$T(z) = T_0 + q_0 \sum_{i=1}^{n} \frac{\Delta z_i}{\lambda_i}. \qquad (13.16)$$

For one single layer with constant heat production rate $H(z) = H_0$ and constant thermal conductivity $\lambda(T) = \lambda_0$, the temperature at depth z equals

$$T(z) = T_0 + \frac{q_0}{\lambda_0} z - \frac{H_0}{2\lambda_0} z^2. \qquad (13.17)$$

For one single layer with constant heat production rate $H(z) = H_0$, but with thermal conductivity that depends on the temperature according to equation (13.7), the temperature at depth z equals

$$T(z) = \frac{1}{C}\left\{ (1 + CT_0)\exp\left[\frac{C}{\lambda_0}\left(q_0 z - \frac{H_0 z^2}{2} \right) \right] - 1 \right\}, \qquad (13.18)$$

where C is the material constant from equation (13.7).

For a single layer were the heat production rate has an exponential depth dependence $H(z) = H_0 \exp(-z/D)$ and constant thermal conductivity $\lambda(T) = \lambda_0$, the temperature at depth z equals

$$T(z) = T_0 + \frac{(q_0 - H_0 D)z}{\lambda_0} + \frac{H_0 D^2}{\lambda_0}(1 - e^{-z/D}). \qquad (13.19)$$

For a single layer were the heat production rate has an exponential depth dependence $(H(z) = H_0 \exp(-z/D))$, but with the thermal conductivity depending on temperature according to equation (13.7), the temperature at depth z equals

$$T(z) = \frac{1}{C}\left[(1 + CT_0) \right.$$
$$\left. \times \exp\left\{ \frac{C}{\lambda_0}[H_0 D^2(1 - e^{-z/D}) - H_0 D_z + q_0 z] \right\} - 1 \right]. \qquad (13.20)$$

For n layers with the heat production H_i and thermal conductivity λ_1 constant across each layer of thickness Δz_i, the temperature at depth z equals

$$T(z) = T_0 + \frac{1}{\lambda_1}\left(q_0 - \sum_{i=1}^{n-1} H_i \Delta z_i \right)\left(z - \sum_{i=1}^{n-1} \Delta z_i \right)$$
$$- \frac{H_n}{2\lambda_1}\left(z - \sum_{i=1}^{n-1} \Delta z_i \right)^2. \qquad (13.21)$$

Thermal conductivity and thermal diffusivity

Thermal conductivity values, λ, of thrustbelt rocks at outcrop vary between 1 and 6 W m^{-1} K^{-1} with few exceptions (Clark, 1966; Kappelmeyer and Haenel, 1974; Čermák and Rybach, 1982; Buntebarth, 1984; Poelchau *et al.*, 1997; Yalcin *et al.*, 1997; Tables 13.1 and 13.2, Fig. 13.1). As is apparent from Tables 13.1 and 13.2 and Fig. 13.1, the defined scalar quantity λ is a material property.

Table 13.1(a). Thermal conductivity for a number of rocks.[a, b, c, d, e, f]

Material	Buntebarth (1984) λ (W m^{-1} K^{-1})	Poelchau *et al.* (1997) λ (W m^{-1} K^{-1})	Yalcin *et al.* (1997) λ (W m^{-1} K^{-1})	Roy *et al.* (1981) λ (W m^{-1} C^{-1})	Schon (1983) λ (W m^{-1} K^{-1})	Ibrmajer *et al.* (1989) λ (W m^{-1} K^{-1})
Air		0.025				
Amphibole				3.6 ± 1.6		
Andesite				2.15		2.1 ± 0.6
Anhydrite				5.1 ± 0.6	4.9–5.75	
Arkose						3.0 ± 0.1
Argilite						2.4 ± 0.2
Basalt				1.65	0.51–2.02 (1.45), 1.84, 2.72	
Bituminous coal	0.26				0.13–0.3 (0.21), 0.2–0.8	
Clay					0.38–3.02 (1.49), 1.0–3.1, 0.9–2.5	2.1 ± 0.4
Claystone						1.9 ± 0.3
Dacite						2.7 ± 0.4
Diabase					2.13–2.90 (2.5), 2.2	
Diorite				2.35	1.38–2.89 (2.20), 2.5–3.4	2.9 ± 0.6
Diorite porphyry						2.6 ± 0.4
Dolomite					1.63–6.5 (3.24), 5.8, 2.7–5.9	3.0 ± 0.5
Gabbro	2.1			2.25, 1.85	1.80–2.83 (2.28), 2.1, 2.5–3.4	
Gneiss	2.7			2.9	0.94–4.86 (2.02), 2.1–2.9 (2.43), 2.13	
Granite	2.6			3.15, 3.05, 3.15, 3.05, 2.85, 3.05	1.34–3.69 (2.64), 2.1–3.8, 2.5–3.8	
Granodiorite				2.6, 3.05		2.6 ± 0.5
Granodiorite porphyry						2.3 ± 0.5
Graywacke						3.5 ± 0.7
Gypsum				2.1 ± 1.8		
Ice		2.1				
Gravel					1.2	
Loess (wet)					0.85–2.1	
Limestone	2.2–2.8	2.5–3.0		3.1, 2.55, 1.8–2.9	0.92–4.4 (2.4), 2.01, 1.9–3.1, 2.5–3.0	2.9 ± 0.6
Marble				2.75	1.59–4.0 (2.56)	
Marlstone						1.9 ± 0.4
Methane		0.033				
Mudstone						2.1 ± 0.5

Table 13.1(a) (cont.)

	Reference					
Material	Buntebarth (1984) λ (W m^{-1} K^{-1})	Poelchau *et al.* (1997) λ (W m^{-1} K^{-1})	Yalcin *et al.* (1997) λ (W m^{-1} K^{-1})	Roy *et al.* (1981) λ (W m^{-1} C^{-1})	Schon (1983) λ (W m^{-1} K^{-1})	Ibrmajer *et al.* (1989) λ (W m^{-1} K^{-1})
Peridotite	3.8				3.78–4.85 (4.37)	
Phyllite					1.67	
Pyroxenite					3.48–5.02 (4.33)	
Quartz diorite					1.98–3.80 (3.00)	2.7 ± 0.5
Quartz diorite porphyry						2.6 ± 0.4
Quartz monzonite				2.95		
Quartzite				5.55	2.68–7.60 (5.26), 2.9, 3.3–7.6, 3.9	
Rock salt	5.5		5.69 (20 °C) 4.76 (100 °C)		5.35–7.22	
Rhyolite				2.25		
Sand, dry clay					0.21–0.85	
Sandstone	3.2	2.5			0.38–5.17 (1.66), 1.26–5.5, 1.2–3.5	2.7 ± 0.5
Sandy shale			2.32 (20 °C) 2.12 (100 °C)			
Schist				3.6		
Serpentinite					2.31–2.87 (2.62), 3.35	
Shale		1.1–2.1	1.98 (20 °C) 1.91 (100 °C)		1.03–4.93 (2.46), 2.3–4.2	2.8 ± 0.6
Slate	2.4					
Syenite					1.74–2.97 (2.26), 2.1, 2.5–2.9	
Tuff				1.95		2.1 ± 0.4
Tuffite						2.0 ± 0.5
Water		0.6				

Table 13.1(b). Thermal conductivity for a number of rocks – continuation.

	Reference			
Material	Clark (1966) (10^{-3} cal cm^{-1} s^{-1} °C^{-1})	Cloetingh *et al.* (1995) λ (W m^{-1} K^{-1})	Hutchison (1985) λ (W m^{-1} K^{-1})	Čermák and Rybach (1982) λ (W m^{-1} K^{-1})
Amphibolite	6.1–9.1 (av. 6.92)			2.46
Basalt		4		
Diorite				2.91
Dolomite and anhydrite	8.9–13.9 (av. 11.93)			
Dolomite	13.2, 9.6–12.0 (av. 11.0)			
Gabbro				2.63
Gneiss	Perp. 4.6–7.7 (av. 6.34), par. 6.0–11.4 (av. 8.9)			2.44
Granite	6.7–8.6 (av. 7.89), 6.2–9.0 (av. 7.77)	3.3		
Granodiorite	6.2–6.9 (av. 6.64), 7.0–8.3 (av. 7.61)			2.65
Granulites				2.4–2.9 (av. 2.6)
Limestone	4.7–7.1 (av. 6.12), 5.2, 6.6–8.0 (av. 7.3)		2.93	
Marl	4.9, 4.2–6.6 (av. 5.22), 2.2–5.3 (av. 3.52)			
Quartzite	14.2–17.6 (av. 16.05), 8.7–19.2 (av. 14.3), 10.4–18.9 (av. 14.5), 7.4–12.7 (av. 10.1)			
Salt			5.86	
Sandstone	3.5–7.7 (av. 4.7), 5.1–10.2 (av. 6.78), 6.0–7.7 (av. 6.62)		4.18	
Shale	4.7–6.9 (av. 5.7), 2.8–4.2 (av. 3.55), 3.2–5.6 (av. 4.2), 3.0–4.3 (av. 3.26)		1.88	
Pore water			0.67	

Table 13.2. Thermal conductivities of several basic rock types[a] versus default values used in Platte River Associates, Inc. (1995).

Rock	Conductivity $(10^{-3}\,\mathrm{cal\,cm^{-1}\,s^{-1}\,{}^{\circ}C})$	Conductivity $(\mathrm{W\,m^{-1}\,K^{-1}})$	Sum	Average conductivity $(\mathrm{W\,m^{-1}\,K^{-1}})$	BasinMod default value $(\mathrm{W\,m^{-1}\,K^{-1}})$
Granite	7.77	3.247 86	15.265 36	3.053 072	2.9
	7.89	3.298 02			
	6.66	2.783 88			
	5.8	2.424 4			
	8.4	3.511 2			
Limestone	6.12	2.558 16	23.600 28	2.622 253 333	2.9
	5.2	2.173 6			
	7.3	3.051 4			
	5.7	2.382 6			
	5.4	2.257 2			
	5.24	2.190 32			
	7.2	3.009 6			
	8.2	3.427 6			
	6.1	2.549 8			
Solomite	11	4.598	15.089 8	5.029 933 333	4.8
	13.2	5.517 6			
	11.9	4.974 2			
Sandstone	4.7	1.964 6	7.565 8	2.521 933 333	4.4
	6.78	2.834 04			
	6.62	2.767 16			
Q sandstone	13.6	5.684 8	11.160 6	5.580 3	4.4
	13.1	5.475 8			
Shale	5.7	2.382 6	6.984 78	1.746 195	1.5
	3.55	1.483 9			
	4.2	1.755 6			
	3.26	1.362 68			
Marl	4.9	2.048 2	5.701 52	1.900 506 667	No
	5.22	2.181 96			
	3.52	1.471 36			
Salt	12.75	5.329 5	30.167 06	6.033 412	5.4
	12.75	5.329 5			
	13.27	5.546 86			
	16.2	6.771 6			
	17.2	7.189 6			

Note:
[a] Clark (1966).

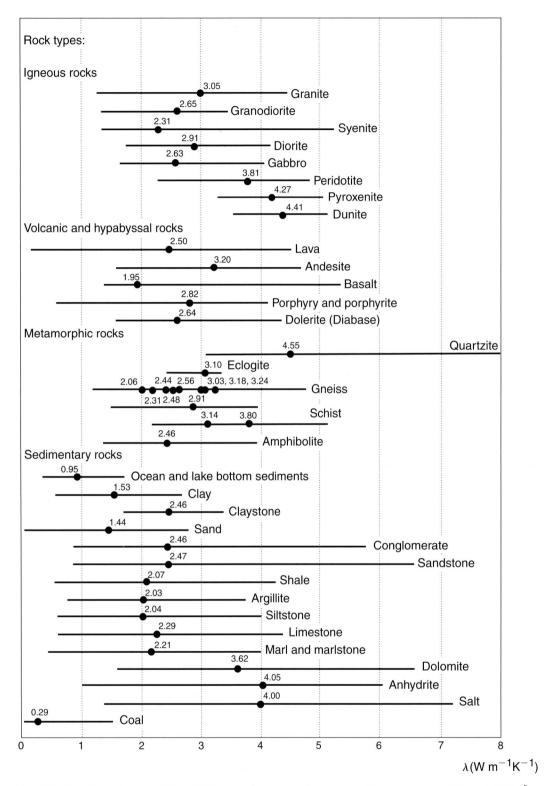

Fig. 13.1. The thermal conductivity of different rocks at room temperature (Kappelmeyer and Haenel, 1974; Čermák and Rybach, 1982). Average values in $W\,m^{-1}\,K^{-1}$ are 2.06 for gneiss perpendicular to schistosity, 2.31 for plagioclase gneiss, 2.44 for all samples of gneiss (randomly oriented), 2.48 for gneiss parallel to schistosity, 2.56 for potassium feldspar gneiss, 3.03 for granite gneiss, 3.18 for amphibolite gneiss, 3.24 for quartz diorite gneiss, 2.91 for schist perpendicular to foliation, 3.14 for all samples of schist (randomly oriented) and 3.80 for schist parallel to foliation.

It is dependent not only on the type of rock or mineral, but also on the crystal structure, which might cause anisotropy in the thermal conductivity (e.g., Schon, 1983; Meissner, 1986). The anisotropy itself causes a diffusion of heat at different rates in different directions (see equations 13.11 and 13.12). The direction of heat flow does not necessarily coincide at any point with the highest temperature gradient. Anisotropy in thermal conductivity λ arises not only from the arrangement of ions in a crystal structure, but also on a macroscopic scale from rocks exhibiting a preferred orientation of individual mineral grains. The most common thrustbelt material, sedimentary rocks, is typically anisotropic. Sedimentary rocks and their minerals have a directional thermal conductivity (equation 13.5). Three components of thermal conductivity parallel to the three perpendicular coordinate directions x, y and z, can be measured either on single crystals or on rocks with distinct structure. Two of these coordinates are parallel to layering and the remaining one is perpendicular. The layering and compositional changes in sedimentary rocks result in a large anisotropy of their thermal conductivity (e.g., Schon, 1983). On the other hand, magmatic rocks, potentially present in basement uplifts, inversional structures and proximal parts of thrustbelts, often show very minor anisotropy, which can generally be ignored in calculations.

The thermal conductivity of a thrustbelt rock can be estimated from the conductivities of the constituent minerals and the fraction of each mineral phase. A maximum conductivity is calculated from the weighted arithmetic mean value $\lambda_{max} = \sum_{i=1}^{n} p_i \lambda_i$, and the minimum conductivity from the harmonical mean value $1/\lambda_{min} = \sum_{i=1}^{n} p_i/\lambda_i$, where p_i is the fraction of the ith mineral having a conductivity λ_i and $\sum_{i=1}^{n} p_i = 1$. Since quartz is a good heat conductor, the rocks bearing this mineral exhibit a strong dependence on the thermal conductivity of the quartz fraction (e.g., Roy *et al.*, 1981; Fig. 13.2a). In the case of granite the thermal conductivity varies from about 2.5 to 4 W m^{-1}K^{-1}, depending on the 20–35% fraction of quartz (Buntebarth, 1984). On the other hand, an increase in plagioclase content, especially of anorthite-rich plagioclase, lowers the thermal conductivity of a rock due to the low heat conductivity of this mineral (e.g., Schon, 1983). The same is true for sandstone, which can be characterized as a randomly inhomogeneous medium with distinct volume fractions of quartz, plagioclase and clay (Palciauskas, 1986).

The grain size also has an effect on the thermal conductivity (Palciauskas, 1986; Fig. 13.2b). Figure 13.2(b) illustrates the thermal conductivity calculated for various grain–water mixtures, which are good analogues for accretionary prism sediments, and which have coarse grains composed of quartz and feldspar and fine grains of clay. The volume fraction of coarse grains is equal to 1 minus the inherent porosity of the coarse-grained sediment ϕ_c. The volume fraction of fine grains is equal to 1 minus the inherent porosity of fine-grained sediment ϕ_f multiplied by the volume occupied by the fines X. The volume of the saturating fluid is (Palciauskas, 1986)

$$\phi = \phi_c - X(1 - \phi_f). \tag{13.22}$$

Of the two end members, the first, with the volume of fines $X = 0$, is a saturated coarse-grained sandstone with a porosity of 0.4 (40%). The second, with $X = 1$, is a clay–water system with a porosity of 0.6 (60%). If sandstone is defined as $X < 0.4$, the calculation shows that large thermal conductivity changes are controlled by both lithology and grain size. A similar effect can be seen by comparing the sandstone and quartz sandstone data from Table 13.2.

The above calculations for effective bulk thermal conductivity have been used in numerous studies and variations (Brigaud *et al.*, 1990; Palciauskas, 1986; Bachu, 1991; Poelchau *et al.*, 1997). Their common characteristic is their use as a simplifying tool, because these mixing formulae are designed for circular particles and infinitely large domains. More heterogeneous rock systems require a different approach.

An indirect method, designed for bounded domains with random shape particles or layers, indicates that the thermal conductivity is controlled by (Bachu, 1991)

(1) the shape and distribution of heterogeneities;
(2) the conductivity contrasts;
(3) the interconnectivity of components; and
(4) the component fraction.

If a sandstone section contains horizontally oriented shale lenses, this sedimentary section will result in the horizontal conductivity being greater than the vertical conductivity. Systems with shale fractions of less than 45–65% have a conductivity controlled mainly by the sand matrix. On the other hand, systems with shale fractions higher than 65% have a conductivity controlled by the shale.

The thermal conductivity of thrustbelt rocks is controlled by two mechanisms: lattice and phonon conductivity by radiation (e.g., Meissner, 1986). Both mechanisms are temperature dependent. Numerous experiments and the resulting functions of thermal conductivity variations with temperature, derived from their results, (e.g., Birch and Clark, 1940; Kappelmeyer and Haenel, 1974; Kopietz and Jung, 1978; Zoth, 1979;

(a)

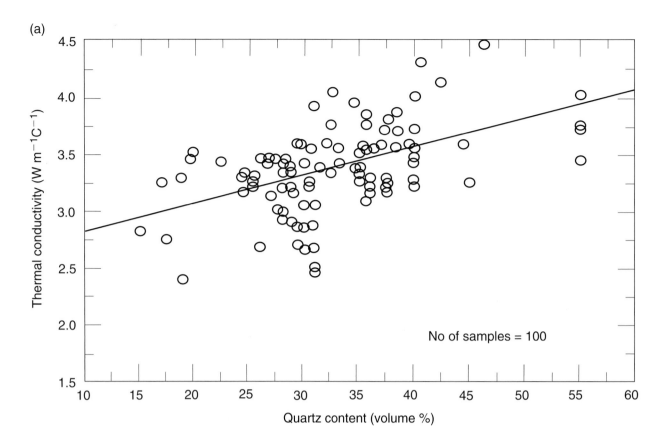

(b)

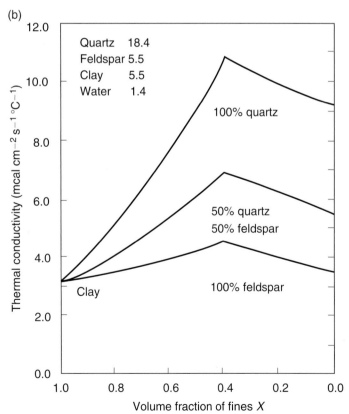

Fig. 13.2. (a) The thermal conductivity dependence on the quartz content of granite and quartz monzonite (Roy *et al.*, 1981). (b) The effective thermal conductivity for clastic rock–water mixtures as a function of the volume X occupied by fine grains (Palciauskas, 1986). The figure also contains thermal conductivities of individual components.

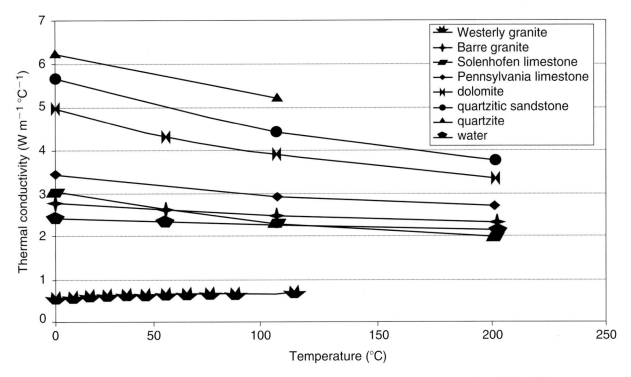

Fig. 13.3. The thermal conductivity temperature dependence of several rock types and water (data from Clark, 1966).

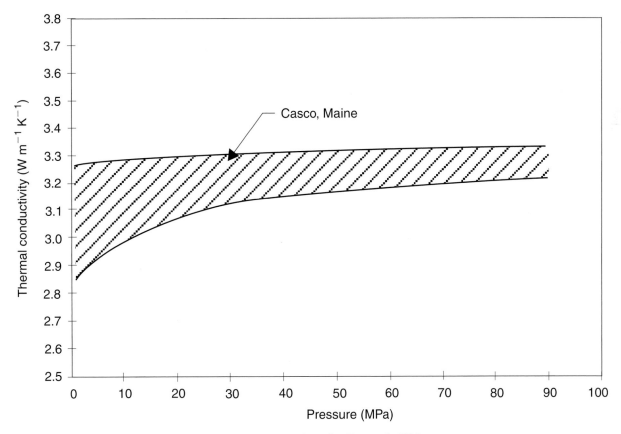

Fig. 13.4. The thermal conductivity dependence on pressure of granite (Roy *et al.*, 1981).

(a)

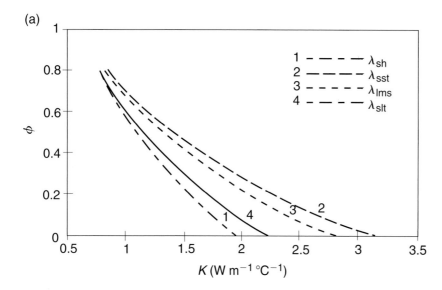

Fig. 13.5. (a) Changes in thermal conductivity (*K*) of sediments (grains + pores) shale (λ_{sh}), sandstone (λ_{sst}), limestone (λ_{lms}) and siltstone (λ_{slt}) due to varying porosities (ϕ) (Yalcin *et al.*, 1997). Thermal conductivities of the pure lithotypes and water (λ_w) used for the calculation are: $\lambda_{sh} = 1.98\,\mathrm{W\,m^{-1}\,{}^\circ C^{-1}}$, $\lambda_{sst} = 3.16\,\mathrm{W\,m^{-1}\,{}^\circ C^{-1}}$, $\lambda_{lms} = 2.83\,\mathrm{W\,m^{-1}\,{}^\circ C^{-1}}$, $\lambda_{slt} = 2.22\,\mathrm{W\,m^{-1}\,{}^\circ C^{-1}}$, $\lambda_w = 0.60\,\mathrm{W\,m^{-1}\,{}^\circ C^{-1}}$. (b) The bulk thermal conductivity dependence on density of several deep-sea sediments (data from Clark, 1966).

(b)

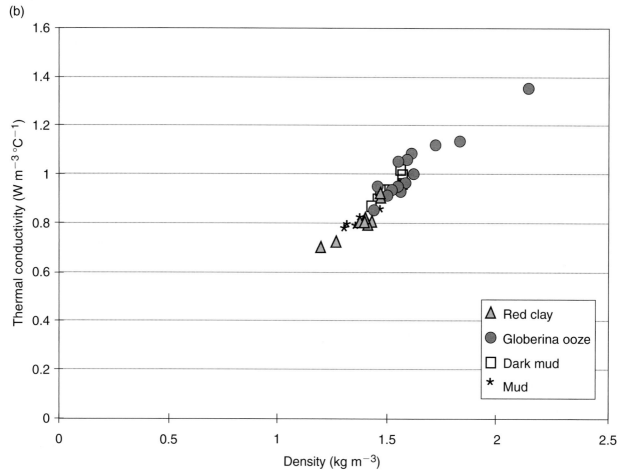

Roy *et al.*, 1981; Čermák and Rybach, 1982; Mongelli *et al.*, 1982; Morin and Silva, 1984; Haenel *et al.*, 1988) show the dependence of the thermal conductivity of salt rocks, limestones and sandstones on temperature. The thermal conductivity decreases with temperature. This function has a different limiting value and gradient for different rocks (Fig. 13.3). The temperature decreases the thermal conductivity of well-conducting lithologies such as sandstone, quartzite and salt, while values for shale with less than 25% porosity are nearly temperature

Table 13.3. Thermal diffusivity for a number of rocks.

Material	Reference	
	Buntebarth (1984) $\kappa\,(10^{-6}\,\mathrm{m^2\,s^{-1}})$	Roy *et al.* (1981) $\kappa\,(10^{-7}\,\mathrm{m^2\,s^{-1}})$
Bituminous coal	0.15	
Dolomite		8.5 ± 2.0
Gneiss	1.2	
Granite	1.4	
Gypsum		3.0 ± 0.3
Limestone	1.1	
Rock salt	3.1	
Slate	1.2	
Sandstone	1.6	
Tuff		6.5 ± 2.5

independent (e.g., Kappelmeyer *et al.*, 1988; Poelchau *et al.*, 1997). This leads to lower thermal conductivity contrasts at higher temperatures.

The influence of pressure on thermal conductivity is the opposite to that of temperature, but is of minor importance (Morin and Silva, 1984). An increase in pressure raises the conductivity (Fig. 13.4), by either compacting the pore volume or deforming the grains.

Under greater pressure the elastic properties of individual crystals influence the thermal conductivity, as a result of deformation of the crystal lattice.

At low pressure all rocks contain porosity consisting of pore spaces between individual mineral grains and microcracks, which occur both between and within the grains. With increasing pressure this porosity gradually decreases. The thermal conductivity, λ, increases with increasing compression, i.e. with decreasing porosity (Fig. 13.5a; Yalcin *et al.*, 1997) and/or with increasing density (e.g. Bachu *et al.*, 1995) (Fig. 13.5b). This increase is linear with pressure, p, up to the elastic limit (Buntebarth, 1984):

$$\lambda = \lambda_0 (1 + ap), \tag{13.23}$$

where a is a constant of the order of magnitude of $1\text{--}5 \times 10^{-5}\,\mathrm{MPa^{-1}}$. It differs for each rock and needs to be determined experimentally.

Even though the porosity of igneous and metamorphic rocks, at about 1%, is not as high as the porosity of sedimentary rocks, their thermal conductivity is considerably altered by about 10% as the pores are closed.

Recent measurements of thermal diffusivity κ (Table 13.3), defined as $\kappa = \lambda/\rho c$, show constant values for

the entire pressure range of 0–300 MPa and can be expressed by the equation (Buntebarth, 1984)

$$\kappa = \kappa_0 (1 + a'p), \tag{13.24}$$

where a' is a material constant in the range $1\text{--}5 \times 10^{-4}\,\mathrm{MPa^{-1}}$ and needs to be determined experimentally for each rock type. The thermal diffusivity decreases with temperature (Fig. 13.6; Roy *et al.*, 1981).

As discussed earlier, porosity plays a major role in controlling the thermal conductivity of sediments (Fig. 13.5a), giving rise to changes by factors of 2–4, depending on the lithotype (Yalcin *et al.*, 1997). For example, sandstones at 1–2 km depth commonly have porosities of up to 15% or more (Wygrala, 1989). Because pores at depth are filled with oil, water or gas, which have lower thermal conductivities than the rock matrix (see Figs. 13.3 and 13.7a), the effective conductivity of the total rock volume can be greatly reduced (Figs. 13.7b–g). For example, the laboratory experiments of Stoll and Bryan (1979) show that gas hydrate, a typical pore constituent in accretionary prisms (Suess *et al.*, 1987; Moore *et al.*, 1990a; Yamano *et al.*, 1990), causes a decrease in the thermal conductivity of the sediment.

Estimates of the thermal conductivity of a porous rock depend on whether the pore spaces represent isolated volumes or form an interconnected network. Models that treat pores as an isolated volume yield maximum values of thermal conductivity and models with interconnected pore volumes yield minimum values. A bulk thermal conductivity calculation for models with isolated pores is given by the equation (Buntebarth, 1984)

$$\lambda = \lambda_{\mathrm{m}} \left[1 - \frac{3\phi(1 - \lambda_{\mathrm{f}}/\lambda_{\mathrm{m}})}{2 + \phi + \lambda_{\mathrm{f}}/\lambda_{\mathrm{m}}} \right], \tag{13.25}$$

where λ_{f} and λ_{m} are the thermal conductivities of the fluid and the rock matrix, respectively, and ϕ is porosity. A bulk thermal conductivity calculation for models with interconnected pore volumes is given by the equation (Buntebarth, 1984)

$$\lambda = \lambda_{\mathrm{m}} \left[1 - \frac{\phi(1 + 2\lambda_{\mathrm{f}}/\lambda_{\mathrm{m}})(1 - \lambda_{\mathrm{f}}/\lambda_{\mathrm{m}})}{\phi(1 - \lambda_{\mathrm{f}}/\lambda_{\mathrm{m}}) + 3\lambda_{\mathrm{f}}/\lambda_{\mathrm{m}}} \right]. \tag{13.26}$$

Factors, such as the fluid pressure, the temperature and diagenetic processes, interactively influence porosity development and compaction (e.g., Chilingarian, 1983; Poelchau *et al.*, 1997 and references therein). Thus overpressure retards porosity reduction and temperature affects porosity reduction by its effect on fluid viscosity, solution and cementation (e.g., Gregory, 1977; Stephenson, 1977). Diagenetic processes can either reduce or enhance porosity. Minerals in contact with the pore fluid are subject to chemical alteration or dissolution, depending on the ion balance in the rock, the fluid

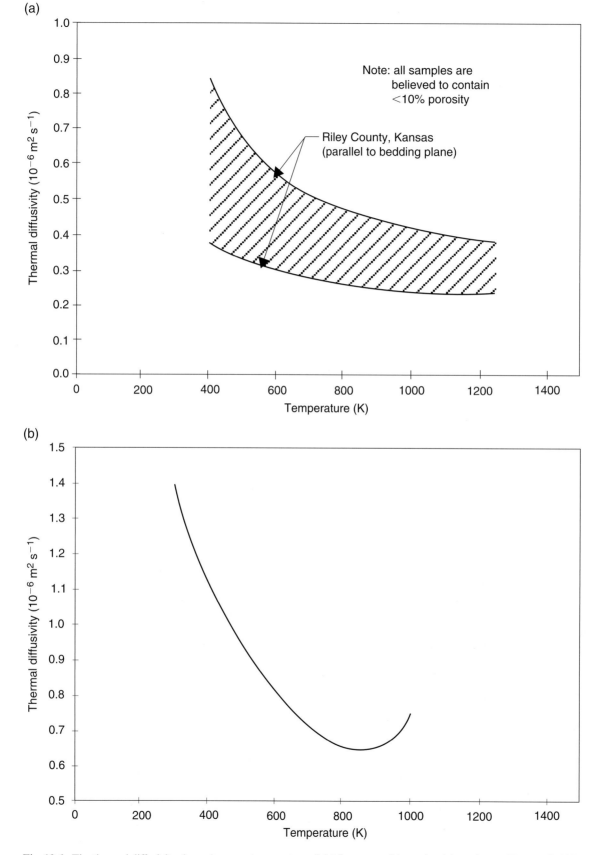

Fig. 13.6. The thermal diffusivity dependence on temperature of (a) limestone, (b) granite, (c) sandstone (Roy *et al.*, 1981).

(c)

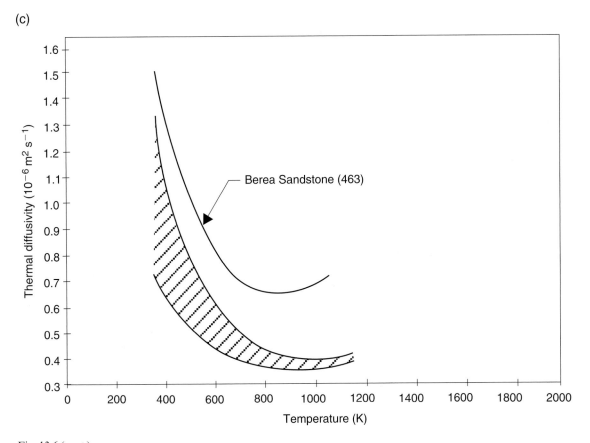

Fig. 13.6 (*cont.*)

and the temperature (Table 13.4). The pore fluid may precipitate certain minerals as cement, depending on the relative saturation of ions needed to form these cements. A critical saturation is of course temperature dependent (Table 13.4; Bjorlykke *et al.*, 1989). The bulk material properties such as the thermal conductivity and the heat capacity are affected by the solid cement properties (e.g. Mann *et al.*, 1997). Some typical cement properties are given in Table 13.5. Therefore, the conductivity of a sedimentary rock becomes a function of porosity, or the state of compaction, as well as the mineralogy and the pore fluid, where $\lambda_{matrix} > \lambda_{wet\ rock} > \lambda_{dry\ rock}$.

All pore fluids have lower thermal conductivities than rocks (Figs. 13.3, 13.5b, 13.7a–g). Porosity, lithology and the pore fluid type are the most important controlling factors, because the thermal conductivity of a saturated sediment can increase by a factor of 2–5 during the course of compaction from its initial to its final porosity value (Woodside and Messmer, 1961; Palciauskas, 1986). Figure 13.7(g) shows the effect on a sandstone of the various overlapping controls that occur with increasing burial and temperature. Both the thermal conductivity of a mineral and a rock matrix decrease with depth as a function of temperature, but

that of the pore fluid increases with temperature. The proportion of pore fluid relative to the rock matrix decreases with compaction, because of the decreasing porosity. The effective thermal conductivity of a sedimentary formation evolves depending on the interaction between the temperature, the pore fluid composition and the porosity. The initial trend is towards increasing conductivity due to rapidly declining porosity (see Fig. 13.8). With increasing burial, i.e. increasing temperature, the effect of decreasing mineral conductivity prevails and the total thermal conductivity shows a reversal towards reduced values. However, when the reservoir is filled with gas, the conductivity increase continues to a greater depth, i.e. the reversal point occurs at a greater depth. This is caused by the rock plus gas conductivity increasing linearly with temperature, while the rock plus water conductivity levels out. This interaction between porosity and temperature dependence is valid for most rocks, including shale, limestone and sandstone (Yalcin *et al.*, 1997). However, in the case of salt and quartzite, with good thermal conductivity, the temperature dependence is dominant, which results in a suppression of the effect from the porosity-dependent reduction (Fig. 13.8; Ungerer *et*

(a)

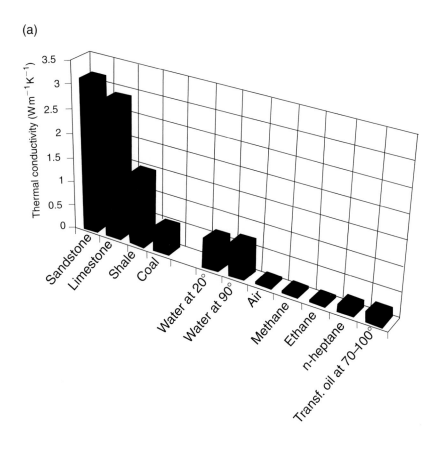

Fig. 13.7. (a) Thermal conductivities of selected rocks and fluids (Poelchau *et al.*, 1997). The values for rocks are matrix conductivities for 0% porosity at 20 °C. Data are from Clark (1966), Weast (1974), Blackwell and Steele (1989) and IES (1993). (b) The bulk thermal conductivity dependence on water content of several deep-sea sediments (data from Clark, 1966). (c) The bulk thermal conductivity of Berea sandstone saturated by water as a function of its porosity. Berea sandstone data (Clark, 1966) are utilized in the equation of Lewis and Rose (1970). (d) The bulk thermal conductivity of quartz-rich sandstone saturated by various saturants as a function of its porosity (data from Clark, 1966). (e) The bulk thermal conductivity of porous quartz-rich sandstones as a function of the saturant (data from Clark, 1966). (f) The thermal conductivity of sandstone as a function of porosity and pore fluid at ambient temperature and pressure (Poelchau *et al.*, 1997). (g) The influence of porosity, temperature, and pore-filling fluids on the effective thermal conductivity of sandstone (Zwach *et al.*, 1994). Note the large difference between sandstone filled with methane and sandstone filled with water. Note also that thermal conductivity of gas-filled sandstones increases to a much greater depth than that of water-filled sandstones.

(b)

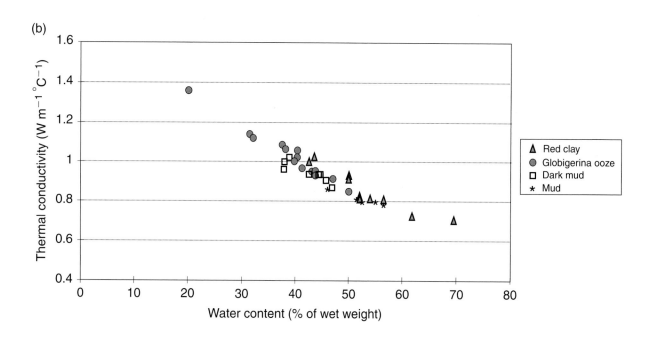

(c)

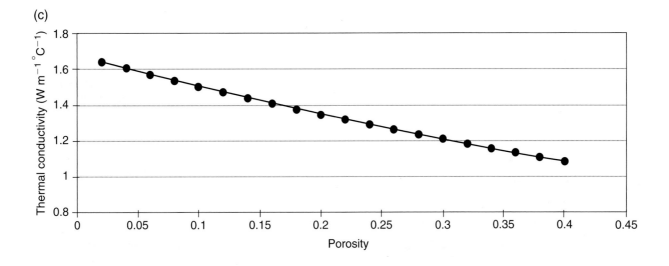

(d)

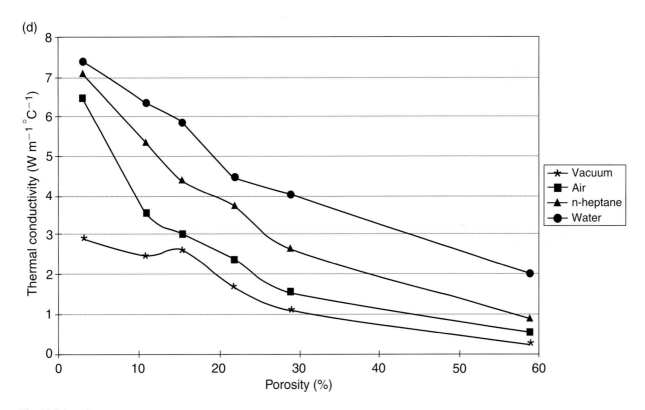

Fig. 13.7 (*cont.*)

(e)

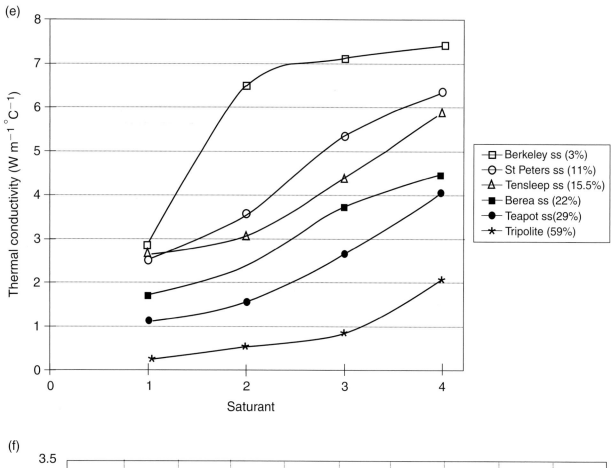

(f)

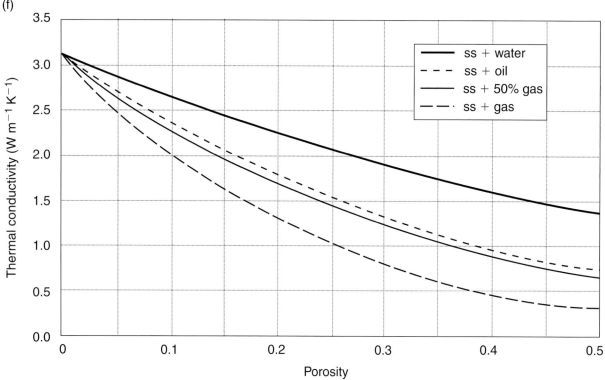

Fig. 13.7 (cont.)

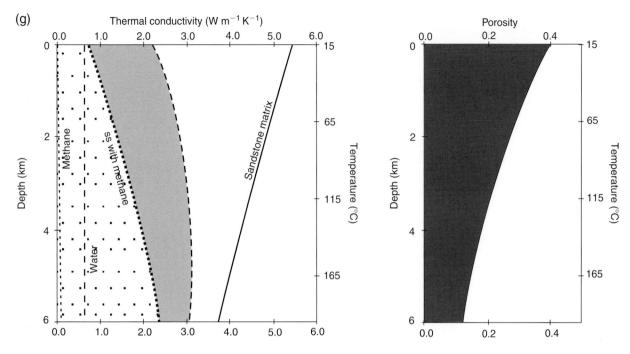

Fig. 13.7 (*cont.*)

al., 1990). Even different types of sandstone have different thermal conductivity–depth profiles, depending on their composition (Palciauskas, 1986). It is interesting to note that the effective conductivity of pure quartz sandstone is approximately constant with depth, because the decrease of the conductivity of the quartz grains due to increasing temperature compensates for the effects of compaction. Feldspathic sandstone and clay behave differently. Feldspar and certain clays do not show a marked temperature effect and thus the increase in the rock thermal conductivity can be more significant than in the case of a quartz sandstone. The behaviour of the two end members, clay–water and feldspar–water mixtures can be described in the following way. The former mixture compacts as shale and the latter mixture compacts as sandstone. Due to the higher initial porosity, the clay–water mixture has a somewhat smaller conductivity than the feldspar–water mixture, but due to its larger compressibility the thermal conductivity increases faster with depth.

Specific heat capacity
The increase of the internal energy ΔQ of a volume element is proportional to its mass m, the temperature and the capability of the element to store heat, the specific heat capacity (see equation 13.1). The specific heat capacity c is a material property (Table 13.6). Rocks that are not porous have a significant specific heat dependence on temperature (e.g., Roy *et al.*, 1981; Mongelli *et al.*, 1982; Figs 13.9a, b). For crystalline rocks, this temperature dependence at constant pressure is given by the equation (Buntebarth, 1984)

$$c = 0.75\left(1 + 6.14 \times 10^{-4}T - 1.928 \times \frac{10^4}{T^2}\right). \qquad (13.27)$$

Sedimentary rocks often have a high porosity and when saturated with water, the corresponding specific heat increases, because of the relatively high specific heat of water (Fig. 13.9c; Shi and Wang, 1986; Yalcin *et al.*, 1997). In this case, the specific heat can be calculated using the volume average from the values of the rock matrix and the pore fluid (e.g., Bear, 1972; van der Kamp and Bachu, 1989; Ferguson *et al.*, 1993):

$$\rho c(z) = \rho_f c_f \phi(z) + \rho_s c_s [1 - \phi(z)], \qquad (13.28)$$

where ρ_f, ρ_s are the fluid and solid grain densities, c_f, c_s are the fluid and solid grain specific heats, ρ, c are the bulk density and bulk specific heat, respectively, and $\phi(z)$ is the depth-dependent porosity.

Heat production rate
Most sedimentary rocks in thrustbelts contain minerals with radioactive elements. All naturally occurring radioactive isotopes generate heat, but only the series of

Table 13.4. List of common diagenetic reactions and their temperature ranges in sandstones.[a]

Source mineral(s)	Pore water	Reaction temperature (°C)	Product mineral(s)
Aragonite and high-Mg calcite		20–50	Low-Mg calcite (sparry)
Feldspar	−Na[+], −K[+]	20–100	Kaolinite
Fine-grained low-Mg calcite		60–70	Low-Mg calcite (poikilotopic)
Amorphous silica and opal CT		60–80	Quartz
Smectite	+K[+]	50–100	Illite + quartz
K-feldspar	+Na[+], −K[+]	>65	Albite
Quartz (pressure solution)		100–150	Quartz
K-feldspar + kaolinite		120–130	Illite + quartz

Note:
[a] Biorlykke *et al.* (1989).

Table 13.5. Thermal conductivity and heat capacity of several cements.[a]

Cement type	Density (kg m^{-3})	Thermal conductivity (W m^{-1} K^{-1})		Heat capacity (kJ kg^{-1} K^{-1})	
		(at 20 °C)	(at 100 °C)	(at 20 °C)	(at 100 °C)
Silica cement	2650	7.70	6.00	0.177	0.212
Calcite cement	2721	3.30	2.70	0.199	0.232
Dolomite cement	2857	5.30	4.05	0.204	0.238
Anhydrite cement	2978	6.30	4.90	0.175	0.193
Halite cement	2150	5.70	4.85	0.206	0.214
Clay cement	2810	1.80	1.60	0.200	0.220

Note:
[a] Mann *et al.* (1997).

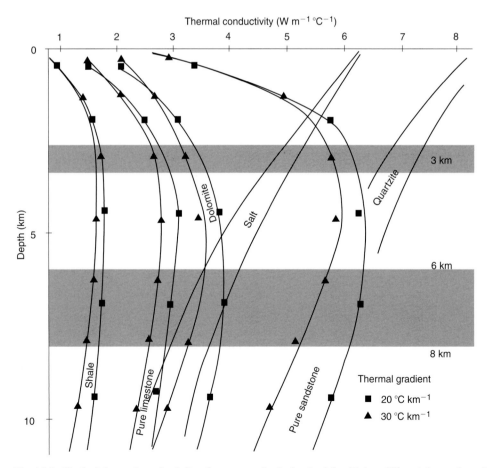

Fig. 13.8. Vertical thermal conductivity of porous rocks during burial, with two different thermal gradients (Ungerer *et al.*, 1990).

Table 13.6. Specific heat for a number of rocks.

Material	Reference				
	Buntebarth (1984) $c \, (\mathrm{W\,g^{-1}\,K^{-1}})$	Yalcin *et al.* (1997) $c \, (\mathrm{cal\,g^{-1}\,°C^{-1}})$	Roy *et al.* (1981) $c \, (\mathrm{J\,kg^{-1}\,C^{-1}})$	Hutchison (1985) $c \, (\mathrm{J\,kg^{-1}\,K^{-1}})$	Schon (1983) $c \, (\mathrm{J\,kg^{-1}\,K^{-1}})$
Basalt					762–2135 (1231)
Bituminous coal	1.26				
Calcareous sandstone	0.84				
Clay	0.86				
Clay/siltstone					753–3546 (1240)
Diabase					791–829 (860)
Diorite					1118–1168 (1136)
Dolomite					648–1465 (1088), 756
Gabbro			775 ± 60		879–1130 (1005)
Gneiss					754–1176 (979)
Granite					741–1548 (946), 672
Gypsum			1010 ± 150		
Ice	2.1				
Gravel					756
Limestone			860 ± 50	1004	753–1712 (887), 714
Marble			883 ± 20		753–879 (857)
Obsidian			920 ± 100		
Oil	2.1				
Peridotite					961–1088 (1005)
Pyroxenite					879–1214 (1005)
Quartz diorite					1214
Quartzite					718–1331 (991)
Salt		0.206 (20 °C)			
		0.212 (100 °C)		854	1474–4651 (2557)
Sandstone	0.71			1088	670–3345 (972)
Sandy shale		0.205 (20 °C)			
		0.248 (100 °C)			
Shale				837	
Serpentinite					963–1130 (1005)
Peat					1758
Shale		0.213 (20 °C)			754–1729 (1096)
		0.258 (100 °C)			
Water	4.2			4180	

(a)

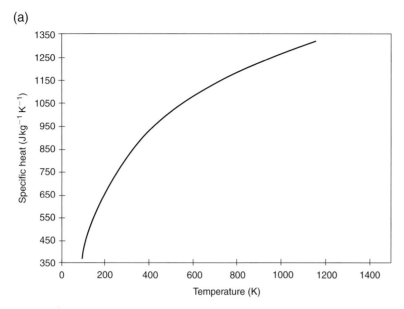

Fig. 13.9. The specific heat dependence
on temperature of (a) calcite,
(b) limestone (Roy *et al.*, 1981). (c) The
bulk specific heat variation with porosity
of sediment saturated by water (data
from Shi and Wang, 1986).

(b)

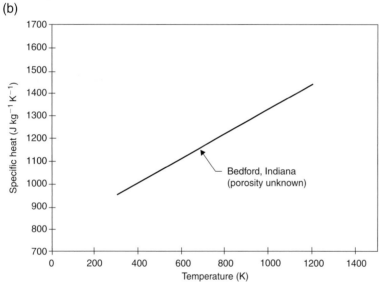

(c)

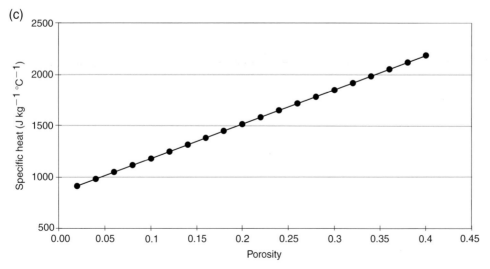

Table 13.7(a). Heat production H in $\mu W\,m^{-3}$ of different rocks and the portions from U, Th and K decay.[a] U and Th concentrations are given in ppm, the K concentration is in per cent.

	Rock						
	U		Th		K		Total
	H	C_U	H	C_{Th}	H	C_K	H
Sandstone, clay, shale	0.69	2.8	0.63	8.9	0.23	1.5	1.55
Granite, liparite	10.90	4.2	1.13	1.6	0.34	3.4	2.56
Plagiogranite	0.52	2.0	0.60	8.5	0.25	2.5	1.37
Diorite, andesite	0.54	2.1	0.52	7.3	0.24	2.4	1.30
Diabase, basalt, gabbro	0.22	0.8	0.18	2.5	0.063	0.6	0.46
Ultrabasite	0.004	0.04	0.003	0.04	0.006	0.06	0.013

Note:

[a] Schon (1983).

Table 13.7(b). Heat production H in $\mu W\,m^{-3}$ of sediments and the portions from U, Th and K decay.[a]

	Rock				
	U (ppm)	Th (ppm)	K (%)	Density (kg m^{-3} $\times 10^3$)	Heat generation ($\mu W\,m^{-3}$)
Black shales	20.2	10.9	2.6	2.4	5.5
Shale/siltstone	3.7	12.0	2.7	2.4	1.8
Sandstones					
Graywacke	2.0	7.0	1.3		0.99
Arkose	1.5	5.0	2.3		0.84
Quartzite	0.6	1.8	0.9		0.32
Deep sea sediments	2.1	11.0	2.5	1.3	0.74
Carbonates				2.6	
Limestone	2.0	1.5	0.3		0.62
Dolomite	1.0	0.8	0.7		0.36
Evaporites					
Anhydrite	0.1	0.3	0.4	2.9	0.090
Salt	0.02	0.01	0.1	2.2	0.012

Note:

[a] Rybach (1986).

uranium, thorium and the potassium isotope ^{40}K contribute significantly to heat generation (Kappelmeyer and Haenel, 1974; Schon, 1983; Rybach, 1986, Table 13.7). In the radioactive decay process, a portion of the mass of each decaying nuclide is converted into energy. Most of this energy is in the form of kinetic energy of emitted particles or of electromagnetic radiation (γ-rays; Van Schmus, 1984). For β^- decays, however, part of the energy is carried away by neutrinos. All of the decay energy other than that carried away by neutrinos is absorbed within the Earth and converted to heat. Table 13.8 summarizes basic data on all four major heat-producing isotopes. Radioactive heat production H is calculated with the help of heat generation constants (amount of heat released per unit mass of U, Th and K per unit time) if the concentrations of uranium C_U, thorium C_{Th} and potassium C_K are known (Schmucker, 1969):

$$H = \rho(3.35C_U + 9.79C_{Th} + 2.64C_K)10^{-5}, \quad (13.29)$$

where H is expressed in $\mu W\,m^{-3}$, ρ is the density, C_K is expressed as a percentage and C_U and C_{Th} as ppm. Rybach (1976, 1986) and Buntebarth (1984) describe slightly different calculations. Rybach (1986), for example, shows that uranium produces about four times more heat than an equivalent amount of thorium. Depending on the concentration, heat production varies with lithology over several orders of magnitude (Tables 13.7 and 13.9). Heat production is generally low in evaporites, except in potassium salt and carbonates, low to medium in sandstone, high in siltstone and shale and highest in black shale (Rybach, 1986). Studies of fine sandy and silty sediments, have found a positive correlation between grain-size fineness and heat production (Pereira *et al.*, 1986). γ-activity has been shown to depend on clay content (Schon, 1983; Fig. 13.10a) and there is a correlation between the concentration of U, Th and K and the amount of quartz in a rock (Meissner, 1986). The density also affects the heat production of each rock type. The greater the rock density is, the smaller the space for large ions of the heat-producing

Table 13.8(a). Decay data on four major heat generating isotopes.[a] Decay mode includes loss of electron (beta particle, β^-), or of alpha particle (α) or electron capture (EC) or spontaneous fission (SF).

Parent isotope	Half-life (yr)	Decay constant (yr^{-1})	Decay mode	Daughter products
^{40}K	1.25×10^9	4.962×10^{-10}	β^-	^{40}Ca (89.5%)
		0.581×10^{-10}	EC	^{40}Ar (10.5%)
^{232}Th	1.40×10^{10}	4.9475×10^{-11}	α, β^-	^{208}Pb + 6 ^{4}He
^{235}U	7.04×10^8	9.8485×10^{-10}	α, β^-	^{207}Pb + 7 ^{4}He
^{238}U	4.47×10^9	1.5513×10^{-10}	α, β^-	^{206}Pb + 6 ^{4}He
		8.46×10^{-17}	SF	Various

Note:
[a] Van Schmus (1984).

elements. This causes an inverse relationship between heat production and density (Buntebarth and Rybach, 1981).

The effect of heat production on the geothermal field can be demonstrated by a simple one-dimensional conductive calculation of temperature for any depth (h_z; Rybach, 1986):

$$T(h_z) = T_0 + h_z \left(\frac{q + Hh}{\lambda} \right) - h_z^2 \left(\frac{H}{2\lambda} \right), \tag{13.30}$$

where T_0 is the surface temperature, q is the basal heat flow, H is the average heat production, h is the thickness and λ is the average thermal conductivity of sediments. Figure 13.10(b) shows the results of this calculation for different values of heat production and thermal conductivity.

Effect of thermal properties on heat distribution
In thrustbelts, the thermal structure is influenced by the thermal conductivity and heat generation of the rock section and heat flow into this section from below. The thermal conductivity and heat capacity of rocks determine to a large degree the distribution of heat in a thrustbelt and the maturation of organic matter. Water has a considerably higher conductivity ($0.6 \, \text{W m}^{-1} \text{K}^{-1}$) than gas ($0.03 \, \text{W m}^{-1} \text{K}^{-1}$), air and even oil (Fig. 13.7a). Therefore, differences in temperature distribution depend on the degree of hydrocarbon saturation of reservoir and source rocks (Poelchau *et al.*, 1997; Figs. 13.7d–g). Even relatively small gas fields have a thermal insulation effect on the sedimentary section, elevating temperatures beneath them (Poelchau *et al.*, 1997).

Table 13.8(b). Energy data on major heat-generating isotopes.

	Element isotope			
	Potassium	Thorium	Uranium	
	^{40}K	^{232}Th	^{235}U	^{238}U
Isotopic abundance (Wt %)	0.0119	100	0.71	99.28
Decay constant (yr^{-1})	5.54×10^{-10}	4.95×10^{-11}	9.85×10^{-10}	1.551×10^{-10}
Total decay energy (MeV/decay)	1.34	42.66	46.40	51.70
Beta decay energy (MeV/decay)	1.19	3.5	3.0	6.3
Beta energy lost as neutrinos (MeV/decay)	0.65	2.3	2.0	4.2
Total energy retained in earth (MeV/decay)	0.69	40.4	44.4	47.5
Specific isotopic heat production (cal g^{-1} yr^{-1})	0.220	0.199	4.29	0.714
Present elemental heat production (cal g^{-1} yr^{-1})	26×10^{-6}	0.199	—	0.740

Note:
[a] Van Schmus (1984).

Most sedimentary rocks are anisotropic with higher horizontal than vertical thermal conductivity values (Gretener, 1981a). However, because the heat flow follows temperature gradients, which have a greatest vertical component, increased horizontal thermal conductivity values do not necessarily have an important effect (Poelchau *et al.*, 1997; Yalcin *et al.*, 1997). Special cases might exist, for example, in the case of fluid flow. In the case of a layered section composed of thermally highly conductive sandstone and low conductivity shale, lateral heat flow can become more significant.

The importance of lithology on the thermal field can be illustrated even with a one-dimensional model. Figure 13.11(a) (Bachu, 1985), based on data from Alberta, Canada, shows a comparison of the average temperature–depth profile in the sedimentary section with the temperature–depth profile through individual layers. This comparison shows the ultimate lithological control. Figures 13.11(b, c) (Bachu, 1985) show the difference of geothermal gradient between shale and sandstone formations in the North Sea,

Table 13.9. Heat production rate for a number of rocks.

Material	Rybach (1986) $H\,(\mu\mathrm{W\,m^{-3}})$	Buntebarth (1984) $H\,(\mu\mathrm{W\,m^{-3}})$	Meissner (1986) $H\,(\mu\mathrm{W\,m^{-3}})$	Pollack and Chapman (1977) $H\,(\mu\mathrm{W\,m^{-3}})$	Hutchison (1985) $H\,(\mu\mathrm{W\,m^{-3}})$	Ibrmajer *et al.* (1989) $H\,(\mu\mathrm{W\,m^{-3}})$
Amphibolite		0.3				
Anhydrite	0.090					
Arkose	0.84					
Black shale	5.5					
Carbonate			0.7			
Chondrite		0.026				
Deep sea sediments	0.74			0.5		
Diorite		1.1				
Dolomite	0.36					
Dunite		0.0042	0.004			
Eclogite-low U		0.034	0.04			
Eclogite-high U		0.15				
Gabbro		0.46				
Gneiss		2.4				
Granite		3.0				4.5
Granodiorite		1.5				
Graywacke	0.99					
Igneous rocks, silicic			2.5			
Igneous rocks, mafic			0.3			
Limestone	0.62					0.84
Mica shist		1.5				
Oceanic lherzolite			0.01			
Olivinfels		0.015				
Peridotite		0.0105				
Quartzite	0.32					
Salt	0.012					0.0
Sand (beach)			1.2			
Sandstone		0.34–1.0				0.84
Shale			2.1			1.05
Shale/siltstone	1.8					
Slate		1.8				
Pore water						0.0

where less conductive shales have a higher temperature gradient.

Because the specific heat gives the energy used or released in a temperature change, it has an obvious effect on heat conduction under transient temperature conditions. It does not play a role under steady-state conditions where temperatures remain fixed.

The effects of heat production are directly proportional to the depth and to the amount of heat generated. As heat generated by internal sources is dissipated by conduction, the temperature increase with time and depth is also controlled by thermal conductivity (Yalcin *et al.*, 1997). Crustal heat sources have a major role in producing the temperature field of the crust. On continents, 20–60% of the surface heat flow originates from crustal radioactivity (Rybach, 1986), and even more in areas of England, Wales, the Central Australian Shield and the French Hercynian Shield (Ungerer *et al.*, 1990; Table 13.10). The effect of heat production on the geothermal field is demonstrated in Fig. 13.10(b). The third term in equation 13.30 controls the curvature of the temperature–depth profile and leads to a decrease of the geothermal gradient with depth. The effect of high heat production is strongest at great depth. Time is also

(a)

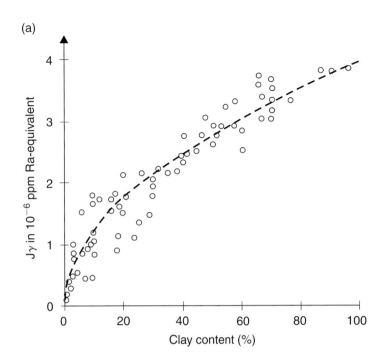

Fig. 13.10. (a) γ-activity as a function of
the clay content (Schon, 1983). (b) The
effect of heat generation in sediment (*H*)
and thermal conductivity (λ) on the
temperature–depth profile (Rybach,
1986). Temperature curves are calculated
with the given λ and *H* values from
equation (13.30) and for a thickness *h* of
6 km, a heat flow *q* of 70 mW m⁻² and a
surface temperature *T* of 10 °C.

(b)

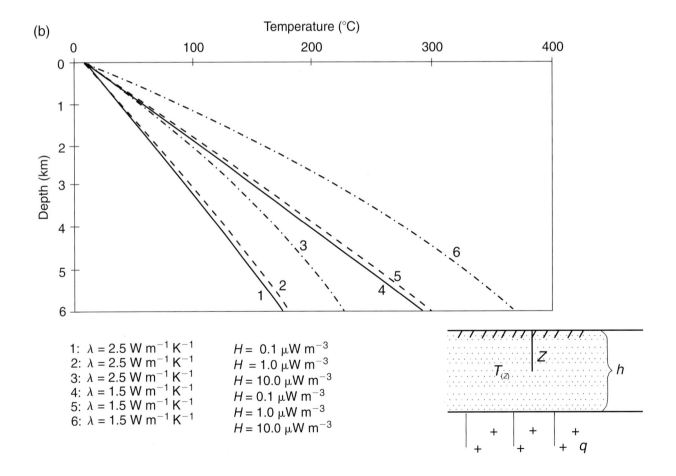

1: $\lambda = 2.5\ \text{W m}^{-1}\text{K}^{-1}$ $H = 0.1\ \mu\text{W m}^{-3}$
2: $\lambda = 2.5\ \text{W m}^{-1}\text{K}^{-1}$ $H = 1.0\ \mu\text{W m}^{-3}$
3: $\lambda = 2.5\ \text{W m}^{-1}\text{K}^{-1}$ $H = 10.0\ \mu\text{W m}^{-3}$
4: $\lambda = 1.5\ \text{W m}^{-1}\text{K}^{-1}$ $H = 0.1\ \mu\text{W m}^{-3}$
5: $\lambda = 1.5\ \text{W m}^{-1}\text{K}^{-1}$ $H = 1.0\ \mu\text{W m}^{-3}$
6: $\lambda = 1.5\ \text{W m}^{-1}\text{K}^{-1}$ $H = 10.0\ \mu\text{W m}^{-3}$

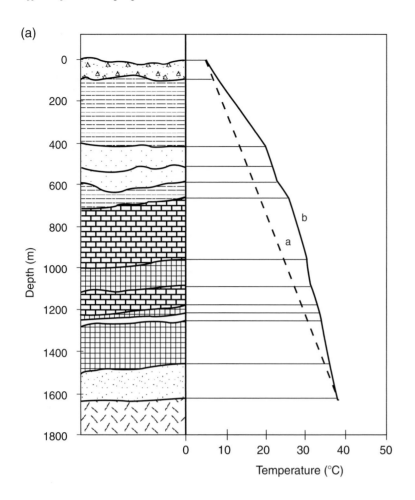

(a)

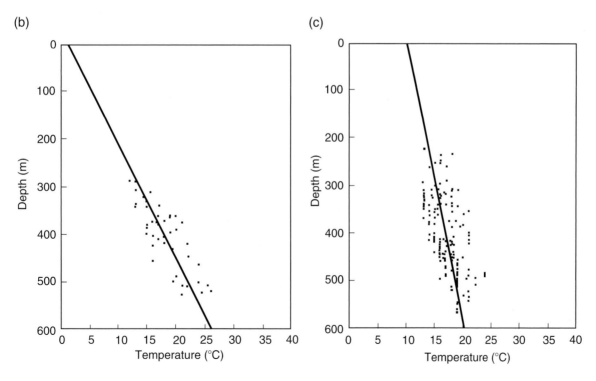

Fig. 13.11. (a) Comparison of the average temperature–depth profile in a sedimentary section (b) with temperature–depth profiles through individual layers (a) (Bachu, 1985). (b), (c) Temperature variation with depth in the shaley Ireton Formation (b) and in Viking Sandstone (c) (Bachu, 1985).

Table 13.10. Contribution of radiogenic heat to the surface heat-flow.[a] A reduced heat-flow approximates the subcrustal flow at the base of the crust, and is obtained by subtracting the estimated radiogenic heat.

Province	Surface heat-flow ($mW\,m^{-2}$)	Reduced heat-flow ($mW\,m^{-2}$)	Reduced/ surface heat-flow (%)
Sierra Nevada	37 ± 13	18 ± 3	49
Eastern USA	57 ± 17	33 ± 4	58
Wales	59 ± 23	23 ± 3	39
Central Australian Shield	83 ± 21	27 ± 6	32
Baltic Shield	36 ± 8	22 ± 6	61
Ukrainian Shield	37	25	67
Superior Province, Canada	34 ± 8	21 ± 1	62
West Australian Shield	39 ± 8	26 ± 6	67
Churchill Province, Canada	44 ± 8	37	84
Indian Archean	49 ± 7	33	67
Indian Proterozoic	76 ± 3	54	71
France Hercynian	68 ± 20	27	38

Note:
[a] Ungerer et al. (1990).

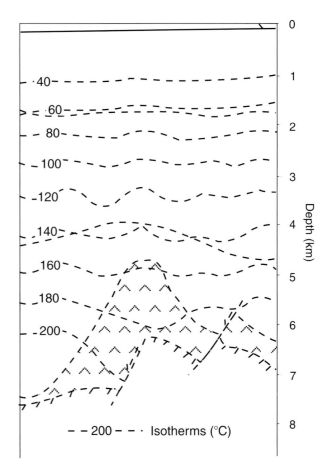

- – 200 – – · Isotherms (°C)

Fig. 13.12. The effect of a salt diapir on the temperature distribution (Yalcin et al., 1997). Note the higher temperature gradient above the salt diapir.

important in heat production as the internal radioactive heat sources of deposited sediments raise the temperature of the sediment column.

As mentioned earlier, conductive heat transfer occurs mainly vertically when the isotherms are horizontal. However, if there are lateral conductivity contrasts within the thrustbelt, heat can also be conducted laterally (Yalcin et al., 1997). Most prominent lateral contrasts in thermal conductivity within thrustbelts are either related to diapiric structures or basement highs. Another possibility is the juxtaposition of distinctly different conductivities by thrusting. Lateral heat conduction leads to a deflection of isotherms that is dependent on the thermal conductivity of the disturbing body. For example, salt diapirs act as heat pipes, carrying positive temperature anomalies over their roofs (Fig. 13.12). The result of anomalous salt conductivity is depressed temperatures in the underlying rock units and increased temperatures in the overlying rock units. In the case of shale diapirs the opposite is observed. Lateral changes in lithofacies or any lateral inhomogeneity in the sedimentary succession, made by thinning or thickening of the strata, or porosity changes, or sat-

uration by hydrocarbons, cause lateral conductivity differences, which result in temperature disturbances. Such conductivity differences exist, for example, in the Brooks Range and its foothills, Alaska (Deming, 1993) and in the Western Canada sedimentary basin (Bachu, 1985). They are caused by a complex fill characterized by thickness and lateral facies changes, resulting in variable concentrations of heat-producing elements and thermal conductivity. Chemical changes brought about by diagenesis and dehydration of clay minerals may provide sources or sinks of heat in the formations in which they occur, normally in young basins (Jessop and Majorowicz, 1994).

Finally, it should be noted that the heat flow in thrustbelts is also affected by paleoclimatic temperature changes, which modify the temperature gradient of the first 100 or more metres in the subsurface (Fig. 13.13).

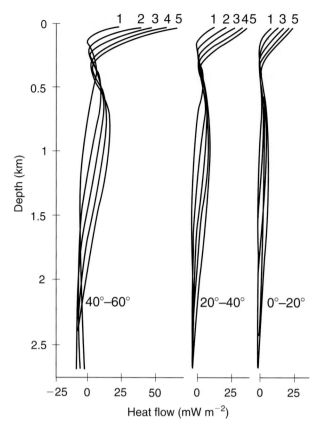

Fig. 13.13. Global averages of the influence of surface paleoclimatic temperature changes on the heat-flow density (Zoth and Haenel, 1988). Values have to be added for the northern latitudes of 0 to 20°, 20 to 40° and 40 to 60°, respectively. Curves are calculated for different thermal conductivities: (1) 1.26, (2) 2.51, (3) 3.77, (4) 5.02 and (5) 6.28 $W\,m^{-1}\,K^{-1}$.

14 Role of syn-orogenic burial and/or uplift and erosion in thermal regimes

The evolution of the overburden affects the thermal structure (Table 14.1) and numerous other factors important to a petroleum system during the development of a thrustbelt. In the case of increasing burial, the underlying sediments subside to greater depths with higher temperatures (Fig. 14.1a). In the case of erosion, the underlying sediments are uplifted to shallower depths with lower temperatures. Whether subsidence or uplift occurs depends on depositional rates and erosional rates specific for different environments.

Because the interplay of deposition and erosion in thrustbelts is complex, this discussion will start by investigating their impact on the thermal structure in simple settings, which have equal depositional or erosional rates over large areas before moving progressively to thrustbelt settings.

Role of deposition on thermal regimes

Assuming no erosion, deposition depresses the heat flow, and the depression persists long after sedimentation ceases (De Bremaeker, 1983; Ungerer *et al.*, 1990). The magnitude of the depression depends on the thermal conductivity of the sediments deposited and the rate and duration of deposition (e.g., Jessop and Majorowicz, 1994; Yalcin *et al.*, 1997).

Insight into the effects of deposition, compaction and related pore fluid movement on the thermal regime in tectonically simple oceanic and deep marginal basins (Hutchison, 1985) allows representation of the sediment system by a two-layer model in which horizontal dimensions extend to infinity (Fig. 14.2a). It allows the application of a simplified one-dimensional solution for vertical heat flow. The rate at which heat $Q(z, t)$ is gained or lost from an element dz (Fig. 14.2a) is given by the difference in heat flow $q(z, t)$ across surfaces Sur_1 and Sur_2, added to the internal heat production $H(z)$ (Hutchison, 1985):

$$\partial_t Q = \left[q - \partial_z q \left(\frac{dz}{2} \right) \right] - \left[q + \partial_z q \left(\frac{dz}{2} \right) \right] + H, \quad (14.1)$$

where the depth and time arguments of the heat flow q, the heat production, etc in this and following equations are omitted for simplicity, or

$$\partial_z(\lambda \partial_z T) - \partial_z \{[\rho_f c_f v_f \phi + \rho_s c_s v_s (1 - \phi)]T\} + H$$
$$= [\rho_f c_f \phi + \rho_s c_s (1 - \phi)] \partial_t T, \quad (14.2)$$

Table 14.1. Factors determining the thermal regime in a sedimentary basin.[a] Limiting deposition rates can be compared with data in Chapter 9.

Factor	Importance (order)	Qualifications
Overburden thickness	First	Always important
Heat-flow	First	Always important
Thermal conductivity	First	Always important
Surface temperature	Second	Always important
Deposition	First	$>0.1\,\mathrm{mm\,yr^{-1}}$
	Second	0.1–$0.01\,\mathrm{mm\,yr^{-1}}$
	Third	$<0.01\,\mathrm{mm\,yr^{-1}}$
Gravity-driven fluid flow	First to second	Foreland basins
Compaction-driven fluid flow	Third	Unless focused
Free convection	Unknown	
Initial thermal event	First (0–20 Ma)	Rift basins only
	Second (20–60 Ma)	
	Third (>60 Ma)	

Note:
[a] Deming (1994c).

where ρ_f, c_f, $v_f(z, t)$ are the pore fluid density, heat capacity and velocity, respectively, ρ_s, c_s, $v_s(z, t)$ are the sediment particle density, heat capacity and velocity, $T(z, t)$ is the temperature at depth z and time t, and $\lambda(z, t)$ is the composite thermal conductivity at depth z. Ongoing deposition modifies the sediment particle velocity, escaping fluid velocity, composite thermal conductivity and porosity. Therefore, the porosity is written as (Hutchison, 1985)

$$\phi(z) = \phi_0 \exp\left(\frac{-z}{C_{\mathrm{const}}} \right), \quad (14.3)$$

where C_{const} is the compaction constant (Rubey and Hubbert, 1960) and ϕ_0 is the surface porosity. The composite thermal conductivity is written as

$$\lambda(\phi) = \frac{-\alpha + (\alpha^2 + 8\lambda_s\lambda_f)^{1/2}}{4}, \quad (14.4)$$

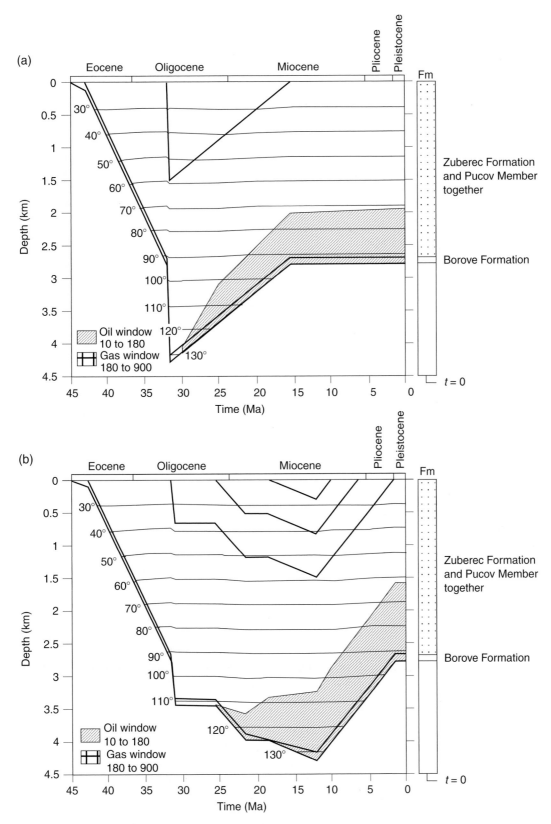

Fig. 14.1. Geohistory curve calculated by Platte River Associates BasinMod 1D for the bore hole Lipany-1 in the Central Carpathian Paleogene basin, West Carpathians (Nemčok *et al.*, 1996): (a) sedimentary–erosional model, (b) sedimentary–erosional model with two thrusting events.

(a)

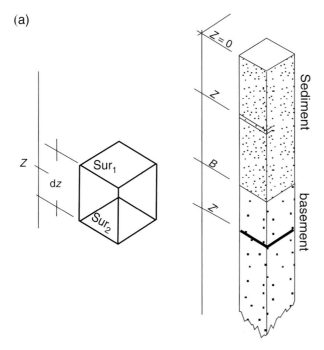

(b)

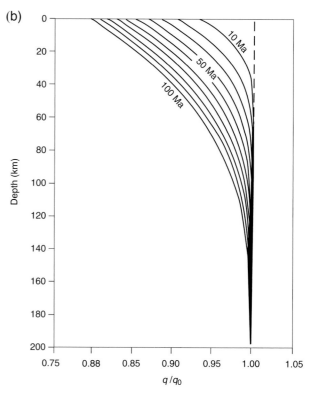

Fig. 14.2. (a) A simplified two-layer model through a sediment–basement column (Hutchison, 1985). Sediment is added at the surface ($z = 0$) with a deposition rate v_0. The porosity is described as a function of depth z ($\phi(z)$). It has the surface value ϕ_0. The basement interface occurs at depth B. Elements of depth dz at Z and Z' within the sediment and basement layers are discussed in the text; the difference in heat flow across Sur_1 (in) and across Sur_2 (out) plus the radioactive heat production H are related to the rate at which heat is gained or lost from dz. (b) The fractional alteration to the heat flow q/q_0 as a function of depth and time (0–100 Ma), calculated from the analytical solutions for the temperatures within a uniform half-space (Hutchison, 1985). q is the heat flow, q_0 is the initial heat flow. Basement with thermal diffusivity of $8 \times 10^{-7}\,\mathrm{m^2\,s^{-1}}$ subsides at velocity 0.1 mm yr^{-1} from depth $z = 0$.

where α equals $3\phi(\lambda_s - \lambda_f) + \lambda_f - 2\lambda_s$ and λ_s and λ_f are sediment and fluid thermal conductivities, respectively. The sediment particle velocity is written as

$$v_s = \frac{v_0(1 - \phi_0)}{1 - \phi}, \tag{14.5}$$

where v_0 is the depositional rate. The fluid velocity is written as

$$v_f = \frac{v_0(1 - \phi_0)}{1 - \phi_0 \exp\left(\dfrac{-B}{C_{\mathrm{const}}}\right)} \exp\left(\frac{z - B}{C_{\mathrm{const}}}\right), \tag{14.6}$$

where B is the depth of the basement and depth z varies between 0 and B. Equation (14.6) indicates that the pore fluid velocity is negligible when the depth of the base-

ment reaches the value of the compaction constant. The porosity function simulates the behaviour of mudstone and fluids are allowed to escape without pressure build-up.

In order to see which component dominates heat-flow advection the fluid and sediment contributions can be compared (Hutchison, 1985):

$$\frac{\text{fluid contribution}}{\text{sediment contribution}} = \frac{\rho_f c_f \phi_0}{\rho_s c_s \left[\exp\left(\dfrac{B}{C_{\mathrm{const}}}\right) - \phi_0\right]}. \tag{14.7}$$

Given typical physical properties of the basement, the sediment and the pore fluid, the fluid contributes the greater heat flow until the basement reaches a depth at which compaction of the sediment cover is constant. The fluid contribution varies between 50 and 80% of that carried by sediment when the basement depth equals this compaction constant. When the basement subsides below the depth equal to the compaction constant of the sediment cover, the fluid contribution becomes negligible.

Before discussing the results of Hutchison (1985), it is useful to check how long thermal transients persist in the continental lithosphere and how far into the lithosphere they penetrate. A useful rule of thumb is provided by Lachenbruch and Sass (1977) and Vasseur and Burrus (1990) who calculate the time required for a

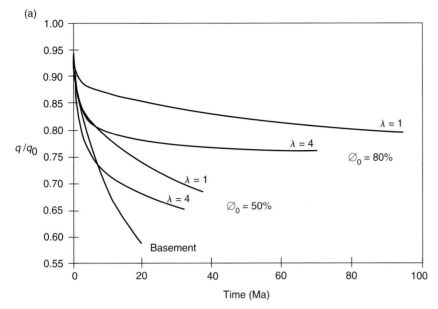

(a)

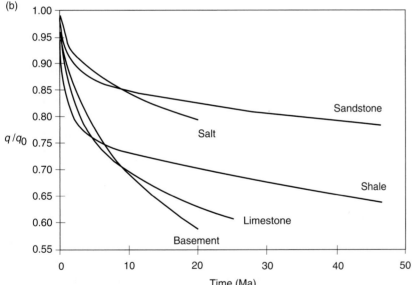

(b)

Fig. 14.3. (a) The dependence of the fractional heat-flow alteration q/q_0 on the compaction constant and the initial sediment porosity (Hutchison, 1985). Deposition rate is $0.5\,\mathrm{mm\,yr^{-1}}$, sediment conductivity is $2.5\,\mathrm{W\,m^{-1}\,K^{-1}}$, heat capacity of sediment particles is $900\,\mathrm{J\,kg^{-1}\,K^{-1}}$ and density of sediment particles is $2700\,\mathrm{kg\,m^{-3}}$. Basement density is $3300\,\mathrm{kg\,m^{-3}}$, thermal conductivity is $3.1\,\mathrm{W\,m^{-1}\,K^{-1}}$, specific heat capacity is $1160\,\mathrm{J\,kg^{-1}\,K^{-1}}$ and heat production is $0\,\mathrm{\mu W\,m^{-3}}$. Pore fluid density is $1030\,\mathrm{kg\,m^{-3}}$, thermal conductivity is $0.67\mathrm{W\,m^{-1}\,K^{-1}}$, specific heat capacity is $4180\,\mathrm{J\,kg^{-1}\,K^{-1}}$ and heat production is $0\,\mathrm{\mu W\,m^{-3}}$. The analytical solution for a basement half-space with the same sedimentation rate is also shown. All curves terminate after deposition of 10 km of sediment. (b) A comparison of the fractional alteration to the surface heat flow due to the deposition of salt, shale, sandstone and limestone for a deposition rate of $0.5\,\mathrm{mm\,yr^{-1}}$ (Hutchison, 1985). The corresponding values for a basement half-space are also shown. Each curve terminates when 10 km of sediment has been deposited. Sandstone matrix density is $2650\,\mathrm{kg\,m^{-3}}$, surface porosity is 62% and compaction constant is 2.78 km. Shale matrix density is $2700\,\mathrm{kg\,m^{-3}}$, surface porosity is 60% and compaction constant is 1.54 km. Limestone matrix density is $2710\,\mathrm{kg\,m^{-3}}$, surface porosity is 24% and compaction constant is 6.25 km. Salt matrix density is $2160\,\mathrm{kg\,m^{-3}}$, surface porosity is 0% and compaction constant is 0 km. Thermal properties of these sediments are listed in the chapter on role of structural and stratigraphic architecture on thermal regime.

thermal disturbance to propagate through the lithosphere. For a 100 km thick continental lithosphere with thermal diffusivity κ of $1.04 \times 10^{-6}\,\mathrm{m^2\,s^{-1}}$, transient thermal events typically have lifetimes of the order of 50–100 Myr (Deming, 1994c). This means that the lithosphere has a relatively high thermal inertia, allowing background thermal states to persist for periods of time comparable to the lifetime of a petroleum system. Calculations of the thermal transient penetration for oceanic lithosphere show that the effects of depositional events shorter than 50 Myr have no effect on the litho-

sphere at depths of 100 km and rapid thermal regime changes due to deposition would only be expected in the upper 10 km of the system (Fig. 14.2b). These changes are controlled by the value of initial sediment porosity, the compaction constant, the heat capacity of sediment particles, the sediment particle density and the depositional rate.

Figure 14.3(a) shows the heat flow q as a proportion of the initial equilibrium value q_0 calculated for fixed sediment thermal conductivity and density, but varying the initial sediment porosity from 50 to 80% and the compaction constant from 1 to 4 km. The magnitude of the heat-flow reduction is primarily determined by the initial porosity. Sediments that are most porous affect the heat-flow least. The compaction constant plays a secondary role and only larger values can cause a slightly increased heat flow.

In addition to the initial porosity and compaction constant, the grain thermal conductivity and heat capacity also vary with sediment type. Therefore, variations of heat flow are dependent on the sedimentary lithology (Fig. 14.3b). The smallest variations are caused by highly conductive salt and sandstone. The largest surface heat-flow reduction results from the deposition of low-porosity, low-conductivity and highly compacted limestone.

Significant surface heat-flow modifications are caused by both the rate and history of the deposition (Figs. 14.4a and b). Figure 14.4(a) shows the effect of the depositional rate of shale on the surface heat flow. The dashed lines indicate the same sediment thickness. The heat-flow alteration is primarily a function of the depositional rate, for rates of from 0 to 0.5 mm yr^{-1}. The only exception is the interval between the onset of deposition and about 10 Ma, when the heat-flow alteration strongly depends on the duration of the depositional event (Fig. 14.4a). The heat-flow alteration is different at sedimentation rates larger than 1 mm yr^{-1}, when significant heat-flow alteration continues through time. Note that even as little as 1 km of sediment, even deposited slowly, causes a significant reduction in the surface heat flow. The heat flow is significantly depressed by deposition rates as high as 0.1 mm yr^{-1} or greater (e.g., Doligez *et al.*, 1986; Deming, 1994c; Fig. 14.5a). The lower the thermal conductivity of these sediments is, the greater the heat-flow reduction at the same deposition rate. If the deposition has a constant rate, but the thermal conductivity of the sediments increases, the heat flow increases. Heat flow is still depressed compared with the initial state of no deposition, but the magnitude of the depression is smaller.

High deposition rates may result in the formation of abnormal pressures and undercompaction. This can retard the heat transfer through sediments by lowering their thermal conductivities (e.g., Yalcin *et al.*, 1997). A case example is the rapidly subsiding Kura basin, located between the Greater Caucasus and the Lesser Caucasus in Azerbaijan (Yukler and Erdogan, 1996). Another example of a high deposition rate lowering the surface heat flow comes from the Absaroka thrust, Wyoming–Utah thrustbelt. One-dimensional conductive modelling of the thermal history reveals that at least half of the heat-flow decrease can be attributed to a transient depression of the heat flow during the Late Cretaceous by high deposition rates (Deming *et al.*, 1989).

Figures 14.4(a) and 14.5(a) show that once the deposition ceases, it may take tens of million years or more for the heat-flow deficit at the surface to be alleviated. For example, if deposition proceeds at 1 mm yr^{-1} for 20 Myr and then stops, it takes 40 Myr for the heat-flow deficit to be reduced to half of its maximum value (Deming, 1994c; Fig. 14.5a). Because the heat-flow depression lasts long after the depositional event, the underlying rock may affect the eventual maturation of organic material in the source rock.

An example of the effect of the deposition history on the modification of heat flow is shown in Fig. 14.4(b). It shows the effect of a shale depositional event, which started at a rate of 0.1 mm yr^{-1} for 10 Myr, then increasing to 0.5 mm yr^{-1} for another 10 Myr before stopping. Figure 14.4(b) shows that the surface heat flow responds quickly to the influence of deposition, but recovers slowly after the end of deposition. The time constant for this recovery is similar to the thermal response of the entire lithosphere (see Fig. 14.2b).

The above calculations of surface heat-flow modification documented in Figs. 14.3 and 14.4) took no account of radiogenic heat production in sediments for simplicity. However, most sediments produce radioactive heat as discussed in Chapter 13. Figure 14.5(b) shows the comparison of modified heat-flow values from Fig. 14.4(a) with those calculated for shale with a heat production of 10^{-6} W m^{-3}. Thus, the curves in Fig. 14.5(b) indicate the range of possible values. The radiogenic contribution is negligible in the case of a very high geothermal flow and the alteration of the surface heat flow follows the lower curves in this case. The radiogenic contribution is important in the case of a low geothermal flow. The effect of the radiogenic contribution can offset the deposition effect by up to 40%.

An excellent example of depositional impact on the thermal regime comes from the initial 35 km of the Barbados accretionary prism, discussed in detail in Chapter 15. One of the principal features of the heat-flow map shown in that chapter (Fig. 15.6a) is a general north to south decrease of 5–10 mW m^{-3} in this zone of the Barbados accretionary prism, related to the effect of the increasing deposition rate (Ferguson *et al.*, 1993). The thickness of incoming sediment carried by the North American plate decreases with increasing distance from the terrigenous sources of the Orinoco and Amazon rivers, located to the south of the complex. The Tiburon Rise, an obliquely subducting basement ridge presents a physical barrier to the northward extent of these deposits. These terrigenous sediments are underlain by a predominantly pelagic and hemipelagic

(a)

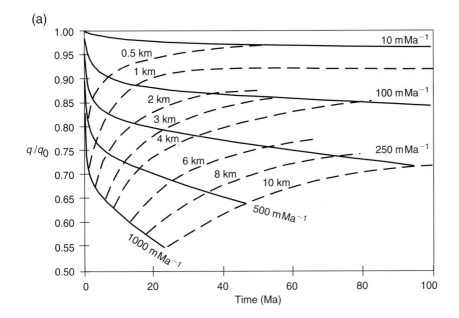

(b)

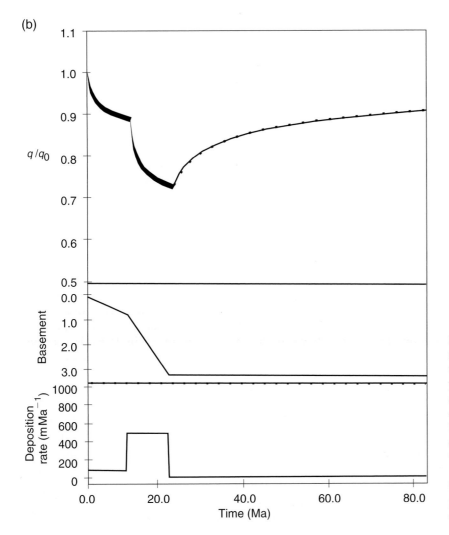

Fig. 14.4. (a) The variation of the fractional heat-flow alteration q/q_0 through time for deposition of shale with a deposition rate of 0.1, 0.25, 0.5 and 1 mm yr^{-1} (Hutchison, 1985). The dashed lines link points of equal basement depth. (b) Example of the changes in surface heat flow q/q_0 due to variable deposition history (Hutchison, 1985). The depth of burial of the basement and the deposition rate are plotted beneath the calculated heat-flow alteration.

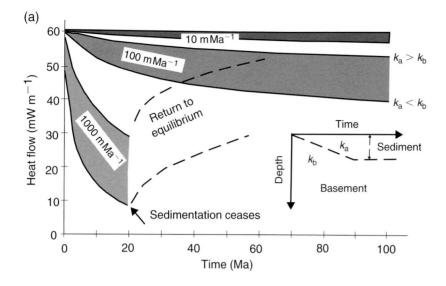

(a)

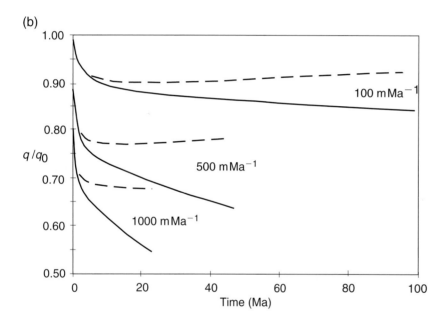

(b)

Fig. 14.5. (a) Effect of the deposition rate on the surface heat flow (Deming, 1994c). (b) The surface heat-flow alteration q/q_0 calculated for shale deposited with a deposition rate of 0.1, 0.5 and 1 mm yr^{-1} (Hutchison, 1985). The solid lines indicate the result ignoring radiogenic heat production. The dashed lines indicate the result with radiogenic heat production of 10^{-6} W m^{-3}. The background heat flow is 50 mW m^{-2}.

sequence of Late Cretaceous to Miocene age, lying on the Late Cretaceous oceanic crust. Ferguson *et al.* (1993) used a finite-element model, described in Chapter 15, to estimate the effect of deposition on heat flow in four areas located along four dip-oriented profiles (Table 14.2). The results of the modelling suggest that the accretion of the Orinoco submarine fan has led to more rapid growth of the southern part of the accretionary prism. From the north to the south of the Tiburon Rise, the width of the accretionary complex changes by 110 km. Thus, since deposition of the fan began, the southern part of the accretionary wedge has grown an additional 110 km further east than the northern part. Deposition related to the Orinoco fan started

during the late Early Miocene and accelerated in the Plio-Pleistocene. This resulted in an increased propagation rate for the southern part of the wedge, relative to the northern part, of 0.6 cm yr^{-1}, but it may have been as much as 1.8 cm yr^{-1} for the Plio-Pleistocene. The propagation rate for the northern part of the wedge is probably less than 0.1 cm yr^{-1}, because of the low thickness of the accreted layer (e.g., Le Pichon *et al.*, 1990b). With a plate convergence rate of 2 cm yr^{-1}, the estimated advance rate for the southern part of the wedge is between 2.7 and 3.9 cm yr^{-1}. Reducing the advance rate in the described profiles reduces the rate of thickening of the wedge at fixed coordinate points, with a consequentially smaller reduction in heat flow. The

Table 14.2. Effect of deposition upon heat-flow in the frontal portion of the Barbados accretionary prism.[a] The range of values represents the variation in heat-flow, obtained using shale and sandstone parameters from Hutchison (1985) listed in Chapter 13 (see also Fig. 14.3b), with and without radiogenic heat production. Pre-Middle-Miocene deposition rates for the three transects north of 11° 35′ N were obtained from Wright (1984). At 11° 35′ N, the greater Pre-middle-Miocene deposition rate reflects early terrigeneous deposition.[b] Deposition rates are for sediments with an initial porosity of 0.7. Compaction follows equation (14.3), where the compaction constant is 1540 m for shale and 2780 m for sandstone. The age of the oceanic lithosphere is 84 Myr.

	Modelled rates of deposition (mm yr^{-1})					
Latitude	Cretaceous–Paleogene 84–53 Ma	L. Eocene 53–49 Ma	M. Eocene–L. Miocene 49–18 Ma	L. Miocene–present 18–0 Ma	Total sediment thickness (km)	Heat flow (mW m^{-3})
15° 30′ N	0.005	0.02	0.015	0.02	0.8	53–55
14° 20′ N	0.005	0.02	0.015	0.35	3.2	44–50
13° 20′ N	0.005	0.02	0.015	0.34	3.1	44–50
11° 35′ N	0.05	0.05	0.05	0.72	6.25	41–52

Notes:
[a] Ferguson *et al.* (1993).
[b] Wu (1990).

difference in heat flow at a location 40 km from the deformation front that would result from halving the advance rate, would be about 9%. However, the advance rate reduction also reduces the frictional heating, that is the principal cause of the arcward recovery in heat flow.

Role of erosion in thermal regimes

The thermal effects of erosion on temperatures in the continental crust in one-dimensional solutions are quantified as (England and Richardson, 1977)

$$\frac{\partial T}{\partial t} = \frac{\partial}{\partial x} \kappa \left(\frac{\partial T}{\partial x} \right) + \frac{H(x,T,t)}{\rho c} + u_x(\partial T \partial x), \qquad (14.8)$$

where T is the temperature, t is the time, κ is the thermal diffusivity, H is the heat production, ρ is the density, c is the specific heat at constant pressure and u_x is the upward movement of the system with time, depending on the erosion rate.

The role of erosion on the two-dimensional thermal structure of an orogenic belt is best discussed in connection with accretion, because it is the interaction between these two processes that changes the velocity at which particles move in both the upper and lower plates and thus affects the advection of heat in both vertical and horizontal directions (Royden, 1993; Fig. 14.6a).

The basic effects of this interaction in orogenic belts, which have reached a steady-state temperature structure, have been studied analytically by Royden (1993). In her models, the detachment fault is located at a depth of 0 m and the accretionary wedge above it has thickness h. The wedge, which has a toe at a distance $x = 0$, is subjected to a surface erosion at rate e, while material from the lower plate is accreted at rate a. A particle in the lower plate moves at rate v relative to the toe of the wedge (Fig. 14.6b). The vertical component of the movement v is the accretion rate at which material is accreted across the detachment fault plane.

When the accretion rate is greater than zero, material is accreted into the wedge (Fig. 14.6a, cases 3 and 4). A particle in the wedge moves relative to the toe of the wedge with horizontal component of velocity u, which depends on the relative rates of erosion and accretion (Fig. 14.6b). If accretion is faster than erosion, the horizontal component of the velocity u is positive and material in the hanging wall moves away from the toe of the wedge (Fig. 14.6a, case 3). If accretion is slower than erosion, the horizontal component of the velocity u is negative and material in the hanging wall moves towards the toe of the wedge (Fig. 14.6a, case 2). If the erosion equals the accretion, particles in the hanging wall remain at a constant distance from the toe of the wedge and move vertically upwards toward the surface (Fig. 14.6a, case 4). Despite the fact that accretion paths in nature are more complex due to the movement of separate thrust sheets (Mulugeta and Koyi, 1992; Fig. 14.7) and that a complexity in erosion paths driven by variable lithology, relief and climate has to be assumed, the special simplified cases in Fig. 14.6(a) are suitable for the illustration of the basic trends.

An example of how the spatial and the temporal distribution of erosion affects the timing and degree of source rock maturation comes from North Alaska

(a)

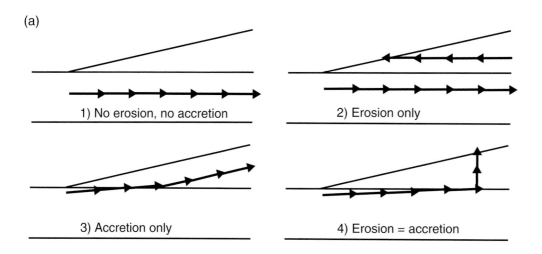

(b)

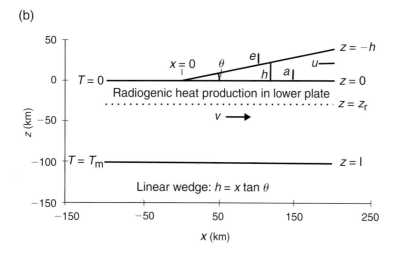

Fig. 14.6. (a) Simplified particle trajectories relative to the toe of the thrust wedge for four special cases of surface erosion and basal accretion: (1) no erosion and no accretion; (2) erosion only; (3) accretion only; and (4) the erosion rate equal to the accretion rate (Royden, 1993). (1) When erosion and accretion rates are zero, particles in the upper plate do not move with respect to the toe of the wedge. (2) When the accretion rate is zero, particles in the upper plate move parallel to the thrust fault and towards the toe of the wedge and particles in the lower plate move parallel to the thrust fault and away from the toe. There is no material transfer from lower to upper plate in either of cases (1) or (2). (3) When the erosion rate is zero, particles in the upper plate move parallel to the surface and away from the toe of the wedge. (4) When erosion equals accretion, particles in the hanging wall remain at a constant distance from the toe of the wedge and move vertically upwards toward the surface. Note that when accretion is faster than erosion, the horizontal component of the velocity u is positive and material of the hanging wall moves away from the toe of the wedge, but that when the accretion is slower than erosion, the horizontal component of the velocity u is negative and material of the hanging wall moves toward the toe of the wedge. (b) A simplified thrustbelt geometry is assumed for a linear wedge where the upper plate thickness h increases linearly with distance x from the wedge toe at $x = 0$, $z = 0$ (Royden, 1993). Surface erosion e and basal accretion rates a are uniform throughout the wedge. T is temperature, T_m is the temperature at base of the lithosphere at depth $z = 1$. Particles in the lower and upper plates move relative to the wedge toe with a horizontal velocity components v and u, respectively. θ is the dip of the detachment relative to the surface.

(Decker *et al.*, 1996), where vitrinite reflectance and apatite fission-track data demonstrate the contributions from short-term compressional and long-term isostatic processes on the distribution of erosion. Erosion associated with folding events of about 5 Myr duration resulted in a removal of 2–3 km thick sections in numer-

ous anticlines at different times. Fission-track data indicate discrete thrusting events affecting different anticlines timed at 60, 45 and 20 Ma. In addition to affecting the local temperature fields of the different thrust sheets, the erosion also affected the thermal regimes of adjacent areas by sourcing synchronous deposition of

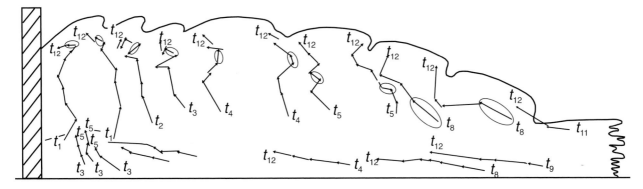

Fig. 14.7. Particle paths of some material points encompassing the entire sand wedge in a sandbox experiment, with respect to a reference frame moving with the rear wall of the sandbox (Mulugeta and Koyi, 1992). Flow paths have been monitored by tracing especially marked sand grains near well-defined and easily distinguishable horizons such as near intersections of passive layering with shear zones and imbricates. Encircled vectors denote particle paths during the episode of shortening.

clastic wedges beyond the deformational fronts (Decker et al., 1996).

Provided erosion is not too high, the hanging walls and footwalls reach a steady-state temperature structure in less than a few tens of Myr (Royden, 1993). Therefore, the insight into the role of erosion and accretion on thermal structure, derived from Royden's (1993) numerical modelling, is suitable for most long-lived orogenic belts and subduction systems. This modelling also assumes homogeneous distribution of heat-producing elements, homogeneous thermal conductivity and a similar temperature distribution in the lower plate beneath the toe of the wedge as that of the equilibrium conditions in the foreland. Figure 14.6(b) shows that the thickness h of the wedge increases linearly with distance x (Royden, 1993):

$$h = x \tan \theta, \tag{14.9}$$

where θ is the dip of the detachment relative to the surface. Therefore, the hanging wall velocity u is related to the erosion and accretion rates e and a, respectively, following (Royden, 1993):

$$ae = u \tan \theta. \tag{14.10}$$

Readers interested in the equations for depth–temperature distributions calculation are referred to Royden (1993).

Cases 2–4 from Fig. 14.6(a) illustrate the end members of the analysis, while intermediate cases where erosion and accretion rates are nonzero and are not equal to one another, produce intermediate steady-state temperature structures. Figures 14.8 and 14.9 show the geotherms for end member cases with erosion and accretion rates of 0–1.5 mm yr^{-1} acting on 5 and 30 km thick thrust plates. An average upper crustal value for the radiogenic heat (Pollack and Chapman, 1977) has been used.

Increasing either the erosion or the accretion rates increases temperatures at depth relative to the case with no erosion or no accretion. Erosion increases the temperatures because it results in advection of hotter material towards the surface and from thicker, hotter parts to thinner, colder parts of the wedge. It mainly affects temperatures inside the wedge. A comparison of Figs. 14.8 and 14.9 shows that, even in the absence of radiogenic heat production, the effects of erosion and accretion are stronger with a detachment fault at a depth of 30 km than at a depth of 5 km. Thus, these effects are more pronounced in thicker parts of the wedge.

In the case of radiogenic heat production in the footwall, erosion increases the contribution of this heat source to the thermal regime of the wedge. However, in the case of radiogenic heat production in the hanging wall, erosion produces a more marked effect than in the case of the footwall and can be dramatic for moderate to large erosion rates. For example, with a detachment fault located at a depth of 30 km, the temperature contribution from hanging wall radiogenic heat production at an erosion rate of 1.5 mm yr^{-1} is three times larger than at a rate of 0 mm yr^{-1} (Royden, 1993; Fig. 14.8).

If the detachment fault is located at a depth of 5 km, neither erosion/accretion nor radiogenic heat production has a very large impact on the shape of the temperature–depth curve in the hanging wall and temperatures increase nearly linearly with depth. In contrast, with a detachment fault depth of 30 km, both erosion and radiogenic heat production cause a significant increase in the hanging wall temperatures, but temperature–depth curves in the upper 5–10 km are still roughly linear (Fig. 14.10a).

The contribution of frictional heating along the fault surface into the thermal regime of the wedge is shown

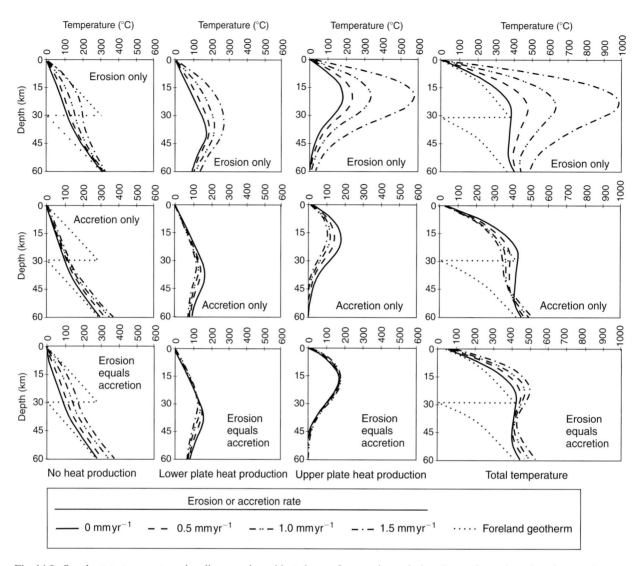

Fig. 14.8. Steady-state temperatures in a linear wedge subjected to surface erosion only, basal accretion only and surface erosion equal to basal accretion (Royden, 1993). Thrust plates are 30 km thick and erosion–accretion varies from 0 to 1.5 mm yr^{-1}. The convergence rate v–u is 20 mm yr^{-1}, fault dip is 11.3°, heat production is 2.5 μW m^{-3}, surface heat flow in the foreland is 60 mW m^{-2} and shear stress on the fault is 0 MPa. See Fig. 14.6(b) for further explanation.

in Fig. 14.10(b), with a shear stress of 100 MPa. Temperatures due to shear heating are always maximal along the detachment fault and the effects of erosion and accretion are similar to their effects on radiogenic production in the footwall (see Figs. 14.8 and 14.9).

A natural example of the erosional impact on the thermal regime of a wedge is discussed by Ferguson *et al.* (1993). It comes from the proximal portion of the Barbados accretionary prism, where the thermal properties of the sedimentary column are predicted from the porosity distribution with depth. The thermal resistance per square metre of a 14 km thick sedimentary column is 5911 °C W^{-1}. If the top 1 km of the column is

removed by erosion and replaced by a 1 km section with the thermal properties of a column predicted for a depth interval of 14–15 km, to keep the total thickness the same, the thermal resistance decreases to 5632 °C W^{-1}. The effect of this replacement is to increase the surface heat flow by 5%, provided the temperature at the base of the wedge is held constant. In fact, this is an overestimate, because the isotherms would migrate downward to increase the thermal gradient in the region beneath the wedge to sustain the increase in heat flow resulting from the reduced thermal resistance of the wedge. Therefore, the temperature at the base of the wedge would be reduced. Similarly, the

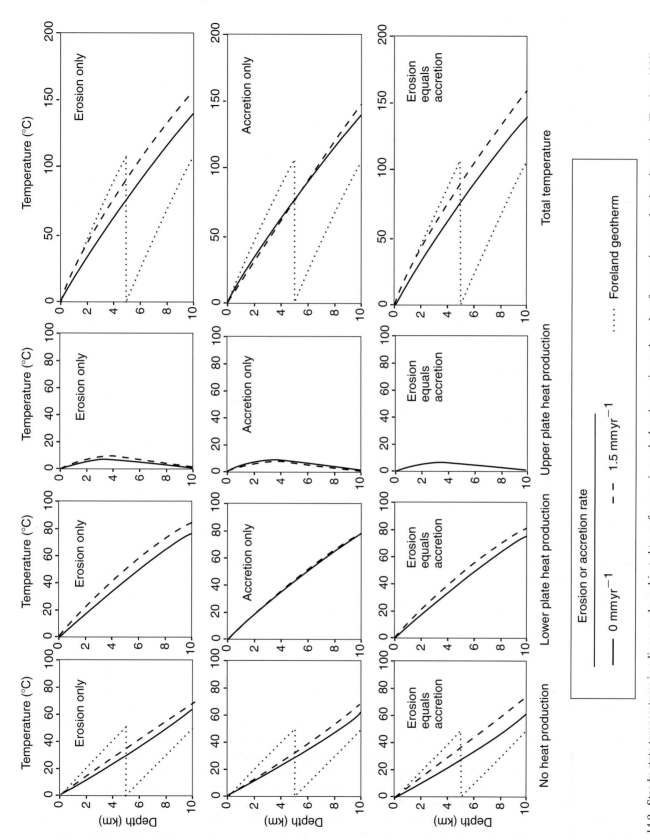

Fig. 14.9. Steady-state temperatures in a linear wedge subjected to surface erosion only, basal accretion only and surface erosion equal to basal accretion (Royden, 1993). Thrust plates are 5 km thick and erosion/accretion varies from 0 to 1.5 mm yr^{-1}. Other parameters are listed in Fig. 14.8.

(a)

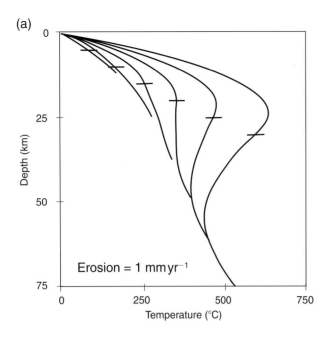

Erosion = 1 mm yr^{-1}

Fig. 14.10. (a) Computed geotherms for the Himalayas assuming a surface erosion rate of 1 mm yr^{-1} and no basal accretion (Royden, 1993). Depth to the basal fault varies from 5 to 30 km in 5 km intervals and short horizontal lines show the fault depth for each geotherm. Other parameters comprise convergence rate of 20 mm yr^{-1}, fault dip of 11.3°, heat production of 2.5 μW m^{-3}. Surface heat flow in the foreland is 60 mW m^{-2} and shear stress on the fault is 0 MPa.
(b) Contribution to steady-state temperature structure in a linear wedge from shear heating along the fault plane (Royden, 1993). Shear stress is 100 MPa, yielding a planar heat production of 63 mW m^{-2} at a slip rate of 20 mm yr^{-1}. Other parameters comprise convergence rate of 20 mm yr^{-1}, erosion–accretion of 0–1.5 mm yr^{-1}, fault dip of 11.3°, no heat production and zero surface heat flow in the foreland.

(b)

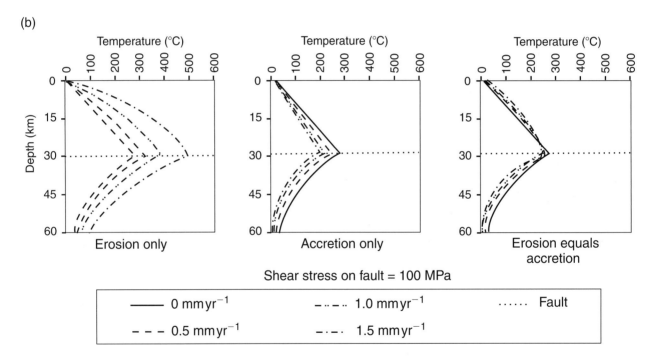

erosional removal of the top 2 km increases the surface heat flow by less than 7.8% and the removal of 3 km increases heat flow by less than 10.2% (Ferguson et al., 1993).

Another natural example of the impact of accretion/erosion on the thermal regime of a wedge is discussed by Barr and Dahlen (1989), whose numerical models focus on the Taiwan accretionary wedge. Their models indicate that reducing the accretion and erosion rates reduces the internal strain heating and shear heating along the décollement by the same fraction. The thermal gradient and surface heat flow are reduced in the case of the total thermal regime, because reduced heating results in a smaller upwarp of the isotherms.

Collision zones, in which a sub-aerial terrain has been developed, consisting of a two-sided orogen with opposite facing, mechanically coupled wedges (e.g., Koons, 1990; Beaumont et al., 1992; Willet et al., 1993) and with

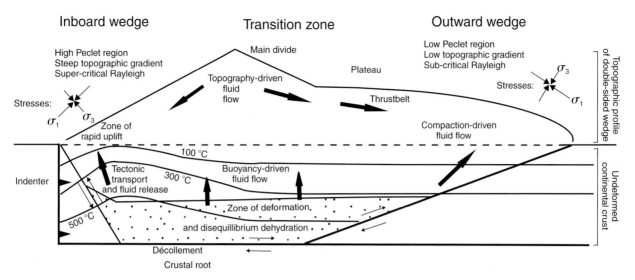

Fig. 14.11. Diagram of a two-sided collisional orogen summarizing the thermal and topographic character of the orogen (Koons and Craw, 1991). The stress orientations for the inboard and outboard wedges are derived from the Mohr circle.

their geometry governed by their position relative to the indenter and by erosion processes on either side of the orogen, have a specific thermal regime. Where the precipitation is orographically controlled, it causes the coexistence of a wet wedge on the windward side, which is rapidly denuded, and a dry wedge on the opposite side, which has a little erosional denudation. If the windward side is adjacent to the indenter, the erosional energy is concentrated in the inboard region. Plate movements are usually sufficiently rapid that Peclet numbers, defined in Chapter 16, are high with the characteristic velocity being the uplift rate and the characteristic length being the depth of the inboard wedge. Under these conditions heat advection dominates and hot rocks are uplifted faster than they can cool (e.g., Allis *et al.*, 1979; Koons, 1987). This produces a pronounced thermal anomaly near the toe of the inboard wedge (Koons and Craw, 1991; Fig. 14.11).

Because erosion is not the only mechanism that brings buried rocks to the surface, it is useful to see how pressure–temperature–time (p–T–t) paths modelled for erosion and tectonic denudation differ from each other. The two pathways were studied on a lithospheric scale by Ruppel and Hodges (1994a), who further divided tectonic denudation into pure and simple shear-controlled denudations. For simplicity, erosion is assumed to affect an area uniformly in their models. Unlike pure shear and simple shear, erosion does not conserve radioactive heat production, because it removes layers with radioactive elements. This difference is the distinguishing factor between tectonic processes and erosion.

Both pure shear thinning and erosional unroofing are treated as one-dimensional processes in these models and do not contribute to the development of lateral thermal gradients. This one-dimensional similarity between pure shear thinning and erosion proves to be more important in controlling p–T–t paths than the differences in how radioactive layers are affected (Fig. 14.12). In Fig. 14.12(a), the comparison of the effect of the various mechanisms on the p–T path is facilitated by the use of a linear initial geotherm A with no radioactive heating (Fig. 14.12c), which eliminates the complicating effect related to any disruption of the radioactive layer. The thermal histories shown in Fig. 14.12(b) correspond to an initial geotherm B, which includes a layer with radioactive heat production in the upper crust (Fig. 14.12c). In the case of both geotherms, simple-shear thinning along a 45° dipping normal fault produces a slightly concave p–T path with nearly constant temperature gradient. Pure shear of the whole lithosphere produces an isothermal decompression during most of the unroofing history in the case of the geotherm A and during the first 10 km of unroofing in the case of the geotherm B. Pure shear affecting only the crust, initially creates an isothermal uplift relative to the syn-orogenic surface and a temperature gradient which becomes steeper with uplift. Erosional unroofing produces a p–T path that lies between paths for uniform pure and simple shear.

The similarity between the nearly isothermal p–T paths, observed in early stages of erosion and pure shear thinning, results from the one-dimensional nature of both mechanisms. Although the mechanisms have important differences, their respective curves shown in Fig. 14.12(b) are separated only by up to 75 °C and 150 MPa. It may make them difficult to distinguish in an actual case. However, in nature, tectonic unroofing tends to be faster than erosion, which makes the separation more pronounced.

(a)

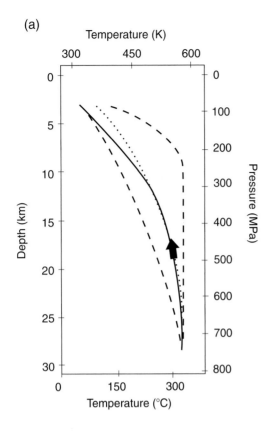

(b)

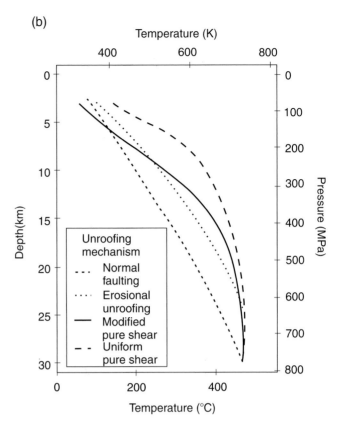

(c)

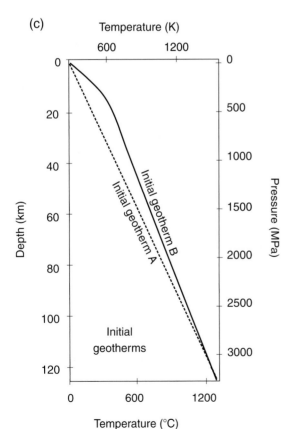

Fig. 14.12. Theoretical post-thrusting *p*–*T* paths for a rock carried from 30 to 3 km by uniform pure shear ($\beta = 10$ throughout the lithosphere), modified pure shear ($\beta = 10$ only in the 30 km thick crust), erosional unroofing at 1 mm yr^{-1} and normal faulting that yields a tectonic denudation rate of 1 mm yr^{-1} (Ruppel and Hodges, 1994a). The scale of figures (a) and (b) is the same to facilitate comparison. (a) With initial geotherm A, pure shear thinning produces nearly linear isothermal decompression, while simple shear (normal faulting) unroofing results in a nearly linear *p*–*T* path. Erosion and modified pure shear, both one-dimensional mechanisms, which primarily affect the crust on the time scale of this test (27 Myr), yield nearly indistinguishable *p*–*T* paths. (b) When radioactive heat is added (initial geotherm B), the pure shear is isothermal only for the first few kilometres of thinning, although the overall morphology of the *p*–*T* path resulting from unroofing by normal fault displacement remains unchanged. In addition, modified pure shear produces a path more clearly distinct from the erosional unroofing than before due to different effects of each mechanism on the configuration of the radioactive layer. (c) The initial geotherms used for models. Initial geotherm A has a constant thermal gradient of 10.4 K km^{-1} between the surface and the base of the lithosphere (125 km) and no crustal radioactive heat. Initial geotherm B is a steady state geotherm calculated by superimposing radioactive heating at a rate of 3 μW m^{-3} in the upper 20 km of the crust on the background gradient of 10.4 K km^{-1} used for geotherm A.

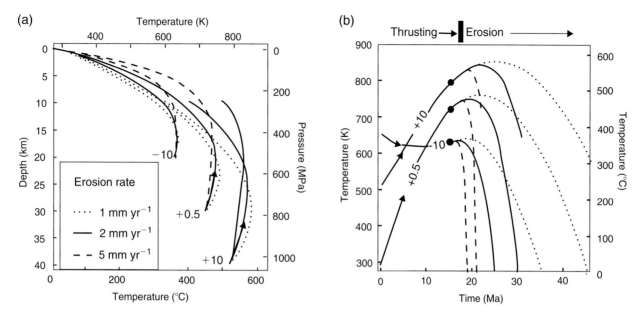

Fig. 14.13. *p–T* and *T–t* paths for thrusting at 2 mm yr⁻¹ along a 45°-dipping fault followed by erosion at rates of 1, 2 and 5 mm yr⁻¹, corresponding to erosional events of duration 30, 15 and 6 Myr, respectively (Ruppel and Hodges, 1994a). The syn-thrusting part of the *p–T* path is not shown. (a) *p–T* paths for rocks located 10 km below, 0.5 km below and 10 km above the fault for the three unroofing rates. Note that both the value of the maximum temperature attained during the rock's thermal history and the depth (time) at which that temperature is reached depend completely on the rate of erosional unroofing. (b) *T–t* paths corresponding to the *p–T* paths of (a). Maximum temperatures are attained shortly after the end of thrusting for the shallowest rocks but not until significantly later for the deeper rocks.

Figure 14.13 shows that the retrograde path depends significantly on the unroofing rate. This is shown for rocks located 10 km below, 0.5 km below and 10 km above the 45° dipping thrust fault that moved at a rate of 2 mm yr⁻¹ for 15 Myr. This thrusting event was followed by the erosional removal of the 30 km thick section at rates of 1, 2 and 5 mm yr⁻¹ over 30, 15 and 6 Myr long periods, respectively. Figure 14.13 indicates that fast unroofing causes the initial decompression phase to be accompanied by very little net heating, producing a nearly isothermal unroofing curve. Because some heating occurs during the unroofing stage for all the examples in Fig. 14.13(a), there is a change in the concavity of the *p–T* curve about an inflection point at the depth of the maximum temperature. At depths below the maximum temperature, the path has a low and relatively constant temperature gradient. After passing through the maximum temperature, the thermal gradient and the rate of cooling increases.

When different burial rates are coupled with similar unroofing rates, temperature differences existing immediately after the end of thrusting are largely eliminated after half of the unroofing (Fig. 14.13b). Figure 14.13(b) indicates the dependence of the shape of the retrograde *p–T* paths on the unroofing rate and not on the burial rate. In the case of faster unroofing, peak temperatures occur at significantly shallower depths. Similar curves characterize different unroofing mechanisms and their general characteristic is the same (Ruppel and Hodges, 1994a). Rapid unroofing produces a distinct nearly isothermal decompression in the early stages and rapid cooling and increasing of the thermal gradient in the later stages of unroofing. If rapid unroofing is characteristic for tectonic denudation, then large thermal gradients and cooling slopes for retrograde portions of *p–T* and *T–t* paths are indicative of tectonic denudation.

Topographic influence on the subsurface temperature field

All the models of thermal regimes of orogenic wedges described above assumed homogeneous erosion, which is a simplification. This section focuses on local erosional effects on the thermal regime.

When a homogeneous rock volume has a flat horizontal surface at constant temperature, its subsurface isotherms are parallel to the surface. No horizontal temperature gradients occur in this case. However, natural erosion and tectonic denudation result in a complex relief that includes flats, hills and valleys. Such a complex relief introduces various perturbations into the thermal regime of a homogeneous rock volume (Buntebarth, 1984). In the case of such a complex topography, the sub-surface isotherms are neither parallel to the surface nor to each other and horizontal gradients occur. The vertical temperature gradient underneath a hill is lower than the gradient under a horizontal surface. The vertical temperature gradient underneath a valley is higher than the gradient under a horizontal surface. Temperatures at the surface are reduced in relation to the surface elevation z above sea level. For example, this reduction has a gradient $(dT/dz)_B$ of $4.5\,°C\,km^{-1}$ in the Alps (Buntebarth, 1984).

The necessary topographic correction of the thermal perturbation is made in two steps. In the first step, the thermal effect of the terrain, which has to be corrected for, is projected onto a horizontal plane. The obtained temperature distribution corresponds to the effect of the uneven topography. The deviation from a constant temperature on the reference plane is then calculated, using the uncorrected measured temperature gradient dT/dz (Buntebarth, 1984):

$$T_B = h\left[\frac{dT}{dz} - \left(\frac{dT}{dz}\right)_B\right].\qquad(14.11)$$

Temperatures T_B on the horizontal plane are computed as average values for concentric circles with radius r around a point P (Fig. 14.14). They are dependent on the average elevation h determined from a topographic map.

In the second step, the temperature T_B distribution from the first step is used as an input for Poisson's integral. It is calculated for the case in which the required temperature gradient $(dT/dz)_c$ at point P of the half-space satisfies Laplace's equation with the boundary function $T_B(r)$ (Buntebarth 1984):

$$\left(\frac{dT}{dz}\right)_c = \int_0^\infty \left(1 - \frac{2z_0^2}{r^2}\right)\left(1 + \frac{z_0^2}{r^2}\right)^{5/2}\left(\frac{T_B(r)}{r^2}\right)dr.\qquad(14.12)$$

The integral formulating the situation illustrated in Fig. 14.14 is solved numerically (Buntebarth, 1984):

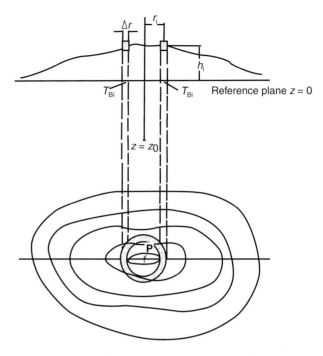

Fig. 14.14. Model of the topographic correction of temperature (Buntebarth, 1984).

$$\left(\frac{dT}{dz}\right)_c \approx \sum_{i=1}^n \left(1 - \frac{2z_0^2}{r_i^2}\right)\left(1 + \frac{z_0^2}{r_i^2}\right)^{5/2}\left(\frac{T_{B,i}}{r_i^2}\right)\Delta r$$
$$= \sum_{i=1}^n \left(1 - \frac{2z_0^2}{r_i^2}\right)\left(1 + \frac{z_0^2}{r_i^2}\right)^{5/2}\left(\frac{h_i\Delta r_i}{r_i^2}\right)\left[\frac{dT}{dz} - \left(\frac{dT}{dz}\right)_B\right].$$
$$(14.13)$$

The corrected temperature gradient $(dT/dz)_0$ at the point P is then given by (Buntebarth, 1984)

$$\left(\frac{dT}{dz}\right)_0 = \frac{dT}{dz} - \left(\frac{dT}{dz}\right)_c.\qquad(14.14)$$

The same method is applied for the correction beneath a valley, with h being negative. If the topographic relief is very large then this approximation can lead to considerable errors.

15 Role of deformation in thermal regimes

The role of deformation on the thermal regime in thrustbelts can be divided into three areas of control:

(1) the velocity at which different thrust sheets are emplaced on top of each other, which controls the rate at which their thermal regimes affect each other;
(2) the heating provided by the internal deformation of the thrust sheet material; and
(3) the heating provided by the friction along the décollement and major thrust faults.

Role of the shortening rate

Shortening brings a hanging wall from depth to rest on top of the footwall, causing warming of the footwall by increased burial and cooling of the hanging wall by placing it into a cooler thermal regime. This concept can be illustrated by petrological data from footwall rocks, which reflect pressure and temperature increase during thrusting, and from hanging wall rocks, which record retrograde reactions (Karabinos, 1984a, b; Chamberlain and Zeitler, 1986; Trzcienski, 1986).

In order to discuss the effect of the velocity at which different thrust sheets are emplaced on top of each other, a set of finite-element models has been designed for various types of thrustbelt structures (Henk and Nemčok, 2000). Because the report in which the results have been described has restricted circulation, they will be discussed here in detail.

Because the shortening rate affects the duration of the heat transfer between thrust sheets moving on top of each other, it is extremely important to understand which rates are realistic for natural cases. While a rate of centimetres per year is appropriate for plate movements, typical shortening rates for local structures in active thrustbelts are in millimetres per year (Table 9.3a). Because initial studies on thermal effects of overthrusting used the former more rapid rates to assume that the thermal structure of both the thrust plate and the underlying sediment were unaffected (e.g., Oxburgh and Turcotte, 1974; England and Richardson, 1977; Angevine and Turcotte, 1983; England and Thompson, 1984), they were unrealistic. The model of Angevine and Turcotte (1983) was an example of this approach, and results in a sawtooth-shaped temperature–depth profile, in which temperatures at the thrust fault reach 90% of the pre-thrusting value at the base of a 4 km thrust sheet after 10 Myr. These models also neglect frictional heating along the bounding thrust faults. Therefore, they provide a simplified insight into the average thermal evolution and only over a time scale an order of magnitude greater than the thrusting event itself (e.g., England and Richardson, 1977; England and Thompson, 1984).

Karabinos and Ketcham (1988) and Wygrala *et al.* (1990), among others, have demonstrated that the previously described instantaneous thrusting is an invalid assumption for models that attempt to simulate the short-term, i.e. 1–10 Myr long, thermal history of rocks in a thrust scenario. They have also shown that the resulting thermal regimes are affected by the thrust ramp dip at a specific slip rate, because the ramp angle controls the time at which the fully grown thrust structure develops (Wygrala *et al.*, 1990). Therefore, because most thrustbelt structures develop over less than 10 Myr (Figs. 15.1a–d and Table 15.1), it is extremely important to provide a correct shortening rate for maturation calculations of the hydrocarbon source rock. Because the geometry inherent for a specific thrustbelt structure controls its thermal regime development, it is important to calculate the thermal regime development of each specific thrustbelt structure and avoid the use of 'approximate' analogues.

The finite-element models of Henk and Nemčok (2000) are based on forward-balanced models of thrustbelt structures (Fig. 15.1) modelled using the programs GeoSec 2D® (Paradigm Geophysical Ltd, Houston, USA) and Move 2D® (Midland Valley Exploration Ltd, Glasgow, UK) from undeformed to the most mature development stages. Different development stages of these structures were used for separate finite-element models. Growth times and slip rates of the structures (Fig. 15.1) are shown in Table 15.1, and most of their physical parameters are listed in Table 15.1. The models are computed in two dimensions, using plane strain conditions, coupling thermal and mechanical calculations and assuming a constant basal heat flow. Thermomechanical coupling is achieved by solving the thermal and mechanical equations successively for two geometrically identical finite-element grids. The computations

(a)

Syn-tectonic deposition case   Syn-tectonic erosion case

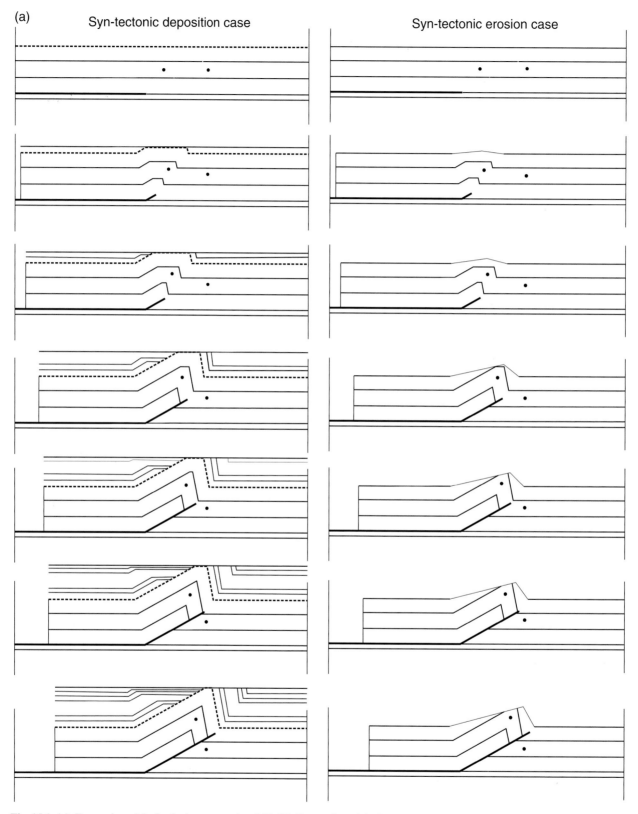

Fig. 15.1. (a) Forward model of a fault-propagation fold. (b) Forward model of a detachment fold. (c) Forward model of a fault-bend fold with a buried duplex. (d) Forward model of an inverted half-graben. For each of the models further explanation is given in the main text.

(b)   Syn-tectonic deposition case   Syn-tectonic erosion case

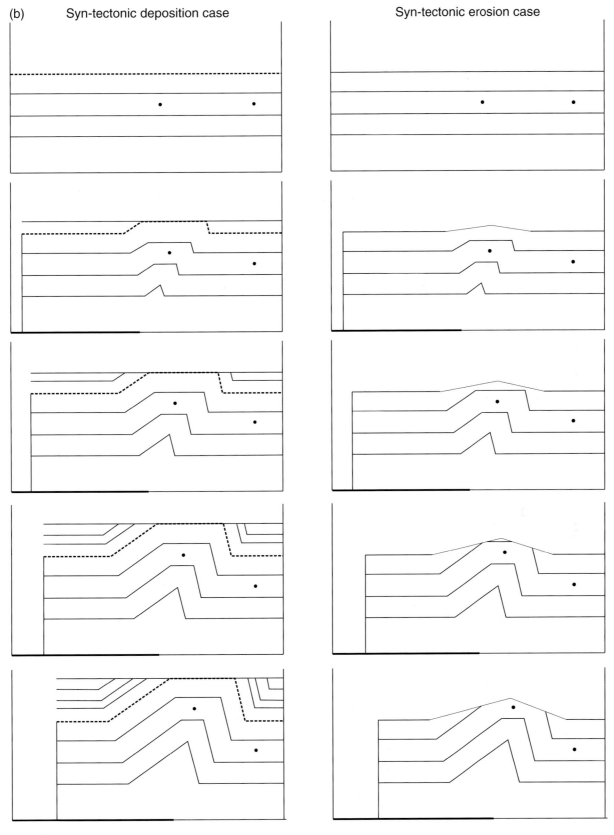

Fig. 15.1 (*cont.*)

(c)   Syn-tectonic deposition case   Syn-tectonic erosion case

(d)   Syn-tectonic deposition case   Syn-tectonic erosion case

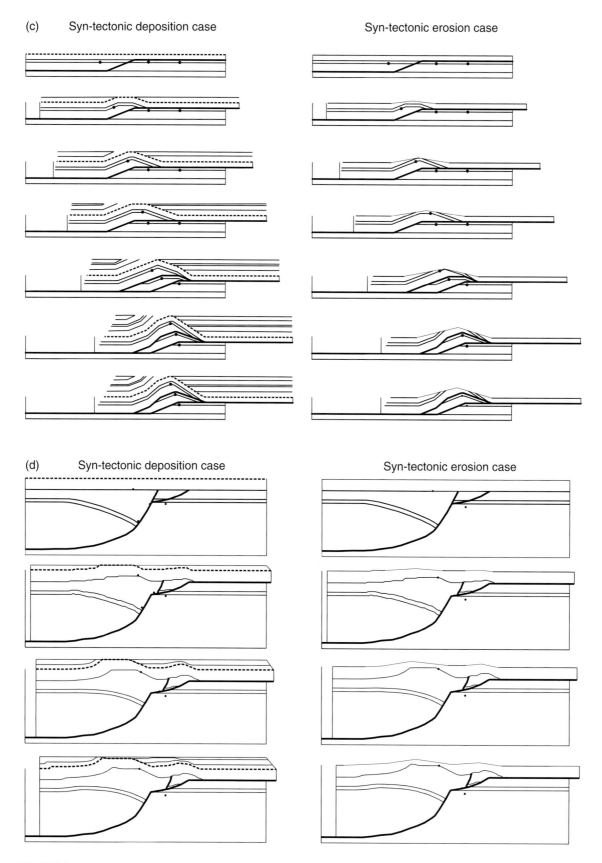

Fig. 15.1 (*cont.*)

Table 15.1. Growth time of the mature structure for each modelled scenario for three chosen slip rates (0.3, 2.3, 4.3 mm yr^{-1}). This is the time when the growing structure actively perturbs the thermal regime. After the growth stops, the thermal perturbation decays with time. Different structures perturb the thermal regime for different times. Therefore they have been separated into categories (long, moderate, short and very short).

Scenario	Structure	Slip rate (mm yr^{-1})	Time of the development to mature stage (Ma)	Category
1	Fault-propagation fold	0.3	6.4	Short
		2.3	0.8	
		4.3	0.4	
2	Passive roof thrust	0.3	11	Moderate
		2.3	1.4	
		4.3	0.8	
3	Detachment fold	0.3	5	Short
		2.3	0.6	
		4.3	0.3	
4	Duplex	0.3	22.7	Long
		2.3	3	
		4.3	1.6	
5	Inverted half-graben	0.3	3.2	Very short
		2.3	0.4	
		4.3	0.2	

are coupled via displacements, temperatures and thermal stresses. The models are sufficiently large to account for transient evolution of the thermal regime during the development of the respective structure. The heat transport is provided by both conduction and advection. No lateral heat flow is assumed through the vertical sides of the models. Physical parameters for each rock type in Table 15.2 are taken from published analogues. Thermal material parameters consist of the temperature-dependent thermal conductivity, the specific heat capacity, the temperature-dependent density and the lithology-dependent radiogenic heat production. It is assumed that the physical properties of layered rock were laterally relatively constant to justify a two-dimensional approach. The calculations utilize the computer program ANSYS® (Ansys Inc., Houston, USA). Each numerical model describes a cross section of the folded sedimentary sequence with isotropic elastic material properties, which is broken by planes of weakness, i.e. pre-existing faults. Four-node isotropic elements are used to describe the material properties of sediment layers and all nodes are in communication during each time-step. Compatibility and equilibrium is

achieved by iteration. Faults in the model are described by so-called contact elements. This approach handles large differential movements between parts of the model, but does not describe the fault propagation itself. The contact element is defined at opposite surfaces of the pre-assigned fault and is capable of describing frictional sliding. Contact stiffnesses at contact elements are used to enforce compatibility between adjacent fault surfaces. The model uses the generalized Hooke's law to relate normal strains to stresses dictated by Young's modulus and Poisson's ratio (e.g., Mandl, 1988).

Since these models use a simplified description of the real physical parameters, they do not attempt to predict absolute values but concentrate on the relative patterns of thermal regimes in relation to controlling factors such as:

(1) the existence of syn-tectonic deposition versus erosion;
(2) initial pre-shortening heat flow;
(3) thermal blanketing of syn-tectonic sediments;
(4) the slip rate;
(5) thrust sheet lithology; and
(6) detachment layer lithology.

The single greatest influence on a system's thermal regime is exerted by whether a growing structure is affected by syn-tectonic deposition or erosion. This control can be illustrated by discussing a model simulating a fault-bend fold with a buried duplex (Fig. 15.1c). This structure deforms a 2 km thick sediment section comprised of three layers. The upper dolomite layer is 663 m thick, the middle limestone layer is 337 m thick and the lower dolomite layer is 1000 m thick. The first alternative represents a structure which undergoes a syn-tectonic shale deposition that keeps pace with the growth of the structure (Fig. 15.1c, case 1; Fig. 15.2a), simulating a marine case. The second alternative represents the development of a structure affected by 50% effective erosion (Fig. 15.1c, case 2; Fig. 15.2b), simulating a continental case. The detachment horizon occurs in shale. The physical properties of the modelled sediments are listed in Table 15.2. The chosen ramp angle has a dip of 23° and the duplex develops beneath the uppermost pre-tectonic layer. The pre-orogenic surface heat flow was selected as 80 mW m^{-2}, from the range of values of 40–80 mW m^{-2}, known to represent compressional basins and described in Chapter 12. The chosen slip rate is 0.3 mm yr^{-1}, the minimum value from the 0.3–4.3 range extracted from Table 15.3a. The two alternative models, described above, include three checkpoints designed to monitor the thermal behaviour of the roof thrust (checkpoint 'hanging wall 1'), duplex (checkpoint 'hanging wall 2') and footwall (checkpoint 'footwall') (Fig. 15.1c).

Table 15.2. Physical properties of rocks used in finite-element models.

Rock	Angle of internal friction (deg)	Friction	Cohesion (MPa)	Young's modulus (GPa)	Poisson's ratio	Thermal conductivity ($W m^{-1} K^{-1}$)	Radiogenic heat production ($W kg^{-1}$)	Density ($kg m^{-3}$)
Sandstone	28.8	0.53	27.2	18.30	0.38	3.0	1.6	2180
Shale	21.0	0.38	0.69	5.52	0.25	2.0	2.1	2470
Limestone	26.4	0.50	26.8	77.40	0.26	2.4	0.7	2720
Dolomite	39.0	0.81	35.9	106.00	0.29	3.0	0.7	2840
Granite	51.0	1.23	55.1	70.60	0.18	3.3	2.5	2640
Salt				28.50	0.22	5.4	3.5	2160

Note:
Values are taken from Clark (1966), Vutukuri *et al.* (1974), Lama and Vutukuri (1978), Čermák and Rybach (1982), Ibrmajer *et al.* (1989) and Cloetingh *et al.* (1995).

Figure 15.2 demonstrates that the alternative with syn-tectonic deposition is much warmer. The observed temperature differences at the checkpoints are as high as several tens of degrees Celsius. This observation holds regardless of the choice of the shortening rate while keeping the other factors fixed (Tables 15.3 and 15.4) or changing the lithology of thrust sheet layers from dolomite–limestone–dolomite to sandstone–shale–sandstone, but holding everything else constant (Table 15.5).

The second largest effect on the thermal regime of a growing thrustbelt structure is exerted by the initial heat-flow value. This impact can be illustrated by using the previously described case of a fault-bend fold with a buried duplex, which detaches along a shale horizon, deforms a carbonate section, starts to grow at an initial heat flow of $80 \, mW \, m^{-2}$, and is associated with deposition of syn-tectonic shale (Fig. 15.2a). Using the same thrusting rate of $0.3 \, mm \, yr^{-1}$, but an initial heat flow of $40 \, mW \, m^{-2}$ results in a much cooler thermal regime for the modelled structure (Fig. 15.2c) than the case with an initial heat flow of $80 \, mW \, m^{-2}$. The observed differences in the temperatures reached at the checkpoints located in the roof thrust, duplex and footwall are of the order of 10–$20 \, °C$ (Table 15.6). Table 15.6 shows that the higher initial heat flow increases the thermal difference between the roof thrust, duplex and footwall.

The same effect can be illustrated in the case of a fault-propagation fold, which deforms a 2 km thick sediment section composed of three layers (Fig. 15.1a), using a model with an upper layer 620 m thick, a middle layer 710 m thick and a lower layer 670 m thick. Two possible thrust sheet lithology alternatives are simulated using dolomite–limestone–dolomite (carbonate) and sandstone–shale–sandstone (siliciclastic) scenarios for the three layers described. Each alternative undergoes folding with 50% erosion affecting the growing structure

(Fig. 15.1a, case 2), in order to illustrate a potential continental setting. The shortening rate is chosen as $0.3 \, mm \, yr^{-1}$, representing the minimum value in Table 15.3a. The thrust ramp angle is fixed at $28°$. Three different pre-tectonic surface heat flows of 40, 60 and $80 \, mW \, m^{-2}$ are chosen to cover the entire interval of reasonable values for compressional basins described in Chapter 12. The computed results are shown in Table 15.7, which shows that the increased value of the initial heat flow for the same shortening causes both a warmer footwall and hanging wall. It also indicates that the temperature difference between the checkpoints in the hanging wall and footwall increases with a faster slip rate within each surface heat-flow category. It is also clear from Table 15.7 that the siliciclastic scenario is warmer than the carbonate scenario for both the footwall and hanging wall, something that will be discussed later.

The third largest impact on the thermal regime of a growing thrustbelt structure is exerted by the blanketing effect of the syn-tectonic sediments. This effect can be documented in the case of a fault-propagation fold that deforms a 2 km thick siliciclastic section, composed of an upper sandstone layer 620 m thick, a middle shale layer 710 m thick and a lower sandstone layer 670 m thick. The growing structure is affected by syn-tectonic deposition of either sandstone (Fig. 15.3a) or shale (Fig. 15.3b) in two alternative models (Fig. 15.1a, case 1). A shortening rate of $0.3 \, mm \, yr^{-1}$ and an initial heat-flow value of $60 \, mW \, m^{-2}$ are used in these two models (Fig. 15.3), and a sensitivity analysis uses the full spectrum of 40, 60 and $80 \, mW \, m^{-2}$ and 0.3, 2.3 and $4.3 \, mm \, yr^{-1}$ for the initial heat-flow value and the slip rate, respectively (Table 15.8). The computed models indicate a temperature difference of up to 10–$15 \, °C$ in the warming of both the footwall and hanging wall between fault-propagation folds affected by shale and sandstone syn-tectonic deposition.

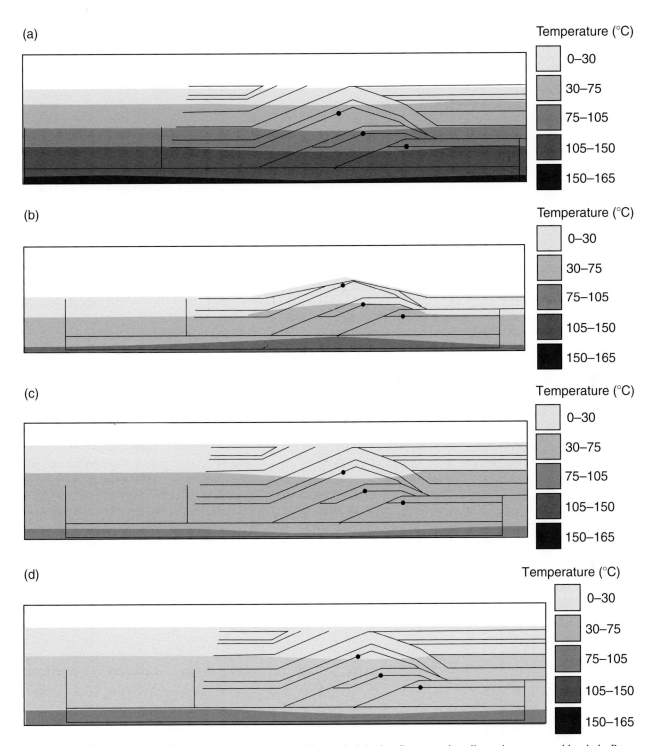

Fig. 15.2. (a) Thermal regime of the fault-bend fold model with a buried duplex. Syn-tectonic sediment is represented by shale. Pre-tectonic section is formed by 663 m thick upper dolomite layer, 337 m thick middle limestone layer and 1000 m thick lower dolomite layer. Detachment horizon occurs in shale, ramp angle is 23°, structure is affected by 50%-effective erosion, pre-orogenic surface heat flow was 80 mW m^{-2}, and slip rate is 0.3 mm/year. Duplex develops beneath the upper dolomite layer. (b) Thermal regime of the fault-bend fold model with a buried duplex. The structure undergoes a syn-tectonic shale deposition that keeps pace with the growth of the structure. Other parameters are the same as in (a). (c) Thermal regime of the fault-bend fold model with a buried duplex. All parameters are as in (a) except for the pre-orogenic surface heat flow, which is 40 mW m^{-2}. (d) Thermal regime of the fault-bend fold model with a buried duplex. All parameters are as in (c) except for the lithology of three pre-tectonic layers, which are formed by sandstone, shale and sandstone.

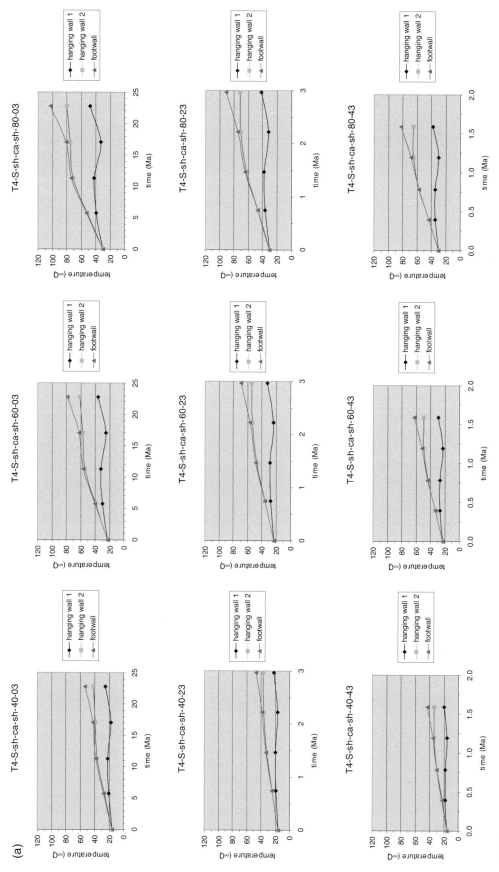

Table 15.3. Temperature history curves for the three checkpoints in the fault-bend fold with a buried duplex shown in Fig. 15.1(c) in the case with syn-tectonic shale deposition (a) and syn-tectonic erosion (b). Further explanation is given in the text.

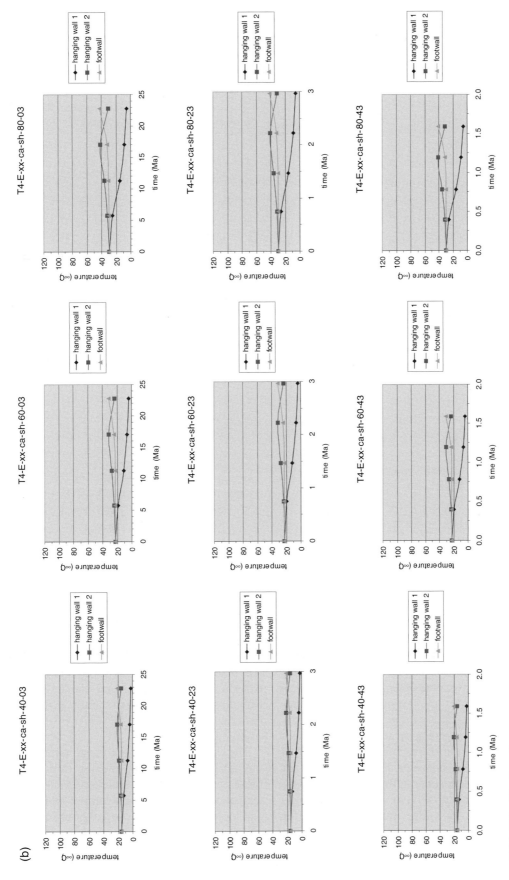

Table 15.3. (Continued)

Table 15.4. Comparison of temperatures at checkpoints located in the roof thrust, duplex and footwall at three different slip rates (0.3, 2.3 and 4.3 mm yr^{-1}) caused by an increase in the initial surface heat-flow value (40, 60 and 80 mW m^{-2}). The deformed section consists of carbonates. The extent that the depositional model is warmer than the erosional model is indicated by the temperature difference in degrees Celsius.

Structure	Slip rate (mm yr^{-1})	Heat flow (mW m^{-2})	Model		Temperature difference (°C)
			Erosional (°C)	Depositional (°C)	
Footwall	0.3	40	21.3	53.6	32.3
		60	31.7	78	46.3
		80	42.1	102.7	60.6
	2.3	40	21	47.2	26.2
		60	31.2	69	37.8
		80	41.4	90.8	49.4
	4.3	40	20.8	42.9	22.1
		60	30.8	62.7	31.9
		80	40.8	82.6	41.8
Duplex	0.3	40	15.6	41.9	26.3
		60	23.3	61.1	37.8
		80	31	80.2	49.2
	2.3	40	15.7	37.2	21.5
		60	23.4	54.3	30.9
		80	31.2	71.3	40.1
	4.3	40	15.8	34	18.2
		60	23.5	49.7	26.2
		80	31.3	65.4	34.1
Roof thrust	0.3	40	2.4	25.1	22.7
		60	3.9	36.4	32.5
		80	5.5	47.8	42.3
	2.3	40	3	21.8	18.8
		60	4	31.8	27.8
		80	5.6	41.7	36.1
	4.3	40	2.5	19.8	17.3
		60	4	28.8	24.8
		80	5.6	37.9	32.3

A similar case is illustrated in Fig. 15.4, which compares two syn-tectonic depositional cases affecting an inverted half-graben (Fig. 15.1d). This model simulates a syn-rift sediment section composed of a limestone layer, which thickens to a maximum thickness of 2 km at the bounding normal fault. It rests on a pre-rift sandstone layer, which lies on crystalline basement characterized as granite. This geological scenario experiences a syn-tectonic deposition keeping pace with the growth of the structure during simulated shortening. Post-rift and syn-shortening deposition in this scenario is simulated by either shale (Fig. 15.4a) or sandstone sedimentation (Fig. 15.4b), providing two alternatives for comparison. The shortening rate is simulated at 0.3 mm yr^{-1} and the applied surface heat-flow value is

60 mW m^{-2}. The physical properties of the modelled sediments are listed in Table 15.2. The maximum angle of dip of the normal fault in its uppermost part is 60°. Alternative thermal models, shown in Fig. 15.4 for comparison, indicate that thermal blanketing of the post-rift and syn-shortening shale results in a warmer model than the model with post-rift and syn-shortening sandstone.

The slip rate provides the fourth largest influence on the thermal regime of a growing thrustbelt structure. Table 15.6 shows that an increasing slip rate raises the temperature difference between the roof thrust, duplex and footwall for the same initial heat flow in the case of a fault-bend fold with a buried duplex (Fig. 15.1c). This is because very similar temperature differences were

Table 15.5. Comparison of temperatures at checkpoints located in the roof thrust, duplex and footwall at three different slip rates (0.3, 2.3 and 4.3 mm yr^{-1}) caused by an increase in the initial surface heat-flow value (40, 60 and 80 mW m^{-2}). Deformed section consists of siliciclastics. The extent that the depositional model is warmer than the erosional model is indicated by the temperature difference in degrees Celsius.

| Structure | Slip rate (mm yr^{-1}) | Heat-flow (mW m^{-2}) | Model | | Difference |
			Erosional	Depositional	
Footwall	0.3	40	24.5	58.8	34.3
		60	35.8	84.6	48.8
		80	47.2	110.3	63.1
	2.3	40	24.1	52.5	28.4
		60	35.3	75.6	40.3
		80	46.5	98.7	52.2
	4.3	40	23.7	48	24.3
		60	34.8	69.2	34.4
		80	45.8	90.5	44.7
Duplex	0.3	40	18.1	46.5	28.4
		60	26.5	66.7	40.2
		80	34.9	86.9	52
	2.3	40	18.1	41.7	23.6
		60	26.6	59.9	33.3
		80	35.1	78.1	43
	4.3	40	18.2	38.5	20.3
		60	28.7	55.3	26.6
		80	35.2	72.2	37
Roof thrust	0.3	40	3.3	28	24.7
		60	5.2	40	34.8
		80	7.1	51.9	44.8
	2.3	40	3.4	24.8	21.4
		60	5.3	35.4	30.1
		80	7.1	46	38.9
	4.3	40	3.4	22.7	19.3
		60	5.3	32.4	27.1
		80	7.2	42.1	34.9

achieved by faster shortening scenarios operating for much shorter times, comparing a time of 22.7 Myr for a slip rate of 0.3 mm yr^{-1} with 3 Myr for 2.3 mm yr^{-2} and 1.6 Myr for 4.3 mm yr^{-1}.

The shortening rate defines the speed at which the hanging wall thrusts over the footwall. A transient thermal perturbation develops immediately following the onset of thrusting and the faster the shortening rate is, the larger the perturbation. The climbing hanging wall cools down and the footwall warms up during shortening. A thermal disequilibrium between the hanging wall and footwall regimes develops during the whole process of thrusting, but undergoes a time-dependent reduction at the same time. This means that the moving base of the hanging wall cools down as it moves upwards and that when thrusting stops the thermal regime of the whole structure has already moved some way towards a thermal equilibrium. The slower the shortening rate, the greater the shift towards equilibrium.

The fifth largest impact on the thermal regime of a growing structure is exerted by the lithology of accreted sediments. This impact can be illustrated by using the previously discussed case of a fault-bend fold with a buried duplex, which detaches along a shale horizon, deforms a carbonate section, starts to grow at an initial heat flow of 40 mW m^{-2} and undergoes syn-tectonic shale deposition (Fig. 15.2c). Given the same thrusting rate of 0.3 mm yr^{-1}, an initial heat flow of 40 mW m^{-2}, and the change of accreted lithologies from carbonate

Table 15.6. Temperature difference between checkpoints in the roof thrust, duplex and footwall increasing with increasing initial surface heat-flow value (40, 60 and 80 mW m^{-2}) for the same slip rate (0.3, 2.3 and 4.3 mm yr^{-1}) for a deformed carbonate section that undergoes syn-tectonic shale deposition.

| Time (Ma) | Slip rate (mm yr^{-1}) | Heat flow (mW m^{-2}) | Temperature | | | Temperature difference | | |
			Roof thrust (°C)	Duplex (°C)	Footwall (°C)	F − RT (°C)	F − D (°C)	D − RT (°C)
22.7	0.3	40	25.1	41.9	53.6	28.5	11.7	16.8
		60	36.4	61.1	78	41.6	16.9	24.7
		80	47.8	80.2	102.7	54.9	22.5	32.4
3	2.3	40	21.8	37.2	47.2	25.4	10	15.4
		60	31.8	54.3	69	37.2	14.7	22.5
		80	41.7	71.3	90.8	49.1	19.5	29.6
1.6	4.3	40	19.8	34	42.9	23.1	8.9	14.2
		60	28.8	49.7	62.7	33.9	13	20.9
		80	37.9	65.4	82.6	44.7	17.2	27.5

to siliciclastic results in a warmer thermal regime for the studied structure (Fig. 15.2d). The differences in temperatures recorded at the checkpoints located in the roof thrust, duplex and footwall for the carbonate and siliciclastic alternatives are of the order of 3–8 °C (Table 15.9). One of the factors behind this lithological control on the thermal regime is the thermal conductivity of the accreted sediments. Lower values of thermal conductivity in the accreted section are the key to preventing vertical heat transfer from being as effective as in sections with higher values of thermal conductivity, and results in the section staying warmer. Another lithological control is the heat production that varies in different lithologies. Yet another factor is the specific heat capacity, which controls the amount of thermal energy necessary to heat the footwall or cool the hanging wall. Since, apart from the grain system, the porosity controls the specific heat capacity, the porosity response to thrusting is important in the lithological control on the syntectonic thermal regime.

The lithology of the detachment horizon plays the least important role in the thermal regime of the growing structure. It can be illustrated using, as an example, a detachment fold that deforms a 2 km thick sediment section composed of three layers; an upper dolomite layer 620 m thick, a middle limestone layer 710 m thick and a lower dolomite layer 670 m thick. The model undergoes syn-tectonic sandstone deposition (Fig.15.1b, case 1). Two alternative models detach along either an 1160 m thick salt unit (Fig. 15.5a) or an 1160 m thick shale horizon (Fig. 15.5b), characterized by different thermal properties. The chosen slip rate and

initial heat-flow values are 0.3 mm yr^{-1} and 60 mW m^{-2}, respectively. The physical properties of the modelled sediments are listed in Table 15.2. The modelled temperature values (Fig. 15.5, Table 15.10) indicate a difference of only 0–2 °C in thermal regimes for these two alternatives. The model with the detachment fold above a salt detachment horizon, which is a better heat conductor, is slightly warmer.

It should be noted that these thermal models involved only a relatively thin layer of shortened sediments. With a thicker sequence of sediments the impact of the relevant factors could be more significant. However, their overall importance for the syn-tectonic thermal regime should be unchanged; the effect of the existence of syn-tectonic deposition versus erosion, the effect of the initial heat-flow value, the effect of potential thermal blanketing, the effect of the slip rate and the effect of the thrust sheet lithology being important for the maturation histories of hydrocarbon source rocks in thrustbelts. The importance increases with the duration over which a specific thrustbelt structure undergoes thermal regime perturbation, since the growth time between different structures varies from 20 Myr to 200 000 yr (Table 15.1).

Since the thermal histories and associated maturation histories of source rocks for initial exploration purposes are determined in either one or two dimensions, it is worth considering the likelihood that these two approaches will establish the syn-tectonic thermal perturbation in a thrustbelt. The thrust displacement of rock layers characterized by different heat production produces lateral (Jaupart, 1983; Furlong and

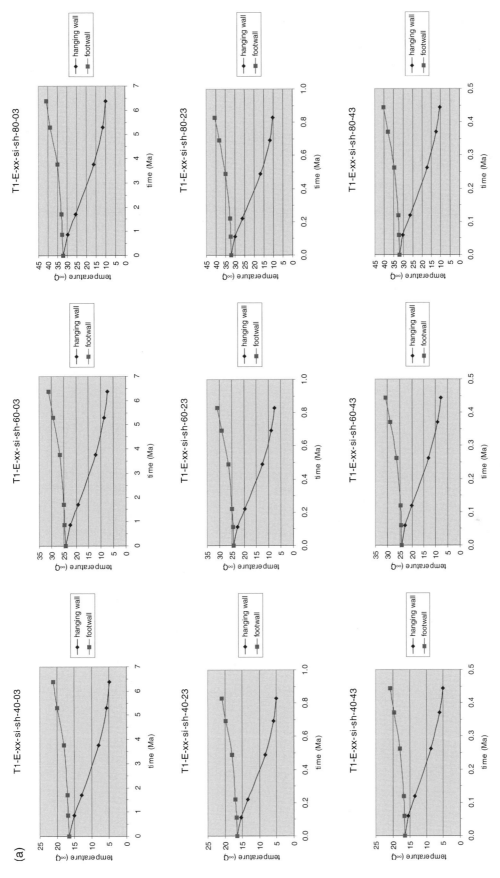

Table 15.7. Temperature history curves for the hanging wall and footwall checkpoints in the fault-propagation fold shown in Fig. 15.1(a), Case 2, in the scenario with deformed siliciclastic section (a) and deformed carbonate section (b). See text for further explanation.

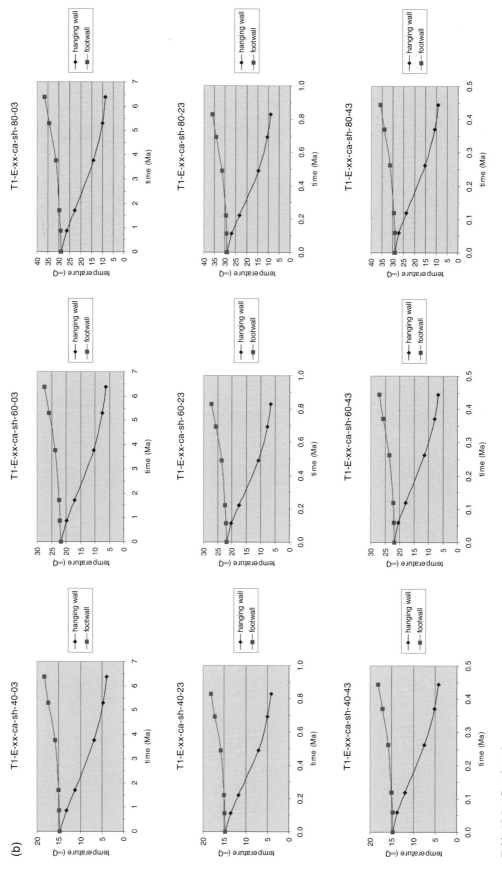

Table 15.7. (Continued)

(a)

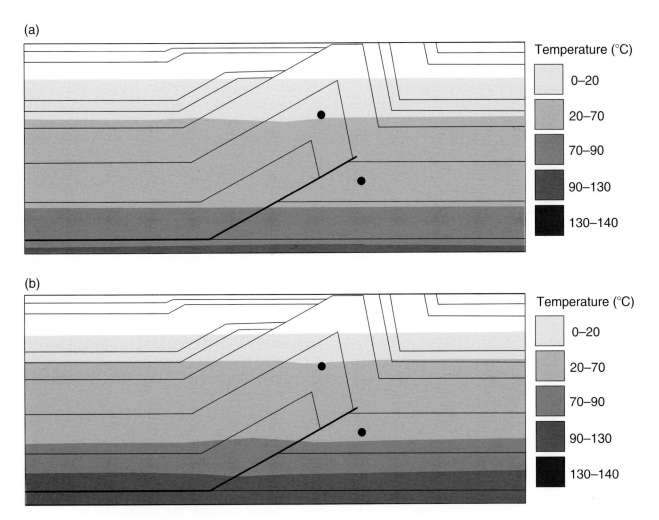

(b)

Fig. 15.3. Thermal regime of a fault-propagation fold. Syn-tectonic sediment are represented by (a) sandstone and (b) shale. The pre-tectonic section is formed from a 620 m thick upper sandstone layer, a 710 m thick middle shale layer and a 670 m thick lower sandstone layer. Ramp angle is 28°, pre-orogenic surface heat flow was 60 mW m^{-2}, and slip rate is 0.3 mm yr^{-1}.

Chapman, 1987; Sonder and Chamberlain, 1992) and vertical thermal gradients, which complicate the thermal regime (e.g., Ruppel and Hodges, 1994a). One-dimensional models of the thermal evolution of active thrustbelts assume that the heat conduction occurs only in a vertical direction. Their failure to include lateral heat conduction results in computations of thermal regimes that are very different from regimes calculated using a two-dimensional conductive approach. With relatively high rates of footwall burial, e.g. 5 mm yr^{-1}, driven by a thrusting rate of 7.07 mm yr^{-1}, the lithospheric thermal regime during shortening and un-roofing is characterized by horizontal temperature gradients similar to or larger than the vertical gradients (Ruppel and Hodges, 1994b). Such high lateral heat conduction results in the relaxation of horizontal

thermal gradients and significantly affects the thermal regime. Therefore, two-dimensional maturation models would be recommended in cases where there is a need to take the development of a syn-tectonic thermal regime into account, because they are capable of eliminating lateral inhomogeneities in the thermal field and reducing the ratio between horizontal and vertical temperature gradients.

For the purposes of the maturity calculation, the maximum temperature T_{max} reached by the source rock during its entire burial history is more important than the temperature reached at the end of the thrustbelt shortening. Therefore, it is far simpler to make a correct one-dimensional calculation of T_{max} in a hanging wall that experiences cooling than in a footwall, which undergoes warming. Ruppel and Hodges (1994b) have

Table 15.8. Warming of the hanging wall and the footwall for three different slip rates (0.3, 2.3, 4.3 mm yr⁻¹) caused by an increase in the initial surface heat-flow value (40, 60 and 80 mW m⁻²), comparing cases with syn-tectonic sandstone versus shale deposition. The temperature difference between syn-tectonic sandstone (ss) and shale (sh) cases indicates how much the section below the shale layer is warmer than the section below the sand layer in degrees Celsius. Calculations were made for deformed siliciclastic sections.

Structure	Slip rate (mm yr⁻¹)	Heat flow (mW m⁻²)	Temperature sediment section ss (°C)	Temperature sediment section sh (°C)	Temperature difference sh − ss (°C)
Footwall	0.3	40	44.2	52.5	8.3
		60	64.1	75.9	11.8
		80	84	99.3	15.3
	2.3	40	36.5	40.4	3.9
		60	53	58.6	5.6
		80	69.5	76.8	7.3
	4.3	40	32.3	34.6	2.3
		60	46.9	50.2	3.3
		80	61.6	65.9	4.3
Hanging wall	0.3	40	24.1	30.5	6.4
		60	34.6	43.8	9.2
		80	45.1	57	11.9
	2.3	40	19.4	22.7	3.3
		60	28	32.7	4.7
		80	36.5	42.7	6.2
	4.3	40	17	19.1	2.1
		60	24.5	27.5	3
		80	32	36	4

further demonstrated that one-dimensional calculations produce errors that decrease with distance from the thrust fault. An additional complexity in such a one-dimensional approach emerges when the internal strain heating and frictional heating along a thrust fault, described in the following section, are included in the calculation. In some orogens an inverted metamorphic field gradient developed within the upper plate (e.g., Le Fort, 1975; Hubbard, 1989), indicating that a sufficiently thick hanging wall can experience such an inversion under slip rates as slow as 1 cm yr⁻¹ (Huerta *et al.*, 1996), which would be relatively high for a thrust structure (Table 15.3a) but is reasonable for an orogenic wedge. The heat production rate in a wedge, the thickness of the wedge containing heat-producing elements and the convergence rate control the steady-state maximum temperatures and the inversion of the geotherms inside the orogenic wedge, as numerically simulated by Huerta *et al.* (1996). The first two factors have the largest impact on the thermal regime. When either the heat production rate or the thickness of the wedge with heat-producing elements reaches a certain upper limit, it triggers a geotherm inversion within the lower plate. An increase in convergence rate would only result in a cooler thermal regime at shallow levels and enhance the geotherm inversion controlled by these two factors (Huerta *et al.*, 1996).

Role of internal strain and frictional heating

Apart from the shortening rate, the role of deformation on the thermal regime of an active thrustbelt includes the effect of internal strain and frictional heating. Case studies of the calculation of frictional heating have been made on the modern accretionary prisms of Taiwan and Barbados (e.g., Barr and Dahlen, 1989; Dahlen, 1990; Ferguson *et al.*, 1993).

Figure 15.6(a) shows a surface heat-flow map of the Barbados accretionary prism (Ferguson *et al.*, 1993). Its characteristic features are:

(1)   a 5–10 mW m⁻² westward decrease in heat flow in a narrow zone located behind the thrust front;
(2)   a 5–10 mW m⁻² southward decrease in heat flow along a zone located within the first 35 km from the thrust front; and
(3)   a westward increase in heat flow in a zone located 100 km westward from the thrust front.

The first of these features corresponds to prism thickening (Fig. 15.6b). The second feature appears to be controlled by a southwards increasing sedimentation rate, as discussed in Chapter 14. In this text the contribution of the frictional heating to the zone of westward increase in heat flow located 100 km west of the thrust front is investigated. Potential mechanisms, which could cause this increase, include:

(1)   an arcward increase in frictional heating along the décollement;
(2)   an arcward decrease in the rate of wedge thickening; and
(3)   erosion of the wedge surface (Langseth *et al.*, 1990).

In order to evaluate their relative roles, and the role of internal heating on the thermal regime, a thermal model of the Barbados accretionary prism has been computed using a finite-element approach (Fig. 15.7; Ferguson *et al.*, 1993). It is based on the assumption that the accretionary prism shape remains constant with time, that it

(a)

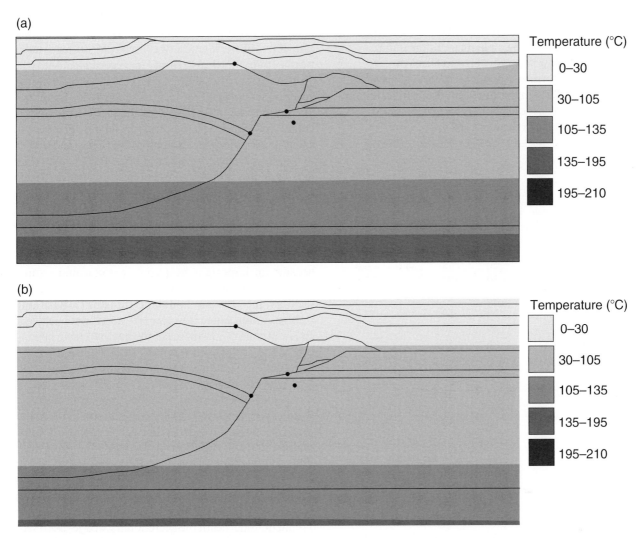

(b)

Fig. 15.4. Thermal regime of an inverted half-graben. The syn-rift sediment section is represented by a limestone layer that thickens to a thickness of 2 km at the bounding normal fault, which has a maximum dip of 60° in its uppermost part. The pre-rift section is represented by a sandstone layer and a granite basement. The post-rift and syn-shortening layers are represented by (a) shale and (b) sandstone. Pre-orogenic surface heat flow was 60 mW m^{-2}, and slip rate is 0.3 mm yr^{-1}.

propagates in a self-similar fashion and that the accreted sediment thickens by horizontal shortening accompanied by a volume reduction and pore fluid expulsion. Accreted sediments are separated from both basement and autochthonous sediments by a décollement (Fig. 15.7). The heat production of the accreted sediments is assumed to be equal to 0.5 μW m^{-3}, which is the average value for oceanic sediments (Pollack and Chapman, 1977). The thickness and lateral velocity of autochthonous sediments and the underlying basement remains constant in the model. The lateral velocity is expressed as the advance rate of the wedge over the subducting plate. The lateral velocity of accreted sediments, however, is progressively slowed as they thicken by shortening.

The shear stress resolved on the décollement plane is calculated as:

$$\tau_b = \mu_b(1-\lambda_b)\rho g H, \tag{15.1}$$

where μ_b is the coefficient of internal friction along the décollement, λ_b is the ratio of the pore fluid pressure to the lithostatic pressure along the décollement, ρ is the average density of the accreted sediment, g is the acceleration due to gravity and H is the wedge thickness. The heat per unit area generated by friction along the décollement is then calculated as a product of the shear stress and the slip velocity. At the décollement level, this heat is added to the heat transported from the section underneath the wedge.

Table 15.9. Warming of the roof thrust, duplex and footwall for three different slip rates (0.3, 2.3 and 4.3 mm yr^{-1}) caused by an increase in the initial surface heat-flow value (40, 60 and 80 mW yr^{-2}), comparing deformed carbonate (ca) and siliciclastic (si) sections. The temperature difference indicates how much the siliciclastic sectioin is warmer than the carbonate section in degrees Celsius.

Structure	Slip rate (mm yr^{-1})	Heat flow (mW m^{-2})	Temperature sediment section ca (°C)	Temperature sediment section si (°C)	Temperature difference ca − si (°C)
Footwall	0.3	40	53.6	58.8	5.2
		60	78	84.6	6.6
		80	102.7	110.3	7.6
	2.3	40	47.2	52.5	5.3
		60	69	75.6	6.6
		80	90.8	98.7	7.9
	4.3	40	42.9	48	5.1
		60	62.7	69.2	6.5
		80	82.6	90.5	7.9
Duplex	0.3	40	41.9	46.5	4.6
		60	61.1	66.7	5.6
		80	80.2	86.9	6.7
	2.3	40	37.2	41.7	4.5
		60	54.3	59.9	5.6
		80	71.3	78.1	6.8
	4.3	40	34	38.5	4.5
		60	49.7	55.3	5.6
		80	64.4	72.2	7.8
Roof thrust	0.3	40	25.1	28	2.9
		60	36.4	40	3.6
		80	47.8	51.9	4.1
	2.3	40	21.8	24.8	3
		60	31.8	35.4	3.6
		80	41.7	46	4.3
	4.3	40	19.8	22.7	2.9
		60	28.8	32.4	3.6
		80	37.9	42.1	4.2

The heating per unit volume produced by the internal strain of the accretionary prism is calculated as a product of the strain rate ε and the shear stress τ:

$$\tau\varepsilon = \mu(1-\lambda)\rho g h(\varepsilon_v + \varepsilon_h), \qquad (15.2)$$

where the effective shear stress at depth h equals:

$$\tau = \mu(1-\lambda)\rho g h, \qquad (15.3)$$

where μ is the coefficient of internal friction for the accreted sediments, λ is the ratio of the pore fluid pressure to the lithostatic pressure of accreted sediments, ε_v is the vertical strain rate and ε_h is the horizontal strain rate. Rates of the internal strain typically decrease landwards from the thrust front, in accordance with the decreasing surface slope and the lateral velocity of accreted sediments (Fig. 15.7). The Barbados accretionary prism has a taper of 3–4° at the thrust front and its lateral velocity reaches a value of about 20 mm yr^{-1}, which places the prism among those that are relatively thin and with slow convergence rates. Its internal strain heating is not an important contributor to its thermal regime (Langseth *et al.*, 1990; Ferguson *et al.*, 1993), as the internal strain heating, calculated along a transect across the Barbados Ridge at 14° 20′ N (Fig. 15.6a) contributes only 1.5 mW m^{-2} to the thermal regime (Langseth *et al.*, 1990). For internal strain heating to become an important heat contributor would require rapidly converging accretionary wedges with large tapers such as the Taiwan fold and thrustbelt (Barr and Dahlen, 1989), where the taper is 9° and the convergence rate is 70 mm yr^{-1}. In this case, the internal strain heating here at the thrust front reaches a value of 10 µW m^{-3}.

Figure 15.8 shows the results of the heat-flow response to various values of initial thickness of accreted sediment, basal pore–fluid pressure ratio and décollement dip, which control the internal strain and frictional heating. The heat-flow values within a narrow zone behind the thrust front decrease landwards with increased thickness of accreted sediments, because their thickening mechanically stretches the geothermal gradient within the accretionary wedge. The heat flow also decreases in authochthonous sediments and oceanic basement underneath the accretionary wedge. Both sediments and basement are cooler than they would be at the same depth in a region with normal steady-state heat flow, unaffected by the advancing wedge. The thickness of the wedge is controlled by the convergence rate. An increased rate causes more rapid sediment thickening. It results in a greater thermal disequilibrium in the material below the accretionary wedge and an enhanced heat-flow decrease.

The convergence rate, wedge taper and initial wedge thickness together control the rate of sediment thickening. Both convergence and taper determine the relative coolness of the material below the wedge. The effective shear stress resolved at the décollement plane and the convergence together determine the amount of heat generated by friction along the décollement, which compensates for the cooling effects of the wedge thickening and underthrusting. A higher convergence rate causes a more rapid reduction of heat flow by tectonic thickening in a narrow zone behind the thrust front and increases the heat-flow landward of this zone by

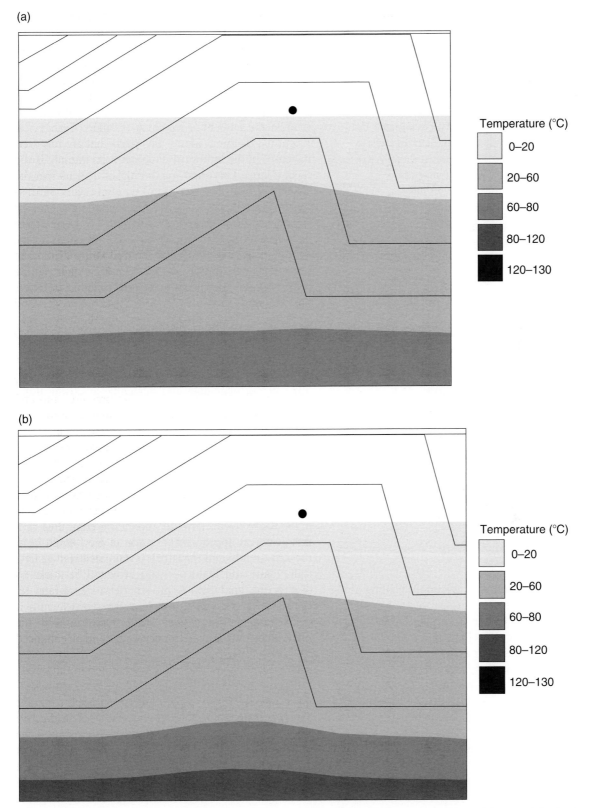

Fig. 15.5. Thermal regime of the detachment fold. The pre-tectonic section is formed from a 620 m thick upper dolomite layer, a 710 m thick middle limestone layer and a 670 m thick lower dolomite layer. The model undergoes syn-tectonic sandstone deposition. Fold detaches along (a) salt unit and (b) shale unit. Pre-orogenic surface heat flow was $60\,\mathrm{mW\,m^{-2}}$, and slip rate is $4.3\,\mathrm{mm\,yr^{-1}}$.

Table 15.10. Warming of the hanging wall and the footwall for three different slip rates (0.3, 2.3, 4.3 mm yr^{-1}) caused by an increase in the initial surface heat-flow value (40, 60 and 80 mW mm^{-2}), comparing a detachment fold deforming a carbonate section detached either along a shale (sh) or salt (sa) horizon. The temperature difference indicates how much the section above the salt layer is warmer than the section above the shale layer in degrees Celsius.

Structure	Slip rate (mm yr^{-1})	Heat flow (mW m^{-2})	Temperature sediment section sh (°C)	Temperature sediment section sa (°C)	Temperature difference sa − sh (°C)
Footwall	0.3	40	36.6	37.3	0.7
		60	53.4	53.7	0.3
		80	70.2	70	−0.2
	2.3	40	29.3	30.2	0.9
		60	42.7	43.6	0.9
		80	56.2	57	0.8
	4.3	40	26.2	27.1	0.9
		60	38.3	39.1	0.8
		80	50.3	51.1	0.8
Hanging wall	0.3	40	15.6	17.1	1.5
		60	22.8	24.5	1.7
		80	29.9	31.9	2
	2.3	40	13	14.1	1.1
		60	18.9	20.4	1.5
		80	24.8	26.6	1.8
	4.3	40	12.2	13.1	0.9
		60	17.7	18.9	1.2
		80	23.3	24.7	1.4

increasing the contribution of frictional heating along the décollement.

As discussed in Chapter 2, a small thickness of accreted sediments increases the strain rates. Therefore, the heat flow decreases more rapidly (Fig. 15.8a). Case C in Fig. 15.8(a) shows that heat-flow curves converge at higher convergence rates, because the frictional heat source is closer to the surface if the thickness of accreted sediments is low, and counteracts the thickening effect.

The basal pore–fluid pressure ratio controls the effective shear stress along the décollement of the accretionary wedge. Keeping the other heat controlling parameters constant, the heat-flow curve attains a distinct minimum even at high convergence rates when the basal pore fluid pressure ratio is low (Fig. 15.8b). This is because a low basal pore fluid pressure ratio increases

the effective stress. The resultant landward increase in heat generated by frictional stress on the décollement overcomes the tendency for the surface heat flow to decrease with wedge thickening.

A large taper angle causes a rapid thickening of the accretionary wedge under high convergence rates. This results in a greater rate of landward heat flow decrease than in the case of a small taper angle (Fig. 15.8c). Frictional heating along the décollement, however, increases in the landward direction more rapidly if the taper is large. This opposing thermal effect decreases the spacing of heat-flow curves in Fig. 15.8(c) with increasing taper angle.

The heat-flow calculations performed by Ferguson *et al.* (1993) along transects through the Barbados accretionary prism shown in Fig. 15.6(a) indicate that the respective contributions of basal frictional heat, radiogenic heat and internal strain heating to the surface heat flow are, respectively:

(1) 44%, 5% and negligible at 200 km along transect 2;
(2) 26%, 8% and negligible at 130 km along transect 3; and
(3) 0.5%, negligible and negligible at 20 km along transect 4.

The contribution of frictional and internal strain heating in the Taiwan fold and thrustbelt (Barr and Dahlen, 1989), are different from those of the Barbados accretionary prism; in the former case the wedge taper is two and half times thicker and the convergence rate is three and half times faster. Because the thermal regime in the Barbados accretionary prism was computed by a finite-element approach (Ferguson *et al.*, 1993) and in the Taiwan fold and thrustbelt it was calculated analytically (Barr and Dahlen, 1989), it would be useful to discuss the analytical approach before analysing its results. It starts with a simple energy equation for a deforming material (Eringen, 1967; Malvern, 1969):

$$\partial_t T + v \cdot \nabla T = \kappa \nabla^2 T + (\rho c_v)^{-1}(H + \sigma : \varepsilon), \qquad (15.4)$$

where T is the temperature, v is the Eulerian velocity, κ is the constant thermal diffusivity, ρ is the constant sediment density, c_v is the constant specific heat capacity, and H is the heat production rate. The stress and strain rate product, $\sigma : \varepsilon$, in the deforming brittle accretionary wedge represents the rate at which mechanical energy dissipates against internal friction, producing heat (Barr and Dahlen, 1989). The energy consumed by fracture propagation and reactivation within the wedge is ignored for simplicity.

In order to further simplify equation (15.4), one can assume that the accretionary wedge advances at a

(a)

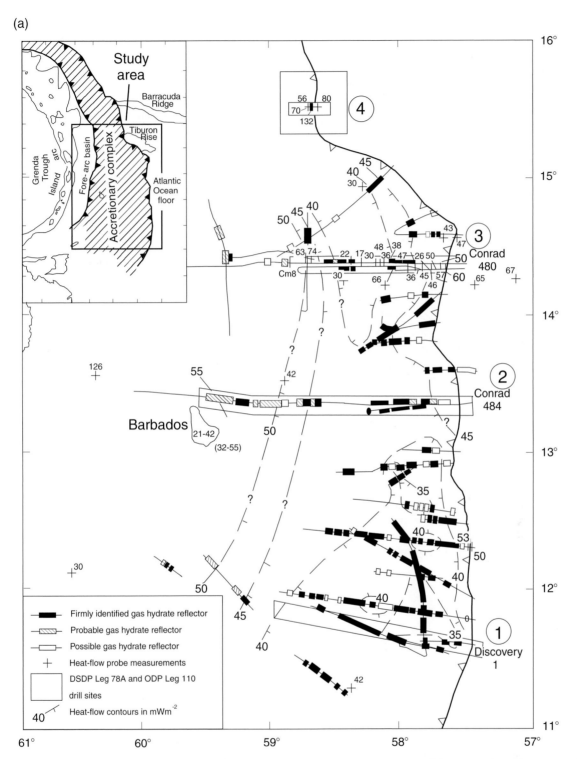

Fig. 15.6. (a) Heat-flow map of the Barbados accretionary prism with contours in m Wm^{-2}, based mostly on depth to gas hydrate reflectors (Ferguson *et al.*, 1993). Location is shown on the inset regional map. Confidence level for the identification of the gas hydrate reflector is based on following. 'Firmly identified' is based on a strong reflector of negative polarity, which crosscuts other reflectors. 'Probable' contains certain ambiguity as to the polarity of a reflector or a lack of clear crosscutting relationships with other reflectors leave some doubt as to the origin of the reflector. 'Possible' is characterized by a reflector, which displays bottom-simulating characteristics, but its polarity cannot be determined and/or it does not crosscut other reflectors. Corridors labelled 1, 2, 3 and 4 are transects along which thermal models of the accretionary wedge were constructed.

(b)

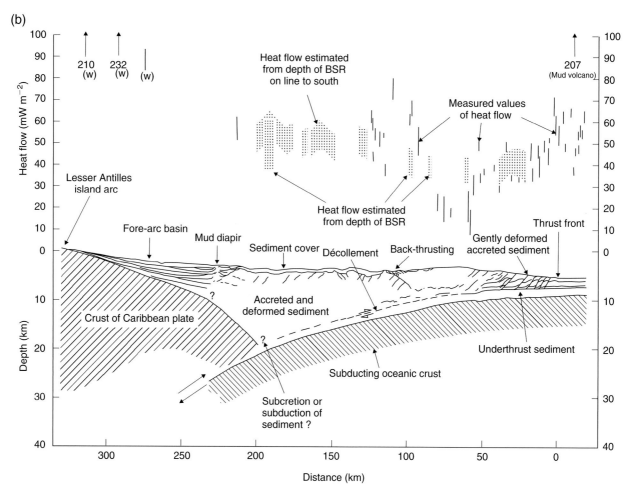

Fig. 15.6 (*cont.*). (b) Transect 3 from (a), which illustrates the relationship between its geological structure and heat flow (Ferguson *et al.*, 1993). The structure of the accretionary prism and forearc basin are interpreted from reflection seismic data. The deep structure between 110 km and 230 km west of the thrust front is derived from gravity modelling. Heat-flow values measured during the cruises of R/V Robert D. Conrad in 1985 and 1988, are shown by vertical bars of length equal to the standard error of the measurement.

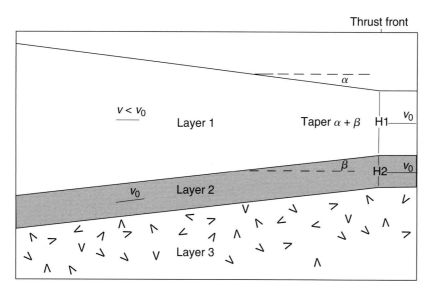

Fig. 15.7. The geometry of the finite-element model of the Barbados accretionary prism (Ferguson *et al.*, 1993). Accreted sediments, shown as layer 1, are slowed on accretion ($v < v_0$) and autochthonous sediments together with basement material, shown as layers 2 and 3, maintain their initial velocity v_0. α is the wedge surface dip, β is the décollement dip, H_1 and H_2 are the initial thicknesses of layers 1 and 2, respectively.

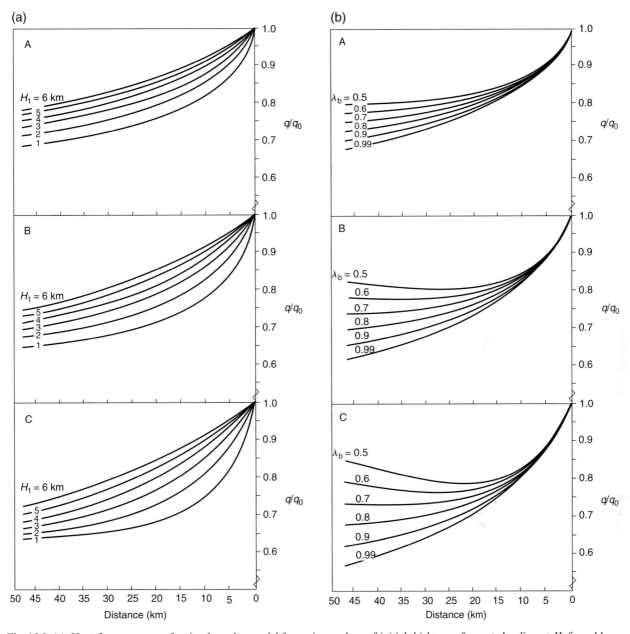

Fig. 15.8. (a) Heat-flow response of a simple wedge model for various values of initial thickness of accreted sediment H_1 from 1 km to 6 km, at three sample velocities v_0 of (A) 20, (B) 50 and (C) 100 mm yr^{-1}, with the other parameters held constant (Ferguson *et al.*, 1993). H_2 is 1 km, α is 2°, β is 1.5°, λ_b is 0.85 and μ_b is 0.85. (b) Heat-flow response of a simple wedge model for various values of the basal pore-fluid pressure ratio λ_b from 0.5 to 0.99, at three example velocities v_0 of (A) 20, (B) 50 and (C) 100 mm yr^{-1}, with the other parameters held constant (Ferguson *et al.*, 1993). H_1 (see Fig. 15.7) is 2 km, H_2 is 1 km, α is 2°, β is 1° and μ_b is 0.85.

constant velocity ($\partial_t v = 0$) and this condition lasts sufficiently long to establish thermal equilibrium ($\partial_t T = 0$). The equation thus reduces to

$$v \cdot \nabla T = \kappa \nabla^2 T + (\rho c_v)^{-1}(H + \sigma : \varepsilon). \qquad (15.5)$$

The constant thermal diffusivity chosen to represent the Taiwan fold and thrustbelt equals 1.2 mm^2 s^{-1}, the

constant sediment density is 2500 kg m^{-3}, the specific heat capacity is 1200 J kg^{-1}°C^{-1}, the heat production rate is 1 μW m^{-3}, the constant thermal conductivity is 3.5 W m^{-1}°C^{-1}, the coefficient of internal friction along the décollement is 0.5 and the coefficient of internal friction of the accreted sediments is 0.7 (Barr and Dahlen, 1989).

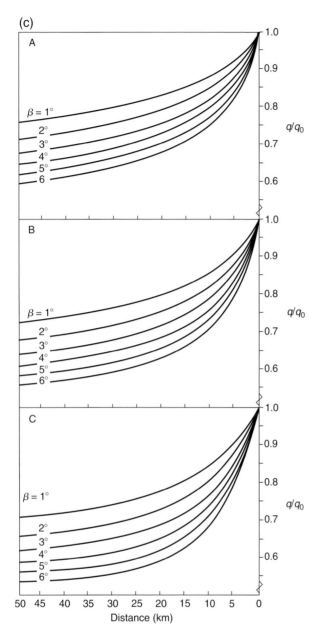

Fig. 15.8 (*cont.*). (c) Heat-flow response of a simple wedge model for various values of décollement dip β, from 1° to 6° at three example velocities v_0 of (A) 20, (B) 50 and (C) 100 mm yr^{-1}, with the other parameters held constant (Ferguson *et al.*, 1993). H_1 is 2 km, H_2 is 1 km, α is 2°, β is 1° and λ_b is 0.85.

The amount of internal strain heating is calculated from (Barr and Dahlen, 1989)

$$\sigma : \varepsilon = \sigma_1 \varepsilon_1 + \sigma_3 \varepsilon_3, \qquad (15.6)$$

where σ_1 and σ_3 are the principal stresses and ε_1 and ε_3 are the principal strain rates. Calculations made for the Taiwan fold and thrustbelt indicate that the internal

strain heating exceeds the radiogenic heating inside the accretionary wedge and reduces to zero within the underlying slab. The distribution of the internal strain heating inside the Taiwan wedge, together with alternatives under modified accretion and erosion rates, is shown in Fig. 15.9(a). It indicates that the internal strain heating is largest in the thrust front area, where the strain rate is largest.

The frictional heating along the décollement, calculated as outlined earlier, increases from zero at the thrust front to its maximum landwards, becoming progressively more important for the thermal regime (Fig. 15.9b).

It is useful to look at a sequential restoration of the thermal regime calculation to appreciate the influence of the controls involved in the Taiwan fold and thrustbelt. Figure 15.10 illustrates such a restoration, progressively adding the effects of sediment accretion and erosion, internal strain heating of accreted sediments and frictional heating along the décollement. A comparison between the undisturbed thermal regime and the regime perturbed by sediment accretion and erosion illustrates how accretion together with erosion reduces the thermal gradient. This cooling is controlled by the wedge shortening and its advance over the cool slab. If one ignored frictional heating, there would be no warming of the thermal regime. However, warming is indicated by the increase in metamorphic grade observed in Taiwan in the direction landward from the thrust front (Liou, 1981; Ernst, 1982). It is a consequence of frictional heating. Part of this heating is produced by the internal strain, which warms up the thickest part of the fold and thrustbelt by about 50–75 °C (Fig. 15.10C). Frictional heating along the décollement appears to be an even more important part of the overall frictional heating. It warms up the thermal regime by more than 150 °C, having a progressively more important effect in the direction away from the frontal thrust (Fig. 15.10D).

When further tested by sensitivity analysis, neither the increase of the mantle heat-flow value nor the thickness of the subducting slab has any noticeable impact on the thermal regime within the Taiwan wedge (Barr and Dahlen, 1989). This lack of influence is caused by the rapid convergence, which prevents the advecting mantle heat from dissipating inside the wedge.

As in the case of the Barbados accretionary prism, which represents the opposite side of the wedge spectrum, one can look at the influence on the frictional heat budget of the coefficient of internal friction and the pore fluid pressure along the décollement, both of which add to the overall thermal regime.

In order to illustrate the effect of décollement friction,

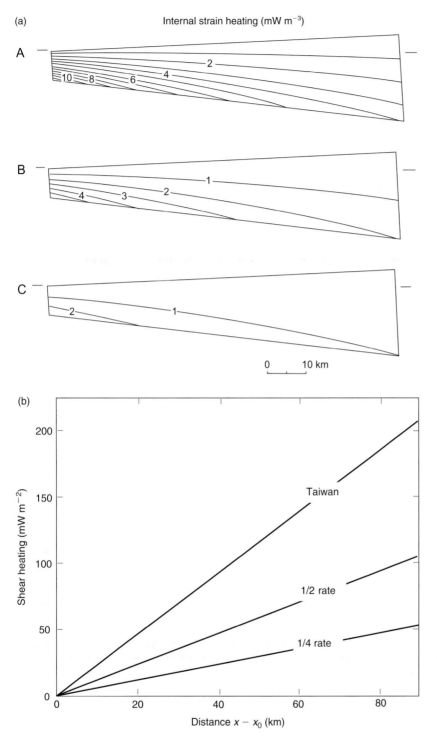

Fig. 15.9. (a) Theoretical east–west cross section through the Taiwan fold and thrustbelt showing the variation of internal strain heating (Barr and Dahlen, 1989). Two cross sections below it show the hypothetical effect of reduced accretion and erosion rates. The quantity $\sigma : \varepsilon$ is contoured at intervals of $1\,\mu W\,m^{-3}$ in each case. In Taiwan the accretionary influx rate is $A = 500\,km^2\,Myr^{-1}$ and the erosion rate is $e_0 = 5.5\,mm\,yr^{-1}$. Reducing the flux rates simply reduces the internal strain heating by the same fraction. The constant slope of the topographic surface is $\alpha = 3°$ and the dip of the basal décollement fault is $\beta = 6°$. The steady-state width of the fold-and-thrustbelt is 90 km and the thickness of incoming sediments at the toe of the wedge is 7 km. The toe of the wedge corresponds to the deformation front of the western edge of the Taiwan fold-and-thrustbelt. A – Real Taiwan accretion and erosion, B – 50% or real rates, C – 25% of real rates. (b) Theoretical variation of shear heating along the décollement fault (Barr and Dahlen, 1989). Distance along the abscissa is measured in the x direction. In Taiwan the accretionary flux rate is $500\,km^2\,Myr^{-1}$, and the erosion rate is $5.5\,mm\,yr^{-1}$. Reducing the flux rates simply reduces the shear heating by the same fraction.

one can vary the associated coefficient of internal friction from 0.3 to 0.9 in 0.2 increments (Fig. 15.11). The minimum value would represent natural shale horizons (e.g., Morrow *et al.*, 1981; Logan and Rauenzahn, 1987), while the maximum value would represent an artificially high value for natural detachment horizons (see Table 2.1). Applying the wedge taper calculations described in Chapter 2 to the Taiwan wedge taper, the chosen coefficients of internal friction along the décollement equal to 0.3, 0.7 and 0.9 are associated with coefficients of internal friction of the accreted sediments equal to 0.4, 0.9 and 1.1, respectively (Barr and

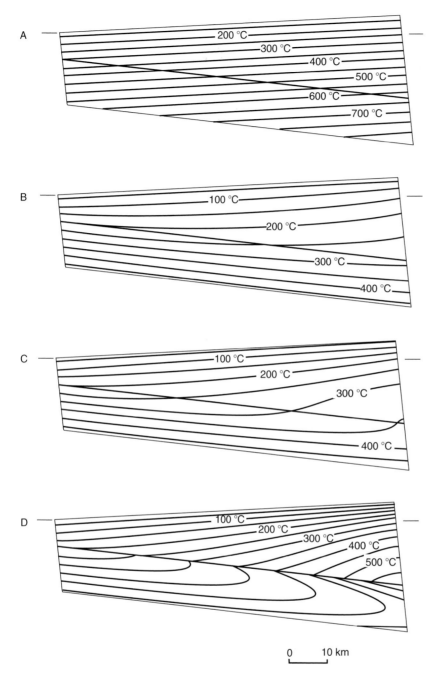

Fig. 15.10. Cross sections of the Taiwan wedge and underlying slab, showing the sequential thermal effects of accretion and erosion, internal strain heating and frictional heating along the décollement (Barr and Dahlen, 1989). The steady-state temperature distribution is contoured at 50° C intervals with the 100° C isotherms labelled. The location of the décollement fault is shown by a solid line. (A) Undisturbed geothermal regime, (B) added effect of accretion and erosion, (C) added effect of internal strain heating, (D) added effect of shear heating along the décollement.

Dahlen, 1989). Figure 15.11 illustrates that the increase in décollement friction causes a significantly warmer thermal regime in the Taiwan wedge partly because of associated frictional heating along the décollement and partly because of associated increased internal strain heating.

The effect of the pore fluid pressure on the thermal regime is associated with the frictional effect, since it modifies both basal and internal friction, reducing them by moving from hydrostatic to lithostatic values. A sen-

sititvity analysis (Barr and Dahlen, 1989) documents that the change of the ratio between pore fluid and lith-ostatic pressures from 0.7 to 0.9 results in a cooling by 100–150 °C in the deepest portion of the Taiwan wedge.

One can investigate further the impact of other factors, such as the initial thermal regime, the accretion rate, the erosion rate and the effect of underplating, on the overall thermal regime of the Taiwan wedge. The effect of the initial thermal regime is shown in Fig. 15.12(a). The initial thermal regime is represented by

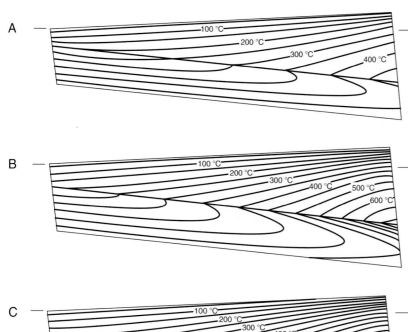

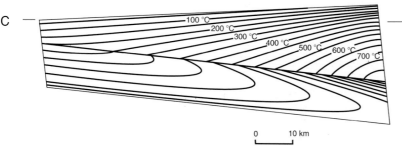

Fig. 15.11. Dependence of the theoretical temperature distribution in Taiwan on the coefficient of basal friction μ_b (Barr and Dahlen, 1989). The pore fluid/lithostatic pressure ratio is kept constant at $\lambda = \lambda_b = 0.7$, and the coefficient of internal friction is varied to fit the observed wedge geometry. (A) $\mu_b = 0.3$, (B) $\mu_b = 0.7$, (C) $\mu_b = 0.9$.

the temperature gradient at the thrust front. An approximately $10\,°C\,km^{-1}$ difference in the initial thermal regime would cause a reduction of about $50–100\,°C$ in the thermal regime in the rear deep portion of the Taiwan accretionary wedge (Barr and Dahlen, 1989).

The effect of the accretion and erosion rates is illustrated in Fig. 15.12(b), where they are represented as fractions of their values determined for the Taiwan fold and thrustbelt. A decrease in these rates results in a difference of tens of degrees Celsius in the thermal regime in the rear and deep part of the wedge (Barr and Dahlen, 1989). Because the slip rate along the décollement and the strain in the accreted sediments are associated with the accretion and erosion rates, a decrease of accretion and erosion would result in a proportional reduction of the frictional heating along the décollement and internal strain heating of the accreted sediments.

The effect of underplating is shown in Fig. 15.12(c). This mechanism is a simple consequence of underplated rock bringing warm thermal regimes into the thermal budget of the accretionary wedge. The temperature difference between the alternative with 50% underplating and the alternative with no underplating is about

$75\,°C$ in the rear and deep part of the wedge (Barr and Dahlen, 1989).

At the end of this discussion of the thermal regime of the Taiwan fold and thrustbelt calculated by Barr and Dahlen (1989) one can look at the contribution of all the discussed heat sources for the surface heat flow Q_T:

$$Q_T = Q_B + Q_A + H_R + H_S, \qquad (15.7)$$

where Q_B is the heat added to the accretionary wedge generated by frictional heating along the décollement, Q_A is the heat flow that enters the accretionary wedge from outside through its boundaries, H_R is the heat produced by radioactive elements included in accreted sediments and H_S is the heat produced by internal strain of the accreted sediments. Figure 15.13 sequentially illustrates these contributions. The first step in this illustration shows that the thermal effect of the radioactive heat production within the accretionary wedge in the surface heat flow is minimal. This is not the case for the remaining factors. The combined effect of accretion and erosion results in an increased surface heat flow, affected by uplifted isotherms in the shallow region and decreased basal heat flow, affected by underthrust cooler material. Both increase and decrease become

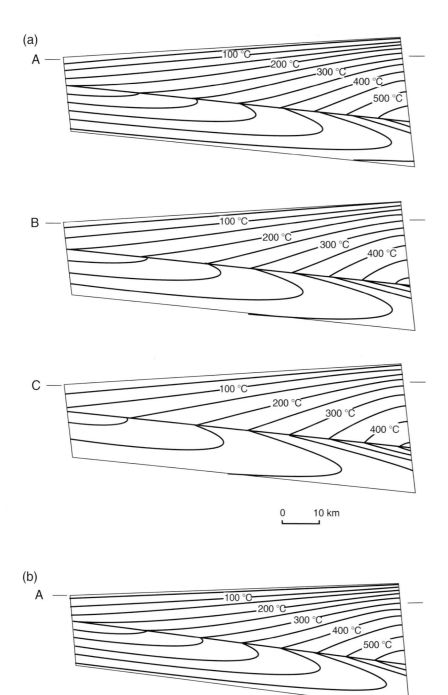

Fig. 15.12. (a) Dependence of the theoretical temperature distribution in Taiwan on the undisturbed thermal gradient T_0 at the toe of the wedge (Barr and Dahlen, 1989). The observed gradient is $T_0 = 27\,°C\,km^{-1}$. (A) $T_0 = 27\,°C\,km^{-1}$, (B) $T_0 = 20\,°C\,km^{-1}$, (C) $T_0 = 15\,°C\,km^{-1}$. (b) Hypothetical effect of accretion and erosion rates on the theoretical temperature distribution of the Taiwan fold and thrustbelt (Barr and Dahlen, 1989). The observed accretionary influx rate is $500\,km^2/Ma$, and the erosion rate is $5.5\,mm\,yr^{-1}$. (A) Real rates, (B) 50% of real rates, (C) 25% of real rates. (c) Hypothetical effect of underplating on the thermal structure of the Taiwan fold and thrustbelt (Barr and Dahlen, 1989). In each case the total accretionary influx rate is $500\,km^2/Ma$, but $R\%$ is underplated beneath the décollement fault and $100\% - R\%$ is accreted at the toe or thrust front. (A) No underplating, (B) 25% of underplating, (C) 50% of underplating.

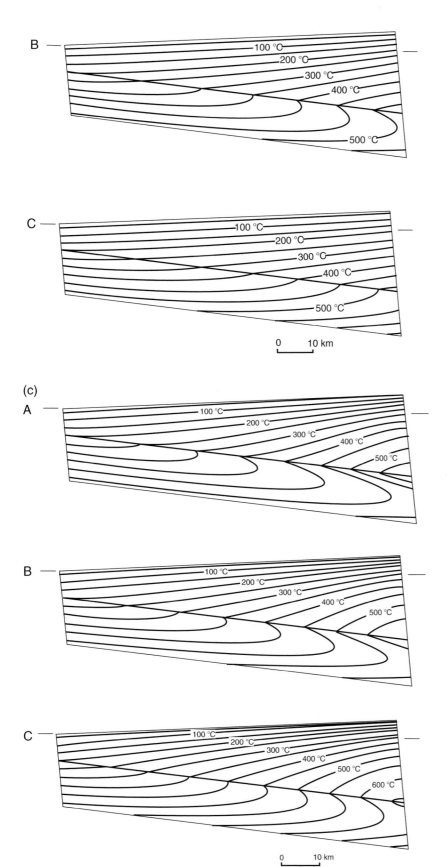

Fig. 15.12 *(cont.)*

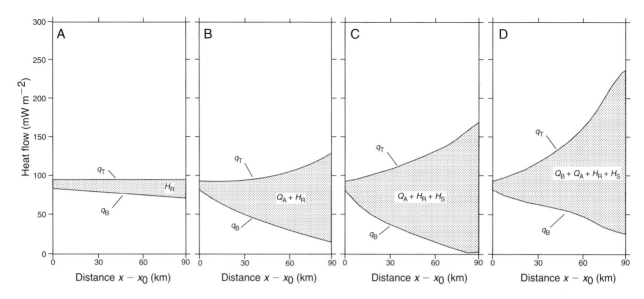

Fig. 15.13. Sequential effects of accretion and erosion, internal strain heating and frictional heating along a décollement fault on the surface heat flow q_T and the basal heat flow into the wedge q_B (Barr and Dahlen, 1989). The corresponding steady-state temperature distributions are shown in Fig. 15.10. Both q_T and q_B are plotted versus distance from the deformation front $x - x_0$. The stippled area denotes the integrated difference $Q_T - Q_B$. (A) Undisturbed geothermal regime, (B) added effect of accretion and erosion, (C) added effect of internal strain heating, (D) added effect of shear heating along the décollement.

more important in the landward direction, enhancing the difference between surface and basal heat flows. Therefore, the second step in the sequential illustration introduces a significant thermal perturbation. The third step of the sequential illustration adds the effect of the internal strain heating into the thermal regime. The heat produced by the mechanical energy dissipated against internal friction is transported in all directions out from the accretionary wedge. This further enhances the landward-increasing difference between the surface and basal heat flow already introduced by the effect of accretion and erosion. The last step in the sequential illustration adds the impact of frictional heat produced along the décollement. It increases both surface and basal heat flow, adding progressively larger amounts of heat in the landward direction from the thrust front.

Recognition of paleothermal perturbations caused by deformation

The recognition of paleothermal perturbations caused by deformation requires using maturity data such as vitrinite reflectance, apatite or zircon fission-track data (e.g., Laslett *et al.*, 1987; Burnham and Sweeney, 1989).

An example when these data show that the deformation did not affect the maturation regime comes from the Fukase thrust of the Shimanto belt, southeast Japan (Ohmori *et al.*, 1997). The Cretaceous–Tertiary Shimanto belt represents a portion of the accretionary

prism located to the NW of the Nankai Trough. The Fukase thrust is located in the Cretaceous Northern Shimanto belt and crops out in the southeast part of Shikoku Island. Vitrinite reflectance and zircon fission-track data generated in the broader Fukase thrust area are characterized by a complex distribution of values. Fission-track ages of partially annealed samples contain two age categories. The first one is syn-depositional and the second one is associated with a younger thermal overprint. Vitrinite reflectance data in the hanging wall of the Fukase ramp indicate that maturity increases up to values of 2.4–2.8% towards the Fukase ramp (Figs. 15.14 and 6.14). Vitrinite reflectance in the footwall zone adjacent to the ramp is characterized by values of 0.6–1.4%. The reflectance distribution in the remaining portion of the footwall indicates a gradual increase away from the Fukase ramp. The ramp itself is associated with a sharp discontinuity between hanging wall and footwall values, documenting an out-of-sequence thrust, which cuts through this portion of the accretionary wedge after it had achieved its thermal maturation structure.

An example of when maturity data show that the deformation affected the maturation regime comes from the Pine Mountain thrust in the Southern Appalachians (Fig. 6.13; O'Hara *et al.*, 1990). Here, the vitrinite reflectance values increase within the hanging wall towards the Pine Mountain thrust to 0.9–1% and drop only to

(a)

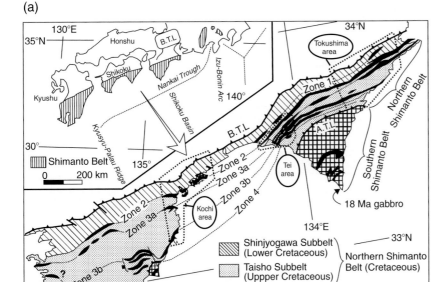

Fig. 15.14. (a) Location of the Shimanto belt in the Shikoku area (Ohmori *et al.*, 1997). Study areas are indicated by dotted lines.
(b) Paleothermal structure indicated by vitrinite reflectance R_0 in the Tokushima area (Ohmori *et al.*, 1997). Temperature is marked every 25 °C from 75 to 225 °C based on the Sweeney and Burnham (1990) equation for conversion from R_0 to maximum paleotemperature. Radial plot (Galbraith, 1990), histograms and age spectrum (Hurford *et al.*, 1984) of fission-track age of zircon are shown. Shaded areas represent depositional age.

(b)

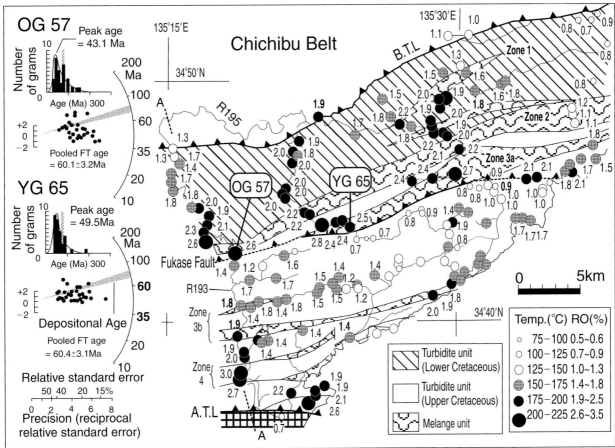

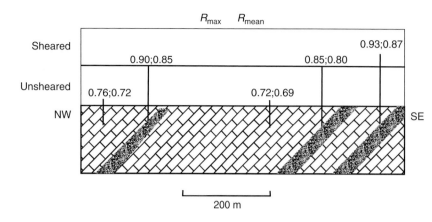

Fig. 15.15. Schematic section through the Cumberland pilot tunnel showing maximum and mean vitrinite reflectance values for sheared and nonsheared Pennsylvanian coals (O'Hara *et al.*, 1990). Strata face to the northwest. Sheared coals display higher maximum and mean reflectance values compared to nonsheared samples regardless of their stratigraphic position.

0.85% in the immediate footwall. This pattern indicates that the thrusting was active during the thermal maturation and the hanging wall affected the thermal regime of the footwall.

Maturity data can also help to determine whether frictional heating has contributed to a paleothermal perturbation recognized in a thrustbelt. Because the organic matter is sensitive to thermal effects, anomalous maturation levels of coal within and near fault zones provide an indication of such frictional heating (e.g., Bustin, 1983). Frictional heating includes two important cases. The first case is the result of steady-state heating during the active history of the thrust. For example, if average shear stresses of about 100 MPa exist over a time period of 1 Myr, frictional heating can result in a temperature increase of 100 °C (Scholz, 1980; Sibson, 1983).

The second case of frictional heating involves short-lived seismic slip events, which result in localized and transient heating on the fault plane. For example, a fault surface, which slips at a rate of $1\,\mathrm{m\,s^{-1}}$ against a shear resistance of 20 MPa for 1 s, undergoes a temperature rise of approximately 100 °C (Sibson, 1980).

In both cases of frictional heating, a decrease in maturation level away from a fault in both the hanging wall and the footwall is good evidence for such heating (Scholz, 1980). A natural example of frictional heating comes from the Cumberland pilot tunnel in the hanging wall of the Pine Mountain thrust in the Appalachians (Fig. 15.15; O'Hara *et al.*, 1990). Here, sheared and slickensided samples of Pennsylvanian coal taken from fault zones penetrated by the tunnel display higher vitrinite reflectance R_{max} values than unsheared samples. R_{mean} values also show the same relationship between sheared and unsheared samples. The absence of chemical alteration of coals along the faults suggests a lack of hydrothermal fluid circulation along the fault planes, which would provide a different method of heating (O'Hara *et al.*, 1990). O'Hara *et al.* (1990) also tried to ensure that they separated the effect of frictional heating on the vitrinite reflectance from the effect of strain. In order to separate the effect of frictional heating from reorientation by strain, O'Hara *et al.* (1990) made a detailed three-dimensional reflectance analysis to show that the reflectance fabric does not show an increase in R_{max}, a decrease in R_{min} and no change in $R_{intermediate}$, which would be caused by a strain effect (see Levine and Davis, 1983). In the case of the Cumberland pilot tunnel, the observation that both R_{mean} and R_{max} display higher values led O'Hara *et al.* (1990) to interpret the effect of frictional heating, which, however, does not completely rule out the impact of strains other than the volume-preserving plane strain.

16 Role of fluid movement in thermal regimes

The magnitude of the thermal anomaly caused by the movement of fluids in thrustbelts depends on the velocity and depth of fluid circulation, and the thermal conductivity of the rock section (e.g., Nunn and Deming, 1991). It requires vertical Darcy velocities of $1\,mm\,yr^{-1}$ and above to make a noticeable impact on the thermal regime of a thrustbelt (e.g., Bredehoeft and Papadopulos, 1965). Lateral movement of pore water has little or no effect on the temperature distribution since the isotherms are almost always parallel to the ground surface.

Numerical simulations of tectonically quiescent sedimentary basins have led to the conclusion that vertical flow velocities generated by normal sediment compaction and the concomitant expulsion of pore fluids are too low (less than $0.1\,mm\,yr^{-1}$) to produce significant thermal perturbations (e.g., Cathles and Smith, 1983; Bethke, 1985). However, thrustbelts involve several other mechanisms for fluid flow in addition to the normal process of compaction. These are described in Chapter 10. The current chapter discusses their impact on the thermal regime.

Conductive and advective heat transport in a porous medium is defined by the diffusion–advection equation (Deming, 1994b):

$$\frac{\partial}{\partial t}(T\rho_b c_b) = \nabla \cdot (\boldsymbol{\lambda} \nabla T) - (\boldsymbol{q} \rho_f c_f \cdot \nabla T) + H, \qquad (16.1)$$

where t is the time, T is the temperature, ρ_b is the bulk density, c_b is the bulk specific heat capacity, λ is the thermal-conductivity tensor characterizing the whole porous rock system, $\boldsymbol{q}$ is Darcy's fluid flow velocity, ρ_f is the fluid density, c_f is the specific heat capacity of the fluid and H is the heat production rate of the porous rock system. For simplicity, one can assume that the heat flow is only vertical, that no heat sources or sinks exist, that Darcy's fluid flow velocity does not vary in space, and that the physical properties of the whole porous rock system, such as the thermal conductivity, density and specific heat capacity change neither spatially nor temporally. It allows a simplification of equation (16.1) to (Deming, 1994b)

$$\rho_b c_b \left(\frac{\partial T}{\partial t}\right) = \lambda_z \left(\frac{\partial^2 T}{\partial z^2}\right) - q_z \rho_f c_f \left(\frac{\partial T}{\partial z}\right), \qquad (16.2)$$

where $(\partial^2 T/\partial z^2)$ is the curvature of the temperature field, λ_z is the thermal conductivity in the z-direction and q_z is Darcy's fluid flow velocity in the z-direction.

Under steady-state conditions ($\partial T/\partial t = 0$), the heat transport across a layer of thickness Δz is (Lachenbruch and Sass, 1977)

$$\frac{\Theta_{top}}{\Theta_{bottom}} = e^{\Delta z/s}, \qquad (16.3)$$

where Θ_{top} is the conductive heat flow at the top surface and Θ_{bottom} is the conductive heat flow at the basal surface, and

$$s = \frac{\lambda}{\rho_f c_f q_z}, \qquad (16.4)$$

where λ is the bulk thermal conductivity, ρ_f is the fluid density, c_f is the specific heat capacity of a fluid and q_z is the vertical Darcy fluid flow velocity. q_z is negative and positive for downward and upward flow, respectively. Equations (16.3) and (16.4) indicate that a downward fluid flow causes a reduction of the geothermal gradient and the total heat flow, because it works against the conductive heat flow. An upward fluid flow increases the geothermal gradient and the total heat flow, because it adds to the conductive heat flow (Fig. 16.1). Because the heat flow depends exponentially on depth (equation 16.3), the already relatively low Darcy fluid flow velocity, of the order of 10^{-3}–$10^{-2}\,m\,yr^{-1}$, can noticeably perturb the thermal regime of a thrustbelt.

Boundary conditions can be set for equation (16.2) by fixing the temperatures for the top and basal surfaces of the assumed layer, which has a thickness of Δz. Assuming steady-state conditions ($\partial T/\partial t = 0$), the equation then can be solved for $T(z)$ (Bredehoeft and Papadopulos, 1965):

$$T(z) = T_0 + \Delta T \frac{e^{(Bz/\Delta z)} - 1}{e^B - 1}, \qquad (16.5)$$

where T_0 is the temperature at the top surface, ΔT is the total temperature change across the layer, z is the depth, and

$$B = \frac{c_f \rho_f q_z \Delta z}{\lambda}. \qquad (16.6)$$

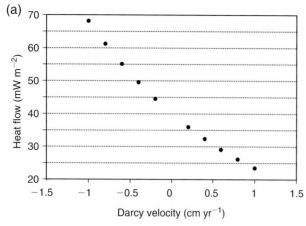

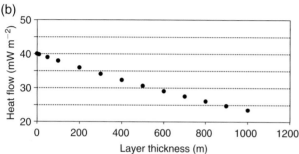

Fig. 16.1. (a) Heat flow at the bottom of a layer with vertically circulating fluid as a function of the Darcy velocity. Darcy velocity is negative for downward flow. The heat flow at the top of the 1 km thick layer is 40 mW m^{-2}, the thermal conductivity of the fluid is 2.5 W m^{-1} K^{-1} and the fluid density is 1000 kg m^{-3}. (b) Heat flow at the bottom of a layer with circulating fluid as a function of the layer thickness. The heat flow at the top of the layer is 40 mW m^{-2}, the thermal conductivity of fluid is 2.5 W m^{-1} K^{-1}, the fluid density is 1000 kg m^{-3} and the fluid has a Darcy velocity of 1 cm yr^{-1}.

When equation (16.5) is doubly differentiated with respect to z, it becomes

$$\frac{\partial^2 T}{\partial z^2} = \frac{\Delta T}{e^B - 1} \frac{B^2}{\Delta z^2} e^{(Bz/\Delta z)}. \qquad (16.7)$$

Equation (16.7) demonstrates that the curvature of the temperature field in a thrustbelt perturbed by fluid flow is exponentially controlled by depth. It also illustrates that it is possible to find almost linear temperature–depth profiles in thrustbelt regions affected by downward fluid flow in cases where the profile depth is shorter than the circulation depth. These profiles give the appearance of conductive temperature profiles leading to the incorrect conclusion that there is no significant fluid flow (Deming, 1994b).

Table 16.1. *Typical behaviour of permeability on production and geological time scales.[a]*

Permeability range	Production time scale	Geologic time scale
>1 d (coarse sand, fractured limestone)	Highly permeable	Highly permeable, hydrostatic pressure gradients
10^{-3} to 1 d (sandstone)	Moderately permeable	Highly permeable, hydrostatic pressure gradients
10^{-3} to 10^{-6} d (cemented sandstone, silt)	Poorly permeable	Moderately permeable, hydrostatic pressure gradients unless rapid burial
10^{-6} to 10^{-9} d (marl, shale)	Practically impermeable	Poorly permeable, often over pressured, sometimes hydraulic fracturing
<10^{-9} d (compacted shale, tight limestone)	Impermeable (measurements difficult)	Practically impervious, frequent hydraulic fracturing, frequently over pressured

Note:
[a] Ungerer *et al.* (1990).

Recognition of the advective part of the heat transport

Recognition based on hydrological calculation

Fluid flow systems in thrustbelts are usually complex, and their understanding requires robust data sets and rigorous numerical analyses. The necessary data and analytical tools are frequently either not available or the cost and time for analyses are beyond the resources of the study. In such cases, simpler but effective methods for the evaluation of the fluid flow effects on the subsurface thermal regime such as a dimensional analysis can be employed.

Domenico and Palciauskas (1973), for example, considered a steady-state fluid flow in an idealized, two-dimensional flow system, characterized by homogeneous and isotropic physical properties. Their dimensional analysis was based on the calculation of the Peclet number, P_c, which served as a relative gauge of the magnitude of convective to conductive heat transport. It is defined as the dimensionless ratio

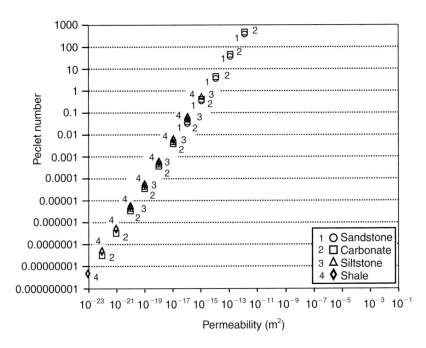

Fig. 16.2. The dependence of the Peclet number on the permeability in a steady-state two-dimensional flow system with homogeneous and isotropic properties. An explanation is given in the text.

$$P_c = \frac{\Delta h\, K\, \Delta z\, \rho_f c_f}{2L\lambda}, \tag{16.8}$$

where Δh is the total head drop across a rock section, K is the hydraulic conductivity, Δz is the depth of the rock section, ρ_f is the fluid density, c_f is the specific heat capacity of the fluid, L is the length of the rock section and λ is the bulk thermal conductivity. One has to bear in mind in this dimensional analysis, when determining the hydraulic conductivity K, that permeability in thrustbelts varies depending on scale. Table 16.1 documents that there is about two to three orders of magnitude difference between production and geological time scales.

Formulating the average head gradient as $\Delta h/L$, equation (16.8) changes to (Deming, 1994b)

$$P_c = \frac{q\, \Delta z\, \rho_f c_f}{2\lambda}, \tag{16.9}$$

where q is Darcy's fluid flow velocity. Using values of the hydraulic head gradient of 0.01, a rock section permeability of $10^{-15}\,\mathrm{m}^2$, a thickness of 5000 m for the rock layer with circulation, a fluid density of 1000 kg m^{-3}, a fluid specific heat of 4200 J kg^{-1} K^{-1} and a thermal conductivity of porous rock of 3.5 W m^{-1} K^{-1} in equation (16.9), Deming (1994b) calculates that the thermal regime of the sedimentary section characterized by an average permeability of about $10^{-15}\,\mathrm{m}^2$ can be perturbed by the fluid flow (Fig. 16.2). It can be implied from this dimensional analysis that a potential thrustbelt section characterized by an average permeability of about $10^{-17}\,\mathrm{m}^2$ or lower would have a particularly low

fluid flow, which would fail in being an important heat transport mechanism causing a significant perturbation of the thermal regime (Deming, 1994b). A threshold Peclet number value of about 0.01 is required before convective heat transfer becomes an important contribution to the overall heat transfer. It has to be stressed that this calculation has been made for the simple case of a homogeneous rock section. The calculation becomes more complex if temperature- and pressure-dependent fluid dynamic viscosities (Fig. 16.3a), temperature-dependent water density and temperature- and pressure-dependent thermal conductivities (Fig. 16.3b) are introduced. Further complexities emerge if one allows for changes of porosity and density with effective mean stress.

Neither of the two described Peclet numbers nor the Peclet numbers introduced later in this chapter are characterized by the existence of a specific critical value below which the thrustbelt thermal regime has a conductive character, as is the case for the Rayleigh number in natural convection (Neild, 1968; Rubin, 1975; Ribando and Torrance, 1976). However, it can be concluded that Peclet numbers significantly lower than 1 indicate that the studied thrustbelt region is conduction-dominated, while values significantly greater than 1 indicate that the system is convection-dominated. Peclet numbers approaching 1 should represent a conductive–convective system.

As mentioned earlier, dimensional analyses utilize a variety of Peclet numbers, each being defined for a different situation. Van der Kamp and Bachu (1989), for

(a)

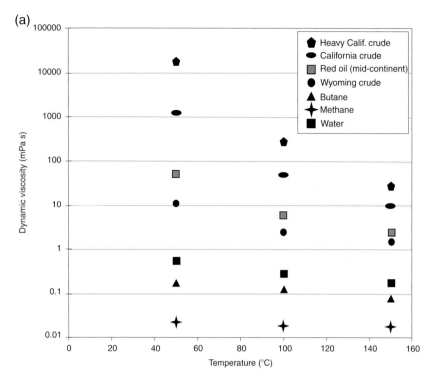

(b)

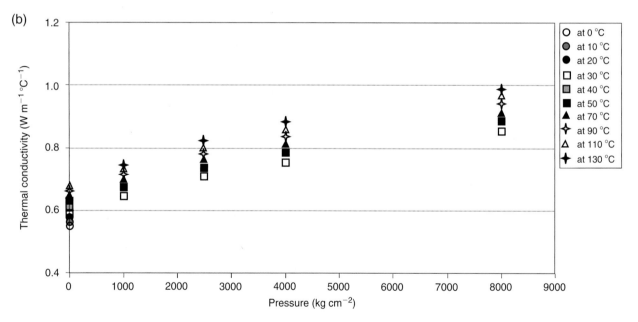

Fig. 16.3. (a) Variation of dynamic viscosity of selected fluids with temperature. Data are from Ungerer *et al.* (1990). (b) Dependence of the thermal conductivity of water on the temperature and the pressure. Data are from Clark (1966).

example, make a dimensional analysis using a representative element of the studied hydrogeological system (Fig. 16.4). For the purpose of easy mathematical formulation, the element is a rectangle. It is characterized by a length L, a height Δz or D, a temperature T_T at the top surface, temperature T_B at the basal surface, a total horizontal fluid flow Q_H and a total vertical fluid flow Q_V. These characteristics help to define the Peclet number as (Van der Kamp and Bachu, 1989)

$$P_c = \frac{\rho_f c_f q_H \Delta z A}{\lambda} = \frac{\rho_f c_f Q_H A}{\lambda}, \qquad (16.10)$$

where $A = \Delta z/L$ or D/L is the aspect ratio of the representative element and $q_H = Q_H/D$ is the horizontal characteristic specific discharge.

The Peclet number used to distinguish between convective and conductive heat transfer in a laminar flow situation, located far from any boundaries, is (Bennett and Myers, 1982)

$$P_c = \frac{v_0 L}{\kappa_f}, \qquad (16.11)$$

where v_0 is the characteristic velocity, L is a characteristic length and κ_f is the thermal diffusivity of the fluid.

The Peclet number designed for identification of the heat transfer in a porous media describes the Darcy fluid flow by a characteristic velocity v_0 (Bachu and Dagan, 1979):

$$P_c = \frac{n\beta v_0 L}{\kappa}, \qquad (16.12)$$

where $\kappa = \lambda/(c_b\rho_b)$ is the thermal diffusivity of the fluid–rock system, c_b is the bulk specific heat capacity, ρ_b is the bulk density and $\beta = \rho_f c_f/c_b\rho_b$ is the ratio of the heat capacity of the fluid to the heat capacity of the fluid-saturated rock.

The relationship between the advective heat transport along a fault and lateral heat loss by conduction from the fault to the wall rocks is evaluated by yet another Peclet number (Wang *et al.*, 1993):

$$P_c = \frac{c_f\rho_f}{c_b\rho_b}\frac{q/L}{\kappa_s/W^2} \approx \frac{qW^2}{\kappa L}, \qquad (16.13)$$

where c_f is the specific heat capacity of the fluid, c_b is the bulk specific heat capacity, ρ_f and ρ_b are the fluid and bulk densities, respectively, q is the Darcy fluid flow velocity along the fault, W is the half-width of the fault, L is the length of the fault and κ_s is the thermal diffusivity of the rocks (Fig. 16.5). Figures 16.5(a–c) indicate that the décollement thickness has the largest impact on the perturbation of the thermal regime. This impact is an order of magnitude larger than the impact of the fluid flow velocity within the décollement zone, which is more important than the role of the décollement length.

Yet another Peclet number uses the geometry of the hydrothermal system described by the aspect ratio A (Van der Kamp and Bachu, 1989):

$$P_c = \beta\left(\frac{q_h D}{\kappa}\right)A \equiv \beta\left(\frac{Q_h A}{\kappa}\right). \qquad (16.14)$$

As indicated by the various dimensional analyses described by equations (16.8)–(16.14), the Peclet number calculation depends on a proper choice of the characteristic data, in order to assess the role of convection in a conductive system (Van der Kamp and Bachu, 1989). The characteristic length L, for example, should be measured parallel to the main heat-flow direction and not parallel to the main fluid flow. In a closed hydrogeological system (Fig. 16.4), the total recharge Q_V equals the total horizontal flow Q_H, which is equal to the total discharge Q_V. Open hydrogeological systems require a different mathematical formulation. Additionally, the lack of precision in determining the characteristic data

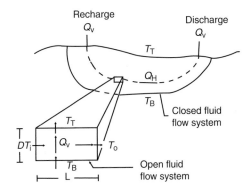

Fig. 16.4. Schematic representation of open and closed hydrogeological systems and of a representative element (Van der Kamp and Bachu, 1989). T_T is the temperature at the top surface, Q_V is the vertical fluid flow, Q_H is the horizontal fluid flow, T_B is the temperature at the basal surface, T_i is the temperature at the inflow, T_0 is the temperature at the outflow, L is the horizontal length of the hydrogeologic system and D is the thickness of the hydrogeological system.

for an open system, in particular permeability, introduces a large degree of freedom in the evaluation of the thermal effects of hydrogeological regimes. The calculation must use the specific discharge q determined according to Darcy's law. Based on the potentiometric distributions and hydraulic parameters, the specific discharge can be calculated as (Bachu, 1985)

$$q = \frac{kgJ}{\eta}, \qquad (16.15)$$

where k is the permeability, g is the acceleration due to gravity, J is the dimensionless hydraulic gradient and η is the fluid viscosity.

Recognition based on maturity data
Another way of recognizing the presence of advective heat transfer in thrustbelts is from maturity data. These data are represented either by vitrinite reflectance (R_0) data (e.g., Burnham and Sweeney, 1989) or by apatite fission-track (AFT) data (e.g., Laslett *et al.*, 1987), which are totally annealed at 0.7% R_0 in more fluorine-rich apatite or at 0.7–0.9% R_0 in more chlorine-rich apatite (Duddy *et al.*, 1991, 1994). If the maturity–depth relationship throughout the potential circulation layer is linear (Fig. 16.6), the thermal regime is probably conductive and not perturbed by advective heat transfer. Maturity data in this situation allow an easy paleogeothermal gradient determination (Duddy *et al.*, 1994). As indicated earlier, if the maturity–depth profiles in the recharge portion of a thrustbelt are shallower than the circulation depth, their maturity–depth relationships

(a)

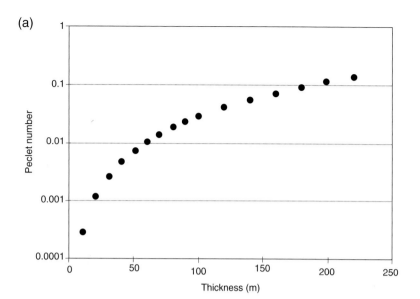

Fig. 16.5. Dependence of the Peclet number designed for the thermal system along a décollement fault on (a) the décollement thickness, (b) the fluid flow velocity and (c) the décollement length. Explanation in text.

(b)

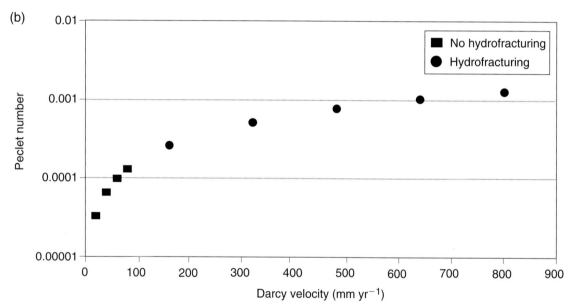

(c)

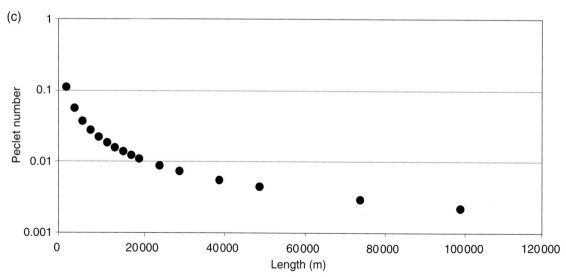

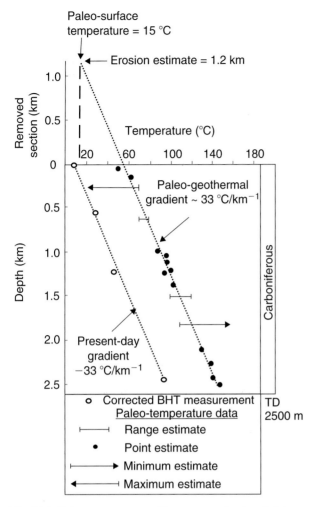

Fig. 16.6. Paleotemperature profile constructed using vitrinite reflectance and apatite fission-track data (Duddy *et al.*, 1994). The amount of uplift and erosion is estimated by construction, which is only appropriate for a pure conductive system.

will give a misleading appearance of a conductive thermal regime (Deming, 1994b).

If the maturity–depth relationship is not linear, a perturbation of the thermal regime by advective heat transfer is suggested. Examples of nonlinear heat-flow–depth and temperature–depth relationships caused by topography-driven fluid flow are shown in Fig. 16.7. The downward-flowing fluid flow in the recharge area depresses the isotherms but raises them in the discharge area, where heated fluids flow laterally and upwards. Therefore, the geothermal gradient in the upper part of the section located in the recharge area, also characterized by the greatest rate of deposition sourced from the uplifted part of the thrustbelt, is lowered. In the discharge area, in contrast, the geothermal gradient is raised. The described perturbations of the temperature profiles by fluid-enhanced lateral heat transfer result in

a low surface heat flow anomaly in the recharge area and a high surface heat flow anomaly in the discharge area, even though the basal heat flow in the region is uniform (Fig. 16.7b).

The flow of heated fluids within one or several aquifers towards the discharge area can result in a complex vertical variation in the paleothermal gradient (Fig. 16.8). The steady-state temperature associated with a single aquifer elevates the temperature gradient above the aquifer (Fig. 16.8a). The background geothermal gradient below the aquifer has the same trend as the gradient prior to the fluid flow, but shifted to higher temperatures. If heated fluids flow in several aquifers, the thermal regime becomes even more complex. If two aquifers containing fluids with similar temperatures are separated by a significantly thick shale aquitard, the outcome can be a zero geothermal gradient through the aquitard (Fig. 16.8b). If local cold recharge interacts with the regional flow of heated fluids, the result can be a hydrogeological situation characterized by a shallow aquifer with fluids hotter than the fluids within a deep aquifer. This situation results in a negative geothermal gradient through the aquitard separating the aquifers (Fig. 16.8c). The described thermal regime perturbations caused by hot aquifers develop over a specific geological time period.

On an immediate time scale, the hot aquifer affects its surroundings by developing a bell-shaped thermal profile. Such a transient profile moves to a steady-state profile with time. In the case of a shallow aquifer, the time scale is rather short (Fig. 16.9). Transient thermal profiles around aquifers are similar phenomena to profiles developed around intrusive sills, although the heating in the latter cases is commonly solely by conduction, as shown by vitrinite reflectance data (e.g., Dow, 1977; Raymond and Murchison, 1988, 1991; Duddy *et al.*, 1994).

The paleothermal regime can be partially or wholly obliterated if the preserved stratigraphic section resides at a higher temperature than the maximum paleotemperature reached during the fluid flow regime (Duddy *et al.*, 1994). Therefore, once the thrustbelt development has ceased, the present-day ability to determine the paleothermal signature varies from location to location.

The probability of observing low paleothermal gradients due to fluid flow in the upper part of a section in the recharge area (Fig. 16.7) is generally low. This is commonly because the largest uplift and erosion, typical of proximal parts of a thrust wedge, remove the section that experienced low geothermal gradients. It is also commonly caused by the cessation of cold-water ingress that is followed by a rapid increase of the geothermal gradient towards the background trend, which erases the older thermal signature (Duddy *et al.*, 1994).

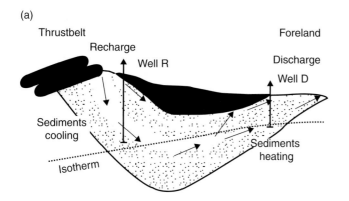

(a)

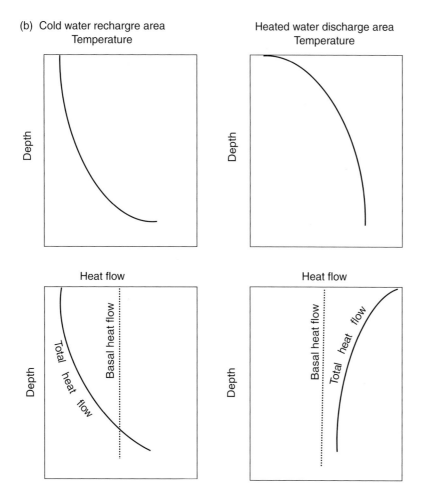

Fig. 16.7. (a) Schematic illustration of fluid movement in an active foreland basin with the location of wells, described in (b) (Majorowicz *et al.*, 1985). Ingress of cold fluid in the elevated thrustbelt results in topography-driven flow discharging heated fluids on the opposite side of the foreland basin. (b) Theoretical temperature and heat-flow distributions with depth for two wells located in the recharge and discharge areas in (a) (Majorowicz *et al.*, 1985).

The chance of observing the paleothermal signature in the former discharge area is much higher, because here smaller amounts of uplift and erosion occur. It is also due to the fact that a cessation of the fluid flow results in the subsequent cooling of the section, but without uplift and erosion the section preserves the vitrinite reflectance paleotemperature profile, characteristic of the former fluid flow regime. The application of apatite fission-track data in this section would also reveal the time of cooling from maximum paleotemperatures providing the time period after the cessation of fluid flow (Duddy *et al.*, 1994).

Role of fluid flow mechanisms in thermal regimes
The perturbation of the regional thermal regime due to groundwater flow near salt domes and the role of thermal expansion was discussed in detail in Chapter 10. Thus, this chapter discusses only the thermal impact

(a)

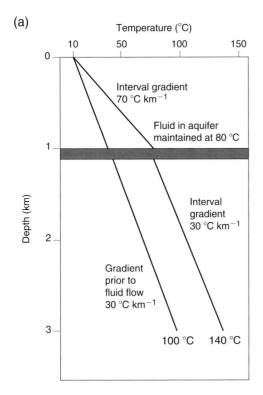

(b)

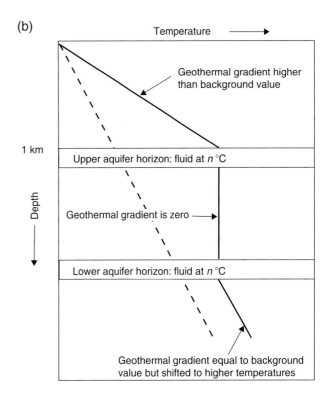

(c)

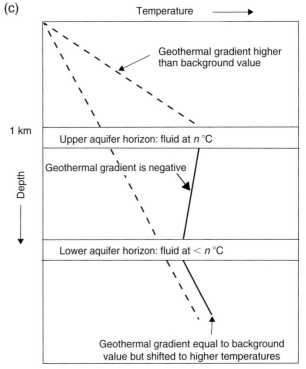

Fig. 16.8. (a) Steady-state temperature profile in the discharge area developed at maximum burial and controlled by a confined aquifer located at a depth of 1 km (Duddy *et al.*, 1994). (b) Temperature profile around two aquifers carrying fluid at the same temperature separated by a thick aquitard (Duddy *et al.*, 1994). (c) Temperature profile around two aquifers with hotter fluid flowing in the shallower aquifer resulting in a negative paleogeothermal gradient in the intervening aquitard (Duddy *et al.*, 1994).

of topography-driven flow compaction-driven flow and the flow of fluids generated by gas hydrate dissociation, clay dehydration and clay transformations.

Role of topography-driven fluid flow in thermal regimes
Among the existing fluid flow mechanisms described in Chapter 10, fluid flow driven by topography or gravity is the most likely mechanism to perturb the thermal regime of a thrustbelt. As demonstrated by radio-carbon, tritium and other measurements, this flow mechanism drives predominately meteoric water, while connate water, expelled by sediment compaction, is of secondary importance (Haenel *et al.*, 1988).

A typical example of topography-driven flow perturbing the thermal regime comes from the Appalachian basin, the thermal anomaly of which is numerically simulated by the brine migration models of Deming and Nunn (1991). The existence of the anomaly was implied from the existence and distribu-

tion of Mississippi Valley-type lead–zinc deposits (Sverjensky, 1986), fluid inclusion studies (Leach, 1979; Leach and Rowan, 1986; Coveney *et al.*, 1987; Hearn *et al.*, 1987; Dorobek, 1989), widespread potassium alteration (Hearn and Sutter, 1985; Elliott and Aronson,

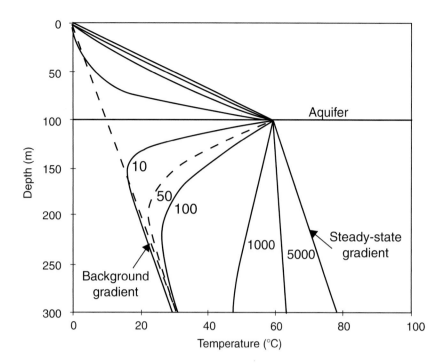

Fig. 16.9. Transition from the background to a steady-state temperature profile around a shallow geothermal aquifer (Ziagos and Blackwell, 1986). Numbers next to transient profiles indicate years after initiation of flow of a 60° C fluid in the aquifer. Note that the dashed background geothermal gradient and the steady-state geothermal gradient are equivalent.

1987; Hearn *et al.*, 1987; Hay *et al.*, 1988), zones of remagnetization (Jackson *et al.*, 1988; McCabe *et al.*, 1989; McCabe and Elmore, 1989; Marshak *et al.*, 1989; Lu *et al.*, 1990), hydrothermal alteration of coal (Daniels *et al.*, 1990) and high thermal maturity, which cannot be explained by burial beneath a section that was subsequently eroded (Johnsson, 1986). Paleomagnetic data (McCabe *et al.*, 1989; McCabe and Elmore, 1989), K–Ar dating (Elliot and Aronson, 1987; Hearn *et al.*, 1987; Hay *et al.*, 1988) and Rb–Sr dating (Hay *et al.*, 1988) indicate that the interpreted brine migration into the foreland was coeval with Late Pennsylvanian to Early Permian orogeny. Oliver (1986), Bethke and Marshak (1990) and Torgersen (1990) linked these flowing brines with fluid sources in the Appalachian and Ouachita orogens. Other examples of topography-driven fluid flow have been described from the Western Canada basin in front of the Rocky Mountains (Garven and Freeze, 1984; Garven, 1985), from the South Wales basin in front of the British Variscan orogen (Gayer *et al.*, 1998) and from the Anadarko basin in front of the Ouachita Mountains (Garven *et al.*, 1993).

The meteoric water enters such a topography-driven fluid system at recharge areas. The water volume is controlled only by the recharge rate. The water flows downward acquiring heat and dissolved salts. The downward flow eventually encounters a permeable aquifer. Within the aquifer the fluid moves laterally and up towards the discharge area where it flows out. While the known fluid discharges typically perturb the thermal regime

only locally, perturbations at recharge areas are more regional (Buntebarth, 1984; Garven, 1995). Typical distances between known recharge and discharge areas fall within the 10–100 km interval. The travel time for the fluid flow starting at the recharge area and finishing at the discharge area spans between 1000 and 100 000 years in active reservoirs (De Marsily, 1981; Smith and Chapman, 1983). The width of the thermal anomaly produced by the fluid flow within the carrier bed is controlled by the thermal diffusivity of the fluid–rock system κ and the circulation time t (Vasseur and Burrus, 1990):

$$x = (\kappa t)^{1/2}. \qquad (16.16)$$

A practically steady-state regime exists when this width reaches the depth of the carrier bed. It happens after a time (Vasseur and Burrus, 1990)

$$t = \frac{\Delta z}{\kappa}, \qquad (16.17)$$

where Δz is the depth of the carrier bed. If one assumes a depth of about 1 km and a thermal diffusivity of about $10^{-6}\,\mathrm{m^2\,s^{-1}}$, the system reaches a steady-state regime after 30 000 years. Such a time is considered instantaneous on a geological time scale. Therefore, numerical simulations of thermal perturbations caused by topography-driven fluid flow typically assume a steady-state regime (Ungerer *et al.*, 1990).

Returning now to interpreted topography-driven flows from the Appalachian and Ouachita orogens, their

(a)

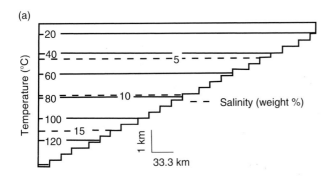

Fig. 16.10. (a) Initial condition temperature in °C and salinity in weight per cent for numerical simulation of topography-driven fluid flow (Deming and Nunn, 1991). Initial temperature is determined by a surface temperature of 10°C, surface heat flow of 60.5 W m⁻², thermal conductivity of 2.5 W m⁻¹ K⁻¹, and radioactive heat production that decreases exponentially with depth. (b) Specific discharge at an elapsed model time of 3 Myr (Deming and Nunn, 1991). Discharge vectors are shown with the same vertical exaggeration (33:1) as the physical dimensions and thus show true direction and magnitude of the fluid flow in the model. Imposed equivalent freshwater head gradient is shown schematically across the top left half.

(b)

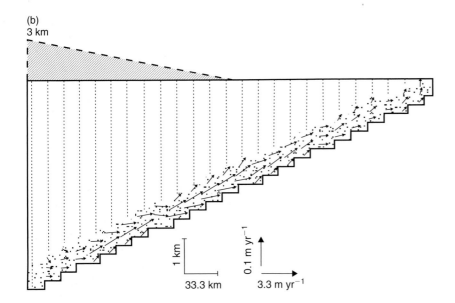

hot saline waters had to be transported for hundreds of kilometres towards the foreland. These hot brines had to produce temperatures of the order of 150 °C at depths around 1.5 km (Deming and Nunn, 1991). However, the pre-existing heat-flow studies in thrustbelts (e.g., Chapman *et al.*, 1984; Majorowicz, 1989) have usually interpreted only modest thermal regime perturbations characterized by a heat flow of 30–40 mW m⁻² in recharge areas and about 100 mW m⁻² in discharge areas.

The brine migration models of Deming and Nunn (1991) assumed a simplified foreland basin scenario. The basin fill included a thick aquitard with horizontal and vertical permeabilities of 10^{-15} and 10^{-16} m², respectively. The aquitard overlays a 500–700 m thick basal aquifer, characterized by horizontal and vertical permeabilities of 10^{-12} and 10^{-13} m², respectively. The aquitard and carrier beds were given distinct permeabilities, which approximated those of a karstic limestone for the aquifer and the most permeable shale for the

aquitard (see Smith and Wiltschko, 1996). The salinity distribution in the fluid system was chosen using the values reported by Hanor (1979) and Ranganathan and Hanor (1987; Fig. 16.10a). The simulated topography was actively undergoing an uplift of 1 mm yr⁻¹ in the proximal portion of the basin due to orogenesis in the hinterland. The growing positive topography was characterized by a thermal gradient of 12 °C km⁻¹. This probably represented an overestimate of the warming beneath the advancing thrustbelt, because the combination of meteoric water recharge, increased deposition and overthrusting was likely to depress the pre-existing thermal gradient.

Figure 16.10(b) shows the direction and magnitude of the specific discharge simulated for a time 3 Myr after the initiation of the topography-driven flow (Deming and Nunn, 1991). This flow pattern developed by a progressive increase of the fluid flow magnitude. As a result of the topography-driven flow, the thermal regime was changed rather dramatically by the heat advection from the stage before the flow initiation, as is apparent by comparing Figs. 16.10(a) and 16.11(a). The upper left-hand portion of the recharge area, where the topography grew to 3 km, underwent an increase in temperature from 10 to 46 °C. However, the left section deeper than 1.5 km experienced a pronounced cooling by the downward-migrating meteoric water. During the initial 3 Myr, the meteoric water became heated and enriched in salts, and flowed within the aquifer into the discharge area. The hot brine warmed the thermal regime in the discharge area. The maximum temperatures in this positive thermal anomaly occur in the portion of the aquifer located at a depth of about 1.5–2 km (Fig. 10.11a).

The higher the pre-circulation surface heat flow in the future discharge area, the more significant will be the positive thermal anomaly during the topography-driven flow. Figure 10.11(b) demonstrates that the anomaly can increase by more than 50% when the initial surface heat flow is 90 mW m^{-2} instead of 60 mW m^{-2} (Deming and Nunn, 1991). The speed of the fluid flow within the aquifer further enhances the steep thermal gradient in the discharge area. It also develops almost isothermal conditions along the aquifer (Deming and Nunn, 1991). Such isothermal conditions have been documented from the zone along the basal detachment of the Sicilian Neogene subduction complex (Larroque *et al.*, 1996).

Numerical simulations of the topography-driven fluid flow of Deming and Nunn (1991) further show that even in the first 1 Myr after the initiation of flow the surface heat flow decreased to 15 mW m^{-2} in the recharge area and increased to 97 mW m^{-2} in the discharge area. Similar increases and decreases of the heat-flow and temperature gradients attributed to topography-driven fluid flow have been reported from the western Canadian, Denver and Uinta basins (Majorowicz and Jessop, 1981; Chapman *et al.*, 1984; Majorowicz *et al.*, 1984, 1985; Gosnold, 1985, 1990). 3 Myr is the time period when the change between the pre-circulation and syn-circulation thermal regimes in thrustbelts becomes significant. The recharge area in the models of Deming and Nunn (1991), for example, underwent a surface heat flow decrease to values below 10 mW m^{-2} and the discharge area experienced an increase to values of around 162 mW m^{-2}.

The sensitivity analysis made by Nunn and Deming (1991) for the recharge area in their models indicates that the predicted temperature and heat flow of their thermal regime simulations are most likely unrealistically low, despite the deliberate overestimate of the thermal gradient for this area. Such an artificially low thermal regime in the recharge area appears to be an inherent feature of the infiltration process, because the meteoric water entering the fluid system is cold and must collect heat on its way down through the positive topographic feature. A further problem in the simulation occurs in the discharge area, where the existence of a positive thermal anomaly is required by studies of lead–zinc deposits, fluid inclusions and coal alteration in the Appalachian foreland (e.g., Hearn *et al.*, 1987; Dorobek, 1989; Daniels *et al.*, 1990) but, given the characteristics of the topography-driven fluid system described above, requires an apparently unrealistically high pre-circulation surface heat flow as high as 100 mW m^{-2} (Nunn and Deming, 1991).

The sensitivity analysis also allows one to map the extent to which the aquifer permeability and its salt content affect the heat transport and to determine the duration of the topography-driven circulation system capable of producing the thermal perturbation. If the foreland basin contained only the pre-circulation sources of salts available for solution, which were not enhanced during the circulation time period, the basin would become essentially flushed by the fresh water after 1 Myr. The fresh water entering the recharge area would flow laterally and up within the carrier bed and leave the circulation system in the discharge area. The only remaining area with higher salinity would be in the area above the aquifer and between the recharge and discharge areas, because here there would be no lateral flow and the vertical flow would be stagnant.

If the topography-driven fluid system were capable of tapping into an unspecified extra source of salt in the recharge area, the distribution of dissolved salts would have a complex syn-circulation history (Fig. 16.12). After 1 Myr of circulation, the discharge area would be characterized by uniformly high salinities, but the supply would start to run out in the upper portion of the recharge section, where the downward migration rates are highest (Fig. 16.12a). After another 0.5 Myr, the recharge area would be stripped of salts and the salinity within the aquifer further forelandward would range from 2% to 14% (Fig. 6.12b). After yet another 0.5 Myr, the sedimentary sequence would be incapable of feeding the fluid in the aquifer with a sufficient amount of salts (Fig. 6.12c). Eventually, after 3 Myr from the initiation of the circulation, fresh water would flow within the carrier bed (Fig. 6.12d).

According to Nunn and Deming (1991), the above described salt migration analysis indicates that the simulated Appalachian foreland basin was most likely not able to supply enough solute to maintain salt concentrations in the topography-driven circulation in the assumed geological situation for longer than 1 Myr. If the fluid flow kept building with the growing topography during the development of the topography-driven

(a)

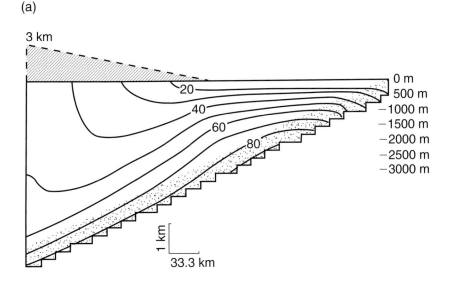

(b)

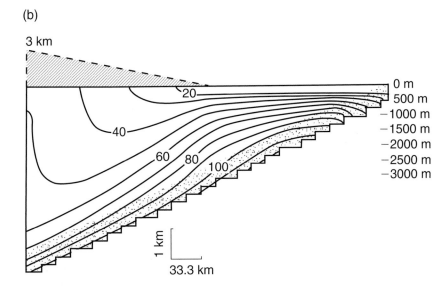

Fig. 16.11. Temperature distribution in °C at an elapsed model time of 3 Myr for an initial surface heat flow of (a) 60.5 mW m^{-2} and (b) 90.5 mW m^{-2} (Deming and Nunn, 1991). On the right half of the model, the surface temperature is 10 °C; on the left half, the surface temperature is 10 °C plus 12 (°C km^{-1}) times the height (km) of simulated topography. Tick marks on right are spaced at 500 m intervals.

circulation, a large portion of salts would have already been removed before the fluid flow could reach a sufficiently high rate to achieve significant heat transport.

The role of permeability on a thermal regime is in controlling the fluid flow rate, as expressed by various forms of Darcy's law described earlier and in Chapter 10, because a flow rate of the order of 10 m yr^{-1} is required to transport sufficient heat to the discharge area (Bethke and Marshak, 1990). Therefore, the described sensitivity analysis of the salt migration and aquifer permeability indicates a problem for models simulating the buildup of the temperature anomaly required for the Mississippi Valley-type lead–zinc deposits (Deming and Nunn, 1991). On the one hand, a significant fluid flow rate would be required to develop

a sufficiently high thermal anomaly to meet the thermal conditions required for the lead-zinc deposits, but on the other hand, the higher the flow velocity, the more quickly the supply of salts in the system is spent. An indication to the solution of this problem comes from models in which the aquifer permeability is lowered by an order of magnitude in both vertical and horizontal directions (Fig. 16.13). In this case, 3 Myr after the initiation of the circulation, the thermal regime at depths greater than 2.5 km is warmer than in the case with a more permeable aquifer (compare Figs. 16.13 and 16.11a). Shallower depth intervals are apparently colder due to the lower permeability aquifer (Fig. 16.13), which prevents advective heat transport from affecting this region as much as in the case of the higher permeability

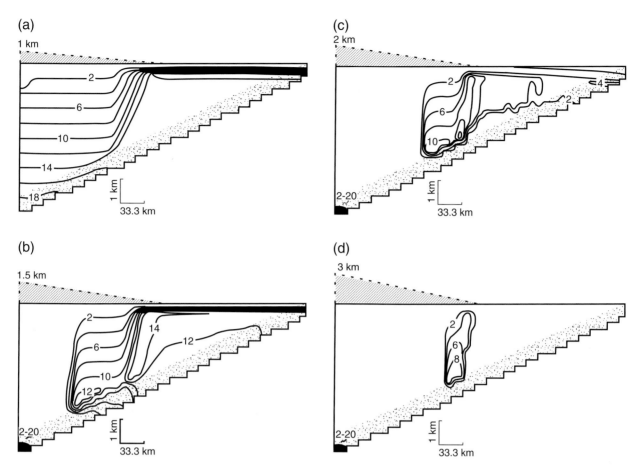

Fig. 16.12. Salinity in weight per cent at elapsed model times of (a) 1, (b) 1.5, (c) 2 and (d) 3 Ma (Deming and Nunn, 1991). Solute source term varies through time to maintain initial condition salinity (see Fig. 16.10a) until an amount of mass equal to 5% of rock matrix per nodal block has been added.

aquifer (Fig. 6.11a). A similar effect also occurs in the forelandward margin of the basin, which does not receive sufficient advective heat.

A satisfactory explanation of the observed Appalachian lead-zinc deposits would combine a reduction in the aquifer permeability, as outlined above, with focused flow of the heated brine to the surface at a rather fast rate (Deming and Nunn, 1991). The simplest case would be the presence of faults in the discharge area, which could act as fluid conduits, as discussed in detail in Chapter 10.

A detailed example of how thermal regime perturbation by topography-driven fluid flow can affect hydrocarbon maturation comes from the Uinta basin within the Colorado Plateau (Willett and Chapman, 1989; Fig. 16.14). Water budget calculations in this region show that only 10% of the total precipitation is available as recharge to any groundwater system (Hood and Fields, 1978). Of this, approximately 40% is discharged by local scale flow systems. The thermal field perturbed by the

associated advective heat transport results in lateral temperature differences of up to 35 °C at the base of the Tertiary section (Fig. 16.15). Vitrinite reflectance data indicate that these temperature variations have existed through the history of the basin (Fig. 16.16). They resulted in the decreased maturity of the organic matter penetrated in wells in the broader recharge area and the increased maturity of the organic matter penetrated in wells in the broader discharge area.

Another example of thermal regime perturbation by topography-driven flow has been described in the South Wales basin in front of the British Variscan orogen (Gayer *et al.*, 1998). Vitrinite reflectance data of coals from the closest vicinity of the recharge area have values as low as 1.1%, while values from the discharge area are as high as 2.5%. The modelled thermal regime, which matches the pattern of coal rank in the basin, indicates that the activity of topography-driven fluid flow lasted for about 1–2 Myr, before erosion destroyed the topographic gradient that drove the fluid flow.

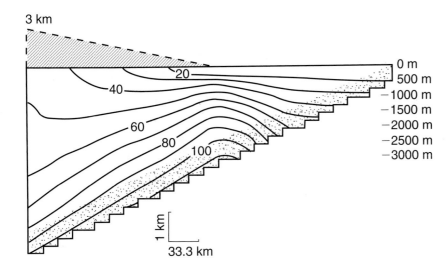

Fig. 16.13. Temperature distribution in °C at elapsed model time of 3 Myr (Deming and Nunn, 1991). Horizontal and vertical permeabilities of basal aquifer have been reduced by a factor of 10 from Fig. 16.11(a), to 10^{-13} m^2 and 10^{-14} m^2, respectively. Initial condition surface heat flow is 60.5 m Wm^{-2}. Tick marks on right are spaced at 500 m intervals.

Role of compaction-driven fluid flow in thermal regimes

Speculations about the effect of compaction-driven fluid flow on the thermal regimes of foreland basins also relied on the same kind of data as for topography-driven flow, such as remagnetization, fluid inclusions, potassium alteration and lead-zinc deposits (e.g., Oliver, 1986). Some studies (e.g., Sharp and Domenico, 1976; Sharp, 1978; Cathles and Smith, 1983; Bethke, 1985) suggested that depositional rates in sedimentary basins are probably not fast enough to produce significant perturbations of a thermal regime unless the conditions are such that geopressured fluids are released. A typical example of a mechanism for the release of geopressured fluids in a thrustbelt scenario would be the advance of the thrustbelt over the proximal portion of the foreland basin. It would significantly enhance both the burial rate and the associated pore collapse leading to fluid expulsion (Deming *et al.*, 1990; Garven *et al.*, 1993; Garven, 1995).

In order to test whether compaction-driven flow is a mechanism capable of perturbing the temperature field, it is necessary to test whether:

(1) it produces sufficient fluid quantity;
(2) it generates a fast enough flow; i.e. about 10 m yr^{-1} (Bethke and Marshak, 1990), to produce relatively high temperatures; i.e. 100–150 °C at shallow depths; i.e. less than 1.5 km, indicated by Mississippi Valley-type lead-zinc deposits;
(3) it produces a sufficiently large vertical component of flow to cause warm enough brines.

Deming *et al.* (1990) tested these requirements using numerical simulations. Their finite-difference models simulated an advancing thrust wedge by nodal compaction at one side of the model and by changing the thermal boundary at the surface (Fig. 16.17). Sediment compaction and fluid expulsion were assumed to be coeval with tectonic loading.

These models indicate that a characteristic distribution of the hydraulic potential (Fig. 16.18a) develops relatively soon after the onset of thrusting and associated compaction of the foreland basin fill. The largest hydraulic potential develops underneath the frontal thrust. The related fluid velocity indicates that impermeable horizons located on the left-hand side of the model and at its bottom, together with a relatively impermeable basal detachment zone, drive the majority of the fluid toward the foreland (Fig. 16.18b). There is also some fluid flow into the thrustbelt itself shown in Fig. 16.18(b), since the thrustbelt is not sealed off completely.

Because sediment porosity decreases with compaction and burial, the amount of pore fluid available for expulsion decreases with depth. The average linear fluid velocity increases with decreasing fluid viscosity (Smith and Chapman, 1983). The maximum Darcy fluid flow velocity calculated in the model of Deming *et al.* (1990) was of the order of 1 cm yr^{-1}, in accordance with a similar model of Ge and Garven (1989), suggesting a reasonable maximum fluid flow value for compaction-driven flow in a typical foreland basin. The fluid flow pattern indicated by the geometry and values from Fig. 16.18(b) does not seem to be sufficiently strong to produce a significant perturbation of the thermal regime. It results in rather a modest heating of the forelandward side of the basin (Figs. 16.18c and d), characterized by a rise of a few degrees Celsius. Figure 16.18(e) illustrates the impact of advective heat transfer alone on a thermal perturbation. A comparison of Fig. 16.18(e)

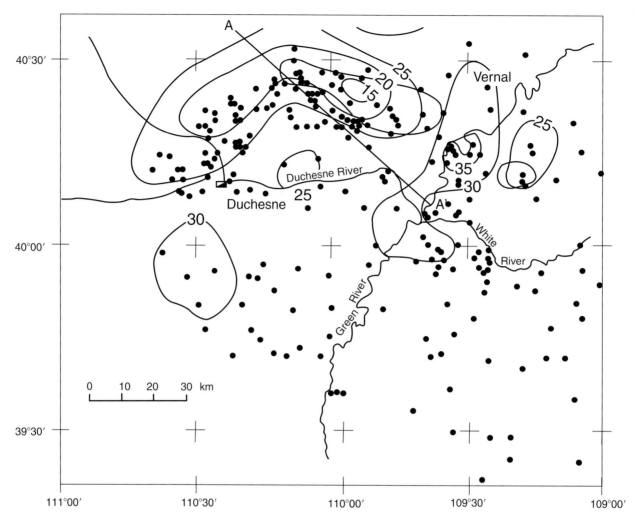

Fig. 16.14. Location map for thermal analysis of the Uinta basin, northern Colorado Plateau, Utah, USA (Willet and Chapman, 1989). Boreholes with bottom hole temperatures are shown by solid dots. Contours in mK m^{-1} indicate a map of thermal gradients within the Uinta Formation deduced from inversion of the bottom hole temperature data. The profile A–A' shows the location of the modelled section.

with Fig. 16.18(d), shows more clearly the impact of advective heat transfer. It is much more effective in transporting heat to the foreland than conductive heat transfer, even when the fluid flow velocity is relatively low. Figure 16.18(e) further illustrates the impact of downward-flowing and upward-flowing cold water beneath the thrust front on the residual temperature under the thrust front, causing negative values that indicate cooling.

Understanding the perturbation of the thermal regime in the fluid discharge area requires an overview of the impact of typical basin fill parameter values. The first to be considered is permeability. Unlike its effect on topography-driven fluid flow (e.g., Smith and Chapman, 1983; Garven and Freeze, 1984a, b; Garven, 1985, 1989;

Bethke, 1986; Deming and Nunn, 1991), it does not seem to have as great an affect on compaction-driven fluid flow, as indicated by the sensitivity analysis of Deming et al. (1990; Figs. 16.19a and b). Two orders of magnitude difference in the permeability of a foreland basin only causes a difference in the thermal regime perturbation of a few degrees Celsius in the discharge area. As documented by Figs. 16.19(c, d), neither does an anisotropy in the permeability appear to have a large impact, although the selected end members simulate contrasting lithologies such as a high percentage of shale in the basin fill, characterized by an average horizontal permeability of 10^{-16} m^2, and a basin fill with a high percentage of permeable sandstone and limestone, characterized by an average horizontal permeability of 10^{-14} m^2.

Another important parameter is the thermal regime in the thrustbelt front. Even if the regime is warmer than in the cases so far discussed, simulated by lowering the thermal gradient from 20 to 10 °C km⁻¹, the discharge area becomes only a few degrees warmer (Deming *et al.*, 1990). This observation of Deming *et al.* (1990) looks counterintuitive and it is probably related to the effect of the thermal regime on water viscosity, which in turn affects the fluid flow pattern in the area most affected by compaction. A cooler thermal regime under the thrustbelt front would cause a higher fluid viscosity, controlling the prominent flush of fluids toward the foreland, while a warmer regime would increase the importance of backflow beneath the thrust wedge.

Differences in the thrusting rate should affect the thermal equilibration. However, these too do not seem to trigger significant differences in perturbed thermal regimes in discharge areas either (Deming *et al.*, 1990). Differences in the fluid budget available for expulsion into the foreland basin would be expected to make a difference, but several tens of per cent difference in porosity only results in a difference of less than 10 °C in the discharge thermal anomaly (Deming *et al.*, 1990).

Because all the above basin characteristics tested by the models of Deming *et al.* (1990) failed by one to two orders of magnitude to produce a thermal anomaly required by the data on remagnetization, fluid inclusions, potassium alteration and lead-zinc deposits located in the Appalachian and Ouachita forelands (e.g., Leach and Rowan, 1986; Sverjensky, 1986; Hearn *et al.*, 1987; Jackson *et al.*, 1988; Dorobek, 1989; Daniels *et al.*, 1990), the fluid flow, as with the case of topography-driven flow, requires some form of enhancement. Possible enhancements of the compaction-driven fluid flow of Oliver (1986) would be either channeling of the fluid flow along aquifers and faults or periodic release of fluids from geopressured zones (Deming *et al.*, 1990; Ungerer *et al.*, 1990). Compaction of sediments controlled by simple vertical loading beneath the thrustbelt front, which drives a continuous fluid expulsion, does not produce a sufficiently fast fluid flow to become a significant heat carrier. A critical reason for the slight warming in the discharge area is the limited pore fluid budget available for expulsion, which implies too small an amount of transported heat to make a significant perturbation of the thermal regime (Nunn and Deming, 1991), since pore fluids make up only a small portion of the volume necessary for the anomaly. Moreover, the pore fluid budget decreases with depth. Since the temperature increases with depth, one is faced with the paradox of having progressively less water tapping into progressively higher heat.

The above-mentioned fluid channeling along aquifers

(a)

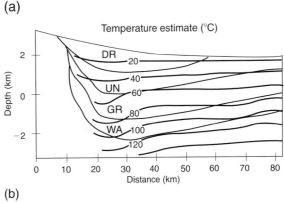

(b)

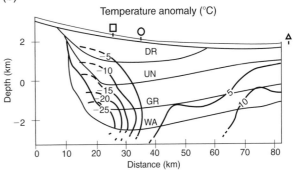

(c)

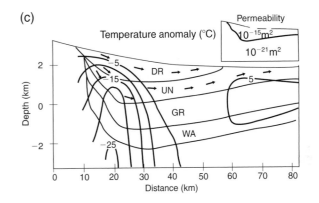

Fig. 16.15. (a) Temperature field for section A–A' from Fig. 16.14 constructed from local formation gradients (Willet and Chapman, 1989). Isotherms are contoured at 20 °C interval. Geometry of Tertiary units of the Uinta basin is shown by thin lines. Symbols refer to formations: Duchesne River (DR), Uinta (UN), Green River, (GR), Wasatch (WA). (b) Uniform gradient temperature anomaly for section A–A'. This anomaly is the difference between the temperature field in (a) and the temperature field that would results if the thermal gradient were everywhere 25 °C mK m⁻¹. Anomaly contours are in °C. Symbols above section represent location of wells where vitrinite reflectance results are available. (c) Uniform gradient temperature anomaly along A–A' for a model simulating both conductive and advective heat transfer. Thermal conductivity varies within and between formations and was constrained by measurements. Permeability is shown in inset; flow in this numeric simulation is restricted to the Duchesne River and Uinta Formations.

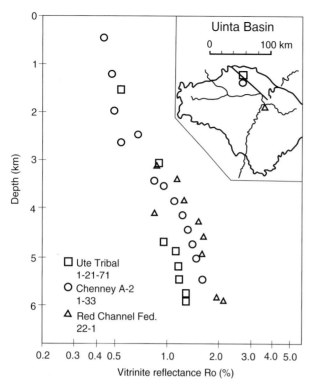

Fig. 16.16. Vitrinite reflectance values versus depth for three wells located in Fig. 16.15b (Willet and Chapman, 1989).

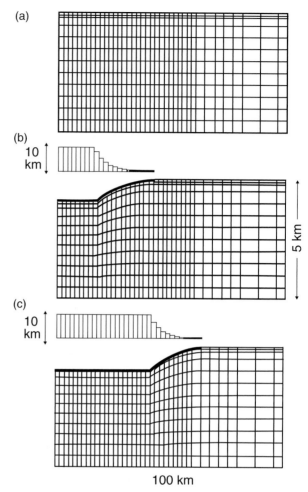

Fig. 16.17. Upper left 5 × 100 km part of the finite-difference model of Deming *et al.* (1990) at: (a) an initial condition with no compaction driven by an advancing thrust wedge; (b) a state of compaction existing after 500 000 yr of thrust movement at a rate of 5 cm yr^{-1}, and (c) a state of compaction existing after 1 000 000 yr of continuous thrust movement at a rate of 5 cm yr^{-1} (Deming *et al.*, 1990). Vertical exaggeration is 10 to 1. Loading of the imaginary thrust wedge is also shown but using a different vertical scale. Maximum thrust wedge height is 10 km. Permeability k_x of top row of nodal blocks, which is darkened, is set to 10^{-19} m^2 ($k_z = k_x/10$) in most model simulations to seal the system underneath the thrust wedge.

can be illustrated by a model of Deming *et al.* (1990), in which the horizontal permeability was 10^{-15} m^2 and the vertical permeability was an order of magnitude smaller, and which contained a basal aquifer with horizontal and vertical permeabilities one order of magnitude higher than the rest of the system. The thrustbelt front, lying above the modelled system, was sealed off from the rest of the fluid system. The calculated fluid flow pattern (Fig. 16.20a) did not differ greatly from the homogeneous permeability model (Fig. 16.18b). The only significant difference is that the fluid flow rate is faster within the aquifer. Since this rate is still only 3.8 cm yr^{-1}, it fails to generate a noticeable thermal anomaly in the discharge area (Fig. 16.20b). Neither does moving the position of the aquifer into intermediate depth levels apparently improve its impact on the thermal regime in the discharge area (Deming *et al.*, 1990).

The previously mentioned channeling of fluid along faults has also been simulated by Deming *et al.* (1990). In the modelled scenario, a vertical permeable fault tapped into the basal aquifer of the above model. The fault permeability of 10^{-13} m^2 caused the fluid flow rate to increase to 5.3 cm yr^{-1}, which resulted in a temperature increase of 37 °C at the centre of the thermal

anomaly in the discharge area. This thermal perturbation could be further considerably enhanced in nature by the presence of a more favourable system of characteristic data involving higher permeabilities, anisotropy, initial surface heat flow, thermal gradient, thrusting rate and sediment porosity (Deming *et al.*, 1990). The discussed simulations imply that effective advective heating of the discharge area in a compaction-driven fluid system requires the existence of permeable faults in

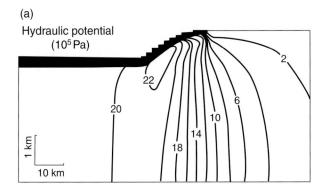

(a)
Hydraulic potential
(10^5 Pa)

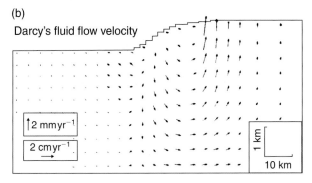

(b)
Darcy's fluid flow velocity

$\uparrow$ 2 mm yr^{-1}

2 cm yr^{-1} $\rightarrow$

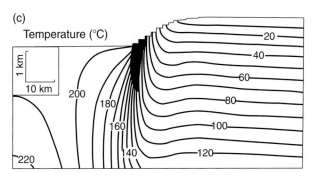

(c)
Temperature (°C)

Fig. 16.18. (a) Hydraulic potential contours after 1 000 000 yr of simulated thrust movement, sediment compaction and fluid flow (Deming *et al.*, 1990). Permeability is homogeneous ($k_x = 10^{-15}$ m^2) and anisotropic ($k_z = 10^{-19}$ m^2), with the exception of the top row of nodal blocks directly underneath the thrust ($k_x = 10^{-19}$ m^2, $k_z = k_x/10$). (b) Fluid velocity calculated using Darcy's law after 1 000 000 yr of simulated thrust movement and sediment compaction in the model introduced in (a). Velocity arrows have the same vertical exaggeration (10:1) as spatial dimensions. Velocity arrows show the Darcy's fluid flow velocity, or specific discharge, which is volume of fluid per unit area per unit time. (c) Temperature contours in °C after 1 000 000 yr of simulated thrust movement, sediment compaction and fluid flow in the model introduced in (a). The model indicates warming of the foreland. (d) Temperature residual. Contours in °C show the temperature field existing after 1 000 000 yr of simulated thrust movement, sediment compaction, and fluid flow in the model introduced in (a), minus the initial thermal condition. Warming of the foreland is less than 5 °C. Large residuals in the compacted region underneath the thrust wedge result from a change in the boundary condition at the surface.
(e) Temperature residual. Contours in °C show the temperature field from (c), minus the temperature field that results from a similar model simulation in which heat transfer is by conduction only (no fluid flow). Positive contours in front of the thrust indicate warming of the foreland due to water being flushed up and out from depth. Negative contours underneath the thrust reflect a cooling caused by fluid being driven down, and back underneath the thrust wedge.

(d)
Temperature residual (°C)

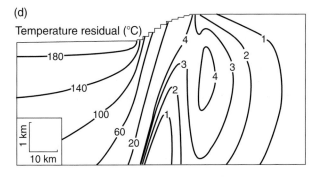

(e)
Temperature residual (°C)

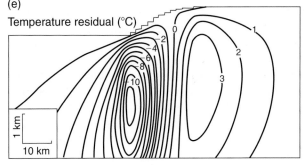

front of the thrustbelt, such as inverted normal faults, flexure-driven normal faults or strike-slip faults. Our own observations from faulted geothermal reservoirs, although in different tectonic settings, for example in west Java, indicate that positive thermal anomalies

along the fault core, damage zone and close surroundings can be significant but volume-restricted. Similarly, Bodner and Sharp (1988) found anomalously warm areas with geothermal gradients of 52–56 °C associated with the Wilcox normal fault in south Texas. In this

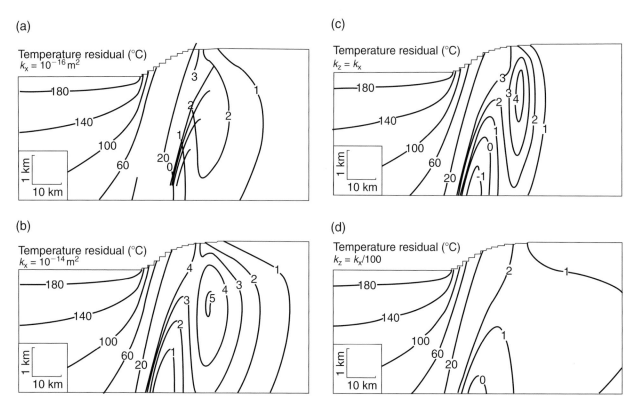

Fig. 16.19. Temperature residuals in °C, resulting from a sensitivity analysis in which parameters from the homogeneous permeability model are varied (Deming *et al.*, 1990). Contours show the temperature field existing after 1 000 000 years of simulated thrusting, sediment compaction and fluid flow, minus the initial thermal condition. Compare to Fig. 16.18(d), in which permeability is homogeneous ($k_x = 10^{-15}\,\mathrm{m}^2$) and anisotropic ($k_z = 10^{-19}\,\mathrm{m}^2$). Here, in (a) the permeability k_x is decreased to $10^{-16}\,\mathrm{m}^2$, in (b) the permeability k_x is increased to $10^{-14}\,\mathrm{m}^2$, in (c) the anisotropy k_x/k_z is decreased from 10 to 1, and in (d) the anisotropy k_x/k_z is increased from 10 to 100.

case, the thermal anomaly was consistent with vertical fluid migration from a geopressured zone. Other examples come from active accretionary prisms, where recent heat-flow surveys found isolated regions of anomalously high heat flow, which are interpreted as resulting from episodic dewatering of sediments channelled upwards from the décollement zone along reverse faults or channelled along the décollement itself (Kinoshita and Yamano, 1986; Davis *et al.*, 1990; Fisher and Hounslow, 1990; Foucher *et al.*, 1990a, b; Henry *et al.*, 1990; Langseth *et al.*, 1990; Yamano *et al.*, 1990, 1992). These examples also document the effect of a third potential fluid flow enhancement mechanism, periodic fluid release. The mechanism for such a transient fluid flow in modern accretionary prisms has been discussed in detail in Chapter 10.

The effect of periodic fluid release on the thermal structure of a thrustbelt and its foreland has not been simulated by the models of Deming *et al.* (1990), but can be discussed qualitatively. As documented by data from accretionary prisms, lower-permeability sediments,

such as shale/silt or well-cemented systems, tend to capture pore fluids in geopressured zones for periods of time. During the periods of entrapment, these fluids become heated. As described in Chapter 10, these fluids periodically migrate to the discharge areas in front of thrust wedges in discrete volumes. The described model simulations, in which fluids flow laterally within a basal aquifer and then upward along a fault, imply that associated advection produces thermal anomalies approaching 100 °C in volume-restricted zones in discharge areas (Deming *et al.*, 1990). Periodically released fluids should increase these thermal anomalies even further, perhaps to 100–150 °C at depths below 1.5 km (Nunn and Deming, 1991).

In thrust wedges that contain an argillite layer, the dewatering is mainly controlled by tectonic conduits, such as faults (e.g. Moore *et al.*, 1990a, b, c), where intense particle realignment (e.g., Arch and Maltman, 1990) or dilative fractures (Brown *et al.*, 1994) are assumed. Numerous lines of evidence also point to the existence of transient fluid flow along basal detach-

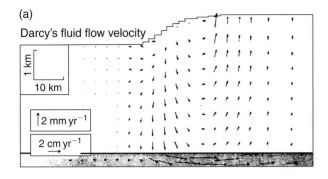

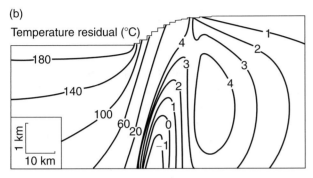

Fig. 16.20. (a) Fluid velocity calculated from Darcy's law for a model simulation with a basal aquifer ($k_x = 10^{-14} \text{m}^2$; $k_z = k_x/10$) after 1 000 000 yr of thrusting and sediment compaction (Deming *et al.*, 1990). Permeability is otherwise homogeneous ($k_x = 10^{-15} \text{m}^2$) and anisotropic ($k_z = k_x/10$), with the exception of the top row of nodal blocks directly underneath the thrust ($k_x = 10^{-19} \text{m}^2$; $k_z = k_x/10$). Velocity arrows have the same vertical exaggeration (10:1) as spatial dimensions. (b) Temperature residual in °C for the model shown in (a).

ments and thrust fault zones ramping from detachments (e.g., Fischer and Hounslow, 1990; Henry and Wang, 1990; Foucher *et al.*, 1992; Knipe, 1993; Le Pichon *et al.*, 1993; Bekins *et al.*, 1994; Brown *et al.*, 1994; Fisher *et al.*, 1996). A detailed fluid inclusion study of such a fluid flow comes from the Pliocene–Pleistocene Sicilian accretionary prism (Larroque *et al.*, 1996). It documents a complex thrust scenario with two detachments on top of each other. The frontal portion of the shallower detachment was used as a flow path by low-temperature fluids. Their temperature was about 60 °C. The frontal portion of the deeper detachment was used for a flow of high-temperature fluids. Their temperature of about 235 °C documents that they must have come from a depth of 6–10 km (Larroque *et al.*, 1996). The regional thermal gradient is two to three times lower than the thermal gradient estimated from fluid inclusion thermometry and the maturation level of the organic matter affected by the heat advection (Guilhaumou *et al.*, 1994; Larroque *et al.*, 1996).

Evidence for fluid flow along detachments and branching faults, which perturbs the regional geothermal regime, comes from direct heat-flow measurements in modern accretionary prisms (Davis *et al.*, 1990; Fisher and Hounslow, 1990; Foucher *et al.*, 1990b, 1992; Langseth *et al.*, 1990; Ferguson *et al.*, 1993; Zwart and Moore, 1993). For example, Ferguson *et al.* (1993) reported heat-flow anomalies in frontal parts of the 15° 30′ N transect across the Barbados accretionary prism. Anomalously high, heat-flow measurements were observed over the interval of 55–218 mW m^{-2}. Geochemical analyses of the pore fluid from cores from ODP Leg 110 provide evidence for low chloride anomalies along faults and permeable sand layers below the décollement, indicating fluid movement (Gieskes *et al.*, 1988; Vrolijk *et al.*, 1991). Observed positive thermal anomalies associated with faults also suggest fluid activity (Fisher and Hounslow, 1990). Not all faults, however, show thermal or geochemical anomalies. Some show one but not the other, suggesting that the fluid flow is episodic and that anomalies decay and are rejuvenated through time (Moore *et al.*, 1988).

Role of combined fluid flow mechanisms in thermal regimes

While the older studies of modern accretionary prisms document the transient fluid flow along faults and permeable horizons within accretionary prisms, which have an impact on thermal anomalies in frontal parts of accretionary wedges, more recent work indicates various fluid sources. For example, Martin *et al.* (1996) indicate that the fluid in their study area was mainly sourced by clay dewatering and clay transformation in combination with gas hydrate dissociation. With the exception of fluids derived from exotic sources or the mentioned thermally activated dehydration processes, fluid expulsion in modern accretionary prisms is largely a consequence of compaction-driven porosity reduction (e.g., Breen and Orange, 1992). These observations raise an important point for advective heat transport in all thrustbelt categories. It becomes apparent that no single fluid flow mechanism usually operates without at least a modest contribution from several others. Therefore, each such contribution further enhances the fluid flow rate affecting the thermal perturbation and the natural examples described later contain the most likely combinations of several fluid flow mechanisms.

The presence of additional fluid flow mechanisms can be indicated by fluid tracers. $\delta^{18}O$ values of water together with Cl^- concentrations provide an excellent tracer for the origin of aqueous fluids, because of the variety of reactions that affect the isotopic and chemical composition of water (Martin *et al.*, 1996). During reactions such as that involved with carbonate diagenesis and the conversion of ash to clay minerals, the $\delta^{18}O$ value of seawater in sediment pores is lowered, but these reactions typically do not alter the Cl^- concentrations. Meteoric water, isotopically lighter than seawater (Craig, 1961; Dansgaard, 1964), when mixed with seawater or marine pore fluids, lowers both their Cl^- concentrations and $\delta^{18}O$ values. In contrast, gas hydrate is enriched in $\delta^{18}O$ because its fractionation factor is similar to that of sea ice or ice made experimentally by freezing water (Craig and Hom, 1968; O'Neil, 1968; Suzuki and Kimura, 1973). Cl^- ions are excluded from its structure simultaneously with this fractionation. Therefore, gas hydrate dissociation produces a low Cl^- concentration fluid that is isotopically heavier than seawater. Because the structural water of clay minerals is typically enriched in $\delta^{18}O$ (Savin and Epstein, 1970), its dehydration and transformation should both lower the Cl^- concentrations and elevate the $\delta^{18}O$ values of pore fluids.

Reports of proven gas hydrate dissociation and of the addition of the products of clay dewatering to the fluid flow system come from the Barbados accretionary prism. Here, fluid flow and mud flow rates measured in mud volcanoes, located either at the wedge front or in front of it, are as high as 1.5 and 60 m yr^{-1}, respectively (Henry *et al.*, 1996), meeting the criteria of successful heat advection of Bethke and Marshak (1990). Additional fluid can be shown to vent from a large area of sediments surrounding the mud volcanoes, but at rates about three orders of magnitude lower than the maximum rates in the mud volcanoes. The oxygen isotopic composition of the fluids together with the Cl^- concentrations suggest that fluids found in the mud volcanoes may be derived from gas hydrate dissociation, or from clay minerals or a combination of these two processes (Vrolijk *et al.*, 1990, 1991; Martin *et al.*, 1996).

Data from the Barbados accretionary prism (Foucher *et al.*, 1990b) show that transient warm fluid flow along the detachment can reach values of 10 m yr^{-1} within a 40 m thick décollement zone (Moore *et al.*, 1987). Such a flow rate meets the criteria of Bethke and Marshak

(1990) to supply sufficient heat to cause a significant thermal regime perturbation, as discussed earlier. Similar values of fluid flow rates of tens of metres per year have been measured directly in the Oregon accretionary prism (Suess *et al.*, 1987) and indirectly, based on temperature gradient studies, in the Nankai accretionary prism (Henry *et al.*, 1990). These flow rates are five to six times larger than the expected steady-state rate caused by local dewatering (Foucher *et al.*, 1990a), indicating an enhancement by transient fluid flow from several different fluid sources. Our own data from the Carpathian thrustbelt show that décollements, or even branching thrust ramps, can form zones several hundred metres thick. This, together with fluid flow and heat-flow data from modern accretionary wedges, would mean that compaction-driven fluid flow, enhanced by gas hydrate dissociation, clay dehydration and clay transformation, could be capable of significant thermal regime perturbations in the frontal parts of accretionary wedges. In the case of sub-aerially exposed thrustbelts, topography-driven fluid flow adds to the fluid flow system.

Further complexities can be involved with the fluid flow system in accretionary wedges if internal fracturing of the sediments occurs. Wang *et al.* (1990) showed that pore pressure and fluid expulsion evolve cyclically with each repeated thrusting event. Once a frontal thrust is initiated, the advancing thrust sheet moves onto porous sediments located seaward of the deformation front. These sediments are pressurized under the increasing overburden and pore fluid is expelled. Unlike the numerical simulations described earlier, a more realistic determination of the flow of the expelled fluid is achieved when the permeability change is also affected by hydrofracturing related to stress perturbations and fluid overpressure (Wang *et al.*, 1993). Permeability enhancement by hydrofracturing can increase the flow rate of expelled fluid by about an order of magnitude. This mechanism results in a transient fluid flow at high rates, exceeding the criteria of Bethke and Marshak (1990) for a sufficient heat supplier.

PART FOUR
Petroleum Systems

17 Hydrocarbons in thrustbelts: global view

Oil and gas deposits are found in a wide range of sedimentary basin settings. These include rifts, such as the Gulf of Suez, rift–sag basin couplets, as in West Siberia or the North Sea, passive margins, such as in the Gulf of Mexico or West Africa, passive margin–foreland basin couplets, as in East Venezuela or Alberta, pull-apart basins, such as in Sumatra and forearc basins, like Sacramento. Portions of these basins at some time in their histories can become subjected to contractional or transpressional structuring creating a thrustbelt in the context of this book. The structuring may be intimately associated with the continuing basin development, or it may be superposed at some later time, being genetically unrelated. Rift basins can be inverted when subjected to far-field intraplate stress (Lowell, 1995; Macgregor, 1995), deforming not just the rift sediment fill, but also any overlying sag and/or foreland basin strata (De Graciansky *et al.*, 1989). Advancing thrust sheets can entrain proximal portions of the passive margin–foreland basin sediments (Suppe, 1987) as a requirement for maintaining a stable tapered wedge (Davis *et al.*, 1983). In turn, the advance of the thrustbelt, together with crustal loading by erosional debris from the rising thrust sheets, helps to maintain accommodation space in the foreland basin and thus also helps its continued development (Johnson and Beaumont, 1995). Small changes in relative plate motion along active margins may shift transtensional pull-apart basins into a transpressional mode resulting in partial basin inversion (Dooley and McClay, 1997).

Hydrocarbons in thrustbelts are tied directly to the history and characteristics of the sedimentary basins in which they are located, thereby forming a continuum in time and space with that basin. The generation, accumulation and preservation of hydrocarbons in the thrustbelt can be fully understood only in the broader context of the sedimentary basin evolution prior to, during and following structuring.

Global distribution of hydrocarbons in thrustbelts

Virtually all thrustbelts in unmetamorphosed sediments contain some organic matter and have the potential for generating liquid hydrocarbons. However, only relatively few host significant commercial accumulations, and fewer still contain an appreciable portion of the world's petroleum resources. As in other petroleum habitats, oil and gas pools are not uniformly distributed spatially within thrustbelts and with respect to their age. With few exceptions, petroleum is found in thrustbelts of Late Mesozoic and Cenozoic age. This is more a factor of preservation than the volume of original petroleum accumulations (Macgregor, 1996). Also with few exceptions, hydrocarbon pools are restricted to the frontal thrusts of thin-skin thrustbelts or specific linear zones of reactivated basement faults in inverted basins. Along the length of any thrustbelt, it is common to find clustering of fields that relates to actual physical conditions leading to the charging, entrapment and preservation of pools, not merely due to contrasting degrees of exploration and development.

Hydrocarbon pools in basins are transient features. This is because the long-term preservation of hydrocarbons in thrustbelt traps appears to be difficult. There are numerous processes that lead to post-entrapment destruction of oil accumulations. They include erosion, fault leakage, gas flushing and biodegradation. Gas and oil will seep from all traps, regardless of seal quality, at some finite rate (Miller, 1992). Of the 350 giant oil fields identified by Macgregor (1996), 64 fields are located in thrustbelt and/or inverted foreland basin settings. 50% of them contain evidence of post-entrapment destruction (Table 1.9). 18% of all giant oil fields had been charged in the Neogene, and nearly all had been charged since the Mid-Cretaceous.

Klett *et al.* (1997) have compiled a ranking of the world's petroleum provinces by known petroleum volume. Within the group of 276 provinces having reserves exceeding 0.1 billion barrels of oil equivalent (BBOE), which collectively contain nearly all of the world's reserves, just 57 are identified as thrustbelt provinces. These are deformed basins or sub-basins in which hydrocarbons occur principally in contractional or transpressional structural traps. These provinces are listed in Table 17.1 and displayed in Fig. 17.1. The total volume of hydrocarbons in all 57 thrustbelt provinces is 364.5 BBOE, which is 12.8% of the total for all 278 provinces. The reader is directed to Fig. 1.1–1.6 for the location of the thrustbelts cited.

The ranking was prepared in 1997 from volumetric data compiled in the mid-1990s. Since that time there

Table 17.1. US Geological Survey ranking of the world's thrustbelt provinces by known hydrocarbon volumes.[a, b]

Province name	Total (BBOE)	Rank	Oil (BB)	Gas (Tcf)	NGL (BB)
Zagros fold belt	189.5	4	121.6	399.4	1.4
Timan–Pechora basin	20.0	27	13.2	36.6	0.7
San Joaquin basin	16.6	30	13.8	12.5	0.7
Middle Caspian basin	14.4	35	9.6	28.7	0.1
Los Angeles basin	10.1	49	8.6	7.0	0.4
East Venezuela thrustbelts	8.3	50	7.1	7.3	0.0
Canadian Foothills belt	7.7	52	1.0	40.0	0.0
Llanos basin	7.3	53	5.4	10.3	0.2
Oriente basin	6.9	56	6.6	1.6	0.0
Neuquen basin	6.2	61	2.4	20.7	0.3
Transylvania	5.1	68	0.1	30.7	0.0
Ventura basin	4.6	72	3.4	5.8	0.2
Santa Cruz–Tarija basin	4.1	75	0.3	18.8	0.6
North Carpathian basin	3.7	82	0.9	16.3	0.1
Assam	3.6	83	2.5	6.5	0.1
Po basin	3.6	84	0.4	18.9	0.1
Indus	3.5	87	0.2	19.6	0.1
Middle Magdalena basin	3.3	89	2.4	5.0	0.1
Big Horn basin	3.0	92	2.7	1.8	0.1
Rocky Mountain deformed belt	2.9	94	0.2	13.8	0.4
North Ustyurt basin	2.8	95	2.4	2.4	0.1
Sulaiman–Kirthar	2.7	99	0.1	15.8	0.1
Southern Alaska	2.6	100	1.3	7.5	0.1
Madre de Dios basin	2.5	101	0.1	10.8	0.7
Irrawaddy	2.5	103	0.7	10.3	0.1
Papua New Guinea fold belt–foreland	2.4	108	0.4	11.0	0.1
Pyrenean Foothills–Ebro basin	2.3	110	0.1	12.6	0.1
Sichuan basin	1.9	117	0.1	10.8	0.1
Santa Maria basin	1.9	119	1.5	1.5	0.1
Taranaki basin	1.7	126	0.2	7.2	0.2
Wyoming thrustbelt	1.5	134	0.2	4.0	0.7
Afgan–Tadjik basin	1.4	137	0.1	7.5	0.1
Cuyo basin	1.4	144	1.3	0.3	0.0
Abu Gharadiq basin	1.3	147	0.5	3.9	0.2
Eastern Cordillera	1.1	152	0.1	5.0	0.3
Wind River basin	1.1	154	0.5	2.8	0.1
Barinas–Apure basin	1.0	158	1.0	0.0	0.0
Upper Magdalena basin	1.0	159	0.9	0.4	0.1
Adriatic basin	0.8	167	0.6	1.0	0.1
Kohat–Potwar	0.7	174	0.3	1.9	0.1
Palmyra Zone	0.6	178	0.1	3.3	0.0
Aquitaine basin	0.4	198	0.4	0.0	0.0
Taiwan thrustbelt	0.4	200	0.0	2.0	0.0
Lesser Antilles deformed belt	0.4	202	0.3	0.1	0.0
Falcon basin	0.3	204	0.2	0.8	0.0
Apennines–Tuscany Latium	0.3	205	0.1	2.0	0.0
Mackenzie thrustbelt	0.3	207	0.2	0.4	0.0
Iberian Cordillera	0.3	208	0.3	0.2	0.0
Lower Magdalena basin	0.3	209	0.1	1.3	0.0
Ucayali basin	0.2	224	0.1	0.9	0.0
Veracruz basin	0.2	227	0.1	0.6	0.0
Zagros thrust zone	0.2	238	0.2	0.0	0.0
South China fold belt	0.1	252	0.1	0.2	0.0

Table 17.1 (cont.)

Province name	Total (BBOE)	Rank	Oil (BB)	Gas (Tcf)	NGL (BB)
Quinling Dabieshan fold belt	0.1	254	0.1	0.0	0.0
Appalachian basin	0.1	254	0.0	0.5	0.0
Lesser Caucasus	0.1	263	0.0	0.3	0.0
Sumatra/Java forearc basin	0.1	268	0.0	0.4	0.0

Notes:

[a] After Klett *et al.* (1997).

[b] Reserves data for the East Venezuela thrustbelts is from Aymard et al. (1990) and for the Western Canadian Foothills belt from Fermor and Moffat (1992) and Newson (2001).

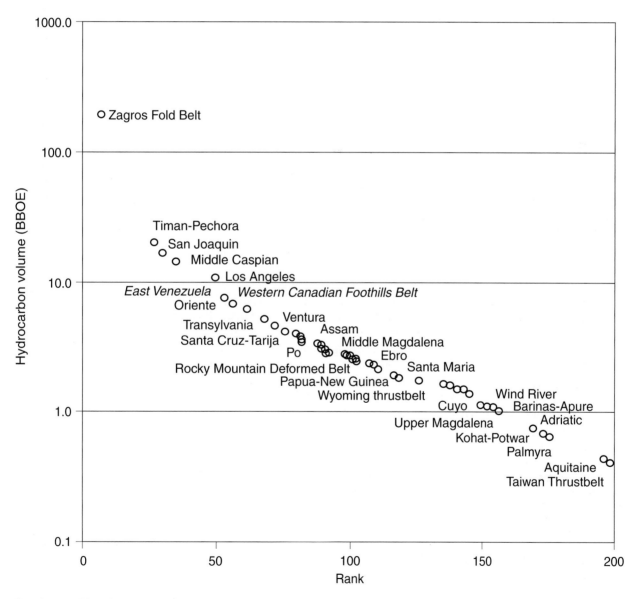

Fig. 17.1. Ranking of the world's thrustbelt petroleum provinces by total known hydrocarbon volumes. Data are from Klett *et al.* (1997).

have been new discoveries and reappraisals in various provinces that have resulted in a slight increase in recognized volumes. For instance, the Ucayali basin ranking of 224 does not include the volume of potential petroleum in the Camisea fields under development, which are estimated to be 13 Tcf and more than 0.5 BBNGL (Martinez *et al.*, 2003). Many provinces also have pools in other than thrustbelt traps. Where it has been possible to separate out the pure thrust-related pools from others, as in the Alberta basin's foothills fold belt and East Venezuela, this has been done. In many other provinces, such as the Llanos basin of Colombia and the Neuquén basin of Argentina, the volumes shown include some nonthrustbelt accumulations. However, single large thrust or inversion traps in an otherwise nonthrustbelt province, such as the Karamay field in the Junggar basin (Zhai and Zhao, 1993), are not included in the summary. The purpose here is to examine the distribution of thrustbelt hydrocarbon accumulations on a global scale, not on a field scale.

The compilation of provinces in Fig. 17.1 highlights the strong asymmetry in global distribution of hydrocarbons in a thrustbelt setting. The Zagros fold belt province with reserves of 189.6 BBOE alone contains more than half of the known petroleum in all thrustbelts worldwide. It is one of just five in the class of provinces with known reserves greater than 100 BBOE located in West Siberia and the Arabian Gulf that together contain more than 40% of the world's hydrocarbons. Only four thrustbelt provinces fall into the 10–100 BBOE class, but together they hold nearly 20% of the thrustbelt reserves. The large group of thrustbelt provinces contains 33 and holds reserves in the range 1.0–10 BBOE. Just 19 thrustbelt provinces fall in the size range 0.1–1.0 BBOE, collectively hosting less than 2% of all thrustbelt reserves.

There may be a tendency, especially among North American explorationists, to think of thrustbelts as dominantly natural gas provinces, but this is not true globally. For all thrustbelt provinces in Fig. 17.1, oil and natural gas liquids comprise nearly two-thirds of the reserves and natural gas just over a third. Nearly the same proportions hold for the Zagros fold belt. In the four provinces in the 10–100 BBOE size class, oil and natural gas liquids constitute more than three-quarters of the total reserves. Only in the smaller provinces with a size less than 10 BBOE are oil and gas reserves about equal. However, the proportion of oil and natural gas liquids relative to natural gas varies considerably from province to province (Fig. 17.2).

The Zagros fold belt, containing 189.5 BBOE, brings together a remarkable assemblage of rocks and physical conditions favourable for the generation and accumulation of petroleum (Bordenave, 2000; Versfelt, 2001). The super-sized oil and gas fields are located in large asymmetric, partially disharmonic anticlines (Fig. 17.3) detached along the horizon of the Infracambrian Hormuz Salt, which is situated at the base of the sedimentary column (Beydoun *et al.*, 1992) resting on a crystalline basement (Jackson *et al.*, 1981). Many of the doubly plunging anticlines exceed 100 km in length and have amplitudes of the order of 1–10 km (Ala, 1990). They contain multiply stacked source rock–reservoir–seal successions (Alsharhan and Nairn, 1997) and most anticlinal traps are filled to spill. The petroleum columns are hundreds of metres thick, or greater. The Gachsaran field has a single oil column 2000 m thick (Hull and Warman, 1970). The greater part of the hydrocarbons is reservoired in fractured Oligocene–Miocene Asmari limestone sealed by thick Miocene Gacharsan evaporites. There is a very large hydrocarbon charge (Bordenave and Burwood, 1994) filling a highly porous fractured reservoir in exceptionally well sealed and very young traps. Consequently, most of the fields have reserves of greater than a billion barrels of oil equivalent, and a few exceed 10 BBOE (Fig. 17.3). Thirty-five of the 64 thrustbelt fields listed in the Macgregor (1996) compilation of giant oil accumulations (Table 1.9) are in the Zagros fold belt.

A large number of the thrustbelt provinces are situated along the leading edge of the Cordilleran complex of western North and South America and the Caribbean. Petroleum-rich segments of this nearly continuous deformed belt extend from the eastern Brooks Range in Alaska, as far south as west-central Argentina. Several segments involve anticlinal traps within thin-skin frontal thrusts. These include the Mackenzie fold-belt containing 0.3 BBOE, the Western Canadian Foothills belt with 7.7 BBOE (Fig. 17.4; Fermor and Moffat, 1992; Newson, 2001), the Rocky Mountain deformed belt with 2.9 BBOE, the Wyoming thrustbelt with 1.5 BBOE, the Eastern Venezuela thrustbelts containing 8.3 BBOE (Fig. 17.5; Aymard *et al.*, 1990; Carnevali, 1992), Madre dos de basin with 2.5 BBOE, Ucayali basin with 0.2 BBOE and Santa Cruz-Tarija basin containing 4.1 BBOE (Dunn *et al.*, 1995; Lindquist, 1998). Other segments contain structural and combination traps within inverted foreland basins flanking the cordillera. These include:

(1)   the Rocky Mountain basins (Hjellming, 1993);
(2)   the basins within and adjacent to the Cordillera Oriental of Colombia, including the Eastern Cordillera with 1.1 BBOE, the Middle Magdalena Valley with 3.3 BBOE, the Upper Magdalena Valley with 1.0 BBOE, the Barinas-

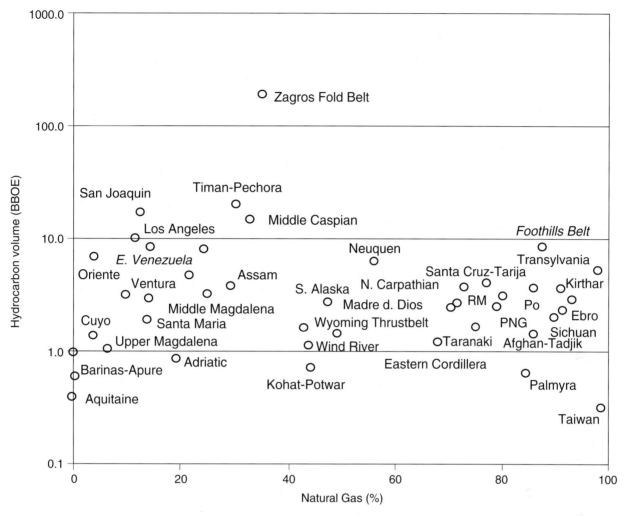

Fig. 17.2. Percentage of natural gas in total hydrocarbon volumes in the world's thrustbelt provinces. Data are from Klett *et al.* (1997). RM – Rocky Mountain deformed belt, PNG – Papua-New Guinea fold belt, Foothills belt – Western Canadian Foothills belt.

Apure basin with 1.0 BBOE and the Llanos basin with 7.3 BBOE);

(3) the Oriente-Marañon basin with 6.9 BBOE;
(4) the Cuyo basin with 1.4 BBOE;
(5) the Neuquén basin with 6.2 BBOE; and
(6) other sub-Andean basins (Dellapé and Hegedus, 1995; Uliana *et al.*, 1995; Urien, 2001).

The remaining segment, the Lesser Antilles deformed belt containing 0.4 BBOE is an accretionary prism within the Caribbean segment of the cordillera.

Four important thrustbelt provinces are located along the San Andreas fault in southern California, on the Pacific active margin. The Los Angeles basin with 10.1 BBOE (Fig. 17.6), the Ventura basin with 4.6 BBOE and the Santa Maria basin containing 1.9 BBOE are partially inverted pull-apart basins (Sylvester and

Brown, 1988; Wright, 1991). In contrast, the San Joaquin basin with 16.6 BBOE is a segment of the Cretaceous Great Valley forearc basin, strongly reactivated in the Late Tertiary along a restraining bend in the San Andreas fault (Webb, 1981; Sylvester, 1988). Transpressional basement-rooted reverse faults and shallow thrusts form the large anticlinal traps of the basin (Harding, 1976; Wentworth and Zoback, 1990). All of these basins owe their very large concentrations of hydrocarbons both to the presence of an exceptional source rock, the Monterey Shale (Graham and Williams, 1985), and also to their young age.

Another group of thrustbelt petroleum provinces is situated along the Late Cretaceous–Tertiary Alpine belts of Europe. The Pyrenean Foothills–Ebro basin province with 2.3 BBOE and Aquitaine basin with 0.4 BBOE (Bourrouilh *et al.*, 1995) are located on the flanks

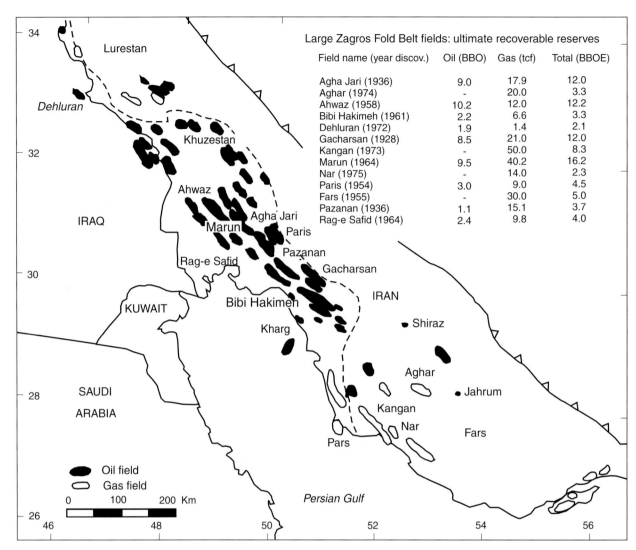

Large Zagros Fold Belt fields: ultimate recoverable reserves			
Field name (year discov.)	Oil (BBO)	Gas (tcf)	Total (BBOE)
Agha Jari (1936)	9.0	17.9	12.0
Aghar (1974)	-	20.0	3.3
Ahwaz (1958)	10.2	12.0	12.2
Bibi Hakimeh (1961)	2.2	6.6	3.3
Dehluran (1972)	1.9	1.4	2.1
Gacharsan (1928)	8.5	21.0	12.0
Kangan (1973)	-	50.0	8.3
Marun (1964)	9.5	40.2	16.2
Nar (1975)	-	14.0	2.3
Paris (1954)	3.0	9.0	4.5
Fars (1955)	-	30.0	5.0
Pazanan (1936)	1.1	15.1	3.7
Rag-e Safid (1964)	2.4	9.8	4.0

Fig. 17.3. Oil and gas fields in the central and southern Zagros fold belt. Field sizes are shown for the larger of the fields in Iran. Reserves data are from Ala (1990) and Iraqi field locations are from Alsharan and Nairn (1997).

of the severely inverted Pyrenees. The Po basin with 3.6 BBOE and the Adriatic province with 0.8 BBOE are situated in the deformed foredeep (Mattavelli *et al.*, 1991, 1993) between the Apennines on their southern and southwestern side and the southern Alps and Dinarides on their northern and northeastern side. The North Carpathian basin with 3.7 BBOE and the Transylvanian basin with 5.1 BBOE provinces are in the frontal and inner zones of the Carpathian mountains, respectively. This group of productive provinces continues eastwards through the Alpine deformed zones of south-central Asia, containing, for example, the Middle Caspian province with 14.4 BBOE and the Afgan–Tadjik province with 1.4 BBOE, and into southeast Asia, containing, for example, the Papua–New Guinea thrustbelt with 2.4 BBOE and the Taiwan thrustbelt with 0.4 BBOE (Suppe, 1987). The Abu Gharadiq basin with 1.3

BBOE and the Palmyra Zone with 0.6 BBOE are both inverted rift systems within the northern African plate (McBride *et al.*, 1990).

It is instructive to consider which major thrustbelts are not included in the above list of important thrustbelt provinces. Prominent among them are the Alps *sensu stricto* and the Himalaya Range, both products of massive continent–continent collision during the Mid- and Late Tertiary. In both instances it would seem that the severity of deformation has worked against the accumulation and preservation of hydrocarbons. Although the frontal thrusts of the Alps have been tested through deep wells into the underlying autochthon and gas and oil shows have been common (Colins *et al.*, 1992; Müller *et al.*, 1992), nowhere has sustained commercial production been established within the allochthon. However, the Tertiary Molasse foredeep

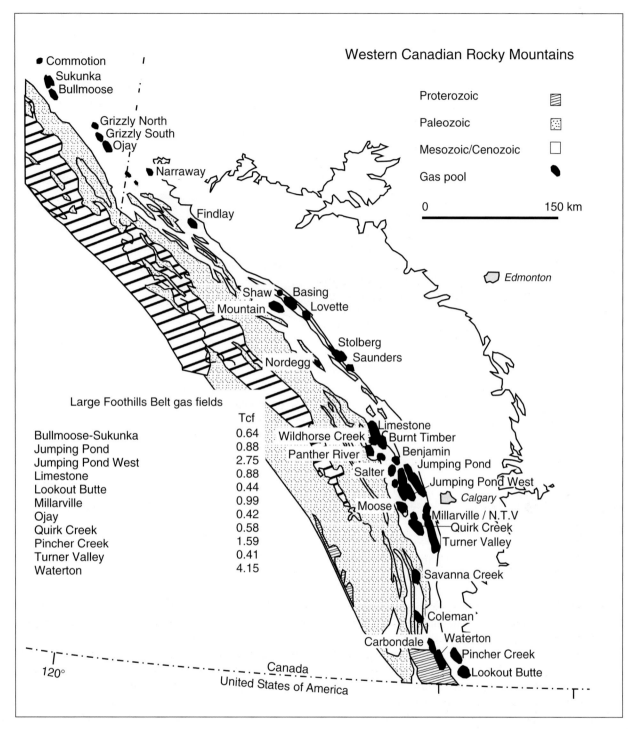

Fig. 17.4. Gas and condensate fields in the Western Canadian Foothills belt. The map identifies the larger fields and their booked reserve sizes (after Fermor and Moffat, 1992).

basin is productive in Austria and southeast Germany from stratigraphic and normal fault traps (Brix and Schultz, 1993). The main portion of the Himalaya Range directly overrides the young Siwalik Group molasse of the foreland basin and is unproductive as it lacks source rock, seals and for the most part, appropriate traps (Burbank *et al.*, 1996; Murphy and Yin, 2003). However, at both ends of the Himalaya Range, where crustal shortening is less severe and source rocks are present, there are the important petroleum provinces of

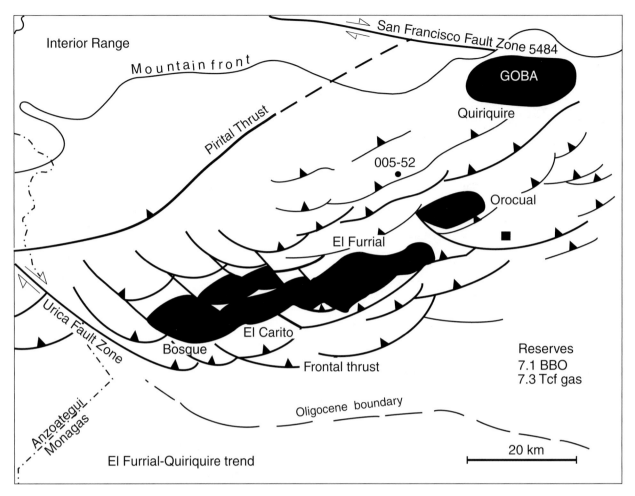

Fig. 17.5. Super-giant oil and gas fields of the El Furrial–Quiriquire belt in the East Venezuela province (after Aymard *et al.*, 1990).

Kohat-Potwar with 0.7 BBOE and the Sulaiman-Kithar with 2.7 BBOE on the west and Assam with 3.6 BBOE and Irrawaddy with 2.5 BBOE on the east.

The latest Cretaceous–Tertiary Atlas Range of North Africa has been the focus of repeated rounds of exploration, but with little success. The oil and gas fields discovered have been very small and short-lived. Yet this complexly inverted system of rift and pull-apart basins (Stets and Wurster, 1982; Lowell, 1995) has good source rock and reservoir successions, and oil seeps are common. It is probable that the basin inversion is simply too extreme to have created adequate traps of any size and integrity (Macgregor, 1995).

In general, the Late Cretaceous–Tertiary sub-Andean thrustbelt provinces of South America are very productive, figuring prominently on the world ranking (Table 17.1). However, south of the northern Neuquén basin of Argentina and through the Austral-Magallenes thrustbelt suitable source rocks and structuring are present (Urien, 2001), but apparently not quality reservoirs. This relatively unexplored array of thrustbelts might still prove productive, but at present it is not.

With a few notable exceptions Paleozoic-age and older thrustbelts are unproductive. This is principally due to erosional stripping of the leading edges where hydrocarbon accumulations are most likely, and tectonic overprinting of otherwise productive zones, allowing the hydrocarbons to escape and degradation. Most prominent in the group of nonpetroliferous thrustbelts are those of central Asia, large portions of the Urals, the Caledonide and Hercynian orogens of Europe, the Anti-Atlas Range of Morocco, the Mauretanide thrustbelt of northwest Africa, the Ouachita–Marathon thrustbelt in southern North America, various Paleozoic thrustbelts in Western North and South America, and the Cape-Ventania fold belt of South Africa and Argentina.

The Paleozoic northern Appalachian basin is now a very mature petroleum province with a relatively low ranking of 254th. But during the period from 1859 to 1993, this province produced a total of 67 BBOE

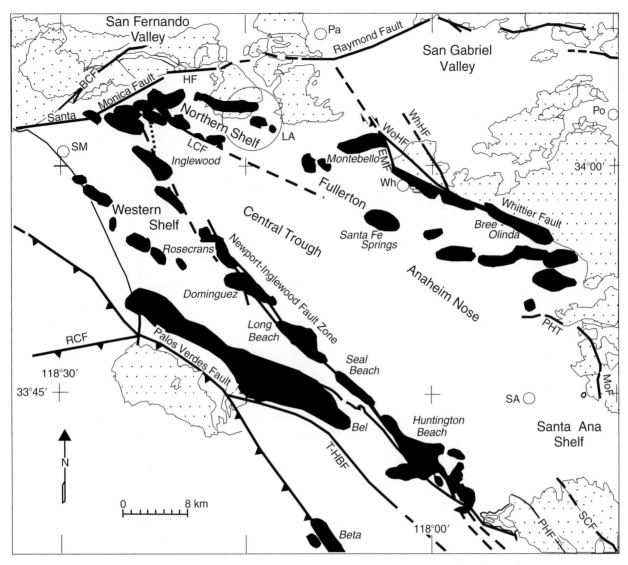

Fig. 17.6. Oil fields of the Los Angeles basin, southern California, resulting from partial inversion of a coastal pull-apart basin (after Wright, 1991).

(Gautier *et al.*, 1996), 60 BBO and 42 Tcf of natural gas. Most of this production was from the gently deformed Allegany Plateau, a large region of the foreland basin detached, faulted and folded above a Silurian salt (Shumaker, 1996). In contrast to the more intensely deformed Appalachians, this region has experienced little erosional stripping, preserved its hydrocarbon accumulations and appears to be generating at least natural gas at present.

In many respects the Timan-Pechora province, ranking 27th with 20 BBOE, stands apart from the other major hydrocarbon-rich thrustbelt provinces. First, it is relatively old, having developed fully prior to the Mid-Jurassic and with many large traps charged before the end of the Paleozoic (Schamel, 1993; Abrams *et al.*, 1999). Secondly, virtually all of the significant fields are within the Northern Urals foreland, located in large doubly plunging folds along the partially inverted flanks of an extensive Mid-Paleozoic aulacogen, in traps associated with reactivated-basement faults, and less commonly in traps on the leading edge of shallow-detached thrust sheets (Dedeev *et al.*, 1994). The only field strictly within the Ural thrustbelt is Vuktyl, a super-giant gas field containing 14.7 Tcf and 0.52 BBNGL (Fossum *et al.*, 2001). Like the Allegany Plateau, this foreland basin has experienced little uplift and erosion, has many excellent seals and probably continues to generate hydrocarbons locally.

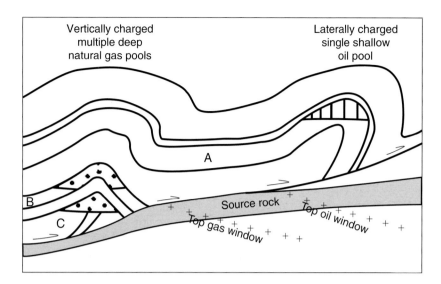

Fig. 17.7. Cross section illustrating hypothetical thrustbelt petroleum systems based on a single source rock unit in the footwall of a thrust sheet carrying three reservoir units, A, B and C. The carrier that allowed the existence of the hanging wall anticlinal traps is the thrust zone itself. It was active at the critical moment in this scenario. Oil migrated laterally to charge reservoir A in the leading anticline via a hanging wall cutoff of that unit. Reservoir B was charged in a similar fashion in the second anticline, but from below the top of the gas window. Reservoir C was charged along small thrust splays off the master thrust. The gas pools were charged vertically from mature source rocks directly beneath the second anticline. The leading anticline is situated in front of the top of the oil window where it intersects the thrust, so lateral migration is required.

Petroleum systems in thrustbelts

Explorationists have long recognized five essential components for the formation of a hydrocarbon play: source rock, hydrocarbon charge, reservoir, effective seal and trap. Implicit in the trap component is timing, the existence of the trap at the time of migration of the hydrocarbon charge through the part of the basin containing the traps. Spatial and temporal coherency is a necessary condition for development of hydrocarbon accumulations. Normally all five conditions must be optimal, but if the charge is very large and migration is very recent then portions of the basin are saturated with hydrocarbons, and seal and trap quality are less critical. The reservoir quality is a measure of both storage of hydrocarbons, the volume in place, and deliverability, production rates in wells completed in the reservoir. A reservoir with relatively high porosity and low permeability can hold large volumes of hydrocarbons, but has low recovery. In all but the youngest oil and gas pools, the question of trap integrity and the preservation of the hydrocarbon is critical. Traps can leak over time, be breached by overprinting structuring or uplift and erosion, or degradation of the oils in the reservoir can occur *in situ*.

The generation, migration and trapping of hydrocarbons in a sedimentary basin involves a sequence of geological processes linking a source rock to one or more reservoir–seal couplets and results in a genetically related family of hydrocarbon accumulations (Biteau *et al.*, 2003). The petroleum system (Perrodon, 1980) integrates the broad range of geological, geodynamical, chemical, thermal and hydrodynamical phenomena that control these genetic relationships. Magoon and Dow (1994) stress the interdependence of elements related to rock strata, such as source, reservoir, seal and overburden,

and the processes of trap formation and generation–migration–accumulation. The petroleum system has a specific duration in geological time and reaches its climax, or critical moment, when generation–migration–accumulation is most active. The geographic extent encompasses the volume of active source rock and all discovered hydrocarbon accumulations, seeps and shows that originate from that source rock volume. To describe hydrocarbons that are probably part of the system, but not yet known, Magoon (1995) coined the term total petroleum system. The petroleum system paradigm, which now guides most exploration activities worldwide, enlarges earlier concepts of oil–source correlation; (i.e. oil system; Dow, 1974) and the spatial–genetic relationships of large hydrocarbon accumulations to proximal generative basins (Demaison, 1984).

Magoon and Dow (1994) formalized the criteria for identifying, naming and mapping petroleum systems. The name assigned identifies the source rock–reservoir pair with a symbolic qualifier added to express the level of certainty in the oil–gas-source correlation supporting the association. Thus, any basin with more than one source rock and/or significant reservoir unit will contain multiple petroleum systems, each of which warrants separate characterization and mapping.

Several graphical devices are recommended for displaying a petroleum system:

(1)   a map and cross section (Fig. 17.7) showing the geographic extent and relative position of the

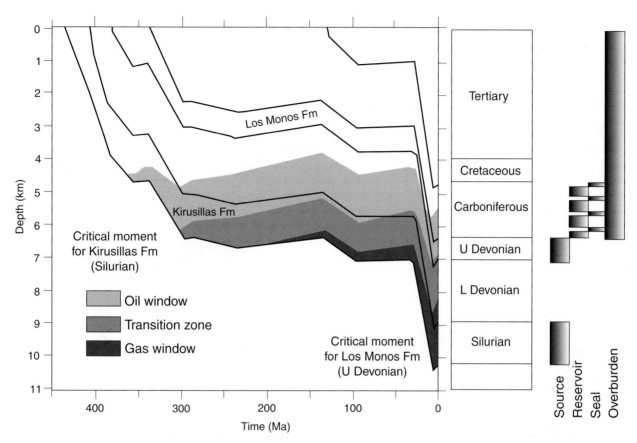

Fig. 17.8. Burial history diagram for the Los Monos–Machareti (!) petroleum system in the Santa Cruz–Tarija province, Bolivia (Dunn *et al.*, 1995, modified by Lindquist, 1998). The diagram represents burial and maturation of two source rock units in a typical synclinal depression between thin-skin anticlinal fields. Modelling shows the Middle Devonian Los Monos source rock entering the oil window immediately before the onset of thrusting and continuing after the end of this event. The Silurian Kirusillas source rock reached maturity too early to have charged the anticlinal traps to any significant extent.

active source rock volume and traps at the critical moment;

(2) a burial history diagram (Fig. 17.8) representing the burial and modelled maturation history of the source rock; and

(3) a petroleum systems chart (Fig. 17.9) that displays the temporal relationships of the elements and processes over the duration of the petroleum system.

The actual operation of a petroleum system in any but the simplest basin is normally very complex and full characterization requires more than just these few charts. Numerous workstation routines are available to model and map the workings of petroleum systems with great specificity, even in the thrustbelt setting.

Demaison and Huizinga (1991) have proposed a useful approach to characterizing key elements of petroleum systems that can assist in screening frontier basins and exploration concepts. Three criteria form the basis of this genetic classification of petroleum systems:

(1) the charge factor (supercharged, normally charged, undercharged) as determined from the richness and volumetrics of mature source rocks in the basin;

(2) the migration drainage style (vertically drained or laterally drained), which is determined from the stratigraphic and structural framework of the basin; seals breached by faults result in vertical drainage, but where unfaulted they force lateral drainage;

(3) the entrapment style (high impedance or low impedance), which integrates the structural framework of the basin with the distribution of quality seals.

Using this terminology, the Zagros fold belt and Los Angeles provinces, which contrast in structural style,

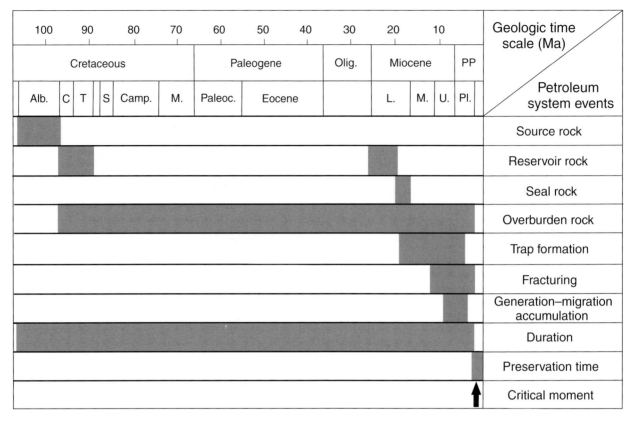

Fig. 17.9. Petroleum systems chart for the central Zagros fold belt within the Dezful Embayment (Bordenave and Burwood, 1994). From this chart it is possible to visualize the ideal temporal relationships that lead to the exceptionally large hydrocarbon accumulations in the large anticlinal traps of the province. Deposition of the principal reservoir, the Oligocene–Miocene Asmari Limestone, overlain and sealed by a thick Miocene Gacharsan evaporite, is followed in rapid succession by trap formation and reservoir enhancement, hydrocarbon generation–migration–accumulation, and a very short preservation time, in this order.

both would be described as supercharged, vertically drained and having high impedance. Many of the world's most productive basins have this set of characteristics.

In basins where the known accumulated petroleum volumes can be compared to an estimate of total volume of hydrocarbons generated, it is possible to determine an efficiency ratio, or petroleum system yield (PSY), which can serve as yet another measure of the exploration potential of a basin (Biteau *et al.*, 2003). However, this parameter must be used with caution. The relatively hydrocarbon-poor Aquitaine basin has a PSY of 12.4%, but ranks as low as 198th in terms of known petroleum volume. In contrast, the petroleum-rich Zagros fold belt that ranks as fourth and the Magdalena provinces of Colombia, ranking as 89th and 159th have PSY values of less than 5%. The richest petroleum provinces are not necessarily the most efficient in retaining hydrocarbons generated (Biteau *et al.*, 2003), but high efficiency can enhance the prospectivity of a basin that is inherently source-rock lean.

In several substantial ways petroleum systems operate differently in the thrustbelt petroleum habitat compared with other basin/tectonic settings. The principal differences include the following.

(1) Hydrocarbon maturation influenced by burial histories changes dramatically during thrusting due to increased rates of burial and/or uplift. The very complex nature of source-rock burial or uplift scenarios in thrustbelts reflects structural styles, as well as whether the thrustbelt is submarine or subaerial, eroded following thrusting, or immediately reburied. Source rocks that are rapidly buried and rapidly uplifted as part of the structural transfer from footwall to hanging wall of an advancing thrust continue to generate hydrocarbons while uplifting, but at greatly diminished rates and yields. Late-orogenic reburial can push the source rocks back into the hydrocarbon generative window.

(2) Generative areas can be highly localized once a portion of a basin is segmented into individual thrust sheets. Whether thrusts detach above or below the major active source-rock intervals is critical to defining the size and maturation levels of individual generative pods. Except where the hydrocarbon kitchens are in the undeformed footwall succession, thrusting has the effect of creating compartmentalized petroleum systems.

(3) Pathways for potential hydrocarbon migration will be controlled by thrust geometries during the time of deformation. The thrust surfaces are likely to be conduits while thrusts are active, aiding in the charging of newly forming structural traps, but serve as fault seals thereafter. Late- and post-orogenic faults normally have a detrimental effect on hydrocarbon accumulations in thrustbelts.

(4) The timing of hydrocarbon generation–migration relative to trap formation and opening of migration conduits is especially critical in thrustbelts as the succession of events tends to be relatively brief.

(5) Stress fields associated with thrusting influence the reaction kinetics of hydrocarbon maturation, early migration of hydrocarbons out of the source rock, the quality of sub-regional seals and the potential for fracture enhancement of reservoirs.

(6) The distribution of source rock types is to some extent determined by thrustbelt evolution. Foreland basins commonly contain early distal oil-prone black shales overlain by late progradational gas-prone carbonaceous shale and coaly successions. In some instances, particularly in multi-phase intracratonic thrustbelts, synorogenic oil-prone lacustrine source rocks are sufficiently buried to charge traps within the thrustbelt. However, in most thrustbelts the principal source rock successions are pre-orogenic.

(7) Preservation of hydrocarbon accumulations can present a special problem in thrustbelts due to uplift and erosion of traps, out-of-sequence faulting breaking earlier traps, and deep circulation of meteoric waters related to topographic elevation.

The salient features of the petroleum system in the thrustbelt setting are discussed in subsequent chapters.

The objective of any exploration endeavour is the identification of probable opportunities for petroleum discoveries assembled in a portfolio of drillable prospects (White, 1993). The specific prospects, in turn, are derived from play concepts that are based on families of geologically related fields, prospects and leads, all of similar geological origin and charged from common petroleum source beds (Rose, 2001). Petroleum systems analysis is now a critical component in developing and risking hydrocarbon plays and prospects (Smith, 1994; Rose, 2001; Naylor and Spring, 2002). Together with commercial parameters, such as product prices and operating costs, and exposure to the exploration and production project to the fiscal, political and business environment (Wood, 2003), assessment of the petroleum system is now just one of many components in the appraisal of prospects and ranking by investment efficiency or profitability (Rose, 2001; Wilson, 2002).

Current methods for risking plays and prospects generally build on a pioneering appraisal system developed by Shell (Sluijk and Nederlof, 1984) for evaluation of specific exploration prospects based on simple geochemical/geological models. This system used Monte Carlo simulations of three factors:

(1) hydrocarbon generation, expulsion and migration leading to the hydrocarbon charge;
(2) trapping and retention of oil and gas; and
(3) recovery.

The first two factors are derived directly from an understanding of the petroleum system, whereas the third relates to the reservoir properties and production technologies. More advanced appraisal systems, such as that described by Naylor and Spring (2002) retain these factors as prospect-specific and add other Monte Carlo simulations related to the state of technology that characterizes the uncertainties in the structural interpretations as a function of seismic data quality and coverage and play- and area-specific risk factors.

Magoon and Dow (1994) cite four levels of hydrocarbon investigation relating sedimentary basins and petroleum systems to plays and prospects. In thrustbelts a fifth level is required. It is a structural analysis following or concurrent with, but separate from, basin analysis (Fig. 17.10). Structural analysis defines the full range of geometric elements and deformation histories that are essential in characterizing the burial history, internal plumbing, and trap styles in the parts of a sedimentary basin overprinted by thrusting (Davison, 1994; Buchanan, 1996). Structural analysis builds on a thorough knowledge of the stratigraphy and stratigraphic variability of the deformed basin, including the influences of stratigraphy on structural style, and in turn is critical to the proper evaluation of the petroleum systems. Thrustbelts commonly combine complex structuring with rugged topography resulting in poor seismic quality compared with other petroleum habitats. Thus, seismic data alone is rarely sufficient to define deep structural features, but must be augmented by the use of a variety of other methods of structural analysis discussed in Chapter 6.

Even at the level of structural analysis, economic considerations should play a role. Given several alternative structural models, each of which is consistent with geological and geophysical constraints, those that can lead to active petroleum systems and productive plays should be given greater consideration in going forward. The importance of economic factors increases in subsequent levels of investigation (Fig. 17. 10) as the focus of investigation becomes more specific.

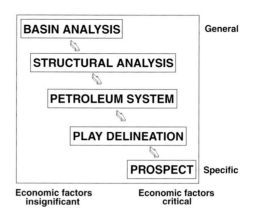

Fig. 17.10. Five levels of petroleum investigation in thrustbelts (modified after Magoon and Dow, 1994).

18 Source rocks in thrustbelt settings

General features of source rocks

Any petroleum system, whether or not in a thrustbelt setting, requires a source rock to be present in the sedimentary section. Without the strata containing at least 0.5–1.0% preserved organic matter, there can be no hydrocarbon charge. Oil and gas form by the thermal breakdown of buried specific organic matter; oil in the temperature range 110–150 °C and natural gas in the temperature range 140–250 °C. However, organic matter starts to undergo chemical alteration from the instant of death of the precursor organism. It progresses from a living organism through a process of diagenesis, to hydrocarbons through catagenesis and finally to a carbon residue through thermal conversion, i.e. metagenesis (Fig. 18.1).

Photosynthesizing aerobic and anaerobic plants and microorganisms convert atmospheric carbon dioxide to organic compounds. Most of these molecules become energy sources that the plants and their predators are able to almost completely break down, converting them back into CO_2 by the process of respiration, thus closing the carbon cycle. Only a very small proportion of organic matter escapes decomposition and is buried in sediments (Hunt, 1979). This proportion represents less than 0.1% by volume of primary productivity and can be considered as a 'leak' in the carbon cycle. This is the organic matter that may eventually be converted to fossil fuels after further burial.

The nature of the organic matter in the source rock and the nature of the initially derived hydrocarbon products are a function of both the precursor organic matter and the depositional environment. The source material for fossil hydrocarbons is derived from four chemically distinct classes of matter: lipids, amino acids, carbohydrates and lignins (Tissot and Welte, 1984; Berkowitz, 1997). By far the most important for oil formation are the lipids, aliphatic hydrocarbons that include fatty acids, waxes, terpenes and steroids. In living organisms the lipids are an energy source, but after death they are converted to carboxyl acids and eventually alkanes by putrefaction and low-temperature diagenesis. Amino acids are the building blocks of proteins, the substance of cells. Carbohydrates are the dominant structural material of plants and include sugars, starches and cellulose. They are hydrogen-poor, oxygen-rich compounds having a characteristic O/C ratio of 2. Lignins first appeared in any significant quantity associated with the development of land plants during and after the Late Silurian (e.g., Glasspool et al., 2004). These complex biopolymers are characterized by an abundance of aromatic units and phenols.

The organic content of recent sediments is determined by the comparative rates of organic and mineral deposition, the availability of oxygen and the level of biological activity feeding on the organic matter (Hunt, 1979). Particularly in coastal areas, the organic matter is produced in a combination of marine and terrestrial environments. Under anoxic conditions marine and lacustrine phytoplankton, such as algae, diatoms and dinoflagellates, are diagenetically altered to sapropel having a high-hydrogen, low-oxygen content. Anaeorobic bacteria associated with the initial decay of the phytoplankton tend to be even more hydrogen-rich. In contrast, terrestrial organic matter rich in cellulose and lignins is altered to a low-hydrogen, high-oxygen organic matter. However, a similar low-hydrogen, high-oxygen organic matter can result from the decay and diagenesis of phytoplankton in a more oxic environment in which various organisms feed on lipids and other hydrogen-rich molecules.

The organic matter within the source rock consists of two principal components: kerogen, i.e. the solid organic matter insoluble in organic solvents, and bitumen, i.e. the soluble liquid hydrocarbon generated from the kerogens. Natural gas, which is also likely to be present in some quantity, only rarely occurs in a separate phase within the organic-rich source rock. Normally, it will be adsorbed on molecular or micro-fracture surfaces within the kerogen or dissolved within the bitumen. If the source rock has not undergone catagenesis, and thus has not entered the maturity window, the bitumen and gas components will be largely absent. However, diagenesis does result in the production of gases such as methane, carbon dioxide and hydrogen sulphide.

Kerogen is a mixture of macerals and reconstituted degraded products of organic matter that preserve something of their original plant structure. The chemical characteristics of isolated kerogen relate to the precursor organisms and the complex diagenetic–catagenic

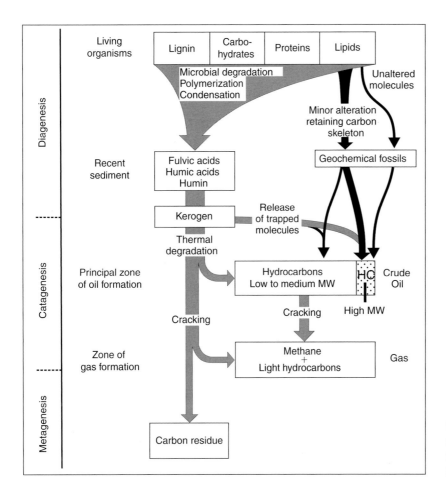

Fig. 18.1. Stages in the transformation of organic matter from living organisms to hydrocarbons and inert carbon (Tissot and Welte, 1984).

pathways leading to the state of the organic matter in the rock. A kerogen may be classified by its hydrogen versus oxygen richness, or a comparison of atomic H/C and O/C ratios. The van Krevelen plot (Fig. 18.2), displaying H/C and O/C ratios, is a common tool for displaying bulk organic matter compositions. Five primary types of kerogen are recognized, the first three of which were identified by Tissot *et al.* (1974). Type I kerogen is rich in hydrogen and strongly oil-prone, having hydrocarbon yields as high as 80 weight %. Type II kerogen is less hydrogen rich and is capable of generating both oil and gas. Its hydrocarbon yields are up to 60 weight %. Type III kerogen is hydrogen-poor and gas-prone. Type IV kerogen is inert residual organic matter incapable of hydrocarbon generation. Its H/C values are less than 0.65. The final type, Type II-S kerogen (Orr, 1986), is a sulphur-enriched form of Type II kerogen that has similar elemental composition but contains at least 6% of organic sulphur. Type II-S kerogens start generating heavy oil at lower temperatures than normal Type II kerogens (Baskin and Peters, 1992).

The three main maceral groups forming coals are vitrinite, the woody and cellulose portions of terrestrial plants, liptinite (exinite), the product of spores, resins and algae, and inertinite (including fusinite), highly oxidized organic matter. Each of these maceral groups has a specific kerogen pathway on the van Krevelen plot (Fig. 18.3).

As organic matter matures and generates hydrocarbons, i.e. bitumen and gas, its composition shifts downward on the van Krevelen plot (Fig. 18.2) to lower values of H/C and O/C. At high degrees of maturity all organic matter takes on the chemical properties of Type III or IV kerogen regardless of its initial immature composition.

A variety of molecular indicators, referred to as biomarkers, point to the precursor organisms, depositional environments and diagenetic pathways leading to the buried organic matter (Philp and Lewis, 1987; Waples and Machihara, 1991; Dahl *et al.*, 1994). In addition, biomarkers have wide application in identifying oil sources and oil–oil correlation (Curiale, 1994). Terrestrial organisms synthesize mainly C_{27}, C_{29} and C_{31} normal paraffins, whereas marine organisms synthesize mainly C_{15}, C_{17} and C_{19} (Hunt, 1979). The relative abundance of C_{27}, C_{28} and C_{29} sterols (Volkman, 1988) can

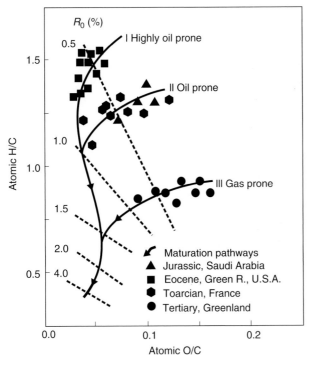

Fig. 18.2. Plot of kerogen atomic H/C and O/C ratios (van Krevelen plot) showing the main hydrocarbon-generative types (Peters, 1986). Thermal maturity of the kerogen, as measured by vitrinite reflectance (R_0), is indicated by the position along the converging maturation pathways.

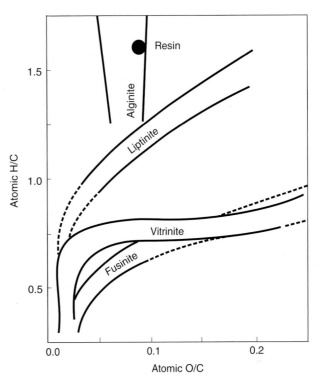

Fig. 18.3. Petrographic components of coal, based on atomic ratios (after Powell and Boreham, 1994).

separate higher land plants, which are C_{29} dominant, from marine phytoplankton, which is C_{27} dominant. The triterpane, gammacerane, is associated with protozoa, and possibly bacteria, and is indicative of a stratified water column, commonly in hypersaline environments (Schoell *et al.*, 1994). Chlorophyll, the molecular basis for photosynthesis in plants, undergoes complex microbial and geochemical transformations in aquatic environments. It transforms to the acyclic isoprenoid pristane (Pr; C_{19}) under oxidizing (terrestrial) conditions or phytane (Ph; C_{20}) under reducing (marine) conditions (Hunt, 1979; Collister *et al.*, 1992). The ratios of Pr/n-C_{17} versus Ph/n-C_{18}, as derived from standard gas chromatography (GC) analysis of crude oil or bitumen extract, can be used to determine whether the oil source is terrestrial kerogen, characterized by higher Pr/n-C_{17}, or marine kerogen, indicated by higher Ph/n-C_{18}. They can also be used for indication of the degree of maturity versus biodegradation (Dydik *et al.*, 1978). As normal n-alkanes form preferentially to and biodegrade more readily than acyclic isoprenoids, such as pristane and phytane, lower values of both ratios characterize higher degrees of thermal maturity, whereas higher values of both indicate biodegradation.

Other useful biomarkers have been identified and are discussed in references cited herein.

The overall quality of a source rock is a function of many factors, including the average organic richness, the type of organic matter and its degree of maturity, and its thickness and spatial distribution. The common measure of richness is total organic content (TOC), reported as weight % of organic carbon. Source rocks that are too lean are simply incapable of generating sufficient quantities of bitumen to break out of the source rock pore space to migrate as liquid hydrocarbons. For most sapropelic kerogens this lower threshold may be about 1.0% TOC (Bissada, 1982). As the barriers to escape are lower for natural gas, it can be generated from kerogens in very low concentrations or from unexpelled bitumens in even the leanest source rocks.

Open system programmed pyrolysis, such as RockEval, has become the industry standard for rapid screening of large numbers of rock samples for petroleum-generative potential and thermal maturity in advance of further analysis (Espitalié *et al.*, 1977; Peters, 1986; Peters and Cassa, 1994). Using this method, products generated during programmed oven heating are swept by a helium carrier gas and are quantified by a flame ionization detector (FID). Pulverized samples are gradually heated in an inert atmosphere

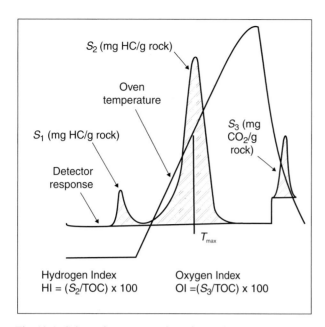

Fig. 18.4. Schematic representation of a RockEval pyrogram showing the evolution of organic compounds from a rock sample during progressive heating (Peters, 1986).

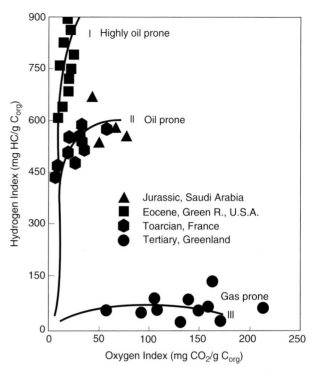

Fig. 18.5. Main hydrocarbon-generative types shown on a plot of the hydrogen index (HI) versus the oxygen index (OI) for the same source rock samples as in the van Krevelen atomic ratios plot of Fig. 18.2 (Peters, 1986).

such that bitumen in the rock is released by distillation and at higher temperatures kerogen is cracked into pyrolytic products. The programmed heating results in a variety of parameters (Fig. 18.4) that collectively characterize the bulk properties of the source rock. The parameters are:

(1) S_1: the hydrocarbons (HC) that can be distilled from the rock (mg HC/g rock);
(2) S_2: the hydrocarbon generated by pyrolytic degradation of kerogen (mg HC/g rock);
(3) T_{max}: the temperature at which the maximum amount of pyrolysate (S2) is generated, which generally increases with depth and may serve as a measure of source rock maturity;
(4) S_3: the carbon dioxide generated during heating up to 390 °C as measured by thermal conductivity detection (TCD; mg CO_2/g C_{org});
(5) production index (PI): the ratio of bitumen to total hydrocarbon or S1/(S1 + S2);
(6) hydrogen index (HI): the quantity of pyrolyzable organic compounds from S2 relative to the quantity of organic carbon (mg HC/g C_{org}); and
(7) oxygen index (OI): the quantity of carbon dioxide from S3 relative to the quantity of organic carbon (mg CO_2/g C_{org}).

HI and OI may serve as proxies for atomic ratios in plotting kerogen compositions in the van Krevelen plot

(Fig. 18.5). Genetic potentials ($S_1 + S_2$) greater than 6 mg HC/g rock and less than 2.5 mg HC/g rock indicate source rocks that are good-to-excellent and poor, respectively. Although vitrinite reflectance (%R_0) is the more common measure of source rock thermal maturity, production index and T_{max} are alternative indicators. Typically, hydrocarbon generation is associated with values of PI and T_{max} in the range 0.2–0.4 and 436–460 °C, respectively. Despite several analytical and interpretative limitations of RockEval methods for characterizing source rocks (e.g., Katz, 1983; Peters, 1986), the method is fast, inexpensive and requires very little sample. These factors enable close-spaced sampling of cores or outcrops and better statistical determinations of the source rock distribution.

Demaison and Huizinga (1991) introduced the source potential index (SPI) as a measure of source rock volumetrics, integrating the average source rock richness and the stratigraphic thickness. The SPI expresses the quantity of hydrocarbons in metric tonnes that can be generated in a column of source rock beneath a square metre of surface area. The genetic potential ($S_1 + S_2$), as determined from RockEval measurements, is used to specify the average source rock richness. The formal expression is

Table 18.1. Average SPI and kerogen types for selected thrustbelt basins.[a]

Basin (country)	Source rock sequence	Kerogen type	Avg. SPI
San Joaquin (USA)	Miocene	II	38
San Joaquin (USA)	Eocene–Oligocene	II to II–III	14
Santa Barbara Channel (USA)	Miocene	II	39
Offshore Santa Marta (USA)	Miocene	II	21
Eastern Venezuela (Venezuela)	Upper Cretaceous	II	27
Maturin (Venezuela)	Upper Cretaceous	II	12
Maracaibo (Venezuela)	Upper Cretaceous	II	10
Plato (Colombia)	Oligocene–Miocene	II–III to III	5
Middle Magdalena (Colombia)	Upper Cretaceous	II	16
Oriente (Ecuador)	Upper Cretaceous	II	6
Metan (Argentina)	Upper Cretaceous	II	1
Tres Cruces (Argentina)	Upper Cretaceous	II	<1
Cuyo (Argentina)	Triassic	I	8

Note:
[a] from Demaison and Huizinga (1991).

$$SPI = \frac{h(S_1 + S_2)d}{1000},\qquad(18.1)$$

where h is the net source rock thickness, $S_1 + S_2$ is the average genetic potential in kg HC/tonne rock and d is the average source rock density, normally about $2.5\,t\,m^{-3}$.

Table 18.1 presents average SPI values and kerogen types published for representative thrustbelt basins in western North America and the Andes (Demaison and Huizinga, 1991). Comparison of listed SPI values with the rankings of the same basins in Table 17.1 shows that basins with large known petroleum reserves are those with rich source rocks as indicated by high SPI.

It needs to be noted that there are several limitations with the SPI approach. The parameter does not capture the variability in source rock quality that is certain to occur within the basin (Katz, 1983). Greater overall source rock thickness by itself does not compensate for inherent leanness. The expulsion efficiency may actually decrease with increasing gross source rock thickness

(Katz, 1995a). However, the presence of thin siltstone and sandstone laminae in the source rock can facilitate migration of hydrocarbons and enhance the expulsion efficiency (Littke, 1994). Examples come from the Carpathians and Balkans where the Oligocene Menilitic and Maykop Formations, respectively, served as excellent source rocks because their dark chocolate shales contain large amounts of thin siltstone or even sandstone horizons, making both source rocks highly efficient in expulsion.

Source rock deposition and preservation

There are two principal requirements for the accumulation of source rocks, bio-productivity and anoxia leading to preservation of organic matter at the sediment–water interface (Katz, 1995a). In nature there are numerous pathways leading to the deposition and preservation of organic-rich sediments. Nevertheless, sediments having sufficient organic matter content to support the generation and expulsion of hydrocarbons are relatively rare and normally have limited temporal and spatial distributions in most sedimentary basins.

With the exception of that accumulating in peat bogs (mires), mangrove swamps and microbial mats, the organic matter in sediments has been transported to its final resting place, in most instances along with the inorganic components (Huc, 1988a). The allochthonous organic matter is represented by

(1) phytoplankton growing in the water mass;
(2) terrigeneous plant matter carried to the sea by rivers; or
(3) wave destruction of coastal mires.

Organic matter of terrestrial origin is identified in marine black shales even hundreds of kilometres offshore (Habib, 1982; Simoneit and Stuermer, 1982). Most coastal and shelf areas have higher than average bio-productivity due to nutrient loading from the land (Kruijs and Barron, 1990; Littke *et al.*, 1997). In all but the lowest-energy seas, currents mix and transport this low-density organic detritus to quiet parts of the basin where it drops out with clay and fine silt. If the bottom waters are anoxic, the organic matter is preserved (Fig. 18.6). This gives rise to the normal association of high organic content with shale and marls. The distribution of organic matter richness in a basin is commonly concentric and mimics the grain-size distribution (Huc, 1988b). The modern Black Sea sediments (Fig. 18.7) display the importance of water mass hydrodynamics in the process of deposition of organic-rich sediments. Their grain-size distribution is concentric around depocentres located in western and eastern parts and anoxic

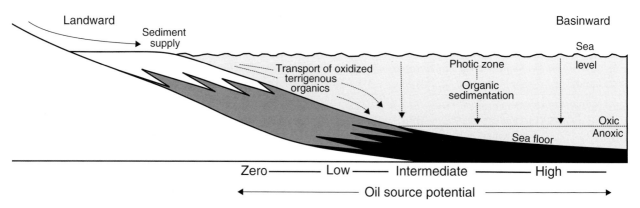

Fig. 18.6. Generalized model of marine oil-prone source rock deposition (Creaney and Allan, 1992).

conditions at the bottom are characteristic for almost the whole basin, with the exception of its relatively small northwestern portion.

The preservation of organic matter in rocks is a function of two competing rates. The first of these is the rate of bio-productivity and accumulation at the sediment–water interface. The second is the rate of destruction by oxidation or by bottom-feeding organisms. The processes are not completely independent in that very high bio-productivity associated with high nutrient supply to the photic zone and in a coastal oceanic current upwelling (Parrish, 1982; Piper and Link, 2002) can lead to oxygen deficiency through decay of the abundant organic matter (Pedersen and Calvert, 1990). In stratified water masses this can create anoxic bottom waters or an oxygen minimum zone within the water mass itself. Some models for source rock deposition suggest the movement of the oxygen minimum zone onto a shelf during a relative sea-level rise to produce temporary basin anoxia (Katz, 1995a). Also rapid burial where sedimentation rates are high has been invoked to help preserve organic matter by isolating it from processes of alteration or destruction at the sediment–water interface.

Certain times in the geological past have been more favourable for source rock deposition and preservation than others. This relates to many factors, including secular changes through time in the world's flora and conditions favouring preservation. More than 90% of the world's petroleum reserves have been generated from source rocks in just six stratigraphic intervals representing only one-third of Phanerozoic time (Ulmishek and Klemme, 1990; Klemme and Ulmishek, 1991). The intervals and their relative contributions to global reserves are: Silurian (9.0%), Upper Devonian–Tournaisian (8.0%), Upper Carboniferous–Lower Permian (8.0%), Upper Jurassic (25.0%), Aptian–Turonian (29.0%), and Oligocene–Miocene (12.5%). Clearly three-quarters of

the source rocks contributing to existing oil and gas deposits were deposited after Middle Jurassic times. Four of the time intervals representing about 80% of the global reserves (Silurian, Upper Devonian–Tournaisian, Upper Jurassic and Aptian–Turonian) coincide with major global transgressions (Sloss, 1979, 1988; Hallam, 1984; Haq *et. al.*, 1987) that resulted in widespread flooding of the continents.

The paleo-relief on flooded continents is shaped by a combination of tectonic activity, often severe, such as that associated with rift and foreland basins, but commonly subtle, as in broad intracratonic depressions, and landscape erosion immediately prior to the transgressive event. The widespread Lower Silurian black shales responsible for nearly all of the Paleozoic-sourced hydrocarbons in North Africa were deposited in the Early Silurian transgression resulting from the melting of the Late Ordovician icecap. The small lateral discontinuities within the Lower Silurian sequences relate principally to paleo-relief on a post-glacial landscape (Lüning *et al.*, 2000). At the other extreme, the very thick, but laterally variable, Late Devonian–Tournaisian source rock succession in the Timan–Pechora basin is the result of marine flooding coinciding with active rifting in the Urals foreland at the early stages of orogeny and foreland basin subsidence (Schamel, 1993). The source rocks are spatially associated with both the early foreland basin and the rifts, thus, conveniently positioned to charge subsequent anticlinal traps of both 'thin-skin' and basin inversion styles.

Conditions suitable for deposition and preservation of organic-rich sediments occur in four main settings of relevance to thrustbelt petroleum resources:

(1)   passive continental margins and epi-continental seas, especially at times of rapid sea-level rise or where ocean-current upwelling periodically occurs;

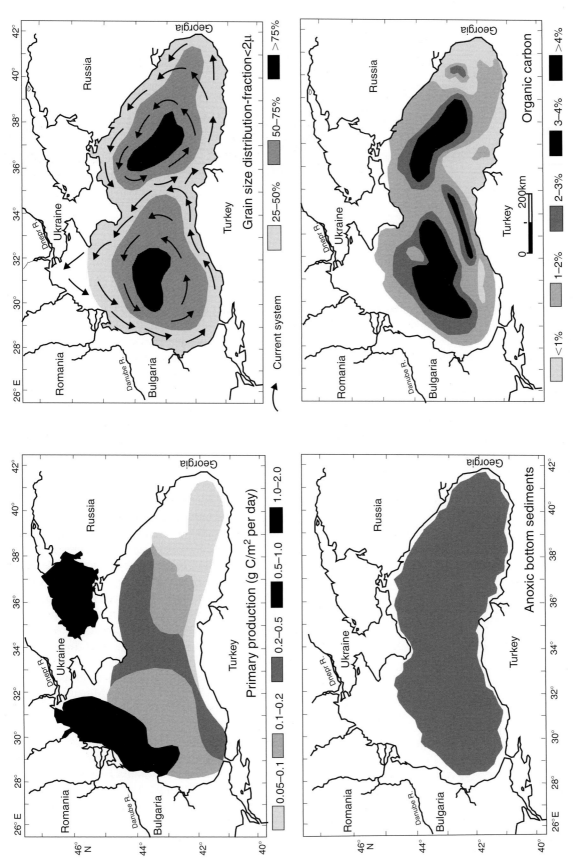

Fig. 18.7. Organic carbon content of modern Black Sea sediments compared to primary organic productivity, area of anoxic bottom sediments, distribution of fine-grained sediments, and water mass current patterns (Huc, 1988a).

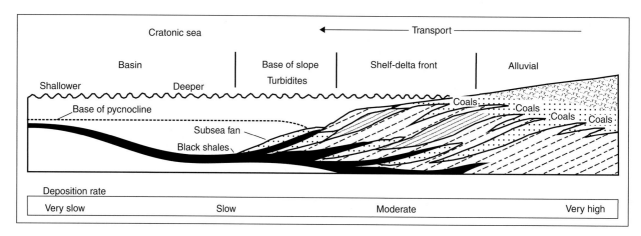

Fig. 18.8. Position of black shale and coal organic facies within an idealized marine foreland basin modelled broadly on the Acadian (Devonian) foredeep in the northern Appalachians (after Kepferle, 1993).

(2) tectonically active silled basins and continental depressions having restricted circulation of marine waters;

(3) peat mires, the site of coal formation; and

(4) lakes, especially large, long-lived and stratified lakes.

Both passive continental margin and silled tectonic basins require water mass stratification for maintenance of anoxic bottom waters and both are subject to loading of nutrients and organic detritus from the adjacent land mass. Lower latitudes are favoured due to higher bio-productivity and greater stability of stratification leading to more effective preservation. The differences are largely related to the thickness and spatial distribution of the source rocks deposited, which are normally thicker, but more laterally restricted, in tectonic basins. Rift basins have a tendency to concentrate high-quality oil-prone source rocks due to their low width–depth ratios and high subsidence rates (Katz, 1995b). Thrustbelts may develop in either of these source-rock settings.

Marine black shales are a relatively deep-water organic-rich facies deposited near the centres of epicontinental depressions and the distal portions of foreland basins during times of rapid relative sea-level rise (Creaney and Passey, 1993). At these times the basin centres are sufficiently deep and isolated to support anoxic bottom waters and inhibit dilution of organic matter by the influx of terrestrial siliciclastic sediments. These thin transgressive black shale units have two possible sequence stratigraphic associations (Wignall and Maynard, 1993).

(1) Basal transgressive black shales are associated with the initial flooding surface at a sequence boundary. They exhibit no lateral facies variation and do not connect laterally with coeval shoreline facies. Rather they pass proximally into sediment-starved facies, such as phosphatic lags, as illustrated by the Duwi–Dakhla source rocks of eastern Egypt (Robinson and Engel, 1993).

(2) Maximum flooding black shales are normally distal condensed sections within a sequence that passes proximally into thicker, coarse-grained nearshore facies.

Black shale successions can form in nearly any basin setting, but in thrustbelts they commonly are developed in the lower and more distal parts of foreland basin successions (Fig. 18.8). The black shale or marl deposition marks the onset of transgressive–regressive cycles during which the foreland depression is deep and anoxic. These may be either basal transgressive or maximum flooding black shales deposited in the deeper parts of the foreland depression. Contemporaneous with the distal black shales, coal may be deposited in the proximal shoreline and coastal plain facies. In the Northern Appalachian basin there are two important source rock successions deposited in foreland basins (Fig. 18.9), the Ordovician Utica–Martinsburg Shale (Wallace and Roen, 1989; Ryder *et al.*, 1998) deposited during the Taconic orogeny and the Middle and Upper Devonian black shales (De Witt *et al.*, 1993; Kepferle, 1993) deposited during the Acadian orogeny.

The precursor to coal deposits is coastal plain peat mires and swamps. Peat preservation leading to coals is very sensitively controlled by water table rise related to relative sea-level rise (McCabe and Parrish, 1992; Petersen *et al.*, 1996). This is accomplished where and when the overall increase in accommodation space

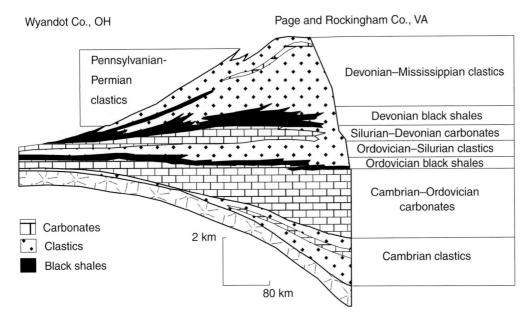

Fig. 18.9. Generalized stratigraphic profile of the northern Appalachian sedimentary basin showing the two principal black shale source-rock successions, the Ordovician Utica–Martinsburg shale and various Middle–Upper Devonian black shales (Wallace and Roen, 1989). Both black shale successions lie at the base of a thick marine foreland basin fill deposited during Taconic (Ordovician) and Acadian (Late Devonian) thrusting in the Appalachians immediately to the east (right) of the profile.

approximately equals the production rate of peat (Bohacs and Suter, 1997). The rates of modern peat production vary between $1\,mm\,yr^{-1}$ in cool temperate regions to about $6\,mm\,yr^{-1}$ in the tropics. At higher rates of relative sea-level rise, the mires become inundated and buried by clastic sediments. At lower rates the mires dry out and are subject to oxidation and erosion. A shallow water table is essential for maintaining the anoxic conditions required for preservation of organic matter in the mire. Although coals can form in coastal plains of virtually any basin setting, the rates of subsidence, or relative sea-level rise, required are closest to those of foreland basins (McCabe, 1991). This may explain why coal deposits worldwide are most commonly associated with foreland basins. Examples come from Carboniferous sections of the Appalachians associated with the Alleghanian orogeny, in Europe associated with the Variscan orogeny, and in the Urals associated with the Uralian orogeny, from the Cretaceous section of the western North American Cordillera, and from the Tertiary section of the Andes.

Over 80% of all coals worldwide are humic and capable of generating only gas (Hunt, 1991), yet oil-prone sapropelic coals do exist and are known oil sources in many basins (Katz *et al.*, 1991; Snowdon, 1991; Philp, 1994; Powell and Boreham, 1994). Macgregor (1994) suggested that significant oil-prone coals are restricted to two paleoclimatic–paleobotanical fairways:

(1) Tertiary angiosperm assemblages within 20° of the paleo-equator and
(2) Late Jurassic–Eocene gymnosperm assemblages formed on the Australian and associated plates.

Petersen *et al.* (1996) noted that coals deposited during rapid water-table rise have an increased average hydrogen index due to better preservation of spores, pollen and algae, and contain organic matter representing kerogens of Types I and II.

Lacustrine source rocks are deposited in tectonic lakes and on flood plains in humid, semi-arid and arid environments. The preservation of organic matter in these lakes depends on the maintenance of anoxic or micro-oxic bottom waters, a condition favoured by thermally or density-stratified lake waters (Powell, 1986). Their preserved organic matter is represented mainly by algae and higher plants, extensively modified by bacteria under slightly oxidizing conditions and supplemented by the remains of the bacteria. Carroll and Bohacs (2001) recognize three end member facies associations that recur in greatly different geographic settings and over a wide range of geological time. These facies associations are fluctuating-profundal, fluvial lacustrine and evaporative. The periodically deep lakes, in which the fluctuating-profundal facies association is deposited, are dominated by algal (Type I) organic matter, resulting in a relatively homogeneous oil-prone source rock. The fluvial lacustrine association is characterized by substantial loading of transported terrestrial organic matter leading to mixed Type I–III kerogens capable of generating waxy oil. This highly heterogeneous facies association contains both oil-prone and gas-prone source rocks. The diversity of source biota is normally restricted in the evaporative facies association and preservation may be enhanced due to more intense water mass stratification and inhibition of bacterial decay (Hite and Anders, 1991; Aizenshtat *et al.*, 1999). Hypersaline, sulphate-rich lakes may preserve a distinctive Type I-S (sulphur-rich) kerogen that is capable of generating oil at thermal maturities as low as 0.45% vitrinite reflectance equivalent (Carroll and Bohacs, 2001). Relative to marine source rocks, lacustrine facies exhibit a high degree of geochemical heterogeneity as is displayed in the dominantly fluvial lacustrine Eocene Green River Formation of the Uinta basin, Utah (Ruble and Philp, 1998; Ruble *et al.*, 2001; Fig. 18.10).

Source rocks in thrustbelts

The relationship to source-rock successions differs between thin-skin thrustbelts developed within passive margin–foreland basin couplets and thick-skin thrustbelts formed along the inverted margins of a rift basin. In the thin-skin thrustbelt habitat there may be some degree of predictability to the distribution of source-rock types in which oil-prone marine (Type II kerogen) source rocks in pre-orogenic strata and early/distal foreland basin shales are overlain by gas-prone coaly (Type III kerogen) late/proximal foreland basin source rocks. Where the foreland succession is very thick near the leading edge of the thrustbelt, pre-orogenic source rocks may be overmature due to deep burial, in which case the source rocks available to charge thrustbelt traps are only those of the foreland succession. The source rocks associated with thick-skin thrustbelts are likely to have been deposited in silled marine or lacustrine basin settings. These can be very rich and oil-prone source rocks that may have reached only a low level of maturity at the onset of thrusting. It is the reason for the very large volumes of hydrocarbons in many thrustbelts of this type.

Source rock successions are not merely passive elements within a thrustbelt. Organic-rich shale and marl, being normally the weaker components in the mechanical stratigraphy, localize detachment surfaces of individual thrust sheets. In addition to being inherently weak, fine-grained source rocks are easily overpressured due to compaction disequilibrium, clay dewatering or active generation of hydrocarbons, as described in Chapter 10. This can have an important impact on petroleum systems in a thrustbelt. The source rock is left

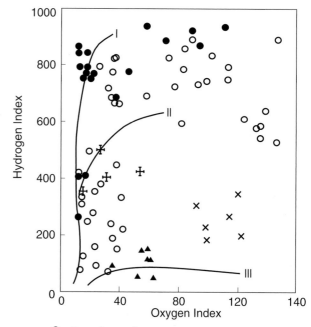

Fig. 18.10. Highly variable lacustrine organic facies within the Eocene Green River Formation of the Uinta basin, Utah. RockEval pyrolysis data are from Anders and Gerrild (1984) and Hunt (1991).

behind in the footwall, where thermal maturation of hydrocarbons can continue. It remains in contact with the thrust surface, which, for as long as the thrust is active, can serve as the migration pathway for hydrocarbons moving upward into hanging wall traps. There are numerous examples of known principal source rocks serving as detachment horizons. Prominent among thrustbelts with source rock basal and/or intermediate detachments are the Wyoming thrustbelt (Frank *et al.*, 1982; Warner, 1982), the sub-Andean Zone of the Santa Cruz–Tarija basin (Belotti *et al.*, 1995b; Dunn *et al.*, 1995) and the thrustbelt developed in the Maturin sub-basin in eastern Venezuela (Prieto and Valdes, 1992; Fig. 17.5).

The precise relationship between thrusting, source rock distribution and maturation history are specific to particular thrustbelts, and even to different sectors of the same thrustbelt. Several examples that follow illustrate the most distinctive of these relationships.

The most prolific of all thrustbelts, the Zagros fold belt, has a variety of important source rock units along its 2000 km length (Fig. 18.11). Nearly everywhere the

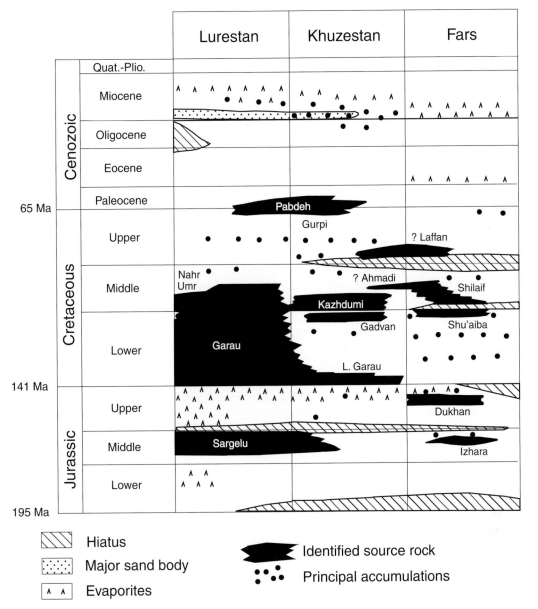

Fig. 18.11. Stratigraphic position of Mesozoic–Cenozoic source rocks in the Zagros fold belt, Iraq and Iran (adapted from Stoneley, 1990).

Lower and Mid-Cretaceous source rocks are significant (Stoneley, 1990). However, by far the most productive source rock is the Albian Kazhdumi Formation. It is this unit that has charged the high concentration of large oil fields of the Dezful embayment in Khuzestan (Fig. 17.3). This single source rock unit, which is less than 300 m thick, is responsible for hydrocarbon accumulations representing over 7.3% of the world's reserves in an area of just 40 000 km² (Bordenave and Burwood, 1994). The unit was deposited in a silled, stratified intra-shelf depression during a period of sea-level rise when organic-rich marls were laid down in a nutrient-charged prodeltaic setting that combined highly productive surface waters with anoxic bottom waters (Figs. 18.12 and 18.13). The unit contains mainly Type II kerogen with TOC as large as 11% and a genetic potential ($S_1 + S_2$) of up to 40 g HC/kg rock. In a relatively small area this highly productive source rock coincides with excellent reservoirs, highly efficient cap rocks, and large amplitude anticlines resulting in the extremely efficient Kazhdumi–Asmari petroleum system. The Dezful depression continued to be active through the

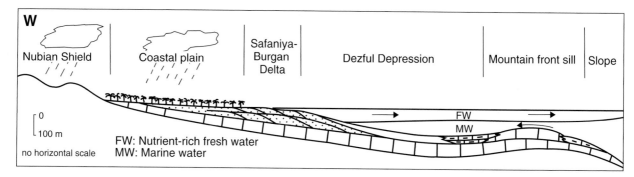

Fig. 18.12. Depositional setting of the Albian Kazhdumi Formation source rock in the Dezful embayment, Iran (Bordenave and Burwood, 1994). The source rock was deposited in a silled intracratonic depression in which a positive water balance maintained water mass stratification.

Tertiary (Fig. 18.13) and controlled the location of both the Paleocene Pabdeh Formation source rock and the Miocene Gacharsan regional evaporite seal to the Asmari carbonate reservoir (Alsharhan and Nairn, 1997). The Paleocene Pabdeh Formation is immature nearly everywhere in the Zagros fold belt except the Dezful embayment, where it is known to have charged several fields (Bordenave and Burwood, 1994).

In the northwest, in Lurestan and Iraq, the Garau and Sargelu formations entered the oil window in the latest Cretaceous–Early Tertiary time (Ala, 1990). They are also productive source rocks, but due to early maturity the oils generated have had a more complex history of migration and trap filling than those of the Dezful embayment (El Zarka, 1993). In the southeast Zagros fold belt, the overmature Silurian Gahkum shale has probably charged the large dry-gas fields in the Permian Dalan carbonate reservoirs of the Fars region (Fig. 17.3). The Mid-Cretaceous Shilaif-Shu'aiba shales are within the oil window, but are relatively lean (Versfelt, 2001). Consequently, the Fars province lacks the very large oil fields so common to the north.

The active, oil-prone source rocks in the thrustbelts flanking the Adriatic, such as the Apennines, Southern Alps and Dinarides, were deposited in small pull-apart basins within a broad carbonate platform during the time of continental break-up forming the Tethyan seaway in the Late Triassic–Early Jurassic (Zappaterra, 1994). Oil fields are associated with each of these oil-generative basins, for example Riva de Soto, Emma, Ionian, Monte Alpi and Noto/Streppenosa. However, the productive areas and reserves are small. Unlike the older source rocks, the thin, deep-water organic-rich marls and shales of mainly Upper Jurassic and Mid-Cretaceous age are widespread, but generally they are immature and have not charged known oil fields. The thick organic-rich turbidite sequences that fill the migrating foreland basins are gas-prone, generating pri-

marily biogenic gases. Only where thrusts impinge on one of the Late Triassic–Early Jurassic pull-apart basins are there oil fields (Mattavelli *et al.*, 1993). In all other parts of the Apennine–Alps–Dinaride system gas fills thrustbelt traps.

The oil and gas fields of the central and southern Canadian Foothills belt lie east of the passive margin hinge line where the pre-orogenic, pre-Cretaceous cratonic succession is thin relative to the foreland basin succession (Thompson, 1982). The fault-propagation folds, duplex and triangle zone structures that characterize this region imbricate Mississippian and Cretaceous reservoirs and Cretaceous source rocks (Fermor and Moffat, 1992). In the northern Foothills belt the large detachment folds hosting the gas fields are developed completely in the passive margin succession west of the hinge line. Both the reservoirs and source rocks are within the passive margin succession. A major change in dominant structural style from south to north relates to changes in the character of the stratigraphic section entrained in the Foothills thrustbelt (McMechan and Thompson, 1989), and that in turn has lead to differences in source rock, maturation–migration scenarios and the nature of the traps (Newson, 2001).

Two thick source rock units are recognized in the Santa Cruz–Tarija basin in Bolivia and Argentina, the Silurian Kirusillas Formation and the Devonian Los Monos Formation (Dunn *et al.*, 1995). Yet, only the Los Monos Formation is likely to have charged the majority of the large anticlinal oil fields in the basin (Lindquist, 1998). The time of entry of the Los Monos into the oil generative window coincided with or post-dated the Late Tertiary time of trap formation (Fig. 17.8), but the deeper Kirusillas shales had been generating hydrocarbons since the Late Paleozoic and were largely overmature, capable of generating only dry gas during the Late Tertiary. The hydrocarbon accumulations in the Tarija

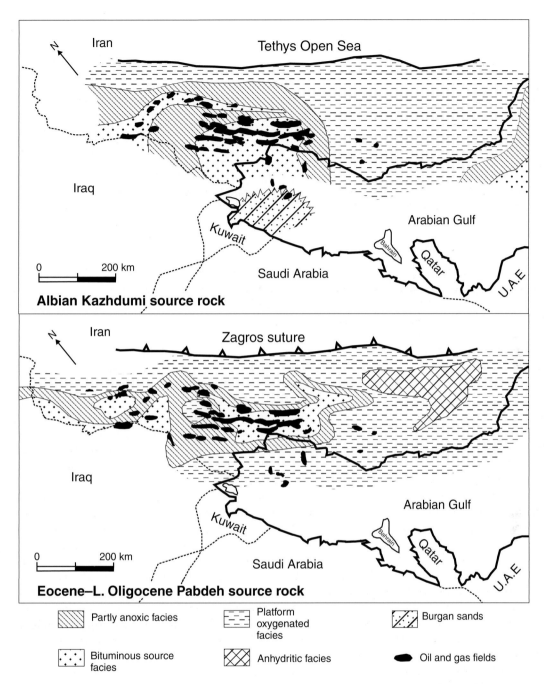

Fig. 18.13. Paleogeographic distribution of bituminous source rocks and associated facies in the Albian Kuzhdumi Formation and the Eocene–Lower Oligocene Pabdeh Formation (after Bordenave and Burwood, 1994 and Alsharhan and Nairn, 1997). Most of the larger oil fields in the central Zagros fold belt are centered on the source rock facies of these two units. Outside of this area, gas fields are more prevalent.

basin are all located inside the erosional limit of the Los Monos Shale (Belotti *et al.*, 1995b).

The highly productive petroleum provinces of southern California are examples of partially inverted pull-apart basins formed adjacent to the San Andreas fault system in the Late Tertiary (Peters *et al.*, 1994).

The principal source rock is the very thick succession of organic-rich shales and diatomites, the Upper Miocene–Lower Pliocene Monterey Shale, deposited in an array of relatively small, silled basins (Isaacs and Garrison, 1983; Graham and Williams, 1985). Coarse siliciclastic sediments entered these basins from adjacent basement highs via alluvial and submarine fans and are intimately intercalated within the Monterey Shale, which is both source rock and seal for these sandstone reservoirs. In the Los Angeles basin, entry of the source rock succession into the oil generative window occurred simultaneously with partial inversion of the rift basin margins in the Late Pliocene to the recent past (Wright, 1991). Virtually every potential structural and combination trap within this and the other basins in the region is charged, normally to spill. In the Ventura–Santa Barbara, Santa Maria and southern San Joaquin basins, even fractured diatomites produce oil (Mero *et al.*, 1992; Reid and McIntyre, 2001).

The oil fields associated with the Naga thrust fault in the Assam province of India are hosted in Tertiary fluvial–lacustrine–paralic sandstone reservoirs. The oil source rocks are carbonaceous shales and sapropelic coals interbedded with the reservoir sandstones. Long-distance migration is not a requirement; the Paleocene–Eocene succession forms one petroleum system and the Oligocene–Miocene succession forms another, separate system (Kent *et al.*, 2002).

The source rocks that charged the thick-skin anticlinal traps of the Cuyo province in western Argentina and the Junggar province in western China are both lacustrine. In the case of the Cuyo basin, the fault-bounded lacustrine basin is formed by rifting pre-dating basin inversion and thrusting (Uliana *et al.*, 1995; Urien, 2001). In contrast, the exceptionally sapropel-rich Upper Permian source rocks of the Junggar basin were deposited in an intramountain basin during the initial stages of the Variscan Karamay thrust belt (Lawrence, 1990). These Upper Permian oil shales are among the richest and thickest source rocks in the world (Carroll *et al.*, 1992) having TOC and S_2 values of up to 34% and 200 mg HC/kg rock, respectively.

19 Maturation and migration in thrustbelts

General statement

Within a sedimentary basin or thrustbelt, petroleum that is generated within a volume of active source rock migrates out of the hydrocarbon kitchen and through the basin along a commonly complex network of carrier beds, faults and fracture arrays. In most instances this carrier network disperses the petroleum to isolated cul de sacs within the sedimentary section or to the ground surface where it forms seeps (Macgregor, 1993). Less commonly, the petroleum is directed towards structural or stratigraphic features that retard its upward migration. Such features are traps, the focus of hydrocarbon exploration and development. Only some of the traps hold a sufficient volume of petroleum to warrant commercial exploitation. At each stage in the expulsion and migration of the petroleum some fraction of the original petroleum generated in the kitchen is dissipated (Fig. 19.1) so that the portion that eventually resides in commercial traps can be quite small. Biteau *et al.* (2003) refer to the efficiency ratio of hydrocarbons accumulated (HCA) to hydrocarbons generated (HCG) as the petroleum system yield (PSY). Different petroleum systems in the same basin or thrustbelt may have different PSY as in the Magdalena basin of Colombia, a thrustbelt setting, where the Villeta-Caballos (!) and Villeta-Monserrate (!) have reported PSYs of 6.0% and 4.9%, respectively (Biteau *et al.*, 2003). In most thrustbelts much lower efficiency

ratios should be expected. The expulsion efficiency of a source rock helps to determine not only the PSY of the petroleum system, but also the composition of accumulated petroleum.

Primary migration or expulsion is the liberation of newly generated oil molecules from the kerogen followed by pressure-driven movement through the fine-grained shale or marl matrix, possibly through transient microfractures (Mann, 1994). Secondary migration refers to long-distance movement through permeable strata or fractures driven mainly by buoyancy resulting from the density contrast of petroleum fluids with respect to coexisting formation water. Oil and gas can become entrained in an externally driven water flux, although, because their solubilities are very low, this is not thought to be an important migration mode (McAuliffe, 1979). Water flow will be more effective at dispersing the lighter, water-soluble components of petroleum, rather than concentrating them in traps. But lighter components of oil can be transported in gaseous solution (Thompson, 1988; Leythaeuser and Poelchau, 1991). There can also be movement along concentration gradients by diffusion or osmosis, but rates are several orders of magnitude lower than for Darcy flow driven by buoyancy (Krooss *et al.*, 1991), and so is of importance only in primary migration. Primary and secondary migration can be treated as multiphase fluid flow driven by petroleum fluid potential gradients (England *et al.*, 1987), but on different scales.

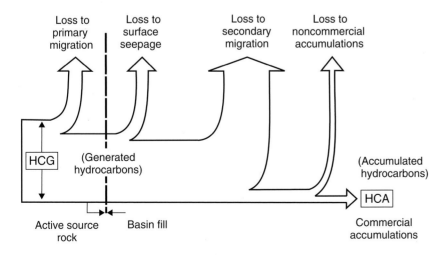

Fig. 19.1. Hydrocarbon loss during migration from an active source rock (England, 1994). Only a small portion of the hydrocarbon generated (HCG) is expelled from the source rock (HCE) to find its way into commercial accumulations (HCA). Most of the expelled petroleum is dispersed in the secondary migration carrier system, lost to surface seeps or accumulates in subcommercial pools.

399

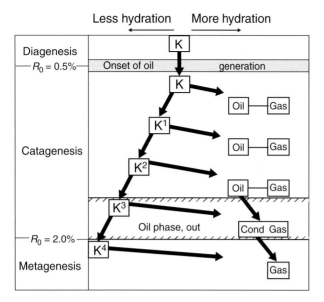

Fig. 19.2. Schematic representation of thermal maturation from diagenesis through catagenesis to metagenesis (Horsfield and Rullkötter, 1994).

Maturation and expulsion of hydrocarbons can be complexly linked processes (Mann *et al.*, 1997). The type of petroleum generated and its composition depends on the type of kerogen within the source rock and the stage of generation or maturity or of the source rock. The nature of the generated petroleum and the rate at which it leaves the source rock depends on the expulsion behaviour of the source rock, which in turn depends on the particular mechanism for expulsion. The efficiency of expulsion influences the maturity level of the hydrocarbon released, and thereby its composition. If the source rock continues to be thermally stressed, oil that is not expelled is ultimately transformed into more easily expelled gas.

Petroleum generation, expulsion and migration differ significantly in thrustbelts from other hydrocarbon settings. There are several reasons for this.

(1)   In thrustbelts rates of syn-tectonic burial and uplift are accelerated due to tectonic loading, rapid sedimentation and isostatic rebound. Thrusting has the effect of compartmentalizing the deformed basin, creating adjacent segments with markedly different burial and thermal histories.

(2)   The compressional stresses associated with thrusting may have more than the normal secondary influence on the generation of petroleum. In addition, they may contribute to the hydrofracturing of maturing source rocks, causing microfractures that aid in migration of petroleum out

of the source rock thereby increasing the expulsion efficiency.

(3)   The complex array of faults and fracture networks associated with thrusting and post-orogenic collapse link deformed carrier beds and modify the efficiency of the carrier beds to form an extremely complicated and ever-changing plumbing system for petroleum migration.

(4)   Topographically driven hydrodynamics can be unusually active in thrustbelts that are sub-aerially exposed.

All of these factors among others make the investigation and modelling of petroleum generation, expulsion and migration especially challenging in thrustbelts, compared with basins in other tectonic settings.

Generation of petroleum

Hydrocarbons, having been derived from organic matter under the influence of temperature, can be considered a form of low-temperature metamorphic fluid (Lawrence and Cornford, 1995). Petroleum generation is the result of a large number of chemical reactions transforming kerogen to liquid and gaseous products of lower molecular weight and to residues of increasing degree of condensation (Schenk *et al.*, 1997). Crude oils show a progressive increase in hydrogen content, API gravity and gas-oil ratio, GOR, with increasing maturity (Fig. 19.2). As oil and associated gas are expelled from the kerogen, its composition shifts to lower hydrogen content. Ultimately oil generation gives way to condensates and finally dry gas. Calculated oil generation occurs in the 0.6–1.1% R_0 window, source-rock gas in the 0.5–2.2% R_0 range and oil cracking in the 1.6–3.3% Ro range (Sweeney and Burnham, 1990).

Most molecular transformations involved in the thermal alteration of sedimentary organic matter to form petroleum during catagenesis involve hemolytic bond breaking, i.e. cracking, and are governed by the laws of chemical kinetics (Vandenbroucke *et al.*, 1999). Most are quasi-first-order reactions in which the concentration of the reacting molecule, *c*, decreases as

$$-\frac{dc}{dt} = kc. \tag{19.1}$$

Consequently, petroleum formation is a function of both time, *t*, and temperature, *T*. The rate of kerogen transformation to petroleum, *k*, is described by the semi-empirical Arrhenius equation:

$$k = Ae^{-E/RT}, \tag{19.2}$$

where *A* is the pre-exponential factor or Arrhenius constant, *E* is the apparent activation energy, *R* is the gas constant and *T* is the absolute temperature.

The higher the temperature is, the higher the rate of

Table 19.1. LLNL kerogen kinetic parameters.

Cracking P-primary S-secondary	Fract. (0–1)	Activation energy (kcal mol^{-1})	Arrhenius constant (1/my)	Reaction constants
Type I				
P	0.07	49.0	1.6×10^{27}	Primary oil = 650
P	0.90	53.0	1.6×10^{27}	Primary gas = 70
P	0.03	54.0	1.6×10^{27}	
S	1.00	54.0	3.2×10^{25}	2ndary gas yield = 50%
Type II				
P	0.05	49.0	9.5×10^{26}	Primary oil = 350
P	0.20	50.0	9.5×10^{26}	Primary gas = 65
P	0.50	51.0	9.5×10^{26}	
P	0.20	52.0	9.5×10^{26}	
P	0.05	53.0	9.5×10^{26}	
S	1.00	54.0	3.2×10^{25}	2ndary gas yield = 50%
Type III				
P	0.04	48.0	5.1×10^{26}	Primary oil = 50
P	0.14	50.0	5.1×10^{26}	Primary gas = 110
P	0.32	52.0	5.1×10^{26}	
P	0.17	54.0	5.1×10^{26}	
P	0.13	56.0	5.1×10^{26}	
P	0.10	60.0	5.1×10^{26}	
P	0.07	64.0	5.1×10^{26}	
P	0.03	68.0	5.1×10^{26}	
S	1.00	54.0	3.2×10^{25}	2ndary gas yield = 50%

kerogen conversion to petroleum. The conversion rate approximately doubles with every 10 °C increase in temperature. The factor RT is a measure of the thermal energy of the system at a given temperature. If the factor RT is large compared with the activation energy, petroleum generation is rapid.

Modern methods of petroleum systems analysis utilize kinetic models based on the Arrhenius relationship (equation 19.2) for simulating petroleum generation within sedimentary basins over geological time. Pyrolysis experiments using immature source rock samples of differing kerogen types are used to determine the values of the kinetic parameters A and E (Tissot and Espitalié, 1975). The range of reactions and processes leading to the thousands of molecular products of petroleum genesis from kerogen is very large and complex. The pyrolysis data attempt to capture, using the bulk kinetic behaviour of a single or limited number of components, the essence of the thousands or even millions of individual molecular transformations that lead to crude oil and gas.

In performing the simulations it is assumed that the same kinetic parameters predict with equal validity the rates of petroleum generation during experimental pyrolysis at temperatures of about 450 °C and in nature at temperatures nearer to 100 °C. That this is not true in every instance is due not only to errors in the measured

kinetic parameters but also to the potential nonuniversality of the underlying assumption. Arguments continue over the applicability of different pyrolysis methods, including open versus closed or hydrous versus anhydrous systems (see Lewan, 1994, 1997; Pepper and Corvi, 1995; Pepper and Dodd, 1995; Schenk and Horsfield, 1998; Vandenbroucke *et al.*, 1999; Henry and Lewan, 2001; Ruble *et al.*, 2003). Catalysis by clays (Kissin, 1987) and by the transition metals (Mango, 1997) has been proposed as an important factor in the generation of light hydrocarbons. The effect of pressure in the transformations is measurable, but it is relatively small compared with the effect of temperature (Schenk *et al.*, 1997). Nevertheless, kinetic modelling remains the most powerful tool currently available for predicting the time, depth and products of petroleum genesis under geological conditions (Waples *et al.*, 1992a, b; Waples, 1994a). There are a variety of commercial software products that facilitate the numerical simulations.

Two widely used kinetic schemes for calculating hydrocarbon generation are those published by the Lawrence Livermore National Laboratory (e.g., Burnham *et al.*, 1995) and the Institut Français du Pétrole (e.g., Tissot *et al.*, 1987; Ungerer and Pelet, 1987). The Lawrence Livermore National Laboratory kinetic schemes are given in Table 19.1. Numerous publications have

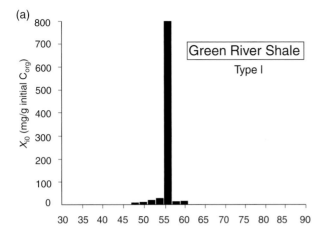

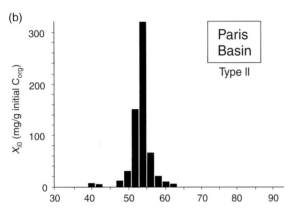

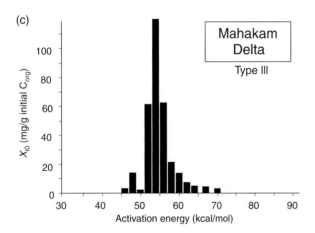

Fig. 19.3. Histograms of activation energies of representative source rocks having differing kerogen types (Tissot and Espitalié, 1975 and Ungerer and Pelet, 1987).

reported kinetic data for source rocks with Type I, II and III kerogens (e.g., Tissot *et al.*, 1987; Lewan, 1993; Tegelar and Noble, 1994; Burnham *et al.*, 1995; Reynolds and Burnham, 1995). Marine Type II and terrigenous

Type III kerogens generally have broad distributions of activation energies (Table 19.1 and Fig. 19.3). In contrast, Type I kerogens have very narrow distributions of activation energies with a slightly higher maximum. Also, maxima for the distributions can be highly variable depending on the sulphur content, the mineralogy of the source rock and numerous other variables (Ungerer, 1990). High sulphur Type II kerogens, such as, for example, the Monterey Formation, California, have particularly low distributions of activation energies because C–S bonds are more readily hydrolyzed than C–C bonds (Tegelar and Noble, 1994; Waples, 1994a).

In thrustbelts, where sedimentary and tectonic burial is followed in rapid succession by uplift, erosion and possibly reburial, the differing kinetic behaviours of kerogen types are especially important to the generation of petroleum. Type II and III kerogens mature over a wide temperature–depth range (Fig. 19.4) and are more likely to continue to generate during multiple episodes of maturation. In contrast, Type I kerogens mature over a very narrow temperature range and expel rapidly (Fig. 19.4) due to their narrow activation energy distributions (Fig. 19.3). They are more likely to exhaust their generative potential in a single burial event.

All kerogen types are capable of generating methane and other natural gases, but sapropelic source rocks, containing Type II-S, II and I kerogens, initially yield larger volumes of liquid hydrocarbons than Type III humic kerogens (Fig. 19.5). They are considered to be oil-prone, whereas humic kerogens are gas-prone by default. Based on hydrous–pyrolysis experiments, Lewan and Henry (2001) report a higher range of gas–oil ratios for humic kerogen than for sapropelic kerogens. Typical ratios for humic and sapropelic kerogens are within the intervals 883–2831 and 382–2381 scf/bbl^{-1}, respectively. Because sapropelic kerogen has a higher elemental concentration of hydrogen at the outset, it will have a larger yield of both oil and methane than humic kerogen (Fig. 19.5). Sapropelic kerogens can generate twice as much hydrocarbon gas per gram of organic matter as humic kerogen (Lewan and Henry, 2001). To a large extent the methane and related hydrocarbon gases from sapropelic kerogen are generated from residual oil (bitumen) remaining in the source rock as it is buried to deeper and hotter levels of the basin. Prolific natural gas provinces are not restricted to coal-bearing basins.

Quigley and Mackenzie (1988) stress the importance of temperature on the maturation of petroleum. Peak maturity is reached in rich oil-prone source rocks between 100 and 150 °C and in lean oil-prone source rocks at 120–150 °C. At temperatures in excess of 150 °C oil retained in the source rock cracks to gas.

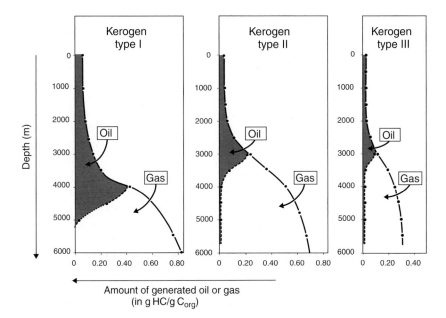

Fig. 19.4. Compared hydrocarbon generation of various kerogen types for a uniform geological subsidence rate and geothermal gradient (Tissot and Espitalié, 1975, reproduced in Robert, 1985).

Generation of liquid hydrocarbons from coals is complex and involves factors such as the type of macerals present, the microstructure of the coals and their constituent macerals controlling petrophysical properties, and hydrocarbon saturation thresholds (Hunt, 1991; Katz *et al.*, 1991; Thompson *et al.*, 1994; Petersen *et al.*, 1996). The pyrolytic products of coals are more dependent on maceral content than age or geographic location (Katz *et al.*, 1991). Coals rich in vitrinite generate predominately aromatic hydrocarbons with minor amounts of n-paraffins. Coals rich in algae and some other exinites (liptinites) generate mainly paraffinic hydrocarbons, whereas coals rich in resins generate naphthenic and aromatic oils.

Methane is formed in sedimentary basins at very shallow depths through microbial breakdown of organic matter, i.e. methanogenesis, and at greater depths by thermal disaggregation of organic molecules containing hydrogen. The microbes responsible for biogenic methane do not survive temperatures above about 40–50 °C (Hunt, 1979; Fig. 19.5), whereas the chemical processes leading to thermogenic methane become significant at temperatures above 135–175 °C (Pepper and Corvi, 1995). There are two main sources of thermogenic gas: maturing source rocks and the in situ cracking of oil in reservoirs (Pepper and Dodd, 1995; Schenk and Horsfield, 1998).

Natural gas of microbial origin (Martini *et al.*, 2003) is nearly pure methane (C_1) with no other hydrocarbon gases, having a low wetness value, and various amounts of carbon dioxide (Fig. 19.5). Owing to the metabolic selectivity of the microbes, biogenic methane is isotopi-

cally lighter than thermogenic methane, having a larger negative $\delta^{13}C$ value. The $\delta^{13}C$ value of microbial methane is typically lighter than $-55‰$ (Mattavelli and Novelli, 1988; Whiticar, 1994). Also low-maturity thermogenic gas produced from primary cracking of kerogen is primarily methane (Seewald *et al.*, 1998). In contrast, mature thermogenic gas associated with the onset of bitumen generation (Fig. 19.5) consists of methane together with other hydrocarbon gases (C_{2+}), resulting in a high wetness value. Its $\delta^{13}C$ values are only slightly lighter than values of the precursor organic matter. It is also associated with nonhydrocarbon gases, such as nitrogen, helium and/or hydrogen sulphide. As the thermal maturity of the natural gas increases, the wetness value again decreases. Thus, shallow microbial and deep, highly mature thermogenic methane are both dry gas, lean in the fraction of other hydrocarbon gases (C_{2+}), with a low wetness value, and only distinguished by their isotopic compositions.

The bulk of C_1–C_4 generation occurs after the peak of oil generation, but CO_2 generation occurs before, during and even after generation of liquid hydrocarbons (Seewald *et al.*, 1998). Type I sapropelic kerogens tend to initiate generation of gas at slightly higher stages of thermal maturity than do either Type II or III kerogens (Fig. 19.4). Gas generation is not restricted to high thermal maturities representing post-oil generation above the $R_0 = 1.2\%$ threshold. Significant amounts of thermogenic gas can be generated together with oil (Schenk and Horsfield, 1998). Approximately 75% of the source rock gas has been generated by the end of oil generation at $R_0 = 1.1\%$. However, bitumen and

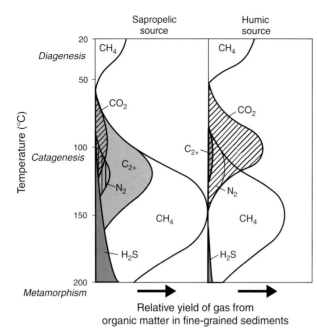

Fig. 19.5. Generation of natural gases with depth of burial
(Hunt, 1979). C_{2+} represents hydrocarbons in the gas phase
heavier than CH_4.

kerogen retained in a maturing source rock that has
passed through the oil window will crack to natural gas
with further thermal stress (Pepper and Dodd, 1995).

In an effort to simplify kinetic modelling, Pepper and
Corvi (1995) have used statistically averaged kinetic data
from many sources to calculate the entry and deadline
temperatures for the oil and gas windows for a reference
heating rate of $2.0\,^{\circ}C\,Myr^{-1}$ (Table 19.2). The temperatures differ with kerogen type and increase by about $5\,^{\circ}C$
with every doubling of heating rate. However, this table
only provides a useful rule of thumb for predicting the
levels of generative windows and is not a satisfactory
substitute for a full numerical simulation.

Migration of petroleum
Secondary migration of petroleum through a carrier
bed or fault network from an active source rock volume
to a reservoir/trap is driven principally by the buoyancy
resulting from the density contrast between petroleum
and formation water (England *et al.*, 1987). A hydrodynamic flow, if present, may deflect migration pathways
by acting either with buoyancy or against it (Hubbert,
1953; Dahlberg, 1995). The main force opposing buoyancy is capillary pressure, which is a function of the
pore-throat radii of the carrier system, the petroleum–water interface tension and wettability (England,
1994).

Under hydrostatic conditions, the driving buoyancy
force, P_b, for a petroleum column or rivulet having a vertical height, h, is

$$P_b = gh(\rho_w - \rho_o), \qquad (19.3)$$

where g is the hydrostatic gradient ($9.795\,kPa\,m^{-1}$ for
pure water) and ρ_w and ρ_o are the densities of formation
water and oil, respectively. Where a vertically rising
petroleum rivulet is deflected along a sloping surface,
the resultant buoyancy force is reduced by the sine of
the angle of dip.

Capillary pressure results from forces acting within
and between liquids and their bounding solids (Vavra
et al., 1992; Vandenbroucke, 1993). Forces between
liquid and liquid are cohesive and forces between liquid
and solid are adhesive. Where adhesive forces are
greater than cohesive forces, the liquid is wetting, otherwise, it is nonwetting. In most porous rocks water is
the wetting phase. The relative wettability of a fluid is
described by the contact angle, θ, which is the angle
between the fluid–fluid interface as measured through
the denser fluid. The capillary pressure, P_c, which is a
function of the interfacial tension between immiscible
fluids and the pore throat size, can be calculated using
the expression

$$P_c = 2\gamma\cos\frac{\theta}{r}, \qquad (19.4)$$

where γ is the interfacial tension between oil and water,
θ is the wettability and r is the average radius of the
interconnected pore space.

Oil or gas as an immiscible phase migrates in rivulets
or filaments upwards through the pore system of the
carrier bed as long as the buoyancy pressure, P_b, is
greater than the capillary entry pressure, P_c, or the resistance of the water in the pore throats, A (Berg, 1975;
Fig. 19.6). For migration to continue when a smaller
pore throat size, B, is encountered, the length of the
rivulet, h_1, must increase until an adequately long
rivulet, h_2, of oil exists for the increased buoyancy, P_b,
to force a breakthrough.

The relative permeability aids migration (Pepper,
1991; Matthews, 1999). As petroleum fills the pore
network, the ability of the network to transport water
decreases and the pore pressure increases, helping to
push the petroleum phase through capillary restrictions.
Migration rates are highly dependent on the migration
process (Thomas and Clouse, 1995; Matthews, 1999).
Typical rates for buoyancy are several $mm\,yr^{-1}$ for oil
and up to metre per day for gas. Hydrodynamic flow
rates are as high as 0.1–$100\,m\,yr^{-1}$. Compaction drive
and diffusion reach 0.001–$1.0\,m\,yr^{-1}$ and 1–$10\,m\,Myr^{-1}$,
respectively.

Table 19.2. Reference oil and gas window thresholds and deadlines for representative organofacies.[a] The calculated window boundaries are for a reference heating rate of 2.0 °C Myr^{-1}. A doubling of the heating rate elevates each of the boundaries by about 5 °C.

Organofacies	Precursor biomass	Environment/ age association	Sulphur incorporation	Kerogen type	Oil window	Gas window
A, Aquatic, marine, siliceous or carbonate/ evaporite	Marine algae and bacteria	Marine, upwelling zones, clastic-starved basins of any age	High	Type II-S	95–135 °C	105–165 °C
B, Aquatic, marine, siliciclastic	Marine algae and bacteria	Marine, clastic basins of any age	Moderate	Type II	105–145 °C	140–210 °C
C, Aquatic, nonmarine, lacustrine	Freshwater algae and bacteria	Tectonic nonmarine basins and coastal plains (Phanerozoic)	Low	Type I	120–140 °C	135–170 °C
D+E, Terrigeneous, nonmarine, waxy	Higher plant cuticule, lignin, bacteria and resin (facies D)	Mesozoic and younger 'ever-wet' coastal plains	Low	Type III	120–160 °C	175–220 °C
F, Terrigeneous, nonmarine, wax-poor	Lignin	Late Paleozoic and younger coastal plains	Low	Type III/IV	145–175 °C	175–220 °C

Note:
[a] Pepper (1991).

Primary migration

Primary migration through a source rock probably occurs initially as diffusion through micropores (Fig. 19.7) connecting with mesopores in to macropores and fractures (Mann, 1994). From there a petroleum bulk phase develops and moves along the macropore and fracture network of the source rock. Overpressure associated with the volume expansion of newly generated petroleum is widely accepted as the principal driving mechanism of expulsion (England *et al.*, 1987; Durand, 1988), but other factors can also play a role.

Primary migration can occur only after the source rock pore space is fully saturated by newly generated hydrocarbons. Mackenzie and Quigley (1988) demonstrate a causal link between expulsion efficiency and the initial petroleum potential of a source rock. The richer the source rock, the larger the excess volume of petroleum that is available for expulsion after the pore and fracture network in the source rock is fully saturated. In rich source rocks, those capable of generating more than 5 kg HC tone rock, the expulsion efficiency is estimated to be of the order of 60–80%. However, in leaner source rocks, the expulsion efficiency is less than 40%.

Pepper (1991) presents a different view of expulsion, arguing that unexpelled petroleum is not retained by capillary effects in the pore network but rather is adsorbed onto organic matter. He proposes that in tight, clay-rich source rocks the relative permeability to oil is actually greater than to water. This is due to bound water adsorbed on hydrophilic/hydrophobic clay particles, affecting the relative permeability so as to favour the Darcy flow of oil. This would explain why Pepper (1991) finds an excellent positive correlation between the expulsion efficiency of a source rock and its hydrogen index, but lower correlations with the total organic carbon and the initial petroleum potential.

Oil expulsion is easiest and most efficient from Type I source rocks typically having a hydrogen index of up to 900 mg (g C$_{org}$)$^{-1}$. Type II source rocks, with a hydrogen index of 500–700 mg (g C$_{org}$)$^{-1}$, exhibit early and relatively efficient expulsion and some sapropellic coals have a similar character of expulsion. Humic coals and Type III source rocks are capable of efficiently expelling only gas/gas-condensate at high levels of thermal maturity. Only at these levels can they expel more than 40% of their potential. The gas expulsion efficiency is greater than 90% regardless of the source rock type (Pepper, 1991).

In very fine-grained source rocks having a matrix permeability less than a microdarcy (10^{-19}–10^{-20} m^2),

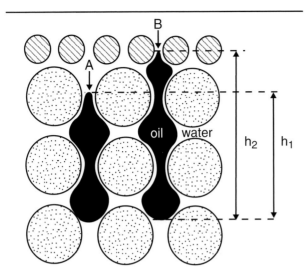

Fig. 19.6. Relationship between buoyancy forces and capillary entry pressure in a porous medium (after Berg, 1975). Explanation in text.

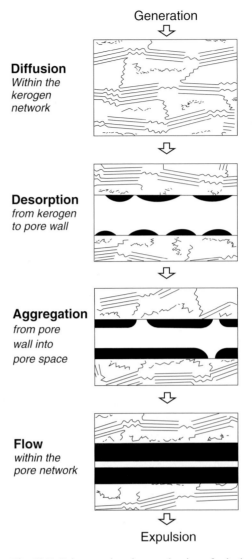

Fig. 19.7. Primary migration mechanisms for hydrocarbons in source rocks (after Mann, 1994).

fracturing may be induced by pore-pressure buildup caused by the conversion of organic matter to less dense oil and gas (Berg and Gangi, 1999). The resulting fractures increase the rock permeability by perhaps as much as six or seven orders of magnitude to 10^{-13} m^2 (Capuano, 1993) and provide pathways for primary migration. Pressure-induced fracturing is more likely to occur in rich source rocks capable of generating large volumes of oil or at elevated levels of maturity where gas is formed (Fig. 19.8). High rates of petroleum generation and/or stress loading of the source rock (Cosgrove, 2001) will further aid this hydraulic fracturing process.

The oil generating potential of sapropelic coal depends on the expulsion efficiency of the coal. Some coal microlithotypes have high microporosity and absorption capacity that trap generated hydrocarbons within the coal structure, thus retarding primary migration (Hunt, 1991). A threshold value of oil saturation of about 30 mg HC (g TOC)$^{-1}$ must be reached before expulsion can occur (Snowdon, 1991). In less oil-prone coals, such threshold levels may not be reached and the oil remains in the coal to be cracked to gas at a still higher grade of maturity.

Secondary migration

Once petroleum is expelled from the source rock it is driven by buoyancy and hydrodynamic flow (England *et al.*, 1987; England, 1994; Schowalter, 1979). The buoyancy force varies with the density difference between the petroleum phase and the formation water. The greater the density difference is, the greater the buoyancy force

for a given height of petroleum column. The tendency will be for the petroleum rivulet to rise vertically moving fastest, and thereby preferentially, along pathways of maximum pore throat size, i.e. maximum permeability. However, in dipping, faulted or fractured strata, the more favoured pathways, those of highest permeability, may be inclined, thus inducing a component of lateral migration. The many small rivulets rising from the interface between a source rock and carrier bed initially form an anastomosing network of filament-size flow paths (England, 1994), but these soon amalgamate into fewer highly efficient rivulets responsible for the bulk of petroleum migration.

The ensuing ultimate flow paths are determined by the morphologies of the more effective carrier beds

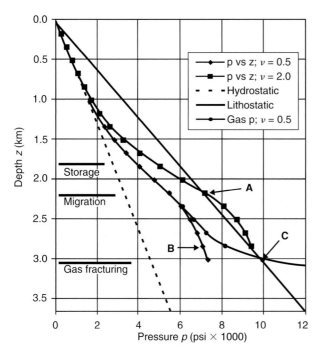

Fig. 19.8. Modelled pressure buildup with depth during oil and gas generation in the Austin Chalk source rock (Berg and Gangi, 1999). Pressure curves are shown for a rich source rock with TOC of about 2.7 wt% and a lean source rock with TOC of 0.65 wt% having ratios of initial kerogen to initial water, ν, of 2.0 and 0.5, respectively. Significant overpressures occur at a depth of 1830 m, just below the top of the oil window and in a region is which generated oil is probably retained in the source rock. For the richer source rock, the fracture pressure is reached at a depth of 2160 m (point A). The leaner source rock, however, does not reach fracture pressure through oil generation alone (B), but rather only after appreciable gas generation (point C) at a depth of 3000 m.

having adjacent top seals (e.g., interbedded sandstones and shales), permeable fault surfaces or fracture networks. The different carrier systems may act alone, in series or in parallel (Fig. 19.9) in complex geometric combinations in different parts of the basin or thrustbelt (e.g., Allan, 1989; Alexander and Handschy, 1998). Continuous seals play an important role in vertical drainage by helping to cause petroleum to converge on faults and other windows through the seals, focusing the vertical drainage into chimneys through which the vertical rise continues until another set of effective top seals is encountered (Nunn and Meulbroek, 2002).

Traps should not be considered final end points for migration, but rather transient storage, through which the petroleum will pass by breaching the top seal when the oil/gas column is sufficiently high, or spilling out the bottom when full (Sales, 1997). Top seal breaching

favours the further migration of gas and condensate over oil (Schowalter, 1979), whereas in trap spilling the oil phase is likely to migrate (Gussow, 1954).

Hydrodynamic flow in a basin or thrustbelt may deflect migration pathways from directions determined by morphology alone. The influence of basin hydrodynamics will be greatest where the buoyancy flow is mostly lateral. Hindle (1997) presents the mathematical foundation for calculating migration pathways under combined buoyancy–hydrodynamic drives.

Laboratory experiments simulating natural petroleum migration (Dembicki and Anderson, 1989; Catalan *et al.*, 1992; Thomas and Clouse, 1995) provide insights into both the configuration and rates of flow paths. After scaling the laboratory model to actual geological conditions, Thomas and Clouse (1995) conclude that most of the carrier bed above an active source rock would be contacted by oil, but at low saturations of about 5–10%. The vertical frontal advance of oil scales to 8 mm yr^{-1}, or 8 km Myr. Once the vertically migrating oil reaches an inclined permeability barrier, it continues to move laterally through only a small portion of the carrier bed just 0.5–1.0 m below the sealing surface. Migration is accomplished through a network of thin petroleum rivulets (Fig. 19.9) that occupy very little of the pore space of the carrier bed, rather than a more pervasive sheet flow. Therefore, the rates of lateral migration are likely to be geologically very rapid and hydrocarbon losses within the migration network low.

The shape of the carrier system is critical to whether migration is dispersive, resulting in the dilution of hydrocarbons away from the petroleum kitchen, or has the tendency to direct and concentrate flow, preferably towards potential traps (Fig. 19.10). There is a tendency for dilution along concave carrier systems and concentration where the system is convex. The distance of the potential trap away from the petroleum kitchen also is a critical factor (Hindle, 1997). Traps located vertically above a kitchen have the greatest potential for filling. Traps which are tens of kilometres away from the kitchen cannot receive more than a small fraction of the oil expelled, even if flow is focused along a convex carrier system. Stratigraphic and structural factors together determine whether migration will be constrained near the petroleum generative kitchen or will be far-ranging (Fig. 19.11). Focused vertical drainage of hydrocarbons is associated with a moderate to high degree of structural deformation resulting in selective breaching and local petroleum leakage through a laterally continuous seal. Basins with laterally drained petroleum systems have several features in common (Demaison and Huizinga, 1991):

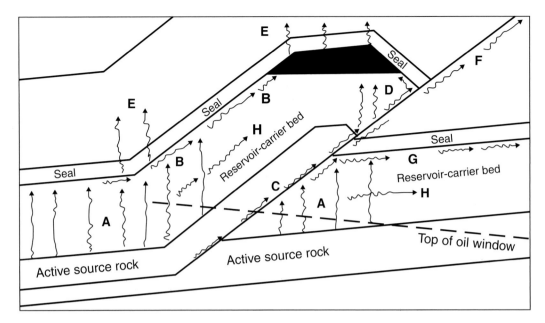

Fig. 19.9. Possible petroleum migration pathways in a simple fault-propagation fold detaching within the active source rock unit. The pathways include: (A) vertical rise of hydrocarbon rivulets through the carrier bed from active source rock areas; (B) migration along the top of the carrier bed immediately below a top seal; (C) migration along the thrust fault and associated fracture system toward a hanging wall trap; (D) migration from a thrust fault zone into the carrier bed at its hanging wall cutoff; (E) migration and dissipation of hydrocarbons along fracture systems breaking the top seal unit; (F) migration along the thrust fault and associated fracture system towards surface or higher traps; (G) migration along top of the carrier bed towards the surface or higher traps; and (H) dissipative migration into more porous zones in the carrier bed. Once thrusting ceases the thrust fault system likely becomes a barrier to flow blocking migration of petroleum from footwall kitchens to the hanging wall trap.

(1)   oil accumulations occur in immature strata at some considerable distance from the hydrocarbon kitchen;

(2)   a single reservoir succession underlying the most effective regional seal is generally host to most of the entrapped hydrocarbons in the petroleum system;

(3)   faulting of the effective regional seal is minor or insignificant.

These are conditions unlikely to be encountered in thrustbelts unless the dominant structural style is detachment folding beneath a highly effective, thick and largely unfaulted top seal. Otherwise, lateral migration will be limited to the width of individual thrust sheets, the footwall block beneath the basal detachment, or the basal detachment itself while active.

Preferential fluid flow through fault zones is well documented, even in the deep crust (MacCaig, 1988; Barton *et al.*, 1995). In sedimentary basins, fault and fracture systems may serve as either migration pathways or barriers to flow, or both at different stages in their histories. During and shortly after faulting/fracturing these are favoured high-permeability conduits for fluid transport (Hooper, 1991; Roberts, 1991; Knipe, 1993; Sibson, 1994a, b, 1996; Dholakia *et al.*, 1998; Connolly and Cosgrove, 1999) and may exhibit a fault valve behaviour as they serve to regulate the release of fluid overpressures (Sibson, 1990a, b, 2003; Grauls and Baleix, 1994; Smith and Wiltschko, 1996; Losh, 1998; Cosgrove, 2001). But once deformation ends they become merely part of the overall porous rock system, equally subject to diagenetic cementation as the matrix pore space (Laubach *et al.*, 2000; Laubach, 2003). Antonellini and Aydin (1994) observe that the capillary pressure within deformation bands in sandstone produced by cataclasis is 10–100 times greater than the surrounding host rock. The deformation bands, rather than acting as subvertical conduits for fluid flow, are highly effective barriers to flow. Stable isotope and fluid-inclusion analysis of vein and cement minerals has proven useful in many thrustbelts in constraining the history of fluid flow in deformed rocks, including the time of petroleum migration (Burruss *et al.*, 1985; Earnshaw *et al.*, 1993; Parris *et al.*, 2003).

The role of hydrodynamics driven by syn-tectonic expulsion of water from the thrustbelt will depend largely on the state of decompaction of the rocks prior

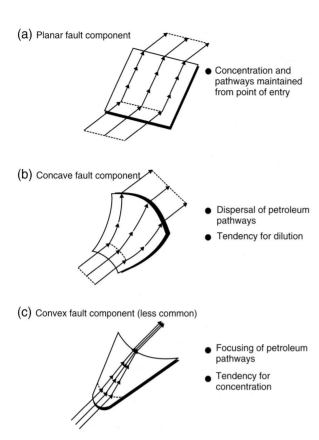

(a) Planar fault component

● Concentration and pathways maintained from point of entry

(b) Concave fault component

● Dispersal of petroleum pathways
● Tendency for dilution

(c) Convex fault component (less common)

● Focusing of petroleum pathways
● Tendency for concentration

Fig. 19.10. Effect of the carrier-bed or the fault-plane shape on the configuration of migration pathways (Hindle, 1997).

to becoming entrained in the thrustbelt. The shallow, largely fine-grained and water-rich sediments that are offscraped into an accretionary prism represent one end member state. Here the expulsion of water during thrusting is large and well documented (Foucher *et al.*, 1990a, b; Speed, 1990; Moore and Vrolijk, 1992; Saffer and Bekins, 2002). As water is tectonically squeezed out of the accretionary prism or tapered wedge of a young thrustbelt, hydrodynamic forces will overwhelm buoyancy as the driving mechanism for hydrocarbon migration (Speed *et al.*, 1991), at least locally. Also high topographic relief related to thrusting coupled with recharge of meteoric waters becomes a possible driving force for large-scale hydrodynamic fluid flow capable of deflecting buoyancy-driven hydrocarbon migration (Hitchon, 1984; Garven, 1989; Ge and Garven, 1989; Montañez, 1994). In contrast, the deeper portions of thrustbelts that imbricate older, well-compacted and largely dewatered sediments may experience only modest hydrodynamic fluid transfer during thrusting. Stable isotopic data on syn-tectonic veins formed along thrusts in thick carbonate successions of the Canadian Front Ranges and Southern Apennines indicate the

movement of only limited amounts of aqueous fluids focused in narrow zones adjacent to the thrust surfaces (Ghisetti *et al.*, 2001; Kirschner and Kennedy, 2001).

Maturation, expulsion and migration in thrustbelts
The generation, expulsion and migration of petroleum in the thrustbelts is inherently different from those in most basinal settings due to the more complex syn-orogenic burial and thermal histories and the more complex array of migration pathways presented by thrust faults and associated fracture systems. The differences occur principally towards the end of burial of potential source rocks, contemporaneous with thrusting, and immediately afterwards as the thrustbelt rebounds isostatically. Prior to the onset of thrusting, the maturity and migration scenarios are characteristic of those of the precursor basin, whatever its type. Even in a thrustbelt setting, substantial generation and migration of hydrocarbons may already have occurred prior to the onset of thrusting and thus influence the nature and amount of petroleum ultimately trapped in the thrust structures.

Petroleum maturation and migration depends to a large extent on the location of the petroleum system within the thrustbelt setting. The generating source rock and traps may lie completely beneath the basal detachment of the thrustbelt (Butler, 1991; Picha, 1996). In this case the role of thrusting is most important in affecting the time and degree of petroleum generation. More likely, petroleum generated in footwall kitchens will migrate upward through and along the basal detachment and other faults to charge traps in the hanging wall (Fig. 19.12). Alternatively, the source rock and reservoirs may be contained completely within the thrust sheets, in which case maturation–migration scenarios can be very complex and site specific. A source rock may generate in one thrust sheet, but not in others, laterally as well as vertically. Migration pathways can be difficult to predict, even if the structural architecture is relatively well constrained. Less commonly, the petroleum system can be fully contained in late- or post-orogenic piggyback or successor basins. If these small basins are sufficiently buried, which normally requires multi-phase deformation in the region (Schamel, 1991), they can generate petroleum that migrates into traps that both pre-date and post-date the basin. The source rock is located commonly within the thrust sheets beneath the floor of the piggyback basin, but in rare instances it can be in the lower part of the basin fill.

The maturation of source rocks, whether those carried within a thrust sheet or charging thrust-related traps from footwall kitchens, will depend on their thermal history (Deming and Chapman, 1989; Waples,

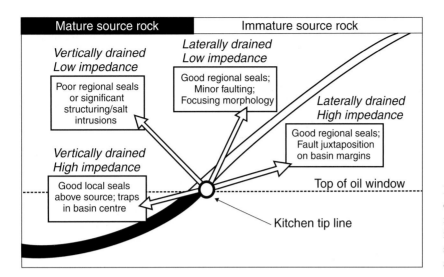

Fig. 19.11. Migration scenarios determined by a combination of regional seal quality and structural styles in a basin (Hindle, 1997). The migration styles named in italics follow Demaison and Huizinga (1991).

1992, 1994b). This, in turn, is dependent on the interplay of the burial history and changing heat-flow regimes in the portion of the basin destined to become entrained in the thrustbelt. The physical factors influencing hydrocarbon maturation within thrustbelts can be understood by considering two contrasting scenarios.

(1) Thin-skin thrustbelt scenario: thin-skin thrustbelts are most likely to develop within a proximal passive margin sedimentary prism and have a thermal history that is both long and multiphase (Fig. 19.13).

    (a) Gradual burial of the source rock as part of a passive margin sedimentary prism with normal or near normal geothermal gradients.

    (b) Acceleration in the rates of burial within a foreland basin setting in front of the advancing thrustbelt and under conditions of slightly less than normal geothermal gradients. Very high rates of sedimentation in the foreland basin can have a cooling effect on heat flows within the basin as a whole.

    (c) As the frontal thrust advances across some portion of the foreland basin that region experiences a tectonic loading. Whether the tectonic loading by the thrust sheet increases the rate of burial of source rocks in the basin depends on the interplay of fault geometry and detachment depths, and relative rates of burial of the basin as a whole. Except where relative sea level rises and the foreland basin becomes

choked with sediment, the loading effect of the repetition of the stratigraphic section by thrusting is decreased by erosion from the top of the advancing thrust sheet. Where sedimentation keeps pace with the vertical rise of the frontal thrust, deep source rock horizons will experience a sudden increase in loading and temperature rise. The arrival of the frontal thrust in some cases causes very little change in the rates of burial and the consequent geothermal gradients. This is due to the interplay of thrust advance and sedimentation/erosion responsible for maintaining the thrustbelt taper.

    (d) With the continued advance of the thrustbelt, a segment of the foreland basin and its underlying passive margin succession is transferred from the autochthonous footwall to the allochthonous hanging wall. As it is very common for source rock successions to serve as detachment horizons, especially if overpressured by rapid sedimentary/tectonic loading or by generation of hydrocarbons, source rocks are commonly left behind in the footwall. In such circumstances, they will continue along their pre-orogenic maturation trajectory, perhaps reaching the hydrocarbon generative window a short time following the cessation of thrusting and trap formation. This is the situation proposed for the Utah–Wyoming Thrustbelt (Warner, 1982), one that appears to be common elsewhere. However, if detached within the frontal

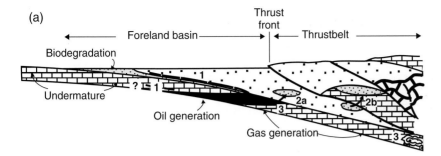

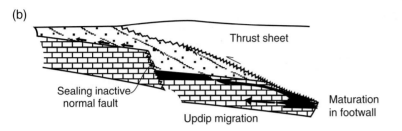

Fig. 19.12. Idealized structural setting of petroleum migration and migration pathways related to footwall petroleum kitchens in the subalpine thrustbelt (Butler, 1991). (A) Oil generation occurs during deposition of the foreland basin sediments in advance of arrival of the thrust sheet loads. Pathway 1 involves long-distance lateral migration along stratigraphically defined conduits within the foreland basin. Pathway 2 involves migration upward into traps within the foreland basin (2a) and overriding thrust sheets (2b). Pathway 3 depicts the overmaturing of source rocks and older oil pools in the footwall block. (B) Oil generation occurs at the time of thrust sheet loading and involves migration into traps within the footwall block.

thrust sheet, the source rock intervals typically will undergo uplift and cooling, perhaps to the point of turning off the petroleum kitchen.

(2) Thick-skin thrustbelt scenario: thick-skin thrustbelts are most likely to develop within or flanking intracratonic rift systems, or along passive margins, formed by wrench faulting. The thrusts may form discrete zones of thrusting or basin inversion, as in the Pyrenees or Caucasus, or they may form large isolated basement uplifts, as in the US Rocky Mountains. The thermal histories for these settings may be both short and complex, especially where plate tectonic-driven transtension evolves into transpression, and they may broadly mirror those of normal passive margins. In either case, the resultant thrust geometries normally result in the source rock and reservoir/trap occupying the same thrust sheet with the consequence that they share similar thermal histories. In contrast to the thin-skin scenario, the hydrocarbon kitchens are within the same thrust sheet as the trap or immediately beneath it. Examples have been described from fields in the Eastern Cordillera of Colombia (Cooper *et al.*, 1995; Cazier *et al.*, 1997), the Los Angeles basin (Wright, 1991) and Papua New Guinea (Buchanan and Warburton, 1996).

Within thick-skin thrustbelts, a four-phase thermal history may be represented (Fig. 19.14).

(a) Rapid burial under conditions of relatively high heat flow during the rifting or wrench-faulting event. Crustal attenuation associated with rifting is responsible for the elevated heat flows and may be associated with volcanic activity if the crust is sufficiently stretched.

(b) Relatively slow rates of sediment accumulation under decreasing heat-flow conditions in the sag stage of the rift basin as it subsides in response to crustal cooling.

(c) Rapid rates of burial with relatively low heat flows within short-lived foredeeps flanking the earliest basement inversions. If the thrust sheet in question is one of the first to invert, it may bypass this stage completely.

(d) Rapid uplift and cooling associated with inversion of half-grabens or basement blocks during thrusting.

If the basement-rooted thrust is essentially an isolated structure, its uplift will probably result in cooling of any potential petroleum kitchens within the thrust sheet. However, if it is part of an array of thrust sheets, an overriding thrust can have the effect of further maturing source rocks in an underlying sheet, especially if the former is developed out of sequence.

Beyond the first or second thrust sheet in a thrustbelt, allochthonous source rocks tend to be capable of generating only gas, if they are not completely barren. This is due to the extreme pre-thrust and syn-thrust burial typical in the more internal portions of a thrustbelt. The degree of maturity reached by the source rocks prior to being overridden or entrained in the frontal thrust sheet will depend,

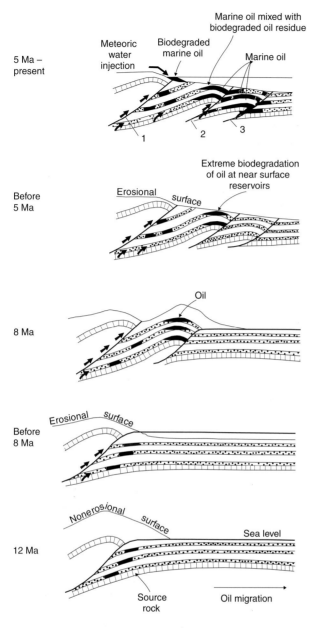

Fig. 19.13. Progression of hydrocarbon generation–migration in an advancing thin-skin thrustbelt modelled on the Maturin sub-basin, eastern Venezuela (Talukdar *et al.*, 1988). Maturation is initiated in the footwall of the leading thrust sheet with migration up-dip within carrier beds in the foreland basin. As the thrust front steps outward, the portion of the foreland basin with an active petroleum system is deformed into an anticlinal trap that fills with petroleum, and the region of newly initiated maturation–migration advances basinward.

in part, on whether the proximal or distal part of the passive margin is incorporated into the thrustbelt.

If the source rock has entered the generative kitchen before being incorporated into the thrustbelt, it is common for the hydrocarbons generated to migrate lat-

erally up-dip along stratigraphic carriers or unconformities either into foreland traps or to the surface (Figs. 19.12 and 19.13). This is the scenario in the Alberta basin (Masters, 1984), Eastern Venezuela (Talukdar *et al.*, 1988) and many other foreland basins worldwide. The arrival of the frontal thrust may drive the kitchen to higher levels of generation, from oil to dry gas, or push it through the generative window completely. In thick-skin thrustbelts, the kitchens will become entrained in the advancing thrust sheets, ultimately to be uplifted and quenched. In many instances, the potential source rocks are already within or beyond the dry gas window prior to becoming entrained in the thrustbelt. For this reason, it is wise to examine the petroleum systems of the inner foreland basin as part of any consideration of the petroleum potential of the adjacent thrustbelt.

Portions of thrustbelts can be subjected to multiple episodes of burial, uplift and reburial. In these circumstances, marine and terrestrial source rocks will generate with each cycle of burial and uplift. They will generate oils that are increasingly more mature and have lower volumetric yields. The original kerogens are compositionally altered to partially depleted kerogen types that progressively resemble Type III kerogen. Owing to the very narrow range of activation energies in Type I kerogens, lacustrine source rocks are likely to exhaust their generative capacity in just a single burial–uplift cycle. Type II and III kerogens have broader activation energy distributions and can support multiphase generation. In principal, source rocks, which are both rapidly buried and rapidly uplifted as an advancing thrust transfers them from a footwall to a hanging wall, can continue to generate hydrocarbons during uplift, but at greatly diminished rates and yields. In contrast, slowly buried source rocks have lost their capacity to continue to generate as they pass quickly back up through the temperature gradient in a cooling hanging wall sheet.

Most of the chemical reactions leading from kerogen to hydrocarbons, especially the cracking of organic molecules, involve volume increase. This would suggest that stress associated with thrusting inherently would retard maturation of organic matter. Pyrolysis experiments support this prediction, but as of yet they do not permit a clear quantification of the significance to the evolution of the petroleum system in actual thrustbelt settings. To the extent that stress can enhance hydrofracturing in an actively generating source rock, primary migration at relatively low hydrocarbon saturations is probably favoured in thrustbelts. Thus, there is likely to be a complex interplay in active thrustbelts between maturation retardation due to stress and the enhancement of petroleum expulsion due to micro-fracturing of the source rock.

1) Oil trapped in Late Cretaceous–Early Tertiary age
extensional footwall structure

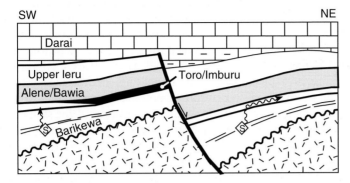

2) Oil remigrates into HW anticline during inversion

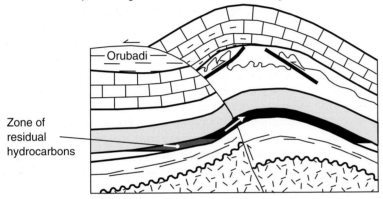

3) HW fold tightens and footwall deformed

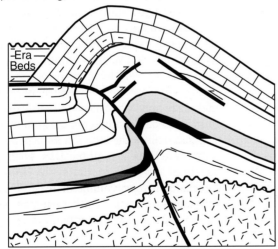

Fig. 19.14. Generalized model for maturation, migration and remigration of hydrocarbons in an inverted rift basin, Papua fold belt (Buchanan and Warburton, 1996). (1) Oil generated from the Barikewa source rock is trapped in early extensional footwall traps during the Late Cretaceous. (2) Oil remigrates across faults from pre-existing footwall traps into hanging wall anticlines during the onset of basin inversion leaving residual hydrocarbons in the footwall reservoir. (3) Transfer across the reverse fault ends when inversion closes off the migration pathway from the footwall to an ever tightening hanging wall anticline.

During thrusting, migration of hydrocarbons is most efficient along and across the active thrust faults. Continued movement on the thrust fault keeps this conduit open for the expulsion of all fluids, including connate waters and hydrocarbons. Thus, the ideal time for the operation of a thrustbelt-related petroleum system is late in the development of the thrustbelt. At this stage in the thrustbelt development unbroken traps are in place or forming, good migration pathways are available, and at this time, prior to isostatic rebound and erosional stripping, footwall hydrocarbon kitchens are likely to be at peak maturities. Traps filled with hydrocarbons prior to thrusting are very likely to be broken or breached by deformation, especially those in the allochthonous section. Pre-thrust hydrocarbon fields in the autochthonous section beneath the thrustbelt might be preserved, however, and are worthy targets for exploration.

Owing to its very high heat capacity, moving water is capable of significant distortions of the conductive geothermal regime of a basin (Nunn and Deming, 1991; Deming, 1994b; Vasseur and Demongodin, 1995) that further complicates maturation histories of involved hydrocarbon kitchens. This can happen when fluid flow crosses isotherms causing lateral fluid flow to be an effective redistributor of heat, as discussed in Chapter 16. Observed basin-scale thermal anomalies are consistent with such fluid-driven heat transport (e.g., Chapman *et al.*, 1991; Jessop and Majorowitz, 1994). In addition, hydrodynamic analysis on the scale of a thrustbelt or basin can be used to search for subtle migration pathways and traps (Davis, 1991).

One-dimensional modelling, using burial history curves alone, can give relatively reliable estimates of the time of entry into petroleum maturation thresholds, especially if well tops and/or quality cross sections are available. Stacked thrusts can be simulated by accelerated burial rates and adjustments can be made to the thermal profile to account for transient perturbation of the geothermal gradient (Furlong and Edman, 1989) due to transfer of deep hot rocks over shallow cooler ones. However, knowledge of the absolute time and duration of thrusting is required. It is also essential to know the history of unroofing and/or post-tectonic burial of the thrust stack. Apatite and zircon fission-track thermochronology methods have been shown to aid in constraining the thermal history of sedimentary basins and thrustbelts (Naeser *et al.*, 1989; Burtner *et al.*, 1994; O'Sullivan *et al.*, 1995). With knowledge of the thickness, richness and kerogen type of the likely source rocks in the vertical profile, it is even possible to predict the petroleum yield as a function of time. Such one-dimensional modelling, despite its limitations discussed in Chapter 15, can be a relatively useful exercise for identifying the location and time of generation/expulsion of petroleum in a thrustbelt. The model results are normally displayed in a burial history diagram (Fig. 17.8), but alternatively may be tabulated or mapped.

However, one-dimensional petroleum maturation models cannot predict migration pathways that ultimately lead to charging of potential traps and charge effectiveness. This requires a much more detailed knowledge of the full three-dimensional structural architecture of the thrustbelt, the fluid transmissivity properties of all potential carrier beds, faults and fracture networks. It also requires a knowledge of the time of movement on all potential carrier faults, and knowledge of the hydrodynamic history of the thrustbelt as these will affect heat flow, geothermal gradients and petroleum migration. It is unusual to have this level of knowledge, even in an intensely explored and mature thrustbelt. However, the two- and three-dimensional software tools that integrate maturation, thermal and fluid flow modelling routines can be helpful for evaluating alternative scenarios for trap filling (Bishop *et al.*, 1984) and thereby further constrain exploration risk assessments. This is a research area still in development and testing.

Roure and Sassi (1995) demonstrate the importance of integrated inverse modelling to restore structure sections to a pre-thrust configuration linked to stepwise forward modelling back to the present configuration in order to capture the intermediate development stages that have controlled maturation and migration in a thrustbelt.

The quality of generation and migration simulations is very dependent on the quality of the structural analysis characterizing the architecture and evolution of the thrustbelt region under investigation. The starting point is the construction of balanced two-dimensional cross sections (e.g., Woodward *et al.*, 1985; Sage *et al.*, 1991; Shaw *et al.*, 1994; Wickham and Moeckel, 1997; Rowan and Linares, 2000), by methods described in Chapter 6, ideally serialized and in time-steps. But this level of analysis is only a bridge to the ultimate requirement of a full three-dimensional representation and reconstruction that defines the location and spatial extent of generative kitchens and potentially filled traps, and the structural elements in the space between them. Yet, there are still a great many thrustbelts in which the most fundamental questions of dominant structural style are unresolved and consequently in which there is a lack of clear understanding of structural architecture and burial histories.

20 Seals and traps in thrustbelts

General statement

The optimum period for a petroleum system to charge a trap is during the late stages of thrustbelt development. Most of the traps are already formed and the chance of their subsequent translation is relatively low. Traps that become filled during earlier stages are likely to be breached by subsequent tectonic events (Clarke and Cleverly, 1991; Macgregor, 1993), except in the case of autochthonous traps buried beneath the thrustbelt, which are likely to escape younger deformation.

Traps in a thrustbelt can also be filled following the cessation of thrusting if the hydrocarbon kitchens continue to operate, but the migration pathways are likely to be complex fracture networks within the volume of the thrust sheets, instead of thrust surfaces themselves. In these situations, vertical migration is favoured. Geological events external to the specific portion of the thrustbelt in consideration are generally required to continue burial and hydrocarbon generation after the close of thrusting. Burial beneath the foredeep debris of another thrust system or beneath a successor basin formed by orogenic collapse are the most common mechanisms for keeping hydrocarbon kitchens active following the end of thrusting. For instance, burial beneath about 2 km of sediments shed into the Green River basin from the flanking Paleocene–Eocene Rocky Mountain uplifts was responsible for the post-thrust generation of hydrocarbons from the Lower Cretaceous foredeep basin source rocks in the Late Cretaceous Wyoming–Utah thrustbelt. Without the superposed post-thrust burial, this extremely prolific thrustbelt would have been oil- and gas-poor. Timing of burial of the source rock before, during and after thrusting is critical.

Although the size and geometries of traps may differ, the trap types that dominate in both thin-skin and thick-skin thrustbelts are the same: hanging wall anticlines and, much less commonly, fault-sealed footwall traps. Very gentle, doubly plunging frontal or inversion anticlines form the largest traps and are by far the most desirable exploration targets. In many of the young thrustbelts, the quality of seal may be very poor, but the active petroleum systems continue to keep the leaky traps filled to spill. However, the presence of gas and condensate in Paleozoic thrustbelts worldwide attests to the possibility of having high-quality shale and diagenetic seals, even in the deformed rocks of a thrustbelt.

All traps are inherently transient and, given sufficient time, will lose their charge through leaks or petroleum degradation. Owing principally to the difficulty of preserving hydrocarbons in reservoirs for long geological periods, more than 80% of the world's discovered oil and gas reserves were generated and trapped since Aptian time, and nearly half since the Oligocene (Ulmishek and Klemme, 1990). Petroleum is a fragile fluid that undergoes compositional changes from the moment of oil separation from bitumen and these changes continue through migration and accumulation (Blanc and Connan, 1994).

Role of seal type and quality in trap integrity

Based on differing modes of failure Watts (1987) defines membrane and hydraulic seals. The capillary entry threshold pressure, P_c, the pressure required for entry through the largest interconnected pore throat of the seal, serves as the dominant control on failure of a membrane seal (Berg, 1975). A hydraulic seal requires that the entry pressure exceeds the rock strength of the seal for seal failure to occur (Berg and Gangi, 1999). The entry pressure results in hydraulic fracturing of the seal rock, thereby opening conduits for fluid flow and pressure release. Seals can be further differentiated according to their position in relationship to the reservoir. Top seals are low-permeability strata that overlie petroleum reservoirs and retard the upward migration of hydrocarbons. Fault seals trap petroleum through a variety of mechanical and diagenetic processes that reduce permeability within a fault zone. Top seals and fault seals exhibit both membrane and hydraulic sealing behaviour (Watts, 1987).

The pressure constrained by the seal is either the buoyancy pressure of a trapped petroleum column or an overpressure resulting from either stress applied to a compressible rock or fluid expansion (Swarbrick et al., 2002). The principal mechanisms for overpressure in the thrustbelts include disequilibrium compaction resulting from sediment or tectonic loading, smectite-to-illite transformation and fluid expansion due to maturation of kerogen to less dense petroleum or cracking of oil to gas, as discussed in Chapter 10. The buoyancy pressure,

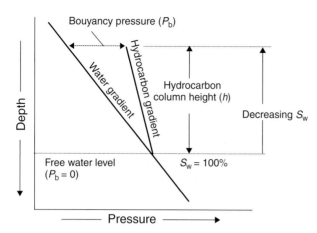

Fig. 20.1. The buoyancy pressure against a top seal is the difference between the water pressure and the hydrocarbon pressure at that surface. As the buoyancy pressure increases upward in the hydrocarbon column and water is displaced from pore spaces in the reservoir, the saturation of water decreases.

P_b, increases with increased height of the static petroleum column (Fig. 20.1) and represents the difference between the petroleum pressure, P_p, and water pressure, P_w, at the same depth level. The pressure is a function of the hydrocarbon column height, h, and the density difference, $\Delta\rho$, between petroleum and water as expressed by

$$P_b = P_p - P_w = \Delta\rho g h, \qquad (20.1)$$

where g is the hydrostatic gradient.

A fault fails as a membrane seal when the buoyancy pressure, P_b, equals the capillary entry threshold pressure, P_c, but requires a larger magnitude of P_b than that to fail by hydraulic fracturing. Owing to upward-increasing buoyancy pressure and relative permeability effects within the water-wet fault zone, a leaky transition zone may exist below the zone of total fault seal failure in which the fault is no longer effective as a membrane seal, but is still acting as a hydraulic seal (Brown, 2003).

The formation or reactivation of brittle faults and fractures within low-permeability rocks capping overpressured reservoirs creates drainage conduits limiting the degree of overpressuring (Sibson, 2003). In the absence of a tectonic stress field and pre-existing fractures, the quality of the hydraulic seal will depend on resistance to fracturing of the cap rock as measured by a material property, such as fracture toughness (Atkinson and Meredith, 1987). Rocks that are either inherently strong, or extremely ductile, are favoured as hydraulic seals. Thus, ductile, low-permeability evaporities and shales are doubly effective as top seals. The maximum sustainable overpressure is affected by the tectonic regime, the local state of differential stress and

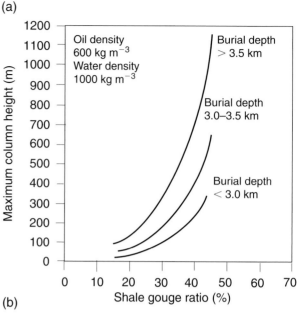

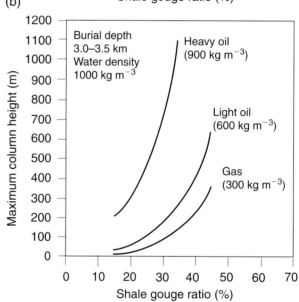

Fig. 20.2. Column heights predicted by Bretan *et al.* (2003) for different burial depths at constant petroleum density (a) and for different petroleum densities at constant burial depth (b).

the presence or absence of inherited brittle architecture within the cap rock (Sibson, 2003).

There are several mechanisms by which fault planes can act as a seal (Watts, 1987; Yielding *et al.*, 1997). These include:

(1)   juxtaposition, in which permeable reservoir sands are juxtaposed against low-permeability shales or other lithologies having a high entry pressure;
(2)   cataclasis, the crushing of sand grains to produce a finer-grained fault gouge resulting in a fault

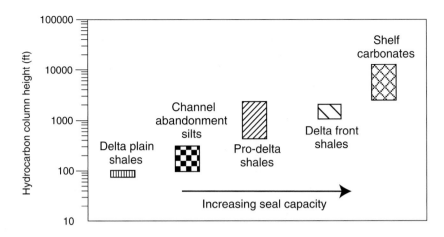

Fig. 20.3. Calculated range of petroleum column heights that could be supported by the Talang Akar Formation seal lithologies (Kaldi and Atkinson, 1997). The calculations are based on measured capillary entry pressures.

with higher capillary entry pressure than the wall rock;

(3) the development of clay smears, resulting from entrainment of shale into the fault plane giving an inherently high capillary entry pressure regardless of the wall rock permeability;

(4) cementation, diagenetic mineralization preferentially within the fault plane that may partially or completely occlude porosity.

The shale gouge ratio, SGR, is one of several useful measures of fault seal quality (Yielding *et al.*, 1997). It is an index based on the volumetric shale fraction of the stratigraphic intervals adjacent to the fault. The larger the shale proportion in the wall rock of the fault plane, the larger the value of SGR. Calibration diagrams based on buoyancy pressures constrained by faults with differing SGR correlate for values between 20 and 40% (Bretan *et al.*, 2003) and permit the use of SGR for estimating the height of oil or gas columns that can be supported by a fault with clay smears (Fig. 20.2). For oil the onset of sealing capacity is at an SGR of about 20% and for gas even lower, but is highly dependent on the depth of burial. The fault plane reaches optimal sealing effectiveness at SGR values of about 40%. Higher values than that do not appear to improve sealing qualities.

Downey (1984, 1994) ranked ductile seal lithologies in order of decreasing capacity: salt, anhydrite, kerogen-rich shale, silty shale, carbonate mudstone and lastly chert. Kaldi and Atkinson (1997) observed that in the upper Oligocene Talang Akar Formation of northwest Java, shales have greatly differing sealing capacities depending on their depositional environment (Fig. 20.3). Those deposited in delta front and prodeltaic settings have capillary entry pressures in the range of 1000–1400 psi, whereas delta plain shales are of the order of 80–90 psi. The shelf carbonates of this formation have the highest capillary entry pressures from all involved facies. Having pressures of 2300–9000 psi, they constitute the most effective top seals. As indicated by the S1/TOC ratio, the poorly sealing delta plain shales are observed to have very high hydrocarbon contents, which result from saturation by migrating petroleum (Noble *et al.*, 1997), whereas the other lithologies do not.

In terms of volume and lateral extent, evaporite successions are virtually insignificant globally; however, they seal more than a third of the world's petroleum resources (Ulmishek and Klemme, 1990). Grunau (1987) observed that one-half of the 25 largest oil fields and more than a third of the 25 largest gas fields worldwide are sealed by evaporites. The remaining fields all have thick shale seals. About half of the known oil and gas reserves in pre-Upper Jurassic reservoirs are sealed by evaporites of primarily Late Permian–Triassic age, mainly in tectonically quiet intracratonic settings (Macgregor, 1996). About a third of the known petroleum reserves in the post-Middle Jurassic reservoirs are sealed by evaporites (Ulmishek and Klemme, 1990). These evaporite seals are located mainly in Tethyan foreland basins where they are commonly involved in thrustbelts, such as the Zagros fold belt (Beydoun *et al.*, 1992). Factors enhancing seal efficiency include weak tectonic deformation and a thickness greater than 20–30 m for evaporites and 50 m for shales (Grunau, 1987).

It is possible to trap a thicker two-phase petroleum column than either oil or gas alone (Watts, 1987). This is due to the fact that the oil column is less buoyant than pure gas, although gas in contact with the seal determines the capillary entry pressure. Sales (1997) distinguishes three broad classes of traps based on seal strength and the type of the entrapped fluid. Class 1 traps are gas-filled, but spill gas and oil (Fig. 20.4 and Table 20.1). Class 2 traps are filled with gas and oil and spill oil, but leak gas. Class 3 traps are oil-filled, but leak gas and minor amounts of oil.

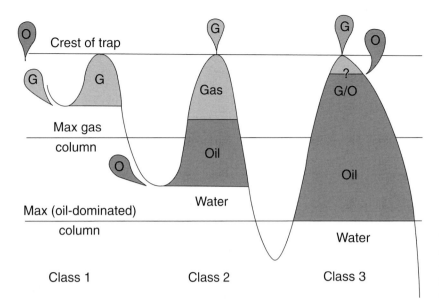

Fig. 20.4. Diagrammatic cross sections showing trap classes and their characteristic fill (Sales, 1997). Bubbles (G = gas, O = oil) indicate the type of fluid that is spilled or leaked from each class of trap.

Trap styles in thrustbelts

The basic requirements for a petroleum trap are that the surface between the reservoir and the seal must have closed contours and an overall convex-up morphology, which are characteristics shown by a very large and diverse array of geological features (Biddle and Wielchowsky, 1994). Traps may be structural, stratigraphic or a combination of both elements. If the trap depends on the flow of water in the reservoir to drive petroleum against the trapping surfaces, it is referred to as a hydrodynamic trap (Hubbert, 1953; Goolsby *et al.*, 1988), otherwise the trap is hydrostatic. Fault traps depend on the sealing qualities of a fault zone, either due to the inherently large capillarity of the sealing fault zone itself or the juxtaposition of the reservoir rocks against a low-permeability rock across the fault.

Milton and Bertram (1992) classify trap styles by the number and types of sealing surfaces that define the trap geometry. Types of sealing surfaces include:

(1)   a conformable bedding surface;
(2)   an unconformity,
(3)   a facies change boundary; and
(4)   a sealing fault.

One-seal traps are those in which closed contours exist on a single sealing surface. In most instances this single top seal is stratigraphic, a bedding or unconformity surface, but in thrustbelts it could equally be a folded sealing fault. Polyseal traps require two or more sealing surfaces to form the closed contours of the trap. They can be combinations of stratigraphic and fault surfaces.

The majority of the discovered oil and gas fields in thrustbelts reside in broad simple anticlines in parts of the thrustbelt where overall shortening and internal strains are relatively small. These are dominantly detachment and fault-propagation folds having a large radius of curvature in cross section. In addition, evaporites are commonly involved in these structures as a basal or intermediate detachment horizon, giving rise to the particular structural style, or as a top seal, or as both detachment and seal. Even where evaporites do not serve as detachment horizons, it is the mechanical stratigraphy of the deforming rock succession, described in Chapter 7, which is the key to predicting structural style. Moderate to weak competency contrasts are more likely to give rise to the detachment and fault-propagation fold styles that are ideal for trapping large petroleum accumulations. Strong mechanical contrasts within the stratigraphic succession are more likely to result in fault-bend folds, duplexes and triangle zones. Whereas each of these structural styles is capable of trapping petroleum, the traps are rarely large.

In thin-skin thrustbelts, the great majority of the known petroleum accumulations reside in hanging wall structures. Footwall traps that depend on fault seal integrity are rare and tend to be relatively small. This may reflect an inherent difficulty in maintaining the quality of a fault seal in a thrustbelt setting, where fault reactivation is easily accomplished, or it may be due to the difficulty of discovering sub-thrust traps. Subtle, poorly imaged structural features in the footwall certainly carry larger exploration risk factors than do large, well-imaged structures in the hanging wall. Also thick relatively ductile stratigraphic seals carry much less risk than do fault seals in a compressional thrustbelt setting (Sibson, 2003). Hydrocarbon accumulations are known to exist in sub-thrust autochthonous and parautochthonous strata beneath the leading edges of thrustbelts

Table 20.1. Trap classes and their characteristics.[a]

Characteristic	Class 1	Class 2	Class 3
Dominant fluid in trap	Gas	Oil and gas	Oil
Trap spills	Oil and gas	Oil	Neither oil nor gas
Trap leaks	Neither oil nor gas	Gas	Oil and gas
Top seal quality	Perfect seal	Leaking gas	Leaking both oil and gas
Trap filled-to-spill	Yes	Yes	No
Factors controlling GOC	Charge volume	Pressure	Pressure
Factors controlling OWC	Spillpoint or charge	Spillpoint	Pressure
Existence of trap up-dip	Gas traps	Oil traps	Water-wet traps
Existence of stratigraphic trap above	Water-wet traps	Gas traps	Both oil and gas traps
Consequences of uplift	Gas flushes	Trap improves	Variable consequences

Note:
[a] Modified after Sales (1997).

(Picha, 1996). Many of these oil and gas fields are as large, or even larger, than the accumulations within the overriding wedge. A few examples of traps in thrustbelts, both thin- and thick-skin, are discussed in the following text to illustrate the diversity of styles.

The large anticlinal traps of the Zagros fold belt are formed by detachment within the Proterozoic Hormuz salt, which rests directly on a crystalline basement (Jackson *et al.*, 1981). They have either a pure detachment fold style or are more likely to be translated detachment folds due to their asymmetry resulting in spill points on their southeast limbs (Hull and Warman, 1970). The most common traps are broad unfaulted anticlines at the level of the Lower Miocene Asmari Limestone, the principal oil reservoir, in structures that can be over 100 km long. Lower–Middle Miocene Lower Fars and Gacharsan evaporites form the highly efficient top seal that itself is a shallow detachment responsible for disharmonic structuring of post-Lower Miocene strata (Beydoun *et al.*, 1992). The fold amplitudes are of the order of 6000 m over a distance of 3–6 km. The vertical distance from the trap spill point to the anticlinal crest is 1200–2100 m. The Gacharsan and Bibi Hakimeh fields in Iran (Fig. 17.3) have single oil columns with heights of 1800 and 900 m, respectively. Most traps are filled to spill. Free gas caps exist in some fields, but others have unsaturated oil. Oil from any particular accumulation is of very uniform composition and pressure due to good fracture connection in the Asmari Limestone. The pressure response to production is nearly immediate across distances of tens of kilometres (Hull and Warman, 1970). There is a strong hydrostatic gradient across the Zagros fold belt. It is related to the elevation difference between the anticlines and the undeformed foreland basin that can produce tilted oil–water contacts in many of the fields. The oil–water contact depths can differ by as much as 150 m on opposite limbs of an anticline.

The Kirkuk field in northeast Iraq is representative of many of the larger Zagros anticlinal oil accumulations (Fig. 20.5). The field and its satellite Alanfal occupy nearly all of the 200 km length of the Kirkuk anticline. The estimated recoverable reserves for the field are 9.85 BBO. In 1975, before disruption of the industry by war and internal strife, the field produced 36–44° API, low GOR oil at the very high rate of 960 000 bopd (Alsharhan and Nairn, 1997). Dunnington (1958) outlined a scenario of trap filling of the southeast Kirkuk field that involves early charging of the Mid-Cretaceous Qamchuqa Group from a coeval basinal source rock succession to the northwest. With further growth of the anticline, the oil migrated through a fracture network vertically into fractured marls of the Upper Cretaceous Shiranish Limestone and higher into the porous and fractured Asmari Limestone, the principal oil reservoir in the Zagros fold belt. Lower Fars and Gacharsan evaporites form the highly efficient top seal to the broad anticlinal trap. A similar trap filling scenario has been proposed in the Ain Zalah field to the north of the Kirkuk field (El Zarka and Ahmed, 1983; El Zarka, 1993).

The Dhulian oil field (Fig. 20.6) in the central Potwar Plateau, northern Pakistan (Jaswal *et al.* 1997) is probably similar in structural style to the Zagros anticlines. As in the Zagros fold belt, a thick, but relatively mechanically uniform, stratigraphic succession is detached on a thick Proterozoic salt succession near the top of the basement to form broad linear fault-propagation anticlines. However, these anticlinal traps are less productive than the Zagros anticlines. This is because the Paleogene carbonate reservoir is located deeper in the core of the structure, resulting in a smaller trap volume, and quality source rocks and top seal are absent. The doubly plunging anticline is also less than 10 km long. Nevertheless, the Dhulian field has produced 41 MMB of 28–32° API oil and 199 bcf of solution gas (Khan and Hasany, 1998).

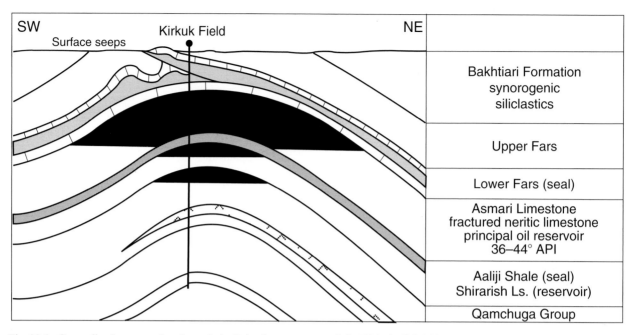

Fig. 20.5. Generalized cross section through the Baba dome segment of the Kirkuk oil field in northeast Iraq (Dunnington, 1958).

The sub-Andean thrustbelt of the Tarija basin, Bolivia is a thin-skin, in-sequence thrust system having two major linked detachment horizons, the Silurian Kirusillas Formation and the Devonian Los Monos Formation. These two thick shale units are known source rocks in the basin (Dunn *et al.*, 1995). In Bolivia, the thrustbelt contains numerous closely spaced anticlines expressed on the surface. They have been developed by shortening that reaches an overall value of about 100 km. This shortening decreases southward towards the southern limit of the thrustbelt in northwestern Argentina. In this region, the surface expression is of a few large north–south-trending symmetric detachment folds developed in Carboniferous–Tertiary strata (Belotti *et al.*, 1995b).

Deep-seated detachment resulted in fault-propagation folds and passive roof duplexes with wedges located beneath the thick and highly ductile Los Monos shale (Fig. 20.7), which is responsible for the shallow detachment folds being distinctly disharmonic. The large Ramos and Aquaragüe fields (Fig. 20.7) produce gas and condensate from quartz-cemented sandstone reservoirs in the Devonian Huamampampa Formation, which is sealed by the Los Monos shale. Favourable production rates are made possible by natural fracture networks that link the relatively low matrix porosity of the sandstones. The Carboniferous Tupambi Formation also produces from shallow box folds. These are very young structures having formed and filled with petroleum since the end of the Miocene (Echavarria *et al.*, 2003). The Ramos field contains

extant reserves of 77.8 B m^3 of gas and 8.5 MMB of oil, whilst the Aquaragüe field has remaining reserves of 57.4 B m^3 of gas and 10.7 MMB of oil (Yrigoyen, 1991). The other fields in the immediate area are quite small.

The Turner Valley field (Figs. 20.8 and 20.9) is located 40 km southwest of Calgary on the leading edge of the outer Rocky Mountains Foothills belt (Fig. 17.4) immediately west of the frontal triangle zone (Newson, 2001). The field is 40 km long and 0.8–5 km wide. It has the distinction of being Canada's first major oil and gas field. The Turner Valley structure is a strongly asymmetric fault-propagation anticline (Gallup, 1954; Mitra, 1990) in Paleozoic strata carried on a major fault, the Turner Valley thrust, which is actually one of many splays off the basal décollement that ends in the triangle zone (Fermor and Moffat, 1992). The thrust fault is arcuate with the tip line having maximum displacement near section B–B′ where the field has maximum height and width. Production is from biostromal limestone and dolomite of the Lower Carboniferous Rundle Group, mainly crinoidal limestone of the Turner Valley and Pekisko Formations (Penner, 1957). The combined thickness of the two pay zones is 45 m. The estimated in-place reserves are 1050 MMB of oil and 0.4 Tcf gas. As in much of the Foothills belt, there is no clear surface expression of the subsurface anticline and it was surface seeps that led to the initial shallow gas field discovery in 1911. Deep step-out drilling in 1936 resulted in the discovery of major oil reserves on the back limb of the fold.

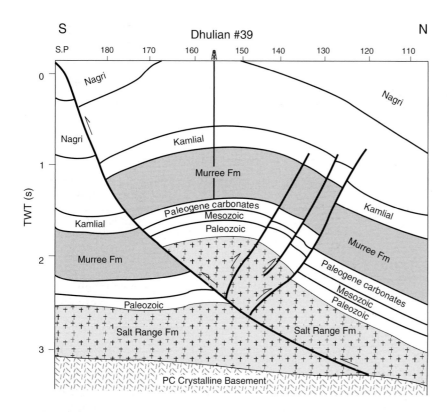

Fig. 20.6. Time–depth cross section of the Dhulian oil and gas field in the central Potwar Plateau, northeast Pakistan (after Khan and Hasany, 1998).

The oil and gas fields of the Wyoming thrustbelt province (Fig. 20.10) in southwest Wyoming and northeast Utah occupy just a small segment of the Late Cretaceous–Paleocene Sevier thrustbelt (Armstrong, 1968), which extends northward to connect with the coeval Montana and Canadian Cordillera–Foothills thrustbelts. Accumulations are all in anticlines associated with Paleozoic or Jurassic cutoffs in the hanging wall of the Absaroka thrust, immediately overlying the frontal Hogsback thrust (Lamerson, 1982). Known mature source rocks occur in the Cretaceous footwall section (Fig. 20.11). The anticlines are supported by hanging wall cutoffs through the Paleozoic carbonates, comprising Ordovician Big Horn and Mississippian Mission Canyon formations, and by the principal reservoir units, the Jurassic Nugget Sandstone and Twin Creek Limestone. Anhydrites of the upper Mission Canyon formation, the Gypsum Spring member of the Twin Creek Formation and the Upper Jurassic Preuss salt play a dual role in serving as both detachments between the ramp cutoffs (West and Lewis, 1982) and highly effective seals in the anticlinal traps (McIntyre, 1988). The two hanging wall cutoff intervals form a double line of anticlinal structures (Fig. 20.11). The eastern folds supported by the Jurassic cutoff host mainly oil pools and the western folds supported by the Paleozoic cutoff host mainly gas and condensate. Owing to a complex hanging wall cutoff geometry, the Anschutz

Ranch East field (Fig. 20.12) and Painter Reservoir field (Fig. 20.13) are paired anticlines separated by a very tight syncline (Frank *et al.*, 1982; White *et al.*, 1990). The matrix porosity of the carbonate and sandstone reservoirs is augmented by fracturing in most of the fields (Nelson, 2001). The doubly plunging anticlinal traps have been charged from active Cretaceous source rocks in the footwall by vertical migration via hanging wall cutoffs of these reservoirs or through fractures in the anticlinal cores. Peak maturity occurred either late syntectonic or immediately post-tectonic (Warner, 1982; Deming and Chapman, 1989). The province has an estimated 1.4 BBOE of recoverable reserves. In 2003, the cumulative production was 1.32 BBOE with 78.7% of the production being gas. The two main producers have been the Anschutz Ranch East gas and oil field containing 0.62 BBOE and Whitney Canyon–Carter Creek gas field with 0.32 BBOE.

The Wilson Creek field is a domal anticline (Fig. 20.14) on an elevated basement block linking the Uinta and White River uplifts along the northeast margin of the Piceance basin, northwest Colorado (Stone, 1990). It belongs to a class of large anticlines described as foreland basement-involved structures (Mitra and Mount, 1998) that form the super-giant Rangely oil field (Hefner and Barrow, 1992) and others in the central Rocky Mountains. An advantage of this style of anticline is that it creates a very large area of four-way

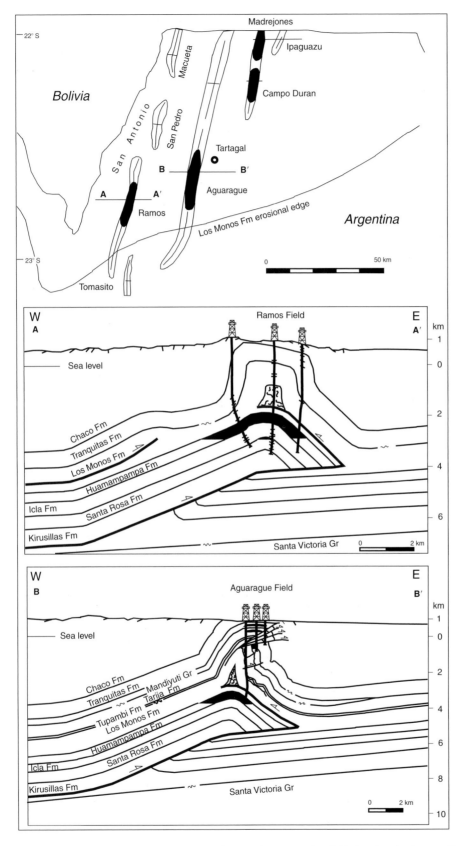

Fig. 20.7. Location map and cross sections of the fault-propagation folds hosting the Ramos (A–A′) and Aguaragüe (B–B′) gas and condensate fields in the southernmost Tarija basin of northwest Argentina (Belotti *et al.*, 1995b).

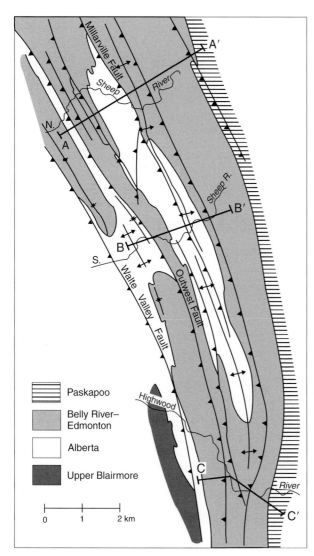

Fig. 20.8. Surface geology of the Turner Valley oil and gas field, southwestern Alberta (Penner, 1957).

of the basin (Dashwood and Abbotts, 1990). The reservoir is the basal Cretaceous Hollin Sandstone and sand intervals intercalated within the overlying Napo Formation, which serves as both source rock and seal to these sandstone reservoirs. Structuring is the result of Late Cretaceous–Cenozoic inversion of Triassic half-grabens and associated normal faults in what becomes the foreland basin to the main Andean ranges immediately to the west of the Oriente basin (Balkwill *et al.*, 1995).

The Los Angeles basin (Fig. 17.6) formed in the Late Neogene through rifting and block rotation along the transform boundary between the Pacific and North American plates. The basin filled with organic-rich marine shale and diatomites of the Monterey Formation into which flowed submarine channels carrying coarse terrigeneous detritus eroded from bordering crystalline basement blocks. Small changes in relative plate motion caused a rapid transition from Early Pliocene extension to Late Pliocene contraction, leading to modest transpressive shortening across older extensional faults. By Late Pliocene times much of the basin was sufficiently buried for the Monterey Formation to have just entered the oil window and to be generating heavy immature oil that leaked upward to fill all available traps and seep to the surface. It is estimated that the known fields in the basin will ultimately yield 10.4 BBOE of petroleum. Approximately 73% of the reserves is trapped in faulted anticlines, 12% is in simple anticlines, 10% is in fault traps and just 5% is in stratigraphic traps (Wright, 1991). These various styles of trapping are observed in the East Beverly Hills and San Vicente fields (Fig. 20.16) on the north edge of the basin south of the Santa Monica fault.

Alteration and destruction of petroleum accumulations
Petroleum undergoes alteration in composition and physical properties from the moment of expulsion from the source rock. This alteration occurs before, during and after entrapment (Blanc and Connan, 1993). The main factors that influence petroleum before trapping are source rock characteristics, such as the kerogen type and the level of maturity, the expulsion efficiency and the secondary migration conditions. The principal influences on oil composition in the reservoir are the temperature and the pressure, which affect the gas–oil ratio, GOR, and related physical properties at the time of accumulation. Unaltered oil and gas in a reservoir normally has an API gravity in the 25–40° range and a GOR of 100–200 $m^3 m^{-3}$. The viscosity is highly dependent on temperature, but at normal reservoir temperatures most mature, nonwaxy and unaltered oils have viscosities of less than 100 mPa s (Mann *et al.*, 1997).

closure with little internal strain of the reservoir units. The Wilson Creek field contains 240 MMB of 47–50° API paraffinic oil in-place in gently arched Jurassic Salt Creek and Entrada Sandstone reservoirs. Although the closing contour of the structure encompasses an area of 107 km^2, only 15% of the potential trap, represented by 107 m of the 305 m vertical closure, is filled and productive. The Jurassic sandstone reservoirs are exposed about 20 km to the south, resulting in a hydrodynamic head across the field and a tilting of the oil–water contact northward at gradient of about 8 m km^{-1}.

The large oil accumulations in the Oriente basin (Fig. 20.15) of Ecuador, particularly the lighter 30–37° API paraffinic oils, are trapped in low-amplitude north–south-trending anticlines that deform mainly the Cretaceous strata resting on the pre-Cretaceous floor

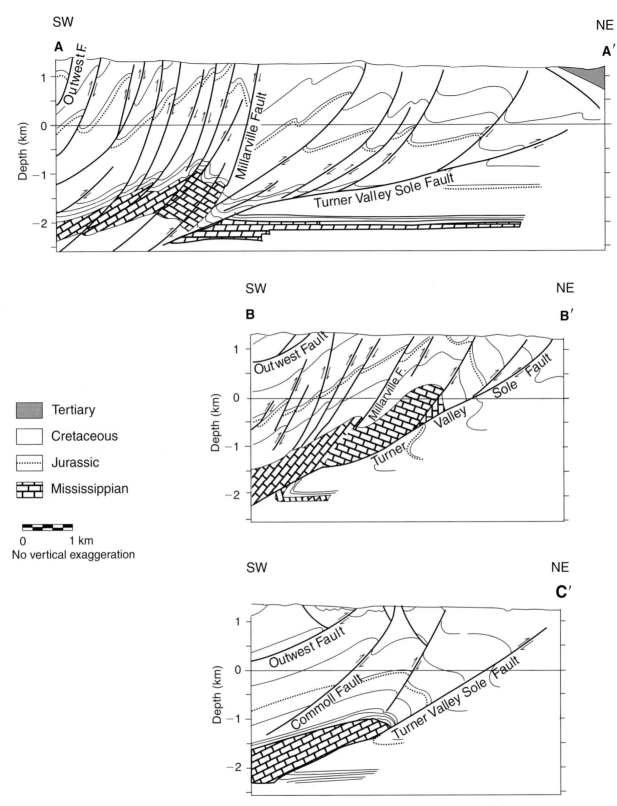

Fig. 20.9. Structural cross sections through the Turner Valley oil and gas field. See Fig. 20.8 for the location of the sections (Penner, 1957).

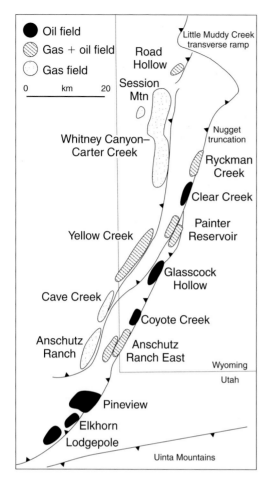

Fig. 20.10. Location of oil and gas fields of the Wyoming Thrustbelt. Constructed from a top Nugget Sandstone structure contour map by Lammerson (1982). The fields are all within hanging wall anticlines of the Absaroka thrust sheet.

The secondary alteration after entrapment is represented by thermal maturation, biological and physical degradation, de-asphalting and dysmigration. Owing to rapid rates of burial, uplift and denudation of topography, thrustbelts are particularly susceptible to secondary alteration and the processes that lead to alteration and degradation of petroleum include:

(1) temperature change related to changes in the depth of burial;
(2) flushing by formation and meteoric waters; and
(3) microbial activity in regions of fresh water circulation.

Temperature increases due to greater burial or geothermal gradient perturbations potentially result in thermal maturation or cracking of reservoired oils. *In situ* cracking within the trap favours formation of light hydrocarbons and gas, an increase in GOR, depletion of polycyclic biomarkers, a decrease in specific gravity, an increase in API gravity, a decrease in viscosity and normally a decrease in sulphur content. Oils in deep reservoirs tend to be depleted in aromatics and consequently enriched in n-alkanes (Vandenbroucke *et al.*, 1999). In the cracking reactions a pyrobitumen residue is formed, which conserves the hydrogen balance. This highly dehydrogenated residue has characteristics that distinguish it from precipitated asphaltines in the reservoir, such as a very high T_{max} with values above 460 °C and vitrinite reflectance values of 1.5–2.5% R_0 (Blanc and Connan, 1994).

The temperature range for onset of oil cracking is dependent on the oil composition (Schenk *et al.*, 1997). However, there remains disagreement on both the kinetic parameters relevant to the process, if controlled

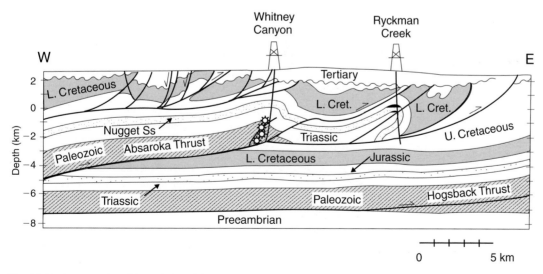

Fig. 20.11. Structural position of the Whitney Canyon and Ryckman Creek fields in hanging wall anticlines of the Absaroka thrust sheet (Warner, 1982).

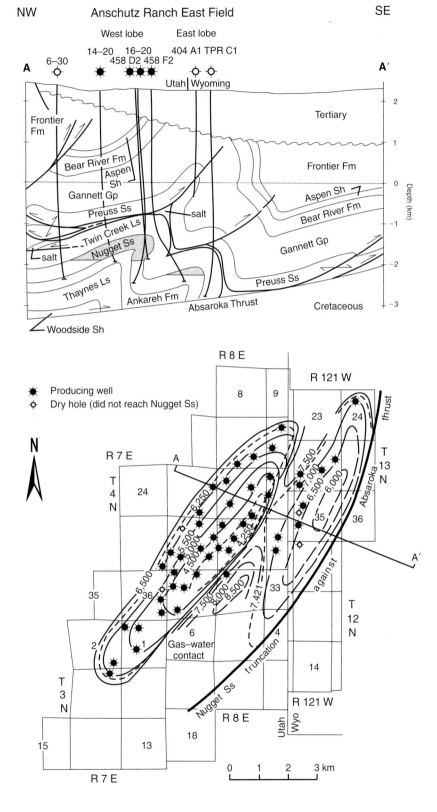

Fig. 20.12. Anschutz Ranch East oil and gas field cross section and structure map on the top of the Jurassic Nugget Sandstone showing the pair of doubly plunging anticlines related to the pair of hanging wall cutoffs above the Absaroka thrust plane. Figures are from Chidsey (1993) and redrawn after Lelek (1982) and West and Lewis (1982).

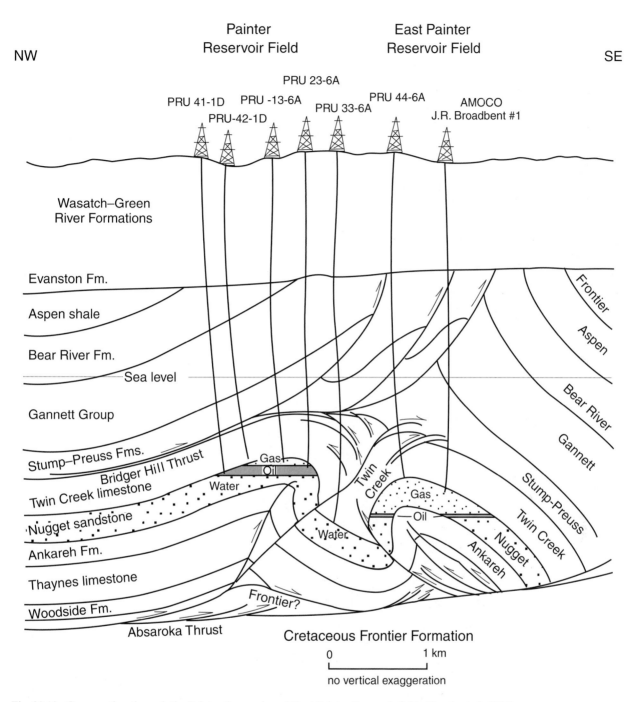

Fig. 20.13. Cross section through the Painter Reservoir and East Painter Reservoir fields (Frank *et al.*, 1982).

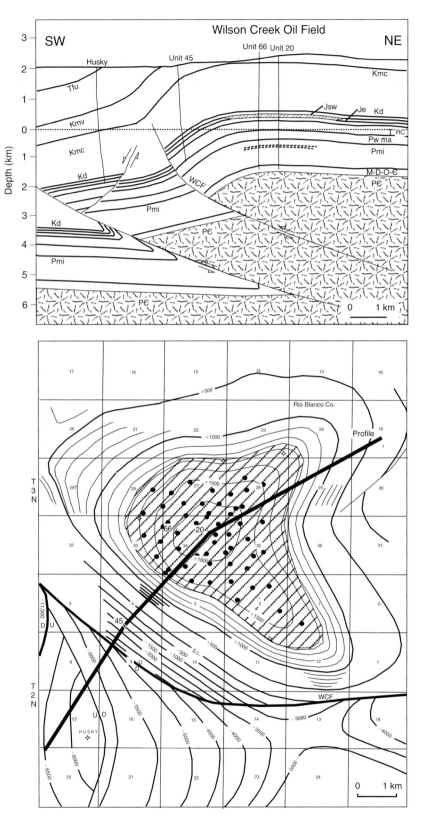

Fig. 20.14. Wilson Creek oil field cross section (A) and structure map (B) on the top of the Salt Wash Sandstone Member of the Jurassic Morrison Formation, the principal productive reservoir unit (Stone, 1990). The seals are formed by shales of the Morrison Formation. Contours are in feet. PC – Precambrian section, M–D–O–C – Mississippian–Devonian–Ordovician–Cambrian section, Pmi – Pennsylvanian Minturn Formation, Pw-ma – Pennsylvanian-Permian Weber-Maroon Formations, Trc – Triassic Chinle Formation, Je – Jurassic Entrada Formation, Jsw – Jurassic Salt Wash Formation, Kd – Cretaceous Dakota Formation, Kmc – Cretaceous Mancos Formation, Kmv – Cretaceous Mesaverde Formation, Tfu – Tertiary Fort Union Formation, WCF – Wilson Creek fault.

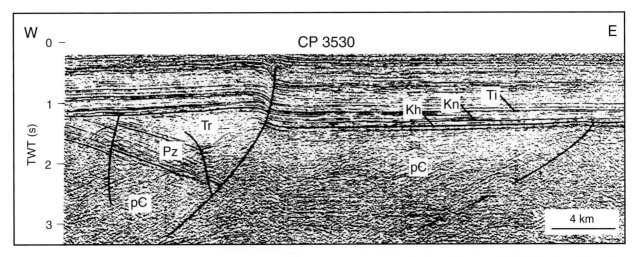

Fig. 20.15. Reflection seismic section from the northeastern Oriente basin, Ecuador, showing the inverted Triassic half-graben style of anticlines typical for the productive structural traps (Balkwill *et al.*, 1995). The basal Cretaceous Hollin Formation (Kh) is the principal sandstone reservoir. pC – Precambrian, Pz – Paleozoic, Tr – Triassic sections, Kn – Cretaceous Napo Formation, Tt – lowermost Tertiary Tena Formation.

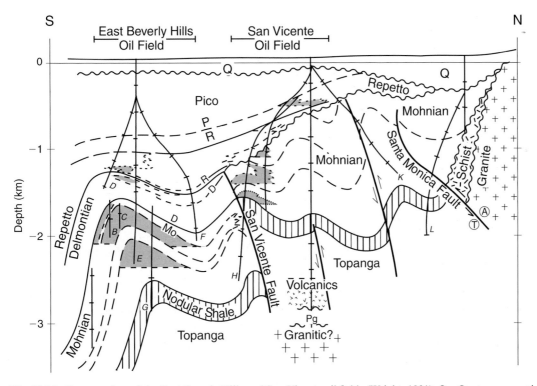

Fig. 20.16. Cross section of the East Beverly Hills and San Vicente oil fields (Wright, 1991). Q – Quaternary section, P – Pico Stage, R – Repetto Stage, D – Delmontian Stage, Mo – Mohnian Stage, Pg – Paleogene section.

Table 20.2. Levels of biodegradation as determined by organic compound classes removed by microbial oxidation or consequently enriched in the residual oil.[a]

Biodegradation level		Compounds removed or enriched
Minor	1	None
	2	Short n-alkanes absent
Moderate	3	>90% of n-alkanes removed
	4	Alkyleyelohexanes absent and isoprenoids reduced
	5	Isoprenoids absent
Extensive	6	Bicyclic alkanes absent
	7	>50% of regular steranes removed
Severe	8	Steranes altered and demethylated hopanes abundant
Extreme	9	Demethylated hopanes predominate, disteranes formed, steranes absent

Note:
[a] After Volkman *et al.* (1983).

by chemical kinetics alone (e.g., Vandenbroucke *et al.*, 1999), or alternatively whether the light hydrocarbons in oils are largely the products of transition-metal catalysis (Mango, 1997) or a complex chemical segregation process during secondary migration (Price and Schoell, 1995). Based on chemical kinetic calculations, Pepper and Dodd (1995) predict that oils rich in n-alkanes will crack relatively slowly over a high, but narrow, temperature range of 155–205 °C. In contrast, low n-alkane oils crack more readily, starting at lower temperatures in the 115–145 °C range. Extreme heating of the trapped oil over 225 °C will result in thermal decomposition to just methane and pyrobitumen.

It is normally difficult to separate the relative influences of water-washing and moderate biodegradation as both processes commonly occur at shallow depths where oil is in contact with meteoric water and shift the composition of the altered oils in the same direction. Water-washing involves the preferential removal of the more water-soluble compounds, principally the aromatics and n-alkanes lighter than C_{15} without altering naphthenes, reflecting their relative aqueous solubilities (Lafargue and Barker, 1988). Inherently, water-washing can occur at any depth by waters of any salinity. However, biodegradation requires temperatures of less than about 80 °C and a continual flow of oxygen-charged fresh or brackish water (Hunt, 1979). Severe biodegradation of petroleum will change both the chemical and physical

properties of normal oil by decreasing the light hydrocarbon content and GOR and increasing the heteroatomic, sulphur content and viscosity (Connan, 1984, 1993; Wenger *et al.*, 2002). Microbial oxidation results in selective removal of alkanes and aromatics and a consequent enrichment of nitrogen–sulphur–oxygen-rich polar compounds. The end product can be a heavy oil with a gravity value in the 8–12° API range and viscosities in the thousands of mPa s, of lower commercial quality and difficult to produce. Normal alkanes are removed first, followed by isoprenoids, regular steranes, disteranes, hopanes and finally neohopanes, in that order (Volkman *et al.*, 1983). The order of removal of compound classes permits the classification of the level of biodegradation (Table 20.2).

The introduction of gas into pooled oil changes the bulk chemistry of the oil, moving it towards a lower average molecular weight and causing the precipitation of a hydrogen-poor asphaltene residue, solid bitumen. The result of deasphalting is a phase separation in the reservoir that leads to oil that is substantially lighter than the original oil. Wet gas formed during the thermal *in situ* cracking of oil causes natural de-asphalting in a process analogous to the propane deasphalting of residuum oil in a refinery (Hunt, 1979). In such cases, deasphalting works with thermal maturation to create lighter oil at greater depths. Gas or condensate may also be introduced into an oil pool through secondary migration (Blanc and Connan, 1994), in which case deasphalting is independent of burial and temperature increase and may be operative in even relatively shallow oil pools.

Depending on the efficiency of the seal, a trap can preferentially leak light fraction hydrocarbons represented by gas and condensate. The resulting evaporative-fractionation results from the differential solution of light hydrocarbons from the oil into the gaseous solution (Thompson, 1987, 1988). The removal of light hydrocarbons in the gas phase causes an increase in content of light aromatic and naphthenic hydrocarbons relative to paraffin in the residual oil. The release of the gas phase may occur due to tectonically induced pressure changes in a gas-saturated oil reservoir (Leythaeuser and Poelchau, 1991). The migrated fraction, if trapped in a higher pool at lower pressure and temperature, will condense as light oil with a gas cap by the process of dysmigration (Blanc and Connan, 1994). This process differs from gas flushing in which the creation of a gas cap in a rising or cooling oil pool will cause some of the pooled oil to spill out of the trap. In gas flushing, it is the oil fraction, made heavier by phase separation of gas and condensate, which migrates away from the original oil pool.

21 Reservoir destruction or enhancement due to thrusting

Rocks with permeabilities of 10^{-13} to 10^{-14} m^2 are considered to be excellent to good reservoirs (Deming *et al.*, 1990; Deming and Nunn, 1991; Thomas and Clouse, 1995). Figure 21.1(a) shows that these values are characteristic of the higher parts of the range of matrix permeability in carbonates and sandstones (e.g., Deming, 1994; Smith and Wiltschko, 1996). Grain size and sorting are the primary control on matrix permeability. For the same lithology, rocks with a larger grain size tend to have a higher permeability than rocks with a smaller grain size (Fig. 21.1b; Prince, 1999). Similarly, well-sorted sedimentary rocks can have a higher permeability than poorly sorted rocks due to a reduction of the pore space in the latter by finer-grained particles occupying the space between larger grains. Matrix permeability is also controlled by pre-orogenic diagenetic processes. An example of contrasting matrix permeability comes from the Painter field in the Absaroka thrust sheet of the Wyoming Thrustbelt. This field contains two facies of the producing Nugget Sandstone; dune and interdune–sand sheet facies (Tillman, 1989). The dune facies, characterized by well-sorted, well-rounded grains and well-connected intergranular porosity, has a porosity, and horizontal and vertical permeabilities of 16.3%, 196 and 13 md, respectively. The interdune–sand sheet facies, represented by alternating fine- and coarse-grained laminae, with poorly sorted, more angular and more tightly packed grains in the finer-grained laminae, has a porosity, and horizontal and vertical permeabilities of 10.2%, 4.5 and 0.67 md, respectively.

However, in thrustbelts, the internal straining that takes place before thrust sheets are developed also influences reservoir permeabilities (e.g., Geiser, 1974; Koyi, 1995, 1997; see Fig. 3.4). Internal deformation that continues to affect an already developed thrust sheet during various stages of its development (see Fig. 3.9) represents an additional control on the reservoir permeability. For example, outcrop observations of the Nugget Sandstone suggest that the finer-grained facies behave in a more brittle fashion than the coarse-grained facies. The brittle behaviour enhances the amount of fracturing, thus counteracting the effects of the finer grain size on permeability. However, diagenetic and chemical processes also affect the rock during thrustbelt development. These processes include dissolution, recrystallization, cementation and mineral transformations. All of these syn-orogenic factors are controlled by the overall energy balance, mechanical stratigraphy of accreted sediments, syn-orogenic deposition, syn-orogenic erosion and syn-orogenic fluid flow.

This chapter will focus on reservoir destruction and enhancement mechanisms in order to develop an understanding of how reservoir permeabilities vary in time and space in thrustbelts.

Reservoir destruction mechanisms

Compaction

As discussed in Chapter 10, reservoir permeability is controlled by factors such as the porosity, the pore geometry, the pore interconnectivity, the pore cementation, the effective stress, the static bulk modulus of the reservoir and the pore compressibility (Zimmerman *et al.*, 1986; Sigal, 1998, personal communication). For simplicity, it will be assumed that a relatively homogeneous system of pores exists and this discussion will focus on the roles of effective stress and static bulk modulus on permeability reduction during the initial compaction. In this case, the permeability decrease is associated with reduction of the pore space due to an increase in the effective stress (Fig. 10.21), as documented by numerous reservoir studies (Wygrala, 1989 and references therein; Fig. 21.2). The rate of porosity reduction with depth, shown in Fig. 21.2 for the case with only overburden load and no tectonic loading, is relatively simple. This figure illustrates the role of the reservoir lithology, represented by the static bulk modulus. Because it is the effective stress that controls porosity reduction, the lithostatic load of the overburden in Fig. 21.2 is reduced by the pore fluid pressure. Unlike tectonically quiescent basins, there are numerous thrustbelt mechanisms that can change the effective stress acting upon thrustbelt reservoirs, such as burial changes driven by syn-tectonic deposition and syn-tectonic erosion, increased burial due to overthrusting and changes in tectonic stress. All of these, except syn-tectonic erosion, have the potential to cause porosity reduction (Fig. 21.3). These mechanisms produce stress paths that are more complex than the burial-loading paths shown in Fig. 21.2. Irrespective of the path, the porosity generally decreases with

(a)

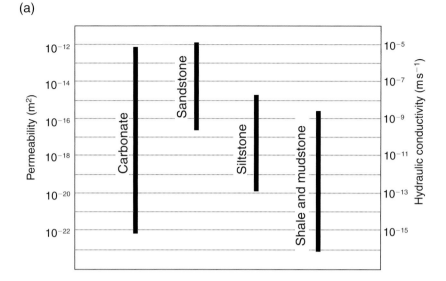

(b)

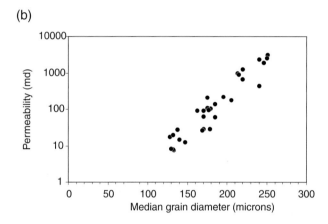

Fig. 21.1. (a) Permeability and equivalent hydraulic conductivity, calculated for water with a viscosity of $\eta = 0.001$ Pa *s and a density of $\rho = 1000$ kg m^{-3}, of common sedimentary rocks measured on laboratory specimens (Smith and Wiltschko, 1996). Data were compiled from Davis (1969), Brace (1980) and Domenico and Schwartz (1990). (b) Permeability dependence on the sandstone grain size (Price, 1999).

increased stresses, which often slows with time. This porosity decrease is caused by progressive grain translation and rotation of grains, resulting in closer grain-to-grain distances and preferential grain orientations (Fig. 21.4). Porosity reduction also changes with time due to progressively increasing creep of the solid reservoir skeleton and the effects of chemical processes.

The importance of the reservoir-loading path cannot be neglected. If the current effective stress is lower than the pre-existing maximum effective stress, which once was loading the reservoir, the porosity will change with the current effective stress at a slower rate than during normal consolidation, i.e. during 'virgin compaction' (e.g., Lambe and Whitman, 1969).

Any prior cementation that has already modified the reservoir before it underwent syn-tectonic compaction will significantly affect its syn-tectonic compaction history. An example of accretionary wedge cementation before the formation of folds and faults comes from the

Alaskan accretionary complex, where calcite and ankerite cementation preceded accretion (Sample, 1990). Here, cementation changed the static bulk modulus and pore compressibility of the reservoir, which in turn affected its compaction history.

The amount of layer parallel shortening, which is one of the important controls of early compaction, depends strongly on whether the rocks are lithified or unlithified. Lithified rocks in accretionary wedges are characterized by smaller changes in their physical and mechanical properties than unlithified rocks during burial, compaction and deformation. For example, the extent of shortening determined from Silurian Skolithos burrows from the Appalachian Valley and Ridge province ranges from 5 to 25% (Geiser, 1974). Internal strains determined in various thrust sheets of the Sevier thrustbelt represent 10–35% of the total amount of shortening (McNaught, 1990; Protzman and Mitra, 1990; Mitra, 1993; Mitra and Sussman, 1997).

(a)

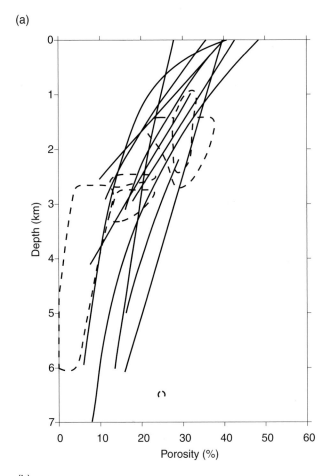

(b)

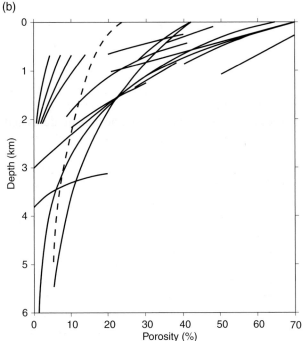

Fig. 21.2. Porosity–depth curves for (a) sandstone and (b) limestone compiled from various sources (Wygrala, 1989).

(a)

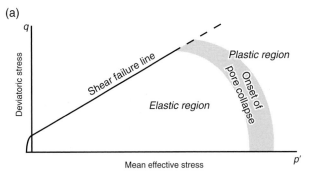

(b)

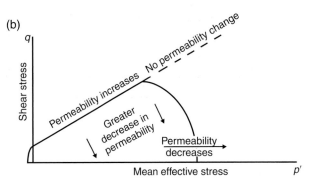

Fig. 21.3. (a) Sediment (chalk) behaviour in stress space (Joint Chalk Research, 1996). The vertical axis indicates the deviatoric stress ($q = \sigma_1 - \sigma_3$) and the horizontal axis indicates the mean effective stress ($p' = (\sigma_1 + \sigma_2 + \sigma_3)/3$). The region of the elastic behaviour of the sediment is bounded by the shear failure line and yield surface (onset of pore collapse surface). The elastic region along the negative mean effective stress line indicates tensile failure. The shear failure line is associated with fracturing and involves a localized distortion and breakdown of the structure along a failure plane. The yield surface is associated with pore collapse and involves pervasive breakdown of the structure due to destruction of cement bonding. The first yield stress is the point where the stress/strain curve departs from linearity. (b) Sediment (chalk) matrix permeability behaviour in stress space (Joint Chalk Research, 1996). Matrix permeabilities change more significantly along loading paths located close to the hydrostatic path, which runs from the origin along the p' axis. The permeability change associated with the loading path passing through the shear failure line depends on whether the path originates in the elastic or plastic region.

Burial, compaction and deformation changes in unlithified accretionary wedges are significant, as discussed in Chapter 10. For example, the average shortening values determined for the Nankai accretionary prism reach 68% (Morgan and Karig, 1995). Compaction-driven porosity reduction in the Barbados accretionary prism is also significant (Bangs *et al.*, 1990; Fig. 21.5a). A closer inspection of Fig. 21.5(a) reveals that the porosity distribution is controlled by the degree of both vertical and horizontal compaction, which

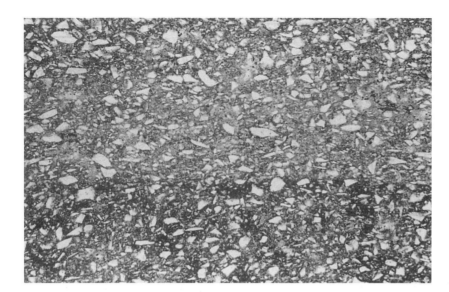

Fig. 21.4. Photomicrograph of a thin section of compacted silty clay illustrating the alignment of the long axes of irregularly shaped silt particles normal to the direction of the applied stress (Dewhurst *et al.*, 1996).

varies from thrust sheet to thrust sheet. Consequently, porosities vary abruptly across the sheet boundaries. A better understanding of the relationship between porosity distributions and deformation is provided by the comparison of the porosity distribution in the Barbados accretionary prism (Fig. 21.5a) with a detailed deformation study of the Nankai accretionary prism by Morgan and Karig (1995; Fig. 21.5b). In the Nankai accretionary prism, deformation starts with pervasive ductile deformation produced by grain boundary sliding, continues with localized ductile failure and formation of deformation bands and ends with brittle fracturing. Since the fracturing develops in zones and clusters along the thrust sheet boundaries (Fig. 21.5b), it is represented in Fig. 21.5(a) by linear zones of abruptly increased porosity along these boundaries. The role of fracturing on reservoir permeability is discussed in more detail below.

As mentioned above, permeability reduction in thrustbelts can be related to straining prior to and during imbrication. Straining prior to imbrication starts with plastic hardening but continues with plastic softening, followed by large-scale faulting and folding, as in the case of the Nankai accretionary prism. The initial straining changes the density and anisotropy of the reservoirs, which influences their subsequent compaction histories. An example of initial layer-parallel straining comes from the Sevier orogenic belt (Fig. 2.10). The effect of lithology on the history and character of straining is documented by examples from the Appalachians (Wojtal and Mitra, 1986; Mitra, 1987).

Plastic softening during initial straining can be caused by several brittle and ductile mechanisms. The brittle mechanisms include shear dilatancy (e.g.,

Handin *et al.*, 1963; Brace *et al.*, 1966; Scholz, 1968; Knipe *et al.*, 1991), cataclastic diminution and rounding of grains (e.g., Aydin, 1978a, b; Antonellini and Aydin, 1994; Antonellini *et al.*, 1994), and growth of microcracks or macroscopic joints (e.g., Dula, 1981; Mitra, 1987; Srivastava and Engelder, 1990; Chester *et al.*, 1991). The ductile mechanisms include, for example, pressure solution and dynamic recrystallization, which are strongly affected by the rate of applied straining, because they are controlled by diffusion (e.g., Elliott, 1976b; Mitra, 1988a; Knipe, 1993; Butler and Bowler, 1995). The effects of various plastic softening mechanisms on reservoir permeability will be discussed below.

Internal straining continues during imbrication in the thrustbelt, competing with new-fault propagation, reactivation of existing faults and other large-scale orogenic mechanisms as a means of releasing accumulated orogenic energy. This straining can react to changes in the thrust sheet dynamics such as those produced by any buttressing forces at its front, changes in its basal friction or changes in gravity resistance against its movement.

Internal straining affects various thrustbelt structures differently. For example, the interlimb and ramp angles in a fault-propagation fold are interrelated with the type of straining in its front limb (see Figs. 4.1 and 4.4). This can be demonstrated using the data from the Northridge earthquake indicating that a fault-bend fold, which grew by movement of the hanging wall from a flatter fault surface onto the base of the thrust ramp, or above the wedge tip, underwent thickening (Treiman, 1995).

A wealth of internal straining examples is provided by studies of natural duplexes. In the Wills Mountain

(a)

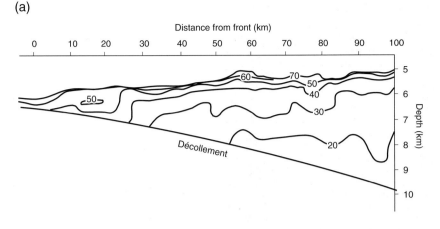

(b)

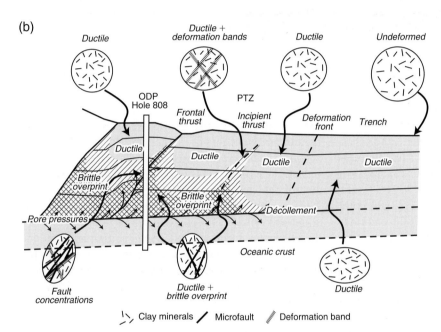

Fig. 21.5. (a) Porosity distributions within the Barbados accretionary prism (after Bangs *et al.*, 1990 and Breen and Orange, 1992). Distances are relative to the thrust front on the left side of the figure. Vertical exaggeration is 6.7*x*. Two ship wide-angle seismic reflection data were inverted for seismic velocity, and empirical relationships were used to infer porosity. See text for further explanation. (b) Distribution of ductile and brittle strains in the front of the eastern Nankai accretionary prism (Morgan and Karig, 1995). The initial tectonic strains are ductile (shaded) near the deformation front. Brittle deformation (cross-hatched) overprints this ductile strain field near the base of the prism, and eventually around propagating or developed thrust faults. These brittle strains may arise from high pore fluid pressures diffusing away from the décollement and the frontal thrust fault. Strain ellipses illustrate the estimated finite strain state, and the relationships between deformation structures and fabrics inferred from diffuse strain calculations, deformation structures observed in ODP drill cores, clay mineral fabrics, and sediment porosities.

Duplex in the Appalachians (Smart *et al.*, 1997), shortening by more than 65% via meso-scale folding and faulting, cleavage, shortening of crinoids and calcite twinning resulted in the strengthening of the roof thrust. In the Variscan orogen in France, the lower Basse Normandie duplex was shortened by about 50% by internal deformation. Here, strengthening of the thrust sheet was a response to increased basal friction along the detachment (Averbuch and Mansy, 1998). Such internal adjustments can result in reservoir permeability reductions.

Natural duplex studies also document that internal straining is controlled by mechanical interactions of various units incorporated in the thrustbelt structure. These interactions are affected by lithology contrasts between the units and their frictional contacts (e.g.,

Morse, 1977; Chester *et al.*, 1991; Riley *et al.*, 1995; Jamison, 1996). As the friction increases, larger stresses are transferred through the contact into the adjacent layer, and greater deformation results.

As discussed in Chapter 4, certain structures have typical locations in thrustbelts. Therefore, the locations of internal straining also can be predicted. Such predictions are useful in estimating reservoir permeability distributions. For example, large areas affected by penetrative straining, which form slip dissipation structures, can be expected in the frontal terminations of thrust systems (Morley, 1986).

Internal straining can also vary spatially and temporally within the same structure, depending on its growth mechanism and the lithologies that are involved. Deformation in cover sequences of basement uplifts

Table 21.1. Stress control of different fracture types.[a] Fluid-pressure criteria for different varieties of macroscopic brittle failure are expressed in terms of rock material properties (the tensile strength T, the cohesive strength for intact rock $C_i \sim 2T$, the coefficient of internal friction for intact rock μ_i, the residual cohesive strength C_r and the residual coefficient of internal friction μ_r. P_f is the pore fluid pressure, σ_1 and σ_3 are the maximum and minimum principal compressional stresses, σ_n and τ are the normal and shear stresses and θ is the angle.

Type of failure	Orientation	Criterion	Conditions
Grain-scale microcracks	$\sim \perp r\, \sigma_3$	—	Grain impingement
Hydraulic extension fractures	$\perp r\, \sigma_3$	$P_f = \sigma_3 + T$	$(\sigma_1 - \sigma_3) < 4T$
Extensional shear fractures (Griffith criterion)	Plane containing σ_2 and at $0 < \theta < \theta_i$ to σ_1	$P_f = \sigma_n + [(4T^2 - \tau^2)/4T]$	$4T < (\sigma_1 - \sigma_3) < 6T$
Shear fractures (Coulomb criterion)	Plane containing σ_2 and at $20° < \theta_i < 30°$ to σ_1	$P_f = \sigma_n + [(2T - \tau)/\mu_i]$	$(\sigma_1 - \sigma_3) > 6T$
Shear reactivation of pre-existing planar anisotropies with residual cohesion C_r	Planes not orthogonal to principal stresses	$P_f = \sigma_n + [(C_r - \tau)/\mu_r]$	—
Stylolitic solution planes	Envelope $\perp r\, \sigma_1$	?	Fine-grained rock matrix

Note:
[a] Secor (1965), Etheridge (1983), Sibson (1996).

frequently contains alternating longitudinal strain minima and maxima (Price, 1988; Means, 1989). They are represented differently in different lithologies. Bioclastic and oolitic limestones are deformed by fracturing at the uplift culmination and by calcite twin gliding and grain boundary sliding on its flanks (Couples and Lewis, 1998). Finely crystalline dolomite is deformed by intense fracturing at the uplift culmination and by periodically occurring fracture clusters on uplift flanks (Cooke and Pollard, 1994, 1997; Couples and Lewis, 1998). A typical example of temporal variations would be the migration of the hanging wall in a fault-bend fold. As the hanging wall moves through the rear syncline and anticline, its internal deformation re-adjusts to the current conditions at each new location.

The preservation of different stages of thrustbelt structures at different places along their strike adds further complexity to the relationship between straining and reservoir permeability. Early stages are preserved at the terminations of the structures and the most mature stages at their culminations. Characteristic patterns of internal straining would then vary with structural maturity along the strike. For example, the producing Turner Valley anticline in the Alberta Foothills becomes tighter as the structure grows (Gallup, 1954; Mitra, 1990). Such tightening also affects fully grown thrustbelt structures if they accommodate further shortening, such that additional compaction commonly occurs in fold crests during subsequent translation.

The existence of large numbers of producing reservoirs in thrustbelts demonstrates that permeability reduction caused by compaction does not rule out the occurrence of good reservoirs. Table 21.4 shows that reasonably good porosities and permeabilities are found in reservoirs within the first 2 km of the sedimentary section in the West Carpathians. There are two main reasons why these reservoirs retain good permeability. As alluded to earlier, increased loading can eventually lead to permeability enhancement by fracturing. This situation will be discussed below under reservoir enhancement mechanisms. The second reason is the existence of numerous factors that minimize the destruction of the pre-orogenic reservoir permeability.

Before discussing these factors, it is useful to describe the special role of large pores in reservoirs. Typical reservoir sediments do not always have homogeneous grain-size distributions. Petrophysical studies have shown that, during deposition, sands tend to self-organize into a mixture of close-packed grain clusters separated by packing flaws (e.g., Ehrlich *et al.*, 1997; Prince, 1999). These flaws propagate through the sediment forming a network of packing flaws; i.e. a system of large pores. Because large pores tend to have the best wetting characteristics of the available pores in the reservoir (Robin *et al.*, 1995; Mann *et al.*, 1997), they are particularly important for hydrocarbon accumulation (e.g., Winter, 1987; Ehrlich *et al.*, 1997; Prince, 1999).

The behaviour of large reservoir pores during permeability destruction in thrustbelts is different from that of the smaller pores. When syn-orogenic compaction, perhaps further enhanced by chemical processes, reduces the porosity of sandstone reservoirs, packing flaws are apparently better protected from destruction than normal pores (e.g., Ehrlich *et al.*, 1997). There are

Table 21.2. Basins with known abnormal pressure compartments.[a]

Location	Basin (part of the basin)	Location	Basin (part of the basin)
Underpressure		African basins	Gabon
	Western Canada basin, central Alberta, Canada		Offshore Sinai, Egypt
			Sirte, Libya
	Gallup Sandstone, San Juan basin, USA	European basins	Lower Saxony trough, Germany
	Keyes dome, Amarillo uplift, USA		Molasse Zone, Germany and Austria
	Mississippian formation, Midland basin, USA		North Sea
	Niobara Chalk, Denver basin, USA		Pannonian basin, Hungary
			Polish trough
	Permian basin, USA		Pre-Caspian
			South Aquitaine, France
Overpressure			South Caspian
Delta	Mackenzie, Canada		Transylvanian, Romania
	Mahakam, Indonesia		Vienna basin, Austria
	Mississippi, USA		Western Siberian basin
	Niger, Africa		Zechstein, Germany and Poland
	Nile, Egypt		
	Po, Italy	Middle Eastern and Asian/Pacific basins	Bengal, Bangladesh
North American and South American basins	Anadarko, USA		Burma, Burma
	Columbus, Trinidad		Cambay, India
	Cook Inlet, USA		Dampier sub-basin, Australia
	Gulf of Paria, Venezuela		Gulf of Bohai, China
	Jeanne D'Arc, Canada		Hsinchu, Taiwan
	Lower Magdalena, Columbia		Jaya, Indonesia
	Mississippi Salt, USA		Mesopotamian, Iraq
	North Ardmore, USA		Nepal, India
	North Slope of Alaska, USA		North Iran
			Northland, New Zealand
	Pacific Northwest, USA		North Sumatra, Sumatra
	Powder River, USA		Northwest shelf, Australia
	Scotian, Canada		Potwar, Pakistan
	Southern Sacramento, USA		Sarawak (offshore)
	Southern San Joaquin, USA		South China Sea
	Texas and Louisiana Gulf Coast		South Papua Coastal, New Guinea
	Uinta, USA		Takzhik, Afghanistan
	Williston, USA		Tertiary reefs, offshore Sumatra
	Wind River, USA		

Note:
[a] Hunt (1990).

several reasons for this. First, wetting phases are preferentially segregated in the close-packed domains, where they remain relatively immobile, counteracting the effects of compaction. Secondly, if the wetting phases are a hydrocarbons, mineral precipitation is slowed. Thirdly, wetting phases with high mobility have the capability of retarding mineral precipitation. Fourthly,

the process of mechanical diagenesis continuously adjusts to increasing stress (Prince, 1999). Associated dewatering accommodates not only the expulsion of locally derived formation fluids, but it also accommodates pore fluid advection from below. Because of this accommodation, the dewatering process allows high-permeability pore throats to remain open, reducing the pore/throat size ratio. Such preferential preservation of the largest pores continues until the porosity decreases to a certain critical value characteristic for each specific sandstone. At that point, the flawed porosity will begin to compact and its proportion of the total porosity will start to decline (Prince, 1999).

As previously mentioned, pore fluid pressure plays a role in permeability preservation because, during dewatering, the driving force must be balanced by the resistive force exerted by the reservoir skeleton upon the moving fluid (e.g., Mandl, 1988). If the fluid cannot be drained away sufficiently quickly to keep up with ongoing compaction, expulsion slows down maintaining porosities larger than those under fully drained conditions. This results in reservoir pore fluid pressures that are higher than hydrostatic. Table 21.2 documents numerous cases of reservoir compartments with abnormal pore fluid pressures from different tectonic settings, some of these are thrustbelts. Especially at greater depths, hydrocarbons are frequently located in seal-bounded horizons, characterized by different internal pressures (Vandenbroucke *et al.*, 1983; Demaison, 1984; Hunt, 1990, 1991; Evans and Battles, 1999). Examples of such well-defined compartments come from the Appalachians (Evans and Battles, 1999), the Alberta basin, Canada (Bachu and Underschultz, 1993; Bachu, 1995) and the Llanos basin, Colombia (Villegas *et al.*, 1994). Figure 10.19 illustrates the effect of porosity preservation due to restricted drainage of the pore fluids from footwalls within the Barbados accretionary prism and from the sedimentary section beneath its décollement.

An additional factor that can potentially minimize the destruction of pre-orogenic reservoir permeability is anisotropy. Under certain conditions, the anisotropy can cause horizons to protect others from greater permeability destruction.

Deformed sedimentary packages are considered to be anisotropic when bedding planes allow for bedding-parallel slip, which does not necessarily occur along each bedding plane in the package. This behaviour can result in independent movement of separate packets of beds during thrustbelt imbrication (Tanner, 1989). It is the friction along layer contacts that determines whether the package behaves anisotropically and which packets or beds will slip 'independently'. This frictional control can be pre-orogenic, determined by primary fabrics or syn-orogenic and driven by initial internal straining (e.g., Knipe, 1985; Mandl, 1988). Primary fabrics of distal turbidites or other planar thin-rhythmic sedimentary sequences cause anisotropic behaviour during subsequent thrusting. These sediments are prone to flexural slip. Proximal turbidites and massive or thick-rhythmic sedimentary sequences have a tendency to behave much less anisotropically than thin-layered sequences, and with less tendency for flexural slip.

An example of syn-orogenic frictional control, driven by initial internal straining, comes from the Double-spring duplex in Idaho (Hedlund *et al.*, 1994). The duplex developed in massive Mississippian limestone, which was then folded. Anisotropic behaviour developed from the initially isotropic rock through the development of a system of parallel shear zones that separated it into layers.

If the anisotropic sedimentary package contains nearly frictionless bedding planes, but the package includes layers of different lithologies, each rheology undergoes different internal deformation during thrusting. In the Appalachians (Mitra, 1987), layers prone to pressure solution underwent shortening forming stylolites, clay-carbon partings and other seams of insoluble residues. This shortening resulted in decreased permeabilities. Argillaceous layers within the package reacted by cleavage development, produced by the preferred alignment of phyllosilicates. Competent beams in the package experienced shortening by extensive faulting and fracturing, which, depending on the ability of the beam material to dilate, enhanced or did not enhance its permeability. The ability of rocks to dilate will be discussed below.

If the anisotropic package contains thick packets of welded layers interspersed with nonwelded layers, the welded packets will undergo extra straining in the fold cores of the thrustbelt structures, while the nonwelded layers, undergoing flexural slip, will be impacted much less. If the anisotropic package contains horizons that are layer-parallel stiff to deformation, the package will be more prone to folding than internal thickening or thinning. This will be the case for an anisotropic package that is layer-parallel soft to deformation. A layer-parallel stiff package would be less likely to suffer syn-thrusting destruction of its initial permeability.

Friction along faults also has the potential to funnel stress through portions of the growing structure, whilst leaving others unaffected. Because stress transfer is utilized by straining it plays a role in controlling reservoir permeability. Thus, when searching for preserved initial permeability in the footwall of an inverted normal fault, it is important to evaluate the stress transfer across the fault. This transfer becomes more efficient with increasing friction along the fault. Consequently, increased straining of the footwall results. Less efficient transfer occurs when the friction is low. In this case the trajectories of the maximum compressive stress become subparallel to the fault and the stress is funnelled primarily through the hanging wall.

(a)

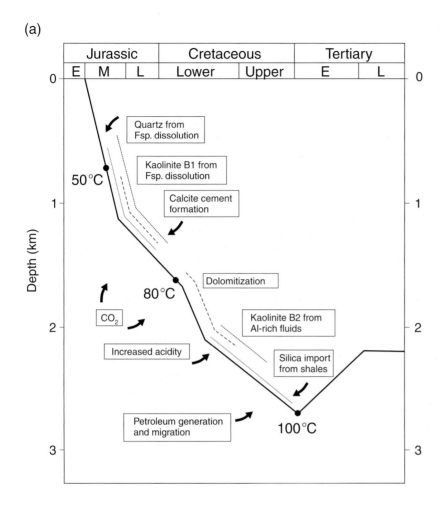

(b)

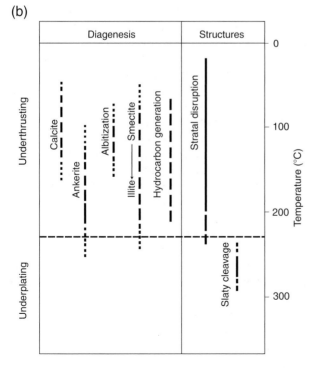

Fig. 21.6. (a) Synthesis of burial, maturation and diagenesis (Schwartzkopf, 1990). (b) Relative timing of cementation in the Alaskan accretionary complex (modified after Sample, 1990). Increasing length of dashes corresponds to increasing confidence in the temperature estimate. Solid lines represent temperatures of stratal disruption and slaty cleavage formation that are constrained by fluid inclusion data and other indicators of metamorphism. The horizontal dashed line is the approximate temperature boundary between deformation dominated by stratal disruption related to underthrusting and deformation dominated by cleavage formation, which is related to underplating.

Cementation

As mechanical compaction continues, pressure solution and cementation join the diagenetic process. Sandstones, for example, compact from about 40% porosity at deposition (Lundegard, 1992) to about 15–25% porosity primarily by mechanical compaction through grain rearrangement and pressure solution (Sclater and Christie, 1980). At 15–25% porosity, the possibility of further reduction in the volume of the porous system by mechanical compaction is essentially exhausted. From this point on, further destruction of reservoir permeability is caused by diagenetic cementation within the pores. While cementation reduces the pore space it can increase pore fluid pressures, which in turn can retard subsequent precipitation.

Apart from the pore fluid pressure, other factors such as increasing temperatures can further effect porosity destruction (e.g., Chilingarian, 1983; Poelchau *et al.*, 1997). Temperature influences porosity reduction through its effects on solution, cementation and fluid viscosity (e.g., Gregory, 1977; Stephenson, 1977). Minerals lining pore walls are subject to chemical alteration, precipitation or dissolution, depending on the temperature, pressure, pore fluid composition and ion balance in the reservoir (Fig. 21.6a, Table 31.4). Table 13.4 and Fig. 21.6(b) show the temperature dependence of common diagenetic minerals (Biorlykke *et al.*, 1989; Schwartzkopf, 1990).

Cement-precipitating fluids, however, do not have to be derived locally from pressure solution. As discussed in Chapters 10 and 16, active major fault systems have a tendency to capture a significant proportion of the fluid flow derived from a variety of sources. These fluids can control the cementation in both the thrustbelt and its foreland, as documented by cement distributions in sandstone reservoirs associated with fluid entries from fault zones (e.g., Flournoy and Ferrell, 1980; Burley *et al.*, 1989). One of the primary influences on the rate of cementation is the rate of fluid supply. Slow cementation rates frequently occur in response to local dissolution, whereas rapid pressure drops following faulting events favour rapid cementation (Knipe, 1993).

Many fluid flow systems associated with thrustbelts also control cementation in the orogenic foreland, as shown by studies of the Canadian Rocky Mountains and the Appalachians (e.g., Sverjensky, 1986; Dorobek, 1989; Qing and Mountjoy, 1994). The dominant fluid flow systems controlling this cementation are topography-driven fluid flow and fluid flow driven by the ongoing compaction inside and under the advancing thrustbelt. This early cementation can hinder later syn-orogenic compaction of cemented reservoirs, increasing the compressibility value of the reservoir that later becomes accreted into the thrustbelt. The distribution of cemented reservoirs in the orogenic foreland is controlled by the geometries of the fluid systems, which are described in detail in Chapter 10.

Potential cements include: anhydrite, ankerite, barite, calcite, dolomite, kaolin, microcrystalline and medium-grained quartz, chert, pyrite, siderite and sphalerite and other less common mineral phases. Carbonates cemented by silica, for example, are known from the North Sea (Hunt, 1990) and diagenetic bands formed by silica, carbonate and chlorite cements are found in the Anadarko basin (Al-Shaieb *et al.*, 1994). Cemented zones cut across lithostratigraphies more commonly in siliciclastic than in carbonate reservoirs. For example, in the Cook Inlet, Alaska, calcite cementation, controlled by the thermal regime, cuts across structures, facies, formations and geological time horizons (Powley, 1980). Depending on their composition and mineralogy, different cements have a variety of roles in reservoir permeability destruction. If the cement is composed of microcrystalline quartz, for example, permeabilities can drop anywhere from 10^{-15} m^2 down to 10^{-19} m^2. The permeability is controlled by the proportion of the highly porous microcrystalline quartz to nonporous chert in the cement.

The extent to which the reservoir cementation is supported by local pressure solution depends on the reservoir lithology. While pressure solution will be discussed in detail later, here we focus on its relationship to cementation in various types of sandstones. In order to compare end members, one can divide sandstones into those containing less than 5% phyllosilicates, 5–15% phyllosilicates and those with a clay mineral content of greater than 15%. The reason for this division is the effect of clay minerals on quartz cementation and pressure solution. Sandstone reservoirs with less than 5% phyllosilicates can undergo significant quartz cementation but not extensive pressure solution.

Sandstones with 5–15% clay minerals can experience both cementation and pressure solution, because small concentrations of clay minerals at quartz grain-to-grain contacts increase the rate of the pressure solution (e.g., Heald, 1955, 1959; Thompson, 1959; De Boer *et al.*, 1977; Chang and Yortos, 1994; Bjorkum, 1996; Oelkers *et al.*, 1996). Fractures in this sandstone type developed during the initial stages of porosity destruction can have a focusing effect on quartz precipitation because the small quantity of clay or other grain surface pollutants such as Al^{3+} (Cecil and Heald, 1971; Iller, 1979) can inhibit quartz cementation. Fracture surfaces contain fewer surface pollutants than quartz grains in the surrounding matrix and provide larger surface areas than the surrounding intact reservoir for quartz deposition.

Therefore, quartz precipitates on the fracture surface faster than in the host sandstone and as a result, most of the quartz cementation will occur along fractures. Even if early fractures are absent, preferential cementation is possible due to the clustered organization of sandstones. This clustered organization can generate networks of higher intergranular volumes around packing flaws, helping to postpone the permeability destruction associated with large pore systems.

Quartz precipitation and pressure solution can be suppressed by hydrocarbons (Griggs, 1940), which lower the rate of silica diffusion from the quartz grain contacts to precipitation sites (Worden *et al.*, 1998).

Sandstones with greater than 15% phyllosilicates behave differently from cleaner sandstones during initial compaction. They can undergo greater porosity destruction because clay minerals lack the stronger compaction resistance of a closely packed quartz grain network. These sandstones are also less prone to cataclasis than clean sandstones (Fisher and Knipe, 1998) because clay-rich sandstones deform by grain sliding, grain rotation and plastic deformation of phyllosilicates without the need for significant dilation and grain crushing. Compaction results in a redistribution of clay minerals. These minerals coat the quartz grains and become more homogeneously distributed with increasing compaction. Rearranged in this way, impure sandstones can experience enhanced pressure solution but are not a good medium for extensive quartz cementation because the coating clay minerals suppress quartz precipitation (Cecil and Heald, 1971; Tada and Siever, 1989) by reducing the reactive quartz area and, therefore, the precipitation rate (Walderhaug, 1996). This inhibition strengthens with the amount of clay minerals, making it difficult for sandstones with more than 25% clay minerals to be affected by cementation and pressure solution.

Cementation and dissolution studies in the North Sea show that up to 30% of the deeply buried reservoirs in this region consist of diagenetic minerals (Giles *et al.*, 1992). Dissolution of minerals in pore walls can increase the pore volume and decrease the pore fluid pressure. The effects of cementation or dissolution on the pore fluid pressure depend on whether the system is open or closed to fluid sources or sinks (Bjorlykke, 1984; Gluyas and Coleman, 1992). The fact that certain diagenetic reactions are controlled in turn by pore fluid pressure makes the process complex. For example, fluid overpressures tend to inhibit quartz cementation in reservoirs where the major source of silica is from local pressure solution. When the pore fluid pressure increases, it decreases the effective stress that loads the reservoir. Closed systems, where cements derived from pressure solution precipitate locally, require compaction equal to the initial porosity for total porosity destruction (Mitra and Beard, 1980).

Pressure solution

Pressure solution affects thrustbelt reservoirs when mechanical compaction is combined with chemical compaction. Chemical compaction, which is represented by dissolution at grain-to-grain contacts, is affected by stress concentration (e.g., Renton *et al.*, 1969; Sprunt and Nur, 1976; Mitra and Beard, 1980; Houseknecht, 1987; Elias and Hajash, 1992). It adds to the overall compaction of the rocks and results in cements that precipitate in the surrounding pore spaces where stresses are lower than those at grain contacts. Together with cementation, pressure solution represents a diffusive mass transfer process that is associated with ductile straining in thrustbelts. It has both pervasive and planar forms (Fletcher and Pollard, 1981). The pervasive form is known to cause reservoir porosity loses as high as several tens of per cent (e.g., Mitra and Beard, 1980; Houseknecht, 1988). The planar form, represented by stylolites and pressure solution seams, acts as a localized permeability barrier due to the decreased permeability in the host rock adjacent to it (Nelson, 1985). Because ductile strain softening is intimately related to brittle strain softening, as discussed in Chapter 3, pressure solution features and fractures are commonly found together in thrustbelts, accommodating contemporaneous shortening and extension in different directions (e.g., Beach, 1977; Mitra, 1988; Fig. 21.7a).

As mentioned earlier, it cannot be anticipated that all of the dissolved material will precipitate locally. The natural rock spectrum provides a whole variety of pressure solution–cementation relationships between end members represented by net cement exporters and importers (Houseknecht, 1988).

Pressure solution is controlled by various factors including stress, temperature, time, lithology, grain size, grain geometry and the presence of catalysts. Increasing differential stress increases the rate of pressure solution, as demonstrated by experimental data on quartz (Elias and Hajash, 1992; Fig. 21.7b). However, pressure solution does not require large differential stresses, as indicated by the sandstone reservoirs of the North Sea, which have undergone pressure solution under mild differential stresses (Bjorkum, 1996). These stresses were typically lower than the differential stresses that drive calcite twinning or quartz cataclasis (Durney, 1972; Groshong, 1975).

Increasing temperature raises the rate of pressure solution, as documented by quartz experiments at

(a)

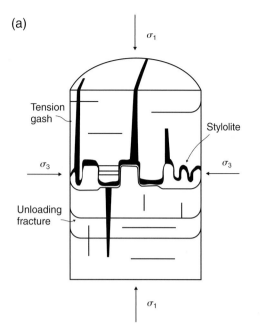

(b)

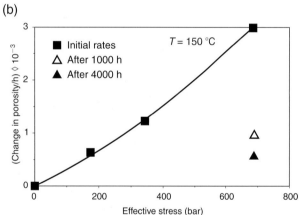

Fig. 21.7. (a) Geometric relationship between stylolites, tension gashes, unloading fractures and stress (Nelson, 1981).
(b) Compaction rate caused by pressure solution as a function of effective stress, determined from experiments on quartz at a temperature of 150 °C (Elias and Hajash, 1992).

Table 21.3. Mineral and lithological controls on pressure solution.[a]

Mineral	Grain-size	Rock fabric
(top, most pressure solution-prone, bottom, least pressure solution-prone)		
Calcite	Fine grained	Uncemented, permeable
Dolomite Quartz Feldspars Micas and clays Pyrite	Coarse grained	Well cemented, impermeable

Note:
[a] Mitra (1988).

sandstone reservoirs show a lower tendency to be affected by this process. For the same lithology, the rate of pressure solution will be highest in fine-grained rocks (Durney, 1972). However, the presence of stylolites in the basal conglomerate of the Central Carpathian Paleogene basin, West Carpathians indicates that coarse-grained facies still undergo pressure solution. Because quartz experiments (Elias and Hajash, 1992) indicate that pressure solution increases with increased differential stress, the grain shape at grain-to-grain contacts should also play a role, because the grain sharpness at these contacts controls local stress concentrations (e.g., Pollard and Aydin, 1988).

Observations of undeformed phyllosilicates penetrating quartz grains (Bjorkum 1996) that have undergone dissolution and re-precipitation, suggest that the clays have catalysed pressure solution of the quartz grains (Walderhaug, 1996; Bjorkum *et al.*, 1998). Other examples are discussed above.

Cataclastic diminution
As the internal straining that precedes the development of thrust sheets progresses, reservoirs in the deforming sedimentary sequence undergo strain hardening followed by strain softening. Strain softening contains both brittle and ductile mechanisms (e.g., Scholz, 1968; Krantz and Scholz, 1977; Lockner *et al.*, 1992). One of the brittle mechanisms, typical of sandstone reservoirs is cataclastic diminution and associated grain rounding (e.g., Aydin, 1978a, b; Antonellini and Aydin, 1994; Antonellini *et al.*, 1994).

Medium- to coarse-grained sandstone typically has porosities higher than other rock types in thrustbelts.

temperatures ranging from 23 to 150 °C (Elias and Hajash, 1992). Similar temperatures are found in hydrocarbon reservoirs in upper parts of thrustbelts. Thus, it can be concluded that pressure solution will be an important permeability control in sandstone-hosted hydrocarbon reservoirs. A sensitivity analysis of quartz experiments indicates that time enhances the effects of lower temperatures on pressure solution.

The tendency of different minerals to undergo pressure solution is shown in Table 21.3. Carbonate reservoirs are most prone to pressure solution. Minerals in

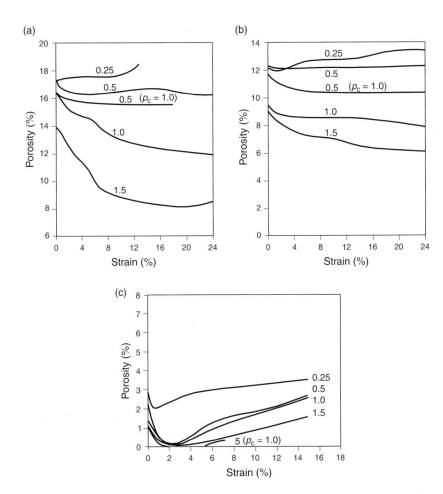

Fig. 21.8. (a) Porosity as a function of total strain in Berea sandstone under different effective confining pressures, p_c, in kilobars at 24 °C (Handin *et al.*, 1963). Curves denoted by $p_c=1$ indicate a confining pressure of 1 kbar. Other curves indicate a confining pressure of 2 kbar. All tests are made under compression. (b) Porosity as a function of total strain in Marianna limestone under different effective confining pressures in kilobars at 24 °C. Curves denoted by $p_c=1$ indicate a confining pressure of 1 kbar. Other curves indicate a confining pressure of 2 kbar. All tests are made under compression. (c) Porosity as a function of total strain in Halsmark dolomite under different effective confining pressures, in kilobars, at 24 °C. All tests are made under compression at a confining pressure of 2 kbar, except for the curve denoted by $p_c=1$.

When these sandstones are loaded by stress, they tend to undergo compression even under very high strains. This behaviour is in sharp contrast with low-porosity rocks, such as dolomite, limestone or granite, which would undergo dilatancy at much smaller strains (e.g., Handin *et al.*, 1963; Fig. 21.8). High-porosity sandstone reservoirs can experience a completely suppressed tendency to dilate (Swanson and Brown, 1972). The likelihood of permeability enhancement in these sandstones due to brittle failure, following initial permeability destruction by compaction, pressure solution and cementation, thus, can be mitigated by permeability destruction by failure (e.g., Edmond and Patterson, 1972; Antonellini and Aydin, 1994).

Faulting studies of medium- to coarse-grained sandstones by Antonellini and Aydin (1994) and Mollema and Antonellini (1996) in Utah demonstrate permeability reduction due to failure. These authors observed that faulting in porous sandstones evolves from a single deformation band, to the creation of deformation band zones and finally to the development of a fault plane (Aydin, 1977, 1978a, b; Aydin and Johnson, 1978, 1983,

Fig. 21.9). Deformation bands are 0.5–2 mm thick fault-like planar structures (Fig. 21.9a), usually with large lateral continuity (Zhao and Johnson, 1991). The pore space within the band may be partially or totally filled with iron oxides, calcite cement or other diagenetic minerals. Deformation bands accommodate a small amount of displacement and are characterized by localized cataclasis and volume changes (Antonellini and Aydin, 1994; Figs. 21.9 and 21.10a). These volume changes produce distinct porosity changes between the deformation band and the host rock (Antonellini and Aydin, 1994; Fig. 21.10b). The porosity inside the deformation band decreases by two to three times and can be accompanied by a permeability decrease of about three orders of magnitude (Antonellini and Aydin, 1994; Fig. 21.10b). Deformation bands can act as localized deformation zones and barriers to fluid flow (Figs. 10.15b and c). The sealing characteristics of a deformation band progressively increases as it evolves into a fault plane.

The reason for the progressive development of a fault plane from a deformation band can be explained using

(a)

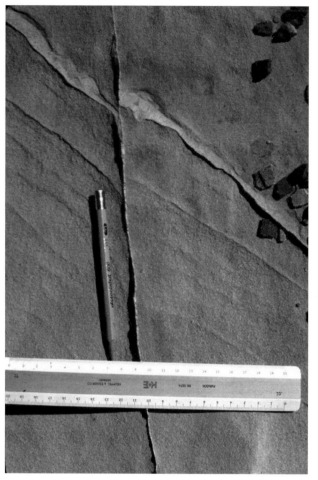

Fig. 21.9. (a) Photomicrograph and outcrop photograph of shear bands formed in the Entrada Sandstone in San Rafael Desert, Utah (Aydin *et al.*, 2004). (b) Sketch of a fault plane in sandstone (Antonellini and Aydin, 1994). Note the numerous bands forming the fault. (c) Photograph of a fault in the Entrada Sandstone from the Delicate Arch viewpoint, Utah.

(b)

(c)

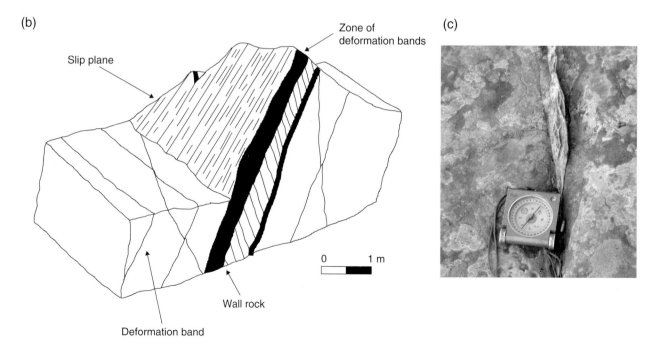

(a)

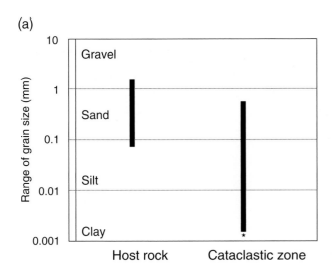

Fig. 21.10. (a) Grain size in a host rock and in the cataclastic zone inside a deformation band in sandstone (Antonellini and Aydin, 1994). Grain sizes of gravel, sand, silt and clay are given by arrows for comparison. The star indicates the limit of resolution using a petrographic microscope. (b) Permeability and porosity profile through the deformation band developed in the Entrada Sandstone (Antonellini and Aydin, 1994).

(b)

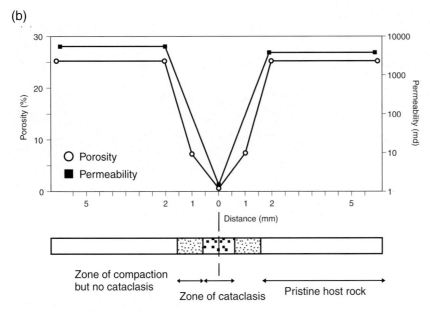

the rock mechanic results of Handin *et al.* (1963). They demonstrate that, as cataclasis along the band proceeds and the average grain size decreases, the extent of grain-to-grain contacts increases, resulting in a frictional increase along the band. At some point the band reaches a stage when the propagation of a new band requires less physical work than continued activity along an existing band. This behaviour is known as strain hardening. Similar observations have been reported from outcrops in basement uplifts around the Uncompahgre Plateau in front of the Rocky Mountains, Colorado where the spacing of deformation bands correlates with the amount of the bulk strain in the outcrop (Jamison and Stearns, 1982).

Because permeability destruction by cataclastic def-ormation bands is not pervasive, determination of its impact on a specific reservoir becomes very difficult to determine, as is the case for compaction bands (Mollema and Antonellini, 1996; Fig. 2.11), formed at higher confining pressures (Wong *et al.*, 2001). These features are difficult to incorporate into any reservoir model.

Reservoir enhancement mechanisms

Fracturing

Thrustbelt reservoirs loaded by triaxial compression undergo volumetric changes, which are negative during early stages of loading. For example, the permeability decrease due to matrix compaction of a sandstone

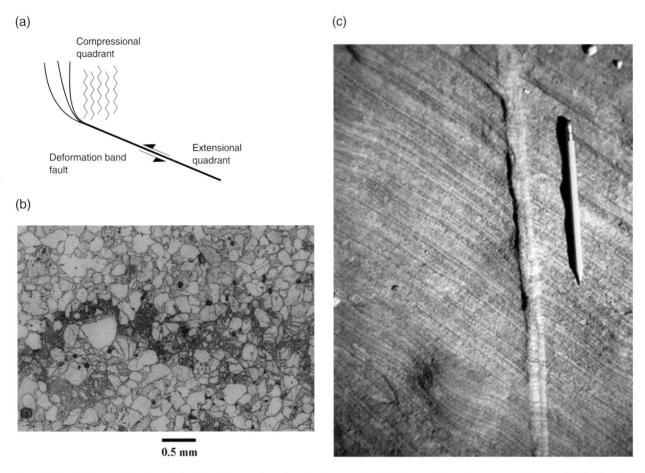

Fig. 21.11. (a) Compaction bands developed in an advancing block near a fault tip in sandstone. They are analogues of stylolites in limestone (Mollema and Antonellini, 1996). (b) Photomicrograph of a compaction band in sandstone (Baud *et al.*, 2004).
(c) Outcrop photograph of a compaction band in sandstone (photograph courtesy of Veronika Vajdová and Kurt Sternlof).

reservoir will generally be less than 50% (Wilhelm and Somerton, 1967; Mordecai and Morris, 1971; Daw *et al.*, 1974). Later in the loading history, the volumetric changes in dilation-prone reservoirs become positive (Vutukuri *et al.*, 1974; Fig. 21.8). This volumetric expansion or dilatancy has long been observed in granular rocks where it is associated with the relative movement of grains or groups of grains in irregularly packed interspaces. Dilatancy develops as a geometrical response to fracturing (Logani, 1973). It is the dominant mechanism in densely packed materials. Fracturing occurs at a differential stress level that is roughly two-thirds of the ultimate rock strength (Zoback and Byrelee, 1976).

As discussed in Chapter 10, the newly produced fracture space has an important impact on the reservoir permeability, increasing the pre-fracturing permeability by several orders of magnitude. An example of such an increase comes from the Tarnawa–Wielopole oil field in the West Carpathians (Table 21.4), where the increase

reaches over two orders of magnitude. Although fractures represent minimal storage capacity in reservoirs, they provide significant transmissibility. As a result, very small increases in fracture porosity can be associated with very large increases in permeability (Nelson, 1985). The largest effect on fracture permeability is imposed by the fracture aperture (Fig. 10.23b). Other factors include the fracture density, the roughness of the fracture surface and the fracture connectivity (e.g., Capuano, 1993). Fracture density is controlled by lithology, grain size, porosity, bed thickness and friction along bedding planes.

Apart from lithology and stress, the confining pressure affects the dilatancy. Porosity changes caused by dilatancy of a variety of rock types document this control. Porosity increases at low effective confining pressures and decreases at high effective confining pressures have been reported by Handin *et al.* (1963). Pore fluid pressure is also an important control on fracturing, because it affects the growth of different types of fractures

Table 21.4. Reservoir porosities in West Carpathian traps in Poland.[a]

Field name	Formation	Lithology	Age	Depth of top (m)	Dominant porosity type	Average porosity (%)	Average permeability (md)
Biecz	Ciezkowice	ss	Eocene	400	Matrix porosity	15.00	
Dabrowka Tuchowska	Istebna	ss (sili)	U. Cretaceous	530	Matrix porosity	14.30	
Gierczyce–Siedlec	Allochthonous molasse	ss	L. Badenian	704	Matrix porosity	19.70	357
Grabownica–Wies	Istebna	ss (sili)	U. Cretaceous	1463	Matrix porosity	10.63	
Jaksmanice–Przemysl	Authochthonous molasse	ss	L. Sarmatian	600	Matrix porosity	19.00	3–243.8
Jaksmanice–Przemysl	Authochthonous molasse	ss	L. Sarmatian	900	Matrix porosity	17.00–19.70	
Jaksmanice–Przemysl	Authochthonous molasse	ss	U. Badenian	1400	Matrix porosity	8.00–21.00	2–916
Jodlowka	Authochthonous molasse	ss	U. Ba-L. Sa	1800	Matrix porosity	5.00–20.00	7–200
Lakta	Authochthonous molasse	ss	L. Badenian	2364	Matrix porosity	8.00–12.00	3
Nosowka	Authochthonous Paleozoic section	carb	Visean	3465	Fract. porosity	17.00 (0.50)[b]	25–27
Osobnica	Ciezkowice	ss	Eocene	400	Matrix porosity	12.00–18.00	500
Osobnica	Istebna	ss (sili)	U. Cretaceous	1060	Matrix porosity	15.00–20.00	
Tarnawa–Wielopole	Lower Krosno	ss (sili)	Oligocene	750	Fr/mat porosity	3.00–20.00	120 (0)
Wola Jasienicka	Weglowka	ss	U. Cretaceous	520	Matrix porosity	13.50	
Zalesie	Authochthonous molasse	ss	U. Badenian	2070	Matrix porosity	12.00	145

Notes:

[a] Data from Karnkowski (1999).

[b] Matrix porosity and permeability in parentheses; ss, sandstone; carb, carbonates; sili, other siliciclastics.

including joints, hydraulic extension fractures, extensional shear fractures, shear fractures and pre-existing planar structures reactivated by shear (e.g., Secor, 1965; Etheridge, 1983; Pollard and Aydin, 1988; Sibson, 1996; Table 21.1). Table 21.1 further indicates that different fracture mechanisms operate under different stress conditions, which has an important implication for the fracture development in specific reservoirs.

Fracturing in thrustbelts can develop either during the internal straining that precedes imbrication or during imbrication. During pre-imbrication straining at differential stresses significantly smaller than those necessary for the main fault propagation, randomly oriented microcracks start to increase in number and form clusters, as shown by rock mechanics tests (e.g., Scholz, 1968; Krantz and Scholz, 1977; Lockner *et al.*, 1992). Microcracks are generally accepted as a precursor to extensional or shear fractures (e.g., Brace *et al.*, 1966; Kranz, 1983; Cox *et al.*, 1991), developing at stresses from 50 to 95% of the fracture strength of the rock (Scholz, 1968). Meso-scale joints also progressively cluster as shown by studies of dolomite reservoirs (Antonellini and Mollema, 2000; Fig. 21.12). In dolomite, which represents one of the most brittle reservoir rock types, small strains are accommodated by pervasive fracturing that affects a large reservoir volume. The appearance of the first small offset faults records more advanced stages of this pervasive fracturing. These

(a)

(b)

(c)

Fig. 21.12. Fault development in brittle carbonate in the frontal thrust sheets of the Dinarides, Dubrovnik, Croatia. (a) Initial stage: development of joints parallel to the maximum principal compressive stress σ_1. The joints are localized in en-echelon fashion with respect to a later fault. Note that a set of cross joints cuts bridges between the main fractures. (b) A more advanced stage in which en-echelon and cross fractures form a breccia zone along the later fault core. Note that the initial en-echelon fracture organization is still visible. (c) The final stage in which the fault core consists of chaotic breccia. The damage zone is formed by a fracture set with increased fracture density.

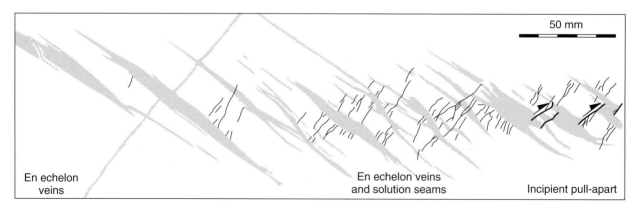

Fig. 21.13. Fault nucleation and growth in limestone, Bristol Channel, Somerset (Willemse *et al.*, 1997). The sketch shows the spatial relationship between en-echelon fractures (left side), en-echelon fractures and solution seams (centre), and incipient pull-aparts (right side) towards more mature stages of fault development. Note the increasing density of solution seams in this direction.

faults have displacements of generally less than 30 mm and are characterized by en-echelon fracture patterns and about 10 mm wide breccia zones. Breccia zones form high permeability paths, with permeabilities of 10^{-14} to 3×10^{-12} m^2, where fractures and high-porosity breccia are concentrated. Further fracture development results in medium-displacement faults. These faults have displacements of 1–10 m and are characterized by 1–2 m wide breccia zones and high-density fracture zones along their brecciated fault cores. The breccias may have porosities as high as 10% and represent zones of preferred fluid flow. The final stage of fracture development is represented by large-displacement faults with displacements greater than 10 m. These faults contain a wide zone of low-porosity breccia in their cores and form permeability barriers. Their fault cores are surrounded by high-permeability fracture zones, with permeabilities reaching values of 10^{-14} to 3×10^{-12} m^2. Similar fracture development can be expected in granitic basement rocks within thick-skin thrustbelt structures. These rocks behave in a similar very brittle fashion as documented by the outcrop fracture studies of Martel *et al.* (1988).

Limestones are not as brittle as dolomites and granites and as a result there are differences in their fracture development. Although the overall process of pervasive fracturing, fracture nucleation and fault development remains roughly similar, the fracture nucleation process proceeds through different deformation mechanisms (e.g., Willemse *et al.*, 1997, Fig. 21.13). Fracture development in limestone reservoirs also starts with en-echelon fractures. However, instead of breaching the bridge between en-echelon fractures by cross joints as in dolomite, two symmetrically arranged zones of solution seams develop in the contractional arches of a bending bridge (Fig. 21.13). The next stage is characterized by incipient pull-aparts, which form because of shear along solution seams. The continued shear along the first generation of solution seams leads to further development of these pull-aparts. Eventually a second generation of solution seams develops, together with tail cracks, which form at the tips of sheared first-generation solution surfaces. The second generation of solution seams, oriented at a high angle to the fault zone, undergoes antithetic slip. Later, a third generation of solution seams, together with tail cracks, forms at the tips of the sheared second-generation solution seams. The result is a complex anastomosing network of discontinuities across the fault zone, which undergoes a block rotation. Solution seams and en-echelon fractures eventually develop in a contractional relay ramp between the two side-stepping fault segments. Ongoing syn- and antithetic slip along the solution seams and en-echelon fractures causes the formation of opening-mode tail cracks and solution seams, fragmenting the relay ramp. A through-going fault eventually develops, linking both fault segments.

Both pervasive fractures and fault systems can act as either fluid conduits or barriers. In either case, these structures can significantly alter the distribution of permeability in a reservoir. The hydrocarbon fields of the Absaroka thrust sheet in the Wyoming thrustbelt including Clear Creek, East Painter and Glasscock Hollow illustrate these effects. All are located in tight anticlinal traps in the hanging wall of the thrust and produce retrograde gas condensates and volatile oils. Production data indicate that faults with low-permeability tectonic gouge along their cores and carbonate-filled fractures restrict hydrocarbon distribution in the Nugget Sandstone reservoir, while discontinuous open fractures enhance the reservoir permeability (Lindquist, 1983; Tillman, 1989). Studies of modern accretionary prisms (e.g., Moore *et al.*, 1990a, b, c), fractured tuff (Davidson *et al.*, 1998) and fractured limestones (Nemčok *et al.*, 2002; Fig. 21.14) indicate that permeability enhancements of up to several orders of magnitude can occur due to fractures capable of focusing fluid flow. If the reservoir contains a system of permeability enhancing faults, an understanding of their damage zone geometries is needed to predict the performance of the field and to map areas of enhanced permeability.

There is a significant difference in fracturing between foreland thrustbelts, which incorporate mostly lithified rocks and accretionary prisms, which accrete mostly unlithified sediments. The initial features that develop in accretionary prisms are mud-filled veins, driven by fluid pressures exceeding the cohesion of the sediment undergoing compaction. The extensional geometry of these veins provides evidence for their origin by disaggregation, dilation and mineral growth. They contain structures indicative of fluid flow (Knipe *et al.*, 1991) that reflect the abrupt enhancement of the reservoir permeability and record temporary fluid storage. Anastomosing cataclastic shear zones, commonly located in partly cemented siltstones and sandstones represent subsequent features that develop in accretionary prisms. They are characterized by a reduction in grain size, fracturing and brecciation and are associated with repeated strain hardening events linked to pore-pressure fluctuations within the fractured material (Knipe *et al.*, 1991).

There have been numerous studies describing fracturing during imbrication in foreland thrustbelts. Tensile fractures, resulting from layer-parallel slip, have been observed in fold forelimbs (Mitra, 1987). Several generations of different fractures were associated with changing local stresses during fault-bend folding in the

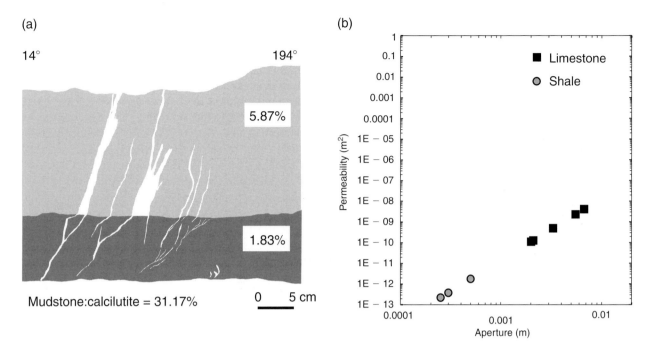

Fig. 21.14. (a) Fracture aperture dependence on rock rheology as indicated by fracture porosity in Liassic sediments, St Donats, Bristol Channel (Nemčok *et al.*, 2002). The light pattern represents bioclastic calcilutite and the dark pattern represents calcareous mudstone. Fracture porosity for each layer and the fracture porosity ratio are expressed as percentages. (b) Variation in permeability (expressed as the coefficient of hydraulic conductivity) with fracture aperture in fractures that propagate from mudstone to calcilutite in (a) (Nemčok *et al.*, 2002). The calculation, which is based on Darcy's law and the Navier–Stokes equation modified by Snow (e.g., Lee and Farmer, 1993), assumes a fluid density, ρ_f, of $1000\,\mathrm{kg\,m^{-3}}$, a dynamic fluid viscosity, η, of $1 \times 10^{-3}\,\mathrm{kg\,m^{-1}\,s^{-1}}$ (Freeze and Cherry, 1979) a hydraulic gradient ratio, I, of 0.06 (Caine, 1999) and a fracture width of 0.173 m. Equations are: $Q = -a^3\rho_\mathrm{f}g/12\eta IW$ and $k = \eta Q/\rho_\mathrm{f}gIA$, where k is the coefficient of hydraulic conductivity, Q is the volumetric flow rate in the direction of decreasing pressure, A is the cross-sectional area across which the perpendicular discharge Q flows, g is the acceleration due to gravity, a is the fracture aperture and W is the fracture width.

Appalachians (Srivastava and Engelder, 1990). Finite-element models, similar to those discussed in Chapter 15, have been designed to illustrate stress control on fracture development (Fig. 21.15) and to study the distribution of dilatant fracturing in various thrustbelt structures (Fig. 21.16) during their growth (Henk and Nemčok, 2000, 2003). The models focus on the influence of specific thrustbelt structures and their lithologies on the development of dilatant fracturing. Growth times and slip rates of the modelled structures (Fig. 15.1) are shown in Table 15.1, and most of their physical parameters are shown in Table 15.2. Each model describes a cross section through a specific thrustbelt structure with isotropic elastic material properties and fault boundaries that were pre-assigned for each model run and simulated by so-called contact elements. Contact elements handle large differential movements along faults but do not allow the simulation of fault propagation. All four-node isotropic elements representing the modelled section are in communication at each time-step. Hooke's law relates normal stresses and strains. The models are based on a simplified simulation of real physical parameters. Their aim is to investigate the major patterns of the fracturing and dilatant strains that develop in response to stress buildup in the thrustbelt structure and the effects of structure geometry, lithology and the presence of the syn-tectonic deposition or erosion.

Figure 21.15 shows the stress buildup in a passive roof duplex, formed by a 2 km thick carbonate in the lower thrust sheet and sandstone in the upper thrust sheet. The carbonate section consists of an upper 620 m thick dolomite layer, a 710 m thick limestone layer and a 670 m thick lower dolomite layer (Fig. 21.15a). The ramp angle is 26°. The model focuses on a single stress buildup cycle. The cycle starts after the last stress release accompanying the last fold growth episode, when the orientation of the maximum principal compressive stress, σ_1, is vertical, indicating no horizontal shortening. Ongoing block movements start a new stress buildup cycle. The very early stage of this buildup is captured in Fig. 21.15(b). This figure shows that several areas of the profile, at this stage, are unaffected by the new tectonic stress cycle. In

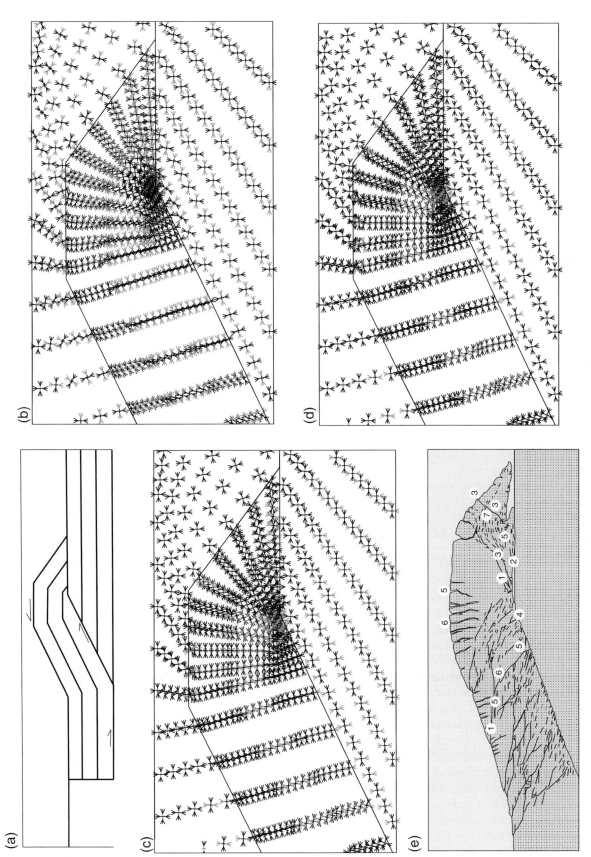

Fig. 21.15. (a) Forward model of a passive roof duplex. The upper thrust is formed by sandstone. The sedimentary section in the lower thrust consists of three layers; an upper 620 m thick dolomite, a 710 m thick limestone and a lower 670 m thick dolomite. The ramp angle is 26°. (b)–(d) Three snapshots of the finite-element model capturing the principal stress perturbations within the passive-roof duplex during the growth stage shown in (a) (Henk and Nemčok, 2000). Stress buildup, recorded in 17 snap shots, was run until failure of the boundary faults occurred. Physical rock properties are shown in Table 14.2. Figures (b–d) show the stress perturbations during the 2nd, 12th and 17th stages of the stress buildup. (e) Simplified interpretation of fracturing succession. See the explanation given in the text.

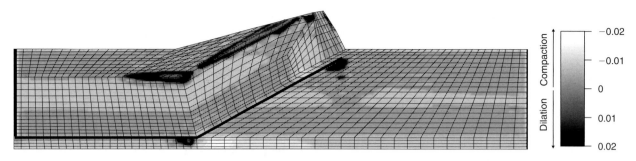

Fig. 21.16. Finite-element model of the dilatancy distribution at the end of a single stress buildup cycle in a mature fault-propagation fold (Henk and Nemčok, 2000). The model shows the deformed section consisting of a 620 m thick upper sandstone layer, a 710 m thick shale layer and a 670 m thick lower sandstone layer. Physical rock properties are shown in Table 14.2. The fold is affected by 50% syn-orogenic erosion. See text for further explanation.

these areas σ_1 remains vertical. Elsewhere σ_1 has been re-oriented towards horizontal. With increasing block movements, the principal stresses become reoriented and differential stresses increase to honor the addition of horizontal loading. An advanced stage of stress buildup is shown in Fig. 21.15(c). Here, the whole profile affected by block movements experiences loading by sub-horizontally and sub-vertically oriented maximum and minimum principal compressional stresses, σ_1 and σ_3, respectively, which are typical of thrusting regimes. Other areas of the profile, however, experience different stress perturbations caused by various factors influencing stress transfer, such as frictional resistance along fault planes, flexure, internal deformation of the thrust sheet and an uneven distribution of gravitational forces. Stress perturbations are not only characterized by changes in the orientations of the principal stresses but also by changes in their magnitudes and the differential stress. Figure 21.15(c) indicates, using diverging arrows to depict σ_3, that this was the stage at which a vertically oriented σ_3 locally switched from compressional to tensile in a small hanging wall area above the upper flat near the ramp. The last stress buildup stage, just before the failure of one of the main thrust faults, is captured in Fig. 21.15(d). The stress regime is similar to that of Fig. 21.15(c) but stress perturbations are more pronounced. This snapshot shows that the upper portion of the anticlinal crest in the lower thrust sheet has already developed a tensile σ_3. Here the horizontally oriented σ_3 became tensile, progressively affecting areas from the centre of the crest to its rear portion.

Differential stresses control the development of fractures, determining when and where each fracture type will develop. The orientation of principal stresses allows a prediction of the fracture orientation. Using these two tools, a simplified interpretation of the succession of fracturing can be made, as, for example, has been carried out for the lower thrust sheet shown in Fig. 21.15(e). This figure shows the fracturing development story of the lower thrust sheet for one stress buildup cycle. Fracturing was initiated simultaneously in two regions of the structure; in the upper dolomite layer in the upper rear limb and in the lower dolomite layer just above the upper flat next to the ramp. Both areas underwent shear fracturing. Back-thrusts propagated in the upper dolomite, whereas thrusts developed in the lower dolomite, both labelled 1 in the figure. Sub-horizontal tensile fractures (2) subsequently developed in the area above the upper flat near the ramp, where Fig. 21.15(c) indicates that σ_3 is vertical and tensile. This was followed by numerous thrusts in the nose of the lower thrust sheet, labelled 3 in the figure. It was only then that the first back-thrusts appeared in the lower dolomite layer in the upper rear limb (4) and propagated towards the back-thrusts in the upper dolomite layer (5). Stage 5 was synchronous with continuous propagation of back-thrusts in the upper dolomite layer. After further stress buildup, the first back-thrust became connected through the limestone layer (6), which behaved in a less brittle fashion than the dolomite layers. Additional back-thrusts connected through the limestone layer progressively affecting areas from the upper part of the rear limb to its lower part. The first sub-vertical tensile fractures appeared in the upper central part of the anti-clinal crest at time 5, roughly synchronously with the development of the first back-thrusts in the lower dolomite layer. These tensile fractures spread towards the rear of the anticlinal crest during time 6. The last fracturing event prior to movement along one of the main faults occurred in the nose of the lower thrust sheet, which underwent some back-thrust development (7). It should be noted that the model simulates newly formed fractures, although in nature numerous fractures would already be present in the reservoir prior to stress build-up. In a natural example, stress build-up could result in either the development of new fractures or the reactivation of pre-existing fractures.

Even though the models represent a simplification of a natural system, they nevertheless illustrate the effects of lithology, friction and flexure on fracturing. The models show that dolomite is more prone to fracturing at lower stress conditions than limestone and that the upper dolomite layer underwent the most intense loading of the layers in the rear limb. Similar rheological effects can commonly be seen in thrustbelt outcrops, for example the frontal thrust sheets of the Dinarides in Croatia (Fig. 21.17a). In the Brooks Range, Alaska, dolomitic mudstone is much more intensively fractured than coarse-grained limestone located in the same structure (Hanks *et al.*, 1997), whereas in the Appalachians the fracture density increases with decreasing grain size (Mitra, 1988b).

The onset of vertical tensile fracturing in the anticlinal crest of the model was apparently controlled by interactions of flexure and friction along the roof thrust plane. If flexure was the prevalent control, the tensile fractures would develop at locations affected by maximum curvature, where the friction along bedding planes is minimal. This is the case in the frontal thrust sheets of the Dinarides, Croatia (Fig. 21.17b).

Because not all fractures significantly enhance reservoir permeability, it is important to study the dilatant portion of the total strain in a thrustbelt structure. Figure 21.16 shows the distribution of dilatancy in a fault-propagation fold affected by syn-orogenic erosion. The numerical model simulates dilatancy in a sedimentary section consisting of an upper 620 m thick sandstone layer, a 710 m thick shale layer and a lower 670 m thick sandstone layer. The ramp angle is 28°. The model is a snapshot at the end of the stress buildup, similar to that discussed above. It indicates that four different areas were affected by dilatancy. The first is where the flat changes to a ramp. This location could be predicted, because it is the likely site for the propagation of a new detachment fault, which would be preceded by intense fracturing. The second location is at the tip of the ramp, where the ramp propagation would potentially start, again preceded by intense fracturing. The two remaining locations affect the upper sandstone layer where the rear limb is most intensely loaded. This layer will undergo a strong fracturing at the bend between the flat and ramp and in the upper portion of the rear limb. Development of thrust faults is most likely in the upper portion of the rear limb because this is where the maximum dilatancy values and dilatant fracture clusters are found.

The models illustrate the effect of lithology on the

(a)

(b)

Fig. 21.17. (a) Tension gashes in the extensional zone of a carbonate layer in an anticlinal crest, frontal Dinaridic thrust sheets, Dubrovnik, Croatia. The tension gash apertures are widest next to an upper bedding plane characterized by low friction. Note how their apertures decrease to zero at the neutral zone. The lowermost zone of the layer is affected by compression. (b) Fracturing in a carbonate layer in a fold limb, frontal Dinaridic thrust sheets, Dubrovnik, Croatia. The layer consists of upper, middle and lower zones with different lithologies. The upper and lower zones are very brittle and deformed by intense fracturing, whereas the middle zone is less brittle and remained intact.

fracture development, among other factors, but the layering is very coarse and relatively strongly coupled. In nature, even fine-scale layering exerts a significant impact on the fracture development in reservoirs. Observations from the Wyoming thrustbelt indicate how bedding plane friction, together with layer thickness, controls the spacing of tensile fractures perpendicular to

(a)

(b)

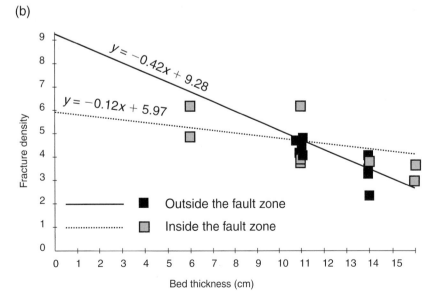

Fig. 21.18. (a) Tensile fractures associated with flexural slip in the Middle Jurassic Twin Creek limestone in the Wyoming thrustbelt, Salt Lake City. See the text for further explanation. (b) Inverse relationship between fracture density and bed thickness for tensile fractures in Liassic limestone, Nash Point, Bristol Channel (Nemčok *et al.*, 1995). See the text for further explanation.

(c)

Fig. 21.18 (*cont.*). (c) En-echelon fractures developed along the contact of Cambrian quartzite and conglomerate in the Utah Thrustbelt, Antelope Island. See the text for further explanation.

bedding planes (Fig. 21.18a). Here, the ends of tensile fractures are filled with fibrous calcite that link with calcite-lined slickensides developed along bedding planes. Slickenside striation vectors are parallel to the fibres in tensile fracture fills, indicating a kinematic linkage of these two types of structures. Each increase in bedding-plane rugosity increases the friction along the associated slip horizon, which reduces the spacing of tensile fractures in adjacent limestone layers. The effect of the layer thickness can be better illustrated using a simpler example from the Bristol Channel, UK (Nemčok *et al.*, 1995). Here, planar bedding planes minimize the effect of the bedding plane rugosity on fracture density (Fig. 21.18b). The relationship between layer thickness and fracture density was examined along pro-

files, some crossing a large-scale fault. Figure 21.18(b) shows that: (1) the thicker the bed is, the stronger the tensile strength; and (2) the stronger the bed is, the larger the fracture spacing. These conclusions are valid for both sedimentary rocks located away from the large-scale fault and rocks located inside the damage zone of the fault. A large difference in fracture density, porosity and permeability between reservoirs away from faults and those located inside damage zones of faults can be documented by fracture porosity measurements from the Bristol Channel (Fig. 21.19; Nemčok *et al.*, 1995). Although the study did not analyse which fractures were open at any particular time in the history of the reservoir, it indicates that the maximum possible fracture porosity decreases away from the damage zone of the fault.

The importance of layer rheology on the development of shear fractures is illustrated in Fig. 21.18(c), which shows propagation of en-echelon shear fractures in adjacent quartzite and conglomerate in the Wyoming thrustbelt. Because the number of grain-to-grain contacts defines the angle of internal friction, which is one of the factors controlling shear strength, fine-grained quartzite behaves in a more brittle fashion than the conglomerate. Therefore, the parts of the shear fractures that developed in quartzite are long and have wide apertures, whereas the parts of the shear fracture that developed in the conglomerate are short and narrow. A similar effect, but for tensile fractures, is shown in Fig. 21.14(a), where fracture apertures are considerably smaller in the brittle limestone compared with the less brittle shale. The smaller apertures in the shale result in a 4% reduction in its fracture porosity, which in turn leads to a difference in permeability between the shale and limestone of three orders of magnitude (Fig. 21.14b).

As discussed in Chapter 3, the thrustbelt structures undergo deformational adjustments in response to changes in the local stress balance. Syn-orogenic extension, for example, may affect the deforming rock section due to changes in the frictional properties along the detachment faults. Syn-orogenic extension and contraction can be controlled by the rheologies of the deforming rocks, as demonstrated by basement uplifts at the eastern side of the Uncompahgre Uplift in Colorado (Stearns, 1978; Couples *et al.*, 1994). Here, the basement is coupled with the sedimentary cover, which can be characterized as being roughly isotropic. The coupling reflects a lack of incompetent Lower Paleozoic strata occurring elsewhere in the region and the presence of Triassic clastic strata in contact with the basement. The isotropic sandstones above the basement–cover contact underwent a thinning from 300 to 50 m in response to only 350 m of uplift of the structure.

(a)   (b)

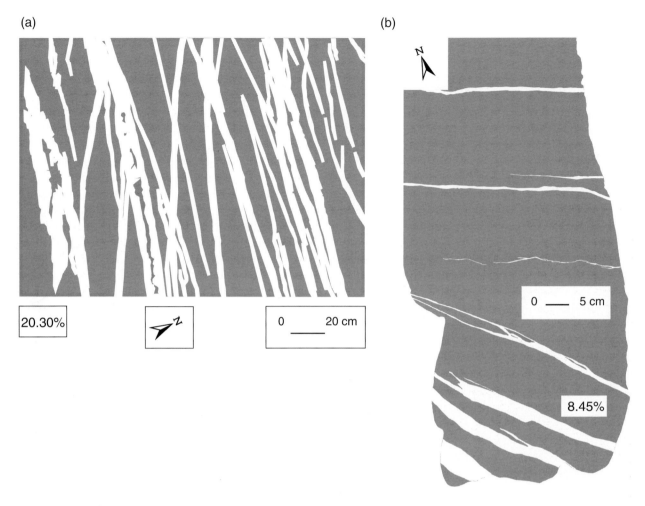

Fig. 21.19. Porosities associated with fractures developed in Liassic limestone: (a) inside a fault damage zone at Kilve, Bristol Channel and (b) outside of a fault damage zone at St Donats, Bristol Channel (Nemčok *et al.*, 1995).

Studies of the Apennines provide an example of extensional adjustment to frictional changes along detachment faults (Bonini *et al.*, 2002; Fig. 4.27). Here, extension developed above the convex-upward fault bend in the frontal portion of a hanging wall that effectively collapsed. Other cases of syn-thrusting extension driven by extensional collapse, determined from earthquake data, have been reported in Algeria (Yielding *et al.*, 1981), Armenia (Philip *et al.*, 1992) and the eastern United States (Kelson *et al.*, 1996).

Another controlling factor in fracture enhancement of reservoir permeability is the timing of fracturing relative to diagenesis. While the early fractures can become cemented during subsequent events, late fractures have a much better chance of remaining open and improving present-day permeabilities. This point introduces a discussion on mechanisms that maintain open fractures once their propagation ceases.

As numerous reservoir studies point out, both tensile

and shear fractures can either remain open or become sealed (Fig. 21.20). There are several mechanisms inhibiting the diagenetic sealing of fractures. One of these is the influx of hydrocarbons that prevent cement growth. Other inhibitors include clay minerals and increased concentrations of Al^{+3} in the fluid. As pointed out by fracture cementation studies in sandstone and carbonate reservoirs (e.g., Fisher and Knipe, 1998; Nemčok *et al.*, 2002), even when fracture cementation occurs, the fracture is unlikely to become homogeneously sealed. Cement distributions are typically uneven, especially in shear fractures, where the tendency to precipitate minerals is greatest in the most dilatant segments of the fracture. This leaves some segments of the fracture permeable (Fig. 21.20b). The likelihood of a fracture remaining partly open increases with increasing aperture (Fig. 21.20a).

As discussed in Chapter 10, fracture permeability is controlled by interactions between stress and fluid pressure (e.g., Carlsson and Olsson, 1979; Sibson, 1990a;

(a)

(b)

Fig. 21.20. (a) Tensile fractures associated with flexural slip in the Middle Jurassic Twin Creek limestone in the Wyoming thrustbelt, Salt Lake City. Their tensile origin is indicated by idiomorphic crystals that grew into open spaces. Fractures that remained open have larger apertures than fractures cemented by calcite. (b) Sinistral shear fracture in Liassic limestone, St Donats, Bristol Channel (Nemčok *et al.*, 2002). The permeability structure of this fracture is heterogeneous, because calcite cement is concentrated only within dilatational jogs. Portions of the fracture that have not experienced extensive dilation remained uncemented.

Fisher *et al.*, 1996; Fig. 10.16). Increased fluid pressure can inhibit fracture collapse. The maintenance of fluid pressure depends on the local draining of fluids or the ability of the surrounding reservoir to feed fluid to the fracture. Thermodynamic arguments further suggest that quartz and calcite are deposited along their flow paths along fractures with decreasing fluid pressure (Bruton and Helgeson, 1983; Capuano, 1990). Therefore, increased fluid pressure would not only keep fractures open, reducing the local stress, but also prevent cement precipitation along them (Roberts *et al.*, 1996). This dynamic control has an important consequence for hydrocarbon production from fractured reservoirs, where production changes can affect the fracture permeability by reducing the fluid pressure.

There are several additional factors that can prevent fractures from total closure; such as the presence of material that props the fracture open (propping material), cement bridges and asperities. Cement bridges develop in cases of partial cementation and prevent total closure with increasing effective stress. The propping material can be formed by breccia, especially in very brittle reservoirs with a tendency to brecciate prior to fault development (Fig. 21.12). Tensile fractures in the Triassic mudstones from the Bristol Channel, which were filled with fluids containing sand from isolated sand bodies, provide an example of sand that acted as a natural propping material (Cosgrove, 2001). Asperities are common in shear fractures. They are formed by juxtaposition of complex fracture wall geometries by strike-parallel displacement. This often results in features with positive relief keeping the fracture open at some residual permeability level.

Dissolution

Dissolution is a reservoir enhancing factor driven by fluid flow in thrustbelts and their forelands. It results in increased intragranular porosity, secondary porosity or karstic features such as cavities, collapse breccias and cave systems. As case studies from the foreland basins of the Appalachians, USA and Canadian Rocky Mountains (e.g., Majorowicz and Jessop, 1981; Chapman *et al.*, 1984; Majorowicz *et al.*, 1984, 1985; Gosnold, 1985, 1990; Deming *et al.*, 1990; Nunn and Deming, 1991) indicate, the solute-starved meteoric waters flow downward on the thrustbelt side of the basin, acquiring heat and dissolved salts. The descending waters eventually encounter a permeable aquifer. Within the aquifer, the fluid moves laterally and up towards the discharge area. The recharge area is the region with the most prominent dissolution in thrustbelts. Compaction-driven fluid flow further contributes to dissolution in a thrustbelt setting, although the ini-

Fig. 21.21. Dissolution of a carbonate layer along fractures in the frontal Dinaridic thrust sheets, Dubrovnik, Croatia.

tially descending fluids, derived in this case from the original pore fluids, have a different chemical character. Solute-poor meteoric waters cause the greatest dissolution in carbonate lithologies (Fig. 21.21), whereas the original pore fluids released from compacting sediments are more important in the dissolution of sandstones (Wright and Smart, 1994). Cases of sandstone dissolution have been reported from the Sulaiman thrustbelt, Pakistan, and the Appalachians (Mitra, 1988b; James, 1995; Sultan and Gipson, 1995), while examples of carbonate dissolution are reported from the US and Canadian parts of the Appalachians and their foreland (Montañez, 1994; Dix *et al.*, 1998).

As indicated by the study of Laramide fluid events in the foreland basin in front of the Richardson and Mackenzie Mountains in Canada (Morris and Nesbitt, 1998), orogenic fluid events occur in groups. Some fluid events result in dissolution, others in mineral precipitation. These events can alternate spatially and temporally. The overall reservoir permeability may not be enhanced if the secondary porosity created is simply redistributed (Giles and de Boer, 1990). Permeability enhancement mitigated by mineral deposition is recognized in the foreland basin in front of the Richardson and Mackenzie Mountains. In the Appalachians, enhanced permeability in Ordovician carbonates was destroyed by subsequent precipitation caused by fluids migrating over long distances (Mussman *et al.*, 1988; Montañez, 1994).

The most typical places for dissolution in thrustbelt–foreland basin systems are in the recharge areas adjacent to the basin or in recharge areas at the forebulge on the landward side of the basin. The Antler orogen in the NW Territories, Canada provides an example of two fluid events; however, only one of them has enhanced permeability (Morris and Nesbitt, 1998).

The first event was related to Silurian–Mississippian diagenesis. During this event very fine crystalline pervasive dolomitization reduced the permeabilities of the reservoir. Fluid flow was partly driven by increased burial and partly by compaction-driven flow from the Late Devonian–Early Carboniferous Antler orogen. Magnesium was provided by residual evaporitic brines and dewatering of shale units. The second event was Devonian–Pennsylvanian post-diagenetic dolomitization characterized by vuggy porosity created by dissolution, which was only partially filled by younger calcite and quartz. Vugs, some as large as 20 cm in diameter, were created. Both topography-driven and compaction-driven fluid flows from the Antler orogen were involved.

The timing of these two events was favourable for reservoir permeability enhancement. It resulted in the formation of the Manetoe facies, a producing reservoir in some of the richest gas fields in Canada (e.g., Morrow *et al.*, 1990). Similar examples of multiple Late Pennsylvanian–Early Permian fluid events have been described from the Appalachians (Oliver, 1986; Bethke

and Marshak, 1990; Torgersen, 1990). Here, dissolution related to some fluid events enhanced the reservoir permeability (Mussman *et al.*, 1988; Montañez, 1994) whereas mineral precipitation during other events resulted in permeability destruction (Sverjensky, 1986; Mussman *et al.*, 1988; Montañez, 1994). Similarly, dissolution of carbonates by meteoric water enhanced fracture permeabilities in the frontal thrust sheets of the Dinarides, Croatia (Fig. 21.21).

In the forebulge area in the NW Territories in Canada, Cretaceous–Early Tertiary fluid deposited large idiomorphic calcite crystals in elongate voids formed by dissolution (Morris and Nesbitt, 1998). Stable isotope compositions of oxygen and carbon indicate that meteoric water entered the recharge area in the forebulge, which is characterized by karstic features and cave systems. Similar examples of dissolution are found in the forebulge area of the Appalachians (Desrochers and James, 1988; James *et al.*, 1989; Dix *et al.*, 1998). Because the forebulge travels in space and time in front of the advancing orogen, the old buried forebulge segments can become potential targets in the search for areas with enhanced reservoir permeability.

22 Remaining petroleum potential of thrustbelts

General statement

Many of the earliest oil and gas provinces of the late nineteenth and early twentieth centuries were in thrustbelts, where discoveries were based on the presence of surface seeps and prominent structures. These include the allochthonous Alleghany Plateau in the Appalachians, the Canadian Foothills belt, the transpressional basins of Southern California, the northern Andes, the Zagros fold belt and the Middle Caspian region. Since these early, relatively easy discoveries, the petroleum exploration in thrustbelts has been driven in cycles in which the sustained product price coincides with technical advances that reduce the generally higher costs and risks of finding and developing hydrocarbons in this setting. The last major cycle of thrustbelt exploration was in the 1970s and 1980s, resulting in the discovery of prolific new provinces such as the Wyoming Thrustbelt, the El Furrial trend in eastern Venezuela, the sub-Andean basins of Peru, Bolivia and northern Argentina, Papua New Guinea, to name just a few. This was also the time of expansion of hydrocarbon plays in the Canadian Foothills belt, the northern Andes and other established thrustbelt provinces. As industry attention shifted to the deep-water continental margins during the 1990s (Esser, 2001), the few new thrustbelt discoveries of significance were all in established petroleum provinces. The sole exception was the discovery of large petroleum accumulations in toe thrusts at the base of the Niger fan (Zafiro field) and the Mississippi fan fold belt beneath the Gulf of Mexico salt canopies (e.g., Mad Dog, Neptune and Atlantis fields). These discoveries point to a whole new class of highly prospective, but high-risk, thrustbelt plays.

Future expansion of petroleum reserves in thrustbelts will be through:

(1) new discoveries in previously under-explored areas of known thrustbelt provinces or previously unexplored thrustbelts, of which few remain;

(2) new discoveries in mature thrustbelt provinces involving both proven play concepts as well as development of new ones;

(3) improved recovery from known petroleum accumulations in established thrustbelt fields.

Technical advances in a wide variety of fields will position thrustbelts for yet another cycle of intensive exploration and development activity. These technical improvements include the following.

(1) Advances in understanding the fundamental geometry of contractional structures and their relationship to the gross and fine-scale rheology of different stratigraphic successions. Modern descriptions extend the older concepts of fault-propagation, fault-bend and detachment folds to a variety of hybrids that more accurately characterize actual structures, such as growth folds (Shaw and Suppe, 1994; Novoa et al., 2000), faulted detachment folds (Mitra, 2002b), fold-accommodation faults (Mitra, 2003) and foreland basement-involved structures (Mitra and Mount, 1998). Building on the technique of axial-surface mapping (Shaw et al., 1994), Rowan and Linares (2000) demonstrate the utility of fold-evolution matrices in examining the influence of independent variables, such as thrust displacement and ramp geometry, on consequent fold configurations and three-dimensional axial surface maps. These and other methods in development will greatly assist in the recognition of plausible subsurface structures where direct control over the geometric configurations is poor.

(2) Improvements in acquisition and processing of two- and three-dimensional seismic surveys that enhance image quality while lowering costs. In thrustbelts, where steep dips and velocity inversions are the norm, pre-stack depth migration and depth imaging of seismic data are proving to be critical for the accurate definition of structural features below the uppermost thrust sheet (e.g., Rigatti et al., 2001; Yan and Lines, 2001). In structurally complex areas, three-dimensional seismic surveys are essential for locating subtle traps and delineating reservoir compartments, thus supporting mature-province exploration and field development. Interpretation of the migrated seismic data is generally improved through the iterative application of forward

seismic modelling (e.g., Lingrey, 1991; Morse *et al.*, 1991), along with other processing and interpretation methods.

(3) Imaging and interpretation workstations that integrate tools for three-dimensional seismic interpretation (e.g., Fagin, 1995; Blangy, 2002), structural visualization (e.g., De Donatis, 2001), palinspastic reconstruction (e.g., Poblet *et al.*, 1998; Guglielmo *et al.*, 2000; Mitra and Leslie, 2003) and form surface analysis to aid in the prediction of natural fracturing of folded strata (e.g., Lisle, 1994; Hennings *et al.*, 2000; Masaferro *et al.*, 2003). Computation tools adequate to accomplish these tasks are now moving out of the visualization centre and onto the desktop (Purday and Sutton, 2004). Accurate characterization of the structural architecture, especially the overall structural style, is critical to all aspects of exploration and development in thrustbelts from petroleum systems analysis to the identification of traps and risks of drillable prospects.

(4) More sophisticated approaches to petroleum systems investigation that recognize the levels of uncertainties inherent in the various integrated components of the analysis. Rigorous petroleum systems analysis involves a complex knowledge chain that includes source rock characterization, burial and thermal history, maturation modelling using kinetic parameters derived from various pyrolysis methods, identification of migration conduits at differing stages of structuring, and reconstruction of the structural and hydrodynamic evolution of the thrustbelt, among other factors. Incomplete knowledge of any component and working in anything less than three-dimensional basin/structural geometry may greatly diminish the validity of the final analysis. Even with the high-level software now available to perform many parts of the analysis, the results at best merely set limits to various trial scenarios and separate the plausible from implausible. Yet, this level of analysis can be extremely useful for identifying new plays or assessing the risks associated with a particular prospect. For example, in the Perijá fold belt of northwest Venezuela, Gallango *et al.* (2002) used detailed structural analysis, oil–oil and oil–source correlations and one-dimensional maturation modelling to support a two-phase scenario for oil generation and migration into differing combinations of trapping structures and reservoirs.

(5) Development of geophysical, statistical and numerical modelling tools for predicting the distribution of natural fracture systems in reservoirs (e.g., Silliphant, 1998; Gray and Head, 2000; Bui *et al.*, 2003; Henk and Nemčok, 2003). A great many of the reservoirs in thrustbelt settings, particularly those deeper than 3000 m, require some degree of natural fracturing to support commercial production rates. In these situations penetration of the part of the structure with dense natural fracturing is essential for a successful well (Cooper, 1991).

(6) Geo-steering and directional drilling methods that permit cost effective penetration of structurally complex, heterogeneous and/or compartmentalized reservoirs, especially in deep traps. In thrustbelts, this approach may be used in initial field development or in field renovation. Horizontal wells have added 2100 bopd of new oil production from Tulare Sands in a 280 hectare portion of the Belridge field, San Joaquin basin (Knight *et al.*, 2001). In the Whitney Canyon gas field (Figs. 20.10 and 20.11) in the Wyoming thrustbelt, BP has drilled a perforated level-4 multilateral well into the bypassed gas-bearing Mission Canyon reservoir resulting in a rate increase from 4 to 16 MMcfpd from the older recompleted well (Harting *et al.*, 2003).

Role of new frontier and mature-field discoveries in thrustbelts

Of the 38 thrustbelt petroleum provinces having known hydrocarbon volumes of 1.0 BBOE or greater (Table 17.1; Klett *et al.*, 1997), less than a third are arguably mature, having little prospects for additional large wildcat discoveries. Of these only six provinces appear firmly in this category. They are the Los Angeles basin, the Oriente basin, the Upper and Middle Magdalena Basins, the Pyrenean Foothills–Ebro basin and the Wyoming thrustbelt. The other possible mature provinces include the Canadian Foothills belt, the North Carpathian basin, the Po basin, the Rocky Mountain Deformed belt in Montana and the Taranaki basin. The other 27 provinces have not been fully explored in their thrustbelt segments and offer an opportunity for substantial additions to reserves through wildcat discoveries.

The Zagros fold belt not only dominates in terms of known petroleum volume, but it is certainly the least intensely explored and developed (Versfelt, 2001) of all of the top 38 thrustbelt provinces. Politically and economically isolated for two decades, yet with large primary production rates from established fields, Iran and Iraq have done little to expand their energy industries with either new field development or renovation of older fields. Exploration drilling all but ended in 1979 due to the Iranian Revolution and subsequent Iran–Iraq War, resuming at a much lower exploration level in Iran only in 1989. Bordenave (2000) has investigated the availability of valid drillable prospects in Iran. He notes that there remain to be tested many tens of structural prospects, principally in Lurestan, in the shallow oil- and gas-bearing Miocene Asmari and Mid-Cretaceous Sarvak limestone reservoirs, some with possible recoverable reserves in excess of 1.0 BBOE. In Pars and Lurestan provinces about 50 such prospects remain to be drilled in the deep gas-bearing Permian–Lower Triassic Dalan and Kangas reservoirs, each with a potential 5–20 Tcf of recoverable reserves. In 1999, the National Iranian Oil Company discovered in the thrustbelt the giant Azadegan field with 5–6 BBOE recoverable (Versfelt, 2001).

The potential for future petroleum development is even greater in the Iraqi sector of the Zagros fold belt. Countrywide, only 17 of 80 discovered fields have been developed and there are few deep wells (EIA, 2003). As in Iran, a very large number of structural prospects remain to be tested. Sanctions and the second Gulf War have delayed projects intended to renovate existing fields. Yet the country has proven reserves of 112 BBO and 110 Tcf and extremely low historic finding costs (EIA, 2003).

The Timan–Pechora basin has a large number of undeveloped or only partially developed oil and gas fields in large anticlines flanking the inverted aulacogens (Dedeev et al., 1994). Smaller pools in stratigraphic traps flanking the anticlines are largely undeveloped and most potential stratigraphic or small structural traps have not even been tested. The foothills of the Urals thrustbelt, with a single known super-giant gas field Vuktyl, remains poorly explored and drilled. The future potential of this province is enormous.

In Southern California, the small Los Angeles basin is fully mature and most fields that are not shut-in are currently undergoing active renovation, despite the highly urban character of the basin. In the offshore portions of the Ventura and Santa Maria basins, environmental concerns have resulted in suspension of exploration and new field development before the basins could be fully tested. There are certainly additional substantial discoveries possible in both basins, and possibly also the small offshore portion of the Los Angeles basin. The shallow oil and gas accumulations of the San Joaquin basin have been in production for many decades and most of these pools are very mature and in secondary or tertiary recovery. However, there are

several deep reservoir targets identified within currently productive anticlines and in the footwall of thrust sheets that have been tested in recent years (Petzet, 2000). The East Lost Hills-1 well came online in February 2000 with 13 MMcfpd gas and 475 bpd of liquids (Fischer, 2001).

In Western Canada, additional frontier opportunities have been identified in new structural and stratigraphic play types in the Foothills belt (Newson, 2001) and the partially deformed Liard basin (Morrow and Potter, 1998). The Madre de Dios (Martinez et al., 2003), Santa Cruz–Tarija and Cuyo–Bolsones (Urien, 2001) Basins are not fully explored and remain prospects for large wildcat discoveries.

In mature petroleum provinces, the future targets for petroleum discoveries will include footwall and stratigraphic traps that have been undetected or undeveloped in past exploration programmes. Improvements in seismic and imaging technology will aid in the recognition of these commonly subtle traps, and improvements in well drilling and completion technology will make them more accessible. Also mature thrustbelts may offer opportunities for development of unconventional oil and gas resources, such as heavy oil and natural gas in fractured low-permeability sandstone, shale and coal-bed reservoirs.

Basin-centred gas systems (Surdam, 1997; Law, 2002), in which low-permeability sandstone reservoirs are saturated with gas that is abnormally pressured and trapped by capillary effects, are known to exist is some thrustbelt settings, and may be much more common than is currently recognized. Overpressured basin-centred gas is produced from the Pinedale anticline and other thrust-related structures of the Greater Green River basin, Wyoming (Kuuskraa and Bank, 2003). Underpressured basin-centred gas is currently produced from the Lower Silurian Medina Sandstone beneath the allochthonous Alleghany Plateau and adjacent foreland basin in western New York and Pennsylvania (Ryder and Zagorski, 2003).

Many thrustbelts have thick organic-rich shale successions that have potential for shale gas development. In the San Joaquin basin, fractured Monterey Formation siliceous shales have produced oil in the Elk Hills (Reid and McIntyre, 2001) and Buena Vista Hills (Montgomery and Morea, 2001) fields. In the allochthonous Alleghany Plateau and foreland basin of the central Appalachians, gas has been produced from Devonian black shales for more than a century (Curtis, 2002). The wells are typically long-term, low-volume producers (Patchen and Hohn, 1993). Yet the estimated volume of gas in-place in the shales is impressive, being as large as 577–1131 Tcf (Charpentier et al., 1993).

Shale gas development has been largely a neglected resource in thrustbelts.

Opportunities for development of coal-bed methane resource in thrustbelts may be limited by structural complexity and hydrodynamics (Freudenberg *et al.*, 1996; Dawson, 1999), inhibiting dewatering of the coals required to stimulate gas production. As the typically weakest lithology in a coal-bearing succession, it is the coal beds that serve as detachments, intensely fractured in a quasi-ductile fashion (Nickelsen, 1979). Whether the deformed coal beds will produce gas will depend, in part, on the relationship between the neotectonic stress regime and the trends of the earlier compressive structures imprinted on the coal beds (Hathaway and Gayer, 1996). It is highly likely that the highly fractured coals in thrustbelts cannot retain gas over geological time and consequently have low gas contents. At present, no significant coal-bed methane production has been developed in a thrustbelt.

Role of improved resource recovery in thrustbelts

Better management of mature fields through improved oil and gas recovery processes has been gaining favour in the petroleum industry as a cost-effective alternative to wildcat drilling (Sneider and Sneider, 2001). Although not yet widespread in thrustbelt fields outside of the USA, there are several major thrustbelt provinces that demonstrate the value of this approach.

The gas–condensate–oil Anschutz Ranch East field (Figs. 20.10 and 20.12) in the Wyoming Thrustbelt was unitized soon after its discovery in 1979 in order to maximize liquid recovery through efficient reservoir pressure management (Welch, 1993). Production-induced pressure depletion of this field would have resulted in retrograde gas condensation in the reservoir and caused reduction in the ultimate condensate recovery, a reduction in gas deliverability and an overall reduction in ultimate gas recovery. Full pressure maintenance of the reservoir above the dew-point pressure of 5080 psi was implemented by injection of a mixture of processed dry gas from the field and nitrogen sufficient to replace the remaining void space of the reservoir. This programme

was continued until 1999, when declining condensate recovery from the field and increasing regional gas price shifted the economics of this reservoir management strategy. Where a favourable market exists for gas sales and the reservoir is appropriate, gravity drainage may be the optimal process for condensate recovery, as was practiced at the Waterton field (Fig. 17.4) in the southern Canadian Foothills belt (Castelijns and Hagoort, 1984).

The domal anticline of the Rangely field in the southeast Uinta uplift of the central Rocky Mountains had been producing oil from the Pennsylvanian Weber Sandstone since the early 1940s. By 1949 the field was fully developed on a 40 acre spacing and already was experiencing declining primary production. In 1958 water flood was initiated to counter declining primary oil production (Hemborg, 1993). That was followed in 1986 by initiation of CO_2 flood. This recovery process is continuing to the present with an anticipated 114 MMBO incremental increase of production from the 1600 MMBO original oil in place. The ultimate recovery expected following CO_2 flood is 939 MMBO (Hefner and Barrow, 1992).

There may exist significant undiscovered petroleum accumulations beneath the leading edge of thin-skin thrustbelts where the trapping mechanism is in no way related to the overlying thrust (Picha, 1996). Key to exploration in the sub-thrust setting is the mapping of the stratigraphy and structure beneath the frontal zone as though the overriding thrustbelt were not present, except as the mechanism for loading the basin and maturing source rocks. In most thrustbelts the complexities of this sub-thrust portion of the foreland basin are poorly known and the potential for hydrocarbon accumulations little appreciated. Picha (1996) describes many smaller sub-thrust oil and gas fields beneath the West Carpathians, including the Ždánice–Krystalinikum oil and gas field producing from Precambrian crystalline basement (Krejčí, 1993). The Cave Gulch gas field in the Wind River basin, Wyoming (Natali *et al.*, 2000; Montgomery *et al.*, 2001) is such a sub-thrust field and several other similar fields probably are yet to be discovered in the central Rocky Mountains.

References

Abbasi, I. A. and McElroy, R. (1991). Thrust kinematics in the Kohat Plateau, Trans-Indus Range, Pakistan. *Journal of Structural Geology*, **13**, 319–27.

Abercombie, R. E. and Leary, P. (1993). Source parameters of small earthquakes recorded at 2.5 km depth, Cajon Pass, southern California; implications for earthquake scaling. *Geophysical Research Letters*, **20**, 1511–14.

Abers, G. and McCaffrey, R. (1988). Active deformation in the New Guinea fold-and-thrust belt; seismological evidence for strike-slip faulting and basement-involved thrusting. *Journal of Geophysical Research B*, **93**, 13 332–54.

Abrams, M. A., Apanel, A. M., Timoshenko, O. M. and Kosenkova, N. N. (1999). Oil families and their potential sources in the northeastern Timan Pechora Basin, Russia. *American Association of Petroleum Geologists Bulletin*, **83**, 553–7.

Adams, J. (1985). Large-scale tectonic geomorphology of the Southern Alps, New Zealand. In *Tectonic Geomorphology. Proceedings of the 15th Annual Binghampton Geomorphology Symposium*, ed. M. Morisawa and J. T. Hack. Boston: Allen and Unwin, pp. 105–28.

Affaton, P., Trompette, R., Uhlein, A. and Boudzoumou, F. (1995). The Panafrican–Brasiliano Aracuai–West-Congo fold belt in the framework of the western Gondwana aggregation at around 600 Ma. In *10th Conference of the Geological Society of Africa: GSA 95 International Conference*, Abstract, Conference Programme, 10. p. 20.

Ahnert, F. (1970). Functional relationships between denudation, relief and uplift in large mid-latitude drainage basins. *American Journal of Science*, **268**, 243–63.

Ahrendt, H., Huhzikker, J. C. and Weber K. (1978). Alterbestimmungen an schwach-metamorphen Gesteinen des Rhenischen Schiefergebirges. *Zeitschrift der Deutschen Geologischen Gesellschaft*, **129**, 229–47.

Aizenshtat, Z., Miloslavski, I., Ashengrau, D. and Oren, A. (1999). Hypersaline depositional environments and their relation to oil generation. In *Microbiology and Biogeochemistry of Hypersaline Environments: the Microbiology of Extreme and Unusual Environments*, ed. A. Oren. Boca Raton: CRC Press, pp. 89–108.

Ala, M. (1990). Seventy-five years of petroleum exploration and production in the Zagros basin of southwest Iran. In *Seventy-five Years of Progress in Oil Field Science & Technology*, ed. M. Ala, H. Hatamian, G. D. Hobson, M. S. King and I. Williamson. Rotterdam: A. A. Balkema, pp. 61–76.

Albouy, E., Casero, P., Eschard, R., Barrier, L. and Rudkiewicz, J. L. (2003). Coupled structural/stratigraphic forward modeling in the Central Apennines. *AAPG Annual Meeting*, Salt Lake City, UT, *Official Program*, vol. 12. Tulsa, OK: American Association of Petroleum Geologists, p. A4.

Allmendinger, R. W., Ramos, V. A., Jordan, T. E., Palma, M. and Isacks, B. L. (1983). Paleogeography and Andean structural geometry, Northwest Argentina. *Tectonics*, **2**, 1–16.

Al-Shaieb, Z., Puckette, J. O., Abdalla, A. A., Tigert, V. and Ortoleva, P. (1994). The banded character of pressure seals. In *Deep Basin Compartments and Seals*, ed. P. Ortoleva. *American Association of Petroleum Geologists Memoir*, vol. 61, pp. 351–67.

Alexander, L. L. and Handschy, J. W. (1998). Fluid flow in a faulted reservoir system: Fault trap analysis for the Block 330 field in Eugene Island, south addition, offshore Louisiana. *American Association of Petroleum Geologists Bulletin*, **82**, 387–411.

Allan, U. S. (1989). Model for hydrocarbon migration and entrapment within faulted structures. *American Association of Petroleum Geologists Bulletin*, **75**, 803–11.

Allen, P. A. and Allen, J. R. (1990). *Basin Analysis: Principles and Applications*. Oxford: Blackwell Scientific Publications.

Allen, M. B., Vincent, S. J. and Wheeler, P. J. (1999). Late Cenozoic tectonics of the Kepingtage thrust zone; interactions of the Tien Shan and Tarim Basin, Northwest China. *Tectonics*, **18**, 639–54.

Allen, M. B., Alsop, G. I. and Zhemchuzhnikov, V. G. (2001). Dome and basin refolding and transpressive inversion along the Karatau fault system, southern Kazakstan. *Journal of the Geological Society of London*, 158, part 1, 83–95.

Allemann, F. (1979). Time of emplacement of the Zhob Valley ophiolites and Bela ophiolites, Baluchistan; preliminary report. In *Geodynamics of Pakistan*, ed. F. Abul Farah and K. A. DeJong. Quetta: Geological Survey of Pakistan.

Allis, R. G., Henley, R. W. and Carman, A. F. (1979). The thermal regime beneath the Southern Alps. In *The Origin of the Southern Alps*, ed. R. I. Walcott and M. M. Cresswell. *Royal Society of New Zealand Bulletin*, **18**, pp. 79–85.

Allmendinger, R. W. (1998). Inverse and forward numerical modeling of trishear fault-propagation folds. *Tectonics*, **17**, 640–56.

Allmendinger, R. W., Brewer, J. A., Brown, L. D., *et al.* (1982).

COCORP profiling across the Rocky Mountain Front in southwestern Wyoming, Part. 2: Precambrian basement structure and its influence on Laramide deformation. *Geological Society of America Bulletin*, **93**, 1253–63.

Allmendinger, R. W., Jordan, T. E., Kay, S. M. and Isacks, B. L. (1997). The evolution of the Altiplano–Puna Plateau of the Central Andes. *Annual Reviews of Earth and Planetary Sciences*, **25**, 139–74.

Almeida, F. F. M., Hasui, Y., Brito Neves, B. B. and Fuck, R. A. (1981). Brazilian structural provinces: an introduction. *Earth Science Reviews*, **17**, 1–29.

Alonso, J. L. and Teixell, A. (1992). Forelimb deformation in some natural examples of fault-propagation folds. In *Thrust Tectonics*, ed. K. R. McClay. London: Chapman and Hall, pp. 175–80.

Alsharhan, A. S. and Nairn, A. E. M. (1997). *Sedimentary Basins and Petroleum Geology of the Middle East*. Amsterdam: Elsevier.

Alvarez-Marron, J. (1995). Three-dimensional geometry and interference of fault-bend folds; examples from the Ponga Unit, Variscan Belt, NW Spain. *Journal of Structural Geology*, **17**, 549–60.

Amato, A. and Cinque, A. (1999). Erosional land surfaces of the Campano–Lucano Apennines (S. Italy); genesis, evolution, and tectonic implications. *Tectonophysics*, **315**, 251–67.

Anders, D. E. and Gerrild, P. M. (1984). Hydrocarbon generation in lacustrine rocks of Tertiary age, Uinta Basin, Utah–organic carbon, pyrolysis yield, and light hydrocarbons. In *Hydrocarbon Source Rocks of the Greater Rocky Mountain Region*, ed. J. Woodward, F. F. Meissner and J. L. Clayton. Denver: Rocky Mountain Association of Geologists, pp. 513–29.

Anderson, G. G. (1956). Subsurface stratigraphy of the pre-Middle Niobara Formations in the Green River Basin, WY. In *Wyoming Stratigraphy*. Casper, WY: Wyoming Geological Association, Nomenclature Committee, pp. 49–68.

Anderson, R. N., Flemings, P. B., Losh, S., Austin, J. and Woodhams, R. (1994). Gulf of Mexico growth fault drilled, seen as oil, gas migration pathway. *Oil and Gas Journal*, **94**, 97–104.

Angelier, J. (1979). Determination of the mean principal directions of stresses for a given fault population. *Tectonophysics*, **56**, T17–26.

Angevine, C. L. and Turcotte, D. L. (1983). Porosity reduction by pressure solution: a theoretical model for quartz arenites. *Geological Society of America Bulletin*, **94**, 1129–34.

Antonellini, M. and Aydin, A. (1994). Effect of faulting on fluid flow in porous sandstones: petrophysical properties. *American Association of Petroleum Geologists Bulletin*, **78**, 355–77.

Antonellini, M., Aydin, A. and Pollard, D. D. (1994). Microstructure of deformation bands in porous sandstones at Arches National Park, Utah. *Journal of Structural Geology*, **16**, 941–59.

Antonellini, M. and Mollema, P. N. (2000). A natural analog for a fractured and faulted reservoir in dolomite: Triassic Sella Group, northern Italy. *American Association of Petroleum Geologists Bulletin*, **84**, 314–44.

Arbenz, J. K. (1989a). The Ouachita system. In *The Geology of North America: an Overview*, ed. A. W. Bally and A. R. Palmer. Boulder: Geological Society of America.

  (1989b). Ouachita thrust belt and Arkoma Basin. In *The Appalachian–Ouachita Orogen in the United States*, ed. R. D. Hatcher, Jr., W. A. Thomas and G. W. Viele. Boulder, CO: Geological Society of America.

Arch, J. and Maltman, A. J. (1990). Anisotropic permeability and tortuosity in deformed wet sediments. *Journal of Geophysical Research*, **95**, 9035–47.

Arch, J., Maltman, A. and Knipe, R. J. (1988). Shear-zone geometries in experimentally deformed clays; the influence of water content, strain rate and primary fabric. *Journal of Structural Geology*, **10**, 91–9.

Armstrong, R. L. (1968). Sevier orogenic belt in Nevada and Utah. *Geological Society of America Bulletin*, **79**, 429–58.

Arrowsmith, J. R. and Strecker, M. R. (1999). Seismotectonic range-front segmentation and mountain-belt growth in the Pamir-Alai region, Kyrgyzstan (India-Eurasia collision zone). *Geological Society of America Bulletin*, **111**, 1665–83.

Asanuma, H., Kizaki, T. and Niitsuma, H. (2002). Analysis of seismic waves propagated through a pressurized fracture by the 3D-TFC method. *SEG Annual Meeting Expanded Technical Program Abstracts with Biographies*, vol. 72, pp. 2385–8.

Athy, L. F. (1930). Density, porosity, and compaction of sedimentary rocks. *American Association of Petroleum Geologists*, **14**, 1–24.

Atkinson, B. K. and Meredith, P. G. (1987). Experimental fracture mechanics data for rocks and minerals. In *Fracture Mechanics of Rock*, ed. B. K. Atkinson. London: Academic Press, pp. 477–525.

Atwood, G. I. and Forman, M. J. (1959). Nature and growth of southern Lousiana salt domes and its effect on petroleum accumulation. *American Association of Petroleum Geologists Bulletin*, **43**, 2595–622.

Audet, D. M. and McConell, J. D. C. (1992). Establishing resolution limits for tectonic subsidence curves by forward basin modeling. *Marine and Petroleum Geology*, **11**, 400–11.

Averbuch, O. and Mansy, J. L. (1998). The 'Basse-Normandie' duplex (Boulonnais, N France); evidence for an out-of-sequence thrusting overprint. *Journal of Structural Geology*, **20**, 33–42.

Averbuch, O., Mattei, M., Kissel, C., Frizon de Lamotte, D. and Speranza, F. (1995). Cinématiques des deformations au sein d'un systeme chevauchant aveugle: l'exemple de la 'Montagna dei Fiori' (front des Apennins centraux, Italie). *Bulletin de la Societé geologiqué de France*, **166**, 451–61.

Avouac, J. P., Tapponier, P., Bai, M., You, H. and Wang, G. (1993). Active thrusting and folding along the

Northern Tien Shan and Late Cenozoic rotation of the Tarim relative to Dzungaria and Kazakhstan. *Journal of Geophysical Research*, **98**, 6755–804.

Aydin, A. (1977). Small faults formed as deformation bands in sandstone. In *Proceedings of Conference II: Experimental Studies of Rock Friction with Application to Earthquake Prediction*, ed. J. F. Evernden. US Geological Survey, Office of Earthquake Studies, Menlo Park, California, pp. 617–53.

(1978a). Faulting in sandstone. Doctoral dissertation, Stanford University, Stanford.

(1978b). Small faults formed as deformation bands in sandstone. *Pure and Applied Geophysics*, **116**, 913–30.

Aydin, A., Borja, R. I. and Eichhubl, P. (2005). Geological and mathematical framework for failure modes in granular rock. *Journal of Structural Geology*, in press.

Aydin, A. and Johnson, A. M. (1978). Development of faults as zones of deformation bands and as slip surfaces in sandstone. *Pure and Applied Geophysics*, **116**, 931–42.

(1983). Analysis of faulting in porous sandstones. *Journal of Structural Geology*, **5**, 19–35.

Aymard, R., Pimentel, L., Eitz, P., Lopez, P., Chaouch, A., Navarro, J., Mijares, J. and Pereira, J. G. (1990). Geological integration and evaluation of northern Monagas, Eastern Venezuelan Basin. In *Classic Petroleum Provinces*, ed. J. Brooks. Bath: Geological Society of London, pp. 37–53.

Babarina, I. I. (1999). Paleozoic deformations in the southern Tamdytau (central Kyzylkum, Uzbekistan). *Geotectonics*, **33**, 234–49.

Baby, P., Herail, G., Salinas, R. and Sempere, T. (1992). Geometric and kinematic evolution of passive roof duplexes deduced from cross section balancing: example from the foreland thrust system of the southern Bolivian subandean zone. *Tectonics*, **11**, 523–36.

Bachu, S. (1985). Influence of lithology and fluid flow on the temperature distribution in a sedimentary basin: a case study from the Cold Lake area, Alberta, Canada. *Tectonophysics*, **120**, 257–84.

(1991). On the effective thermal and hydraulic conductivity of binary heterogeneous sediments. *Tectonophysics*, **190**, 299–314.

(1995). Synthesis and model of formation-water flow, Alberta Basin, Canada. *American Association of Petroleum Geologists Bulletin*, **79**, 1159–78.

Bachu, S. and Dagan, G. (1979). Stability of displacement of a cold fluid by a hot fluid in a porous medium. *Physics of Fluids*, **22**, 54–9.

Bachu, S., Ramon, J. C., Villegas, M. E. and Underschultz, J. R. (1995). Geothermal regime and thermal history of the Llanos Basin, Colombia. *American Association of Petroleum Geologists Bulletin*, **79**, 116–29.

Bachu, S. and Underschultz, J. R. (1993). Hydrogeology of formation waters, northeastern Alberta basin. *American Association of Petroleum Geologists Bulletin*, **77**, 1745–68.

Badley, M. E., Price, J. D. and Backshall, L. C. (1989). Inversion, reactivated faults and related structures: seismic examples from the southern North Sea. In *Inversion Tectonics*, ed. M. A. Cooper and G. D. Williams. *Geological Society of London Special Publication*, vol. 44, pp. 201–19.

Bagnall, W. D., Beardsley, R. W. and Drabish, R. A. (1979). The Keyser gas field, Mineral County, West Virginia. Morgantown: *West Virginia Geological and Economical Survey*.

Balkwill, H. R. (1978). Evolution of Sverdrup Basin, Arctic Canada. *AAPG Bulletin*, **62**, 1004–28.

Balkwill, H. R., Rodrigue, G., Paredes, F. L. and Almeida, J. P. (1995). Northern part of Oriente basin, Ecuador: reflection seismic expression of structures. In *Petroleum Basins of South America*, ed. A. J. Tankard, R. Suárez-Soruco and H. J. Welsink. *American Association of Petroleum Geologists Memoir*, vol. 62, pp. 559–71.

Balla, Z. (1984). The Carpathian loop and the Pannonian basin: a kinematic analysis. *Geophysical Transactions*, **30**, 313–53.

Bally, A. W. (1983). Seismic expression of structural styles: a picture and work atlas. *American Association of Petroleum Geologists Studies in Geology*, vol. 15, three separate atlases.

Bangs, N. L. B. and Westbrook, G. K. (1991). Seismic modeling of the basal décollement at the base of the Barbados Ridge accretionary complex. *Journal of Geophysical Research*, **96**, 3853–66.

Bangs, N. L. B., Westbrook, G. K., Ladd, J. W. and Buhl, P. (1990). Seismic velocities from the Barbados Ridge complex: indicators of high pore fluid pressures in an accretionary complex. *Journal of Geophysical Research*, **95**, 8767–82.

Banks, C. J., and Warburton, J. (1986). 'Passive-roof' duplex geometry in the frontal structures of the Kirthar and Sulaiman belts, Pakistan. *Journal of Structural Geology*, **8**, 229–37.

Banner, J. L., Wasserburg, G. J., Dobson, P. F., Carpenter, A. B. and Moore, C. H. (1989). Isotopic and trace element constraints on the origin and evolution of saline groundwaters from central Missouri. *Geochimica et Cosmochimica Acta*, **53**, 383–98.

Barazangi, M. and Isacks, B. L. (1979). Subduction of the Nazca Plate beneath Peru; evidence from spatial distribution of earthquakes. *Geophysical Journal of the Royal Astronomical Society*, **57**, 537–55.

Barker, C. (1987). Development of abnormal and subnormal pressures in reservoirs containing bacterially generated gas. *American Association of Petroleum Geologists Bulletin*, **71**, 1404–13.

Barnes, P. M. (1996). Active folding of Pleistocene unconformities of the edge of the Australian–Pacific plate boundary zone, offshore North Canterbury, New Zealand. *Tectonics*, **15**, 623–40.

Barr, T. D. and Dahlen, F. A. (1989). Brittle frictional mountain building; 2, Thermal structure and heat budget. *Journal of Geophysical Research*, **94**, 3923–47.

Barton, C. A., Zoback, M. D. and Moos, D. (1995). Fluid flow along potentially active faults in crystalline rock. *Geology*, **23**, 683–6.

Baskin, D. K. and Peters, K. E. (1992). Early generation characteristics of a sulfur-rich Monterey kerogen. *American Association of Petroleum Geologists Bulletin*, **76**, 1–13.

Baud, P., Klein, E. and Wong, T. F. (2004). Compaction localization in porous sandstones: spatial evolution of damage and acoustic emission activity. *Journal of Structural Geology*, **26**, 603–24.

Bazhenov, M. L., Burtman, V. S. and Dvorova, A. V. (1999). Permian paleomagnetism of the Tien Shan fold belt, Central Asia; post-collisional rotations and deformation. *Tectonophysics*, **312**, 303–29.

Beach, A. (1977). Vein arrays, hydraulic fractures and pressure-solution structures in a deformed flysch sequence, S.W. England. *Tectonophysics*, **40**, 201–25.

Bear, J. (1972). *Dynamics of Fluids in Porous Media*. New York: Elsevier.

Beauchamp, W., Barazangi, M., Demnati, A. and El Alji, M. (1996). Intracontinental rifting and inversion; Misour Basin and Atlas Mountains, Morocco. *American Association of Petroleum Geologists Bulletin*, **80**, 1459–82.

Beaumont, J. (1981). Foreland basins. *Geophysical Journal of Royal Astronomic Society*, **65**, 291–329.

Beaumont, C., Fullsack, P. and Hamilton, J. (1992). Erosional control of active compressional orogens. In *Thrust Tectonics*, ed. K. R. McClay. London: Chapman and Hall, pp. 1–18.

(1994). Styles of crustal deformation in compressional orogens caused by the subduction of the underlying lithosphere. *Tectonophysics*, **232**, 119–32.

Bebout, G. E. (1991). Geometry and mechanics of fluid flow at 15 to 45 km depths in an early Cretaceous accretionary complex. *Geophysical Research Letters*, **18**, 923–6.

Beer, J. A., Allmendinger, R. W., Figueroa, D. E. and Jordan, T. E. (1990). Seismic stratigraphy of a Neogene piggyback basin, Argentina. *American Association of Petroleum Geologists Bulletin*, **74**, 1183–202.

Begin, S. B., Meyer, D. F. and Schumm, S. A. (1981). Development of longitudinal profiles of alluvial channels in response to basement lowering. *Earth Surface Processes and Landforms*, **6**, 49–68.

Behrmann, J. H. (1991). Conditions for hydrofracture and the fluid permeability of accretionary wedges. *Earth and Planetary Science Letters*, **107**, 550–8.

Bekins, B. A., McCaffrey, A. M. and Dreiss, S. J. (1994). The influence of kinetics on the smectite to illite transition in the Barbados accretionary prism. *Journal of Geophysical Research*, **99**, 18 145–58.

(1995). Episodic and constant flow models for the origin of low-chloride waters in a modern accretionary complex. *Water Resources Research*, **31**, 3205–15.

Belitz, K. and Bredehoeft, J. D. (1988). Hydrodynamics of Denver Basin: explanation of subnormal fluid pressures. *American Association of Petroleum Geologists Bulletin*, **72**, 1334–59.

Belotti, H. J., Rebori, L. O. and Ibanez, G. H. (1995a). Igneous reservoirs in triangle zones, Neuquen Basin, Argentina. In *American Association of Petroleum Geologists 1995 Annual Convention. Annual Meeting Abstracts*. American Association of Petroleum Geologists and Society of Economic Paleontologists and Mineralogists, p. 8.

Belotti, H. J., Saccavino, L. L. and Schachner, G. A. (1995b). Structural styles and petroleum occurrence in the sub-Andean fold and thrust belt of northern Argentina. In *Petroleum Basins of South America*, ed. A. J. Tankard, R. Suárez-Soruco and H. J. Welsink. *American Association of Petroleum Geologists Memoir*, vol. 62, pp. 545–55.

Bennett, C. O. and Myers, J. E. (1982). *Momentum, Heat and Mass Transfer*. New York: McGraw-Hill.

Bentham, P. A., Burbank, D. W. and Puidefabregas, C. (1992). Temporal and spatial controls on the alluvial architecture of an axial drainage system: Late Eocene Escanilla Formation, southern Pyrenean foreland basin, Spain. *Basin Research*, **4**, 335–52.

Bentham, P. A., Talling, P. J. and Burbank, D. W. (1993). Braided stream and flood-plain deposition in a rapidly aggrading basin; the Escanilla Formation, Spanish Pyrenees. In *Braided Rivers*, ed. J. L. Best and C. S. Bristow, *Geological Society Special Publications*, vol. 75. Oxford: Blackwells, pp. 177–94.

Berg, R. R. (1962). Mountain flank thrusting in Rocky Mountain foreland, Wyoming and Colorado. *American Association of Petroleum Geologists Bulletin*, **46**, 2019–32.

Berg, R. R. (1975). Capillary pressure in stratigraphic traps. *American Association of Petroleum Geologists Bulletin*, **59**, 939–56.

Berg, R. R. and Gangi, A. F. (1999). Primary migration by oil-generation microfracturing in low-permeability source rocks: application to the Austin Chalk, Texas. *American Association of Petroleum Geologists Bulletin*, **83**, 727–56.

Bergamin, J. F., De Vicente, G., Tejero, R., Sanchez Serrano, F., Gomez, D., Munoz Martin, A. and Perucha, M. A. (1996). Cuantification del desplazamiento dextrose alpino en la Cordillera Iberica a partir de datos gravimetricos. Quantification of Alpine dextral displacement in the Iberian Cordillera from gravimetric data. 4. *Congreso Nacional de Geolica. Fourth National Geological Congress*, vol. 20. Madrid: Sociedad Geologica de España. pp. 917–20.

Berger, P. and Johnson, A. M. (1980). First order analysis of deformation of a thrust sheet moving over a ramp. *Tectonophysics*, **70**, T9–24.

(1982). Folding of passive layers and forms of minor structures near terminations of blind thrust faults –

application to the Central Appalachian blind thrust. *Journal of Structural Geology*, **4**, 343–53.

Berggren, W. A., Kent, D. V., Flynn, J. J. and van Couvering, J. A. (1985). Cenozoic geochronology. *Geological Society of America Bulletin*, **96**, 1407–18.

Bergh, S. G., Braathen, A. and Andresen, A. (1997). Interaction of basement-involved and thin-skinned tectonism in the Tertiary fold-thrust belt of central Spitsbergen, Svalbard. *American Association of Petroleum Geologists Bulletin*, **81**, 637–61.

Berkowitz, N. (1997). *Fossil Hydrocarbons: Chemistry and Technology*. San Diego: Academic Press.

Bethke, C. M. (1985). A numerical model of compaction-driven groundwater flow and heat transfer and its application to the paleohydrology of intracratonic basins. *Journal of Geophysical Research*, **90**, 6817–28.

(1986). Hydrologic constraints on the genesis of the Upper Mississippi Valley district from Illinois Basin brines. *Economic Geology*, **81**, 233–49.

Bethke, C. M. and Marshak, S. (1990). Brine migrations across North America – the plate tectonics of groundwater. *Annual Review of Earth and Planetary Sciences*, **18**, 287–315.

Betzler, C. (1989). The Upper Paleocene to mid-Eocene between the Rio Segre and the Rio Llobnegat (eastern south Pyrenees): facies stratigraphy and structural evolution. *Tubinger Geowissenschaftliche Arbeiten*, vol. 2.

Beydoun, Z. R., Clarke, M. W. H. and Stonely, R. (1992). Petroleum in the Zagros basin: a late Tertiary foreland basin overprinted onto the outer edge of a vast hydrocarbon-rich Paleozoic–Mesozoic passive-margin shelf. In *Foreland Basin and Fold Belts*, ed. R. W. Macqueen and D. A. Leckie. *American Association of Petroleum Geologists Memoir*, vol. 55, pp. 309–39.

Biddle, K. T. and Wielchowski, C. C. (1994). Hydrocarbon traps. In *The Petroleum System – From Source to Trap*, ed. L. B. Magoon and W. G. Dow. *American Association of Petroleum Geologists Memoir*, vol. 60, pp. 219–35.

Bilotti, F. D. and Shaw, J. H. (2001). Modeling the compressive toe of the Niger Delta as a critical taper wedge. *Annual Meeting Expanded Abstracts*. American Association of Petroleum Geologists, pp. 18–19.

Biot, M. A. (1941). General theory of three dimensional consolidation. *Journal of Applied Physics*, **12**, 155–64.

Birch, F. and Clark, H. (1940). The thermal conductivity of rocks and its dependence upon temperature and composition. *American Journal of Science*, **238**, 529–58, 613–35.

Birkenland, P. W. (1984). *Soils and Geomorphology*. New York: Oxford University Press.

Bischke, R. E. and Suppe, J. (1990). Calculating uplift rates on the Zayantes Fault utilizing well log correlations. *EOS, Transactions, American Geophysical Union*, **71**, 555.

Bishop, R. S., Gehman, J. H. M. and Young, A. (1984). Concepts for estimating hydrocarbon accumulation and dispersion. In *Petroleum Geochemistry and Basin Evaluation*, ed. G. Demaison and R. J. Murris. *American Association of Petroleum Geologists Memoir*, vol. 35, pp. 41–52.

Bissada, K. K. (1982). Geochemical constraints on petroleum generation and migration – a review. *Proceedings of the Second ASCOPE Conference*, Manilla, October 1981, pp. 69–87.

Biteau, J. J., Choppin de Janvry, G. and Perrodon, A. (2003). How the petroleum system relates to the petroleum province. *Oil and Gas Journal*, **101**, 46–9.

Bjorkum, P. A. (1996). How important is pressure in causing dissolution of quartz in sandstones? *Journal of Sedimentary Research*, **66**, 147–54.

Bjorkum, P. A., Oelkers, E. H., Nadeau, P. H., Walderhaug, O. and Murphy, W. M. (1998). Porosity prediction in quartzose sandstones as a function of time, temperature, depth, stylolite frequency, and hydrocarbon saturation. *American Association of Petroleum Geologists Bulletin*, **82**, 637–48.

Bjorlykke, K. (1984). Secondary porosity: how important is it? In *Clastic Diagenesis*, ed. D. A. McDonald and R. C. Surdam. *American Association of Petroleum Geologists Memoir*, vol. 37, pp. 277–86.

Bjorlykke, K., Ramm, M. and Saigal, G. C. (1989). Sandstone diagenesis and porosity modification during basin evolution. In *Geologic Modeling: Aspects of Integrated Basin Analysis and Numerical Simulation*, ed. H. S. Poelchau and U. Mann. *Geologische Rundschau*, vol. 78, pp. 243–68.

Black, R. and Fabre, J. (1983). A brief outline of the geology of West Africa. In *West Africa: Geological Introduction and Stratigraphic Terms*, ed. J. Fabre. Oxford: Pergamon Press, pp. 17–26.

Blackstone Jr., D. L. (1983). Laramide compressional tectonics, southern Wyoming. *University of Wyoming Contributions to Geology*, **19**, 105–22.

Blackwell, D. D. and Steele, J. L. (1989). Thermal conductivity of sedimentary rocks: measurement and significance. In *Thermal History of Sedimentary Basins – Methods and Case Histories*, ed. N. D. Naeser and T. H. McCulloh. Heidelberg: Springer, pp. 14–36.

Blair, T. C. and Bilodeau, W. L. (1988). Development of tectonic cyclothems in rift, pull-apart, and foreland basins: sedimentary response to episodic tectonism. *Geology*, **16**, 517–20.

Blanc, G., Gieskes, J. M., Vrolijk, P. J., *et al.* (1988). Advection de fluides interstitiels dans les series sedimentaires du complexe d'accretion de la Barbade (leg 110 ODP). *Bulletin de la Societé Géologique de France, Huitième Serie*, **4**, 453–60.

Blanc, P. and Connan, J. (1993). Crude oils in reservoirs: the factors influencing their composition. In *Applied Petroleum Geochemistry*, ed. M. L. Bordenave. Paris: Technip, pp. 151–74.

(1994). Preservation, degradation, and destruction of trapped oil. In *The Petroleum System – From Source to Trap*, ed. L. B. Magoon and W. G. Dow. *American*

Association of Petroleum Geologists Memoir, vol. 60, pp. 237–47.

Blangy, J.-P. (2002). Target-oriented, wide-patch, 3-D seismic yields trap definition and exploration success in the sub-Andean thrust belt Devonian gas play, Tarija Basin, Argentina. *The Leading Edge*, **21**, 142–51.

Blanpied, M. L., Lockner, D. A. and Byerlee, J. (1995). Frictional slip of granite at hydrothermal conditions. *Journal of Geophysical Research*, **100**, 13 045–64.

Blay, P., Cosgrove, J. W. and Summers, J. M. (1977). An experimental investigation of the development of structures in multilayers under the influence of gravity. *Journal of the Geological Society of London*, **133**, 329–42.

Blenkinsop, T. G. and Rutter, E. H. (1986). Cataclastic deformation of quartzite in the Moine Thrust Zone. *Journal of Structural Geology*, **8**, 669–82.

Blythe, A. E. Sugar, A. and Phipps, S. P. (1988). Structural profiles of the Ouachita Mountains, western Arkansas. *AAPG Bulletin*, **72**, 810–19.

Bocharova, N. Y., Golonka, J. and Meisling, K. E. (1995). Tectonic evolution and sedimentary basins of the eastern regions of Russia. American Association of Petroleum Geologists Pacific Section meeting abstracts. *American Association of Petroleum Geologists Bulletin*, **79**, 580.

Bodenhausen, J. W. A. and Ott, W. F. (1981). Habitat of the Rijswijk oil province, onshore, the Netherlands. In *Petroleum Geology of the Continental Shelf of Northwest Europe, 2nd Conference*, ed. L. V. Illing and G. D. Hobson. London: Graham and Trotman, pp. 301–9.

Bodenlos, A. J. (1970). Cap rock development and salt stock movement. In *The Geology and Technology of Gulf Coast Salt, Symposium Proceedings*, ed. D. H. Kupfer. Baton Rouge: Louisiana State University, p. 192.

Bodner, D. P. and Sharp, J. M. (1988). Temperature variations in south Texas subsurface. *American Association of Petroleum Geologists Bulletin*, **72**, 21–32.

Bohacs, K. and Suter, J. (1997). Sequence stratigraphic distribution of coaly rocks: fundamental controls and paralic examples. *American Association of Petroleum Geologists Bulletin*, **81**, 1612–39.

Bombolakis, E. G. (1994). Applicability of critical-wedge theories to foreland belts. *Geology*, **22**, 535–8.

Bonini, M., Sokoutis, D., Mulugeta, G. and Katrivanos, E. (2002). Modelling hanging wall accommodation above rigid thrust ramps. *Journal of Structural Geology*, **22**, 1165–79.

Bonini, M., Sokoutis, D., Talbot, C. J., Boccaletti, M. and Milnes, A. G. (1999). Indenter growth in analogue models of Alpine-type deformation. *Tectonics*, **18**, 119–28.

Booker, J. R. (1974). Time dependent strain following faulting of a porous medium. *Journal of Geophysical Research*, **79**, 2037–44.

Bordenave, M. L. (2000). Zagros domain of Iran. *Oil & Gas Journal*, **98**, 36–8.

Bordenave, M. L. and Burwood, R. (1994). The Albian Kazdhumi Formation of the Dezful Embayment, Iran. In *Petroleum Source Rocks*, ed. B. Katz. New York: Springer-Verlag, pp. 183–208.

Bosworth, W. (1984). Foreland deformation in the Appalachian Plateau, central New York: the role of small-scale detachment structures in regional overthrusting. *Journal of Structural Geology*, **6**, 73–81.

Bott, M. H. P. (1959). The mechanics of oblique slip faulting. *Geological Magazine*, **96**, 109–17.

Bott, M. H. P. and Dean, D. S. (1973). Stress diffusion from plate boundaries. *Nature*, **243**, 339–41.

Bouchon, M. (1997). The state of stress on some faults of the San Andreas system as inferred from near-field strong motion data. *Journal of Geophysical Research, B, Solid Earth and Planets*, **102**, 11 731–44.

Bouchon, M., Sekikuchi, H., Irikura, K. and Iwata, T. (1998). Some characteristics of the stress field of the 1995 Hyogo-ken Nanbu (Kobe) earthquake. *Journal of Geophysical Research*, **103**, 24 271–82.

Boudjema, A. (1987). *Evolution Structurale du Bassin Petrolier Triassique du Sahara Nord–Oriental (Algerie)*. Paris: Universite Paris-Sud.

Boulegue, J., Charlou, J. L. and Jedwab, J. (1986). Fluids from subduction zones off Japan. *EOS Transactions, American Geophysical Union*, **67**, 1204.

Bourrouilh, R., Richert, J.-P. and Zolnai, G. (1995). The North Pyrenean Aquitaine Basin, France: evolution and hydrocarbons. *American Association of Petroleum Geologists Bulletin*, **79**, 831–53.

Bowler, S. (1987). Duplex geometry: an example from the Moine thrust belt. *Tectonophysics*, **135**, 25–35.

Boyer, S. E. (1986). Styles of folding within thrust sheets: examples from the Appalachian and Rocky Mountains of the USA and Canada. *Journal of Structural Geology*, **3/4**, 325–39.

(1995). Sedimentary basin taper as a factor controlling the geometry and advance of thrust belts. *American Journal of Science*, **295**, 1220–54.

Boyer, S. E. and Elliott, D. (1982). Thrust systems. *American Association of Petroleum Geologists Bulletin*, **66**, 1196–230.

BP (2004). *BP Statistical Review of World Energy*. Sunbury: BP plc.

Braathen, A., Bergh, S. G. and Maher Jr., H. D. (1999). Application of a critical wedge taper model to the Tertiary transpressional fold-thrust belt on Spitsbergen, Svalbard. *Geological Society of America Bulletin*, **111**, 1468–85.

Brace, W. F. (1961). Dependence of fracture strength of rocks on grain size. *Proceedings of the 4th Symposium on Rock Mechanics*. Reston, VA: US Geological Survey, pp. 99–103.

(1980). Permeability of crystalline and argillaceous rocks: a review. *International Journal of Rock Mechanics and Mining Sciences & Geomechanics Abstracts*, **17**, 241–51.

Brace, W. F. and Byerlee, J. D. (1966). Stick-slip as a mechanism for earthquakes. *Science*, **153**, 990–2.

Brace, W. F. and Kohlstedt D. L. (1980). Limits on lithospheric stress imposed by laboratory experiments. *Journal of Geophysical Research*, B, **85**, 6248–52.

Brace, W. F., Paulding, B. W. and Scholz, C. (1966). Dilatancy in the fracture of crystalline rocks. *Journal of Geophysical Research*, **71**, 3939–53.

Bradbury, H. J. and Woodwell, G. R. (1987). Ancient fluid flow within foreland terrains. In *Fluid Flow in Sedimentary Basins and Aquifers*, ed. J. C. Goff and B. P. J. Williams. *Geological Society of London Special Publication*, vol. 34, pp. 87–102.

Bradley, D. C. and Bradley, L. M. (1988). Early Acadian west-directed thrusting in the Connecticut Valley–Gaspe synclinorium, northern Maine, and its bearing on northern Appalachian plate kinematics. *Geological Society of America, Abstracts with Programs*, **20**, 9.

(1994). Geometry of an outcrop-scale duplex in Devonian flysch, Maine. *Journal of Structural Geology*, **16**, 371–80.

Brantley, S. L., Evans, B., Hickman, S. H. and Crear, D. A. (1990). Healing of microcracks in quartz: implications for fluid flow. *Geology*, **18**, 136–9.

Bray, C. E., and. Karig, D. E. (1985). Porosity of sediments in accretionary prisms and some implications for dewatering processes. *Journal of Geophysical Research*, **90**, 768–78.

Bredehoeft, J. D., Belitz, K. and Sharp-Hansen, S. (1992). The hydrodynamics of the Big Horn Basin: a study of the role of faults. *American Association of Petroleum Geologists Bulletin*, **76**, 530–46.

Bredehoeft, J. D., Djevanshir, R. D. and Belitz, K. R. (1988). Lateral fluid flow in a compacting sand-shale sequence: South Caspian Basin. *American Association of Petroleum Geologists Bulletin*, **72**, 416–24.

Bredehoeft, J. D., Neuzil, C. E. and Milly, P. C. D. (1983). Regional flow in the Dakota aquifer: a study of the role of confining layers. *US Geological Survey Water-Supply Paper*, **2237**, 45.

Bredehoeft, J. D. and Papadopulos, I. S. (1965). Rates of vertical groundwater movement estimated from the Earth's thermal profile. *Water Resources Research*, **1**, 325–8.

Breen, N. A. and Orange, D. L. (1992a). The effects of fluid escape on accretionary wedges 1. Variable porosity and wedge convexity. *Journal of Geophysical Research*, **97**, 9265–75.

(1992b). The effects of fluid escape on accretionary wedges 2. Seepage force, slope failure, headless submarine canyons, and vents. *Journal of Geophysical Research*, **97**, 9277–95.

Bretan, P., Yielding, G. and Jones, H. (2003). Using calibrated shale gouge ratio to estimate hydrocarbon column heights. *American Association of Petroleum Geologists Bulletin*, **87**, 397–413.

Brewer, J. A., Smithson, S. B., Oliver, J. E., Kaufman, S. and Brown, L. D. (1980). The Laramide Orogeny: evidence from COCORP deep crustal seismic profiles in the Wind River Mountains, Wyoming. *Tectonophysics*, **62**, 165–89.

Brigaud, F., Chapman, D. S. and Le Douaran, S. (1990). Estimating thermal conductivity in sedimentary basins using lithologic data and geophysical well logs. *American Association of Petroleum Geologists Bulletin*, **74**, 1459–77.

Briggs, D. J. (1980). An investigation into variations in mineralogical and mechanical properties of coal measure strata. Ph.D. thesis, Cardiff: University of Wales.

Brinkman, R. and Logters, H. (1968). Diapirs in western Pyrenees and foreland, Spain. In *Diapirism and Diapirs: a Symposium*, ed. J. Braunstein and G. D. O'Brien, *Geology Memoir*, vol. 8. Tulsa, OK: American Association of Petroleum Geologists, pp. 275–93.

Brix, F. and Schultz, O. (1993). *Erdöl und Erdgas in Österreich, Neue Folge 19*. Vienna: Naturhistorisches Museum Wien.

Brookfield, M. E. (2000). Geological development and Phanerozoic crustal accretion in the western segment of the southern Tien Shan (Kyrgyzstan, Uzbekistan and Tajikistan). *Tectonophysics*, **328**, 1–14.

Brookfield, M. E. and Hashmat, A. (2001). The geology and petroleum potential of the North Afghan Platform and adjacent areas (northern Afghanistan, with parts of southern Turkmenistan, Uzbekistan and Tajikistan). *Earth Science Reviews*, **55**, 41–71.

Brown, A. (2003). Capillary effects on fault-fill sealing. *American Association of Petroleum Geologists Bulletin*, **87**, 381–95.

Brown, K. M. (1995). The variation of the hydraulic conductivity structure of an overpressured thrust zone with effective stress. *Proceedings of the Ocean Drilling Program, Scientific Results*, **146**, 281–9.

Brown, K. M. and Behrmann, J. H. (1990). Genesis and evolution of small scale structures in the toe of the Barbados Ridge accretionary wedge. *Ocean Drilling Programme, Scientific Results*, **110**, 229–43.

Brown, K. M., Bekins, B. A., Clennell, B., Dewhurst, D. and Westbrook, G. K. (1994). Heterogeneous hydrofracture development and accretionary fault dynamics: reply. *Geology*, **22**, 1053–4.

Brown, R., Carr, S. D., Johnson, B. J., *et al.* (1992). The Monashee décollement of the southern Canadian Cordillera: a crustal-scale shear zone linking the Rocky Mountain foreland belt to lower crust beneath accreted terranes. In *Thrust Tectonics*, ed. K. R. McClay. London: Chapman & Hall.

Brown, R. A. C. (1952). Carbonate stratigraphy and paleontology in the Mt. Greenock area, Alberta. *Geological Survey of Canada Memoir*, **264**, 119.

Brown, S. P. and Spang, J. H. (1978). Geometry and mechanical relationship of folds to thrust fault propagation using a minor thrust in the Front Ranges of the Canadian Rocky Mountains. *Bulletin of Canadian Petroleum Geology*, **26**, 551–71.

Brown, W. G. (1984a). A reverse fault interpretation of Rattlesnake Mountain anticline, Big Horn Basin, Wyoming. *Mountain Geologist*, **21**, 31–5.

(1984b). Basement involved tectonics: foreland areas. *American Association of Petroleum Geologists Course Notes*, vol. 26. Tulsa, OK: American Association of Petroleum Geologists.

(1988). Deformational style of Laramide uplifts in the Wyoming foreland. In *Interaction of the Rocky Mountains Foreland and the Cordilleran Thrust Belt*, ed. C. J. Schmidt and W. J. Perry, Jr. *Geological Society of America Memoir*, vol. 171, pp. 1–25.

Bruce, C. H. (1984). Smectite dehydration – its relation to structural development and hydrocarbon accumulation in northern Gulf of Mexico Basin. *American Association of Petroleum Geologists Bulletin*, **68**, 673–83.

Brun, J.-P. and Nalpas, T. (1996). Graben inversion in nature and experiments. *Tectonics*, **15**, 677–87.

Bruton, D. J. and Helgeson, H. C. (1983). Calculation of the chemical and thermodynamic consequences of differences between fluid and geostatic pressure in hydrothermal systems. *American Journal of Science*, **283A**, 540–88.

Buchanan, P. G. (1996). The application of cross-section construction and validation within exploration and production: a discussion. In *Modern Developments in Structural Interpretation, Validation and Modeling*, ed. P. G. Buchanan and D. A. Nieuwland. *Geological Society of London Special Publication*, vol. 99, pp. 41–50.

Buchanan, P. G. and McClay, K. R. (1991). Sandbox experiments of inverted listric and planar fault systems. *Tectonophysics*, **188**, 97–115.

(1992). Experiments on basin inversion above reactivated domino faults. *Marine and Petroleum Geology*, **9**, 486–500.

Buchanan, P. G. and Warburton, J. (1996). The influence of pre-existing basin architecture in the development of the Papuan Fold and Thrust Belt: implications for petroleum prospectivity. In *Proceedings of the Third PNG Petroleum Convention*: Port Moresby, ed. P. G. Buchanan. Port Moresby: PNG Chamber of Mines and Petroleum, pp. 89–131.

Bui, T. D., Brinton, J., Karpov, A. V., Hanks, C. L. and Jensen, J. L. (2003). Evidence and implication for significant late and post-fold fracturing on detachment folds in the Lisburne Group of the northeastern Brooks Range. *SPE Reservoir Evaluation & Engineering*, **6**, 197–205.

Buntebarth, G. (1984). *Geothermics*. New York: Springer-Verlag.

Buntebarth, G. and Rybach, L. (1981). Linear relationships between petrophysical properties and mineralogical constitution; preliminary results. *Tectonophysics*, **75**, 41–6.

Burbank, D. W. (1992). Causes of recent Himalayan uplift deduced from deposited patterns in the Ganges basin. *Nature*, **357**, 680–2.

Burbank, D. W., Beck, R. A. and Mulder, T. (1996). The Himalayan foreland basin. In *Tectonic Evolution of Asia*, ed. A. Yin and T. M. Harrison. New York: Cambridge University Press, pp. 149–88.

Burbank, D. W., Derry, L. A. and France-Lanord, C. (1993). Reduced Himalayan sediment production 8 Myr ago despite an intensified monsoon. *Nature*, **364**, 48–50.

Burbank, D. W., Meigs, A. and Brozovic, N. (1996). Interactions of growing folds and coeval depositional systems. *Basin Research*, **8**, 199–223.

Burbank, D. W. and Verges, J. (1994). Reconstruction of topography and related depositional systems during active thrusting. *Journal of Geophysical Research*, **99**, 20 281–97.

Burbank, D. W., Verges, J., Munoz, J. A. and Bentham, P. (1992). Coeval hindward- and forward-imbricating thrusting in the south-central Pyrenees, Spain: timing and rates of shortening and deposition. *Geological Society of America Bulletin*, **104**, 3–17.

Burchfiel, B. C., Brown, E. T., Deng, Q., Feng, X., Li, J., Molnar, P., Shi, J., Wu, Z. and You, H. (1999). Crustal shortening on the margins of the Tien Shan, Xinjiang, China. *International Geology Review*, **41**, 665–700.

Burley, S. D., Mullis, J. and Matter, A. (1989). Timing of diagenesis in the Tartan Reservoir (UK North Sea): constraints from combined cathodoluminescence microscopy and fluid inclusion studies. *Marine and Petroleum Geology*, **6**, 98–120.

Burnham, A. K., Schmidt, B. J. and Braun, B. L. (1995). A test of the parallel reaction model using kinetic measurements on hydrous pyrolysis residues. *Organic Geochemistry*, **23**, 931–9.

Burnham, A. K., and Sweeney, J. J. (1989). A chemical kinetic model of vitrinite maturation and reflectance. *Geochimica et Cosmochimica Acta*, **53**, 2649–57.

Burrus, J., Schneider, F. and Wolf, S. (1994). Modeling overpressures in sedimentary basins: consequences for permeability and rheology of shales, and petroleum expulsion efficiency. *American Association of Petroleum Geologists Bulletin*, **78**, 1137.

Burruss, R. C., Cercone, K. R. and Harris, P. M. (1985). Timing of hydrocarbon migration: evidence from fluid inclusions in calcite cements, tectonics and burial history. In *Carbonate Cements Revisited*, ed. P. M. Harris and N. M. Schneidermann. *SEPM Special Publication*. Tulsa, OK: Society for Sedimentary Geology, pp. 277–89.

Burtman, V. S. (2000). Cenozoic crustal shortening between the Pamir and Tien Shan and a reconstruction of the Pamir-Tien Shan transition zone for the Cretaceous and Palaeogene. *Tectonophysics*, **319**, 69–92.

Burtner, R. L., Nigrini, A. and Donelick, R. A. (1994). Thermochronology of Lower Cretaceous source rocks in the Idaho–Wyoming thrust belt. *American Association of Petroleum Geologists Bulletin*, **78**, 1613–36.

Busk, H. G. (1929). *Earth Flexures*. Cambridge: Cambridge University Press.

Bustin, R. M. (1983). Heating during thrust faulting in the Rocky Mountains: friction or fiction? *Tectonophysics*, **95**, 309–28.

Butler, R. W. H. (1985). The restoration of thrust systems and displacement continuity around the Mont Blanc massif, NW external Alpine thrust belt. *Journal of Structural Geology*, 7, 569–82.

(1989). The geometry of crustal shortening in the western Alps. In *Tectonic Evolution of the Tethyan Region*, ed. A. M. C. Sengor. Dordrecht: Kluwer Academic Publishers, pp. 43–76.

(1991). Hydrocarbon maturation, migration and thrust sheet loading in the Western Alpine foreland thrust belt. In *Petroleum Migration*, ed. W. A. England and A. J. Fleet. *Geological Society of London Special Publication*, vol. 59, pp. 227–44.

(1992). Thrusting patterns in the NW French subalpine chains. *Annales Tectonicae*, 6, 150–72.

Butler, R. W. H. and Bowler, S. (1995). Local displacement rate cycles in the life of a fold-thrust belt. *Terra Nova*, 7, 408–16.

Butler, K. and Schamel, S. (1988). Structure along the eastern margin of the Central Cordillera, Upper Magdalena Valley, Colombia, *Journal of South American Earth Sciences*, 1, 109–20.

Buxtorf, A. (1916). Prognosen und befunde beim Hauenstembasis-und Grenchenberg-Tunnel und die Bedeutung der latzeren fur die Geologie des Juragebirges. *Verhandlungen der Naturforschenden Gesellschaft in Basel*, 27, 181–205.

Byerlee, J. D. (1968). Brittle–ductile transition in rocks. *Journal of Geophysical Research*, 73, 4741–50.

(1978). Friction of rocks. *Pure and Applied Geophysics*, 116, 615–26.

(1993). Model for episodic flow of high-pressure water in fault zones before earthquakes. *Geology*, 21, 303–6.

Byrne, D. E., Davis, D. M. and Lynn, R. S. (1988). Loci and maximum size of thrust earthquakes and the mechanics of the shallow region of subduction zones. *Tectonics*, 7, 22–61.

Byrne, T. and Fisher, D. M. (1990). Evidence for a weak and overpressured décollement beneath sediment-dominated accretionary prisms. In *Special Section on the Role of Fluids in Sediment Accretion, Deformation, Diagenesis, and Metamorphism in Subduction Zones*, ed. M. G. Langseth and J. C. Moore. *Journal of Geophysical Research*, 95, 9081–97.

Caine, J. S. (1999). *The Architecture and Permeability Structure of Brittle Fault Zones*. Doctoral dissertation, University of Utah, Salt Lake City.

Caine, J. S., Evans, J. P. and Forster, C. B. (1996). Fault zone architecture and permeability structure. *Geology*, 24, 1025–8.

Calassou, S., Laroque, C. and Malavieille, J. (1993). Transfer zones of deformation in thrust wedges: an experimental study. *Tectonophysics*, 221, 325–44.

Campbell, C. J. and Burgl, H. (1965). Section through the Eastern Cordillera of Colombia, South America. *GSA Bulletin*, 76, 567–90.

Caprarelli, G. and Leitch, E. C. (1998). Magmatic changes during the stabilisation of a cordilleran fold belt; the Late Carboniferous-Triassic igneous history of eastern New South Wales, Australia. *Lithos*, 45, 413–30.

Capuano, R. M. (1990). Hydrochemical constraints on fluid–mineral equilibria during compaction diagenesis of kerogen-rich geopressured sediments. *Geochimica et Cosmochimica Acta*, 54, 1283–99.

Capuano, R. M. (1993). Evidence of fluid flow in microfractures in geopressured shales. *American Association of Petroleum Geologists Bulletin*, 77, 1303–14.

Carlsson, A. and Olsson, T. (1979). Hydraulic conductivity and its stress dependence. *Proceedings, Workshop on Low-Flow, Low Permeability Measurements in Largely Impermeable Rocks*, Paris. pp. 249–59.

Carminati, E. and Siletto, G. B. (1997). The effects of brittle–plastic transitions in basement-involved foreland belts: the central Southern Alps case (N Italy). In *Thermal and Mechanical Interactions in Deep-seated Rocks*, ed. K. Schulmann. *Tectonophysics*, 280, 107–23.

Carnevali, J. O. (1992). Monagas thrust-fold belt in the eastern Venezuela Basin: anatomy of a giant discovery of the 1980s. In *Exploration and Production, Proceedings – World Petroleum Congress, Actes et Documents, Congrès Mondial du Pétrole*, vol. 13. Chichester: John Wiley & Sons, pp. 47–58.

Carroll, A. R. and Bohacs, K. M. (2001). Lake-type controls on petroleum source rock potential in nonmarine basins. *American Association of Petroleum Geologists Bulletin*, 85, 1033–53.

Carroll, A. R., Brassell, S. C. and Graham, S. A. (1992). Upper Permian lacustrine oil shales, southern Junggar basin, northwest China. *American Association of Petroleum Geologists Bulletin*, 76, 1874–902.

Carson, B., Suess, E. and Strasser, J. C. (1990). Fluid flow and mass flux determinations at vent sites on the Cascadia margin accretionary prism. *Journal of Geophysical Research*, 95, 8891–9.

Carter, N. L. and Tsenn, M. C. (1987). Flow properties of continental lithosphere. *Tectonophysics*, 136, 27–63.

Castelijns, J. H. P. and Hagoort, J. (1984). Recovery of retrograde condensate from naturally fractured gas condensate reservoirs. *Journal of the Society of Petroleum Engineers*, 24, 707–17.

Catalan, L., Xiaowen, F., Chatzis, I. and Dullien, F. A. I. (1992). An experimental study of secondary oil migration. *American Association of Petroleum Geologists Bulletin*, 76, 638–50.

Cathles III., L. M. (1990). Scales and effects of fluid flow in the upper crust. *Science*, 248, 323–9.

Cathles, L. M. and Smith, A. T. (1983). Thermal constraints on the formation of Mississippi Valley-type lead-zinc deposits and their implication for episodic basin dewatering and deposit genesis. *Economic Geology*, 78, 983–1002.

Caus, E. (1973). Aportaciones al conocimiento del Eoceno del anticlinal de Oliana (prov. de Lerida). The Eocene of the Oliana Anticline, Lerida, Spain. *Acta Geologica Hispanica*, 8, 7–10.

Cazier, E. C., Cooper, M. A., Eaton, S. G. and Pulham, A. J. (1997). Basin development and tectonic history of the

Llanos Basin, Eastern Cordillera, and Middle Magdalena Valley, Colombia: reply. *American Association of Petroleum Geologists Bulletin*, **81**, 1332–5.

Cecil, C. B. and Heald, M. T. (1971). Experimental investigation of the effects of grain-coatings on quartz over-growth. *Journal of Sedimentary Petrology*, **41**, 582–4.

Cello, G. and Nur, A. (1988). Emplacement of foreland thrust systems. *Tectonics*, **7**, 261–71.

Čermák, V. and Bodri, L. (1986). Temperature structure of the lithosphere based on 2-D temperature modeling, applied to Central and Eastern Europe. In *Thermal Modeling in Sedimentary Basins*, ed. J. Burrus. Paris: Editions Technip, pp. 7–32.

Čermák, V. and Haenel, R. (1988). Geothermal maps. In *Handbook of Terrestrial Heat-flow Density Determination; With Guidelines and Recommendations of the International Heat Flow Commission*, ed. R. Haenel, L. Rybach and L. Stegena. Dordrecht: Kluwer Academic Publishers.

Čermák, V. and Hurtig, E. (1977). *Preliminary Heat Flow Map of Europe, 1:5,000,000; Explanatory Text*. Potsdam: IASPEI, International Heat Flow Communication.

Čermák, V. and Rybach, L. (1982). Thermal conductivity and specific heat of minerals and rocks. In *Landolt–Bornstein Numerical Data and Functional Relationships in Science and Technology: New Series*, ed. G. Angenheister. vol. 16, Berlin: Springer-Verlag.

Cerveny, P. F., Naeser, N. D., Naeser, C. W. and Johnson, N. M. (1988). History of uplift, relief and provenance of the Himalaya during the past 18 m.y.: evidence from fission track ages of detrital zircons from Siwalik group sediments. In *New Perspectives in Basin Analysis*, ed. K. L. Kleinspehn and C. Paola. New York: Springer-Verlag, pp. 43–61.

Chamberlain, C. P. and Zeitler, P. (1986). Pressure–temperature–time paths in the Nanga Parbat massif: constraints on the tectonic development of the northwest Himalayas. *Geological Society of America, Abstracts with Programs*, Boulder, CO: Geological Society of America, vol. 18, p. 561.

Chamberlin, R. T. (1910). The Appalachian folds of central Pennsylvania. *Journal of Geology*, **18**, 228–51.

Chang, J. and Yortsos, Y. C. (1994). Lamination during silica diagenesis – effects of clay content and Oswald ripening. *American Journal of Science*, **294**, 137–72.

Chapman, D. S., Keho, T. H., Bauer, M. S. and Picard, M. D. (1984). Heat flow in the Uinta basin determined from bottom hole temperature (BHT) data. *Geophysics*, **49**, 453–66.

Chapman, D. S., Willett, S. D. and Clauser, C. (1991). Using thermal fields to estimate basin-scale permeabilities. In *Petroleum Migration*, ed. W. A. England and A. J. Fleet. *Geological Society of London Special Publication*, vol. 59, pp. 123–5.

Chapple, W. M. (1978). Mechanics of thin-skinned fold-and-thrust belts. *Geological Society of America Bulletin*, **93**, 1189–98.

Charlton, T. R. (1988). Tectonic erosion and accretion in steady-state trenches. *Tectonophysics*, **149**, 233–43.

Charollais, R. J., Pairis, J. L. and Rosset, J. (1977). Compte rendu de l'excursion de la Societé Géologique Suisse en Haute-Savoie (France) du 10 au 12 Octobre 1976. *Eclogae Geologicae Helvetiae*, **70**, 253–85.

Charpentier, R. R., de Witt Jr., W., Claypool, G. E., Harris, L. D., Mast, R. F., Megeath, J. D., Roen, J. B. and Schmoker, J. W. (1993). Estimates of unconventional natural gas resources of the Devonian shales of the Appalachian Basin. In *Petroleum Geology of the Devonian and Mississippian Black Shale of Eastern North America*, ed. J. B. Roen and R. C. Kepferle. US Geological Survey Bulletin 1909, pp. N1–20.

Chester, F. M. and Logan, J. M. (1986). Composite planar fabric of gouge from the Punchbowl fault, California. *Journal of Structural Geology*, **9**, 621–34.

Chester, J. S. (1988). Geometry and fracture distribution in fault-propagation folds in nature and experiments. *American Association of Petroleum Geologists Bulletin*, **72**, 171.

(1992). Role of mechanical anisotropy in the internal evolution of a thrust sheet. Ph.D. thesis, College Station: Texas A&M University.

Chester, J. S. and Chester, F. M. (1990). Fault-propagation folds above thrusts with constant dip. *Journal of Structural Geology*, **12**, 903–10.

Chester, J. S., Logan, J. S. and Spang, J. H. (1991). Influence of layering and boundary conditions on fault-bend and fault-propagation folding. *Geological Society of America Bulletin*, **103**, 1059–72.

Chidsey Jr., T. C. (1993). Thrustbelt structure plays; Jurassic–Triassic Nugget Sandstone. In *Atlas of Major Rocky Mountain Gas Reservoirs*, ed. C. A. Hjellming. Socorro, NM: New Mexico Bureau of Mines & Mineral Resources, pp. 77–9.

Chigne, N., Loureiro, D., Cabrera, E. and Osuna, S. (1996). Tectonostratigraphic evolution and petroleum systems of the Barinas–Apure Basin and surrounding areas. *American Association of Petroleum Geologists International Conference, Abstracts, American Association of Petroleum Geologists Bulletin*, **80**, 1281.

Chilingarian, G. V. (1983). Compactional diagenesis. In *Sediment Diagenesis, Reidel*, ed. A. Parker and B. W. Sellwood. Dordrecht: Reidel Publishing Co., pp. 57–168.

Choukroune, P. and ECORS Team (1989). The ECORS Pyrenean deep seismic profile reflection data and the overall structure of an orogenic belt. *Tectonics*, **8**, 23–39.

Clark Jr., S. P. (1966). *Handbook of Physical Constants. Geological Society of America Memoir*, vol. 97. New York: Geological Society of America.

Clarke, R. H. and Cleverly, R. W. (1991). Petroleum seepage and post-accumulation migration. In *Petroleum Migration*, ed. W. A. England and A. J. Fleet. *Geological Society of London Special Publication*, vol. 59, pp. 265–71.

Clauer, N., Caby, R., Jeannette, D. and Trompette, R. (1982). Geochronology of sedimentary and metasedimentary Precambrian rocks of the West African Craton. In *Geochronological Correlation of Precambrian Sediments and Volcanics in Stable Zones: Precambrian Research*, vol. 18, ed. M. G. Bonhomme. Amsterdam: Elsevier, pp. 53–71.

Cloetingh, S. A. P. L., Zoetemeijer, R. and van Wees, J. D. (1995). Tectonics I. Tectonics and basin formation in convergent settings; thermo-mechanical evolution of the lithosphere and basin evolution in compressive tectonic regimes. *Short Course*. Amsterdam: Vrije University.

Coates, J. S. (1945). The construction of geological sections. *Quarterly Journal of the Geological, Mining and Metallurgical Society of India*, **17**, 1–11.

Cobbold, P. R. (1976). Fold shapes as functions of progressive strain. *Philosophical Transactions of the Royal Society of London, Series A: Mathematical and Physical Sciences*, **283**, 129–38.

Cobbold, P. R., Cosgrove, J. W. and Summers, J. M. (1971). Development of internal structures in deformed anisotropic rocks. *Tectonophysics*, **12**, 23–53.

Cobbold, P. R., Thomas, J. C., Gapais, D. and Sadybakasov, E. (1993). Cenozoic deformation of the western Tien-Shan. In *Seventh Meeting of the European Union of Geosciences*, Strasbourg, *Terra Abstracts*, vol. 5, Suppl. 1. Oxford: Blackwell Scientific Publications, p. 256.

Cobbold, P. R., Szatmari, P., Demercian, S. L., Coelho, D. and Rossello, E. A. (1995). Seismic and experimental evidence for thin-skinned horizontal shortening by convergent radial gliding on evaporites, deep-water Santos Basin, Brazil. In *Salt Tectonics: a Global Perspective*, ed. M. P. A. Jackson, D. G. Roberts and S. Snelson. *American Association of Petroleum Geologists Memoir*, vol. 65, pp. 305–21.

Cobbold, P. R., Meisling, K. E. and Van Mount, S. (2001). Reactivation of an obliquely rifted margin, Campos and Santos basins, southeastern Brazil. *AAPG Bulletin*, **85**, 1925–44.

Coffin, M. F. and Rabinowitz, P. D. (1988). Evolution of the conjugate East African–Madagascaran margins and the western Somali Basin. *Geological Society of America Special Paper*, **226**, 78.

Colins, E., Hamilton, W. and Schmidt, F. (1992). The hydrocarbon potential of the Alpine subthrust and overthrust, Austria. In *Generation, Accumulation and Prediction of Europe's Hydrocarbons* vol. II, ed. A. M. Spencer. Berlin: Springer-Verlag, pp. 193–9.

Colletta, B., Letouzey, J., Pinedo, R., Ballard, J. F. and Balle, P. (1991). Computerized X-ray tomography analysis of sandbox models: examples of thin-skinned thrust systems. *Geology*, **9**, 1063–7.

Collister, J. W., Simmons, R. E., Lichtfouse, E. and Hayes, J. M. (1992). An isotopic biochemical study of the Green River oil shale. *Organic Geochemistry*, **19**, 265–76.

Colombo, F. and Verges, J. (1992). Geometria del margen S.E. de la cuenca del Ebro; discordancias progresivas en el Grupo Scala Dei, Serra de la Llena (Tarragona). Geometry of the southeastern margin of the Ebro Basin; progressive discordancies in the Scala Dei Group, Serra de la Llena, Tarragona. *Acta Geologica Hispanica*, **27**, 33–53.

Colten-Bradley, V. A. C. (1987). Role of pressure in smectite dehydration – effects on geopressure and smectite-to-illite transition. *American Association of Petroleum Geologists Bulletin*, **71**, 1414–27.

Connan, J. (1984). Biodegradation of crude oils in reservoirs. In *Advances in Petroleum Geochemistry*, vol. 1, ed. J. Brooks and D. Welte. London: Academic Press, pp. 299–335.

(1993). Origin of severely biodegraded oils: a new approach using biomarker pattern of asphaltene pyrolysates. In *Applied Petroleum Geochemistry*, ed. M. L. Bordenave. Paris: Technip, pp. 457–63.

Connelly, J. N. and Mengel, F. C. (2000). Evolution of Archean components in the Paleoproterozoic Nagssugtoqidian Orogen, West Greenland. *Geological Society of America Bulletin*, **112**, 747–63.

Connolly, P. and Cosgrove, J. (1999). Prediction of fracture-induced permeability and fluid flow in the crust using experimental stress data. *American Association of Petroleum Geologists Bulletin*, **83**, 757–77.

Constenius, K. N. (1996). Late Paleogene extensional collapse of the Cordilleran foreland fold and thrust belt. *Geological Society of America Bulletin*, **108**, 20–39.

Contreras, J. and Suter, M. (1997). A kinematic model for the formation of duplex systems with a perfectly planar roof thrust. *Journal of Structural Geology*, **19**, 269–78.

Coogan, J. C. (1992a). Structural evolution of piggyback basins in the Wyoming–Idaho–Utah thrust belt. In *Regional Geology of Eastern Idaho and Western Wyoming, Boulder, Colorado*, ed. P. K. Link, M. A. Kuntz and L. B. Platt. *Geological Society of America Memoir*, vol. 179, pp. 55–81.

(1992b). Thrust systems and displacement transfer in the Wyoming–Idaho–Utah thrust belt. Ph.D. dissertation, Laramie: University of Wyoming.

Coogan, J. C. and Royse Jr., F. (1990). Overview of recent developments in thrust belt interpretation. *Public Information Circular, Geological Survey of Wyoming*, **29**, 89–124.

Cook, D. G. (1992). MacKenzie Mountains, Franklin Mountains, and Coleville Hills. In *Geology of Cordilleran Orogen in Canada*, ed. H. Gabrielse and C. J. Yorath. Ottawa: Geological Survey of Canada, pp. 642–6.

Cook, D. G. and Aitken, J. D. (1976). Two cross-sections across selected Franklin Mountain structures and their implications for hydrocarbon exploration. *Geological Survey of Canada, Report of Activities*, vol. 76-1B. Ottawa: Geological Survey of Canada, pp. 315–22.

Cook, T. D. and Bally, A. W., eds. (1975). *Stratigraphic Atlas of North and Central America*. Princeton, NJ: Princeton University Press.

Cook, D. G. and MacLean, B. C. (1999). The Imperial anticline: a fault-bend fold above bedding-parallel thrust ramp, Northwest Territories, Canada. *Journal of Structural Geology*, **21**, 215–28.

Cooke, M. L., Mollema, P., Pollard, D. D. and Aydin, A. (1998). Interlayer slip and joint localization in East Kaibab monocline, Utah: field evidence and results of numerical modeling. In *Forced (Drape) Folds and Associated Fractures*, ed. J. W. Cosgrove and M. S. Ameen. *Geological Society of London Special Publication*, vol. 169, pp. 23–49.

Cooke, M. L. and Pollard, D. D. (1994). Development of bedding plane faults and fracture localization in a flexed multilayer: a numerical model. In *Rock Mechanics, Models and Measurements, Challenges from Industry: Industry Proceedings of First North American Rock Mechanics Symposium*, ed. P. P. Nelson and S. E. Laubach. Rotterdam: Balkema, pp. 131–8.

(1997). Bedding plane in initial stages of fault-related folding. *Journal of Structural Geology*, **19**, 567–81.

Cooper Jr., H. H. (1966). The equation of groundwater flow in fixed and deforming coordinates. *Journal of Geophysical Research*, **71**, 4785–90.

Cooper, M. (1991). The analysis of fracture systems in subsurface thrust structures from the Foothills of Canadian Rockies. In *Thrust Tectonics*, ed. K. R. McClay. London: Chapman & Hall, pp. 391–405.

Cooper, M. A., Addison, F. T., Alvarez, R., Hayward, A. B., Howe, S., Pulham, A. J. and Taborda, A. (1995). Basin development and tectonic history of the Llanos Basin, Colombia. In *Petroleum Basins of South America*, ed. A. J. Tankard, R. S. Soruco and H. J. Welsink. *American Association of Petroleum Geologists Memoir*, vol. 62, pp. 659–65.

Cooper, M. A., Garton, M. R. and Hossack, J. R. (1983). The origin of the Basse-Normandie duplex, Boulonnais, France. *Journal of Structural Geology*, **5**, 139–52.

Copper, J. A. Wells, A. T. and Nicholas, T. (1971). Dating of glauconite from the Ngalia Basin, Northern Territory, Australia. *Journal of the Geological Society of Australia*, **18**, 97–106.

Cosgrove, J. W. (2001). Hydraulic fracturing during the formation and deformation of a basin: a factor in the dewatering of low-permeability sediments. *American Association of Petroleum Geologists Bulletin*, **85**, 737–48.

Couples, G. D. (1986). *Kinematic and Dynamic Considerations in the Forced Folding Process as Studied in the Laboratory (Experimental Models) and in the Field (Rattlesnake Mountain, Wyoming)*. Texas: A&M University, College Station.

Couples, G. D. and Lewis, H. (1998). Lateral variations of strain in experimental forced folds. *Tectonophysics*, **295**, 79–91.

Couples, G. D., Stearns, D. L. and Handin, J. W. (1994). Kinematics of experimental forced folds and their relevance to cross-section balancing. *Tectonophysics*, **233**, 193–213.

Coutand, I., Strecker, M. R., Arrowsmith, J. R., *et al.* (2002) Late Cenozoic tectonic development of the intramontane Alai Valley (Pamir-Tien Shan region, Central Asia); an example of intracontinental deformation due to the Indo-Eurasia collision. *Tectonics*, **21**, 19.

Couzens, B. A. and Dunne, W. M. (1994). Displacement transfer at thrust termination: the Saltville thrust and Sinking Creek anticline, Virginia, USA. *Journal of Structural Geology*, **16**, 781–93.

Couzens, B. A. and Wiltschko, D. V. (1996). The control of mechanical stratigraphy on the formation of triangle zones. In *Triangle Zones and Tectonic Wedges*, ed. P. A. MacKay, J. L. Varsek, T. E. Kubli, R. G. Dechesne, A. C. Newson and J. P. Reid. *Society of Canadian Petroleum Geologists: Bulletin of Canadian Petroleum Geology*, vol. 44, pp. 165–79.

Couzens, B. A., Wiltschko, D. V. and Vendeville, B. C. (1997). Experimental investigation of duplex and triangle-zone development. *Annual Meeting Abstracts*, 6. Tulsa, OK: American Association of Petroleum Geologists and Society of Economic Paleontologists and Mineralogists, p. 23.

Couzens-Schultz, B. A. and Wiltschko, D. V. (2000). The control of the smectite–illite transition on passive-roof duplex formation; Canadian Rockies foothills, Alberta. *Journal of Structural Geology*, **22**, 207–30.

Coveney, R. M., Goebel Jr., E. D. and Ragan, V. M. (1987). Pressures and temperatures from aqueous fluid inclusions in sphalerite from midcontinent country rocks. *Economic Geology*, **82**, 740–51.

Covey, M. (1986). The evolution of foreland basins to steady state: evidence from the Western Taiwan foreland basin. *Special Publication of International Association of Sedimentologists*, vol. 8, pp. 77–90.

Cowan, D. S. and Silling, R. M. (1978). A dynamic, scaled model of accretion at trenches and its implications for the tectonic evolution of subduction complexes. *Journal of Geophysical Research, A, Space Physics*, **83**, 5389–96.

Coward, M. P. (1984). A geometrical study of the Arnaboll and Heilam thrust sheets, NW of Ben Arnaboll, Sutherland. *Scottish Journal of Geology*, **20**, 87–106.

Coward, M. P. and Dietrich, D. (1989). Alpine tectonics: an overview. In *Alpine Tectonics*, ed. M. P. Coward, D. Dietrich and R. G. Park. *Geological Society of London Special Publication*, vol. 45, pp. 1–29.

Coward, M. P., Enfield, M. A. and Fischer, M. W. (1989). Devonian basins of northern Scotland; extension and inversion related to Late Caledonian–Variscan tectonics. In *Inversion Tectonics Meeting*, ed. M. A. Cooper and G. D. Williams. *Geological Society of London Special Publication*, vol. 44, pp. 275–308.

Coward, M. P. and Kim, J. H. (1981). Strain within thrust sheets. In *Thrust and Nappe Tectonics*, ed. K. R. McClay and N. J. Price. *Geological Society of London Special Publication*, vol. 9, pp. 275–92.

Cowgill, E. and Kapp, P. (2001). Did Tarim (North China) collide with Qiangtang (South China) in both the Devonian and the Triassic? In *Geological Society of America, Cordilleran Section, 97th Annual Meeting, Abstracts with Programs*, vol. 33. Boulder, CO: Geological Society of America, p. 42.

Cox, S. J. D., Meredith, P. G. and Stuart, C. E. (1991). Microfracturing during brittle rock failure: a model for the Kaiser effect including sub-critical crack growth. *Seventh International Congress on Rock Mechanics*, pp. 703–7.

Craddock, J. P., Jackson, M., van der Pluijm, B. A. and Versical, R. T. (1993). Regional shortening fabrics in eastern North America: far-field stress transmission from the Appalachian–Ouachita orogenic belt. *Tectonics*, **12**, 257–64.

Craig, H. (1961). Isotopic variations in meteoric waters. *Science*, **133**, 1702–3.

Craig, H. and Hom, B. (1968). Relationship of deuterium, oxygen-18 and chlorinity in the formation of sea ice. *EOS, Transactions, American Geophysical Union*, **49**, 217–18.

Crane, R. C. (1987). Use of fault cut-offs and bed travel distance in balanced cross sections. *Journal of Structural Geology*, **9**, 243–7.

Creaney, S. and Allan, J. (1992). Petroleum systems in the foreland basin of western Canada. In *Foreland Basin and Fold Belts*, ed. R. W. Macqueen and D. A. Leckie. *American Association of Petroleum Geologists Memoir*, vol. 55, pp. 279–308.

Creaney, S. and Passey, Q. R. (1993). Recurring patterns of total organic carbon and source rock quality within a sequence stratigraphic framework. *American Association of Petroleum Geologists Bulletin*, **77**, 386–401.

Criss, R. E. and Hofmeister, A. M. (1991). Application of fluid dynamics principles in tilted permeable media to terrestrial hydrothermal systems. *Geophysical Research Letters*, **18**, 199–202.

Cruikshank, K. M. and Johnson, A. M. (1993). High-amplitude folding of linear-viscous multilayers. *Journal of Structural Geology*, **15**, 79–94.

Csontos, L., Nagymarosy, A., Horváth, F. and Kováč, M. (1992). Tertiary evolution of the intracarpathian area: a model. *Tectonophysics*, **208**, 221–41.

Cunningham, D. Dijkstra, A., Howard, J. and Badarch, G. (2000). Crustal architecture of the northwestern Mongolian Altai. *EOS, Transactions*, **81**, 1164.

Curiale, J. A. (1994). Correlation of oils and source rocks: a conceptual and historical perspective. In *The Petroleum System – From Source to Trap*, ed. L. B. Magoon and W. G. Dow. *American Association of Petroleum Geologists Memoir*, vol. 60, pp. 251–60.

Currie, J. B., Patnode, H. W. and Trump, R. P. (1962). Development of folds in sedimentary strata. *Geological Society of America Bulletin*, **73**, 655–73.

Curtis, J. B. (2002). Fractured shale-gas systems. *American Association of Petroleum Geologists Bulletin*, **86**, 1921–38.

Czassny, B., Young, E. M., Arrowsmith, J. R. and Strecker, M. R. (1999). Stratigraphic and structural evidence of late Paleogene to early Neogene deformation in the southwestern Tien Shan, Pamir-Alai region, Kyrgyzstan. *EOS, Transactions, American Geophysical Union*, **80**, 1016–17.

Dahl, J. E. P., Moldowan, J. M., Teerman, S. C., McCaffrey, M. A., Sundararaman, P. and Stelting, C. E. (1994). Source rock quality determination from oil biomarkers I: a new geochemical technique. *American Association of Petroleum Geologists Bulletin*, **78**, 1507–26.

Dahlberg, E. C. (1995). *Applied Hydrodynamics in Petroleum Exploration*. New York: Springer-Verlag.

Dahlen, F. A. (1984). Noncohesive critical Coulomb wedges: an exact solution. *Journal of Geophysical Research, B*, **89**, 10 125–33.

(1990). Critical taper model of fold-and-thrust belts and accretionary wedges. *Annual Review of Earth and Planetary Sciences*, **8**, 55–99.

Dahlen, F. A. and Barr, T. D. (1989). Brittle frictional mountain building; 1: Deformation and mechanical energy budget. *Journal of Geophysical Research*, **94**, 3906–22.

Dahlen, F. A. and Suppe, J. (1988). Mechanics, growth, and erosion of mountain belts. In *Processes in Continental Lithospheric Deformation*, ed. S. P. J. Clark, B. C. Burchfiel and J. Suppe. *Geological Society of America Special Paper*, vol. 218, pp. 161–78.

Dahlen, F. A., Suppe, J. and Davis, D. (1984). Mechanics of fold-and-thrust belts and accretionary wedges: cohesive Coulomb theory. *Journal of Geophysical Research, B*, **89**, 10 087–101.

Dahlstrom, C. D. A. (1969). Balanced cross sections. *Canadian Journal of Earth Sciences*, **6**, 743–57.

(1970). Structural geology in eastern margin of Canadian Rocky Mountains. *The American Association of Petroleum Geologists Bulletin*, **54**, 843.

Daniels, E. J., Altaner, S. P., Marshak, S. and Eggleston, J. R. (1990). Hydrothermal alteration in anthracite from eastern Pennsylvania: implications for mechanisms of anthracite formation. *Geology*, **18**, 247–50.

Dansgaard, W. (1964). Stable isotopes in precipitation. *Tellus*, **XVI**, 436–68.

Darcy, J. (1856). *Les Fontaines Publiques de la Ville de Dijon*. Paris: Dalmont.

Dashwood, M. F. and Abbotts, I. L. (1990). Aspects of the petroleum geology of the Oriente basin, Ecuador. In *Classic Petroleum Provinces*, ed. J. Brooks. *Geological Society of London Special Publication*, vol. 50, pp. 89–107.

Davidson, J. (1975). Jumping Pound and Sarcee gas fields. In *Structural Geology of the Foothills between Savanna*

Creek and Panther River, S. W. Alberta, Canada, ed. H. J. Evers and J. E. Thorpe. Canadian Society of Petroleum Geologists, Calgary, Alta, Canada, pp. 30–4.

Davidson, D. W., El-Defraway, M. K., Fuglem, M. D. and Judge, A. S. (1978). Natural gas hydrates in northern Canada. *International Conference on Permafrost: Proceedings*, **1**, pp. 937–43.

Davidson, G. R., Bassett, R. L., Hardin, E. L. and Thompson, D. L. (1998). Geochemical evidence of preferential flow of water through fractures in unsaturated tuff, Apache Leap, Arizona. *Applied Geochemistry*, **13**, 185–95.

Davies, G. R. (1977). Carbonate-anhydrite facies relationships, Otto Fiord Formation (Mississippian-Pennsylvanian), Canadian Arctic Archipelago. *Studies in Geology (Tulsa)*, **5**, 145–67.

Davies, V. M. (1982). Interaction of thrusts and basement faults in the French external Alps. *Tectonophysics*, **88**, 325–31.

Davis, D. and Engelder, T. (1985). The role of salt in fold-and-thrust belts. *Tectonophysics*, **119**, 67–88.

Davis, D., Suppe, J. and Dahlen, F. A. (1983). Mechanics of fold and thrust belts and accretionary wedges. *Journal of Geophysical Research*, **88**, 1153–72.

Davis, D. M. and Lillie, R. J. (1994). Changing mechanical response during continental collision; active examples from the foreland thrust belts of Pakistan. *Journal of Structural Geology*, **16**, 21–34.

Davis, E. E., Hyndman, R. D. and Villinger, H. (1990). Rates of fluid expulsion across the northern Cascadia accretionary prism: constraints from new heat flow and multichannel seismic reflection data. *Journal of Geophysical Research*, **95**, 8869–91.

Davis, G. A., Zheng, Y., Wang, C., Darby, B. J., Zhang, C. and Gehrels, G. (2001). Mesozoic tectonic evolution of the Yanshan fold and thrust belt, with emphasis on Hebei and Lianoning provinces, northern China. In *Paleozoic and Mesozoic Tectonic Evolution of Central Asia; From Continental Assembly to Intracontinental Deformation*, ed. M. S. Hendrix and G. A. Davis, *Geological Society of America Memoir*, vol. 194. Boulder, CO: Geological Society of America, pp. 171–97.

Davis, P. N. (1983). Gippsland Basin, southeastern Australia. In *Seismic Expression of Structural Styles: a Picture and Work Atlas*, vol. 3, ed. A. W. Bally. *American Association of Petroleum Geologists Studies in Geology*, vol. 15, pp. 3.3-19–24.

Davis, R. W. (1991). Integration of geological data into hydrodynamic analysis of hydrocarbon movement. In *Petroleum Migration*, ed. W. A. England and A. J. Fleet. *Geological Society of London Special Publication*, vol. 59, pp. 127–35.

Davis, S. N. (1969). Porosity and permeability of natural materials. In *Flow Through Porous Media*, ed. R. J. M. DeWiest. New York: Academic Press, pp. 54–89.

Davis, T. L., Namson, J. and Yerkes, R. F. (1989). A cross section of the Los Angeles area: seismically active fold and thrust belt, the 1987 Whitier Narrows earthquake and earthquake hazard. *Journal of Geophysical Research*, **94**, 9644–64.

Davison, I. (1994). Linked fault systems: extensional, strike-slip and contractional. In *Continental Deformation*, ed. P. L. Hancock. Oxford: Pergamon Press, pp. 121–42.

Davy, P. and Cobbold, P. R. (1988). Indentation tectonics in nature and experiment, 1. Experiments scaled for gravity. *Bulletin of the Geological Institutions of the University of Uppsala*, **14**, 129–41.

Daw, G. P., Howell, F. T. and Woodward, F. A. (1974). The effect of applied stress upon the permeability of some Permian and Triassic sandstones of northern England: advances in rock mechanics. *Proceedings of the Third International Congress of Rock Mechanics*, vol. IIA, 537–42.

Dawson, F. M. (1999). Coalbed methane exploration in structurally complex terrain: a balance between tectonics and hydrogeology. In *Coalbed Methane: Scientific, Environmental and Economic Evaluation*, ed. M. Mastalerz, M. Glikson and S. D. Golding. Dordrecht: Kluwer Academic Publishers, pp. 111–21.

Day, S. M., Yu, G. and Wald, D. J. (1998). Dynamic stress changes during earthquake rupture. *Bulletin of the Seismologic Society of America*, **88**, 512–22.

De Azevedo, R. P. (1991). Tectonic evolution of Brazilian equatorial continental margin basins. Ph.D. thesis, Imperial College London.

De Boer, R. D., Nagtegaal, P. J. C. and Duyvus, E. M. (1977). Pressure solution experiments on quartz sand. *Geochimica et Cosmochimica Acta*, **41**, 249–56.

De Bremaeker, J. C. (1983). Temperature, subsidence and hydrocarbon maturation in extensional basins: a finite element model. *American Association of Petroleum Geologists Bulletin*, **67**, 1410–14.

de Cserna, Z. (1971). Development and structure of the Sierra Madre Oriental of Mexico. *Abstracts with Programs, Geological Society of America*, **3**, 377–8.

De Donatis, M. (2001). Three-dimensional visualization of the Neogene structures of an external sector of the Northern Apennines, Italy. *American Association of Petroleum Geologists Bulletin*, **85**, 419–31.

De Donatis, M., Mazzoli, S., Nesci, O., Santini, S., Savelli, D., Tramontana, M. and Veneri, F. (2001). Recent tectonics of the external zones of the Northern Apennines: evidence from the northern Marche foothills (Italy). *Biuletyn Panstwowego Instytutu Geologicznego*, **396**, 38–9.

De Graciansky, P. C., Dardeau, G., Lemoine, M. and Tricart, P. (1989). The inverted margin of the French Alps and foreland basin inversion. In *Inversion Tectonics*, ed. M. A. Cooper and G. D. Williams. *Geological Society of London Special Publications*, vol. 44, pp. 87–104.

De Jager, J., Doyle, M., Grantham, P. and Mabillard, J. (1993). Hydrocarbon habitat of the West Netherlands Basin.

American Association of Petroleum Geologists Bulletin, **77**, 1618.

De Marsily, G. (1981). *Hydrogeologie Quantitative*. Paris: Mason.

De Paor, D. G. (1988). Balanced section in thrust belts, Part 1: Construction. *American Association of Petroleum Geologists Bulletin*, **72**, 73–90.

De Paor, D. G. and Bradley, D. C. (1988). Balanced sections in thrust belts, Part 2: Computerized line and area balancing. *Geobyte*, **3**, 33–7.

De Witt Jr., W., Roen, J. B. and Wallace, L. G. (1993). Stratigraphy of Devonian black shales and associated rocks in the Appalachian basin. In *Petroleum Geology of the Devonian and Mississippian Black Shale of Eastern North America*, ed. J. B. Roen and R. C. Kepferle. *US Geological Survey Bulletin*, **1909**, B1–47.

Debelmas, J. (1989). On some key features of the evolution of the Western Alps. *NATO ASI Series, Series C: Mathematical and Physical Sciences*, vol. 259, pp. 23–42.

DeCelles, P. G. (1994). Late Cretaceous–Paleocene synorogenic sedimentation and kinematic history of the Sevier thrust belt, Northeast Utah and Southwest Wyoming. *Geological Society of America Bulletin*, **106**, 32–56.

DeCelles, P. G., Gray, M. B., Ridgway, R. B., Cole, P., Srivastava, P., Pequera, P. and Pivnik, D. A. (1991). Kinematic history of a foreland uplift from Paleocene synorogenic conglomerate, Beartooth Range, Wyoming and Montana. *Geological Society of America Bulletin*, **103**, 1458–75.

DeCelles, P. G. and Mitra, G. (1995). History of the Sevier orogenic wedge in terms of critical taper models, northeast Utah and southwest Wyoming. *Geological Society of America Bulletin*, **107**, 454–62.

Decker, J., Corrigan, J. and Bergman, S. (1996). Brookian maturation and erosion framework of North Alaska. *Annual Meeting Abstracts, American Association of Petroleum Geologists and Society of Economic Paleontologists and Mineralogists*, vol. 5, p. 34.

Dedeev, V., Aminov, L. and Schamel, S. (1994). Stratigraphic distribution of oil and gas resources of the Timan–Pechora basin. *International Geology Review*, **36**, 24–32.

Dellapé, D. and Hegedus, A. (1995). Structural inversion and oil occurrence in the Cuyo basin of Argentina. In *Petroleum Basins of South America*, ed. A. J. Tankard, R. Suárez-Soruco and H. J. Welsink. *American Association of Petroleum Geologists Memoir*, vol. 62, pp. 359–76.

Delville, N., Arnaud, N., Montel, J.-M., *et al.* (2001). Paleozoic to Cenozoic deformation along the Altyn Tagh Fault in the Altun Shan Massif area, eastern Qilian Shan, northeastern Tibet, China. In *Geological Society of America Memoir*, ed. M. S. Hendrix and G. A. Davis. Boulder, CO: Geological Society of America, vol. 194, pp. 269–92.

Demaison, G. (1984). The generative basin concept. In *Petroleum Geochemistry and Basin Evaluation*, ed. G. Demaison and R. J. Murris. *American Association of Petroleum Geologists Memoir*, vol. 35, pp. 1–14.

Demaison, G. and Huizinga, B. J. (1991). Genetic classification of petroleum systems. *American Association of Petroleum Geologists Bulletin*, **75**, 1626–43.

Dembicki Jr., H. and Anderson, M. L. (1989). Secondary migration of oil experiments supporting efficient movement of separate, buoyant oil phase along limited conduits. *American Association of Petroleum Geologists Bulletin*, **73**, 1018–21.

Demercian, S., Szatmari, P. and Cobbold, P. R. (1993). Style and pattern of salt diapirs due to thin-skinned gravitational gliding, Campos and Santos basins, offshore Brazil. *Tectonophysics*, **228**, 393–433.

Deming, D. (1993). Regional permeability estimates from investigations of coupled heat and groundwater flow, North Slope of Alaska. *Journal of Geophysical Research*, **98**, 16 271–86.

(1994a). Factors necessary to define a pressure seal. *American Association of Petroleum Geologists Bulletin*, **78**, 1005–9.

(1994b). Fluid flow and heat transport in the upper continental crust. In *Geofluids; Origin, Migration and Evolution of Fluids in Sedimentary Basins*, ed. J. Parnell. *Geological Society of London Special Publication*, vol. 78, pp. 27–42.

(1994c). Overburden rock, temperature, and heat flow. In *The Petroleum System – From Source to Trap*, ed. L. B. Magoon and W. G. Dow. *American Association of Petroleum Geologists Memoir*, vol. 60, pp. 165–86.

Deming, D. and Chapman, D. S. (1989). Thermal histories and hydrocarbon generation: example from Utah–Wyoming thrust belt. *American Association of Petroleum Geologists Bulletin*, **73**, 1455–71.

Deming, D. and Nunn, J. A. (1991). Numerical simulations of brine migration by topographically driven recharge. *Journal of Geophysical Research*, **96**, 2485–99.

Deming, D., Nunn, J. A. and Chapman, D. S. (1989). Thermal history of north-central Utah thrust belt; significance of sedimentation, thrusting, and compaction-driven groundwater flow. In *American Association of Petroleum Geologists Rocky Mountain Section Meeting, Abstracts. American Association of Petroleum Geologists Bulletin*, **73**, 1153.

Deming, D., Nunn, J. A. and Evans, D. G. (1990). Thermal effects of compaction-driven groundwater flow from overthrust belts. *Journal of Geophysical Research*, **95**, 6669–83.

Deming, D., Sass, J. H., Lachenbruch, A. H. and De Rito, R. F. (1992). Heat flow and subsurface temperature as evidence for basin-scale ground-water flow, North Slope of Alaska. *Geological Society of America Bulletin*, **104**, 528–42.

Dengo, C. A. and Covey, M. C. (1993). Structure of the Eastern Cordillera of Colombia; implications for trap styles and regional tectonics. *AAPG Bulletin*, **77**, 1315–37.

Dennison, J. M. (1971). Petroleum related to Middle and Upper Devonian deltaic facies in the central Appalachians. *AAPG Bulletin*, **55**, 1179–93.

Dercourt, J. L., Zonenshain, P., Ricou, L. E., *et al.* (1986). Geological evolution of the Tethys belt from the Atlantic to the Pamirs since the Lias. In *Evolution of the Tethys*, ed. J. Aubouin, X. Le Pichon and A. S. Monin. *Tectonophysics*, **123**, 241–315.

Dercourt, J., Cotiereau, N. and Vrielynck, B. (1993). Reconstruction of Tethys from Permian to Recent; implications for sedimentary facies distribution and oceanic circulation. In *American Association of Petroleum Geologists 1993 Annual Convention*, New Orleans, LA, *Annual Meeting Abstracts*, vol. 1993. Tulsa, OK: American Association of Petroleum Geologists and Society of Economic Paleontologists and Mineralogists, p. 91.

Desegaulx, P., Koyi, H., Cloetingh, S. and Moretti, I. (1991). Consequences of foreland basin development on thinned continental lithosphere: application to the Aquitaine basin (SW France). *Earth Planetary Science Letters*, **106**, 116–32.

Desrochers, A. and James, N. P. (1988). Early Paleozoic surface and subsurface paleokarst: Middle Ordovician carbonates, Mingan Islands, Quebec. In *Paleokarst*, ed. N. P. James and P. W. Choquette. New York: Springer-Verlag, pp. 183–210.

De Vera, J., McClay, K. R., Young, L. E., King, A. R. and Clark, J. L. (2001). Structural setting of the Red Dog District, western Brooks Range, Alaska. In *Geological Society of America, 2001 Annual Meeting*, Boston, MA, *Abstracts with Programs*, vol. 33. Boulder, CO: Geological Society of America, p. 272.

Dewey, J. F. (1993). Basins and granites in the Caledonides. *Seventh Meeting of the European Union of Geosciences, Abstract Supplement*, Terra Abstracts, **5**, Suppl. 1, p. 22.

Dewhurst, D. N., Brown, K. M., Clennell, M. B. and Westbrook, G. K. (1996). A comparison of the fabric and permeability anisotropy of consolidated and sheared silty clay. *Engineering Geology*, **42**, 253–67.

Dholakia, S. K., Aydin, A., Pollard, D. D. and Zoback, M. D. (1998). Fault-controlled hydrocarbon pathways in the Monterey Formation, California. *American Association of Petroleum Geologists Bulletin*, **82**, 1551–74.

Di Croce, J. (1995). Eastern Venezuela Basin: sequence stratigraphy and structural evolution. PhD Dissertation, Houston, TX: Rice University, Texas.

Dibblee, T. W., Jr (1982). Regional geology of the Transverse Ranges Province of Southern California. In *Geology and Mineral wealth of the California Transverse Ranges; Mason Hill volume*, ed. D. L. Fife and J. A. Minch. Santa Ana, CA: South Coast Geological Society, pp. 7–26.

Dickinson, G. (1953). Geological aspects of abnormal reservoir pressures in Gulf Coast, Louisiana. *American Association of Petroleum Geologists*, **37**, 410–32.

Diecchio, R. J. (1986). Taconian clastic sequence and general geology in the vicinity of the Allegheny Front in Pendleton County, West Virginia. In *Southeastern Section of the Geological Society of America Centennial Field Guide*, vol. 6. Boulder, CO: Geological Society of America, pp. 85–90.

Diegel, F. A. (1986). Topological constraints on imbricate thrust networks: examples from the Mountain City window, Tennessee, USA. *Journal of Structural Geology*, **8**, 269–79.

Dinares, J., McClelland, E. and Santanach, P. (1992). Contrasting rotations within thrust sheets and kinematics of thrust tectonics as derived from palaeomagnetic data: an example from the southern Pyrenees. In *Thrust Tectonics*, ed. K. R. McClay. London: Chapman and Hall, pp. 265–75.

Dix, G. R., Robinson, G. W. and McGregor, D. C. (1998). Paleokarst in the Lower Ordovician Beekmantown Group, Ottawa Embayment: structural control inboard of the Appalachian orogen. *Geological Society of America Bulletin*, **110**, 1046–59.

Dixon, J. S. (1982). Regional structural synthesis, Wyoming Salient of the Western Overthrust Belt. *American Association of Petroleum Geologists Bulletin*, **66**, 1560–80.

Dixon, J. S. and Liu, S. (1991). Centrifuge modeling of the propagation of thrust faults. In *Thrust Tectonics*, ed. K. McClay. London: Chapman and Hall, pp. 53–70.

Doglioni, C. (1992). The Venetian Alps thrust belt. In *Thrust Tectonics*, ed. K. R. McClay. London: Chapman & Hall, pp. 319–24.

Doglioni, C. (1993a). Geological evidence for a global tectonic polarity. *Journal of the Geological Society of London*, **150**, 991–1002.

(1993b). Some remarks on the origin of foredeeps. In *Crustal Controls on the Internal Architecture of Sedimentary Basins*, ed. R. A. Stephenson. Amsterdam: Elsevier, vol. 228, pp. 1–20.

Dolan, J. F., Sieh, K., Rockwell, T. R., Yeats, R. S., Shaw, J., Suppe, J., Huftile, G. J. and Gath, E. M. (1995). Prospects of larger or more frequent earthquakes in the Los Angeles Metropolitan Region, California. *Science*, **267**, 195–205.

Doligez, B., Bessi, F., Burrus, J., Ungerer, P. and Chenet, P. Y. (1986). Integrated numerical simulation of the sedimentation heat transfer, hydrocarbon formation and fluid migration in a sedimentary basin: the Themis model. In *Thermal Modeling in Sedimentary Basins*, ed. J. Burrus. Paris: Technip, pp. 173–95.

Domenico, P. A. and Palciauskas, V. V. (1973). Theoretical analysis of forced convective heat transfer in regional ground-water flow. *Geological Society of America Bulletin*, **84**, 3803–14.

Domenico, P. A. and Schwartz, F. W. (1990). *Physical and Chemical Hydrogeology*. New York: John Wiley and Sons.

Domingues, S., Lallemand, S. E. and Malavieille, J. (1994). New results from sandbox modeling of seamount subduction and possible applications. *EOS, Transactions, American Geophysical Union*, **75**, Suppl., 671.

Donath, F. A. (1961). Experimental study of shear failure in anisotropic rocks. *Geological Society of America Bulletin*, **76**, 985–99.

Donnell, J. R. (1961). Tripartition of the Wasatch Formation near de Beque in northwestern Colorado. In *Sphort Papers in the Geologic and Hydrologic Sciences. US Geological Survey Professional Paper*, **424-B**, 147.

Dooley, T. and McClay, K. (1997). Analog modeling of pull-apart basins. *American Association of Petroleum Geologists Bulletin*, **81**, 1804–26.

Dorobek, S. (1989). Migration of the orogenic fluid through the Siluro-Devonian Helderberg Group during late Paleozoic deformation: constraints on fluid sources and implications for thermal histories of sedimentary basins. *Tectonophysics*, **159**, 25–45.

Dover, J. H. (1994). Geology of part of east-central Alaska. In *The Geology of North America*, G. Plafker and H. C. Berg. Boulder, CO: Geological Society of America.

Dow, W. G. (1974). Application of oil-correlation and source-rock data to exploration in Williston basin. *American Association of Petroleum Geologists Bulletin*, **58**, 1253–62.

(1977). Kerogen studies and geological interpretations. *Journal of Geochemical Exploration*, 7, 79–99.

Downey, M. W. (1984). Evaluating seals for hydrocarbon accumulations. *American Association of Petroleum Geologists Bulletin*, **68**, 1752–63.

(1994). Hydrocarbon seal rocks. In *The Petroleum System – From Source to Trap*, ed. L. B. Magoon and W. G. Dow. *American Association of Petroleum Geologists Memoir*, vol. 60, pp. 159–64.

Drahovzal, J. A., Hertig, S. P. and Thomas, W. A. (1984). Basement faults and cover tectonics in southernmost Appalachians. *American Association of Petroleum Geologists Bulletin*, **68**, 471–2.

Drewes, H. (1988). Development of the foreland zone and adjacent terranes of the Cordilleran orogenic belt near the US–Mexican border. In *Interaction of the Rocky Mountain Foreland and the Cordilleran Thrust Belt*, ed. C. J. Schmidt and W. J. Perry Jr. *Geological Society of America Memoir*, vol. 171, pp. 447–63.

Dreyer, W. (1972). *The Science of Rock Mechanics: Series on Rock and Soil Mechanics*, vol. 1. Bay Village, Ohio: Trans. Tech. Publications.

Dron, D., Boulegue, J. and Mariotti, A. (1986). Biogeochemical constraints and fluid circulation in subduction zones off Japan. *EOS, Transactions, American Geophysical Union*, **67**, 1218.

Drucker, D. C. and Prager, W. (1952). Soil mechanics and plastics analysis for limited design. *Quarterly of Applied Mathematics*, **10**, 157–65.

Duddy, I. R., Green, P. F., Bray, R. J. and Hegarty, K. A. (1994). Recognition of the thermal effects of fluid flow in sedimentary basins. In *Geofluids: Origin, Migration and Evolution of Fluids in Sedimentary Basins*, ed. J. Parnell. *Geological Society of London Special Publication*, vol. 78, pp. 325–45.

Duddy, I. R., Green, P. F., Hegarty, K. A. and Bray, R. J.

(1991). Reconstruction of thermal history in basin modeling using apatite fission track analysis: what is really possible? *Offshore Australia Conference Proceedings*, vol. 1, pp. III-49–61.

Dula, W. F. (1981). Correlation between deformation lamellae, microfractures, macrofractures, and *in situ* stress measurements, White River Uplift, Colorado. *Geological Society of America Bulletin*, Part I, **92**, 37–46.

Dunlap, W. J. (1993). Direct dating of thrusting events: thermal, geochemical, and microstructural constraints. *EOS, Transactions, American Geophysical Union*, **74**, 547.

Dunn, J. F., Hartshorn, K. G. and Hartshorn, P. W. (1995). Structural styles and hydrocarbon potential of the sub-Andean thrust belt of southern Bolivia. In *Petroleum Basins of South America*, ed. A. J. Tankard, R. Suárez-Soruco and H. J. Welsink. *American Association of Petroleum Geologists Memoir*, vol. 62, pp. 523–43.

Dunne, W. M. and Ferrill, D. A. (1988). Blind thrust systems. *Geology*, **16**, 33–6.

Dunnington, H. V. (1958). Generation, migration, accumulation and dissipation of oil in Northern Iraq. In *Habitat of Oil: a Symposium, Tulsa, OK*. American Association of Petroleum Geologists, ed. G. L. Weeks, pp. 1194–251.

Durand, B. (1988). Understanding of HC migration in sedimentary basins (present state of knowledge). *Organic Geochemistry*, **13**, 445–59.

Durney, D. W. (1972). Solution-transfer: an important geological deformation mechanism. *Nature*, **235**, 315–17.

Dydik, B. M., Simoneit, B. R. T., Brassell, S. C. and Eglinton, G. (1978). Organic geochemical indicators of palaeoenvironmental conditions of sedimentation. *Nature*, **272**, 216–22.

Earnshaw, J. P., Hogg, A. J. C., Oxtoby, N. H. and Cawley, S. J. (1993). Petrographic and fluid inclusion evidence for the timing of diagenesis and petroleum entrapment in the Papuan Basin. In *Petroleum Exploration, Development and Production in Papua New Guinea: Proceedings of the Second PNG Petroleum Convention 31st May–2nd June*, ed. G. J. Carman and Z. Carman. Port Moresby: Petroleum Exploration and Development, pp. 459–75.

Echavarria, L., Hernández, R., Allmendinger, R. and Reynolds, J. (2003). Subandean thrust and fold belt of northwestern Argentina: geometry and timing of the Andean evolution. *American Association of Petroleum Geologists Bulletin*, **87**, 965–85.

Edmond, J. M. and Patterson, M. S. (1972). Volume changes during the deformation of rocks at high pressures. *International Journal of Rock Mechanics and Mining Sciences*, **9**, 161–82.

Edmund, W. and Eggleston, J. (1989). Characteristics of the Mid-Carboniferous boundary and associated coal-bearing rocks in the Appalachian basin. *28th International Geological Congress Fieldtrip Guidebook*, vol. T352. Washington, DC: American Geophysical Union, p. 118.

Ehrlich, R., Prince, C. and Carr, M. B. (1997). Sandstone reservoir assessment and production is fundamentally affected by properties of a characteristic porous microfabric. *SPE* **38 712**, 591–9.

EIA (2003). Iraq, U.S. Energy Information Agency Country Analysis Briefs.

Einsele, G., Ratschbacher, L. and Wetzel, A. (1996). The Himalaya–Bengal Fan denudation–accumulation system during the past 20 Ma. *Journal of Geology*, **104**, 163–84.

Eisbacher, G. H. (1976). Sedimentology of the Dezadeash Flysch and its implications for strike-slip faulting along the Denali Fault, Yukon Territory and Alaska. *Canadian Journal of Earth Sciences*, **13**, 1495–513.

Ekstroem, G. and England, P. (1989). Seismic strain rates in regions of distributed continental deformation. *Journal of Geophysical Research*, **94**, 10 231–57.

El Zarka, M. H. (1993). Ain Zalah field – Iraq Zagros Folded Zone, northern Iraq. In *Structural Traps*, vol. VIII, ed. N. H. Foster and E. A. Beaumont. *American Association of Petroleum Geologists Atlas of Oil and Gas Fields*, pp. 57–65.

El Zarka, M. H. and Ahmed, W. A. M. (1983). Formational water characteristics as an indicator for the process of oil migration and accumulation at the Ain Zalah field, northern Iraq. *Journal of Petroleum Geology*, **6**, 165–78.

Elias, B. P. and Hajash, A. (1992). Changes in quartz solubility and porosity due to effective stress: an experimental investigation of pressure solution. *Geology*, **20**, 451–4.

Elliot, W. C. and Aronson, J. L. (1987). Alleghenian episode of K-bentonite illitization in the southern Appalachian basin. *Geology*, **15**, 735–9.

Elliott, D. (1976a). The energy balance and deformation mechanisms of thrust sheets. *Philosophical Transactions of the Royal Society of London*, **A283**, 183–92.

(1976b). The motion of thrust sheets. *Journal of Geophysical Research*, **81**, 949–63.

(1977). Some aspects of the geometry and mechanics of thrust belts. *Canadian Society of Petroleum Geology, 8th Annual Seminar Publication*. Notes, Continuing Education Dept., University of Calgary, vol. 1, p. 2.

(1983). The construction of balanced cross-sections. *Journal of Structural Geology*, **5**, 101.

Elliott, D. and Johnson, M. R. W. (1980). Structural evolution in the northern part of the Moine thrust belt, NW Scotland. *Transactions of the Royal Society of Edinburgh: Earth Sciences*, **71**, 69–96.

Engelder, T. (1979). Mechanisms for strain within the upper Devonian clastic sequence of the Appalachian plateau, western New York. *American Journal of Science*, **279**, 527–42.

Engelder, T. and Engelder, R. (1977). Fossil distortion and décollement tectonics of the Appalachian plateau. *Geology*, **5**, 457–60.

England, P. and Molnar, P. (1990). Surface uplift, uplift of rocks, and exhumation of rocks. *Geology*, **18**, 1173–7.

England, P. C. and Richardson, S. W. (1977). The influence of erosion upon mineral facies of rocks from different metamorphic environments. *Journal of the Geological Society of London*, **134**, 201–13.

England, P. C. and Thompson, A. B. (1984). Pressure–temperature–time paths of regional metamorphism, I: Heat transfer during the evolution of regions of thickened continental crust. *Journal of Petrology*, **25**, 894–928.

England, W. A. (1994). Secondary migration and accumulation of hydrocarbons. In *The Petroleum System – From Source to Trap*, ed. L. B. Magoon and W. G. Dow. *American Association of Petroleum Geologists Memoir*, vol. 60, pp. 211–17.

England, W. A., Mackenzie, A. S., Mann, D. M. and Quigley, T. M. (1987). The movement and entrapment of petroleum fluids in the subsurface. *Journal of the Geological Society of London*, **144**, 327–47.

Englund, K. J., Arndt, H., Schweinfurth, S. and Gillespie, W. (1986). Pennsylvanian System stratotype sections, West Virginia. In *Southeastern Section of the Geological Society of America Centennial Field Guide*, vol. 6. Boulder, CO: Geological Society of America, pp. 59–68.

Epard, J. L. and Groshong, R. H. J. (1995). Kinematic model of detachment folding including limb rotation, fixed hinges and layer-parallel strain. In 30 years of Tectonophysics: a special volume in honour of Gerhard Oertel, ed. T. Engelder. *Tectonophysics*, **247**, 85–103.

Epstein, A. G., Epstein, J. B. and Harris, L. D. (1976). Conodont color alteration: an index to organic metamorphism. *US Geological Survey Professional Paper*, **995**, 1–27.

Erickson, S. G. (1995). Mechanics of triangle zones and passive-roof duplexes: implications for finite-element models. *Tectonophysics*, **245**, 1–11.

Eringen, A. C. (1967). *Mechanics of Continua*. New York: John Wiley & Sons.

Ernst, W. G. (1982). Mountain building and metamorphism: a case history from Taiwan. In *Mountain Building Processes*, ed. K. J. Hsu. San Diego: Academic Press, pp. 247–56.

Erslev, E. A. (1991). Trishear fault-propagation folding. *Geology*, **19**, 617–20.

Erslev, E. A. and Mayborn, K. R. (1997). Multiple geometries and modes of fault-propagation folding in the Canadian thrust belt. *Journal of Structural Geology*, **19**, 321–35.

Erslev, E. A. and Rogers, J. L. (1993). Basement-cover geometry of Laramide fault-propagation folds. *Geological Society of America Special Paper*, **280**, 125–46.

Erslev, E. A., Rogers, J. L. and Harvey, M. (1988). The northeastern Front Range revisited: horizontal compression and crustal wedging in a classic locality for vertical tectonics. In *Geological Society of America Field Trip Guide*, ed. G. S. Holden. *Colorado School of Mines Professional Contributions*, vol. 12, pp. 122–33.

Espitalié, J., Madec, M., Tissot, B., Menning, J. J. and Leplat, P. (1977). Source rock characterization method for petroleum exploration. *Ninth Annual Offshore Technology Conference*, pp. 439–48.

Esser, R. (2001). Discoveries of the 1990's: were they significant? In *Petroleum Provinces of the Twenty-first Century*, ed. M. W. Downey, J. C. Tereet and W. A. Morgan. *American Association of Petroleum Geologists Memoir*, vol. 74, pp. 35–43.

Etheridge, M. A. (1983). Differential stress magnitudes during regional deformation and metamorphism: upper bounds imposed by tensile fracturing. *Geology*, **11**, 231–4.

Eubank, R. and Makki, A. C. (1981). Structural geology of the central Sumatra back-arc basin. *Oil and Gas Journal*, **79**, 200–6.

Eva, A. N., Burke, D., Mann, P. and Wadge, G. (1989). Four-phase tectonostratigraphic development of the southern Caribbean. *Marine and Petroleum Geology*, **6**, 9–21.

Evans, B. and Dresen, G. (1991). Deformation of Earth materials: six easy pieces, U.S. Nat. Rep. Int. Union Geod. Geophys. 1987–1990. *Reviews of Geophysics*, **29**, 823–43.

Evans, D. G. and Nunn, J. A. (1989). Free thermohaline convection in sediments surrounding a salt column. *Journal of Geophysical Research*, **94**, 12413–22.

Evans, D. G., Nunn, J. A. and Hanor, J. S. (1991). Mechanisms driving groundwater flow near salt domes. In *Crustal-scale Fluid Transport; Magnitude and Mechanisms, Geophysical Research Letters*, vol. 18, ed. T. Torgensen, pp. 927–30.

Evans, J. P., Yang, J. and Martel, S. J. (1996). Three-dimensional structure, mineralization, and fluid flow along small strike-slip faults in granite. *Geological Society of America 28th Annual Meeting, Abstracts*, vol. 28, p. 254.

Evans, M. A. and Battles, D. A. (1999). Fluid inclusion and stable isotope analyses of veins from the central Appalachian Valley and Ridge province: Implications for regional synorogenic hydrologic structure and fluid migration. *Geological Society of America Bulletin*, **111**, 1841–60.

Evers, H. J. and Thorpe, J. E. eds. (1975). Structural geology of the Foothills between Savanna Creek and Panther River, SW Alberta, Canada. *Canadian Society of Petroleum Geologists and Canadian Society of Exploration Geophysicists Convention Guidebook*.

Fagin, S. (1995). Depth imaging's role in structural exploration. *World Oil*, **116**, 93–6.

Faill, R. T. (1987). The Birmingham window; Alleghanian decollement tectonics in the Cambrian–Ordovician succession of the Apppalachian Valley and Ridge Province, Birmingham, Pennsylvania. In *Northeastern Section of the Geological Society of America Centennial Field Guide*, vol. 5 Boulder, CO: Geological Society of America, pp. 37–42.

Fang, H., Yongchaun, S., Sitian, L. and Qiming, Z. (1995). Overpressure retardation of organic matter and petro-leum generation: a case study from the Yinggehai and Qwiongdongnan basins, South China Sea. *American Association of Petroleum Geologists Bulletin*, **79**, 551–62.

Farhoudi, G. (1978). A comparison of Zagros geology to island arcs. *Journal of Geology*, **86**, 323–34.

Ferguson, A. and McClay, K. R. (2000). Detached inversion: a new structural style applied to the Mahakam fold belt, Kutai Basin, East Kalimantan, Indonesia. *American Association of Petroleum Geologists Bulletin*, **84**, 1425.

Ferguson, I. J., Westbrook, G. K., Langseth, M. G. and Thomas, G. P. (1993). Heat flow and thermal models of the Barbados Ridge accretionary complex. *Journal of Geophysical Research*, **98**, 4121–42.

Fermor, P. R. and Moffat, I. W. (1992). Tectonics and structure of the Western Canadian Foreland Basin. In *Foreland Basin and Fold Belts*, ed. R. W. Macqueen and D. A. Leckie. *American Association of Petroleum Geologists Memoir*, vol. 55, pp. 81–105.

Fermor, P. R. and Price, R. A. (1976). Imbricate structures in the Lewis thrust sheet around the Cate Creek and Haig Brook windows, southeastern British Columbia. *Geological Survey of Canada Special Paper*, **76–1B**, Report of activities, Part B, 7–10.

(1987). Multiduplex structure along the base of the Lewis thrust sheet in the southern Canadian Rockies. *Bulletin of Canadian Petroleum Geology*, **35**, 159–85.

Ferrill, D. A. and Dunne, W. M. (1989). Cover deformation above a blind duplex: an example from West Virginia, U.S.A. *Journal of Structural Geology*, **11**, 421–31.

Ferrill, D. A., Stamatakos, J. A. and Sims, D. (1999). Normal fault corrugation: implications for growth and seismicity of active normal faults. *Journal of Structural Geology*, **21**, 1027–38.

Feybesse, J. and Milesi, J. (1994). The Archaean/Proterozoic contact zone in West Africa: a mountain belt of décollement thrusting and folding on a continental margin related to 2.1 Ga convergence of Archaean cratons. *Precambrian Research*, **69**, 199–227.

Finkbeiner, T., Barton, C. A. and Zoback, M. D. (1997). Relationships among *in-situ* stress, fractures and faults, and fluid flow: Monterey Formation, Santa Maria Basin, California. *American Association of Petroleum Geologists Bulletin*, **81**, 1975–99.

Fischer, D. M. and Anastasio, D. J. (1994). Kinematic analysis of large scale leading edge fold, Lost River Range, Idaho. *Journal of Structural Geology*, **16**, 337–54.

Fischer, D. M. and Byrne, T. (1990). The character and distribution of mineralized fractures in the Kodiak Formation, Alaska: implications for fluid flow in an underthrust sequence. In *Special Section on the Role of Fluids in Sediment Accretion, Deformation, Diagenesis, and Metamorphism in Subduction Zones*, ed. M. Langseth and J. C. Moore. *Journal of Geophysical Research*, **95**, pp. 9069–80.

Fischer, P. A. (2001). What's happening in production. *World Oil*, **222**, 27.

Fisher, A. T. and Hounslow, M. (1990). Transient fluid flow through the toe of the Barbados accretionary complex: constraints from Ocean Drilling Program Leg 110 heat flow studies and simple models. *Journal of Geophysical Research*, **95**, 8845–59.

Fisher, A. T., Iturrino, G., Cochrane, G., Langseth, M. and Hobart, M. (1992). Heat flow measurements along the Oregon accretionary margin: where's the fluid flow? *EOS, Transactions, American Geophysical Union*, **3**, 294.

Fisher, A. T. and Zwart, G. (1996). Permeability, fluid pressure, and effective stress in an active plate boundary fault zone: observations and models. *Annual Meeting Abstracts – American Association of Petroleum Geologists and Society of Economic Paleontologists and Mineralogists*, vol. 5, p. 45.

Fisher, A. T., Zwart, G., Shipley, T., *et al.* (1996). Relation between permeability and effective stress along a plate-boundary fault, Barbados accretionary complex. *Geology*, **24**, 307–10.

Fisher, Q. J. and Knipe, R. J. (1998). Fault sealing processes in siliciclastic rocks. In *Faulting, Fault Sealing and Fluid Flow in Hydrocarbon Reservoirs*, ed. G. Jones, Q. J. Fisher and R. J. Knipe. *Geological Society of London Special Publication*, vol. 147, pp. 117–34.

Flemings, P. B. and Jordan, T. E. (1989). A synthetic stratigraphic model of foreland basin development. *Journal of Geophysical Research*, **94**, 3851–66.

(1990). Stratigraphic modeling of foreland basins: interpreting thrust deformation and lithosphere rheology. *Geology*, **18**, 430–4.

Fletcher, R. C. (1974). Wavelength selection in the folding of a single layer with power-law rheology. *American Journal of Science*, **274**, 1029–43.

Fletcher, R. C. and Pollard, D. D. (1981). Anticrack model for pressure solution surfaces. *Geology*, **9**, 419–24.

Fletcher, R. C. and Sherwin, J. (1978). Arc lengths of single layer folds: a discussion of the comparison between theory and observation. *American Journal of Science*, **278**, 1085–98.

Flournoy, L. A. and Ferrell, R. E. (1980). Geopressure and diagenetic modifications of porosity in the Lirette field area, Terrebonne Parish, Louisiana. *Gulf Coast Geological Association Transactions*, **30**, 341–5.

Folk, R. L. (1960). Petrography and origin of the Tuscarora, Rose Hill and Keefer formations, Lower and Middle Silurian of eastern West Virginia. *Journal of Sedimentary Petrology*, **30**, 1–58.

Fossum, B. J., Schmidt, W. J., Jenkins, D. A., Bogatsky, V. I. and Rappoport, B. I. (2001). New frontiers for hydrocarbon production in the Timan–Pechora basin, Russia. In *Petroleum Provinces of the Twenty-first Century*, ed. M. W. Downey, J. C. Tereet and W. A. Morgan. *American Association of Petroleum Geologists Memoir*, vol. 74, pp. 259–79.

Foster, D. A. and Gray, D. R. (2000). Evolution and structure of the Lachlan fold belt (orogen) of eastern Australia. *Annual Review of Earth and Planetary Sciences*, **28**, 47–80.

Foster, D. A., Gray, D. R. and Offler, R. (1996). The western subprovince of the Lachlan Fold Belt. Victoria: structural style, geochronology, metamorphism, and tectonics. *Geological Society of Australia, Specialist Group in Geochemistry, Mineralogy, and Petrology Field Guide*, vol. 1, p. 89.

Foucher, J. P. and Tisseau, C. (1984). Thermal regime of Atlantic type continental margins: Bay of Biscay and Gulf of Lion. In *Thermal Phenomena in Sedimentary Basins*, ed. B. Durand. Paris: Editions Technip, pp. 221–5.

Foucher, J. P., Henry, P., Le Pichon, X. and Kobayashi, K. (1992). Time-variations of fluid expulsion velocities at the toe of the eastern Nankai accretionary complex. *Earth Planetary Science Letters*, **109**, 373–82.

Foucher, J. P., Henry, P., Sibuet, M., Kobayashi, K. and Le Pichon, X. (1990a). Heat flow and fluid budget at the toe of the Nankai accretionary prism near 138 E. *International Conference: Fluids in Subduction Zones, Abstracts*.

Foucher, J. P., Le Pichon, X., Lallemant, S., Hobart, M. A., Henry, P., Benedetti, J., Westbrook, G. K. and Langseth, M. G. (1990b). Heat flow, tectonics, and fluid circulation at the toe of the Barbados ridge accretionary prism. *Journal of Geophysical Research*, **95**, 8859–967.

Fox, F. G. (1959). Structure and accumulation of hydrocarbons in southern foothills, Alberta, Canada. *American Association of Petroleum Geologists Bulletin*, **43**, 992–1025.

Fraissinet, C., El Zouine, M., Morel, J.-L., Poisson, A., Andrieux, J. and Faure-Muret, A. (1988). Structural evolution of the southern and northern central High Atlas in Paleogene and Mio-Pliocene times. In *Lecture Notes in Earth Sciences*, vol. 15. Berlin: Springer, pp. 273–91.

France-Lanord, C., Derry, L. and Michard, A. (1993). Evolution of the Himalaya since Miocene time: isotopic and sedimentological evidence from Bengal fan. In *Himalayan Tectonics*, ed. P. J. Treloar and M. Searle. *Geological Society of London Special Publication*, vol. 74, pp. 603–21.

Frank, J. R., Cluff, S. and Bauman, J. M. (1982). Source and time of generation of hydrocarbons in the Fossil Basin, western Wyoming Thrust Belt. In *Geologic Studies of the Cordilleran Thrust Belt*, ed. R. B. Powers, Denver: Rocky Mountain Association of Geologists, vol. 2, pp. 601–11.

Franke, W. (1989). Tectonostratigraphic units in the Variscan belt of central Europe. *Geological Society of America Special Paper*, vol. 230, pp. 67–90.

Frape, S. K. and Fritz, P. (1987). Geochemical trends for groundwaters from the Canadian Shield. In *Saline Water and Gases in Crystalline Rocks*, ed. P. Fritz and S. K. Frape. *Geological Association of Canada Special Papers*, vol. 33, pp. 19–38.

Fraser, A. J., Nash, D. F., Steele, R. P. and Ebdon, C. C. (1990). A regional assessment of the intra-Carboniferous play of Northern England. In *Classic Petroleum Provinces*, ed. J. Brooks. *Geological Society of London Special Publication*, vol. 50, pp. 417.

Frebold, H. (1957). The Jurassic Fernie Group in the Canadian Rocky Mountains and Foothills. *Geological Survey of Canada Memoir*, **287**, 197.

Fredrich, J., McCaffrey, R. and Denham, D. (1988). Source parameters of seven large Australian earthquakes determined by body wave inversion. *Geophysical Journal of the Royal Astronomical Society*, **95**, 1–13.

Freeze, R. A. and Cherry, J. A. (1979). *Groundwater*. Englewood Cliffs, New Jersey: Prentice-Hall.

Freudenberg, U., Lou, S., Schlüter, R., Schütz, K. and Thomas, K. (1996). Main factors controlling coalbed methane distribution in the Ruhr District, Germany. In *Coalbed Methane and Coal Geology*, ed. R. Gayer and J. Harris. *Geological Society of London Special Publication*, vol. 109, pp. 67–88.

Friedman, M., Handin, J., Logan, J. M., Min, K. D. and Stearns, D. W. (1976a). Experimental folding of rocks under confining pressure, Part III: Faulted drape folds in multilithologic layered specimens. *Geological Society of America Bulletin*, **87**, 1049–66.

Friedman, M., Hugman, R. H. H. and Handin, J. (1980). Experimental folding of rocks under confining pressure, Part VIII: Forced folding of unconsolidated sand and lubricated layers of limestone and sandstone. *Geological Society of America Bulletin*, **91**, 307–12.

Friedman, M., Teufel, L. W. and Morse, J. D. (1976b). Strain and stress analyses from calcite twin lamellae in experimental buckles and faulted drape folds. *Philosophical Transactions of the Royal Society of London*, **4283**, 87–107.

Frisch, W., Kuhlemann, J., Dunkl, I. and Brugel, A. (1998). Palinspastic reconstruction and topographic evolution of the Eastern Alps during late Tertiary tectonic extrusion. *Tectonophysics*, **297**, 1–15.

Frizon de Lamotte, D., Mercier, E., Outtani, F., Addoum, B., Ghandriche, H., Ouali, J., Bouazziz, S. and Andrieux, J. (1998). Structural inheritance and kinematics of folding and thrusting along the front of the eastern Atlas Mountains, Algeria and Tunisia. In *Peri-Tethys Memoir 3; Stratigraphy and Evolution of Peri-Tethyan Platforms*, ed. S. Crasquin-Soleau and E. Barrier. *Memoires du Museum National d'Histoire Naturelle*, vol. 177, pp. 237–52.

Froitzheim, N., Stets, J. and Wurster, P. (1988). Aspects of western High Atlas tectonics. In *Lecture Notes in Earth Sciences*, vol. 15. Berlin: Springer, pp. 219–44.

Froitzheim, N., Müntener, O., Puschnig, A., Schmid, S. M. and Trommsdorff, V. (1996). Der penninisch-ostalpine Grenzbereich in Graubünden und in der Val Malenco; Bericht über die gemeinsame Exkursion der Schweizerischen Geologischen Gesellschaft, der Schweizerischen Mineralogischen und Petrographischen Gesellschaft und der Schweizerischen. *Fachgruppe der Geophysider*, vol. 8, *Eclogae Geologicae Helvetiae*, **89**, 617–34.

Fuis, G. S. and Kohler, W. M. (1984). Crustal structure and tectonics of the Imperial Valley region, California. *Field Trip Guidebook – Pacific Section: Society of Economic Paleontologists and Mineralogists*, vol. 40, pp. 1–13.

Furlong, K. P. and Chapman, D. S. (1987). Crustal heterogeneities and the thermal structure of the continental crust. *Geophysical Research Letters*, **14**, 314–17.

Furlong, K. P. and Edman, J. D. (1989). Hydrocarbon maturation in thrust belts: thermal considerations. In *Origin and Evolution of Sedimentary Basins and Their Energy and Mineral Resources: Geophysical Monograph*, ed. R. A. Price. Washington, DC: American Geophysical Union, vol. 48, pp. 137–44.

Furlong, K. P., Chapman, D. S. and Alfeld, P. W. (1982). Thermal modeling of the geometry of subduction with implications for the tectonics of the overriding plate. *Journal of Geophysical Research* B, **87**, 1786–802.

Galbraith, R. F. (1990). The radial plot: graphical assessment of spread in ages. *Nuclear Tracks*, **17**, 207–14.

Gallango, O., Novoa, E. and Bernal, A. (2002). The petroleum system of the central Perijá fold belt, western Venezuela. *American Association of Petroleum Geologists Bulletin*, **86**, 1263–84.

Gallup, W. B. (1954). Geology of the Turner Valley oil and gas field, Alberta, Canada. In *Western Canada Sedimentary Basins*, ed. L. M. Clarke. *American Association of Petroleum Geologists Bulletin*, 397–414.

Gao, J., Li, M., Xiao, X., Tang, Y. and He, G. (1998). Paleozoic tectonic evolution of the Tianshan Orogen, northwestern China. *Tectonophysics*, **287**, 213–31.

Garde, A. A., Chadwick, B., Grocott, J., Hamilton, M. A., McCaffrey, K. J. W. and Swager, C. P. (2002). Mid-crustal partitioning and attachment during oblique convergence in an arc system, Palaeoproterozoic Ketilidian Orogen, southern Greenland. *Journal of the Geological Society of London*, **159**, 247–61.

Garven, G. (1985). The role of regional fluid flow in the genesis of the Pine Point Deposit, western Canada sedimentary basin. *Economic Geology*, **80**, 307–24.

(1989). A hydrogeologic model for the formation of the giant oil sand deposits of the western Canada sedimentary basin. *American Journal of Science*, **289**, 105–66.

(1995). Continental-scale groundwater flow and geologic processes. *Annual Review of Earth and Planetary Sciences*, **23**, 89–117.

Garven, G. and Freeze, R. A. (1984a). Theoretical analysis of the role of groundwater flow in the genesis of stratabound ore deposits, 1, Mathematical and numerical model. *American Journal of Science*, **284**, 1085–124.

(1984b). Theoretical analysis of the role of groundwater flow in the genesis of stratabound ore deposits, 2, Quantitative result. *American Journal of Science*, **284**, 1125–74.

Garven, G., Ge, S., Person, M. A. and Sverjensky, D. A. (1993). Genesis of stratabound ore deposits in the Midcontinent Basins of North America, 1: The role of regional groundwater flow. *American Journal of Science*, **293**, 497–568.

Gaullier, V., Vendeville, B., Loncke, L., Maillard, A. and Mascle, J. (2003). Role of basement architecture during salt tectonics in the northwest and southeast Mediterranean: comparison between the Rhone and Nile deep-sea fans. *American Association of Petroleum Geologists Annual Meeting, Salt Lake City 2003, Official Program*, vol. 12, pp. A60–1.

Gaullier, V., Vendeville, B. C., Loncke, L. and Mascle, J. (2000). Gravity-driven tectonics in salt provinces with implications for thin-skinned vs. thick-skinned deformation in the Mediterranean. *EOS, Transactions, American Geophysical Union*, **81**, Suppl., 1224.

Gautier, D. L., Dolton, G. L., Takahashi, K. I. and Varnes, K. L. eds. (1996). *1995 National Assessment of United States Oil and Gas Resources: Results, Methodology, and Supporting Data*. Washington, DC: US Geological Survey Digital Data Series DDS-30, Release 2.

Gayer, R. A., Garven, G. and Rickard, D. (1998). Fluid migration and coal-rank development in foreland basins. *Geology*, **26**, 679–82.

Gayer, R. A., Greiling, R. O., Hecht, C. and Jones, J. A. (1993). Comparative evolution of Variscan coal-bearing foreland basins. In *Rhenohercynian and Subvariscan Folds Belts*, ed. R. A. Gayer, R. O. Greiling and A. K. Vogel. Braunschweig: Vieweg, pp. 47–82.

Gayer, R. A., Hathaway, T. and Nemčok, M. (1998). Transpressionally driven rotation in the external orogenic zones of the Western Carpathians and the SW British Variscides. *Geological Society Special Publications*, vol. 135, pp. 253–66.

Gayer, R. A. and Nemčok, M. (1994). Transpressionally driven rotation in the external Variscides of South-west Britain. *Proceedings of the Ussher Society*, **9**, 317–20.

Ge, S. and Garven, G. (1989). Tectonically induced transient groundwater flow in foreland basins. In *Origin and Evolution of Sedimentary Basins and Their Energy and Mineral Resources: Geophysical Monograph*, ed. R. A. Price. Washington, DC: American Geophysical Union, vol. 48, pp. 145–57.

(1992). Hydromechanical modeling of tectonically driven groundwater flow with application to the Arkoma foreland basin. *Journal of Geophysical Research*, **97**, 9119–44.

Geiser, P. A. (1974). Cleavage in some sedimentary rocks of the central Valley and Ridge province, Maryland. *Geological Society of America Bulletin*, **85**, 1399–412.

(1988a). Mechanisms of thrust propagation: some examples and implications for the analysis of overthrust terranes. *Journal of Structural Geology*, **10**, 829–45.

(1988b). The role of kinematics in the reconstruction and analysis of geological cross sections in deformed terranes. In *Geometries and Mechanism of Thrusting, with Special Reference to the Appalachians*, ed. G. Mitra and S. F. Wojtal. *Geological Society of America Special Paper*, vol. 222, pp. 47–76.

Geiser, P. A. and Engelder, T. (1983). The distribution of layer parallel shortening fabrics in the Appalachian foreland of New York and Pennsylvania: evidence of toe noncoaxial phases of Alleghanian orogeny. In *Contributions to the Tectonics and Geophysics of Mountain Chains*, ed. R. Hatcher, H. Williams, I. Zeitz. *Geological Society of America Memoir*, vol. 158, pp. 161–75.

Ghisetti, F., Kirschner, D. L., Vezzani, L. and Agosta, F. (2001). Stable isotope evidence for contrasting paleo-fluid circulation in thrust faults and normal faults of the central Apennines, Italy. *Journal of Geophysical Research*, **106**, 8811–25.

Gibbs, A. D. (1987). Linked tectonics of the northern North Sea basins. In *Sedimentary Basins and Basin-forming Mechanisms*, ed. C. Beaumont and A. J. Tankard. *Atlantic Geoscience Society Special Publication*, vol. 5, pp. 163–71.

Gieskes, J. M., Vrolijk, P. and Blanc, G. (1990). Hydrogeochemistry of the northern Barbados accretionary complex transect. Ocean Drilling Project leg 101. *Journal of Geophysical Research*, **95**, 8809–18.

Gieskes, J. M., Blanc, G., Vrolijk, P. *et al.* (1988). Hydrogeochemistry in the Barbados Accretionary Complex. *Ocean Drilling Program: Palaeogeography, Palaeoclimatology, Palaeoecology*, **71**, 83–96.

Giles, M. R. and de Boer, R. B. (1990). Origin and significance of redistributional secondary porosity. *Marine and Petroleum Geology*, **7**, 378–97.

Giles, M. R., Stevenson, S., Martin, S. V., Cannon, S. J. C., Hamilton, P. J., Marshall, J. D. and Samways, G. M. (1992). The reservoir properties and diagenesis of the Brent Group: a regional perspective. In *Geology of the Brent Group*, ed. A. C. Morton, R. S. Haszeldine, M. R. Giles and S. Brown. Bath: Geological Society of London, pp. 289–327.

Gill, W. D. (1953). Construction of geological sections of folds with steep-limb attenuation. *American Association of Petroleum Geologists Bulletin*, **37**, 2389–406.

Gilotti, J. A. (1989). Reaction progress during mylonitization of basaltic dikes along the Sarv thrust, Swedish Caledonides. *Contributions to Mineralogy and Petrology*, **101**, 30–45.

Girdler, R. W. (1970). A review of Red Sea heat flow. *Philosophical Transactions of the Royal Society of London*, **A267**, 191–203.

Glasspool, I. J., Edwards, D. and Axe, L. (2004). Charcoal in the Silurian as evidence for the earliest wildfire. *Geology*, **32**, 381–3.

Glen, R. A. (1990). Formation and thrusting in some Great Valley rocks near the Franciscan complex, California, and implications for the tectonic wedging hypothesis. *Tectonics*, **9**, 1451–77.

Gluyas, J. G. and Coleman, M. L. (1992). Material flux and porosity changes during sandstone diagenesis. *Nature*, **356**, 52–4.

Gobson, D. W. (1985). Stratigraphy, sedimentology and depositional environments of the coal-bearing Jurassic–

Cretaceous Kootnay Group, Alberta and British Columbia. *Geological Survey of Canada Bulletin*, **357**, 108.

Goddard, J. V. and Evans, J. P. (1995). Chemical changes and fluid-rock interaction in faults of crystalline thrust sheets, northwestern Wyoming, U.S.A. *Journal of Structural Geology*, **17**, 533–47.

Goetze, C. and Evans, B. (1979). Stress and temperature in the bending lithosphere as constrained by experimental rock mechanics. *Geophysical Journal of Royal Astronomical Society*, **59**, 463–78.

Goetze, G. B. (1978). The mechanisms of creep in olivine. *Philosophical Transactions of the Royal Society of London*, **288**, 99–119.

Goff, D. F. and Wiltschko, D. V. (1992). Stresses beneath a ramping thrust sheet. *Journal of Structural Geology*, **14**, 437–49.

Goff, D. F., Wiltschko, D. V. and Fletcher, R. C. (1996). Decollement folding as a mechanism for thrust-ramp spacing. *Journal of Geophysical Research*, **101**, 11 341–52.

Goguel, J. (1962). *Tectonics*. San Francisco: W. H. Freeman and Co.

Golonka, J. (2000). *Cambrian–Neogene Plate Tectonic Maps*. Krakow: Wydawnictwo Universitetu Jagiellonskiego.

Gonevchuk, V. G., Seltmann, R. and Gonevchuk, G. A. (2000). Tin mineralization and granites of the main ore districts of central Amur region, Russian Far East. In *Ore-bearing Granites of Russia and Adjacent Countries*, ed. A. A. Kremenetsky, B. Lehmann and R. Seltmann. Moscow: Institute of Mineralogy, Geochemistry and Crystal Chemistry of Rare Elements.

Goolsby, S. M., Druyff, L. and Fryt, M. S. (1988). Trapping mechanisms and petrophysical properties of the Permian Kaibab Formation, south-central Utah. In *Properties of Carbonate Reservoirs in the Rocky Mountain Region Occurrence and Petrophysical*, ed. S. M. Goolsby and M. W. Longman. Denver: Rocky Mountain Association of Geologists 1988 Guidebook, pp. 193–210.

Gordy, P. L., Frey, F. R. and Norris, D. K. (1977). *Geological Guide for CSPG 1977 Waterton-Glacier Park Field Conference*. Calgary: Canadian Society of Petroleum Geologists, p. 93.

Gordy, P. L., Frey, F. R. and Norris, D. K. (1982). Geology of the Waterton area, Alberta, Calgary, AB, Canada. *Canadian Society of Petroleum Geologists Field Trip No 2, American Association of Petroleum Geologists Annual Convention*.

Gosnold, Jr., W. D. (1985). Heat flow and groundwater flow in the Great Plains of the United States. *Journal of Geodynamics*, **4**, 247–64.

   (1990). Heat flow in the Great Plains of the United States. *Journal of Geophysical Research*, **95**, 353–74.

Graham, S. A. and Williams, L. A. (1985). Tectonic, depositional, and diagenetic history of Monterey Formation (Miocene), central San Joaquin basin, California.

American Association of Petroleum Geologists Bulletin, **69**, 385–411.

Grantz, A., Dinter, D. A. and Culotta, R. C. (1987). Geology of the continental shelf north of the Arctic National Wildlife Refuge, northeastern Alaska. In *Field Trip Guidebook – Pacific Section, Society of Economic Paleontologists and Mineralogists*, ed. I. L. Tailleur and P. Weimer. Los Angeles, CA: Society of Economic Paleontologists and Mineralogists, Pacific Section, vol. 50, pp. 759–63.

Grantz, A., May, S. D. and Hart, P. E. (1990). Geology of the Arctic continental margin of Alaska. In *The Arctic Ocean Region, The Geology of North America*, ed. A. Grantz, L. Johnson and J. F. Sweeney. Boulder: Geological Society of America, pp. 257–88.

   (1994). Geology of Arctic continental margin of Alaska. In *The Geology of Alaska*, ed. G. Plafker and H. C. Berg. Boulder: Geological Society of America, pp. 17–48.

Grauls, D. J. and Baleix, J. M. (1994). Role of overpressures and *in situ* stresses in fault-controlled hydrocarbon migration: a case study. *Marine and Petroleum Geology*, **11**, 734–42.

Gray, F. D. and Head, K. J. (2000). Fracture detection in the Manderson field: a 3D AVAZ case history. *The Leading Edge*, **19**, 1214–21.

Gregory, A. R. (1977). Aspects of rock physics from laboratory and log data that are important to seismic interpretation. In *Seismic Stratigraphy, Applications to Hydrocarbon Exploration*, ed. C. E. Payton. *American Association of Petroleum Geologists Memoir*, vol. 26, pp. 15–46.

Gretener, P. E. (1981a). Geothermics: using temperature in hydrocarbon exploration. *American Association of Petroleum Geologists Short Course Notes*, vol. 17, pp. 1–156.

   (1981b). Pore pressure, discontinuities, isostasy and overthrusts. In *Thrust and Nappe Tectonics*, ed. K. R. McClay and N. J. Price. *Geological Society of London Special Publications*, vol. 9, pp. 33–9.

Grette, J. F. and Coney, P. J. (1974). Absolute motion of the Eurasian Plate: a problem in vector geometry. *Geology*, **2**, 527–8.

Griboulard, R., Faugeres, J. C., Blanc, G., Gontier, E. and Vernette, G. (1989). Nouvelles evidences sedimentologiques et géochimiques de l'activité actuelle du prisme Sud Barbade. *Comptes Rendus de l'Académie des Sciences*, Ser. 2, **308**, 75–87.

Griggs, D. T. (1940). Experimental flow of rocks under conditions favouring recrystallisation. *Geological Society of America Bulletin*, **51**, 1001–22.

Grocott, J. and Pulverlaft, T. C. R. (1990). The early Proterozoic Rinkian Belt of central West Greenland. *Geological Association of Canada Special Paper*, vol. 37, pp. 443–63.

Groshong, Jr., R. H. (1975). Strain, fractures, and pressure solution in natural single-layer folds. *Geological Society of America Bulletin*, **86**, 1363–76.

(1988). Low temperature deformation mechanisms and their interpretation. *Geological Society of America Bulletin*, **100**, 1329–60.

Groshong, Jr., R. H. and Epard, J. L. (1992). New excess-area and depth-to-detachment relationship for fold-thrust structures. *American Association of Petroleum Geologists Annual Convention 1992*, Abstracts, p. 49.

(1994). The role of strain in area-constant detachment folding. *Journal of Structural Geology*, **16**, 613–18.

Groshong, Jr., R. H. and Usdansky, S. I. (1986). Deformation in thrust ramp anticlines and duplexes: implications for geometry and porosity. Unpublished Ph.D. thesis, Université de Grenoble.

Grunau, H. R. (1987). A world-wide look at the cap-rock problem. *Journal of Petroleum Geology*, **10**, 245–66.

Guglielmo, Jr., G., Vendeville, B. C. and Jackson, M. P. A. (2000). 3-D visualization and isochore analysis of extensional diapirs overprinted by compression. *American Association of Petroleum Geologists Bulletin*, **84**, 1095–108.

Guilhaumou, N., Larroque, C., Nicot, E., Roure, F. and Stephan, J. F. (1994). Mineralized veins resulting from fluid flow in décollement zones of the Sicilian prism: evidence from fluid inclusions. *Bulletin de la Société Géologique de France*, **165**, 425–36.

Guimera, J. (1983). Evolution de la déformation Alpine dans le N.E. de la chaine Iberique et dans la chaine côtière Catalane. *Comptes Rendus de l'Académie des Sciences Paris*, **297**, 425–30.

Guiraud, R. and Bellion, Y. (1996). Late Carboniferous to recent geodynamic evolution of the West Gondwanian cratonic Tethyan margins. In *The Tethys Ocean*, ed. A. E. M. Nairn, L.-E. Ricou, B. Vrielynck and J. Dercourt. New York: Plenum Press.

Guliyev, I. S., Kadirov, F. A., Reilinger, R. E., Gasanov, R. I. and Mamedov, A. R. (2002). Active tectonics in Azerbaijan based on geodetic, gravimetric, and seismic data. *Transactions of the Russian Academy of Sciences, Earth Science Section*, **383**, 174–7.

Gussow, W. C. (1954). Differential entrapment of gas and oil: a fundamental principle. *American Association of Petroleum Geologists Bulletin*, **38**, 816–53.

Gutscher, M. A., Kukowski, N., Malavieille, J. and Lallemand, S. (1996). Cyclical behaviour of thrust wedges: insights from high basal friction sandbox experiments. *Geology*, **24**, 135–38.

Gwin, V. E. (1964). Thin-skinned tectonics in the Plateau and northwestern Valley and Ridge provinces of the Central Appalachians. *Geological Society of America Bulletin*, **75**, 863–900.

Habib, D. (1982). Sediment supply origin of Cretaceous black shales. In *Nature and Origin of Cretaceous Carbon-rich Facies*, ed. S. O. Schlanger and M. B. Cita. London: Academic Press, pp. 113–27.

Hack, J. T. (1960). Interpretation of erosional topography in humid temperate climate. *American Journal of Science*, **258**, 80–97.

Haenel, R., Rybach, L. and Stegena, L., eds. (1988). *Handbook of Terrestrial Heat-flow Density Determination*. Dordrecht: Kluwer Academic Publishers.

Haeussler, P. J., Bruhn, R. L. and Pratt, T. L. (2000). Potential seismic hazards and tectonics of the upper Cook Inlet basin, Alaska: based on analysis of Pliocene and younger deformation. *Geological Society of America Bulletin*, **112**, 1414–29.

Hale-Erlich, W. and Coleman, J. L. (1993). Ouachita–Appalachian juncture: a Paleozoic transpressional zone in the southwestern USA. *AAPG Bulletin*, **77**, 552–68.

Hall, R. (1997). Cenozoic plate tectonic reconstructions of SE Asia. In *Petroleum Geology of Southeast Asia*, ed. A. J. Fraser, S. J. Matthews and R. W. Murphy. *Geological Society Special Publications*. London: Geological Society of London, vol. 126, pp. 11–23.

Hallam, A. (1984). Pre-Quaternary sea-level changes. *Annual Review of Earth and Planetary Science*, **12**, 205–43.

Hamburger, M. W., Sarewitz, D. R., Pavlis, T. L. and Popandopulo, G. A. (1992). Structural and seismic evidence for intracontinental subduction in the Peter the First Range, Central Asia. *Geological Society of America Bulletin*, **104**, 397–408.

Hamilton, W. B. (1978). Mesozoic tectonics of the western United States. *Pacific Coast Paleogeography Symposium*, **2**, 33–70.

(1988). Laramide crustal shortening. In *Interaction of the Rocky Mountain Foreland and the Cordilleran Thrust Belt*, ed. C. J. Schmidt and W. J. Perry Jr. *Geological Society of America Memoir*, vol. 171, pp. 27–39.

Han, M. W. and Suess, E. (1989). Subduction induced pore fluid venting and the formation of authigenic carbonates along the Cascadia continental margin: implications for the global Ca cycle. *Palaeogeography, Palaeoclimatology, Palaeoecology*, **71**, 97–118.

Hancock, P. L. (1985). Brittle microtectonics: principles and practice. *Journal of Structural Geology*, **7**, 437–57.

Handin, J., Hager, R. V., Friedman, M. and Feather, J. N. (1963). Experimental deformation of sedimentary rocks under confining pressure: pore pressure tests. *American Association of Petroleum Geologists Bulletin*, **47**, 717–55.

Hanks, C. L. (1993). Rapid evaluation of regional geometry and shortening of a fold and thrust belt: an example from near Brooks Range, Alaska, *AAPG Bulletin*, **77**, 19–28.

Hanks, C. L., Lorenz, J., Teufel, L. and Krumhardt, A. P. (1997). Lithologic and structural controls on natural fracture distribution and behavior within the Lisburne Group, Northeastern Brooks Range and North Slope subsurface, Alaska. *American Association of Petroleum Geologists Bulletin*, **81**, 1700–20.

Hanks, T. C., Bucknam, R. C., Lajoie, K. R. and Wallace, R. E. (1984). Modification of wave-cut and faulting-controlled landforms. *Journal of Geophysical Research*, **89**, 5771–90.

Hanor, J. S. (1979). Sedimentary genesis of hydrothermal fluids. In *Geochemistry of Hydrothermal Ore Deposits*, ed. H. L. Barnes. New York: John Wiley, pp. 137–68.

Haq, B. U., Hardenbol, J. and Vail, P. R. (1987). Chronology of fluctuating sea levels since the Triassic. *Science*, **235**, 1156–67.

Hardcastle, K. C. and Hills, S. L. (1991). BRUTE3 and SELECT: Quickbasic 4 programs for determination of stress tensor configurations and separation of heterogeneous populations of fault-slip data. *Computers & Geosciences*, **17**, 23–43.

Harding, T. P. (1976). Tectonic significance and hydrocarbon trapping consequences of sequential folding synchronous with San Andreas faulting, San Joaquin Valley, California. *American Association of Petroleum Geologists Bulletin*, **69**, 582–600.

(1985a). Seismic characteristics and identification of negative flower structures, positive flower structures, and positive structural inversion. *American Association of Petroleum Geologists Bulletin*, **69/4**, 582–600.

(1985b). Structural styles, plate tectonic settings, and hydrocarbon traps of divergent (transtensional) wrench faults. In *Strike Slip Deformation, Basin Formation, and Sedimentation*, ed. K. T. Biddle and N. Christie-Blick. *Society of Economic Paleontologists and Mineralogists. Special Publication*, vol. 37, pp. 51–77.

Hardy, S. and Ford, M. (1997). Numerical modeling of trishear fault propagation folding. *Tectonics*, **16**, 841–54.

Hardy, S., Poblet, J., McClay, K. R. and Waltham, D. (1996). Mathematical modeling of growth strata associated with fault-related fold structures. In *Modern Developments in Structural Interpretation, Validation and Modeling*, ed. P. G. Buchanan and D. A. Nieuwland. *Special Publication of Geological Society of London*, vol. 99, pp. 265–82.

Harp, E. L., Wells II., W. G. and Sarmiento, J. G. (1990). Pore pressure response during failure in soils. *Geological Society of America Bulletin*, **102**, 428–38.

Harris, L. D. and Milici, R. C. (1977). Characteristics of thin-skinned style of deformation in the southern Appalachians, and potential hydrocarbon traps. *USGS Professional Paper*, **1018**, 1–40.

Harrison, J. C. (1993). Salt involved tectonics of a foreland folded belt, Arctic Canada (abstract). *AAPG Hedgerg Conference on Salt Tectonics*. Tulsa, OK: American Association of Petroleum Geologists.

(1995). Tectonics and kinematics of a foreland folded belt influenced by salt, Arctic Canada. In *Salt Tectonics: a Global Perspective*, ed. M. P. A. Jackson, D. G. Roberts and S. Snelson. *American Association of Petroleum Geologists Memoir*, vol. 65, pp. 379–412.

Harrison, J. C., Fox, F. G. and Okalitch, A. V. (1991). Late Devonian–Early Carboniferous deformation of the Parry Island and Canrobert Hills fold belts, Bathurst and Melville Islands. In *Geology of the Innuitian Orogen and Arctic Platform of Canada*, ed. H. P. Trettin. Ottawa: Geological Survey of Canada, pp. 321–33.

Harrison, T. M., Copeland, P., Hall, S. A., Quade, J., Burner, S., Ojha, T. P. and Kidd, W. S. F. (1993). Isotopic preservation of Himalayan/Tibetan uplift, denudation, and climatic histories of two molasse deposits. *Journal of Geology*, **101**, 157–75.

Harry, D. L., Oldow, J. S. and Sawyer, D. S. (1995). The growth of orogenic belts and the role of crustal heterogeneities in décollement tectonics. *Geological Society of America Bulletin*, **107**, 1411–26.

Harting, T. A., Anderson, T., Birse, D. and Seale, R. A. (2003). World's first perforated Level-4 multilateral completion. *World Oil*, **224**, 27–32.

Hatcher, Jr., R. D. and Hooper, R. J. (1992). Evolution of crystalline thrust sheets in the internal parts of mountain chains. In *Thrust Tectonics*, ed. K. R. McClay. London: Chapman & Hall.

Hathaway, T. M. and Gayer, R. A. (1996). Thrust-related permeability in the South Wales Coalfield. In *Coalbed Methane and Coal Geology*, ed. R. Gayer and J. Harris. *Geological Society of London Special Publication*, vol. 109, pp. 121–32.

Hay, R. L., Lee, M., Kolata, D. R., Matthews, J. C. and Morton, J. P. (1988). Episodic potassic diagenesis of Ordovician tuffs in the Mississippi Valley area. *Geology*, **16**, 743–7.

Hayward, A. B. and Graham, R. H. (1989). Some geometrical characteristics of inversion. *Geological Society Special Publications*, **44**, 17–39.

Heald, M. T. (1955). Stylolites in sandstones. *Journal of Geology*, **63**, 101–14.

(1959). Significance of stylolites in permeable sandstones. *Journal of Sedimentary Petrology*, **29**, 251–3.

Hearn, P. P. J. and Sutter, J. F. (1985). Authigenic potassium feldspar in Cambrian carbonates: evidence of Alleghanian brine migration. *Science*, **228**, 1529–31.

Hearn, P. P. J., Sutter, J. F. and Belkin, H. E. (1987). Evidence for late-Paleozoic brine migration in Cambrian carbonate rocks of the central and southern Appalachians: implications for Mississippi Valley type sulfide mineralization. *Geochimica et Cosmochimica Acta*, **51**, 1323–34.

Hedberg, H. D. (1936). Gravitational compaction of clays and shales. *American Journal of Science*, **31**, 241–87.

Hedlund, C. A., Anastasio, D. J. and Fisher, D. M. (1994). Kinematics of fault-related folding in a duplex, Lost River Range, Idaho, U.S.A. *Journal of Structural Geology*, **16**, 571–84.

Heezen, B. C., MacGregor, I. D., Foreman, H. P. *et al.* (1973). Initial reports of the Deep Sea Drilling Project, covering Leg 20 of the cruises of the drilling vessel Glomar Challenger; Yokohama, Japan to Suva, Fiji, September–November 1971. In *Initial Reports of the Deep Sea Drilling Project*, vol. 20, ed. A. G. Kaneps, pp. 955–8.

Hefner, T. A. and Barrow, K. T. (1992). Rangely Field, U.S.A., Uinta/Piceance basins, Colorado. In *Structural Traps*, vol. VII, ed. E. A. Beaumont and N. H. Foster. *American Association of Petroleum Geologists Treatise* of Petroleum Geology, Atlas of Oil and Gas Fields, vol. A-25, pp. 29–56.

Heim, A. (1919). *Geologie der Schweiz*. Leipzig: Touchnitz.

Heling, D. and Teichmuller, M. (1974). The transition zone between montmorillonite and mixed-layer minerals and its relation to coalification in the Graue Beds of the Oligocene in the Upper Rhine Graben. *Fortschritte in der Geologie von Rheinland und Westfalen*, **24**, Inkohlung und Erdoel, 113–28.

Hemborg, H. T. (1993). PC-6: Weber Sandstone. In *Atlas of Major Rocky Mountain Gas Reservoirs*, ed. C. A. Hjellming. Socorro, NM: New Mexico Bureau of Mines & Mineral Resources, p. 104.

Henk, A. and Nemčok, M. (2000). Finite element models of the role of deformation on the thermal regime of thrustbelts. In *Systematics of Hydrocarbon Exploration and Production in Thrustbelts. EGI Technical Report* 50500459-3-00, ed. M. Nemčok, J. Collister, R. James and S. Schamel. Salt Lake City: Archive of Energy and Geoscience Institute at University of Utah.

(2003). Stress perturbations and strain localization in evolving fold-and-thrust structures: insights from numerical models (abstract). *American Association of Petroleum Geologists Program with Abstracts*, Salt Lake City, Utah, May 11–14, 2003, pp. A75.

Hennessey, W. J. (1975). A brief history of the Savanna Creek gas field. In *Structural Geology of the Foothills between Savanna Creek and Panther River, S.W. Alberta, Canada*, ed. H. J. Evers and J. E. Thorpe. Calgary, Alta: Canadian Society of Petroleum Geologists, pp. 18–21.

Hennings, P. H., Olson, J. E. and Thompson, L. B. (2000). Combining outcrop data and three-dimensional structural models to characterize fractured reservoirs: an example from Wyoming. *American Association of Petroleum Geologists Bulletin*, **84**, 830–49.

Henry, A. A. and Lewan, M. D. (2001). Comparison of kinetic-model prediction of deep gas generation. In *Geologic Studies of Deep Natural Gas Resources*, ed. T. S. Dyman and V. A. Kuuskraa. *US Geological Survey Digital Data Series*, vol. 67, pp. D1–25.

Henry, P., Foucher, J. P., Le Pichon, X., Lallemant, S. and Chamot-Rooke, N. (1990). Thermal modeling of clam colonies: fluid flow velocity estimates from Kaiko-Nankai thermal data. *International Conference: Fluids in Subduction Zones*, Abstracts.

Henry, P., Le Pichon, X., Lallemant, S., *et al.* (1996). Fluid flow in and around a mud volcano field seaward of the Barbados accretionary wedge: results from Manon cruise. *Journal of Geophysical Research*, **101**, 20 297–323.

Henry, P. and Wang, C.-Y. (1990). Modeling of fluid flow and pore pressure at the toe of the Oregon and Barbados accretionary wedges. *EOS, Transactions, American Geophysical Union*, **71**, 1576.

Herzer, R. H. (1981). Late Quaternary stratigraphy and sedimentation of the Canterbury continental shelf, New Zealand. *New Zealand Oceanographic Institute Memoir*, **89**, 1–71.

Hewitt, D. F. (1920). Measurement of folded bends. *Economic Geology*, **15**, 367–85.

Heyman, O. G. (1983). Distribution and structural geometry of faults and folds along the northwestern Uncompahgre Uplift, western Colorado and eastern Utah. In *Northern Paradox Basin–Uncompahgre Uplift*, ed. W. R. Averett. Grand Junction Colorado: Grand Junction Geological Society, pp. 45–57.

Hickman, S. H. and Evans, B. (1987). Diffusional crack healing in calcite. The influence of crack geometry upon healing rate. *Physical Chemical Minerology*, **15**, 91–102.

Hickman, S., Sibson, R. H. and Bruhn, R. (1995). Introduction to special section: Mechanical involvement of fluids in faulting. *Journal of Geophysical Research*, **100**, 12 831–40.

Hill, K. C. (1991). Structure of the Papuan fold belt, Papua New Guinea. *AAPG Bulletin*, **75**, 857–872.

Hindle, A. D. (1997). Petroleum migration pathways and charge concentration: a three-dimensional model. *American Association of Petroleum Geologists Bulletin*, **81**, 1451–81.

Hirono, T. and Ogawa, Y. (1998). Duplex arrays and thickening of accretionary prisms: an example from Boso Peninsula, Japan. *Geology*, **26**, 779–82.

Hitchon, B. (1984). Geothermal gradients, hydrodynamics, and hydrocarbon occurrences, Alberta, Canada. *American Association of Petroleum Geologists Bulletin*, **68**, 713–43.

Hite, R. J. and Anders, D. E. (1991). Petroleum and evaporites. In *Evaporites, Petroleum and Mineral Resources*, ed. J. L. Melvin. Amsterdam: Elsevier, pp. 349–411.

Hjellming, C. A., ed. (1993). *Atlas of Major Rocky Mountain Gas Reservoirs*. Socorro, NM: New Mexico Bureau of Mines & Mineral Resources.

Hobson, D. M. (1986). A thin-skinned model for the Papuan thrust belt and some implications for hydrocarbon exploration. *Australian Petroleum Exploration Association Journal*, **26**, 214–24.

Hoholick, J. D., Metarko, T. and Potter, P. E. (1984). Regional variation of porosity and cement: St. Peter and Mount Simon sandstones in Illinois Basin. *American Association of Petroleum Geologists Bulletin*, **68**, 753–64.

Holl, J. E. and Anastasio, D. J. (1993). Paleomagnetically derived folding rates, southern Pyrenees, Spain. *Geology*, **21**, 271–4.

Homza, T. X. and Wallace, W. K. (1997). Detachment folds with fixed hinges and variable detachment depth, northeastern Brooks Range, Alaska. *Journal of Structural Geology*, **19**, 337–54.

Hood, J. W. and Fields, F. K. (1978). Water resources of the northern Uinta Basin Area – Utah and Colorado – with special emphasis on ground-water supply. State of Utah Department of Natural Resources.

Hooper, E. C. D. (1991). Fluid migration along growth faults in compacting sediments. *Journal of Petroleum Geology*, **14**, 161–80.

Horsfield, B. and Rullkötter, J. (1994). Diagenesis, catagenesis, and metagenesis of organic matter. In *The Petroleum System – From Source to Trap*, ed. L. B. Magoon and W. G. Dow. *American Association of Petroleum Geologists Memoir*, vol. 60, pp. 189–99.

Hoshino, K. (1972). Brittle fracturing of non-foliated rocks. *International Geological Congress*, Abstracts, *Congrès Géologique Internationale*, Resumés, vol. 24, pp. 81–2.

Hoshino, K., Koide, H., Inami, K., Iwamura, S. and Mitsui, S. (1972). Mechanical properties of Japanese Tertiary sedimentary rocks under high confining pressures. *Geological Survey of Japan*, **244**, 200.

Hossack, J. R. (1979). The use of balanced cross-sections in the calculation of orogenic contraction: a review. *Journal of the Geological Society of London*, **136**, 705–11.

(1983). A cross-section through the Scandinavian Caledonides constructed with the aid of branchline maps. *Journal of Structural Geology*, **5**, 103–11.

House, W. M. and Gray, D. R. (1982). Cataclasites along the Saltville thrust, U.S.A. and their implications for thrust-sheet emplacement. *Journal of Structural Geology*, **4**, 257–69.

Houseknecht, D. W. (1987). Assessing the relative importance of compaction processes and cementation to reduction of porosity in sandstones. *American Association of Petroleum Geologists Bulletin*, **71**, 633–42.

(1988). Intergranual pressure solution in four quartzose sandstones. *Journal of Sedimentary Petrology*, **58**, 228–46.

Housen, B. A., Tobin, H. J., Labaume, P., *et al.* (1996). Strain decoupling across the décollement of the Barbados accretionary prism. *Geology*, **24**, 127–30.

Hrouda, F. (1982). Magnetic anisotropy of rocks and its application in geology and geophysics. *Geophysical Surveys*, **5**, 37–82.

Hubbard, M. S. (1989). Thermobarometric constraints on the thermal history of the Main Central thrust zone and Tibetan Slab, eastern Nepal Himalaya. *Journal of Metamorphic Geology*, **7**, 19–30.

Hubbard, M. S., Grew, E. S., Hodges, K. V., Yates, M. G. and Pertsev, N. N. (1999). Neogene cooling and exhumation of upper-amphibolite-facies 'whiteschists' in the Southwest Pamir Mountains, Tajikistan. In *Tectonics of Continental Interiors*, ed. S. Marshak, B. A., van der Pluijm and M. Hamburger. *Tectonophysics*, **305**, 325–37.

Hubbert, M. K. (1940). The theory of ground-water motion. *Journal of Geology*, **48**, 785–944.

(1953). Entrapment of petroleum under hydrodymanic conditions. *American Association of Petroleum Geologists Bulletin*, **37**, 1954–2026.

(1969). *The Theory of Ground-water Motion and Related Papers*. New York: Hafner Publishing Company.

Huc, A. Y. (1988a). Aspects of depositional processes of organic matter in sedimentary basins. In *Advances in Organic Geochemistry 1987, Part I, Organic Geochemistry in Petroleum Exploration: Proceedings of the 13th International Meeting on Organic Geochemistry*, ed. L. Mattavelli and L. Novelli. *Organic Geochemistry*, vol. 13, pp. 263–72.

(1988b). Sedimentology of organic matter. In Humic substances and their role in the environment: report of the Dahlem Workshop, ed. F. H. Frimmel and R. F. Christman. *Life Sciences Research Reports*, **41**, 215–43.

Huchon, P. and Le Pichon, X. (1984). Sunda Strait and Central Sumatra Fault. *Geology*, **12**, 668–72.

Hudleston, P. J. (1973a). Fold morphology and some geometrical implications of theories of fold development. *Tectonophysics*, **16**, 1–46.

(1973b). The analysis and interpretation of minor folds developed in the Moine rocks of Monar, Scotland. *Tectonophysics*, **17**, 89–132.

Hudleston, P. J. and Lan, L. (1993). Information from fold shapes. *Journal of Structural Geology*, **15**, 253–64.

(1994). Rheological control on the shapes of single-layer folds. *Journal of Structural Geology*, **16**, 1007–21.

Huerta, A. D., Royden, L. and Hodges, K. (1996). The interdependence of deformational and thermal processes in mountain belts. *Science*, **273**, 637–9.

Hull, C. E. and Warman, H. R. (1970). Asmari oil fields of Iran. In *Geology of Giant Petroleum Fields*, ed. M. T. Halbouty. *American Association of Petroleum Geologists Memoir*, vol. 14, pp. 428–37.

Humayon, M., Lillie, R. J. and Lawrence, R. D. (1991). Structural interpretation of the eastern Sulaiman foldbelt and foredeep, Pakistan. *Tectonics*, **10**, 299–324.

Hunt, J. M. (1979). *Petroleum Geochemistry and Geology*. San Francisco: W. H. Freeman & Co.

(1990). Generation and migration of petroleum from abnormally pressured fluid compartments. *American Association of Petroleum Geologists Bulletin*, **74**, 1–12.

(1991). Generation of gas and oil from coal and other terrestrial organic matter. *Organic Geochemistry*, **17**, 673–80.

Hunter, R. B. (1988). Timing and structural interaction between the thrust belt and foreland Hoback Basin, Wyoming. In *Interaction of Rocky Mountain Foreland and the Cordilleran Thrust Belt*, ed. C. J. Schmidt and W. J. Perry, *GSA Memoir*, vol. 171. Boulder, CO: Geological Society of America, pp. 367–94.

Hunziker, J. C., Frey, H., Clauer, N., Dallmayer, R. D., Freidrichsen, H., Roggwiller, P. and Schwander, H. (1986). The evolution of illite to muscovite: mineralogical and isotopic data from the Glarus Alps, Switzerland. *Contributions in Mineral Petrology*, **92**, 157–80.

Hurford, A. J., Fitch, F. J. and Clarke, A. (1984). Resolution of the age structure of the detrital zircon populations of two Lower Cretaceous sandstones from the Weald of England by fission track dating. *Geological Magazine*, **121**, 269–396.

Hurford, A. J., Flisch, M. and Jager, E. (1989). Unraveling the thermo-tectonic evolution of the Alps: a contribution from fission-track analysis and mica-dating. In *Alpine*

Tectonics, ed. M. P. Coward, D. Dietrich and R. G. Park. *Special Publication of Geological Society of London*, vol. 45, pp 369–98.

Hutchinson, I. (1985). The effects of sedimentation and compaction on oceanic heat flow. *Geophysical Journal of the Royal Astronomic Society*, **82**, 439–59.

Ibraham, S. S. M. (1977). Stratigraphy of Pakistan. In *Geological Survey of Pakistan Memoir*, vol. 12. Quetta: Geological Survey of Pakistan, p. 138.

Ibrmajer, J., Suk, M. *et al.* (1989). *Geophysical Image of ČSSR*. Prague: ÚÚG (in Czech).

IES (1993). In Poelchau, H. S., Baker, D. R., Hantschel, T., Horsfield, B. and Wygrala, B. (1997). Basin simulation and the design of the conceptual basin model. In *Petroleum and Basin Evolution: Insights from Petroleum Geochemistry, Geology and Basin Modeling*, ed. D. H. Welte, B. Horsfield and D. R. Baker. Berlin: Springer, pp. 3–70.

Iller, R. K. (1979). *The Chemistry of Silica*. New York: John Wiley and Sons.

Inners, J. D. (1987). Upper Paleozoic stratigraphy along the Allegeny topographic front at the Horseshoe Curve, west-central Pennsylvania. In *Northeastern Section of the Geological Society of America Centennial Field Guide*, vol. 5. Boulder, CO: Geological Society of America, pp. 29–36.

Isaacs, C. M. and Garrison, R. E., eds. (1983). *Petroleum Generation and Occurrence in the Miocene Monterey Formation, California. Guidebook.* Los Angeles: SEPM, Pacific Section.

Jackson, J. and Fitch, T. (1981). Basement faulting and the focal depths of the larger earthquakes in the Zagros Mountains (Iran). *Geophysical Journal of the Royal Astronomical Society*, **64**, 561–86.

Jackson, J. A. (1980). Reactivation of basement faults and crustal shortening in orogenic belts. *Nature*, **283**, 343–6.

Jackson, J. A., Fitch, T. J. and McKenzie, D. P. (1981). Active thrusting and the evolution of the Zagros Fold Belt. In *Thrust and Nappe Tectonics*, ed. K. R. McClay and N. J. Price. *Geological Society of London Special Publication*, vol. 9, pp. 371–80.

Jackson, M., McCabe, C., Ballard, M. M. and Van der Voo, R. (1988). Magnetite authigenesis and diagenetic paleotemperatures across the northern Appalachian Basin. *Geology*, **16**, 592–5.

Jacob, H. and Kuckelhorn, K. (1977). The coalification profile of the Miesbach 1 well and its interpretation for oil geology. *Oil and Gas Magazine*, **93**, 115–23.

Jacobeen, Jr., F. and Kanes, W. H. (1974). Structure of the Broadtop synclinorium and its implications for Appalachian structural style. *American Association of Petroleum Geologists Bulletin*, **58**, 362–75.

(1975). Structure of the Broadtop synclinorium, Wills Mountain anticlinorium, and Allegheny frontal zone. *American Association of Petroleum Geologists Bulletin*, **59**, 1136–50.

Jadoon, I. A. K., Lawrence, R. D. and Lillie, R. J. (1992). Balanced and retrodeformed geological cross-section from the frontal Sulaiman Lobe, Pakistan; duplex development in thick strata along the western margin of the Indian Plate. In *Thrust Tectonics 1990*, Egham, ed. K. R. McClay. London: Chapman & Hall.

Jadoon, I. A. K., Lawrence, R. D. and Lillie, R. J. (1994). Seismic data, geometry, evolution, and shortening in the active Sulaiman fold-and-thrust belt of Pakistan, southwest of the Himalayas. *American Association of Petroleum Geologists Bulletin*, **78**, 758–74.

Jaeger, J. C. (1960). Shear failure of anisotropic rocks. *Geological Magazine*, **97**, 65–72.

Jaeger, J. C. and Cook, N. G. W. (1976). *Fundamentals of Rock Mechanics*. London: Chapman and Hall.

Jaillard, E., Herail, G., Monfret, T. and Worner, G. (2002). Andean geodynamics: main issues and contributions from the 4th ISAG, Gottingen. *Tectonophysics*, **345**, 1–15.

James, N. P., Stevens, R. K., Barnes, C. R. and Knight, I. (1989). Evolution of a Lower Paleozoic continental-margin carbonate platform, Northern Canadian Appalachians. In *Controls on Carbonate Platform and Basin Development: Special Publication*, vol. 44, ed. P. D. Crevello, J. L. Wilson, J. F. Sarg and J. F. Read. Tulsa: Society of Economic Paleontologists and Mineralogists, pp. 123–46.

James, R. A. (1995). Pore structure and petroleum reservoir characteristics of arcosic sandstones with consideration of the tectonic context of diagenesis: Examples from Pennsylvania and West Siberia. Ph.D. thesis, Columbia: University of South Carolina, SC.

Jamieson, R. A. and Beaumont, C. (1988). Orogeny and metamorphism: a model for deformation and pressure–temperature–time paths with application to the central and southern Appalachians. *Tectonics*, **7**, 417–45.

(1989). Deformation and metamorphism in convergent orogens: a model for uplift and exhumation of metamorphic terrains. *Geological Society Special Publications*, vol. 43, pp. 117–29.

Jamison, J. W. (1987). Geometric analysis of fold development in overthrust terranes. *Journal of Structural Geology*, **9**, 207–19.

Jamison, W. R. (1979). Laramide deformation of the Wingate Sandstone, Colorado National Monument: a study of cataclastic flow. Ph.D. thesis, Texas: A&M University, College Station.

(1992). Stress controls on fold thrust style. In *Thrust Tectonics*, ed. K. R. McClay. London: Chapman and Hall, pp. 155–64.

(1996). Mechanical models of triangle zone evolution. In *Triangle Zones and Tectonic Wedges*, ed. P. A. MacKay, J. L. Varsek, T. E. Kubli, R. G. Dechesne, A. C. Newson and J. P. Reid. *Bulletin of Canadian Petroleum Geology*, vol. 44, pp. 180–94.

(1997). Quantitative evaluation of fractures on Monkswood Anticline: a detachment fold in the foothills of Western Canada. *American Association of Petroleum Geologists Bulletin*, **81**, 1110–32.

Jamison, W. R. and Pope, A. (1996). Geometry and evolution of a fault-bend fold, Mount Bertha Anticline. *Geological Society of America Bulletin*, **108**, 208–24.

Jamison, W. R. and Stearns, D. W. (1982). Tectonic deformation of Wingate Sandstone, Colorado National Monument. *American Association of Petroleum Geologists Bulletin*, **66**, 2584–608.

Jaswal, T. M., Lillie, R. J. and Lawrence, R. D. (1997). Structure and evolution of the northern Potwar deformed zone, Pakistan. *American Association of Petroleum Geologists Bulletin*, **81**, 308–28.

Jaupart, C. (1983). Horizontal heat transfer due to radioactivity contrasts: causes and consequences of the linear heat flow relation. *Geophysical Journal of the Royal Astronomic Society*, **75**, 411–35.

Jessop, A. M. and Majorowicz, J. A. (1994). Fluid flow and heat transfer in sedimentary basins. In *Geofluids: Origin, Migration and Evolution of Fluids in Sedimentary Basins*, ed. J. Parnell. *Geological Society of London Special Publications*, vol. 78, pp. 43–54.

Ji, J., Zhong, D., Sang, H. and Zhang, L. (2000). The western boundary of extrusion blocks in the southeastern Tibetan Plateau. *Chinese Science Bulletin*, **45**, 876–81.

Jiang, Z., Oliver, N. H. S., Barr, T. D., Power, W. L. and Ord, A. (1997). Numerical modeling of fault-controlled fluid flow in the genesis of tin deposits of the Malage ore field, Gejiu mining district, China. *Economic Geology and the Bulletin of the Society of Economic Geologists*, **92**, 228–47.

Johnson, A. M. and Page, B. M. (1976). A theory of concentric, kink and sinusoidal folding and of monoclinal flexuring of compressible, elastic multilayers VII: Development of folds within Huasna syncline, San Luis Obispo County, California. *Tectonophysics*, **33**, 97–143.

Johnson, B. D., Powell, C. M. and Veevers, J. J. (1976). Spreading history of the eastern Indian Ocean and Greater India's northward flight from Antarctica and Australia. *Geological Society of America Bulletin*, **87**, 1560–6.

Johnson, D. D. and Beaumont, C. (1995). Preliminary results from a planform kinematic model of orogen evolution, surface processes and the development of clastic foreland basin stratigraphy. In *Stratigraphic Evolution of Foreland Basins: Special Publications*, vol. 12, ed. S. L. Dorobek and G. M. Ross. Tulsa: Society for Sedimentary Geology, pp. 3–24.

Johnson, K. M. and Johnson, A. M. (2002a). Mechanical models of trishear-like folds. *Journal of Structural Geology*, **24**, 277–87.

(2002b). Mechanical analysis of the geometry of forced-folds. *Journal of Structural Geology*, **24**, 401–10.

Johnson, N. M., Jordan, T. E., Johnson, P. A. and Naeser, C. W. (1986). Magnetic polarity stratigraphy, age and tectonic setting of fluvial sediments in an eastern Andean foreland basin, San Juan Province, Argentina. In *Foreland Basins: International Association of Sedimentologists Special Publication*, vol. 8, ed. P. Allen and P. Homewood, pp. 63–75.

Johnson, N. M., Stix, J., Tauxe, L., Cerveny, P. F. and Tahirkheli, R. A. K. (1985). Paleomagnetic chronology, fluvial deposits, and tectonic implications of the Siwalik deposits near Chinji Village, Pakistan. *Journal of Geology*, **93**, 27–40.

Johnsson, M. J. (1986). Distribution of maximum burial temperatures across northern Appalachian basin and implications for Carboniferous sedimentation patterns. *Geology*, **14**, 384–7.

Joint Chalk Research (1996). Geology, rock mechanics, rock properties and improved oil recovery in chalk of the Danish and Norwegian sectors of the North Sea. *Joint Chalk Research, Phase IV Monograph*.

Jolivet, M., Brunel, M., Seward, D., *et al.* (2001). Mesozoic and Cenozoic tectonics of the northern edge of the Tibetan Plateau; fission-track constraints. *Tectonophysics*, **343**, 111–34.

Jones, P. B. (1982). Oil and gas beneath east-dipping underthrust faults in the Alberta Foothills, Canada. In *Geologic Studies of the Cordilleran Thrust Belt*, ed. R. B. Powers. Denver: Rocky Mountains Association of Geologists, pp. 61–74.

Jones, P. B. and Linsser, H. (1986). Computer synthesis of balanced structural cross-sections by forward modeling. *American Association of Petroleum Geologists Bulletin*, **65**, 605.

Jordan, T. E. (1995). Retroarc foreland and related basins. In *Tectonics of Sedimentary Basins*, ed. C. J. Busby and R. V. Ingersoll. Cambridge: Blackwell Science, Inc., pp. 331–62.

Jordan, T. E., Allmendinger, P. W., Damanti, J. F. and Drake, R. E. (1993). Chronology of motion in a complete thrust belt: the Precordillera, 30–31 deg. S, Andes Mountains. *Journal of Ecology*, **101**, 135–56.

Jordan, T. E. and Flemings, P. B. (1991). Large-scale stratigraphic architecture, eustatic variation, and unsteady tectonism: a theoretical evaluation. In *Special Section on Long-term Sea Level Changes*, ed. S. A. P. L. Cloetingh. *Journal of Geophysical Research*, **96**, pp. 6681–99.

Jowett, E. C., Cathles III., L. M. and Davis, B. W. (1993). Predicting depths of gypsum dehydration in evaporitic sedimentary basins. *American Association of Petroleum Geologists Bulletin*, **77**, 402–13.

Joyce, J. E., Tjalsma, L. R. C. and Prutzman, J. M. (1990). High-resolution planktic stable isotope record and spectral analysis for the last 5.35 m.y.: Ocean drilling Program Site 625, northeast Gulf of Mexico. *Paleoceanography*, **5**, 507–29.

Julivert, M. (1970). Cover and basement tectonics in the Cordillera Oriental of Colombia, South America, and a comparison with some other folded chains. *Geological Society of America Bulletin*, **81**, 3623–46.

Kaldi, J. G. and Atkinson, C. D. (1997). Evaluating seal potential: example from the Talang Akar Formation, offshore northwest Java, Indonesia. In *Seals, Traps, and the Petroleum System*, ed. R. C. Surdam. *American Association of Petroleum Geologists Memoir*, vol. 67, pp. 85–101.

Kalsbeek, F., Pulvertaft, T. C. R. and Nutman, A. P. (1998). Geochemistry: age and origin of metagreywackes from the Palaeoproterozoic Karrat Group, Rinkian Belt, West Greenland. *Precambrian Research*, **91**, 383–99.

Kapp, P., Murphy, M. A., Yin, A., Harrison, T. M., Ding, L. and Guo, J. (2003). Mesozoic and Cenozoic tectonic evolution of the Shiquanhe area of western Tibet. *Tectonics*, **22**, in press.

Kappelmeyer, O. and Haenel, R. (1974). *Geothermics with Special Reference to Application*. Berlin: Gebrueder Borntraeger.

Karabinos, P. (1984a). Deformation and metamorphism on the east side of the Green Mountain massif in southern Vermont. *Geological Society of America Bulletin*, **95**, 584–93.

(1984b). Polymetamorphic garnet zoning from southwestern Vermont. *American Journal of Science*, **95**, 584–93.

Karabinos, P. and Ketcham, R. (1988). Thermal structure of active thrust belts. *Journal of Metamorphic Geology*, **6**, 559–70.

Karig, D. E. (1986). Physical properties and mechanical state of accreted sediments in the Nankai Trough, Southwest Japan Arc. In *Structural Fabric in Deep Sea Drilling Project Cores from Forearcs*, ed. J. C. Moore. *Geological Society of America Memoir*, vol. 166, pp. 117–36.

(1990). Experimental and observational constraints on the mechanical behaviour in the toes of accretionary prisms. In *Deformation Mechanisms, Rheology and Tectonics*, ed. R. J. Knipe and E. H. Rutter. *Geological Society of London, Special Publications*, vol. 54, pp. 383–98.

Karig, D. E. and Morgan, K. (1990). A dynamically sealed décollement: Nankai Prism. *EOS, Transactions, American Geophysical Union*, **71**, 1626–7.

Karner, G. D. (2000). Rifts of the Campos and Santos basins, southeastern Brazil: distribution and timing. In *Petroleum Systems of South Atlantic Margins*, ed. M. R. Mello and B. J. Katz. *American Association of Petroleum Geologists Memoir*, vol. 73, pp. 301–15.

Karner, G. D., Driscoll, N. W. and Weissel, J. K. (1993). Response of the lithosphere to in-plane force variations. *Earth and Planetary Science Letters*, **114**, 397–416.

Karnkowski, P. (1999). *Oil and Gas Deposits in Poland*. Cracow: The Geosynoptics Society, University of Mining and Metallurgy.

Katz, B. J. (1983). Limitations of 'Rock-Eval' pyrolysis for typing organic matter. *Organic Geochemistry*, **4**, 195–9.

(1995a). A survey of rift basin source rocks. In *Hydrocarbon Habitat in Rift Basins*, ed. J. J. Lambiase. *Geological Society of London Special Publication*, vol. 80, pp. 213–42.

(1995b). Petroleum source rocks: an introductory overview.

In *Petroleum Source Rocks*, ed. B. J. Katz. Berlin: Springer-Verlag, pp. 1–8.

Katz, B. J., Kelley, P. A., Royle, R. A. and Jorjorian, T. (1991). Hydrocarbon products of coals as revealed by pyrolysis-gas chromatography. *Organic Geochemistry*, **17**, 711–22.

Kazantsev, Y. V. and Kamaletdinov, M. A. (1977). Salt tectonics in the southern part of the Cis-Uralian foredeep and its connection with thrusts. *Geotectonics*, **11**, 294–8.

Keen, C. E. and Potter, D. P. (1995). Formation and evolution of the Nova Scotia rifted margin: evidence from deep seismic reflection data. *Tectonics*, **14**, 918–32.

Keith, C. M., Simpson, D. W. and Soboleva, O. V. (1982). Induced seismicity and style of deformation at Nurek Reservoir, Tadjik SSR. *Journal of Geophysical Research B*, **87**, 4609–24.

Kelley, J. S. and Foland, R. L. (1987). Structural style and framework geology of the coastal plain and adjacent Brooks Range. In *Petroleum Geology of the Northern Part of the Arctic National Wildlife Refuge, Northeastern Alaska*, ed. K. J. Bird and L. B. Magoon. *US Geological Survey Bulletin*, Reston, VA: US Geological Survey, vol. B 1778, pp. 255–70.

Kellogg, J. N. and Bonini, W. E. (1982). Subduction of the Caribbean Plate and basement uplifts in the overriding South American Plate. *Tectonics*, **1**, 251–76.

Kelson, K. I., Simpson, G. D., Van Arsdale, R. B., Haraden, C. C. and Lettis, W. R. (1996). Multiple Late Holocene earthquakes along the Reelfoot fault: Central New Madrid seismic zone. *Journal of Geophysical Research*, **101**, 6151–70.

Kemp, A. E. S. (1990). Fluid flow in 'vein structures' in Peru forearc basins: evidence from back-scattered electron microscope studies. *Proceedings, Scientific reports, ODP, Leg 112, Peru continental margin*, pp. 33–41.

Kent, W. N., Hickman, R. G. and Dasgupta, U. (2002). Application of a ramp/flat-fault model to interpretation of the Naga thrust and possible implications for petroleum exploration along the Naga thrust front. *American Association of Petroleum Geologists Bulletin*, **86**, 2023–45.

Kenyon, P. M. and Turcotte, D. L. (1985). Morphology of a delta prograding by bulk sediment transport. *Geological Society of America Bulletin*, **96**, 1457–65.

Kepferle, R. C. (1993). A depositional model and basin analysis for the gas-bearing black shale (Devonian and Mississippian) in the Appalachian basin. In *Petroleum Geology of the Devonian and Mississippian Black Shale of Eastern North America*, ed. J. B. Roen and R. C. Kepferle. *US Geological Survey Bulletin*, **1909**, pp. F1–23.

Kerr, J. W. (1977). Cornwallis fold belt and the mechanism of basement uplift. *Canadian Journal of Earth Sciences*, **14**, 1374–401.

Khan, F. A. and Hasany, S. T. (1998). Dhulian oilfield: a case study. *Pakistan Petroleum Convention*, pp. 1–20.

King, G. C. P. and Vita-Finzi, C. (1981). Active folding in the

Algerian earthquake of 10 October 1980. *Nature*, **292**, 22–6.

Kinoshita, H. and Yamano, M. (1986). The heat flow anomaly in the Nankai Trough area. *Initial Report of the Deep Sea Drilling Program*, **87**, 737–43.

Kirby, S. H. (1983). Rheology of the lithosphere. *Reviews of Geophysics*, **21**, 1458–87.

Kirby, S. H. and Kronenberg, A. K. (1987). Rheology of the lithosphere: selected topics. *Reviews of Geophysics*, **25**, 1219–44.

Kirschner, D. L. and Kennedy, L. A. (2001). Limited syntectonic fluid flow in carbonate-hosted thrust faults of the Front Ranges, Canadian Rockies, inferred from stable isotope data and structures. *Journal of Geophysical Research*, **106**, 8827–40.

Kissin, Y. V. (1987). Catagenesis and composition of petroleum: origin of n-alkanes and isoalkanes in petroleum crudes. *Geochimica et Cosmochimica Acta*, **51**, 2445–57.

Klemme, H. D. and Ulmishek, G. F. (1991). Effective petroleum source rocks of the world: Stratigraphic distribution and controlling depositional factors. *American Association of Petroleum Geologists Bulletin*, **75**, 1809–51.

Klett, T. R., Ahlbrandt, T. S., Schmoker, J. W. and Dolton, G. L. (1997). Ranking of the world's oil and gas provinces by known petroleum volumes. *US Geological Survey Open File Report* 97-463.

Kley, J. and Eisbacher, G. H. (1999). How Alpine or Himalayan are the Central Andes? *International Journal of Earth Sciences*, **88**, 175–89.

Kligfield, R., Geiser, P. and Geiser, J. (1986). Construction of geologic cross-sections using microcomputer system. *Geobyte*, **1**, 60–6.

Klinger, R. E. and Rockwell, T. K. (1989). Flexural-slip folding along the Eastern Elmore Ranch Fault in the Superstition Hills Earthquake sequence of November 1987. *Bulletin of Seismological Society of America*, **79**, 297–303.

Knight, L., Martin, J. and McAlee, J. (2001). Horizontal wells extend the Tulare Sands play in Belridge field, California. *World Oil*, **222**, 93–4.

Knipe, R. J. (1985). Footwall geometry and the rheology of thrust sheets. *Journal of Structural Geology*, **7**, 1–10.

(1986). Faulting mechanisms in slope sediments: examples from deep sea drilling project cores. In *Structural Fabric in Deep Sea Drilling Project Cores from Forearcs*, ed. J. C. Moore. *Geological Society of America Memoir*, vol. 166, pp. 45–54.

(1989). Deformation mechanisms: recognition from natural tectonites. *Journal of Structural Geology*, **11**, 127–46.

(1993). The influence of fault zone processes and diagenesis on fluid flow. In *Diagenesis and Basin Development*, ed. A. D. Horbury and A. Robinson. *American Association of Petroleum Geologists Studies in Geology*, vol. 36, pp. 135–51.

Knipe, R. J., Agar, S. M. and Prior, D. J. (1991). The microstructural evolution of fluid flow paths in semi-lithified sediments from subduction complex. In *The Behaviour*

and Influence of Fluids in Subduction Zones, ed. J. Tarney, K. T. Pickering, R. J. Knipe and J. F. Dewey. *Philosophical Transactions of the Royal Society of London*, **335**, 261–73.

Kong, F. (1998). Continental margin deformation analysis and reconstruction; evolution of the East China Sea basin and adjacent plate interaction. Ph.D Thesis, Austin, TX: University of Texas.

Koons, P. O. (1987). Some thermal and mechanical consequences of rapid uplift: an example from the Southern Alps, New Zealand. *Earth Planetary Science Letters*, **86**, 307–19.

(1990). The two-sided orogen: collision and erosion from the sand box to the Southern Alps, New Zealand. *Geology*, **18**, 679–82.

Koons, P. O. and Craw, D. (1991). Evolution of fluid driving forces and composition within collisional orogens. *Geophysical Research Letters*, **18**, 935–8.

Koopman, A., Speksnijder, A. and Horsfield, W. T. (1987). Sandbox model studies of inversion tectonics. *Tectonophysics*, **137**, 379–88.

Kopietz, J. and Jung, R. (1978). Geothermal *in situ* experiments in the Asse salt-mine. In *Seminar on In Situ Heating Experiments in Geological Formations*. Paris: OECD Publications.

Kopp, M. L. (1997). *Struktury Lateral'nogo Vyzhimaniya v Al'piysko-Gimalayskom Kollizionnom Poyase. Structures of Lateral Escape in the Alpine-Himalayan Collision Belt*. Moscow: Nauka.

Kováč, M., Baráth, I. and Nemčok, M. (1992). Bericht 1991 ueber geologische Aufnahmen im Quartaer und Tertiaer im Suedoestlichen Teil des Wiener Beckens auf Blatt 77 Eisenstadt. *Jahrbuch der Geologischen Bundensanstalt*, vol. 135, pp. 701–3.

Koyi, H. (1995). Mode of internal deformation in sand wedges. *Journal of Structural Geology*, **17**, 293–300.

(1997). Analogue modeling: from a qualitative to a quantitative technique, a historical outline. *Journal of Petroleum Geology*, **20**, 223–38.

Kranz, R. L. (1983). Microcracks in rocks: a review. In *Continental Tectonics; Structure, Kinematics and Dynamics*, ed. M. Friedman and M. N. Toksoez. *Tectonophysics*, **100**, 449–80.

Kranz, R. L. and Scholz, C. H. (1977). Critical dilatant volume of rocks at the onset of tertiary creep. *Journal of Geophysical Research*, **82**, 4893–8.

Krejčí, J. J. (1993). Ždánice-Krystalinikum field, Czechoslovakia, Carpathian foredeep, Moravia. In *Structural Traps*, ed. N. N. Foster and E. A. Beaumont. *American Association of Petroleum Geologists Atlas of Oil and Gas Fields*, vol. VIII, pp. 153–75.

Krooss, B. M., Brothers, L. and Engel, M. H. (1991). Geochromatography in petroleum migration: a review. In *Petroleum Migration*, ed. W. A. England and A. J. Fleet. *Geological Society of London Special Publication*, vol. 59, pp. 149–63.

Kruijs, E. and Barron, E. (1990). Climatic model prediction of paleoproductivity and potential source-rock distribution. In *Deposition of Organic Facies*, ed. A. Y. Huc. *American Association of Petroleum Geologists Studies in Geology*, vol. 30, pp. 195–216.

Krus, S. and Šutora, A., eds. (1986). Geophysical–geological atlas of the Alpine–Carpathian mountain system. *Report, Archive Geofyzika Brno*.

Krzywiec, P. (1997). Large-scale tectono-sedimentary middle Miocene history of the central and eastern Polish Carpathian foredeep basin: results of seismic data interpretation. In *Dynamics of the Pannonian–Carpathian–Dinaride System*; PANCARDI 97, Przeglad Geologiczny, **45**, Part 2, ed. M. Krobicki and W. Zuchiewicz, pp. 1039–53.

Kukal, Z. (1990). The rate of geological processes. *Earth Science Reviews*, vol. 28. Amsterdam: Elsevier.

Kukowski, N., Lallemand, S. E., Malavieille, J., Gutscher, M. A. and Reston, T. (2002). Mechanical decoupling and basal duplex formation observed in sandbox experiments with application to the western Mediterranean Ridge accretionary complex. In *The Accretionary Complex of the Mediterranean Ridge; Tectonics, Fluid Flow, and the Formation of Brine Lakes: Marine Geology*, vol. 186, ed. G. K. Westbrook and T. Reston, pp. 29–42.

Kukowski, N., Von Huene, R. R., Malavieille, J. and Lallemand, S. E. (1994). Sediment accretion against a buttress beneath the Peruvian continental margin at 12 degrees S as simulated with sandbox modeling. *Geologische Rundschau*, **83**, 822–31.

Kulander, B. R. and Dean, S. L. (1986). Structure and tectonics of Central and Southern Appalachian Valley and Ridge and Plateau Provinces, West Virginia and Virginia. *American Association of Petroleum Geologists Bulletin*, **70**, 1674–84.

Kulik, D. M. and Schmidt, C. J. (1988). Region of overlap and styles of interaction of Cordilleran thrust belt and Rocky Mountain Foreland. *Geological Society of America Memoir*, vol. 171, pp. 75–98.

Kulm, L. D. and Suess, E. (1990). Relationship between carbonate deposits and fluid venting: Oregon accretionary prism. Special section on the Role of fluids in sediment accretion, deformation, diagenesis, and metamorphism in subduction zones. *Journal of Geophysical Research*, **95**, 8899–915.

Kuuskraa, V. A. and Bank, G. C. (2003). Gas from tight sands, shales a growing share of US supply. *Oil & Gas Journal*, **101**, 34–43.

Kvenvolden, K. A. and McMeanamin, M. A. (1980). Hydrates of natural gas: a review of their geologic occurrence. *US Geological Survey Circular*, vol. 825, pp. 1–11.

Lachenbruch, A. H. and Sass, J. H. (1977). Heat flow in the United States and the thermal regime of the crust. In *The Earth's Crust, its Nature and Physical Properties*, ed. J. G. Heacock. *American Geophysical Union, Geophysical Monograph*, vol. 20, pp. 626–75.

Lafargue, E. and Barker, C. (1988). Effect of water washing on crude oil compositions. *American Association of Petroleum Geologists Bulletin*, **72**, 263–76.

Lallemand, S. E. (1995). High rates of arc consumption by subduction processes: some consequences. *Geology*, **23**, 551–4.

Lallemand, S. E., Gutscher, M. A., Kukowski, N. and Malavieille, J. (1995). Material transfer, accretion and deformation observed in sandbox experiments including weak horizons. *EOS, Transactions, American Geophysical Union*, **76**, Suppl., 534.

Lallemand, S. E. and Malavieille J. (1992). L'erosion profonde des continents (Deep erosion of continents). *La Recherche*, 23, 1388–97.

Lallemand, S. E., Mallavieille J. and Calassou, S. (1992). Effects of oceanic ridge subduction on accretionary wedges: experimental modeling and marine observations. *Tectonics*, **11**, 1301–13.

Lallemand, S. E., Schnuerle, P. and Malavieille, J. (1994). Coulomb theory applied to accretionary and nonaccretionary wedges: possible causes for tectonic erosion and/or frontal accretion. *Journal of Geophysical Research*, **99**, 12033–55.

Lallemant, S. J., Chamot, R. N., Henry, P., Le Pichon, X., Lallemand, S. E. and Cadet, J. P. (1990a). Tectonic context of fluid venting at the toe of the eastern Nankai accretionary prism (Kaiko–Nankai diving cruise). *EOS, Transactions, American Geophysical Union*, **71**, 1627.

Lallemant, S. J., Henry, P., Le Pichon, X. and Foucher, J. P. (1990b). Detailed structure and possible fluid paths at the toe of the Barbados accretionary wedge (ODP Leg 110 area). *Geology*, **18**, 854–7.

Lama, R. D. and Vutukuri, V. S. (1978). *Handbook on Mechanical Properties of Rocks, Series on Rock and Soil Mechanics*, vol. 2. Bay Village: Trans Tech Publications.

Lamb, S., Hoke, L., Kennan, L. and Dewey, J. (1997). Cenozoic evolution of the Central Andes in Bolivia and Northern Chile. In *Orogeny Through Time*, ed. J.-P. Burg and M. Ford. *Geological Society of London Special Publication*, vol. 121, pp. 237–64.

Lamb, M. A., Hanson, A. D., Graham, S. A., Badarch, G. and Webb, L. E. (1999). Left-lateral sense offset of upper Proterozoic to Paleozoic features across the Gobi Onon, Tost, and Zuunbayan faults in southern Mongolia and implications for other Central Asian faults. *Earth and Planetary Science Letters*, **173**, 183–94.

Lambe, T. W. and Whitman, R. V. (1969). *Soil Mechanics*. New York: John Wiley.

Lambeck, K., McQueen, H. W. S., Stephenson, R. A. and Denham, D. (1984). The state of stress within the Australian continent. *Annales Geophysicae*, **2**, 723–41.

Lamerson, P. R. (1982). The Fossil basin and its relationship to the Absaroka thrust fault system, Wyoming and Utah. In *Geologic Studies of the Cordilleran Thrust Belt*, vol. 1, ed. R. B. Powers. Rocky Mountain Association of Geologists, pp. 279–340.

Lane, L. S. (1996). Geometry and tectonics of early Tertiary triangle zones, northeastern Eagle Plain, Yukon

Territory. In *Triangle Zones and Tectonic Wedges*, ed. P. A. MacKay, T. E. Kubli, A. C. Newson, J. L. Varsek, R. G. Dechesne and J. P. Reid. *Bulletin of Canadian Petroleum Geology*, **44**, 337–48.

Langseth, M. G. and Moore, J. C. (1990). Introduction to special section on the role of fluids in sediment accretion, deformation, diagenesis, and metamorphism in subduction zones. *Journal of Geophysical Research*, **95**, 8737–41.

Langseth, M. G., Westbrook, G. K. and Hobart, M. (1990). Contrasting geothermal regimes of the Barbados ridge accretionary complex. *Journal of Geophysical Research*, **95**, 1049–61.

Larroque, C., Guilhaumou, N., Stephan, J. F. and Roure, F. (1996). Advection of fluids at the front of the Sicilian Neogene subduction complex. *Tectonophysics*, **254**, 41–55.

Larsen, S. C. and Reilinger, R. (1991). Age constraints for the present fault configuration in the Imperial Valley, California: evidence for northwestward propagation of the Gulf of California rift system. *Journal of Geophysical Research, B, Solid Earth and Planets*, **96**, 10 339–46.

Laslett, G. M., Green, P. F., Duddy, I. R. and Gleadow, A. J. W. (1987). Thermal annealing of fission tracks in apatite, 2: A quantitative analysis. *Chemical Geology (Isotope Geoscience Section)*, **65**, 1–13.

Laubach, S. E. (2003). Practical approaches to identifying sealed and open fractures. *American Association of Petroleum Geologists Bulletin*, **87**, 561–79.

Laubach, S. E., Marrett, R. and Olson, J. (2000). New directions in fracture characterization. *The Leading Edge*, **19**, 704–11.

Laubscher, H. P. (1972). Some overall aspects of Jura dynamics. *American Journal of Science*, **272**, 293–304.

  (1986). The eastern Jura: relations between thin skinned and basement tectonics, local and regional. *Geologische Rundschau*, **73**, 535–53.

  (1992). The Alps: a transpressive pile of peels. In *Thrust Tectonics*, ed. K. R. McClay. London: Chapman and Hall, pp. 277–85.

Law, B. E. (2002). Basin-centered gas systems. *American Association of Petroleum Geologists Bulletin*, **86**, 1891–919.

Lawrence, D. T., Doyle, M. and Aigner, T. (1990). Stratigraphic simulation of sedimentary basins: concepts and calibrations. *American Association of Petroleum Geologists Bulletin*, **74**, 273–95.

Lawrence, S. R. (1990). Aspects of the petroleum geology of the Junggar basin, Northwest China. In *Classic Petroleum Provinces*, ed. J. Brooks. *Geological Society of London Special Publication*, vol. 50, pp. 545–57.

Lawrence, S. R. and Cornford, C. (1995). Basin geofluids. *Basin Research*, **7**, 1–7.

Lawton, T. F., Boyer, S. E. and Schmit, J. G. (1994). Influence of inherited taper on structural variability and conglomerate distribution: Cordilleran fold and thrust belt, western United States. *Geology*, **22**, 339–42.

Lawton, T. F. and Giles, K. A. (2000). Southwestern ancestral Rocky Mountains province, Arizona, New Mexico, Sonora and Chihuahua: broken foreland or continental borderland? *Geological Society of America Annual Meeting 2000*, Abstracts with programs, vol. 32, pp. 466–7.

Lawton, T. F., Guimera, J. and Roca, E. (1994). Kinematics of detachment folding from conglomeratic growth strata: linking zone, northeastern Spain (Catalunya). *Geological Society of America*, Abstracts with Programs, vol. 26, p. 315.

Le Fort, P. (1975). Himalayas: the collided range, present knowledge of the continental arc. *American Journal of Science*, **275A**, 1–44.

Le Pichon, X., Huchon, P., Angelier, J. *et al.* (1982). Subduction in the Hellenic Trench; probable role of a thick evaporitic layer based on Seabeam and submersible studies. In *Trench-Forearc Geology; Sedimentation and Tectonics on Modern and Ancient Active Plate Margins, Conference*, London, June 23–25, 1980, ed. J. K. Leggett, *Geological Society of London Special Publication*, vol. 10. London: Geological Society, pp. 319–33.

Le Pichon, X., Foucher, J. P., Boulegue, J., Henry, P., Lallemant, S., Benedetti, F., Avedik, F. and Mariotti, A. (1990a). Mud volcano field seaward of the Barbados accretionary complex: a submersible survey. *Journal of Geophysical Research*, **95**, 8931–45.

Le Pichon, X., Henry, P. and Lallemant, S. (1990b). Water flow in the Barbados accretionary complex. In *Special Section on the Role of Fluids in Sediment Accretion, Deformation, Diagenesis, and Metamorphism in Subduction Zones*, ed. M. G. Langseth and J. C. Moore. *Journal of Geophysical Research*, **95**, 8945–67.

  (1993). Accretion and erosion in subduction zones: the role of fluids. *Annual Reviews of Earth and Planetary Science*, **21**, 307–31.

Le Pichon, X., Lallemant, S. and Lallemand, S. (1986). Tectonic context of fluid venting along Japanese trenches. *EOS, Transactions, American Geophysical Union*, **67**, 1204.

Leach, D. L. (1979). Temperature and salinity of the fluids responsible for minor occurrences of sphalerite in the Ozark region of Missouri. *Economic Geology*, **74**, 931–7.

Leach, D. L. and Rowan, E. L. (1986). Genetic link between Ouachita foldbelt tectonism and the Mississippi Valley lead-zinc deposits of the Ozarks. *Geology*, **14**, 931–5.

Lebel, D., Langenberg, W. and Mountjoy, E. W. (1996). Structure of the central Canadian Cordilleran thrust-and-fold belt, Athabasca–Brazeau area, Alberta: a large, complex intercutaneous wedge. In *Triangle Zones and Tectonic Wedges*, ed. P. A. MacKay, T. E. Kubli, A. C. Newson, J. L. Varsek, R. G. Dechesne and J. P. Reid. *Canadian Society of Petroleum Geologists: Bulletin of Canadian Petroleum Geology*, vol. 44, pp. 282–98.

Lecorche, J. P., Dallmeyer, R. D. and Villeneuve, M. (1989). Definition of tectonostratigraphic terranes in the Mauritanide, Bassaride, and Rokelide orogens, West Africa. *Geological Society of America Bulletin*, **230**, 131–44.

Lee, C. H. and Farmer, I. (1993). *Fluid Flow in Discontinuous Rocks*. London: Chapman & Hall.

Lee, T.-Y. and Lawver, L. A. (1994). Cenozoic plate reconstruction of the South China Sea region. *Tectonophysics*, **235**, 149–80.

Lehner, F. K. (1986). Comments on 'Noncohesive critical Coulomb wedges: an exact solution' by F. A. Dahlen. *Journal of Geophysical Research*, **91**, 793–6.

Leith, W. and Alvarez, W. (1985). Structure of the Vakhsh fold-and-thrust belt, Tadjik SSR; geologic mapping on a Landsat image base. *Geological Society of America Bulletin*, **96**, 875–85.

Leith, W. S. (1984). The Tadjik Depression, USSR; geology, seismicity and tectonics. Thesis, New York: Columbia University, Teachers College.

Lelek, J. J. (1982). Anschutz Ranch East field, northeast Utah and southwest Wyoming. In *Geologic Studies of the Cordilleran Thrust Belt*, vol. 2, ed. R. B. Powers. Denver: Rocky Mountain Association of Geologists, pp. 619–31.

Lepvrier, C., Maluski, H., Nguyen, V. V., Roques, D., Axente, V. and Rangin, C. (1997). Indosinian NW-trending shear zones within the Truong Son Belt (Vietnam); ^{40}Ar- ^{39}Ar Triassic ages and Cretaceous to Cenozoic overprints. *Tectonophysics*, **283**, 105–27.

Letouzey, J. (1990). Petroleum and tectonics in mobile belts. *Proceedings of the IFP Exploration and Production Research Conference*. Paris: Technip.

Letouzey, J., Colletta, B., Benard, F., Sassi, W. and Bale, P. (1990). Fault reactivation and structural inversion physical models analyzed with X-ray scanner and seismic examples. In *AAPG Annual Convention with DPA/EMD Divisions and SEPM, an Associated Society*: technical program with abstracts. *American Association of Petroleum Geologists Bulletin*, **74**, 703–4.

Letouzey, J., Sherkati, S. and Motiei, H. (2003). Salt tectonics and compressive structures in the Central Zagros fold and thrust belt (Iran). *AAPG Annual Meeting*, Salt Lake City, Utah, *Official Program*, vol. 12. Tulsa, OK: American Association of Petroleum Geologists, p. A103.

Levine, J. R. and Davis, A. (1983). Tectonic history of coal-bearing sediments in eastern Pennsylvania using coal reflectance anisotropy. Pennsylvania St University Special Research Report, SR-118, 1–314.

Lewan, M. D. (1993). Laboratory simulation of petroleum formation: hydrous pyrolysis. In *Organic Geochemistry*, ed. M. H. Engel and S. A. Macko. New York: Plenum Press, pp. 419–42.

(1994). Assessing natural oil expulsion from source rocks by laboratory pyrolysis. In *The Petroleum System – From Source to Trap*, ed. L. B. Magoon and W. G. Dow. *American Association of Petroleum Geologists Memoir*, vol. 60, pp. 201–10.

(1997). Experiments on the role of water in petroleum formation. *Geochimica et Cosmochimica Acta*, **61**, 3691–723.

Lewan, M. D. and Henry, A. A. (2001). Gas: oil ratios for source rocks containing type-I, -II, -IIS, and -III kerogens as determined by hydrous pyrolysis. In *Geologic Studies of Deep Natural Gas Resources*, ed. T. S. Dyman and V. A. Kuuskraa. *US Geological Survey Digital Data Series*, vol. 67, E1–9.

Lewis, B. T. R. and Cochrane, G. C. (1990). Relationship between the location of chemosynthetic benthic communities and geologic structure on the Cascadia subduction zone. *Journal of Geophysical Research*, **95**, 8783–93.

Lewis, C. R. and Rose, S. C. (1970). A theory relating high temperatures and overpressures. *Journal of Petroleum Technology*, **22**, 11–16.

Lewis, P. D. and Ross, J. V. (1991). Mesozoic and Cenozoic structural history of the central Queen Charlotte Islands, British Columbia. In *Evolution and Hydrocarbon Potential of the Queen Charlotte Basin, British Columbia*, ed. G. J. Woodsworth. *Geological Survey of Canada, Report*, vol. 90-10, pp. 31–50.

Leythaeuser, D. and Poelchau, H. S. (1991). Expulsion of petroleum from type III kerogen source rocks in gaseous solution: modeling of solubility fractionation. In *Petroleum Migration*, ed. W. A. England and A. J. Fleet. *Geological Society of London Special Publication*, vol. 59, pp. 33–46.

Li, D. (1991). *Tectonic Types of Oil and Gas Basins in China*. Beijing: China Petroleum Industry Press.

Liechti, P., 1968. Salt features of France. *GSA Special Paper*, **88**, 83–106.

Lillie, R. J. (1984). Tectonic implications of subthrust structures revealed by seismic profiling of Appalachian–Ouachita orogenic belt. *Tectonics*, **3**, 619–46.

Lindquist, S. J. (1983). Nugget formation reservoir characteristics affecting production in the overthrust belt of Southwestern Wyoming. *Journal of Petroleum Technology*, **35**, 1355–65.

(1998). The Santa Cruz–Tarija province of central South America: Los Monos-Machareti(!) petroleum system. *US Geological Survey Open-File Report*, vol. 99-50-C, 16.

Lingrey, S. (1991). Seismic modeling of an imbricate thrust structure from the Foothills of the Canadian Rocky Mountains. In *Seismic Modeling of Geologic Structures: Applications to Exploration Problems*, ed. S. W. Fagin. *Society of Exploration Geophysicists Geophysical Development Series*, pp. 111–25.

Linzer, H. G., Moser, F., Nemes, F., Ratschbacher, L. and Sperner, B. (1997). Build-up and dismembering of a classical fold-thrust belt, from non-cylindrical stacking to lateral extrusion in the eastern Alps. *Tectonophysics*, **272**, 97–124.

Liou, J. G. (1981). Recent high CO_2 activity and Cenozoic progressive metamorphism in Taiwan. *Geological Society of China Memoir*, vol. 4, pp. 551–81.

Liro, L. M. and Coen, R. (1995). Salt deformation history and postsalt structural trends, offshore southern Gabon, West Africa. In *Salt Tectonics: a Global Perspective*, ed. M. P. A. Jackson, D. G. Roberts and S. Snelson. *American Association of Petroleum Geologists Memoir*, vol. 65, pp. 323–31.

Lisle, R. J. (1994). Detection of zones of abnormal strains in structures using Gaussian curvature analysis. *American Association of Petroleum Geologists Bulletin*, **78**, 1811–19.

Littke, R. (1994). *Deposition, Diagenesis, and Weathering of Organic Matter-rich Sediments*. Heidelberg: Springer-Verlag.

Littke, R., Baker, D. R. and Rullkötter, J. (1997). Deposition of petroleum source rocks. In *Petroleum and Basin Evolution: Insights from Petroleum Geochemistry, Geology and Basin Modeling*, ed. D. H. Welte, B. Horsfield and D. R. Baker. Berlin: Springer-Verlag, pp. 273–333.

Liu, H., McClay, K. R. and Powell, D. (1992). Physical models of thrust wedges. In *Thrust Tectonics*, ed. K. R. McClay. London: Chapman & Hall, pp. 71–81.

Liu, S. and Dixon, J. M. (1991). Centrifuge modeling of thrust faulting: structural variations along strike in fold-thrust belts. *Tectonophysics*, **188**, 39–62.

Liu, X., Fu, D., Yao, J., *et al.* (1996). Tectonic evolution of Tarim Plate and surrounding areas since late Paleozoic. *Continental Dynamics*, **1**, 109–22.

Lockner, D. A., Moore, D. E. and Reches, Z. (1992). Microcrack interaction leading to shear fracture. *33rd Symposium on Rock Mechanics*, pp. 807–16.

Logan, J. M., Friedman, M. and Stearns, M. T. (1978). Experimental folding of rocks under confining pressure, Part VI: Further studies of faulted drape folds. In *Laramide Folding Associated with Basement Block Faulting in the Western United States*, ed. V. Matthews. *Geological Society of America Memoir*, vol. 151, pp. 79–99.

Logan, J. M. and Rauenzahn, K. M. (1987). Frictional dependence of gouge mixtures of quartz and montmorillonite on velocity, composition and fabric. *Tectonophysics*, **144**, 87–108.

Logani, K. (1973). *Dilatancy Model for the Failure of Rocks*. Iowa State University.

Longley I. M. (1997). The tectonostratigraphic evolution of SE Asia. In *Petroleum Geology of Southeast Asia, Geological Society Special Publications*, vol. 126. London: Geological Society of London, pp. 311–39.

Lorenz, J. C., Lawrence, W. T. and Warpinski, N. R. (1991). Regional fractures 1: a mechanism for the formation of regional fractures at depth in flat lying reservoirs. *American Association of Petroleum Geologists Bulletin*, **75**, 1714–37.

Losh, S. (1998). Oil migration in a major growth fault: structural analysis of the Pathfinder core, south Eugene Island block 330, offshore Louisiana. *American Association of Petroleum Geologists Bulletin*, **82**, 1694–710.

Lowell, J. D. (1995). Mechanics of basin inversion from world-wide examples. In *Basin Inversion*, ed. J. G. Buchanan and P. G. Buchanan. *Geological Society of London Special Publication*, vol. 88, pp. 39–57.

Lowell, R. P. (1990). Thermoelasticity and the formation of black smokers. *Geophysical Research Letters*, **17**, 709–12.

Lu, G., Marshak, S. and Kent, D. V. (1990). Characteristics of magnetic carriers responsible for the late Paleozoic remagnetization in carbonate strata of the midcontinent, U.S.A. *Earth and Planetary Science Letters*, **99**, 351–61.

Lundegard, P. D. (1992). Sandstone porosity loss – a 'big picture' view of the importance of compaction. *Journal of Sedimentary Petrology*, **62**, 250–60.

Lundberg, N. and Moore, J. C. (1986). Macroscopic structural features in Deep Sea Drilling Project cores from forearc regions. In *Structural Fabrics in Deep Sea Drilling Project Cores from Forearcs*, ed. J. C. Moore. *Geological Society of America Memoir*, vol. 166, pp. 13–44.

Lüning, S., Craig, J., Loydell, D. K., Storch, P. and Fitches, B. (2000). Lower Silurian 'hot shales' in North Africa and Arabia: regional distribution and depositional model. *Earth Science Reviews*, **49**, 121–200.

Lucazeau, F. (1986). The post rift evolution of the Massif Central (France). In *Thermal Modeling in Sedimentary Basins*, ed. J. Burrus. Paris: Editions Technip, pp. 75–89.

Lucazeau, F. and Le Douaran, S. (1984). Numerical model of sediment thermal history comparison between the Gulf of Lion and the Viking Graben. In *Thermal Phenomena in Sedimentary Basins*, ed. B. Durand. Paris: Editions Technip, pp. 211–18.

Lyberis, N. and Manby, G. (1999). Oblique to orthogonal convergence across the Turan Block in the post-Miocene. *AAPG Bulletin*, **83**, 1135–60.

MacCaig, A. M. (1988). Deep fluid circulation in fault zones. *Geology*, **16**, 867–70.

Macedo, J. and Marshak, S. (1999). Controls on the geometry of fold-thrust belt salients. *Geological Society of America Bulletin*, **111**, 1808–22.

Macgregor, D. S. (1993). Relationships between seepage, tectonics and subsurface petroleum reserves. *Marine and Petroleum Geology*, **10**, 606–19.

(1994). Coal-bearing strata as source rock: a global overview. In *Coal and Coal-bearing Strata as Oil-prone Source Rocks*, ed. A. C. Scott and A. J. Fleet. *Geological Society of London Special Publication*, vol. 77, pp. 107–16.

(1995). Hydrocarbon habitat and classification of inverted rift basins. In *Basin Inversion*, ed. J. G. Buchanan and P. G. Buchanan. *Geological Society of London Special Publication*, vol. 88, pp. 83–93.

(1996). Factors controlling the destruction or preservation of giant light oilfields. *Petroleum Geoscience*, **2**, 197–217.

Mack, C. and Jerzykiewicz, T. (1989). Provenance of post-

Wapiabi sandstones and its implications for Campanian to Paleocene tectonic history of the southern Canadian Cordillera. *Canadian Journal of Earth Sciences*, **26**, 665–77.

MacKay, M. E., Jarrard, R. D., Westbrook, G. K. and Hyndman, R. D. (1994). Origin of bottom-simulating reflectors: geophysical evidence from the Cascadia accretionary prism. *Geology*, **22**, 459–62.

MacKay, M. E., Moore, G. F., Klaeschen, D. and Von Huene, R. (1995). The case against porosity change: seismic velocity decrease at the toe of the Oregon accretionary prism. *Geology*, **23**, 827–30.

MacKay, P. A., Varsek, J. L., Kubli, T. E., Dechesne, R. G., Newson, A. C. and Reid, J. P. (1996). Triangle zones and tectonic wedges: an introduction. *Bulletin of Canadian Petroleum Geology*, **44**, I.1–5.

Mackenzie, A. S. and Quigley, T. M. (1988). Principles of geochemical prospect appraisal. *American Association of Petroleum Geologists Bulletin*, **72**, 399–415.

Magara, K. (1976). Water expulsion from elastic sediments during compaction: directions and volumes. *American Association of Petroleum Geologists Bulletin*, **60**, 543–53.

Magoon, L. B. (1995). The play that complements the petroleum system: a new exploration equation. *Oil and Gas Journal*, **93**, 85–7.

Magoon, L. B. and Dow, W. G. (1994). The petroleum system. In *The Petroleum System – From Source to Trap*, ed. L. B. Magoon and W. G. Dow. *American Association of Petroleum Geologists Memoir*, vol. 60, pp. 3–24.

Majorowicz, J. A. (1989). The controversy over the significance of the hydrodynamic effect on heat flow in the Prairies Basin. In *Hydrological Regimes and Their Subsurface Thermal Effects*, ed. A. E. Beck. Washington, DC: American Geophysical Union, pp. 101–5.

Majorowicz, J. A. and Jessop, A. M. (1981). Regional heat flow patterns in the western Canadian sedimentary basin. *Tectonophysics*, **74**, 209–38.

Majorowicz, J. A., Jones, F. W., Lam, H. L. and Jessop, A. M. (1984). The variability of heat flow both regional and with depth in southern Alberta, Canada: effect of groundwater flow? *Tectonophysics*, **106**, 1–29.

Majorowicz, J. A., Rahman, M., Jones, F. W. and McMillen, N. J. (1985). The paleogeothermal and present thermal regimes of the Alberta Basin and their significance for petroleum occurrences. *Bulletin of Canadian Petroleum Geology*, **33**, 12–21.

Malavieille, J. (1984). Experimental model for imbricated thrusts: comparison with thrust-belts. *Bulletin de la Société Géologique de France*, **26**, 129–38.

Malavieille, J., Larroque, C. and Calassou, S. (1993). Modelisation experimentale des relations tectoniques/sedimentation entre bassin avant-arc et prisme d'accretion. *Comptes Rendus de l'Academie des Sciences*, Serie 2, **316**, 1131–7.

Malin, P. E. (1994). The seismology of extensional hydrothermal systems. *Geothermal Resources Council Transactions*, **18**, 17–22.

Mallory, W. W. (1972). Pennsylvanian arkose and the Ancestral Rocky Mountains. In *Geologic Atlas of the Rocky Mountain Region*, Denver, CO: Rocky Mountain Association of Geologists, pp. 131–2.

Maltman, A. J. (1994). Prelithification deformation. In *Continental Deformation*, ed. P. L. Hancock. Tarrytown: Pergamon Press, pp. 143–58.

Maltman, A. J., Byrne, T., Karig, D. E. and Lallemant, S. (1992). Structural geological evidence from ODP Leg 131 regarding fluid flow in the Nankai Prism, Japan. In *Fluids in Convergent Margins: Earth and Planetary Science Letters*, vol. 109, ed. M. Kastner and X. Le Pichon, pp. 463–8.

(1993). Deformation at the toe of an active accretionary prism; synopsis of results from ODP Leg 131, SW Japan. *Journal of Structural Geology*, **15**, 949–64.

Malvern, L. E. (1969). *Introduction to the Mechanics of a Continuous Medium*. Englewood Cliffs: Prentice-Hall.

Mandl, G. (1988). *Mechanics of Tectonic Faulting: Models and Basic Concepts*. Amsterdam: Elsevier.

Mandl, G. and Shippam, G. K. (1981). Mechanical model of thrust sheet gliding and imbrication. In *Thrust and Nappe Tectonics*, ed. K. R. McClay and N. J. Price. *Geological Society of London Special Publication*, vol. 9, pp. 79–98.

Mango, F. D. (1997). The light hydrocarbons in petroleum: a critical review. *Organic Geochemistry*, **26**, 417–40.

Mann, D. M. and Mackenzie, A. S. (1990). Prediction of pore fluid pressures in sedimentary basins. *Marine and Petroleum Geology*, **7**, 55–65.

Mann, U. (1994). An integrated approach to the study of primary petroleum migration. In *Geofluids: Origin, Migration and Evolution of Fluids in Sedimentary Basins*, ed. J. Parnell. *Geological Society of London Special Publication*, vol. 78, pp. 233–60.

Mann, U., Hantschel, T., Schaefer, R. G., Krooss, B., Leythaeuser, D., Littke, R. and Sachsenhofer, R. F. (1997). Petroleum migration: mechanisms, pathways, efficiencies and numerical simulations. In *Petroleum and Basin Evolution: Insights from Petroleum Geochemistry, Geology and Basin Modeling*, ed. D. H. Welte, B. Horsfield and D. R. Baker. Berlin: Springer-Verlag, pp. 405–520.

Marker, M., Whitehouse, M. J., Scott, D. J., Stecher, O., Bridgewater, D. and Van Gool, J. A. M. (1999). Deposition, provenance and tectonic setting for metasediments in the Palaeoproterozoic Nagssugtoqidian Orogen, West Greenland: a key for understanding crustal collision. *Journal of Conference Abstracts*, **4**, 128.

Markevich, P. V., Zyabrev, S. V., Filippov, A. N. and Malinovskiy, A. I. (1996). Vostochnyy flang Kiselevsko-Manominskogo terreyna: fragment ostrovnoy dugi v akkretsionnoy prizme (Severnyy Sikhote-Alin). Eastern flank of the Kiselyevsko-Manominskiy terrane; an island-arc fragment in accretionary wedge, northern Sikhote-Alin. *Tikhookeanskaya Geologiya, Pacific Geology*, **15**, 70–98.

Marques, F. O. and Cobbold, P. R. (2002). Topography as a

major factor in the development of arcuate thrust belts: insights from sandbox experiments. *Tectonophysics*, **348**, 247–68.

Marshak, S. (1992). Relationships between brine-migration pulses and deformation processes in the continental interior, USA. *Geological Society of America, Abstracts with Programs*, **24**, 111.

Marshak, S., Lu, G. and Kent, D. V. (1989). Reconnaissance investigation of remagnetization in Paleozoic limestones from the mid-continent of North America. *EOS, Transactions, American Geophysical Union*, **70**, 310.

Marshak, S. and Wilkerson, M. S. (1992). Effect of overburden thickness on thrust belt geometry and development. *Tectonics*, **11**, 560–6.

Marshak, S., Wilkerson, M. S. and Hsui, A. T. (1992). Generation of curved fold-thrust belts: insight from simple physical and analytical models. In *Thrust Tectonics*, ed. K. R. McClay. London: Chapman & Hall, pp. 83–92.

Martel, S. J., Pollard, D. D. and Segall, P. (1988). Development of simple strike-slip fault zones, Mount Abbot Quadrangle, Sierra Nevada, California. *Geological Society of America Bulletin*, **100**, 1451–65.

Martin, J. B., Kastner, M., Henry, P., Le Pichon, X. and Lallemant, S. (1996). Chemical and isotopic evidence for sources of fluids in a mud volcano field seaward of the Barbados accretionary wedge. *Journal of Geophysical Research*, **101**, 20 325–45.

Martinez, A., Rivero, L. and Casas, A. (1997). Integrated gravity and seismic interpretation of duplex structures and imbricate thrust systems in the southeastern Pyrenees (NE Spain). *Tectonophysics*, **282**, 303–29.

Martinez, E., Fernandez, J., Calderon, Y. and Galdos, C. (2003). Reevaluation defines attractive areas in Peru's Ucayali-Ene basin. *Oil and Gas Journal*, **101**, 32–8.

Martini, A. M., Walter, L. M., Ku, T. C. W., Budai, J. M., McIntosh, J. C. and Schoell, M. (2003). Microbial production and modification of gases in sedimentary basins: a geochemical case study from a Devonian shale gas play, Michigan basin. *American Association of Petroleum Geologists Bulletin*, **87**, 1355–75.

Márton, E. and Fodor, L. (1995). Combination of paleomagnetic and stress data: a case study from North Hungary. *Tectonophysics*, **242**, 99–114.

Marton, L. G., Tari, G. C. and Lehmann, C. T. (2000). Evolution of the Angolan passive margin, West Africa, with emphasis on post-salt structural styles. In *Atlantic Rifts and Continental Margins*, ed. W. U. Mohriak and M. Talwani. *Geophysical Monograph*, vol. 115, pp. 129–49.

Masaferro, J. L., Bulnes, M., Poblet, J. and Casson, N. (2003). Kinematic evolution and fracture prediction of the Valle Morado structure inferred from 3-D seismic data, Salta province, northwest Argentina. *American Association of Petroleum Geologists Bulletin*, **87**, 1083–104.

Mascle, A. and Moore, J. C. (1990). ODP Leg 110: Tectonic and hydrogeologic synthesis. *Proceedings of Ocean Drilling Program, Scientific Results*, vol. 110, 409–22.

Mascle, A., Moore, J. C. Taylor, E. *et al.* (1988). Synthesis of shipboard results; Leg 110 transect of the northern Barbados Ridge. *Proceedings of the Ocean Drilling Program, Part A: Initial Reports*, vol. 110, pp. 577–91.

Mascle, A., Huyghe, P. and Deville, E. (2003). Deep marine tectonic, mud diapirism and sedimentary processes at the southern edge of the Barbados Ridge and eastern margins of Trinidad and Venezuela. *AAPG Annual Meeting*, Salt Lake City, UT, *Official Program*, vol. 12. Tulsa, OK: American Association of Petroleum Geologists, p. A114.

Masek, J. G., Isacks, B. L., Gubbels, T. L. and Fielding, E. J. (1994). Erosion and tectonics at the margins of continental plateaus. *Journal of Geophysical Research*, **99**, 13 941–56.

Massari, F., Grandesso, P., Stefani, P. and Jobstraibizer, J. B. (1986). A small polyhistory foreland basin evolving in a context of oblique convergence: the Venetian Basin (Chattian to Recent, Southern Alps, Italy). *International Association Special Publication*, vol. 8, pp. 141–68.

Masters, J. A. (1984). Lower Cretaceous oil and gas in western Canada. In *Elmworth: Case Study of a Deep Basin Gas Field*, ed. J. A. Masters. *American Association of Petroleum Geologists Memoir*, vol. 38, pp. 1–33.

Mathews, M. (1986). Logging characteristics of methane hydrate. *The Log Analyst*, **27**, 26–63.

Mathur, L. P. and Evans, P., eds. (1964). Oil in India: 22nd Session. *International Geological Congress Proceedings*.

Mattavelli, L. and Novelli, L. (1988). Geochemistry and habitat of natural gases in Italy. *Organic Geochemistry*, **13**, 1–13.

Mattavelli, L., Novelli, L. and Anelli, L. (1991). Occurrence of hydrocarbons in the Adriatic Basin. In *Generation, Accumulation, and Production of Europe's Hydrocarbons*, ed. A. M. Spencer. *Germany: Special Publication of the European Association of Petroleum Geoscientists*, vol. 1, pp. 369–80.

Mattavelli, L., Pieri, M. and Groppi, G. (1993). Petroleum exploration in Italy: a review. *Marine and Petroleum Geology*, **10**, 410–25.

Mattern, F. and Schneider, W. (2000). Suturing of the Proto- and Paleo-Tethys oceans in the western Kunlun (Xinjiang, China). *Journal of Asian Earth Sciences*, **18**, 637–50.

Matthews, M. D. (1999). Migration of petroleum. In *Exploring for Oil and Gas Traps: American Association of Petroleum Geologists Treatise of Petroleum Geology*, ed. E. A. Beaumont and N. H. Foster, pp. 7-31–8.

Matviyevskaya, N. D., Ivantsov, Y. F. and Yaralov, B. A. (1986). New petroleum: promising objectives and methods for prospecting in the Timan–Pechora oil province. *Geologiya Nefti i Gaza*, **1986**, 1–4.

McAuliffe, C. D. (1979). Oil and gas migration: chemical and physical constraints. *American Association of Petroleum Geologists Bulletin*, **73**, 1455–71.

McBride, J. H., Barazangi, M., Best, J., Al-Saad, D., Sawaf, T., Al-Otri, M. and Gebran, A. (1990). Seismic reflection structure of intracratonic Palmyride fold-thrust belt and surrounding Arabian platform, Syria. *American Association of Petroleum Geologists Bulletin*, **74**, 238–59.

McCabe, C. and Elmore, R. D. (1989). The occurrence and origin of Late Paleozoic remagnetization in the sedimentary rocks of North America. *Reviews of Geophysics*, **27**, 471–94.

McCabe, C., Jackson, M. and Saffer, B. (1989). Regional patterns of magnetite authigenesis in the Appalachian Basin: implications for the mechanism of late Paleozoic remagnetization. *Journal of Geophysical Research*, B, *Solid Earth and Planets*, **94**, 10 429–43.

McCabe, P. J. (1991). Tectonic controls on coal accumulations. *Société Géologique de France Bulletin*, **162**, 277–82.

McCabe, P. J. and Parrish, J. T. (1992). Tectonic and climatic controls on the distribution and quality of Cretaceous coals. In *Controls on the Distribution and Quality of Cretaceous Coals*, ed. P. J. McCabe and J. T. Parrish. *Geological Society of America Special Paper*, vol. 267, pp. 1–15.

McClay, K. R. (1995). The geometries and kinematics of inverted fault systems: a review of analogue model studies. In *Basin Inversion*, ed. J. G. Buchanan and P. G. Buchanan. *Geological Society of London Special Publication*, vol. 88, pp. 97–118.

McClay, K. R., Dooley, T., Ferguson, A. and Poblet, J. (2000). Tectonic evolution of the Sanga Sanga Block, Mahakam Delta, Kalimantan, Indonesia. *American Association of Petroleum Geologists Bulletin*, **84**, 765–86.

McClay, K. R. and Insley, M. W. (1986). Duplex structures in the Lewis thrust sheet, Crowsnest Pass, Rocky Mountains, Alberta, Canada. *Journal of Structural Geology*, **8**, 911–22.

McDougal, J. W. and Hussain, A. (1991). Fold and thrust propagation in western Himalayas based on balanced cross-section of the Surshar Range. *AAPG Bulletin*, **75**, 463–78.

McGarr, A. (1980). Some constraints on levels of shear stress in the crust from observations and theory. In *Magnitude of Deviatoric Stresses in the Earth's Crust and Uppermost Mantle*, ed. T. C. Hanks and C. B. Raleigh. *Journal of Geophysical Research*, **85**, 6231–8.

McIntyre, J. F. (1988). Presence and control of evaporite top seals on occurrence and distribution of hydrocarbon traps: main fairway, central overthrust belt, Wyoming and Utah (abstract). *American Association of Petroleum Geologists Bulletin*, **72**, 221.

McLaughlin, D. H., Jr (1972). Evaporite deposits of Bogota area, Cordillera Oriental, Columbia. *GSA Bulletin*, **56**, 2240–59.

McNaughton, D. A., Quinlan, T., Hopkins, R. M., Jr and Wells, A. T. (1968). Evolution of salt anticlines and salt domes in the Amadeus Basin, central Australia, *Geological Society of America Special Paper*, vol. 88. Boulder, CO: Geological Society of America, pp. 229–47.

McLean, B. C. and Cook, D. G. (1999). Salt tectonism in the Fort Norman area, Northwest Territories, Canada. *Bulletin of Canadian Petroleum Geology*, **47**, 104–35.

McMechan, M. E. (1985). Low-taper triangle-zone geometry: an interpretation for the Rocky Mountain foothills, Pine Pass-Peace River area, British Columbia. *Bulletin of Canadian Petroleum Geology*, **33**, 31–8.

McMechan, M. E. and Thompson, R. I. (1989). Structural style and history of the Rocky Mountain fold and thrust belt. In *Western Canadian Sedimentary Basin: a Case Study*, ed. B. D. Ricketts. Calgary: Canadian Society of Petroleum Geologists, pp. 47–72.

McNaught, M. A. (1990). The use of retrodeformable cross sections to constrain the geometry and interpret the deformation of the Mead thrust sheet, southeastern Idaho and northern Utah. Ph.D. thesis, New York: University of Rochester.

Means, W. D. (1989). Stretching faults. *Geology*, **17**, 893–6.

Medd, D. M. (1996). Triangle zone deformation at the leading edge of the Papuan fold belt. In *Petroleum Exploration, Development and Production in Papua New Guinea: Proceedings of the Third Papua New Guinea Petroleum Convention*, ed. P. G. Buchanan, pp. 217–29.

Medwedeff, D. A. (1989). Growth fault-bend folding at Southeast Lost Hills, San Joaquin Valley, California. *American Association of Petroleum Geologists Bulletin*, **73**, 54–67.

  (1992). Geometry and kinematics of an active, laterally propagating wedge thrust, Wheeler Ridge, California. In *Structural Geology of Fold and Thrust Belts*, ed. S. Mitra and G. W. Fisher. Baltimore: John Hopkins University Press, pp. 3–28.

Meghraoui, M., Jaegy, R., Lammali, K. and Albarede, F. (1988). Late Holocene earthquake sequences on the El Asnam (Algeria) thrust fault. *Earth and Planetary Science Letters*, **90**, 187–203.

Meier, B., Schwander, M. and Laubscher, H. P. (1987). The tectonics of Tachira; a sample of North Andean tectonics. In *The Anatomy of Mountain Ranges*, ed. J.-P. Schaer and J. Rodgers. Princeton, NJ: Princeton University Press, pp. 229–37.

Meigs, A. J. and Burbank, D. W. (1997). Growth of the South Pyrenean orogenic wedge. *Tectonics*, **16**, 239–58.

Meigs, A. J., Verges, J. and Burbank, D. W. (1996). Ten-million-year history of a thrust sheet. *Geological Society of America Bulletin*, **108**, 1608–25.

Meissner, F. F. (1978). Petroleum geology of the Bakken Formation, Williston Basin, North Dakota and Montana. *Proceedings of 1978 Williston Basin Symposium*, September 24–27. Billings: Montana Geological Society, pp. 207–27.

Meissner, R. (1986). *The Continental Crust: a Geophysical Approach*. San Diego: Academic Press.

Mellon, G. B. (1967). Stratigraphy and petrology of the Lower Cretaceous Blairmore and Mannville Groups, Alberta Foothills and Plains. *Research Council of Alberta Bulletin*, **21**, 270.

Merewether, E. A. (1971). Geology of the Knoxville and Delaware quadrangles. Johnson and Logan cities and vicinity, Arkansas. *US Geological Survey Professional Paper*, **657-B**, 18.

Merewether, E. A. and Haley, B. R. (1969). Geology of the Hartman and Clarksville quadrangles, Johnson city and vicinity, Arkansas. *US Geological Survey Professional Paper*, **536-C**, 27.

Mero, W. E., Thurston, S. P. and Kropschot, R. E. (1992). The Point Arguello field. In *Giant Oil and Gas Fields of the Decade 1978–1988*, ed. M. T. Halbouty. *American Association of Petroleum Geologists Memoir*, vol. 5, pp. 3–25.

Mertosono, S. (1975). Geology of Pungut and Tandun oil fields, Central Sumatra. *Proceedings of the Annual Convention: Indonesian Petroleum Association*, vol. 4, pp. 165–79.

Meyer, B., Avouac, J. P., Tapponnier, P. and Meghraoui, M. (1990). Topographic measurements of the southwestern segment of the El Asnam fault zone and mechanical interpretation of the relationship between reverse and normal faults. *Bulletin de la Société Géologique de France*, Huitieme Serie, **6**, 447–56 (in French).

Meyer, T. J. and Dunne, W. M. (1990). Multi-stage deformation of cover limestones above a blind thrust system: an example from central Appalachians. *Journal of Geology*, **98**, 108–17.

Midland Valley (2000). Modeling software goes global with BP Amoco. *First Break*, **18**, 50.

Miller, R. G. (1992). The global oil system: the relationship between oil generation, loss, half-life, and the world crude oil resource. *American Association of Petroleum Geologists Bulletin*, **76**, 489–500.

Miller, R. M. (1983). A possible model for the Damara Orogen in the light of recent data. In *Geodynamics of Orogenic Belts: Geodynamics Series*, vol. 10, ed. N. Rast and F. N. Delany, pp. 31–4.

Milnes, A. G., Wennberg, O. P., Skar, O. and Koestler, A. G. (1997). Contraction, extension and timing in the South Norwegian Caledonides: the Sognefjord transect. In *Orogeny Through Time*, ed. J-P. Burg and M. Ford. *Geological Society of London Special Publication*, vol. 121, pp. 123–48.

Milton, N. J. and Bertram, G. T. (1992). Trap styles: a new classification based on sealing surfaces. *American Association of Petroleum Geologists Bulletin*, **76**, 983–99.

Mitra, G. (1994). Strain variation in thrust sheets across the Sevier fold-and-thrust belt (Idaho–Utah–Wyoming): implications for section restoration and wedge taper evolution. *Journal of Structural Geology*, **16**, 585–602.

(1997). Evolution of salients in a fold-and-thrust belt: the effects of sedimentary basin geometry, strain distribution and critical taper. In *Evolution of Geological Structures in Micro- to Macro-Scales*, ed. S. Singupta. London: Chapman and Hall, pp. 59–90.

Mitra, G. and Beard, W. (1980). Theoretical models of porosity reduction by pressure solution for well-sorted sandstones. *Journal of Sedimentary Petrology*, **50**, 1347–60.

Mitra, G. and Boyer, S. E. (1986). Energy balance and defor-

mation mechanisms of duplexes. *Journal of Structural Geology*, **8**, 291–304.

Mitra, G., Hull, J. M., Yonkee, W. A. and Protzman, G. M. (1988). Comparison of mesoscopic and microscopic deformational styles in the Idaho-Wyoming thrust belt and the Rocky Mountain foreland. In *Interaction of the Rocky Mountain Foreland and the Cordilleran Thrust Belt*, ed. C. J. Schmidt and W. J. Perry Jr. *Geological Society of America Memoir*, vol. 171, pp. 119–41.

Mitra, G. and Sussman, A. J. (1997). Structural evolution of connecting splay duplexes and their implications for critical taper: an example based on geometry and kinematics of the Canyon Range culmination, Sevier Belt, central Utah. *Journal of Structural Geology*, **19**, 503–21.

Mitra, G. and Wojtal, S., eds. (1988). Geometries and mechanisms of thrusting. *Geological Society of America Special Paper*, 222.

Mitra, S. (1984). Brittle to ductile transition due to large strains along the White Rock thrust, Wind River Mountains, Wyoming. *Journal of Structural Geology*, **6**, 51–61.

(1986). Duplex structures and imbricate thrust systems: geometry, structural position, and hydrocarbon potential. *American Association of Petroleum Geologists Bulletin*, **70**, 1087–112.

(1987). Regional variations in deformation mechanisms and structural styles in the central Appalachian orogenic belt. *Geological Society of America Bulletin*, **98**, 569–90.

(1988a). Three-dimensional geometry and kinematic evolution of the Pine Mountain thrust system, Southern Appalachians. *Geological Society of America Bulletin*, **100**, 72–95.

(1988b). Effects of deformation mechanisms on reservoir potential in Central Appalachian overthrust belt. *American Association of Petroleum Geologists Bulletin*, **72**, 536–54.

(1990). Fault-propagation folds: geometry, kinematic evolution, and hydrocarbon traps. *American Association of Petroleum Geologists Bulletin*, **74**, 921–45.

(1992). *Balanced Structural Interpretations in Fold and Thrust Belts*, ed. S. Mitra and G. W. Fisher. Baltimore, MD: Johns Hopkins University Press.

(1993). Geometry and kinematic evolution of inversion structures. *American Association of Petroleum Geologists Bulletin*, **77**, 1159–91.

(2002a). Fold-accommodation faults. *American Association of Petroleum Geologists Bulletin*, **86**, 671–93.

(2002b). Structural models of faulted detachment folds. *American Association of Petroleum Geologists Bulletin*, **86**, 1673–94.

(2003). Fold-accommodation faults. *American Association of Petroleum Geologists Bulletin*, **86**, 671–93.

Mitra, S. and Leslie, W. (2003). Three-dimensional structural model of the Rhourde el Baguel field, Algeria. *American Association of Petroleum Geologists Bulletin*, **87**, 231–50.

Mitra, S. and Mount, V. S. (1998). Foreland basement-involved structures. *American Association of Petroleum Geologists Bulletin*, **82**, 70–109.

Mitra, S. and Namson, J. (1989). Equal-area balancing. *American Journal of Science*, **289**, 563–99.

Mollema, P. N. and Antonellini, M. A. (1996). Compaction bands: a structural analog for anti-mode I cracks in aeolian sandstone. *Tectonophysics*, **267**, 209–28.

Molnar, P. and Chen, W.-P. (1983). Focal depths and fault plane solutions of earthquakes under the Tibetan Platau. *Journal of Geophysical Research B*, **88**, 1180–96.

Molnar, P. and Lyon-Caen, H. (1988). Some simple physical aspects of the support, structure, and evolution of mountain belts. In *Processes in Continental Lithospheric Deformation*, ed. S. P. Clark, B. C. Burchfiel and J. Suppe. *Geological Society of America Special Paper*, vol. 218, pp. 179–207.

Mongelli, F., Loddo, M. and Tramacere, A. (1982). Thermal conductivity, diffusivity and specific heat variation of some Travale field (Tuscany) rocks versus temperature. *Tectonophysics*, **83**, 33–43.

Monger, J. W. H., Price, R. A. and Tempelman-Kluit, D. J. (1982). Tectonic accretion and the origin of the two major metamorphic and plutonic welts in the Canadian Cordillera. *Geology*, **10**, 70–5.

Montanari, A., Stewart, K., Bice, D. and Alwarez, W. (1983). Apennine fold-belt tectonics 3: reactivation of fault-block mosaic. *EOS, Transactions, American Geophysical Union*, **64**, 861.

Montañez, I. P. (1994). Late diagenetic dolomitization of Lower Ordovician, Upper Knox carbonates: a record of the hydrodynamic evolution of the Southern Appalachian Basin. *American Association of Petroleum Geologists Bulletin*, **78**, 1210–39.

Montgomery, D. R. (1994). Valley incision and the uplift of mountain peaks. *Journal of Geophysical Research*, **99**, 13 913–21.

Montgomery, S. L., Barrett, F., Vickery, K., *et al.* (2001). Cave Gulch field, Natrona County, Wyoming: large gas discovery in the Rocky Mountain foreland, Wind River basin. *American Association of Petroleum Geologists Bulletin*, **85**, 1543–64.

Montgomery, S. L. and Morea, M. F. (2001). Antelope shale (Monterey Formation), Buena Vista Hills field: advanced reservoir characterization to evaluate CO_2 injection for enhanced oil recovery. *American Association of Petroleum Geologists Bulletin*, **85**, 561–85.

Moore, G. F., Shipley, T. H., Stoffa, P. L., Karig, D. E., Taira, A., Kuramoto, S., Tokuyama, H. and Suyehiro, K. (1990a). Structure of the Nankai Trough accretionary zone from multichannel seismic reflection data. In *Special Section on the Role of Fluids in Sediment Accretion, Deformation, Diagenesis, and Metamorphism in Subduction Zones*, ed. M. G. Langseth and J. C. Moore. *Journal of Geophysical Research*, **95**, 8753–65.

Moore, J. C., Mascle, A. and the ODP Leg 110 Scientific Party. (1987). Expulsion of fluids from depth along a subduction-zone décollement horizon. *Nature*, **326**, 785–7.

Moore, J. C., Mascle, A. *et al.* (1988). Tectonics and hydrogeology of the northern Barbados ridge: results from Ocean Drilling Program Leg 110. *Geological Society of America Bulletin*, **100**, 1578–93.

Moore, J. C., Orange, D. L. and Kulm, L. D. (1990b). Interrelationship of fluid venting and structural evolution, Oregon margin. *Journal of Geophysical Research*, **95**, 8795–808.

Moore, J. C., Orange, D., Stiles, S. and Geddes, D. (1990c). Fluid and hydrocarbon migration at active margins: a conceptual synthesis of results from the Barbados Ridge and Pacific Northwest. *American Association of Petroleum Geologists Bulletin*, **74**, 724–5.

Moore, T. E., Fuis, G. S., O'Sullivan, P. B. and Murphy, J. M. (1994). Evidence of Laramide age deformation in the Brooks Range, Alaska. In *Geological Society of America, 1994 Annual Meeting*, Seattle, WA, *Abstracts with Programs*, vol. 26. Boulder, CO: Geological Society of America, p. 383.

Moore, J. C., Shipley, T. H., Goldberg, D. *et al.* (1995). Abnormal fluid pressures and fault-zone dilation in the Barbados accretionary prism: evidence from logging while drilling. *Geology*, **23**, 605–8.

Moore, J. C. and Vrolijk, P. (1992). Fluids in accretionary prisms. *Reviews of Geophysics*, **30**, 130–5.

Moore, J. N. and Adams, M. C. (1988). Evolution of the thermal cap in two wells from the Salton Sea geothermal system, California. *Geothermics*, **17**, 695–710.

Morand, V. J. (1993). Structure and metamorphism of the Calliope Volcanic Assemblage; implications for Middle to Late Devonian orogeny in the northern New England fold belt. *Australian Journal of Earth Sciences*, **40**, 257–70.

Mordecai, M. and Morris, L. H. (1971). An investigation into the changes of permeability occurring in a sandstone when failed under triaxial stress condition. In *Dynamic Rock Mechanics, Proceedings of the 12th Symposium of Rock Mechanics*. American Institute of Mining, Metallurgical and Petroleum Engineers, ed. D. B. Clark, pp. 221–40.

Moretti, I. and Larrere, M. (1989). LOCACE: Computer-aided construction of balanced geological cross-section. *Geobyte*, **16**, 24.

Moretti, I. and Turcotte, D. L. (1985). A model for erosion, sedimentation, and flexure with application to New Caledonia. *Journal Geodynamics*, **3**, 155–68.

Morgan, J. K. and Karig, D. E. (1995). Kinematics and a balanced and restored cross-section across the toe of the eastern Nankai accretionary prism. *Journal of Structural Geology*, **17**, 31–45.

Morgan, P. (1984). The thermal structure and thermal evolution of the continental lithosphere. *Physics and Chemistry of the Earth*, **15**, 107–93.

Morin, R. and Silva, A. J. (1984). The effects of high pressure and high temperature on physical properties of ocean sediments. *Journal of Geophysical Research*, **89**, 511–26.

Morley, C. K. (1986). A classification of thrust fronts.

American Association of Petroleum Geologists Bulletin, **70**, 12–25.

(1987). Origin of a major cross-element, Moroccan Rif. *Geology*, **15**, 761–4.

(1992). Tectonic and sedimentary evidence for synchronous and out-of-sequence thrusting, Larache-Acilah area, Western Moroccan Rif. *Journal of Geological Society*, **149**, 39–49.

(1993). Discussion of origins of hinterland basins to the Rif-Betic Cordillera and Carpathians. In *The Origin of Sedimentary Basins – Inferences from Quantitative Modelling and Basin Analysis*, ed. S. Cloetingh, W. Sassi and F. Horváth. *Tectonophysics*, **226**, 359–76.

(1994). Fold-generated imbricates: examples from the Caledonides of Southern Norway. *Journal of Structural Geology*, **16**, 619–31.

Morris, G. A. and Nesbitt, B. E. (1998). Geology and timing of palaeohydrogeological events in the Mac-Kenzie Mountains, Northwest Territories, Canada. *Geological Society Special Publication*, vol. 144, pp. 161–72.

Morrow, C. A., Shi, L. Q. and Byerlee, J. D. (1981). Permeability and strength of San Andreas fault gouge under high pressure. *Geophysical Research Letters*, **8**, 325–8.

Morrow, D. W., Cumming, G. L. and Aulstead, K. L. (1990). The gas-bearing Devonian Manetoe facies, Yukon and Northwest Territories. *Bulletin of the Geological Survey of Canada, Report* 400.

Morrow, D. W. and Potter, J. (1998). Internal stratigraphy, petrography and porosity development of the Manetoe Dolomite in the region of the Pointed Mountain and Kotaneelee gas fields. In *Oil and Gas Pools of the Western Canada Sedimentary Basin*, ed. J. R. Hogg. *Canadian Society of Petroleum Geologists Special Publication*, vol. S-51, pp. 137–60.

Morse, J. D. (1977). Deformation in the ramp regions of over-thrust faults: experiments with small-scale rock models. *Wyoming Geological Association Guidebook*, pp. 457–70.

Morse, P. F., Purnell, G. W. and Medwedeff, D. A. (1991). Seismic modeling of fault-related folds. In *Seismic Modeling of Geologic Structures: Applications to Exploration Problems*, ed. S. W. Fagin. *Society of Exploration Geophysicists Geophysical Development Series*, vol. 2, pp. 127–52.

Mosar, J. and Suppe, J. (1992). Role of shear in fault-propagation folding. In *Thrust Tectonics*, ed. K. McClay. London: Chapman and Hall, pp. 123–32.

Mou, D. and Su, F. (2003). Tectonic evolution and its control on hydrocarbon generation and migration, northern Taiwan thrust belt. In *AAPG Annual Meeting*, Salt Lake City, UT, *Official Program*, vol. 12. Tulsa, OK: American Association of Petroleum Geologists, p. A124.

Mount, V. S., Suppe, J. and Hook, S. (1990). A forward modeling strategy for balancing cross sections. *American Association of Petroleum Geologists Bulletin*, **74**, 521–31.

Mueller, K. and Suppe, J. (1997). Growth of Wheeler Ridge anticline, California: geomorphic evidence for fault-bend folding behavior during earthquakes. In *Special Issue: Fault-related Folding*, ed. D. J. Anastasio, E. A. Erslev, D. M. Fisher and J. P. Evans. *Journal of Structural Geology*, **19**, 383–96.

Mugnier, J. L. and Vialon, P. (1986). Deformation and displacement of the Jura cover in its basement. *Journal of Structural Geology*, **8**, 373–88.

Müller, M., Nieburding, F. and Wanninger, A. (1977). Tectonic style and pressure distribution at the northern margin of the Alps between Lake Constance and the River Inn. *Geologische Rundschau*, **77**, 787–96.

Müller, M., Nieberding, F. and Weggen, K. (1992). Hindelang 1 (Bavarian Alps): a deep wildcat with implications for future exploration in the alpine thrust belt. In *Generation, Accumulation and Production of Europe's Hydrocarbons*, vol. II, ed. A. M. Spencer. Berlin: Springer-Verlag, pp. 193–9.

Mulugeta, G. (1988). Squeeze box in a centrifuge. *Tectonophysics*, **148**, 323–35.

Mulugeta, G. and Koyi, H. (1987). Three-dimensional geometry and kinematics of experimental piggyback thrusting. *Geology (Boulder)*, **15**, 1052–6.

(1992). Episodic accretion and strain partitioning in a model sand wedge. *Tectonophysics*, **202**, 319–33.

Munoz, J. A. (1992). Evolution of a continental collision belt: ECORS–Pyrenees crustal balanced cross-section. In *Thrust Tectonics*, ed. K. R. McClay. London: Chapman & Hall.

Murphy, M. A. and Yin, A. (2003). Structural evolution and sequence of thrusting in the Tethyan fold-thrust belt and Indus–Yalu suture zone, southwest Tibet. *Geological Society of America Bulletin*, **115**, 21–34.

Murphy, M. A., Yin, A., Harrison, T. M. *et al.* (1997). Did the Indo-Asian collision alone create the Tibetan Plateau? *Geology*, **25**, 719–22.

Mussman, W. J., Montanez, I. P. and Read, J. F. (1988). Ordovician Knox Paleokarst Unconformity, Appalachians. In *Paleokarst*, ed. N. P. James and P. W. Choquette. New York: Springer-Verlag, pp. 211–28.

Myers, R. E., McCarthy, T. S. and Stanistreet, I. G. (1989). A tectono-sedimentary reconstruction of the development and evolution of the Witwatersrand Basin, with particular emphasis on the Central Rand Group. *Information Circular, University of the Witwatersrand, Economic Geology Research Unit*, Report: vol. 216.

Nadai, A. (1963). *Theory of Flow and Fracture of Soil*, vol. 22. New York: McGraw-Hill.

Naeser, N. D., Naeser, C. W. and McCulloh, T. H. (1989). The application of fission-track dating to depositional and thermal history of rocks in sedimentary basins. In *Thermal History of Sedimentary Basins*, ed. N. D.

Naeser and T. H. McCulloh. New York: Springer-Verlag, pp. 157–80.

Nalivkin, D. V. (1973). *Geology of the USSR*. Edinburgh: Oliver & Boyd.

Nalpas, T., Douaran, S., Le Brun, J. P., Unternehr, P. and Richert, J. P. (1995). Inversion of the Broad Fourteens Basin (offshore Netherlands): a small-scale model investigation. *Sedimentary Geology*, **95**, 237–50.

Namson, J. S. and Davis, T. L. (1988). Seismically active fold and thrust belt in the San Joaquin Valley, central California. *Geological Society of America Bulletin*, **100**, 257–73.

(1990). Late Cenozoic fold and thrust belt of the Southern Coast Ranges and Santa Maria Basin, California. *American Association of Petroleum Geologists Bulletin*, **74**, 467–92.

Narr, W. (1993). Deformation of basement-involved, compressive structures. In *Laramide Deformation in the Rocky Mountain Foreland of the Western United States*, ed. C. J. Schmidt, R. B. Chase and E. A. Erslev. *Geological Society of America Special Paper*, vol. 280, pp. 107–24.

Narr, W. and Suppe, J. (1994). Kinematics of basement-involved compressive structures. *American Journal of Science*, **294**, 802–60.

Nash, D. B. (1980). Morphologic dating of degraded normal fault scarps. *Journal of Geology*, **88**, 353–60.

Natali, S., Roux, R., Dea, P. and Barrett, F. (2000). Cave Gulch 3-D survey, Wind River Basin, Wyoming. *The Mountain Geologist*, **37**, 3–13.

Natalin, B. A. and Chernysh, S. G. (1992). Tipy i istoriya deformatsiy osadochnogo vypolneniya i fundamenta Sredneamurskoy vpadiny. Types and history of sedimentary cover and basement deformation of the Middle Amur Basin. *Tikhookeanskaya Geologiya, Pacific Geology*, **1992**, 43–61.

Navarro, V. D. (1991). Hercynian structures in the Herrera unit, eastern branch of the Paleozoic massif in the Iberian Cordillera. *Boletin Geologico y Minero*, **102**, 830–37.

Naylor, M. A. and Spring, L. Y. (2002). Exploration strategy development and performance management: a portfolio-based approach. *The Leading Edge*, **21**, 159–67.

Neild, D. A. (1968). Onset of thermohaline convection in a porous medium. *Water Resource Research*, **4**, 553–60.

Nelson, R. A. (1981). Significance of fracture sets associated with stylolite zones. *American Association of Petroleum Geologists Bulletin*, **65**, 2417–25.

(1985). *Geologic Analysis of Naturally Fractured Reservoirs*. Houston: Gulf Professional Publishing.

Nemčok, M., Coward, M. P., Sercombe, W. J. and Klecker, R. A. (1999). Structure of the West Carpathian accretionary wedge: insights from cross section construction and sandbox validation. *Physics and Chemistry of the Earth*, **24**, 659–65.

Nemčok, M., Gayer, R. A. and Miliorizos, M. (1995). Structural analysis of the inverted Bristol Channel Basin: implications for the geometry and timing of the fracture porosity. In *Basin Inversion*, ed. J. G. Buchanan and P. G. Buchanan. *Geological Society of London Special Publication*, vol. 82, pp. 355–92.

Nemčok, M., Henk, A., Gayer, R. A., Vandycke, S. and Hathaway, T. M. (2002). Strike-slip fault bridge fluid pumping mechanism: insights from field-based palaeostress analysis and numerical modeling. *Journal of Structural Geology*, **24**, 1885–901.

Nemčok, M., Hók, J., Kováč, P., Marko, F., Coward, M. P., Madarás, J., Houghton, J. J. and Bezák, V. (1998a). Tertiary extension development and extension/compression interplay in the West Carpathian mountain belt. *Tectonophysics*, **290**, 137–67.

Nemčok, M., Houghton, J. J. and Coward, M. P. (1998b). Strain partitioning along the western margin of the Carpathians. *Tectonophysics*, **292**, 119–43.

Nemčok, M. and Kantor, J. (1990). Movement study in the selected area of the Velky Bok Unit. *Regionálna Geológia Západných Karpát*, 75–83 (in Slovak).

Nemčok, M., Keith Jr., J. F. and Neese, D. G. (1996). Development and hydrocarbon potential of the central Carpathian Paleogene Basin, West Carpathians, Slovak Republic. *Memoires du Museum National d'Histoire Naturelle*, **170**, 321–42.

Nemčok, M., Nemčok, J., Wojtaszek, M., Ludhová, L., Klecker, R. A., Sercombe, W. J., Coward, M. P. and Keith, Jr. J. F. (2000). Results of 2D balancing along 20° and 21°30' longitude and pseudo-3D in the Smilno Tectonic Window: implications for shortening mechanisms of the West Carpathian accretionary wedge. *Geologica Carpathica*, **51**, 281–300.

Nemčok, M., Nemčok, J., Wojtaszek, M. *et al.* (2001). Reconstruction of Cretaceous rifts incorporated in the Outer West Carpathian wedge by balancing. In *The Hydrocarbon Potential of the Carpathian–Pannonian Region: Marine and Petroleum Geology*, vol. 18, ed. S. Cloetingh, M. Nemčok, F. Neubauer, F. Horváth and P. Seifert, pp. 39–63.

Nemčok, M., Pospíšil, L., Lexa, J. and Donelick, R. A. (1998c). Tertiary subduction and slab break-off model of the Carpathian–Pannonian region. *Tectonophysics*, **295**, 307–40.

Neurath, C. and Smith, R. B. (1982). The effect of material properties on growth rates of folding and boudinage: experiments with wax models. *Journal of Structural Geology*, **4**, 215–29.

Newman, J. and Mitra, G. (1994). Fluid-influenced deformation and recrystallization of dolomite at low temperatures along a natural fault zone, Mountain City Window, Tennessee. *Geological Society of America Bulletin*, **106**, 1267–80.

Newson, A. C. (2001). The future of natural gas exploration in the foothills of the Western Canadian Rocky Mountains. *The Leading Edge*, **20**, 74–9.

Nickelsen, R. P. (1979). Sequence of structural stages of the Alleghany orogeny at the Bear Valley Strip Mine,

Shamokin, Pennsylvania. *American Journal of Science*, **279**, 225–71.

(1986). Cleavage duplexes in the Marcellus Shale of the Appalachian foreland. *Journal of Structural Geology*, **8**, 361–71.

Nicol, A., Alloway, B. and Tonkin, P. (1994). Rates of deformation, uplift, and landscape development associated with active folding in the Waipara area of north Canterbury, New Zealand. *Tectonics*, **13**, 1327–44.

Nieuwland, D. A. and Saher, M. H. (2002). *Punctuated Equilibrium in Fold-and-Thrust Belts: New Insights from Analogue Modeling*. Amsterdam: Vrije Universiteit, de Boelelaan 1085, 1081 HV.

Niggli, P. (1948). *Gesteine und Minerallagerstaetten: Erster Band, Allgemeine Lehre von den Gesteinen und Minerallagerstaetten*. Basel: Verlag Birkhaeuser.

(1958). *Gesteins- und Minerallagerstatten*, vol. 1. Basel: Verlag Birkhauser.

Nikishin, A. M., Ziegler, P. A., Stephenson, R. A. *et al.* (1996). Late Precambrian to Triassic history of the East European craton: dynamics of sedimentary basin evolution. In *Europrobe: Interpolate Tectonics and Basin Dynamics of the Eastern European Platform*, ed. R. A. Stephenson, M. Wilson, H. de Boorder and V. I. Starostenko. *Tectonophysics*, **268**, 23–63.

Nizamuddin, M. (1997). Structural traps in the Potwar Basin, Pakistan. In *AAPG International Conference and Exhibition; Abstracts, AAPG Bulletin*, vol. 81. Tulsa, OK: American Association of Petroleum Geologists, pp. 1401–2.

Noble, R. A., Kaldi, J. G. and Atkinson, C. D. (1997). Oil saturations in shales: applications in seal evaluation. In *Seals, Traps, and the Petroleum System*, ed. R. C. Surdam. *American Association of Petroleum Geologists Memoir*, vol. 67, pp. 13–29.

Nokleberg, W. J., Plafker, G. and Wilson, F. (1994). Geology of south-central Alaska. In *The Geology of North America*, ed. G. Plafker and H. C. Berg. Boulder, CO: Geological Society of America.

Novoa, E., Suppe, J. and Shaw, J. H. (2000). Inclined-shear restoration of growth folds. *American Association of Petroleum Geologists Bulletin*, **84**, 787–804.

Nunn, J. A. and Deming, D. (1991). Thermal constraints on basin-scale flow systems. In *Crustal-Scale Fluid Transport – Magnitude and Mechanisms*, ed. T. Torgensen. *Geophysical Research Letters*, **18**, 967–70.

Nunn, J. A. and Meulbroek, P. (2002). Kilometer-scale upward migration of hydrocarbons in geopressured sediments by buoyancy-driven propagation of methane-filled fractures. *American Association of Petroleum Geologists Bulletin*, **86**, 907–18.

Nur, A. and Booker, J. R. (1972). Aftershocks caused by pore-fluid flow? *Science*, **175**, 885–87.

Oelkers, E. H., Bjorkum, P. A. and Murphy, W. M. (1996). A petrographic and computational investigation of quartz cementation and porosity reduction in the North Sea sandstones. *American Journal of Science*, **296**, 420–52.

O'Hara, I., Hower, J. G. and Rimmer, S. M. (1990). Constraints on the emplacement and uplift history of the Pine Mountain thrust sheet, eastern Kentucky: evidence from coal rank trends. *Journal of Geology*, **98**, 43–51.

Ohmori, K., Taira, A., Tokuyama, H., Sakaguchi, A., Okamura, M. and Aihara, A. (1997). Paleothermal structure of the Shimanto accretionary prism, Shikoku, Japan: role of an out-of-sequence thrust. *Geology*, **25**, 327–30.

Ojeda, H. A. O. (1982). Structural framework, stratigraphy, and evolution of Brazilian marginal basins. *American Association of Petroleum Geologists Bulletin*, **66**, 732–49.

Okulitch, A. V., Packard, J. J. and Zolnai, A. I. (1986). Evolution of the Boothia Uplift, Arctic Canada. *Canadian Journal of Earth Sciences*, **23**, 350–8.

Oldow, J. S., Bally, A. W., Avé Lallemant, H. G. and Leeman, W. P. (1989). Phanerozoic evolution of the North American Cordillera. In *The Geology of North America: an Overview*, ed. A. W. Bally and A. R. Palmer. Geological Society of America, pp. 139–232.

Oliver, J. (1986). Fluids expelled tectonically from orogenic belts: their role in hydrocarbon migration and other geologic phenomena. *Geology*, **14**, 99–102.

Olovyanishnikov, V. G., Siedlecka, A. and Roberts, D. (1997). Aspects of the geology of the Timans, Russia, and linkages with Varanger Peninsula, NE Norway. *Bulletin – Norges Geologiske Undersokelse*, **433**, 28–9.

O'Neil, J. R. (1968). Hydrogen and oxygen isotope fractionation between ice and water. *Journal of Physical Chemistry*, **72**, 3683–4.

Ord, A. and Hobbs, B. E. (1989). The strength of the continental crust, detachment zones and the development of plastic instabilities. *Tectonophysics*, **158**, 269–89.

Ori, G. G. and Friend, P. F. (1984). Sedimentary basins formed and carried piggyback on active thrust sheets. *Geology*, **12**, 475–78.

Ori, G. G., Roveri, M. and Vannoni, F. (1986). Plio-Pleistocene sedimentation in the Apenninic–Adriatic foredeep (Central Adriatic Sea, Italy). *International Association of Sedimentologists Special Publication*, vol. 8, 183–98.

Orr, W. L. (1986). Kerogen/asphaltene/sulfur relationships in sulfur-rich Monterey oils. *Organic Geochemistry*, **10**, 499–506.

Osborne, M. J. and Swarbrick, R. E. (1997). Mechanisms for generating overpressures in sedimentary basins: a reevaluation. *American Association of Petroleum Geologists Bulletin*, **81**, 1023–41.

O'Sullivan, P. B., Hanks, C. I., Wallace, W. K. and Green, P. F. (1995). Multiple episodes of Cenozoic denudation in the northeastern Brooks Range: fission-track data from the Okpilak batholith, Alaska. *Canadian Journal of Earth Sciences*, **32**, 1106–18.

O'Sullivan, P. B., Foster, D. A., Kohn, B. P. and Gleadow, A. J. W. (1996). Multiple postorogenic denudation events;

an example from the eastern Lachlan fold belt, Australia. *Geology*, **24**, 563–6.

O'Sullivan, P. B., Belton, D. X. and Orr, M. (2000). Post-orogenic thermotectonic history of the Mount Buffalo region, Lachlan fold belt, Australia; evidence for Mesozoic to Cenozoic wrench-fault reactivation. *Tectonophysics*, **317**, 1–26.

Otto, S. C., Tull, S. J., Macdonald, D., Voronova, L. and Blackbourn, G. (1997). Mesozoic–Cenozoic history of deformation and petroleum systems in sedimentary basins of Central Asia; implications of collisions on the Eurasian margin. In *Thematic Set; Habitat of Oil and Gas in the Former Soviet Union. Petroleum Geoscience*, London: Geological Society, vol. 3, pp. 327–41.

Oxburgh, E. R. and Turcotte, D. L. (1974). Thermal gradients and regional metamorphism in overthrust terrains with special reference to the Eastern Alps. *Schweizerische mineralogische und petrographische Mitteilungen*, **54**, 614–22.

Palciauskas, V. V. (1986). Models for thermal conductivity and permeability in normally compacting basins. In *Thermal Modeling of Sedimentary Basins*, ed. J. Burrus. Paris: Technip, pp. 323–36.

Paradigm Geophysical (1999). *GeoSec 3D 2.0 Overview*.

Paraschiv, D. and Olteanu, Gh. (1970). Oil fields in Miocene–Pliocene zone of eastern Carpathians (district of Ploiesti). *American Association of Petroleum Geologists Memoir*, vol. 14. Tulsa, OK: American Association of Petroleum Geologists, pp. 399–427.

Parfenov, L. M., Natapov, L. M., Sokolov, S. D. and Tsukanov, N. V. (1993). Terranes and accretionary tectonics of northeastern Asia. *Geotectonics*, **27**, 62–72.

Parris, T. M., Burruss, R. C. and O'Sullivan, P. B. (2003). Deformation and the timing of gas generation and migration in the eastern Brooks Range foothills, Arctic National Wildlife Refuge, Alaska. *American Association of Petroleum Geologists Bulletin*, **87**, 1823–46.

Parrish, J. T. (1982). Upwelling and petroleum source beds with reference to the Paleozoic. *American Association of Petroleum Geologists Bulletin*, **66**, 750–74.

Patchen, D. G. and Hohn, M. E. (1993). Production and production controls in Devonian shales, West Virginia. In *Petroleum Geology of the Devonian and Mississippian Black Shale of Eastern North America*, ed. J. B. Roen and R. C. Kepferle. *US Geological Survey Bulletin*, **1909**, L1–28.

Paterson, M. S. (1978). *Experimental Rock Deformation: the Brittle Field*. Berlin: Springer-Verlag.

(1987). Problems in the extrapolation of laboratory rheological data. *Tectonophysics*, **133**, 33–43.

Patriat, P., Segoufin, J., Schlich, R. *et al.* (1982). Les mouvements relatifs de l'Inde, de l'Afrique et de l'Eurasie. *Bulletin de la Societe Geologique de France*, **24**, 363, 73.

Patton, T. L., Logan, J. M. and Friedman, M. (1998). Experimentally generated normal faults in single- and multilayer limestone beams at confining pressure. *Tectonophysics*, **295**, 53–77.

Pavlis, T. L. and Bruhn, R. L. (1983). Deep-seated flow as mechanism for the uplift of broad fore arc ridges and its role in the exposition of high *P/T* metamorphic terranes. *Tectonics*, **2**, 473–97.

Pedersen, T. F. and Calvert, S. E. (1990). Anoxia vs. productivity: what controls the formation of organic-carbon-rich sediments and sedimentary rocks? *American Association of Petroleum Geologists Bulletin*, **74**, 454–66.

Penner, D. G. (1957). Turner Valley oil and gas field. *7th Annual Field Conference, Alberta Society of Petroleum Geologists Guidebook*, pp. 131–7.

Pennock, E. S. (1989). Structural interpretation of seismic reflection data from eastern Salt Range and Potwar Plateau, Pakistan. *American Association of Petroleum Geologists Bulletin*, **80**, 841–57.

Pennock, E. S., Lillie, R. J., Zaman, A. S. H. and Yousaf, M. (1989). Structural interpretation of seismic reflection data from eastern Salt Range and Potwar Plateau, Pakistan. *American Association of Petroleum Geologists Bulletin*, **73**, 841–57.

Peper, T. (1993). *Tectonic Control on the Sedimentary Record in Foreland Basins, Inferences from Quantitative Subsidence Analyses and Stratigraphic Modeling*. Amsterdam: Vrije University.

Pepper, A. S. (1991). Estimating the petroleum expulsion behavior of source rocks: a novel quantitative approach. In *Petroleum Migration*, ed. W. A. England and A. J. Fleet. *Special Publication of Geological Society of London*, vol. 59, pp. 9–31.

Pepper, A. S. and Corvi, P. J. (1995). Simple kinetic models of petroleum formation. Part I: Oil and gas generation from kerogen. *Marine and Petroleum Geology*, **12**, 291–319.

Pepper, A. S. and Dodd, T. A. (1995). Simple kinetic models of petroleum formation. Part II: Oil-gas cracking. *Marine and Petroleum Geology*, **12**, 321–40.

Pereira, E. B., Hamza, V. M., Furtado, V. V. and Adams, J. A. S. (1986). U, Th and K content, heat production and thermal conductivity of Sao Paulo, Brazil, continental shelf sediments: a reconnaissance work. *Chemical Geology*, **58**, 217–26.

Pereira, M. J. and Macedo, J. M. (1990). A Bacia de Santos: perspectivas de uma nova provincia petrolifera na plataforma continental Sudeste Brasileira. Santos Basin: the outlook for a new petroleum province on the southeastern Brazilian continental shelf. *Boletim de Geociencias da PETROBRAS*, **4**, 3–11.

Perrodon, A. (1980). Géodynamique pétrolière. *Genése et Répartition des Gisements d'Hydrocarbures*. Paris: Masson Elf-Aquitaine.

Perry, W. J. (1975). Tectonics of the western Valley and Ridge fold-belt, Pendleton County, West Virginia: a summary report. *USGS Journal of Research*, **3**, 583–8.

(1978). Sequential deformation in the Central Appalachians. *American Journal of Science, Appalachian Geodynamic Research*, **278**, 518–42.

Persson, K. S. (2001). Effective indenters and the development of double-vergent orogens: insights from analogue sand models. In *Tectonic Modeling: a Volume in Honor of Hans Ramberg*, ed. H. Koyi and N. S. Mancktelow. *Geological Society of America Memoir*, vol. 193, pp. 191–206.

Persson, K. S. and Soukotis, D. (2002). Analogue models of orogenic wedges controlled by erosion. *Tectonophysics*, **356**, 323–36.

Peters, K. E. (1986). Guidelines for evaluating petroleum source rock using programmed pyrolysis. *American Association of Petroleum Geologists Bulletin*, **70**, 318–29.

Peters, K. E. and Cassa, M. R. (1994). Applied source rock geochemistry. In *The Petroleum System – From Source to Trap*, ed. L. B. Magoon and W. G. Dow. *American Association of Petroleum Geologists Memoir*, vol. 60, pp. 93–120.

Peters, K. E., Pytte, M. H., Elam, T. D. and Sundararaman, P. (1994). Identification of petroleum systems adjacent to the San Andreas fault, California. In *The Petroleum System – From Source to Trap*, ed. L. B. Magoon and W. G. Dow. *American Association of Petroleum Geologists Memoir*, vol. 60, pp. 423–36.

Petersen, H. I., Rosenberg, P. and Andsbjerg, J. (1996). Organic geochemistry in relation to the depositional environments of Middle Jurassic coal seams, Danish Central Graben, and implications for hydrocarbon generative potential. *American Association of Petroleum Geologists Bulletin*, **80**, 47–62.

Petroy, D. E. and Wiens, D. A. (1989). Historical seismicity and implications for diffuse plate convergence in the Northeast Indian Ocean. *Journal of Geophysical Research*, B, *Solid Earth and Planets*, **94**, 12 301–19.

Petzet, A. (2000). Ekho deep test is spud amid San Joaquin Valley uncertainty. *Oil & Gas Journal*, **98**, 71–2.

Pfiffner, O. A. (1986). Evolution of the North Alpine foreland basin in the Central Alps. In Foreland Basins, ed. P. A. Allen and P. Homewood, *International Association of Sedimentologists and Society of Economic Paleontologists and Mineralogists; Foreland basins*, Fribourg, Switzerland, 1986, *Special Publication of the International Association of Sedimentologists*, vol. 8. Oxford: Blackwells, pp. 219–28.

Pfiffner, O. A. and Ramsay, J. G. (1982). Constraints on geological strain rates: arguments from finite strain states of naturally deformed rocks. *Journal of Geophysical Research*, **87**, 311–21.

Philip, H., Rogozhin, E., Cisternas, A., Bosquet, J. C., Borisov, B. and Karakhanian, A. (1992). The Armenian Earthquake of 1988 December 7: faulting and folding, neotectonics and palaeoseismicity. *Geophysical Journal International*, **110**, 141–58.

Philp, R. P. (1994). Geochemical characteristics of oils derived predominantly from terrigenous source materials. In *Coal and Coal-bearing Strata as Oil-prone Source Rocks*, ed. A. C. Scott and A. J. Fleet. *Geological Society of London Special Publication*, vol. 77, pp. 71–91.

Philp, R. P. and Lewis, C. A. (1987). Organic geochemistry of

biomarkers. *Annual Review of Earth and Planetary Science*, **15**, 363–95.

Pialli, G. and Alvarez, W. (1997). Tectonic setting of the Miocene Northern Apennines: the problem of contemporaneous compression and extension. *Developments in Palaeontology and Stratigraphy*, **15**, 167–85.

Pícha, F. J. (1996). Exploring for hydrocarbons under thrust belts: a challenging new frontier in the Carpathians and elsewhere. *American Association of Petroleum Geologists Bulletin*, **80**, 1547–64.

Pieri, M. (1989). Three seismic profiles through the Po Plain. In *Atlas of Seismic Stratigraphy*, ed. A. W. Bally. *American Association of Petroleum Geologists Studies*, vol. 27, pp. 90–110.

Pieri, M. and Groppi, G. (1981). Subsurface geological structure of the Po plain, Italy. Italy: C.N.R., Publication 414, Program Final Geodinamica, pp. 1–13.

Pillevuit, A., Marcoux, J., Stampfli, G. M. and Baud, A. (1997). The Oman exotics; a key to the understanding of the Neotethyan geodynamic evolution. *Geodinamica Acta*, **10**, 209–38.

Pinet, P. and Souriau, M. (1988). Continental erosion and large-scale relief. *Tectonics*, **7**, 563–82.

Piper, D. Z. and Link, P. K. (2002). An upwelling model for the Phosphoria sea: a Permian, ocean-margin sea in the northwest United States. *American Association of Petroleum Geologists Bulletin*, **86**, 1217–35.

Plafker, G. and Berg, H. C., eds (1994). Overview of the geology and tectonic evolution of Alaska. In *The Geology of Alaska*. Boulder, CO: Geological Society of America.

Plašienka, D. (1999). Definition and correlation of tectonic units with a special reference to some central Western Carpathian examples. *Mineralia Slovaca*, **31**, 3–16.

Platt, J. P. (1986a). Dynamics of orogenic wedges and the uplift of high-pressure metamorphic rocks. *Geological Society of America Bulletin*, **97**, 1037–53.

(1986b). Mechanics of frontal imbrication: a first-order analysis. *Geologische Rundschau*, **77**, 577–89.

Platt, J. P. and Leggett, J. K. (1986). Stratal extension in thrust footwalls, Makran accretionary prism: implications for thrust tectonics. *American Association of Petroleum Geologists Bulletin*, **70**, 191–203.

Platte River Associates (1995). BasinMod® 1-D for Windows™. Basin Modeling System. Document Version: 5.0.

Plesch, A. and Oncken, O. (1999). Orogenic wedge growth during collision: constraints on mechanics of a fossil wedge from its kinematic record (Rhenohercynian FTB, Central Europe). *Tectonophysics*, **309**, 117–39.

Poblet, J., McClay, K. R., Storti, F. and Muñoz, J. A. (1997). Geometries of syntectonic sediments associated with single-layer detachment folds. In *Special Issue, Fault-Related Folding*, ed. D. J. Anastasio, E. A. Erslev, D. M. Fisher and J. P. Evans. *Journal of Structural Geology*, **19**, pp. 369–81.

Poblet, J., Muñoz, J. A., Travé, A. and Serra-Keil, J. (1998). Quantifying the kinematics of detachment folds using the 3D geometry: the example of the Mediano anticline (Pyrenees, Spain). *Geological Society of America Bulletin*, **110**, 111–25.

Poelchau, H. S., Baker, D. R., Hantschel, T., Horsfield, B. and Wygrala, B. (1997). Basin simulation and the design of the conceptual basin model. In *Petroleum and Basin Evolution. Insights from Petroleum Geochemistry, Geology and Basin Modeling*, ed. D. H. Welte, B. Horsfield and D. R. Baker. Berlin: Springer, pp. 3–70.

Pogue, K. R., Hylland, M. D., Yeats, R. S., Khattak, W. U. and Hussain, A. (1999). Stratigraphic and structural framework of Himalayan foothills, northern Pakistan. In *Himalaya and Tibet; Mountain Roots to Mountain Tops*, ed. A. Macfarlane, R. B. Sorkhabi and J. Quade, *Geological Society of America Special Paper*, vol. 328. Boulder, CO: Geological Society of America, pp. 257–74.

Pollack, H. N. and Chapman, D. S. (1977). On the regional variation of heat flow, geotherms, and lithospheric thickness. *Tectonophysics*, **38**, 279–96.

Pollard, D. D. and Aydin, A. (1988). Progress in understanding jointing over the past century. *Geological Society of America Bulletin*, **100**, 1181–204.

Pollitz, F. F. (1986). Pliocene change in Pacific-plate motion. *Nature*, **320**, 738–41.

Potrel, A., Peucat, J. J. and Fanning, C. M. (1998). Archean crustal evolution of the West African Craton: example of the Amsaga area (Reguibat Rise). U-Pb and Sm-Nd evidence for crustal growth and recycling. *Precambrian Research*, **90**, 107–17.

Powell, T. G. (1986). Petroleum geochemistry and depositional setting of lacustrine source rocks. *Marine and Petroleum Geology*, **3**, 200–19.

Powell, T. G. and Boreham, C. J. (1994). Terrestrially sourced oils: where do they exist and what are our limits of knowledge? A geochemical perspective. In *Coal and Coal-bearing Strata as Oil-prone Source Rocks*, ed. A. C. Scott and A. J. Fleet. *Geological Society of London Special Publication*, vol. 77, pp. 11–29.

Powers, M. K. (1989). Magnetostratigraphy and rock magnetism of Eocene foreland basin sediments, Esera and Isabena valleys, Tremp-Graus Basin, southern Pyrenees, Spain. MSc Thesis, Los Angeles: University of South California.

Powley, D. E. (1980). Pressures, normal and abnormal. *American Association of Petroleum Geologists Advanced Exploration Schools Unpublished Lecture Notes*.

Price, L. C. (1973). The solubility of hydrocarbons and petroleum in water as applied to the primary migration of petroleum. Ph.D. thesis, Riverside: University of California.

(1981). Aqueous solubility of crude oil to 400 °C and 2,000 bars pressure in the presence of gas. *Journal of Petroleum Geology*, **4**, 195–223.

Price, L. C. and Schoell, M. (1995). Constraints on the origins of hydrocarbon gas from compositions of gases at their site of origin. *Nature*, **378**, 368–71.

Price, L. C. and Wenger, L. M. (1992). The influence of pressure on petroleum generation and maturation as suggested by aqueous pyrolysis. *Organic Geochemistry*, **19**, 141–59.

Price, R. A. (1973). Large-scale gravitational flow of supracrustal rocks, Southern Canadian Rockies. In *Gravity and Tectonics*, ed. K. A. de Jong and R. Scholten. New York: John Wiley, pp. 491–502.

(1986). The southeastern Canadian Cordillera: thrust faulting, tectonic wedging, and delamination of the lithosphere. *Journal of Structural Geology*, **8**, 239–54.

(1988). The mechanical paradox of large overthrusts. *Geological Society of America Bulletin*, **100**, 1898–1908.

Prieto, R. and Valdes, G. (1992). El Furrial oil field. In *Giant Oil and Gas Fields of the Decade: 1978–1988*, ed. M. T. Halbouty. *American Association of Petroleum Geologists Memoir*, vol. 54, pp. 155–61.

Prince, C. M. (1999). Textural and diagenetic controls on sandstone permeability. In *Forty-ninth Annual Convention of the Gulf Coast Association of Geological Societies, American Association of Petroleum Geologists Regional Meeting and the 46th Annual Convention of the Gulf Coast Section SEPM*, ed. W. C. Terrell and L. Czernikowski. *Transactions, Gulf Coast Association of Geological Societies*, **49**, 42–53.

Prior, D. J. and Behrmann, J. H. (1990). Thrust related mudstone fabrics from the Barbados forearc: a backscattered electron microscope study. *Journal of Geophysical Research*, **85**, 9055–67.

Protzman, G. and Mitra, G. (1990). Strain fabric associated with the Meade thrust sheet: implications for cross-section balancing. *Journal of Structural Geology*, **12**, 403–17.

Prucha, J. J., Graham, J. A. and Nickelsen, R. P. (1965). Basement-controlled deformation in Wyoming Province of Rocky Mountains Foreland. *American Association of Petroleum Geologists Bulletin*, **49**, 966–92.

Puchkov, V. N. (1991). The Paleozoic of the Uralo–Mongolian fold system. *Occasional Publications ESRI*, New Series, vol. 7, p. 69.

(1997). Structure and geodynamics of the Uralian Orogen. In *Orogeny Through Time*, ed. J.P. Burg and M. Ford. *Geological Society Special Publications*, vol.121. London: Geological Society of London, pp. 201–36.

Puigdefabregas, C., Munoz, J. A. and Verges, J. (1992). Thrusting and foreland basin evolution in the southern Pyrenees. In *Thrust Tectonics*, ed. K. R. McClay. London: Chapman and Hall, pp. 247–54.

Purday, N. and Sutton, J. (2004). Volume interpretation, visualization mature. *Hart's E&P*, **77**, 36–40.

Qing, H. and Mountjoy, E. W. (1994). *Formation of Coarsely Crystalline, Hydrothermal Dolomite Reservoirs in the Presqu'ile Barrier, Western Canada Sedimentary Basin. American Association of Petroleum Geologists Bulletin*, **78**, 55–77.

Quigley, T. M. and Mackenzie, A. S. (1988). The temperatures of oil and gas formation in the subsurface. *Nature*, **333**, 549–52.

Radney, B. and Byerlee, J. D. (1988). Laboratory studies of the shear strength of montmorillonite and illite under crustal conditions. *EOS, Transactions, American Geophysical Union*, **69**, 1463.

Ramos, V. A. (1989). Andean foothills structures in northern Magallanes Basin, Argentina. *AAPG Bulletin*, **73**, 887–903.

Ramsay, J. G. (1967). *Folding and Fracturing of Rocks*. New York: McGraw-Hill.

Ramsay, J. G. and Huber, M. I. (1983). *The Techniques of Modern Structural Geology*, vol. 1 *Strain Analysis*. London: Academic Press.

  (1987). *The Techniques of Modern Structural Geology*, vol. 2 *Fractures*. London: Academic Press.

Ranalli, G. and Murphy, D. C. (1987). Rheological stratification of the lithosphere. *Tectonophysics*, **132**, 281–95.

Ranganathan, V. and Hanor, J. S. (1987). A numerical model for the formation of saline waters due to diffusion of dissolved NaCl in subsiding sedimentary basins with evaporites. *Journal of Hydrology*, **92**, 97–102.

Rankin, D. W., Drake, A. A., Jr., Glover, L., III *et al.* (1989). Pre-orogenic terranes. In *The Appalachian–Ouachita Orogen in the United States*, ed. R. D. Hatcher Jr., W. A. Thomas and G. W. Viele. Geological Society of America, *Geology of North America*, vol. F-2, pp. 7–100.

Raoult, J. F. and Meilliez, F. (1987). The Variscan front and the Midi Fault between the Channel and the Meuse River. *Journal of Structural Geology*, **9**, 473–9.

Ratschbacher, L., Frisch, W. Linzer, H. G. and Merle, O. (1991). Lateral extrusion in the Eastern Alps, part 2, structural analysis. *Tectonics*, **10**, 257–71.

Raymond, A. C. and Murchison, D. G. (1988). Development of organic maturation in the thermal aureoles of sills and its relation to sediment compaction. *Fuel*, **67**, 1599–608.

  (1991). Short paper: the relationship between organic maturation, the widths of thermal aureoles and the thicknesses of sills in the Midland Valley of Scotland and Northern England. *Journal of the Geological Society of London*, **148**, 215–18.

Reches, Z. and Johnson, A. M. (1978). Theoretical analysis of monoclines. In *Laramide Folding Associated with Basement Block Faulting in the Western United States*, ed. V. Mathews III. *Geological Society of America Memoir*, vol. 151, pp. 273–331.

Reid, S. A. and McIntyre, J. L. (2001). Monterey Formation porcelanite reservoirs of the Elk Hills field, Kern County, California. *American Association of Petroleum Geologists Bulletin*, vol. 85, 169–89.

Renton, J. J., Heald, M. T. and Cecil, C. B. (1969). Experimental investigation of pressure solution of quartz. *Journal of Sedimentary Petrology*, **39**, 1107–17.

Reynolds, J. G. and Burnham, A. K. (1995). Comparison of kinetic analysis of source rocks and kerogen concentrates. *Organic Geochemistry*, **23**, 11–19.

Ribando, R. J. and Torrance, K. E. (1976). Natural convection in a porous medium: effects of confinement, variable permeability and thermal boundary conditions. *Journal of the Heat Transfer, Transactions of the ASME*, **98**, 42–8.

Ricci Lucchi, F. (1986). The Oligocene to Recent foreland basins of the Northern Apennines. In *Foreland Basins*, ed. P. A. Allen and P. Homewood, *Conference of the International Association of Sedimentologists and Society of Economic Paleontologists and Mineralogists; Foreland Basins*, Fribourg, Switzerland, 1986, *Special Publication of the International Association of Sedimentologists*, vol. 8. Oxford: Blackwells, pp. 105–39.

Rich, J. L. (1934). Mechanics of low angle overthrust faulting as illustrated by the Cumberland thrust block, Virginia, Kentucky, Tennessee. *American Association of Petroleum Geologists Bulletin*, **18**, 1584–96.

Ricou, L. E. (1996). *The Plate Tectonic History of the Past Tethys Ocean*. New York: Plenum Press.

Ridley, J. and Casey, M. (1989). Numerical modeling of folding in rotational strain histories: strain regimes expected in thrust belts and shear zones. *Geology*, **17**, 875–8.

Riffault, R. (1969). *Catalogue des Characteristiques Géologiques et Mécaniques de Quelques Roches Française*. Laboratoire Central des Ponts et Chaussées.

Rigatti, V., Fox, A., Roden, R. *et al.* (2001). 3-D PSDM case history in a thrust belt: Quiriquire block, Eastern Venezuela Basin. *The Leading Edge*, **20**, 514–18.

Riley, D. J., Sanderson, D. J. and Harkness, R. M. (1995). Mechanics of duplex development: a numerical model. *EOS, Transactions, American Geophysical Union*, **76**, 574–5.

Robert, P. (1988). *Organic Metamorphism and Geothermal History*. Dordrecht: Elf-Aquitaine and Reidel Publishing.

Roberts, D. G. (1989). *Basin Inversion In and Around the British Isles. Geological Society of London Special Publication*, vol. 44, pp. 3–16.

Roberts, G. P. (1991). Structural controls on fluid migration through the Rencurel thrust zone, Vercors, French Sub-Alpine chains. In *Petroleum Migration*, ed. W. A. England and A. J. Fleet. *Geological Society of London Special Publication*, vol. 59, pp. 245–62.

Roberts, S. J. and Nunn, J. A. (1995). Episodic fluid expulsion from geopressured sediments. *Marine and Petroleum Geology*, **12**, 195–204.

Roberts, S. J., Nunn, J. A., Cathles, L. and Cipriani, F. D. (1996). Expulsion of abnormally pressured fluids along faults. *Journal of Geophysical Research*, **101**, 28 231–52.

Robertson, A. H. F., Clift, P. D., Degnan, P. J. and Jones, G. (1991). Palaeogeographic and palaeotectonic evolution of the eastern Mediterranean Neotethys. In *Palaeogeography and Palaeoceanography of Tethys: Palaeogeography, Palaeoclimatology, Palaeoecology*, vol. 87, ed. J. E. T. Channell, E. L. Winterer and L. F. Jansa, pp. 289–343.

Robertson, P. and Burke, K. (1989). Evolution of southern Caribbean Plate boundary, vicinity of Trinidad and Tobago, *AAPG Bulletin*, **73**, 490–509.

Robin, M., Rosenberg, E. and Fassi-Fihri, O. (1995). Wettability studies at the pore level: a new approach by use of cryo-SEM. *SPE Formation Evaluation*, 11–19.

Robinson, V. D. and Engel, M. H. (1993). Characterization of the source horizons within the Late Cretaceous transgressive sequence of Egypt. In *Source Rocks in a Sequence Stratigraphic Framework*, ed. B. J. Katz and L. M. Pratt. *American Association of Petroleum Geologists Studies in Geology*, vol. 37, pp. 101–17.

Roca, E. and Desegaulx, P. (1989). *Geodynamic Evolution of the Valencia Trough from Mesozoic Extensional Basin to the Early Miocene Foredeep of the Pyrenees*. Minneapolis: UMSI, 89/59.

Rockwell, T. K., Keller, E. A. Clark, M. N. and Johnson, D. L. (1984). Chronology and rates of faulting of Ventura River terraces, California. *Geological Society of America Bulletin*, **95**, 1466–74.

Rockwell, T. K., Keller, E. A. and Dembroff, G. R. (1988). Quaternary rate of folding of the Ventura anticline, western Transverse Ranges, southern California. *American Geological Society Bulletin*, **100**, 850–8.

Rodgers, J. (1963). Mechanics of Appalachian foreland folding in Pennsylvania and West Virginia. *AAPG Bulletin*, **47**, 1527–36.

(1970). The Valley and Ridge and Appalachian Plateau; structure and tectonics; the Pulaski Fault, and the extent of Cambrian evaporites in the central and southern Appalachians. In *Studies of Appalachian Geology, Central and Southern*, ed. K. N. Weaver. New York: Interscience, pp. 175–8.

(1987). Chains of basement uplifts within cratons marginal to orogenic belts. *American Journal of Science*, **287**, 661–92.

(1995). Lines of basement uplifts within the external parts of orogenic belts. *American Journal of Science*, **295**, 455–87.

Rogers, C. L., De Cserna, Z., Ojeda Rivera, J., Tavera Amezcua, E. and Van Vloten, R. (1962). Tectonic framework of an area within the Sierra Madre Oriental and adjacent Mesa Central, north central Mexico. *US Geological Society Professional Papers*, **450-C**, article 68, C21–4.

Rose, P. R. (2001). *Risk Analysis and Management of Petroleum Exploration Ventures. Methods in Exploration Series*, vol. 12.

Rossetti, F., Faccenna, C., Ranalli, G., Funiciello, R. and Storti, F. (2001). Modeling of temperature-dependent strength in orogenic wedges: first results from a new thermomechanical apparatus. *Geological Society of America Memoir*, vol. 193, pp. 253–9.

Rossetti, F., Faccenna, C., Ranalli, G. and Storti, F. (2000). Convergence fate-dependent growth of experimental viscous orogenic wedges. *Earth and Planetary Science Letters*, **178**, 367–72.

Rossetti, F., Faccenna, C., Ranalli, G., Storti, F. and Funiciello, R. (1999). Modeling the evolution of viscous orogenic wedges using paraffin as an analogue material. *European Union of Geosciences, 10th Meeting*, Strasbourg, France. *Journal of Conference Abstracts*, **4**, 607.

Rossetti, F., Ranalli, G. and Faccenna, C. (1999). The rheology of paraffin as an analogue material for viscous crustal deformation. *Journal of Structural Geology*, **21**, 413–7.

Roure, F., Choukroune, P., Berastegui, X. and Munoz, J. A. (1989). ECORS deep seismic data and balanced cross sections, geometric constraints on the evolution of the Pyrenees. *Tectonics*, **8**, 41–50.

Roure, F., Roca, E. and Sassi, W. (1993). The Neogene evolution of the outer Carpathian flysch units (Poland, Ukraine and Romania): kinematics of a foreland/fold-and-thrust belt system. *Sedimentary Geology*, **86**, 177–201.

Roure, F. and Sassi, W. (1995). Kinematics of deformation and petroleum system appraisal in Neogene foreland fold-and-thrust belts. *Petroleum Geoscience*, **1**, 253–69.

Rowan, M. G. (1997). Three-dimensional geometry and evolution of a segmented detachment fold, Mississippi Fan foldbelt, Gulf of Mexico. In *Special Issue: Fault-related Folding*, ed. D. J. Anastasio, E. A. Erslev, D. M. Fisher and J. P. Evans. *Journal of Structural Geology*, **19**, 463–80.

Rowan, M. G. and Linares, R. (2000). Fold-evolution matrices and axial-surface analysis of fault-bend folds: application to the Medina anticline, Eastern Cordillera, Colombia. *American Association of Petroleum Geologists Bulletin*, **84**, 741–64.

Rowan, M. G., Peel, F. J. and Vendeville, B. C. (2004). Gravity-driven foldbelts on passive margins. In *Thrust Belts and Petroleum Systems: AAPG Memoir* (in Press).

Rowan, M. G., Trudgill, B. D. and Fiduk, J. C. (2000). Deep-water, salt-cored foldbelts: lessons from the Mississippi Fan and Perdido foldbelts, northern Gulf of Mexico. In *Atlantic Rifts and Continental Margins*, ed. W. U. Mohriak and M. Talwani. *Geophysical Monograph*, vol. 115, pp. 173–91.

Rowlands, D. and Kanes, W. H. (1972). The structural geology of a portion of the Broadtop synclinorium, Maryland and south-central Pennsylvania. In *Appalachian Structures – Origin, Evolution, and Possible Potential for New Exploration Frontiers*, ed. W. H. Kanes. West Virginia University and West Virginia Geological and Economic Survey, pp. 195–225.

Roy, R. F., Beck, A. E. and Toulokian, Y. S. (1981). Thermophysical properties of rocks. In *Physical Properties of Rocks and Minerals*, ed. Y. S. Toulokian, W. R. Judd and R. F. Roy. New York: McGraw-Hill, pp. 409–88.

Royden, L. (1986). A simple method for analyzing subsidence and heat flow in extensional basins. In *Thermal Modeling in Sedimentary Basins*, ed. J. Burrus. Paris: Editions Technip. pp. 49–73.

(1993). The steady state thermal regime of eroding orogenic

belts and accretionary prisms. *Journal of Geophysical Research*, **98**, 4487–507.

Royse, F. (1993). An overview of the geologic structure of the thrust belt in Wyoming, northern Utah, and eastern Idaho. In *Geology of Wyoming*, ed. A. Snoke, J. Steidtmann and S. Roberts. *Geological Survey of Wyoming Memoir*, vol. 5, pp. 273–311.

Royse, F, Warner, M. A. and Reese, E. L. (1975). Thrust belt of Wyoming, Idaho and northern Utah: structural geometry and related stratigraphic problems. In *Deep Drilling Frontiers of the Central Rocky Mountains*, ed. D. W. Boylard. Rocky Mountain: Association of Geologists Guidebook, pp. 41–50.

Rubey, W. W. and Hubbert, M. K. (1960). Role of fluid pressure in mechanics of overthrust faulting, II: Overthrust belt in geosynclinal area of western Wyoming in light of fluid pressure hypothesis. *Geological Society of America Bulletin*, **60**, 167–205.

Rubin, C. M. and Saleeby, J. B. (1992). Thrust tectonics and Cretaceous intracontinental shortening in Southeast Alaska. In *Thrust Tectonics*, ed. K. R. McClay. London: Chapman & Hall, pp. 407–17.

Rubin, H. (1975). On the analysis of cellular convection in porous media. *International Journal of Heat and Mass Transfer*, **18**, 1483–6.

Ruble, T. E., Lewan, M. D. and Philp, R. P. (2001). New insights on the Green River petroleum system in the Uinta basin from hydrous pyrolysis experiments. *American Association of Petroleum Geologists Bulletin*, **85**, 1333–71.

(2003). New insights on the Green River petroleum system in the Uinta basin from hydrous pyrolysis experiments. Reply. *American Association of Petroleum Geologists Bulletin*, **87**, 1535–41.

Ruble, T. E. and Philp, R. P. (1998). Stratigraphy, depositional environments and organic geochemistry of source-rocks in the Green River petroleum system, Uinta basin, Utah. In *Modern and Ancient Lake Systems: New Problems and Perspectives*, ed. J. K. Pitman and A. R. Carroll. Salt Lake City, Utah: Geological Association Guidebook 26, pp. 289–328.

Rumelhart, P. E., Yin, A., Butler, R., *et al.* (1997). Oligocene initiation of deformation of northern Tibet, evidence from the Tarim Basin, northwestern China. In *Geological Society of America, 1997 Annual Meeting*, Salt Lake City, UT, *Abstracts with Programs*, vol. 29. Boulder, CO: Geological Society of America, p. 143.

Ruppel, C. and Hodges, K. V. (1994a). Pressure–temperature–time paths from two-dimensional thermal models: prograde, retrograde, and inverted metamorphism. *Tectonics*, **13**, 17–44.

(1994b). Role of horizontal thermal conduction and finite time thrust emplacement in simulation of pressure–temperature–time paths. *Earth and Planetary Science Letters*, **123**, 49–60.

Rutter, E. H. and Brodie, K. H. (1991). Lithosphere rheology: a note of caution. *Journal of Structural Geology*, **13**, 363–7.

Rutter, E. H., Maddock, R. H., Hall, S. H. and White, S. H. (1986). Comparative microstructures of natural and experimentally produced clay-bearing fault gouges. *Pure and Applied Geophysics*, **124**, 3–30.

Rybach, L. (1976). Radioactive heat production in rocks and its relation to other petrophysical parameters. *Pure and Applied Geophysics*, **114**, 309–17.

(1986). Amount and significance of radioactive heat sources in sediments. In *Thermal Modeling in Sedimentary Basins*, ed. J. Burrus. Paris: TECHNIP, pp. 311–22.

Ryder, R. T., Burruss, R. C. and Hatch, J. R. (1998). Black shale source rocks and oil generation in the Cambrian and Ordovician of the central Appalachian Basin, USA. *American Association of Petroleum Geologists Bulletin*, **82**, 412–41.

Ryder, R. T. and Zagorski, W. A. (2003). Nature, origin, and production characteristics of the Lower Silurian regional oil and gas accumulation, central Appalachian basin, United States. *American Association of Petroleum Geologists Bulletin*, **87**, 847–72.

Saez, A. (1987). *Estratigrafía y Sedimentología de las Formaciones Lacustres del Tránsito Eoceno–Oligoceno del NE de la Cuenca del Ebro*. Ph.D. Thesis, Barcelona: University of Barcelona.

Saffer, D. M. and Bekins, B. A. (2002). Hydrologic controls on the morphology and mechanics of accretionary wedges. *Geology*, **30**, 271–4.

Sage, L., Mosconi, A., Moretti, I., Riva, E. and Roure, F. (1991). Cross section balancing in the central Apennines: an application of LOCACE. *American Association of Petroleum Geologists Bulletin*, **75**, 832–44.

Saint Bezar, B., Frizon, D., Frizon de Lamotte, D., Morel, J. L., Mercier, E. (1998). Kinematics of large scale tip line folds from the High Atlas thrust belt, Morocco. *Journal of Structural Geology*, **20**, 999–1011.

Sales, J. K. (1997). Seal strength vs. trap closure: a fundamental control on the distribution of oil and gas. In *Seals, Traps, and the Petroleum System*, ed. R. C. Surdam. *American Association of Petroleum Geologists Memoir*, vol. 67, pp. 57–83.

Sammis, C. G., Osborne, R. H., Anderson, J. L., Banerdt, M. and White, P. (1986). Self-similar cataclasis in the formation of fault gouge. *Pure and Applied Geophysics*, **124**, 191–213.

Sample, J. C. (1990). The effect of carbonate cementation of underthrust sediments on deformation styles during underplating. *Journal of Geophysical Research*, **95**, 9111–21.

Sanderson, D. J. (1982). Models of strain variation in nappes and thrust sheets: a review. *Tectonophysics*, **88**, 201–33.

Sanlav, F., Tolgay, M. and Genca, M. (1963). Geology, geophysics, and production history of the Garzan–Germik field, Turkey. *6th World Petroleum Congress Proceedings*, section 1, pp. 749–71.

Sarewitz, D. (1988). High rates of late Cenozoic crustal shortening in the Andean Foreland, Mendoza Province, Argentina. *Geology (Boulder)*, **16**, 1138–42.

Sassi, W., Guiton, M. L. E., Daniel, J. M., Faure, J. L., Mengus, J. M., Schmidtz, J., Delisle, S., Leroy, Y. M. and Massot, J. (2003). Mechanical reconstruction of fracture development in Weber Sandstone Formation, Split Mountain (Utah). *American Association of Petroleum Geologists Annual Meeting, Salt Lake City, 2003*, Official Program, 12, pp. A151–2.

Saunders, I. and Young, A. (1983). Rates of surface processes on slopes, slope retreat and denudation. *Earth Surface Processes and Landforms*, **8**, 473–501.

Savin, S. M. and Epstein, S. (1970). The oxygen and hydrogen isotope geochemistry of clay minerals. *Geochimica et Cosmochimica Acta*, **51**, 1727–41.

Schamel, S. (1991). Middle and Upper Madgalena basins, Colombia. In *Active Margin Basins*, ed. K. T. Biddle. *American Association of Petroleum Geologists Memoir*, vol. 52, pp. 283–301.

(1993). *Hydrocarbon Prospects in the Pechora Basin: ESRI Technical Report 92–09–373(1)*. Columbia, SC: University of South Carolina, Earth Science & Resources Institute.

Schedl, A. and Wiltschko, D. V. (1987). Possible effects of pre-existing basement topography on thrust fault ramping. *Journal of Structural Geology*, **9**, 1029–37.

Schelling, D. D. (1999). Frontal structural geometries and detachment tectonics of the northeastern Karachi Arc, southern Kirthar Range, Pakistan. In *Himalaya and Tibet; Mountain Roots to Mountain Tops*, ed. A. Macfarlane, R. B. Sorkhabi and J. Quade, *Geological Society of America Special Paper*, vol. 328. Boulder, CO: Geological Society of America, pp. 287–302.

Schelling, D. D., Pilifosov, V. and Vasilyev, B. (1998). Tectonic framework and evolution of West Kazakhstan. In *American Association of Petroleum Geologists 1998 Annual Meeting*, Salt Lake City, UT, *Annual Meeting Expanded Abstracts*, vol. 1998. Tulsa, OK: American Association of Petroleum Geologists and Society of Economic Paleontologists and Mineralogists.

Schelling, D. D., Wawrek, D. A., Mirzoev, D., Dashtiyev, Z. and Shlygin, D. (2003). Structural geology and petroleum systems of the Dagestan fold-thrust belt and adjacent Terek-Caspian foredeep. *AAPG Annual Meeting*, Salt Lake City, UT, *Official Program*, vol. 12. Tulsa, OK: American Association of Petroleum Geologists, p. A152.

Schenk, H. J. and Horsfield, B. (1998). Using natural maturation series to evaluate the utility of parallel reaction kinetics models: an investigation of Toarcian shales and Carboniferous coals, Germany. Organic Geochemistry, **29**, 137–44.

Schenk, H. J., Horsfield, B., Krooss, B., Schaefer, R. G. and Schwochau, K. (1997). Kinetics of petroleum formation and cracking. In *Petroleum and Basin Evolution: Insights from Petroleum Geochemistry, Geology and Basin Modeling*, ed. D. H. Welte, B. Horsfield and D. R. Baker. Berlin: Springer-Verlag, pp. 233–69.

Schlunegger, F., Leu, W. and Matter, A. (1997). Sedimentary sequences, seismic facies, subsidence analysis, and evolution of the Burdigalian Upper Marine Molasse Group, central Switzerland. *American Association of Petroleum Geologists Bulletin*, vol. 81, pp. 1185–207.

Schmid, S. M. (1982). Microfabric studies as indicators of deformation mechanisms and flow laws operative in mountain building. In *Mountain Building Processes*, ed. K. J. Hsu. New York: Academic Press, pp. 95–110.

Schmidt, C. J. and Garihan, J. M. (1983). Laramide tectonic development of the Rocky Mountain foreland of southwestern Montana. *Rocky Mountain Association of Geologists Field Conference*, 1983, pp. 271–94.

Schmidt, C. J., O'Neill, J. M. and Brandon, W. C. (1988). Influence of Rocky Mountain foreland uplifts on the development of the frontal fold and thrust belt, south-western Montana. In Interaction of the Rocky Mountain Foreland and the Cordilleran thrust belt, ed. C. J. Schmidt and W. J. Perry Jr. *Geological Society of America Memoir*, vol. 171, pp. 171–201.

Schmoker, J. W. and Halley, R. B. (1982). Carbonate porosity versus depth: a predictable relation for South Florida. *American Association of Petroleum Geologists Bulletin*, **66**, 2561–70.

Schmucker, U. (1969). Conductivity anomalies, with special reference to the Andes. In *The Application of Modern Physics to the Earth and Planetary Interiors*, ed. S. K. Runcorn. New York: Wiley-Interscience, pp. 125–38.

Schoell, M., Hwang, R. J., Carlson, R. M. K. and Welton, J. E. (1994). Carbon isotopic composition of individual biomarkers in gilsonites. *Organic Geochemistry*, **21**, 673–83.

Scholl, D. W., Christensen, M. N., Huene, R. and Marlow, M. S. (1970). Peru–Chile Trench sediments and sea floor spreading. *Geological Society of America Bulletin*, **61**, 1339–60.

Scholz, C. H. (1968). Microfracturing and the inelastic deformation of rock in compression. *Journal of Geophysical Research*, **73**, 1417–32.

(1980). Shear heating and the state of stress on faults. *Journal of Geophysical Research*, **85**, 6174–84.

(1990). *The Mechanics of Earthquakes and Faulting*. New York: Cambridge University Press.

Schon, J. (1983). *Petrophysik: Physikalische Eigenschaften von Gesteinen und Mineralen*. Stuttgart: Ferdinand Enke Verlag.

Schonborn, G. (1992) Kinematics of a transverse zone in the southern Alps. In *Thrust Tectonics*, ed. K. R. McClay. London: Chapman and Hall, pp. 299–310.

Schowalter, T. T. (1979). Mechanics of secondary hydrocarbon migration and entrapment. *American Association of Petroleum Geologists Bulletin*, **63**, 723–60.

Schwartzkopf, T. A. (1990). Relationship between petroleum generation, migration and sandstone diagenesis, Middle Jurassic, Gifhorn Trough, N. Germany. *Marine and Petroleum Geology*, **7**, 153–68.

Sclater, J. G. and Christie, P. A. F. (1980). Continental stretching: an explanation of the post mid-Cretaceous subsidence of the central North Sea basin. *Journal of Geophysical Research*, **85**, 3711–39.

Screaton, E. J., Wuthrich, D. R. and Dreiss, S. J. (1990). Permeabilities, fluid pressures and flow rates in the Barbados Ridge Complex. *Journal of Geophysical Research*, **95**, 8997–9008.

Sebrier, M., Mercier, J. L., Machare, J., Bonnot, D., Cabrera, J. and Blanc, J. L. (1988). The Andes of central Peru. *Tectonics*, **7**, 895–928.

Secor, D. T. (1965). Role of fluid pressure in jointing. *American Journal of Science*, **263**, 633–46.

Seewald, J. S. (1994). Evidence for metastable equilibrium between hydrocarbons under hydrothermal conditions. *Nature*, **370**, 285–87.

Searle, M. P. (1996). Geological evidence against large-scale pre-Holocene offsets along the Karakoram Fault; implications for the limited extrusion of the Tibetan Plateau. *Tectonics*, **15**, 171–86.

Seewald, J. S., Benitez-Nelson, B. C. and Whelan, J. K. (1998). Laboratory and theoretical constraints on the generation and composition of natural gas. *Geochimica et Cosmochimica Acta*, **62**, 1599–617.

Seguin, M. K. (1982). Geophysics of the Quebec Appalachians. *Tectonophysics*, **81**, 1–50.

Selley, R. C. and Stoneley, R. (1987). Petroleum habitat in south Dorset. In *Petroleum Geology of Northwest Europe*, ed. J. Brooks and K. W. Glennie. London: Graham and Trotman, pp. 623–32.

Sengor, A. M. C. and Natalin, B. A. (1996). Paleotectonics of Asia: fragments of a synthesis. In *The Tectonic Evolution of Asia*, ed. A. Yinand and M. Harrison. New York: Cambridge University Press, pp. 486–640.

Serra, S. (1977). Styles of deformation in the ramp regions of overthrust faults. *Wyoming Geological Association Handbook, 29th Annual Field Conference*, pp. 487–98.

Shanmugam, G., Moiola, R. J., Sales, J. K. (1988). Duplex-like structures in submarine fan channels, Ouachita Mountains, Arkansas. *Geology*, **16**, 229–32.

Sharp, J. M. (1978). Energy and momentum transport model of the Ouachita Basin and its possible impact on formation of economic mineral deposits. *Economic Geology*, **73**, 1057–68.

Sharp, J. M. and Domenico, P. A. (1976). Energy transport in thick sequences of compacting sediment. *Geological Society of America Bulletin*, **87**, 390–400.

Sharp Jr., J. M. (1983). Permeability controls on aquathermal pressuring. *American Association of Petroleum Geologists Bulletin*, **67**, 2057–61.

Shaw, J. H., Bilotti, F. and Brennan, P. A. (1999). Patterns of imbricate thrusting. *Geological Society of America Bulletin*, **111**, 1140–54.

Shaw, J. H., Hook, S. C. and Suppe, J. (1994). Structural trend analysis by axial surface mapping. *American Association of Petroleum Geologists Bulletin*, **78**, 700–21.

Shaw, J. H. and Suppe, J. (1994). Active faulting and growth folding in the eastern Santa Barbara Channel, California. *Geological Society of America Bulletin*, **106**, 607–26.

Shi, Y. and Wang, C. Y. (1986). Pore pressure generation in sedimentary basins: overloading versus aquathermal. *Journal of Geophysical Research*, **91**, 2153–62.

Shipley, T. H., Houston, M. H., Buffler, R. T., Shaub, F. J., McMillen, K. J., Ladd, J. W. and Worzel, J. L. (1979). Seismic evidence for widespread possible gas hydrate horizons on continental slopes and rises. *American Association of Petroleum Geologists Bulletin*, **63**, 2204–13.

Shipley, T. H., Moore, G. F., Bangs, N. L., Stoffa, P. L. and Moore, J. C. (1993). Seismically inferred spatial pattern of fluid content of the northern Barbados Ridge décollement: implications for fluid migration and fault strength. *EOS, Transactions, American Geophysical Union*, **74**, 579.

Shipley, T. H., Ogawa, J. J., Blum, P. *et al.*, eds. (1995). *Proceedings of the Ocean Drilling Program*, Initial results, 156. Texas: College Station, pp. 13–27.

Shipley, T. H., Stoffa, P. L. and Dean, D. F. (1990). Underthrust sediments, fluid migration paths and mud volcanoes associated with the accretionary wedge off Costa Rica: Middle America Trench. *Journal of Geophysical Research*, **95**, 8743–52.

Shu, L., Charvet, J., Guo, L., Lu, H. and Laurent-Charvet, S. (1999). A large-scale Palaeozoic dextral ductile strike-slip zone; the Aqqikkudug-Weiya Zone along the northern margin of the central Tien Shan belt, Xinjiang, NW China. *Acta Geologica Sinica*, **73**, 146–62.

Shumaker, R. C. (1992). Paleozoic structure of the Central Basin Uplift and adjacent Delaware Basin, west Texas. *American Association of Petroleum Geologists Bulletin*, **76**, 1804–24.

   (1996). Structural history of the Appalachian basin. In *The Atlas of Major Appalachian Gas Plays*, ed. J. B. Roen and B. J. Walker. *West Virginia Geological and Economic Survey, Publication* V-25, pp. 8–21.

Sibson, R. H. (1980). Power dissipation and stress levels on faults in the upper crust. *Journal of Geophysical Research*, **85**, 6239–47.

   (1981a). Controls on low-stress hydro-fracture dilatancy in thrust, wrench and normal fault terrains. *Nature*, **289**, 665–7.

   (1981b). Fluid flow accompanying faulting: field evidence and models. In *Earthquake Prediction: an International Review*, ed. D. W. Simpson and P. G. Richards. *American Geophysical Union, Maurice Ewing Series*, vol. 4, pp. 593–603.

   (1983). Continental fault structure and the shallow earthquake source. *Journal of Geological Society of London*, **140**, 741–67.

   (1986). Brecciation processes in fault zones: Inferences from earthquake rupturing. *Pure and Applied Geophysics*, **124**, 159–75.

   (1989). Earthquake faulting as a structural process. *Journal of Structural Geology*, **11**, 1–14.

   (1990a). Conditions for fault-valve behaviour. In *Deformation Mechanisms, Rheology and Tectonics*, ed.

R. J. Knipe and E. H. Rutter. *Geological Society of London Special Publication*, vol. 54, pp. 15–28.

(1990b). Faulting and fluid flow. In *Fluids in Tectonically Active Regimes of the Continental Crust*, ed. B. E. Nesbitt. *Mineralogical Association of Canada Short Course*, vol. 18, pp. 93–132.

(1992a). Implications of fault-valve behaviour for rupture nucleation and recurrence. In *Earthquake Source Physics and Earthquake Precursors*, ed. T. Mikumo, K. Aki, M. Ohnaka, P. K. P. Spudich and L. J. Ruff. *Tectonophysics*, **211**, 283–93.

(1992b). Fault-valve behavior and the hydrostatic-lithostatic fluid pressure interface. In *Metamorphic Fluids*, ed. W. S. Fyfe. *Earth Science Reviews*, **32**, 141–4.

(1994a). Crustal stress, faulting and fluid flow. In *Geofluids: Origin, Migration and Evolution of Fluids in Sedimentary Basins*, ed. J. Parnell. *Geological Society of London Special Publication*, vol. 78, pp. 69–84.

(1994b). Hill fault/fracture meshes as migration conduits for overpressured fluids. In *Proceedings of Workshop LXIII, US Geological Survey Red-Book Conference on the Mechanical Involvement of Fluids in Faulting*, ed. H. Hickman, R. H. Sibson, R. L. Bruhn and M. L. Jacobson. *Open-File Report* OF 94–0228, pp. 224–30.

(1995). Selective fault reactivation during basin inversion: potential for fluid redistribution through fault-valve action. In *Basin Inversion*, ed. J. G. Buchanan and P. G. Buchanan. *Geological Society of London Special Publication*, vol. 88, pp. 3–19.

(1996). Structural permeability of fluid-driven fault-fracture meshes. *Journal of Structural Geology*, **18**, 1031–42.

(2003). Brittle-failure controls on maximum sustainable overpressure in different tectonic regimes. *American Association of Petroleum Geologists Bulletin*, **87**, 901–8.

Sibson, R. H., Robert, F. and Poulsen, K. H. (1988). High angle reverse faults, fluid pressure cycling and meso-thermal gold-quartz deposits. *Geology*, **16**, 551–5.

Sibuet, M., Fiala-Medioni, A., Foucher, J. P. and Ohta, S. (1990). Spatial distribution of clam colonies at the toe of the Nankai accretionary prism near 138 E. *International Conference: Fluids in Subduction Zones.*

Silliphant, L. J. (1998). Regional and fold-parallel joint distribution at Split Mountain anticline, Utah: a study of the relationship between joint development and 2-D curvature in Laramide folds. M.S. Thesis, Pennsylvania: Pennsylvania State University.

Silver, E. A. and Reed, D. L. (1988). Back thrusting in accretionary wedges. *Journal of Geophysical Research*, **93**, 3116–26.

Simon Gomez, J. L. and Cardona, P. J. (1988). Sobre la compression neogena en la Cordillera Iberica. Neogene compression in the Iberian Cordillera. *Estudios Geologicos*, **44**, 271–83.

Simoneit, B. R. T. and Stuermer, D. H. (1982). Organic geochemical indicators for sources of organic matter and paleoenvironmental conditions in Cretaceous oceans. In *Nature and Origin of Cretaceous Carbon-rich Facies*, ed. S. O. Schlanger and M. B. Cita. London: Academic Press, pp. 145–63.

Sinclair, H. D. (1997). Tectonostratigraphic model for under-filled peripheral foreland basins: an Alpine perspective. *Geological Society of America Bulletin*, **109**, 324–46.

Sinclair, H. D. and Allen, P. A. (1992). Vertical versus horizontal motions in the Alpine orogenic wedge: stratigraphic response in the foreland basin. *Basin Research*, **4**, 215–32.

Sinclair, H. D., Coakley, B. J., Allen, P. A. and Watts, A. B. (1991). Simulation of foreland basin stratigraphy using a diffusion model of mountain belt uplift and erosion: an example from the Central Alps, Switzerland. *Tectonics*, **10**, 599–620.

Sinclair, H. D., Juranov, S. G., Georgiev, G., Byrne, P. and Mountney, N. P. (1997). The Balkan thrust wedge and foreland basin of eastern Bulgaria; structural and stratigraphic development. In *Regional and Petroleum Geology of the Black Sea and Surrounding Region*, ed. A. G. Robinson, *AAPG Memoir*, vol. 68. Tulsa, OK: American Association of Petroleum Geologists, pp. 91–114.

Singh, R. N. (1976). Measurement and analysis of strata deformation around mining excavations. Ph.D. thesis, Cardiff: University of Wales.

Skilbeck, C. G. and Cawood, P. A. (1994). Provenance history of a Carboniferous Gondwana margin forearc basin, New England fold belt, eastern Australia; modal and geochemical constraints. *Sedimentary Geology*, **93**, 107–33.

Skipp, B. (1987). Basement thrust sheets in the Clearwater orogenic zone, central Idaho and western Montana. *Geology*, **15**, 220–4.

Sleep, N. H. and Blanpied, M. L. (1992). Creep, compaction and the weak rheology of major faults. *Nature*, **359**, 687–92.

Sloss, L. L. (1979). Global sea level change: a view from the craton. *American Association of Petroleum Geologists Memoir*, vol. 29, pp. 461–7.

Sloss, L. L. (1988). Forty years of sequence stratigraphy. *Geological Society of America Bulletin*, **100**, 1661–5.

Slotboom, R. T., Lawton, D. C. and Spratt, D. A. (1996). Seismic interpretation of the triangle zone at Jumping Pound, Alberta. *Bulletin of Canadian Petroleum Geology*, **44**, 233–43.

Sluijk, D. and Nederlof, M. H. (1984). Worldwide geological experience as a systematic basis for prospect appraisal. In *Petroleum Geochemistry and Basin Evaluation*, ed. G. Demaison and R. J. Murris. *American Association of Petroleum Geologists Memoir*, vol. 35, pp. 15–26.

Smart, K. J., Dunne, W. M. and Krieg, R. D. (1997). Roof sequence response to emplacement of the Wills Mountain duplex: the roles of forethrusting and scales of deformation. *Journal of Structural Geology*, **19**, 1443–59.

Smith, J. E. (1973). Shale compaction. *Society of Petroleum Engineers Journal*, **13**, 12–22.

Smith, J. T. (1994). The petroleum system logic as an exploration tool in a frontier setting. In *The Petroleum System – From Source to Trap*, ed. L. B. Magoon and W. G. Dow. *American Association of Petroleum Geologists Memoir*, vol. 60, pp. 25–49.

Smith, L. and Chapman, D. S. (1983). On the thermal effects of groundwater flow, 1: Regional scale systems. *Journal of Geophysical Research*, **88**, 593–608.

Smith, L., Forster, C. B. and Evans, J. P. (1990). Interaction of fault zones, fluid flow, and heat transfer at the basin scale. In *Hydrogeology of Low Permeability Environments*, ed. S. P. Newman and I. Neretnieks. International Association of Hydrogeologists, *Selected Papers*, **2**, pp. 41–67.

Smith, R. B. (1975). Unified theory of the onset of folding, boudinage, and mullion structure. *Geological Society of America Bulletin*, **86**, 1601–9.

(1977). Formation of folds, boudinage and mullions in non-Newtonian materials. *Geological Society of America Bulletin*, **88**, 312–20.

Smith, R. E. and Wiltschko, D. V. (1996). Generation and maintenance of abnormal fluid pressures beneath a ramping thrust sheet: isotropic permeability experiments. *Journal of Structural Geology*, **18**, 951–70.

Smith, W. H. F. and Sandwell, D. T. (1997). Global sea floor topography from satellite altimetry and ship depth soundings. *Science Magazine*, **227**, 5334.

Sneider, R. M. and Sneider, J. S. (2001). New oil in old places: the value of mature field redevelopment. In *Petroleum Provinces of the Twenty-first Century*, ed. M. W. Downey, J. C. Tereet and W. A. Morgan. *American Association of Petroleum Geologists Memoir*, vol. 74, pp. 63–84.

Snowdon, L. R. (1991). Oil from type III organic matter: Resinite revisited. *Organic Geochemistry*, **17**, 743–7.

Sobornov, K. O. (1992). Blind duplex structure of the North Urals thrust belt front. *Journal of Geodynamics*, **15**, 1–11.

(1994). Structure and petroleum potential of the Dagestan thrust belt, northeastern Caucasus, Russia. *Bulletin of Canadian Petroleum Geology*, **42**, 352–64.

(1996). Lateral variations in structural styles of tectonic wedging in the northeastern Caucasus, Russia. In *Triangle Zones and Tectonic Wedges*, ed. P. A. MacKay, T. E. Kubli, A. C. Newson, J. L. Varsek, R. G. Dechesne and J. P. Reid. *Bulletin of Canadian Petroleum Geology*, **44**, 385–99.

Solomon, S. C., Sleep, N. H. and Jurdy, D. M. (1977). Mechanical models for absolute plate motions in the early Tertiary. *Journal of Geophysical Research*, **82**, 203–12.

Sonder, L. J. and Chamberlain, C. P. (1992). Tectonic controls of metamorphic field gradients. *Earth Planetary Science Letters*, **111**, 517–36.

Soper, N. J. and Higgins, A. K. (1987). A shallow detachment beneath the North Greenland fold belt: implications for sedimentation and tectonics. *Geological Magazine*, **124**, 441–50.

(1990). Models for Ellesmerian mountain thrust front in North Greenland: a basin margin inverted by basement uplift. *Journal of Structural Geology*, **12**, 83–98.

Speed, R. (1990). Volume loss and defluidization history of Barbados. *Journal of Geophysical Research*, **95**, 8983–96.

Speed, R. C., Barker, L. H. and Payne, P. L. B. (1991). Geologic and hydrocarbon evolution of Barbados. *Journal of Petroleum Geology*, **14**, 323–42.

Sperner, B., Ratschbacher, L. and Nemčok, M. (2002). Interplay between subduction retreat and lateral extrusion: tectonics of the Western Carpathians. *Tectonics*, **21**, 1051–75.

Spicher, A. (1980). *Carte Tectonique de la Suisse*. Bern: Comm. Geol. Suisse.

Spraggins, S. A. and Dunne, W. M. (2002). Deformation history of the Roanoke Recess, Appalachians, USA. *Journal of Structural Geology*, **24**, 411–33.

Spratt, D. and Lawton, D. C. (1996). Variations in detachment levels, ramp angles and wedge geometries along the Alberta thrust front. *Bulletin of Canadian Petroleum Geology*, **44**, 313–23.

Sprunt, E. S. and Nur, A. (1976). Reduction of porosity by pressure solution: experimental verification. *Geology*, **4**, 463–6.

Srivastava, D. C. and Engelder, T. (1990). Crack-propagation sequence and pore-fluid conditions during fault-bend folding in the Appalachian Valley and Ridge, central Pennsylvania. *Geological Society of America Bulletin*, **102**, 116–28.

St.-Julien, P., Slivitsky, A. and Feininger, T. (1983). A deep structural profile across the Appalachians of southern Quebec. In *Contributions to the Tectonics and Geophysics of Mountain Chains*, ed. R. D. Hatcher, H. Williams and I. Zietz. *Geological Society of America Memoir*, vol. 158, pp. 103–11.

Stearns, D. W. (1971). Mechanisms of drape folding in the Wyoming province. *Wyoming Geological Association 23rd Annual Field Conference Guidebook*, pp. 125–44.

(1978). Faulting and forced folding in the Rocky Mountains foreland. In *Laramide Folding Associated with Basement Block Faulting in the Western United States*, ed. V. Matthews. *Geological Society of America Memoir*, vol. 151, pp. 1–37.

Stearns, D. W. and Weinberg, D. M. (1975). A comparison of experimentally created and naturally formed drape folds. *Wyoming Geological Association Guidebook*, vol. 27, pp. 159–66.

Steidtmann, J. R. and Schmitt, J. G. (1988). Provenance and dispersal of tectogenic sediments in thin-skinned, thrusted terrains. In *New Perspectives in Basin Analysis*, ed. K. L. Kleinspehn and C. Paola. New York: Springer Verlag, pp. 353–66.

Stendal, H. and Frei, R. (2000). Gold occurrences and lead isotopes in Ketilidian mobile belt, South Greenland. In *Exploration in Greenland: Discoveries of the 1990s*, ed. H. Stendal. *Transactions of the Institution of*

Mining and Metallurgy, Section B, *Applied Earth Science*, **109**, 6–13.

Stephenson, L. P. (1977). Porosity dependence on temperature: limits on maximum possible effect. *American Association of Petroleum Geologists Bulletin*, **61**, 407–15.

Stets, J. and Wurster, P. (1982). Atlas and Atlantic: Structural relations. In *Geology of the Northwest African Continental Margin*, ed. U. von Rad, K. Hinz, M. Sarntein and E. Seibold. Berlin: Springer-Verlag, pp. 69–85.

Stewart, A. J. (1979). A barred-basin marine evaporite in the upper Proterozoic of the Amadeus Basin, central Australia. *Sedimentology*, **26**, 33–62.

Stocklin, J. (1968). Structural history and tectonics of Iran: a review. *American Association of Petroleum Geologists Bulletin*, **52**, 1229–58.

Stockmal, G. S. (1983). Modeling of large-scale accretionary wedge deformation. *Journal of Geophysical Research*, B, **88**, 8271–87.

Stoll, R. D. and Bryan, G. M. (1979). Physical properties of sediments containing gas hydrates. *Journal of Geophysical Research*, **84**, 1629–34.

Stone, D. S. (1984). The Rattlesnake Mountain, Wyoming debate: a review and critique of models. *Mountain Geology*, **21**, 37–46.

(1990). Wilson Creek Field U.S.A., Piceance Basin, northern Colorado. In *Structural Traps III: Tectonic Fold and Fault Traps*, ed. E. A. Beaumont and N. H. Foster. *American Association of Petroleum Geologists Treatise of Petroleum Geology: Atlas of Oil and Gas Fields*, vol. A-019, pp. 57–101.

(2002). Morphology of the Casper Mountain uplift and related, subsidiary structures, central Wyoming: implications for Laramide kinematics, dynamics, and crustal inheritance. *American Association of Petroleum Geologists Bulletin*, **86**, 1417–40.

Stoneley, R. (1990). The Middle East basin: a summary overview. In *Classic Petroleum Provinces*, ed. J. Brooks. *Geological Society of London Special Publication*, vol. 50, pp. 293–8.

Storti, F. and McClay, K. R. (1995). Influence of syntectonic sedimentation on thrust wedges in analogue models. *Geology*, **23**, 999–1002.

Storti, F., Salvini, F. and McClay, K. R. (1997). Fault-related folding in sandbox analogue models of thrust wedges. *Journal of Structural Geology, Special Issue: Fault-related Folding*, **19**, 583–602.

(2000). Synchronous and velocity-partitioned thrusting and thrust polarity reversal in experimentally produced, doubly-vergent thrust wedges: implications for natural orogens. *Tectonics*, **19**, 378–96.

Strayer, L. M. (1998). Controls on duplex formation: results from numerical models. Geological Society of America, Abstracts with Programs, Session 38, Regional-Scale Faulting.

Strayer, L. M. and Hudleston, P. J. (1997). Numerical modeling of fold initiation at thrust ramps. *Journal of Structural Geology*, **19**, 551–66.

Strayer, L. M., Hudleston, P. J. and Lorig, L. J. (2001). A numerical model of deformation and fluid-flow in an evolving thrust wedge. *Tectonophysics*, **335**, 121–45.

Strecker, M. R., Frisch, W., Hamburger, M. W. *et al.* (1995). Quaternary deformation in the eastern Pamirs, Tadzhikistan and Kyrgyzstan. *Tectonics*, **14**, 1061–79.

Strecker, M. R., Hilley, G. E., Arrowsmith, J. R. and Coutand, I. (2003). Differential structural and geomorphic mountain-front evolution in an active continental collision zone; the Northwest Pamir, southern Kyrgyzstan. *Geological Society of America Bulletin*, **115**, 166–81.

Suess, E., Kulm, L. D., Carson, B. and Whiticar, M. J. (1987). Fluid flow and methane fluxes from vent sites at the Oregon subduction zone. *EOS, Transactions, American Geophysical Union*, **68**, 1487.

Sultan, M. and Gipson, M. (1995). Reservoir potential of the Maastrichtian Pab Sandstone in the eastern Sulaiman fold-belt, Pakistan. *Journal of Petroleum Geology*, **18**, 309–28.

Suppe, J. (1980). Imbricated structure of western foothills belt, south-central Taiwan. *Petroleum Geology of Taiwan*, **17**, 1–16.

(1981). Mechanics of mountain-building and metamorphism in Taiwan. In *ROC-USA Seminar on Plate Tectonics and Metamorphic Geology, Memoir of the Geological Society of China*, vol. 4. Taipei: Chung Kuo Ti Chih Hseuh Hui, pp. 67–89.

(1983). Geometry and kinematics of fault-bend folding. *American Journal of Science*, **283**, 684–721.

(1985). *Principles of Structural Geology*. Englewood Cliffs: Prentice Hall.

(1987). The active Taiwan mountain belt. In *The Anatomy of Mountain Belts*, ed. J.-P. Schaer and J. Rodgers. New Jersey: Princeton University Press, pp. 277–93.

Suppe, J. and Connors, C. (1992). Critical taper wedge mechanics of fold-and-thrust belts on Venus: initial results from Magellan. *Journal of Geophysical Research*, E, *Planets*, **97**, 13 545–61.

Suppe, J., Chou, G. T. and Hook, S. C. (1992). Rates of folding and faulting determined from growth strata. In *Thrust Tectonics*, ed. K. R. McClay. London: Chapman and Hall, pp. 105–21.

Suppe, J. and Hardy, S. (1997). Deformation of the upper crust by rigid-block adjustment, fragmentation and healing. *Geological Society of America, Abstracts with Programs*, vol. 29, p. 44.

Suppe, J. and Medwedeff, D. A. (1984). Fault-propagation folding. *Geological Society of America Bulletin*, **16**, 670.

(1990). Geometry and kinematics of fault-propagation folding. *Eclogae Geologica Helvetica*, **83**, 909–54.

Suppe, J. and Namson, J. S. (1979). Fault-bend origin of frontal folds of the western Taiwan fold-and-thrust belt. *Petroleum Geology of Taiwan*, **16**, 1–18.

Surdam, R. C. (1997). A new paradigm for gas exploration in anomalously pressured 'tight gas sands' in the Rocky Mountain Laramide basins. In *Seals, Traps, and the*

Petroleum System, ed. R. C. Surdam. *American Association of Petroleum Geologists Memoir*, vol. 67, pp. 283–98.

Suzuki, T. and Kimura, T. (1973). D/H and $^{18}O/^{16}O$ fractionation in ice–water systems. *Mass Spectroscopy*, **21**, 229–33.

Sverjensky, D. A. (1986). Genesis of Mississippi Valley-type lead-zinc deposits. *Annual Review of Earth and Planetary Sciences*, **14**, 177–99.

Swanson, S. R. and Brown, W. S. (1972). The influence of state of stress on the stress-strain behaviour of rocks. *Transactions of the American Society of Mechanical Engineers, Journal of Basic Engineering*, **94**, 238–42.

Swarbrick, R. E., Osborne, M. J. and Yardley, G. S. (2002). Comparison of overpressure magnitude resulting from the main generating mechanisms. In *Pressure Regimes in Sedimentary Basins and Their Prediction*, ed. A. R. Huffman and G. L. Bowers. *American Association of Petroleum Geologists Memoir*, vol. 76, pp. 1–12.

Sweeney, J. J. and Burnham, A. K. (1990). Evaluation of a simple model of vitrinite reflectance based on chemical kinetics. *American Association of Petroleum Geologists Bulletin*, **74**, 1559–70.

Sylvester, A. G. (1988). Strike-slip faults. *Geological Society of America Bulletin*, **100**, 1666–703.

Sylvester, A. G. and Brown, G. C., eds. (1988). Santa Barbara and Ventura Basins. In *Coast Geological Society Guidebook*, vol. 64. Ventura, CA: Coast Geological Society.

Sylvester, A. G. and Smith, R. R. (1979). Structure section across the San Andreas fault zone, Mecca Hills In *Tectonics of the Juncture between the San Andreas Fault System and the Salton Trough, Southeastern California; a Guidebook*, ed. J. C. Crowell and A. G. Sylvester. Santa Barbara, CA: University of California, Department of Geological Sciences, pp. 125–39.

Tada, R. and Siever, R. (1989). Pressure solution during diagenesis. *Annual Reviews of Earth and Planetary Sciences*, **17**, 89–118.

Talukdar, S., Gallango, O. and Ruggiero, A. (1988). Generation and migration of oil in the Maturin subbasin, Eastern Venezuela basin. *Organic Geochemistry*, **13**, 537–47.

Tankard, A. J., Suarez, S. R. and Welsink, H. J. (1995). Petroleum basins of South America. *American Association of Petroleum Geologists Memoir*, vol. 62, pp. 1–780.

Tankard, A. J., Uliana, M. A., Welsink, H. J. *et al.* (1995). Structural and tectonic controls of basin evolution in Southwestern Gondwana during the Phanerozoic. In *Petroleum Basins of South America*, ed. A. J. Tankard, S. R. Suarez and H. J. Welsink. *American Association of Petroleum Geologists Memoir*, vol. 62, pp. 5–52.

Tanner, P. W. G. (1989). The flexural-slip mechanism. *Journal of Structural Geology*, **11**, 635–55.

(1992). The duplex model: implications from a study of flex-

ural-slip duplexes. In *Thrust Tectonics*, ed. K. R. McClay. London: Chapman & Hall, pp. 201–8.

Tapponnier, P. and Molnar, P. (1979). Active faulting and Cenozoic tectonics of the Tien Shan, Mongolia, and Baykal regions. *Journal of Geophysical Research*, **84**, 3425–59.

Tari, G. (1991). Multiple Miocene block rotation in the Bakony Mountains, Transdanubian Central Range, Hungary. *Tectonophysics*, **199**, 93–108.

Tari, G., Dicea, O., Faulkerson, J., Georgiev, G., Popov, S., Stefanescu, M. and Weir, G. (1997). Cimmerian and Alpine stratigraphy and structural evolution of the Moesian Platform (Romania/Bulgaria). In *Regional and Petroleum Geology of the Black Sea and Surrounding Region*, ed. A. G. Robinson. *American Association of Petroleum Geologists Memoir*, vol. 68, pp. 63–90.

Tari, G. C., Ashton, P. R., Coterill, K. L., Molnar, J. S., Sorgenfrei, M. C., Thompson, P. W. A., Valasek, D. W. and Fox, J. F. (2002). Are West Africa deepwater salt tectonics analogous to the Gulf of Mexico? *Oil and Gas Journal*, **100**, 73–74, 76–82.

Tarling, D. H. and Hrouda, F. (1993). *The Magnetic Anisotropy of Rocks*. London: Chapman & Hall.

Tavarnelli, E. (1994). Evidences for fault-propagation folding in the Umbria–Marche–Sabina Apennines (Central Italy). *Annales Tectonicae*, **7**, 87–99.

Taylor, D. H. and Cayley, R. A. (2000). Character and kinematics of faults within the turbidite-dominated Lachlan Orogen; implications for tectonic evolution of eastern Australia; discussion. *Journal of Structural Geology*, **22**, 523–8.

Taylor, E. and Leonard, J. (1990). Sediment consolidation and permeability at the Barbados forearc. In *Proceedings of the Ocean Drilling Program*, ed. J. C. Moore, A. Mascle *et al.* College Station: ODP, Scientific Results, **110**, pp. 129–40.

Tchalenko, J. S. (1970). Similarities between shear zones of different magnitudes. *Geological Society of America Bulletin*, **81**, 1625–40.

Tegelar, E. W. and Noble, T. A. (1994). Kinetics of hydrocarbon generation as a function of the molecular structure of kerogen as revealed by pyrolysis-gas chromatography. *Organic Geochemistry*, **22**, 543–74.

Teichmuller, M. and Teichmuller, R. (1975). The geological basis of coal formation. In *Stach's Textbook of Coal Petrology*, 2nd edn., ed. E. Stach, G. H. Taylor, M.-Th. Mackowsky, D. Chandra, M. Teichmuller and R. Teichmuller. Stuttgart: Gebruder Borntraeger, pp. 5–54.

Teng, L. S. (1990). Geotectonic evolution of late Cenozoic arc–continent collision in Taiwan. *Tectonophysics*, **183**, 57–76.

Tessensohn, F. and Piepjohn, K. (1998). Eocene compressive deformation in Arctic Canada, North Greenland and Svalbard and its plate tectonic causes. In *Third International Conference on Arctic Margins*, ed. N. W. Roland and F. Tessensohn. *Polarforschung*, vol. 68, pp. 121–4.

Thanh, T.-D. (1998). Paleontology, stratigraphy and geology of central Viet Nam. *Journal of Geology*, B, **11–12**, 168.

Theil, K. (1995). *Wlasciwosci Fizyko-mechaniczne i Modele Masywow Skalnych Polskich Karpat Fliszowych*. Gdansk: Institut Budownictwa Wodnego, Polska Akademia Nauk.

Thomas, M. M. and Clouse, J. A. (1995). Scaled physical model of secondary oil migration. *American Association of Petroleum Geologists Bulletin*, **79**, 19–29.

Thomas, W. A. (1983). Basement-cover relations in the Appalachian fold and thrust belt. *Geological Journal*, **18**, 267–76.

(1990). Controls on locations of transverse zones in thrust belts. *Eclogae Geologicae Helvetiae*, **83**, 727–44.

Thompson, A. (1959). Pressure solution and porosity. In *Silica in Sediments*, ed. H. A. Ireland. *Society of Economic Paleontologists and Mineralogists Special Publication*, vol. 7, pp. 92–110.

Thompson, A. H., Katz, A. J. and Krohn, C. E. (1987). The microgeometry and transport properties of sedimentary rock. *Advances in Physics*, **36**, 625–94.

Thompson, K. F. M. (1987). Fractionated aromatic petroleums and the generation of gas-condensates. *Organic Geochemistry*, **18**, 573–90.

(1988). Gas-condensate migration and oil fractionation in deltaic systems. *Marine and Petroleum Geology*, **5**, 237–46.

Thompson, R. I. (1979). A structural interpretation across part of northern Rocky Mountains, British Columbia, Canada. *Canadian Journal of Earth Sciences*, **16**, 1228–41.

(1981). The nature and significance of large 'blind' thrusts within the northern Rocky Mountains of Canada. In *Thrust and Nappe Tectonics*, ed. K. R. McClay and N. J. Price, *Geological Society of London Special Publication*, vol. 9. London: Geological Society, pp. 449–62.

(1982). The nature and significance of large 'blind thrusts' within the northern Rocky Mountains of Canada. In *Geologic Studies of the Cordilleran Thrust Belt*, ed. R. B. Powers. *Rocky Mountain Association of Petroleum Geologists*, vol. 1, pp. 47–59.

Thompson, S., Cooper, B. S. and Barnard, P. C. (1994). Some examples and possible explanations for oil generation from coals and coaly sequences. In *Coal and Coal-bearing Strata as Oil-prone Source Rocks*, ed. A. C. Scott and A. J. Fleet. *Geological Society of London Special Publication*, vol. 77, pp. 119–37.

Tillman, L. E. (1989). Sedimentary facies and reservoir characteristics of the Nugget Sandstone (Jurassic), Painter Reservoir Field, Uinta County, Wyoming. In *Petrogenesis and Petrophysics of Selected Sandstone Reservoirs of the Rocky Mountain Region*, ed. E. B. Coalson, S. S. Kaplan, C. W. Keighin, C. A. Oglesby and J. W. Robinson. Rocky Mountain Association of Geologists, pp. 97–108.

Timoshenko, S. (1936). *Theory of Elastic Stability*. New York: McGraw-Hill.

Tissot, B. P., Durand, B., Espitalié, J. and Combaz, A. (1974). Influence of nature and diagenesis of organic matter in formation of petroleum. *American Association of Petroleum Geologists Bulletin*, **58**, 499–506.

Tissot, B. P. and Espitalié, J. (1975). The thermal evolution of organic matter in sediments: application of a mathematical model simulation. *Revue de l'Institut Français du Pétrole*, **30**, 743–77.

Tissot, B. P., Mattavelli, L. and Brosse, E. (1990). Trends in organic geochemistry and petroleum exploration in Italy. In Deposition of Organic Facies, ed. A. Y. Huc. *American Association of Petroleum Geologists Studies in Geology*, vol. 30, pp. 161–79.

Tissot, B. P., Pelet, R. and Ungerer, P. (1987). Thermal history of sedimentary basins, maturation indices, and kinetics of oil and gas generation. *American Association of Petroleum Geologists Bulletin*, **71**, 1445–66.

Tissot, B. P. and Welte, D. H. (1984). *Petroleum Formation and Occurrence*. New York: Springer-Verlag.

Tobin, H. J., Moore, J. C. and Moore, G. F. (1994). Seismic velocity as an indicator of high fluid pressure in thrust faults of the Oregon accretionary prism: laboratory and modeling results. *EOS, Transactions, American Geophysical Union*, **75**, 323.

Toksoz, M. N. (1977). Planetary seismology. *Trends and Opportunities in Seismology*.

Torgersen, T. (1990). Crustal-scale fluid transport. *EOS, Transactions, American Geophysical Union*, **71**, 4–13.

Torres, M. A. (1997). Onshore South Caspian Basin oil fields in western Turkmenistan; the Keimir joint venture as an example. In *1997 AAPG International Conference and Exhibition*, Vienna, Austria, *AAPG Bulletin*, vol. 81, no.8. Tulsa, OK: American Association of Petroleum Geologists, p. 1417.

Torrini, R., Jr and Speed, R. C. (1989). Tectonic wedging in the fore-arc basin – accretionary prism transition, Lesser Antilles forearc. *Journal of Geophysical Research*, **94**, 10 549–84.

Torsvik, T. H., Smethurst, M. A., Meert, J. G., Van der Voo, R., McKerrow, W. S., Brasier, M. D., Sturt, B. A. and Walderhaug, H. J. (1996). Continental break-up and collision in the Neoproterozoic and Palaeozoic: a tale of Baltica and Laurentia. *Earth Science Reviews*, **40**, 229–58.

Tortochaux, F. (1978). Occurrence and structure of evaporites in North Africa. *GSA Special Paper*, **88**, 107–38.

Towsend, C., Roberts, B., Rice, A. H. N. and Gayer, R. A. (1986). The Gaissa Nappe, Finnmark, North Norway: an example of a deeply eroded external imbricate zone within the Scandinavian Caledonides. *Journal of Structural Geology*, **8**, 431–40.

Treiman, J. E. (1995). Surface faulting near Santa Clarita. *Californian Division of Mines and Geology Special Publication*, vol. 116, pp. 103–10.

Treloar, P. J., Coward, M. P., Chambers, A. F., Izatt, C. N. and Jackson, K. C. (1992). Thrust geometries, interferences and rotations in the Northwest Himalaya. In *Thrust Tectonics 1990*. London: Chapman & Hall.

Trettin, H. P. (1989). The Arctic Islands. In *The Geology of North America: an Overview*, ed. A. W. Bally and A. R. Palmer. Geological Society of America, pp. 349–70.

Trettin, H. P. and Balkwill, H. R. (1979). Contributions to the tectonic history of the Innuitian Province, Arctic Canada. *Canadian Journal of Earth Sciences, Journal Canadien des Sciences de la Terre*, **16**, 748–69.

Tribble, J. E. (1990). Clay diagenesis in the Barbados Ridge accretionary prism: potential impact on hydrogeology and subduction dynamics. In *Proceedings of the Ocean Drilling Program*, ed. J. C. Moore and A. Mascle. College Station: ODP, Scientific results, **110**, pp. 97–110.

Trzcienski, W. E. (1986). Disequilibrium textures: petrologic indicators of convergent P–T paths, Shickshock Mountains, northern Gaspe, Quebec, Canada. *Geological Society of America, Abstracts with Programs*, vol. 18, p. 73.

Tullis, J. and Yund, R. A. (1985). Diffusion creep in feldspar aggregates: Experimental evidence. *Journal of Structural Geology*, **13**, 987–1000.

Turcotte, D. L. and Schubert, G. (1982). *Geodynamics Applications of Continuum Physics to Geological Problems*. New York: John Wiley & Sons.

Turner, F. J. and Weiss, L. E. (1963). *Structural Analysis of Metamorphic Tectonites*. New York: McGraw-Hill.

Tysdal, R. G. (1986). Thrust faults and backthrusts in the Madison Range of southwestern Montana foreland. *AAPG Bulletin*, **70**, 360–76.

Ujeta, G. L. (1969). Salt in the Eastern Cordillera of Colombia. *GSA Bulletin*, **80**, 2317–20.

Uliana, M. A., Arteaga, M. E., Legarreta, L., Cerdan, J. J. and Perono, G. O. (1995). Inversion structures and hydrocarbon occurrence in Argentina. In *Basin Inversion*, ed. J. G. Buchanan and P. G. Buchanan. *Geological Society of London Special Publication*, vol. 88, pp. 211–33.

Ulmishek, G. F. (1982). Petroleum geology and resource assessment of the Timan-Pechora Basin, USSR, and the adjacent Barents–northern Kara Shelf. *Argonne National Laboratory, Applied Geoscience and Engineering Group Report* ANL/EES-TM-199.

Ulmishek, G. F., Bogino, V. A., Keller, M. B. and Poznyakevich, Z. L. (1994). Structure, stratigraphy, and petroleum geology of the Pripyat and Dnieper–Donets basins, Byelarus and Ukraine. In *Interior Rift Basins*, ed. S. M. Landon. *American Association of Petroleum Geologists Memoir*, vol. 59, pp. 125–56.

Ulmishek, G. F. and Klemme, H. D. (1990). Depositional controls, distribution, and effectiveness of world's petroleum source rocks. *US Geological Survey Bulletin, Report* B, 1931.

Ungerer, P. (1990). State of the art of research in kinetic modeling of oil formation and expulsion. *Organic Geochemistry*, **16**, 1–25.

Ungerer, P., Behar, E. and Discamps, D. (1983). Tentative cal-

culation of the overall volume expansion of organic matter during hydrocarbon genesis from geochemistry data: implications for primary migration. In *Advances in Organic Geochemistry 1981*, ed. M. Bioroy *et al.* Chichester: John Wiley and Sons, pp. 129–35.

Ungerer, P., Burrus, J., Doligez, B., Chenet, P. Y. and Bessis, F. (1990). Basin evaluation by integrated two-dimensional modeling of heat transfer, fluid flow, hydrocarbon generation, and migration. *American Association of Petroleum Geologists Bulletin*, **74**, 309–35.

Ungerer, P. and Pelet, R. (1987). Extrapolation of the kinetics of oil and gas formation from laboratory to sedimentary basin. *Nature*, **327**, 52–4.

Unruh, J. R. and Moores, E. M. (1992). Quaternary blind thrusting in the southwestern Sacramento Valley, California. *Tectonics*, **11**, 192–203.

Unruh, J. R., Ramirez, V. R., Phipps, S. P. and Moores, E. M. (1991). Tectonic wedging beneath fore-arc basins: ancient and modern examples from California and the Lesser Antilles. *GSA Today*, September.

Urai, J. (1985). Water-enhanced dynamic recrystallization and solution transfer in experimentally deformed carnallite. *Tectonophysics*, **120**, 285–317.

Urai, J., Spiers, C. J., Zwart, H. J. and Lister, G. S. (1986). Weakening of rock salt by water during long-term creep. *Nature*, **324**, 554–7.

Urien, C. M. (2001). Present and future petroleum provinces of southern South America. In *Petroleum Provinces of the Twenty-first Century*, ed. M. W. Downey, J. C. Threet and W. A. Morgan. *American Association of Petroleum Geologists Memoir*, vol. 74, pp. 373–402.

Ustaomer, T. and Robertson, A. (1997). Tectonic-sedimentary evolution of the North Tethyan margin in the central Pontides of northern Turkey. In *Regional and Petroleum Geology of the Black Sea and Surrounding Region*, ed. A. G. Robinson, *AAPG Memoir*, vol. 68. Tulsa, OK: American Association of Petroleum Geologists, pp. 255–90.

Valasis, A. G. and Gornostay, B. A. (1989). Problem of Baykalides and the Riphean Complex of the Kanin Peninsula and northern Timan. *Geotectonics*, **23**, 103–12.

Van der Kamp, G. and Bachu, S. (1989). Use of dimensional analysis in the study of thermal effects of various hydrogeological regimes. In *Hydrogeological Regimes and Their Subsurface Thermal Effects*, ed. A. E. Beck, G. Garven and L. Stegena. *Geophysical Monograph*, vol. 47, IUGG Vol. 2, pp. 23–8.

Van Schmus, W. R. (1984). Radioactivity properties of minerals and rocks. In *CRC Handbook of Physical Properties of Rocks*, ed. R. S. Carmichael. Boca Raton: CRC Press, pp. 281–93.

Vandenbroucke, M. (1993). Migration of hydrocarbons. In *Applied Petroleum Geochemistry*, ed. M. L. Bordenave. Paris: Technip, pp. 125–48.

Vandenbroucke, M., Behar, F. and Rudkiewicz, J. L. (1999). Kinetic modeling of petroleum formation and cracking: implication from the high pressure/high temperature Elgin Field (UK, North Sea). *Organic Geochemistry*, **30**, 1105–25.

Vandenbroucke, M., Durand, B. and Oudin, J. L. (1983). Detecting migration phenomena in a geological series by means of C_1–C_{35} hydrocarbon amounts and distributions. In *Advances in Organic Geochemistry 1981*, ed. M. Bjoroey *et al.* Chichester: John Wiley and Sons, pp. 147–55.

Vann, I. R., Graham, R. H. and Hayward, A. B. (1986). The structure of mountain fronts. *Journal of Structural Geology*, **8**, 215–27.

Vasseur, G. and Burrus, J. (1990). Contraintes hydrodynamiques et thermiques sur la genese des gisements stratiformes a plomb-zinc. In *Mobilité et Concentration des Métaux de Base dans les Couvertures Sédimentaires, Manifestations, Mécanismes, Prospection; Actes du Colloque International*, ed. H. Pelissonnier and J. F. Sureau. Documents, B. R. G. M., **183**, pp. 305–54.

Vasseur, G. and Demongodin, L. (1995). Convective and conductive heat transfer in sedimentary basins. *Basin Research*, **7**, 67–79.

Vavra, C. L., Kaldi, J. G. and Sneider, R. M. (1992). Geological applications of capillary pressure: a review. *American Association of Petroleum Geologists Bulletin*, **76**, 840–50.

Verges, J., Burbank, D. W. and Meigs, A. (1996). Unfolding: an inverse approach to fold kinematics. *Geology*, **24**, 175–78.

Verges, J. and Munoz, J. A. (1990). Thrust sequences in the southern Central Pyrenees. *Bulletin de la Société Géologique Française*, **8**, 265–71.

Verges, J., Munoz, J. A. and Martinez, A. (1992). South Pyrenean fold and thrust belt: The role of foreland evaporitic levels in thrust geometry. In *Thrust Tectonics*, ed. K. R. McClay. London: Chapman and Hall, pp. 255–64.

Vernikovskiy, V. A. (1995). Geodinamicheskaya evolyutsiya Taymyrskoy skladchatoy oblasti. Geodynamic evolution of Taymyr folded area. *Tikhookeanskaya Geologiya, Pacific Geology*, **14**, 71–80.

(1996). Geodinamicheskaya evolyutsiya Taymyrskoy skladchatoy oblasti. Geodynamic evolution of the Taymyr fold region. In *Trudy Rossiyskoy Akademii Nauk Sibirskoye Otdeleniye, Obiyedinennyy Institut Geologii, Geofiziki i Mineralogii*, vol. 831.

Verschuren, M., Nieuwland, D. A. and Gast, J. (1996). Multiple detachment levels in thrust tectonics: sandbox experiments and palinspastic reconstruction. In *Modern Developments in Structural Interpretation, Validation and Modeling*, ed. P. G. Buchanan and D. A. Nieuwland. *Geological Society of London Special Publication*, vol. 99, pp. 227–34.

Versfelt, Jr., P. L. (2001). Major hydrocarbon potential in Iran. In *Petroleum Provinces of the Twenty-first Century*, ed. M. W. Downey, J. C. Tereet and W. A. Morgan. *American Association of Petroleum Geologists Memoir*, vol. 74, pp. 417–27.

Viele, G. W. and Thomas, W. A. (1989). Tectonic synthesis of the Ouachita orogenic belt. In *The Appalachian–Ouachita Orogen in the United States*, ed. R. D. Hatcher Jr., W. A. Thomas and G. W. Viele. *Geology of North America*, vol. F-2, Geological Society of America. pp. 695–728.

Villegas, M. E., Bachu, S., Ramon, J. C. and Underschultz, J. R. (1994). Flow of formation waters in the Llanos basin, Colombia. *American Association of Petroleum Geologists Bulletin*, **78**, 1843–62.

Vissers, R. L. M., Platt, J. P. and Van der Wal, D. (1995). Late orogenic extension of the Betic Cordillera and the Alboran domain: a lithospheric view. *Tectonics*, **14**, 786–803.

Volkman, J. K. (1988). Biological marker compounds as indicators of the depositional environments of petroleum source rocks. In *Lacustrine Petroleum Source Rocks*, ed. A. J. Kelts and M. R. Talbot. *Geological Society of London Special Publication*, vol. 40, pp. 103–22.

Volkman, J. K., Alexander, R., Kagi, P. I. and Woodhouse, G. W. (1983). Demethylated hopanes in crude oils and their applications in petroleum geochemistry. *Geochimica et Cosmochimica Acta*, **47**, 785–94.

Von Gosen, W. (1992). Structural evolution of the Argentine Precordillera, the Rio San Juan section. *Journal of Structural Geology*, **14**, 643–67.

Von Herzen, R. P. and Helwig, J. A. (1984). Geothermal heat flux determined from Cost wells on the Atlantic continental margin. In *Thermal Phenomena in Sedimentary Basins*, ed. B. Durand. Paris: Editions Technip. pp. 219–20.

Von Huene, R. (1984). Tectonic processes along the front of modern convergent margins: research of the past decade. *Earth and Planetary Science Reviews*, **12**, 351–81.

Von Huene R. and Scholl, D. W. (1991). Observations at convergent margins concerning sediment subduction, subduction erosion, and the growth of continental crust. *Reviews of Geophysics*, **29**, 279–316.

(1993). The return of sialic material to the mantle indicated by terrigenous material subducted at convergent margins. *Tectonophysics*, **219**, 163–75.

Von Huene R., Suess, E. and the Leg 112 shipboard scientists. (1988). Results of Leg 112 drilling, Peru continental margin, Part 1: Tectonic history. *Geology*, **16**, 934–38.

Vrielynck, B., Odin, G. S. and Dercourt, J. (1997). Miocene palaeogeography of the Tethys ocean; potential global correlations in the Mediterranean. In *Miocene Stratigraphy; an Integrated Approach*, ed. A. Montanari, G. S. Odin and R. Coccioni. *Developments in Palaeontology and Stratigraphy*, **15**, 157–65.

Vrolijk, P. J. (1987). Tectonic driven fluid flow in the Kodiak accretionary complex, Alaska. *Geology*, **15**, 466–9.

Vrolijk, P., Chambers, S. R., Gieskes, J. M. and O'Neil, J. R. (1990). Stable isotope ratios of interstitial fluids from the northern Barbados accretionary prism, ODP Leg 110. In *Proceedings of the Ocean Drilling Program*, ed. J. C. Moore, A. Mascle *et al.* College Station: ODP, Scientific Results, 110, pp. 189–205.

Vrolijk, P., Fisher, A. and Gieskes, J. M. (1991). Geochemical and geothermal evidence for fluid migration in the Barbados accretionary prism, ODP Leg 110. In *Crustal-scale Fluid Transport: Magnitude and Mechanisms*, ed. T. Torgersen. *Geophysical Research Letters*, **18**, 947–50.

Vrolijk, P. and Sheppard, S. M. F. (1991). Syntectonic carbonate veins from the Barbados accretionary prism (ODP Leg 110): record of palaeohydrology. *Sedimentology*, **38**, 671–90.

Vutukuri, V. S., Lama, R. D. and Saluja, S. S. (1974). *Handbook on Mechanical Properties of Rocks: Testing Techniques and Results*, vol. 1. Bay Village: Trans Tech Publications.

Walderhaug, O. (1996). Kinetic modeling of quartz cementation and porosity loss in deeply buried sandstone reservoirs. *American Association of Petroleum Geologists Bulletin*, **80**, 731–45.

Wallace, L. G. and Roen, J. B. (1989). Petroleum source rock potential of the Upper Ordovician black shale sequence, northern Appalachian basin. *US Geological Survey, Open-File Report* 89–488.

Wallace, R. E. (1951). Geometry of shearing stress and relation to faulting. *Journal of Geology*, **59**, 118–30.

Wallace, W. K. (2003). Geometry and evolution of detachment folds in deformed foreland basin deposits of the Brooks Range Foothills, Northern Alaska. In *AAPG Annual Meeting*, Salt Lake City, UT, *Official Program*, vol. 12. Tulsa, OK: American Association of Petroleum Geologists, p. A177.

Wallace, W. K. and Hanks, C. L. (1990a). Cenozoic thrust emplacement of a Devonian batholith, northeastern Brooks Range: involvement of crystalline rocks in a foreland fold-and-thrust belt. *Geology*, **18**, 395–8.

(1990b). Structural provinces of the northeastern Brooks Range, Arctic National Wildlife Refuge, Alaska. *American Association of Petroleum Geologists Bulletin*, **74**, 1100–18.

Walsh, J. B. (1981). Effect of pore pressure and confining pressure on fracture permeability. *International Journal of Rock Mechanics and Mining Sciences*, **18**, 429–35.

Walsh, J. J. and Watterson, J. (1990). New methods of fault projection for coalmine planning. *Proceedings of the Yorkshire Geological Society*, **48**, 209–19.

Wang, C. Y., Liang, G. and Shi, Y. (1993). Heat flow across the toe of accretionary prisms: the role of fluid flux. *Geophysical Research Letters*, **20**, 659–62.

Wang, C. Y., Shi, Y., Hwang, W. and Chen, H. (1990). Hydrogeological processes in the Oregon–Washington accretionary complex. *Journal of Geophysical Research*, **95**, 9009–24.

Wang, W. H. and Davis, D. M. (1996). Sandbox model simulation of forearc evolution and non-critical wedges. *Journal of Geophysical Research*, **101**, 11 329–39.

Wang Y., Yuan X., Mooney, W. D. and Coleman, R. G. (1999). Crustal structure of Northwest China. *EOS, Transactions*, **80**, 1019–20.

Waples, D. W. (1992). Future developments in basin modeling.

29th International Geological Congress, Abstracts, *Congrès Géologique Internationale*, Resumes, **29**, Vol. 3, p. 812.

(1994a). Maturity modeling: thermal indicators, hydrocarbon generation, and oil cracking. In *The Petroleum System – From Source to Trap*, ed. L. B. Magoon and W. G. Dow. *American Association of Petroleum Geologists Memoir*, vol. 60, pp. 285–306.

(1994b). Modeling of sedimentary basins and petroleum systems. In *The Petroleum System – From Source to Trap*, ed. L. B. Magoon and W. G. Dow. *American Association of Petroleum Geologists Memoir*, vol. 60, pp. 307–22.

Waples, D. W., Kamata, H. and Suizu, M. (1992a). The art of maturity modeling, Part 1: Finding a satisfactory geologic model. *American Association of Petroleum Geologists Bulletin*, **76**, 31–46.

(1992b). The art of maturity modeling, Part 2: Alternative models and sensitivity analysis. *American Association of Petroleum Geologists Bulletin*, **76**, 47–66.

Waples, D. W. and Machihara, T. (1991). *Biomarkers for Geologists. American Association of Petroleum Geologists Methods in Exploration Series*, vol. 9.

Warner, M. A. (1982). Source and time of generation of hydrocarbons in the Fossil Basin, western Wyoming Thrust Belt. In *Geologic Studies of the Cordilleran Thrust Belt*, ed. R. B. Powers. *Rocky Mountain Association of Geologists*, vol. 2, pp. 805–15.

Waschbusch, P. J. and Royden, L. H. (1992). Episodicity in foredeep basins. *Geology*, **20**, 915–18.

Watanabe, Y., Turmagnai, D., Byambasuren, D., Oyuchimeg, G., Tsedenbaljir, Y. and Sato, Y. (1999). Geology and K-Ar ages of the South, Huh Bulgiin Hundii, Saran Uul, Taats Gol and Han Uul deposits in the Bayankhongor region, Mongolia. *Resource Geology*, **49**, 123–30.

Watts, N. L. (1987). Theoretical aspects of cap-rock and fault seals for single- and two-phase hydrocarbon columns. *Marine and Petroleum Geology*, **4**, 274–307.

Weast, R. C. (1974). *CRC Handbook of Chemistry and Physics*. Cleveland: CRC Press.

Webb, G. W. (1981). Stevens and earlier Miocene turbidite sandstones, southern San Joaquin Valley, California. *American Association of Petroleum Geologists Bulletin*, **65**, 438–65.

Web, J. B. and Hertlein, L. G. (1934). Zones in the Alberta shale in foothills of southwestern Alberta. *AAPG Bulletin*, **18**, 1387–416.

Webel, S. (1987). Significance of backthrusting in the Rocky Mountain thrust belt. 1987 *Wyoming Geological Association Guidebook*, pp. 37–53.

Weinberg, D. M. (1978). Some two-dimensional kinematic analyses of the drape-fold concept. *Geological Society of America Memoir*, vol. 151, pp. 51–78.

(1979). Experimental folding of rocks under confining pressure, Part VII: Partially scaled models of drape folds. *Tectonophysics*, **54**, 1–24.

Weislogel, A. L. (1998). Influence of salt diapirism versus Sierra Madrean shortening on deposition of the

Cretaceous (Maastrichtian) Muerto Formation, La Popa Basin, northeastern Mexico. *American Association of Petroleum Geologists Bulletin*, Foundation grants-in-aid abstracts, **83**, 1898.

Welch, V. (1993). EE-4. Maximizing condensate recovery in a rich gas reservoir. In *Atlas of Major Rocky Mountain Gas Reservoirs*, ed. C. A. Hjellming. Socorro, NM: New Mexico Bureau of Mines & Mineral Resources, pp. 178–9.

Wellborn, R. (2000). Crooks Gap Field: limitations of 3-D seismic interpretation. *American Association of Petroleum Geologists Explorer*, **November 2000**, 33–5.

Wells, D. L. and Coppersmith, K. J. (1994). Updated empirical relationships between magnitude, rupture length, rupture area and surface displacement. *Bulletin of Seismological Society*, **84**, 972–1002.

Wenger, L. M., Davis, C. L. and Isaksen, G. H. (2002). Multiple controls on petroleum biodegradation and impact on oil quality. *Society of Petroleum Engineers Reservoir Formation Evaluation and Engineering*, **5**, 375–83.

Wenjiao, X.; Windley, B. F.; Chen, H., Zhang, G. and Li, J. (2002). Carboniferous–Triassic subduction and accretion in the western Kunlun, China; implications for the collisional and accretionary tectonics of the northern Tibetan Plateau. *Geology*, **30**, 295–8.

Wentworth, C. M. and Zoback, M. D. (1990). Structure of the Coalinga area and thrust origin of the earthquake. In *The Coalinga, California, Earthquake of May 2, 1983*, ed. M. J. Rymer and W. L. Ellsworth. *US Geological Survey Professional Paper*, vol. 1487, pp. 41–68.

West, J. and Lewis, H. (1982). Structure and palinspastic reconstruction of the Absaroka Thrust, Anschutz Ranch area, Utah and Wyoming. In *Geologic Studies of the Cordilleran Thrust Belt*, ed. R. B. Powers. Rocky Mountain Association of Geologists, pp. 633–9.

Westbrook, G. K., Ladd, J. W., Buhl, P. and Tiley, G. J. (1988). Cross section of an accretionary wedge: Barbados Ridge Complex. *Geology*, **16**, 631–5.

White, D. A. (1993). Geologic risking guide for prospects and plays. *American Association of Petroleum Geologists Bulletin*, **77**, 2048–64.

White, R. R., Alcock, T. J. and Nelson, R. A. (1990). Anschutz Ranch East Field U.S.A., Utah–Wyoming thrust belt. In *Structural Traps III, Tectonic Fold and Fault Traps*, ed. E. A. Beaumont and N. H. Foster. *American Association of Petroleum Geologists Treatise of Petroleum Geology: Atlas of Oil and Gas Fields*, vol. A-019, pp. 31–55.

Whiticar, M. J. (1994). Correlation of natural gases with their sources. In *The Petroleum System – From Source to Trap*, ed. L. B. Magoon and W. G. Dow. *American Association of Petroleum Geologists Memoir*, vol. 60, pp. 261–83.

Wibberley, C. A. J. (1997). A mechanical model for the reactivation of compartmental faults in basement thrust sheets, Muzelle region, Western Alps. *Journal of the Geological Society of London*, **154**, 123–8.

Wickham, J. and Moeckel, G. P. (1997). Restoration of structural cross-sections. *Journal of Structural Geology*, **19**, 975–86.

Wiens, D. A. and Stein, S. (1983). Age dependence of oceanic intraplate seismicity and implications for lithospheric evolution. *Journal of Geophysical Research*, **88**, 6455–68.

Wignall, P. B. and Maynard, J. R. (1993). The sequence stratigraphy of transgressive black shales. In *Source Rocks in a Sequence Stratigraphic Framework*, ed. B. J. Katz and L. M. Pratt. *American Association of Petroleum Geologists Studies in Geology*, vol. 37, pp. 35–47.

Wilhelm, B. and Somerton, W. H. (1967). Simultaneous measurement of pore and elastic properties of rocks under triaxial stress condition. *Journal of the Society of Petroleum Engineers*, **7**, 283–94.

Willemse, E. J. M., Peacock, D. C. P. and Aydin, A. (1997). Nucleation and growth of strike-slip faults in limestones from Somerset, UK. *Journal of Structural Geology*, **19**, 1461–77.

Willett, S. D., Beaumont, C. and Fullsack, P. (1993). Mechanical model for the tectonics of doubly vergent compressional orogens. *Geology*, **21**, 371–4.

Willett, S. D. and Chapman, D. S. (1989). Temperatures, fluid flow and heat transfer mechanisms in the Uinta Basin. In *Hydrogeological Regimes and Their Subsurface Thermal Effects*, ed. A. E. Beck, G. Garven and L. Stegena. *Geophysical Monograph*, vol. 47, IUGG Vol. 2, pp. 29–33.

Williams, C. A., Connors, C., Dahlen, F. A., Price, E. J. and Suppe, J. (1994). Effect of the brittle–ductile transition on the topography of compressive mountain belts on Earth and Venus. *Journal of Geophysical Research*, **99**, 19 947–74.

Williams, G. D. (1985). Thrust tectonics in the south central Pyrenees. *Journal of Structural Geology*, **7**, 11–17.

Williams, G. D. and Chapman, T. J. (1983). Strain developed in the hanging-wall of thrusts due to their slip/propagation rate: a dislocation model. *Journal of Structural Geology*, **5**, 563–71.

Williams, G. D. and Fischer, M. W. (1984). A balanced section across the Pyrenean orogenic belt. *Tectonics*, **3**, 773–80.

Williams, P. R., Johnston, C. R. Almond, R. A. and Simamora, W. H. (1988). Late Cretaceous to early Tertiary structural elements of West Kalimantan. *Tectonophysics*, **148**, 279–97.

Williams, P. R., Supriatna, S., Johnston, C. R., Almond, R. A. and Simamora, W. H. (1989). A Late Cretaceous to early Tertiary accretionary complex in West Kalimantan. *Bulletin of the Geological Research and Development Centre*, **13**, 9–29.

Williams, R. T. (1987). Energy balance for large thrust sheets and fault-bend folds. *Journal of Structural Geology*, **9**, 375–9.

Williams, W. D. and Dixon, J. S. (1985). Seismic interpretation of the Wyoming overthrust belt. In *Seismic Exploration of Rocky Mountain Region*, ed. R. R.

Gries and R. C. Dyer. *Rocky Mountain Association of Geologists*, vol. 4, pp. 13–22.

Willis, B. and Willis, R. (1934). *Geological Structures*. New York: McGraw-Hill.

Willis, G. C. (1999). The Utah thrust system: an overview. *Utah Geological Association Publication*, **27**, 1–10.

Wilson, D. L. (2002). Planning for success: profit vs. reserves. *Oil & Gas Journal*, **100**, 30–1.

Wiltschko, D. V. (1979a). A mechanical model for thrust sheet deformation at a ramp. *Journal of Geophysical Research*, **84**, 1091–104.

(1979b). Partitioning of energy in a thrust sheet and implications concerning driving forces. *Journal of Geophysical Research*, **84**, 6050–8.

(1981). Thrust sheet deformation at a ramp: summary and extensions of an earlier model. In *Thrust and Nappe Tectonics*, ed. K. R. McClay and N. J. Price. *Geological Society of London Special Publication*, vol. 9, pp. 55–63.

Wiltschko, D. V. and Chapple, W. M. (1977). Flow of weak rocks in the Appalachian Plateau folds. *American Association of Petroleum Geologists Bulletin*, **61**, 653–70.

Wiltschko, D. V. and Dorr, J. A. (1983). Timing of deformation in overthrust belt and foreland of Idaho, Wyoming and Utah. *American Association of Petroleum Geologists Bulletin*, **67**, 1304–22.

Wiltschko, D. V. and Eastman, D. (1983). Role of basement warps and faults in localizing thrust fault ramps. In *Contributions to the Tectonics and Geophysics of Mountain Chains*, ed. R. D. Hatcher, H. Williams and I. Zietz. *Geological Society of America Memoir*, vol. 158, pp. 177–90.

Winter, A. (1987). Percolative aspects of hydrocarbon migration. In *Migration of Hydrocarbons in Sedimentary Basins*, ed. B. Doligez. Paris: Editions Technip, pp. 237–55.

Withjack, M. O., Olson, J. and Peterson, E. (1990). Experimental models of extensional forced folds. *American Association of Petroleum Geologists Bulletin*, **74**, 1038–54.

Wojtal, S. and Mitra, G. (1986). Strain hardening and strain softening in fault zones from foreland thrusts. *Geological Society of America Bulletin*, **97**, 674–87.

Wong, T., Baud, P. and Klein, E. (2001). Localized failure modes in a compactant porous rock. *Geophysical Research Letters*, **28**, 2521–4.

Wood, D. (2003). More aspects of E&P asset and portfolio risk analysis. *Oil & Gas Journal*, **101**, 28–32.

Woodside, W. and Messmer, J. H. (1961). Thermal conductivity of porous media I: Unconsolidated sand. *Journal of Applied Geophysics*, **32**, 1688–706.

Woodward, N. B. (1985). Valley and ridge thrust belt: balanced structural sections, Pennsylvania to Alabama. *University of Tennessee Studies in Geology*, vol. 12.

(1992). Deformation styles and geometric evolution of some Idaho–Wyoming thrust belt structures. In *Structural Geology of Fold and Thrust Belts*, ed. S.

Mitra and G. W. Fisher. Baltimore: Johns Hopkins University Press, pp. 191–206.

Woodward, N. B., Boyer, S. E. and Suppe, J. (1985). An outline of balanced cross-sections. *Studies in Geology*, vol. 11. Knoxville, TN: University of Tennessee, Department of Geological Sciences.

Woodward, N., Gray, D. and Spears, D. (1986). Including strain data in balanced cross sections. *Journal of Structural Geology*, **8**, 313–24.

Worden, R. H., Oxtoby, N. H. and Smalley, P. C. (1998). Can oil emplacement stop quartz cementation in sandstones? *Petroleum Geoscience*, **4**, 129–38.

Worrall, D. M. and Snelson, S. (1989). Evolution of the northern Gulf of Mexico, with emphasis on Cenozoic growth faulting and the role of salt. In *The geology of North America: an Overview*, ed. A. W. Bally and A. R. Palmer. Boulder, CO: Geological Society of America, pp. 97–138.

Wright, A. (1984). Sediment accumulation rates of the Lesser Antilles intraoceanic island arc. *Initial Report of the Deep Sea Drilling Project*, 78A, 301–24.

Wright, L. D. and Coleman, J. M. (1974). Mississippi River mouth processes: effluent dynamics and morphological development. *Journal of Geology*, **82**, 751–8.

Wright, T. L. (1991). Structural geology and tectonic evolution of the Los Angeles basin, California. In *Active Margin Basins*, ed. K. T. Biddle. *American Association of Petroleum Geologists Memoir*, vol. 52, pp. 35–135.

Wright, V. P. and Smart, P. L. (1994). Paleokarst (dissolution diagenesis): its occurrence and hydrocarbon exploration significance. *Developments in Sedimentology*, **51**, 477–517.

Wu, F. T. (1979). Recent tectonics of Taiwan. In *Geodynamics of the Western Pacific*, ed. S. Uyeda, R. W. Murphy and K. Kobayashi. *Advances in Earth and Planetary Sciences*, vol. 6, Tokyo: Center for Academic Publications, pp. 265–99.

Wu, F. T., Yeh, Y. H. and Tsai, Y. B. (1979). Seismicity in the Tsengwen Reservoir area, Taiwan. *Bulletin of the Seismological Society of America*, **69**, 1783–96.

Wu, Q. (1990). *Growth of the Orinoco Deep-Sea Fan in Front of the Barbados Ridge Accretionary Complex*. Ph.D. Thesis, Birmingham: University of Birmingham.

Wu, S., Bally, A. and Cramez, C. (1990). Allochthonous salt, structure and stratigraphy of the north-eastern Gulf of Mexico, Part II: Structure. *Marine and Petroleum Geology*, **7**, 334–70.

Wygrala, B. P. (1989). *Integrated Study of an Oil Field in the Southern Po Basin, Northern Italy*. Ph.D. Thesis, Koln: University of Koln.

Wygrala, B. P., Yalcin, M. N. and Dohmen, L. (1990). Thermal histories and overthrusting: application of numerical simulation technique. In *Advances in Organic Geochemistry 1989; Part I: Organic Geochemistry in Petroleum Exploration*, ed. B. Durand and F. Behar. *Organic Geochemistry*, vol. 16, pp. 267–85.

Yalcin, M. N., Littke, R. and Sachsenhofer, R. F. (1997).

Thermal history of sedimentary basins. In *Petroleum and Basin Evolution: Insights from Petroleum Geochemistry, Geology and Basin Modeling*, ed. D. H. Welte, B. Horsfield and D. R. Baker. Berlin: Springer, pp. 71–167.

Yamano, M., Fisher, A., Kinoshita, M., Foucher, J. P., Hyndman, R. D. and the ODP Leg 131 Shipboard Scientific Party (1990). Heat flow at the toe of the Nankai accretionary prism. *EOS, Transactions, American Geophysical Union*, **71**, 1627.

Yamano, M., Foucher, J. P., Kinoshita, M. *et al.* (1992). Heat flow and fluid flow regime in the western Nankai accretionary prism. In *Fluids in Convergent Margins*, ed. M. Kastner and X. Le Pichon. *Earth and Planetary Science Letters*, **109**, 451–62.

Yan, L. and Lines, L. (2001). Imaging of an Alberta foothills seismic survey. *The Leading Edge*, **20**, 80–4.

Yang, W. (1985). Daqing oil field, People's Republic of China: a giant field with oil of non-marine origin. *American Association of Petroleum Geologists Bulletin*, **69**, 1101–11.

Yeats, R. S. and Lawrence, R. D. (1985). Tectonics of the Himalaya thrust belt in northern Pakistan. *Proceedings of US-Pakistan Workshop on Marine Science in Pakistan*, Karachi, Pakistan, November 11–16, 1982.

Yeh, M. and Yang, C. Y. (1990). A stratigraphic study of upper Miocene–Pliocene series in the Tainan–Kaohsiung Foothills area. In *Central Geological Survey Publication*, vol. 4. Taipei: Central Geological Survey, pp. 147–76.

Yielding, G., Freeman, B. and Needham, D. T. (1997). Quantitative fault seal prediction. *American Association of Petroleum Geologists Bulletin*, **81**, 897–917.

Yielding, G., Jackson, J. A., King, G. C. P., Sinval, H., Vita-Finzi, C. and Wood, R. M. (1981). Relations between surface deformation, fault geometry, seismicity and rupture characteristics during the El Asnam (Algeria) earthquake of 10 October, 1980. *Earth and Planetary Science Letters*, **56**, 287–304.

Yilmaz, P. O., Norton, I. O., Leary, D. and Chuchla, R. J. (1996). Tectonic evolution and paleogeography of Europe. In *Peri-Tethys Memoir 2: Structure and Prospects of Alpine Basins and Forelands*, ed. P. A. Ziegler and F. Horváth. *Memoires du Museum National d'Histoire Naturelle*, vol. 17, pp. 47–60.

Yin, A., Gehrels, G., Chen, X., Wang, X.-F. and Harrison, T. M. (2000). Normal-slip motion on the northern Altyn Tagh Fault. *EOS, Transactions, American Geophysical Union*, **81**, 1092.

Yin, A. and Harrison, T. M. (2000). Geologic evolution of the Himalayan-Tibetan orogen. *Annual Review of Earth and Planetary Sciences*, **28**, 211–80.

Yin, A., Harrison, T. M., Murphy, M. A. *et al.* (1999). Tertiary deformation history of southeastern and southwestern Tibet during the Indo-Asian collision. *Geological Society of America Bulletin*, **111**, 1644–64.

Yonkee, A. W. (1992). Basement–cover relations, Sevier orogenic belt, northern Utah. *Geological Society of America Bulletin*, **104**, 280–302.

Yrigoyen, M. R. (1991). Hydrocarbon resources of Argentina. Buenos Aires: Argentine Petroleum Institute, vol. 23, pp. 38–54.

Yukler, M. A. and Erdogan, T. L. (1996). Effects of geological parameters on temperature histories of basins. *Annual Meeting Abstracts, American Association of Petroleum Geologists and Society of Economic Paleontologists and Mineralogists*, vol. 5, pp. 157–8.

Zanchi, A., Poli, S., Fumagalli, P. and Gaetani, M. (2000). Mantle exhumation along the Tirich Mir fault zone, NW Pakistan; pre-mid-Cretaceous accretion of the Karakoram terrane to the Asian margin. In *Tectonics of the Nanga Parbat Syntaxis and the Western Himalaya*, ed. M. A. Khan, P. J. Treloar, M. P. Searle and M. Q. Jan, *Geological Society Special Publication*, vol. 170. London: Geological Society of London, pp. 237–52.

Zappaterra, E. (1994). Source-rock distribution model of the Periadriatic region. *American Association of Petroleum Geologists Bulletin*, **78**, 333–54.

Zehnder, A. T. and Allmendinger, R. W. (2000). Velocity field for the trishear model. *Journal of Structural Geology*, **22**, 1009–14.

Zepeda, R. L., Keller, E. A., Rockwell, T. K. and Ku, T. L., Dinklage, W. S. (1998). Active tectonics and soil chronology of Wheeler Ridge, Southern San Joaquin Valley, California. *American Geological Society Bulletin*, **110**, 298–310.

Zhai, G. and Zhao, W. (1993). Kelamayi field – People's Republic of China, Zhungeer basin, Xinjiang Province. In *Structural Traps*, vol. VIII, ed. N. N. Foster and E. A. Beaumont. *American Association of Petroleum Geologists: Atlas of Oil and Gas Fields*, pp. 175–92.

Zheng H., Powell, C. McA., An, Z., Zhou, J. and Dong, G. (2000). Pliocene uplift of the northern Tibetan Plateau. *Geology*, **28**, 715–18.

Zhao, G. and Johnson, A. M. (1991). Sequential and incremental formation of conjugate faults. *Journal of Structural Geology*, **13**, 887–96.

Zhao, W. L., Davis, D. M., Dahlen, F. A. and Suppe, J. (1986). Origin of convex accretionary wedges: evidence from Barbados. *Journal of Geophysical Research*, **91**, 10 246–58.

Ziagos, J. P. and Blackwell, D. D. (1986). A model for the transient temperature geothermal systems. *Journal of Volcanology and Geothermal Research*, **27**, 371–97.

Ziegler, P. A. (1988). Evolution of the Arctic-North Atlantic and the Western Tethys. *American Association of Petroleum Geologists Memoir*, vol. 43, p. 198.

(1989). Geodynamic model for Alpine intra-plate compressional deformation in Western and Central Europe. In *Inversion Tectonics Meeting*, ed. M. A. Cooper and G. D. Williams. *Geological Society of London Special Publication*, vol. 44, pp. 63–85.

Zimmerman, R. W., Somerton, W. H. and King, M. S. (1986). Compressibility of porous rocks. *Journal of Geophysical Research*, **91**, 12 765–77.

Zoback, M. D. and Byerlee, J. D. (1976). Effect of high-pressure deformation on permeability of Ottawa Sand. *American Association of Petroleum Geologists Bulletin*, **60**, 1531–42.

Zoback, M. D., Zoback, M. L., Mount, V. S. *et al.* (1987). New evidence on the state of stress of the San Andreas fault system. *Science*, **238**, 1105–11.

Zoetemeijer, R. (1993). *Tectonic Modeling of Foreland Basins: Thin-skinned Thrusting, Syntectonic Sedimentation and Lithospheric Flexure*. P.h.D. Thesis. Amsterdam: Vrije University.

Zoetemeijer, R., Cloetingh, S., Sassi, W. and Roure, F. (1993). Modelling of piggyback-basin stratigraphy: record of tectonic evolution. *Tectonophysics*, **226**, 253–69.

Zonenshain, L. P., Kuzmin, M. I. and Natapov, L. M. (1990). *Geology of the USSR: a Plate-tectonic Synthesis. American Geophysical Union, Geodynamics Series*, vol. 21, p. 242.

Zorin, Yu A. (1999). Geodynamics of the western part of the Mongolia-Okhotsk collisional belt, Trans-Baikal region (Russia) and Mongolia. *Tectonophysics*, **306**, 33–56.

Zorin, Yu. A., Zorina, L. D., Spiridonov, A. M. and Rutshtein, I. G. (2001). Geodynamic setting of gold deposits in eastern and central Trans-Baikal (Chita region, Russia). *Ore Geology Reviews*, **17**, 215–32.

Zoth, G. (1979). Temperaturmessungen in der Bohrung Hanigsen sowie Warmeleitfahigkeitbestimmungen von Gesteinproben. *Report* BGR/NLfB Hannover, Archive No. 81 828.

Zoth, G. and Haenel, R. (1988). Appendix. In *Handbook of Terrestrial Heat-flow Density Determination*, ed. R. Haenel, L. Rybach and L. Stegena. Dordrecht: Kluwer Academic Publishers, pp. 453–63.

Zwach, C., Poelchau, H. S., Hantschel, T. and Welte, D. H. (1994). Simulation with contrasting pore fluids: can we afford to neglect hydrocarbon saturation in basin modeling? In *Basin Modeling – What Have We Learned?*, ed. S. Dueppenbecker and J. Iliffe. *Proceedings of Basin Modeling Conference*, 1.–2.11. 1994, Petroleum Group, British Geological Survey, p 4.

Zwart, G. and Moore, J. C. (1993). Variations in temperature gradient of the Oregon accretionary prism. *EOS, Transactions, American Geophysical Union*, **74**, 369.

Index